	Design or Purpose	Measurement Variables	Ranked Variables	Attributes
1 variable 2 samples only	Independent observations	Back-to-back stem-and-leaf display, Section 2.5; Single-classification anova; unequal sample sizes, Box 9.1; t-test comparing two means or other statistics of location, Box 9.4; Test of equality of two variances, Box 8.1; Testing difference between two means with unequal variances, Box 13.3	Mann-Whitney U-test, Box 13.8; Kolmogorov–Smirnov two-sample test, Box 13.10	Testing difference between two percentages, Box 17.6; Testing difference between two counts, Section 13.7; Sample size required to detect a given difference in percentages, Box 17.17, Box 17.18; Log-odds ratios, Section 17.6
	Paired observations	Paired comparisons test, Box 11.6	Wilcoxon's signed ranks test, Box 13.12	Sign test, Section 13.11; Repeated testing of the same individuals, Section 17.16
2 variables 1 sample	Prediction or functional relationship	Regression statistics: computation, Box 14.1; standard errors, Box 14.2; confidence limits, Box 14.3; outliers and residuals, Section 14.10; regression through the origin, Section 14.4; Estimation of X from Y, Box 14.6; Polynomial regression, Box 16.4; Model II regression, Box 14.10	Kendall's robust line fit method, Box 14.9; Ordering test, Section 14.11	Runs-up and -down test for trends, Box 18.8
	Association	Computation of correlation coefficient, Box 15.1; Hypothesis tests and confidence limits for correlation coefficients, Box 15.2; required sample size for correlation coefficient, Box 15.7; Principal axes, Box 14.10; confidence regions for bivariate scattergram, Box 15.6; Olmstead and Tukey's corner test for association, Box 15.5; Mantel test, Box 18.10	Coefficient of rank correlation: Kendall's, Box 15.4; Spearman's, Section 15.7	2×2 test for independence: by G-test, Box 17.6; by chi-square test, Box 17.5; by Fisher's exact test, Box 17.7; The phi coefficient, Section 17.4; $R \times C$ test for independence, Box 17.8; Unplanned tests of homogeneity of $R \times C$ tables, Section 17.4
2 variables ≥ 2 samples	Prediction or functional relationship	Comparison of regression coefficients, Box 14.7; unplanned comparison by T and GT2, Box 14.8; Analysis of covariance, Section 16.8		
	Association	Tests of homogeneity of correlation coefficients, Box 15.3; Power and sample size, Box 15.7		The Mantel-Haenszel procedure, Box 17.12
2 variables ≥ 3 samples	Prediction or functional relationship	Each sample represents a different value of X: regression with more than one value of Y per value of X, Box 14.4; standard errors, Box 14.2; confidence limits, Box 14.5, also Section 14.9; Test for curvilinear regression by orthogonal polynomials, Box 16.5; Effect size, Box 16.7; confidence limits and sample size, Box 16.8; Isotonic regression, Box 18.9	Page's L-test for ordered alternatives, Section 14.11	
≥ 3 variables 1 sample	Prediction or functional relationship	Multiple regression: computation, Box 16.1; standard errors, Box 16.3; hypothesis testing and confidence limits, Box 16.2; Path coefficients, Table 16.2; structural equation modeling, Section 16.3		
	Association	Coefficients of partial and multiple correlation, Section 16.4	Kendall's coefficient of concordance, Section 15.7	Log-linear model analysis of a three-way table, Box 17.10, Box 17.11
	Design or Purpose	**Measurement Variables**	**Ranked Variables**	**Attributes**

Biometry FOURTH EDITION

Biometry

The Principles and Practice of Statistics in Biological Research

FOURTH EDITION

Robert R. Sokal and **F. James Rohlf**

Stony Brook University

RELEASE

W.H. Freeman and Company
New York

Publisher: Peter Marshall
Acquisitions Editor: Jerry Correa
Marketing Manager: Debbie Clare
Project Editor: Marni Rolfes
Art Director: Diana Blume
Project management, Illustrations, and Composition: MPS Limited, a Macmillan
  Company
Production Coordinator: Lawrence Guerra
Printing and Binding: RR Donnelley

Library of Congress Control Number: 2010939805

ISBN-13: 978-0-7167-8604-7
ISBN-10: 0-7167-8604-4

Printed in the United States of America

First printing

W.H. Freeman and Company
41 Madison Avenue
New York, NY 10010
Houndmills, Basingstoke RG21 6XS, England
www.whfreeman.com

To our wives
Julie and Janice

Contents

Preface

The success of the first three editions of *Biometry* among teachers, students, and others who use biological statistics has encouraged us to prepare this extensively revised Fourth Edition.

We wrote and have now revised this book three times because we feel that there is a need for an up-to-date text aimed primarily at the academic biologist—a text that develops the subject from an elementary introduction up to the advanced methods necessary nowadays for biological research and for an understanding of the published literature. Personal inclination and the nature of the institutions at which we have taught cause us to address ourselves to general biologists, evolutionists, ecologists, geneticists, physiologists, and other biologists working largely on nonapplied subjects in universities, research institutes, and museums. Some of our presentation will deal with agricultural experimentation. More broadly, the statistical methods treated here are useful in many applied fields, including medicine and allied health sciences. Because it is a well-known pedagogical dictum that people learn best by familiar examples, we have endeavored to make our examples as pertinent as possible for our readers.

Much biological work is by its very nature experimental. This book, while furnishing ample directions for the analysis of experimental work, also stresses descriptive and analytical statistical study of biological phenomena. These powerful methods are often overlooked and sometimes, by implication, the validity of non-experimental biological work is put in doubt. We think that descriptive, analytical, and experimental approaches are all of value, and we have tried to strike a balance among them. This approach will be appreciated by workers in applied fields such as the health sciences where ethical, financial, or other considerations may prevent free use of direct experimentation.

The readers we hope to interest are graduate students in biological departments who require a knowledge of biometry as part of their professional training and professional biologists in universities, museums, and research positions who for one reason or another did not receive biometric training during their student careers—that is, people who need to acquire this knowledge largely in an autodidactic way. This book has been designed to serve not only as a text to accompany a lecture course but also as a complete volume for self-study.

We aim to instill in our readers an ability to think through biological research problems in such a way as to grasp the essentials of the experimental or analytical setup, to know which types of statistical tests to apply in a given case, and to carry out the computations required. A modicum of statistical theory has to be presented to introduce the subject matter, but once this is done, further knowledge of statistical theory is acquired as part of the learning process in solving statistical problems, rather than as a preamble to their solution. Because of limitations of time and in the mathematical preparation of most students, discussion of theory must be curtailed and some methods are presented almost entirely in "cookbook" style.

Over the years, we have found that students with very limited mathematical backgrounds can handle biometry excellently. There is remarkably little correlation between innate mathematical ability and capacity to understand the ordinary biometric methodologies. By contrast, in our experience a very high correlation exists between success in the biometry course and success in the chosen field of biological specialization. Teachers should encourage students who at the start have little confidence in their mathematical capabilities, because it does not apply to the study of applied statistics.

We must make our apologies to the mathematical statisticians who may chance on this volume. Compromises had to be made on occasion between precise definitions as required by mathematical statistics and formulations of theory that teaching experience has shown to be most easily understood by beginning classes. In judging our effort, allowance should be made for such changes and circumscriptions.

To make our book more useful to the autodidactic reader, we have employed a more direct and personal style than is customary in textbooks of this sort. No book can replace a dedicated and able teacher, but we have tried to make our written approach as similar as we can to our classroom manner.

Arrangement of This Book

Because this book deviates in several important ways from the customary statistics textbook, we would like to point out the differences so that the prospective reader may derive the maximum benefit from its use. Subject matter is arranged in chapters and sections, numbered by the conventional decimal numbering system: The number before the decimal point refers to the chapter, the number after the point to the section. Topics can be referred to in the table of contents as well as in the index. The tables are also numbered with the decimal system: The first number denotes the chapter, the second the number of the table within the chapter.

Certain special tables are designated "boxes" and are numbered as such, again using the decimal system. The boxes serve a double function. First, they illustrate computational methods for solving various types of biometric problems and can therefore be used as convenient patterns for computation. The boxes usually contain all the steps necessary—from the initial setup of the problem to the final result; hence to students familiar with the book, the boxes can serve as quick summary reminders of the technique. A second important use of the boxes relates to their origin as mimeographed sheets handed out in a biometry course at the University of Kansas. In the allowed lecture time, the instructor would not have been able to convey even as much as half the subject matter covered in the course (or in this book) if the material contained in the sheets had to be put on the chalkboard. Asking the students to refer in class to their biometry sheets (now the boxes in this book) saved much time, and, incidentally, students were able to devote their attention to understanding the content of the sheets rather than to copying them. Instructors who employ this book as a text may wish to use the boxes in a similar manner.

Figures are numbered by the same decimal system as are the boxes. A similar numbering system is also used for (mathematical) expressions, and (homework) exercises.

Because the emphasis in this book is on practical applications of statistics to biology, discussions of statistical theory are deliberately kept to a minimum. However, we feel that some exposure to statistical theory is a good idea for any biologist, so we have provided some in various places. Derivations are given for a number of formulas, but these are consigned to Appendix A, where they should be studied and reworked by the student. A new Appendix B has been added to provide an overview or review of matrix algebra. This is needed for an understanding of some of the material in Chapter 16.

In this book, we present what we consider the minimum statistical knowledge required currently for a PhD in the biological sciences. This material is of equal importance for basic research and for applications in areas such as medicine and allied health sciences. The most important topics treated are the simple distributions (binomial, Poisson, and normal), simple statistical tests (including nonparametric tests), analysis of variance through factorial analysis, analysis of frequencies, simple and multiple regression, and correlation. This is more material than is covered in most elementary texts, yet even this much is usually not sufficient to treat even the common problems of everyday biological research. To enable the reader to become acquainted with some advanced methods, we have therefore introduced at appropriate places throughout the book so-called signpost paragraphs, set in a smaller type and bordered by a gray bar to the left. Signpost paragraphs explain some of the advanced methods and their application and refer the reader to books and papers that we have found to be most useful in explaining a given technique in simple and instructive fashion. We hope that these signposted discussions will open up the wider field of statistics for the readers of this book and permit them at the very least to consult professional statisticians intelligently.

Homework exercises for students reading this book in connection with a biometry course (or studying on their own) are given at the end of each chapter. In keeping with our own convictions, these are largely real research problems. Some of them, therefore, require a fair amount of computation for their solution.

We have been dissatisfied with the customary placement of statistical tables at the end of textbooks of biometry and statistics. Users of these books and tables are constantly inconvenienced by having to turn back and forth between the text material on a certain method and the table necessary for hypothesis testing or for some other computational step. Occasionally, the tables are interspersed throughout a textbook at sites of their initial application; they are then difficult to locate, and turning back and forth in the book is not avoided. Constant users of statistics, therefore, generally use one or more sets of statistical tables, not only because such handbooks usually contain more complete and diverse statistical tables than do the textbooks but also because it avoids the problem of constantly turning pages in the latter.

When we first planned this textbook, we thought we would eliminate statistical tables altogether, asking readers to furnish their own from those available or use probability values provided by statistical software. However, for pedagogical reasons, it was found desirable to refer to a standard set of tables, and we consequently undertook to furnish them, bound separately from the text. The forthcoming *Statistical Tables*, Fourth Edition (Rohlf and Sokal, W. H. Freeman and Company), are identified by boldface letters (for example, Statistical Table **A**).

When a textbook is published over a considerable span of time in several editions, some of the helpful work by former readers and reviewers lives on in later editions, having become part of the immanent (*pace* Kant) spirit of the book. It therefore behooves us to recall those early contributors here. They were Professor R. C. Lewontin (Harvard University), the late Professor K. R. Gabriel (University of Rochester), Dr. Michel Kabay (then of the Université de Moncton in New Brunswick, Canada), Professor F. J. Sonleitner (University of Oklahoma), Professor Theodore J. Crovello (then of the University of Notre Dame), Professor Albert J. Rowell (University of Kansas), Mrs. Barbara A. Thomson (then at Stony Brook University), and Mrs. Donna DiGiovanni (currently at Stony Brook University). All these persons, as well as many students in more than 59 biometry classes contributed comments on several versions of the text. Because not all suggestions were accepted, the reviewers named here cannot be held responsible for errors and other failings in the book.

Turning now to the Fourth Edition, we are greatly indebted to Professor Dean C. Adams of Iowa State University at Ames, Iowa, who in his capacity as "fact checker" fine combed the text and in the process found errors as well as unfortunately worded passages. Preparation of the Fourth Edition was greatly aided by the help of Leone Brown of our home institution, who converted ASCII text files from the Third Edition into formatted MS Word files that we could edit for the Fourth Edition. She also helped with a number of other tasks including the index.

We are obligated to the many generations of biometry students who have enabled us to try out our ideas for teaching biometry and whose responses and criticism have guided us in developing our course. It has been a never-ending source of satisfaction to see these students appropriate biometric techniques as their own and employ them actively in their researches.

RRS is deeply grateful to Dr. Philip Schulman and his colleagues of the Memorial Sloan-Kettering Cancer Center, whose skillful and dedicated care enabled him to complete this book.

To Julie we owe a continuing debt of gratitude for her forbearance during the long periods of gestation of four editions of this book. Janice gracefully survived her initiation to biometry book widowhood with the Fourth Edition. Authors as well as readers are indebted to both of these ladies for their support and encouragement during the several years of preparation of this edition.

Notes on the Fourth Edition

W e trust that the changes in the Fourth Edition of *Biometry*, outlined below, will substantially enhance the book's utility.

When the First Edition was published in 1969, most calculations in biometry were carried out on mechanical desk calculators. Electronic desk calculators were just beginning to make an appearance. The availability of handheld pocket calculators was still several years away. Heavy, repetitive calculations were carried out on large mainframe computers. By now, most statistical computations, except for very large tasks, are carried out on personal computers or workstations, making obsolete the desk calculator formulas we furnished in the text and in the boxes of earlier editions of the book. In the Third Edition, we began to replace these formulas with simpler structural formulas that readers find easier to understand because they furnish direct expressions of the nature of the computation. In the Fourth Edition, we have completed this task for all the boxes in the book. We have also added more complex methods that are only practical when done by computer, and in some cases we left out the computational details on the assumption that appropriate software would be available.

For the benefit of those readers who are familiar with earlier editions of this book, we list below the more important new topics in the sequence of their appearance in the book. Additional homework exercises have also been included for the new methods.

Chapter 3, on the handling of data, has been rewritten to keep abreast of the developments in computing hardware and software. Some of the material in Chapters 5 and 6 (binomial, Poisson, and normal distribution) has been rewritten; we hope it is now easier to understand. Chapter 7, estimation and hypothesis testing, has been revised extensively. We now begin with an introduction to hypothesis testing based on permutation test methods. There is now a separate section on the power of statistical tests. The section on the jackknife and bootstrap methods have been moved here from Chapter 18.

Changes in Chapter 9, single-classification anova, included a new method for estimating variance components and their confidence intervals. The discussion of methods for testing differences among means has been restructured extensively. Several new boxes have been added, including one for Dunnett's test. Chapter 10, nested anova, has been revised to include more recent methods for the estimation of variance components and their confidence intervals. The discussion of optimal allocation of resources in a nested design has been moved to the end of a new Chapter 12. Old Chapters 11 and 12 have been merged and revised extensively. There is now a new box on multiple comparisons among cell means in a two-way anova when interaction is present.

Chapter 12 is almost entirely new. It includes a discussion of effect sizes and their confidence intervals making use of the noncentral t- and F-distributions. There are also new sections on sample size estimation and post-hoc power analyses. We introduce Levene's test for homogeneity of variances in Chapter 13 as well as the Shapiro–Wilk test for normality. Chapter 14, linear regression, has a number of important changes. This section on comparing regression lines was simplified because of the introduction of more comprehensive methods for this in Chapter 16. The section on the analysis of covariance was also moved to Chapter 16. The discussion of model II regression has been expanded and updated with the inclusion of a new box.

The major change in Chapter 15, correlation, is the inclusion of a new section on effects size, power, and sample size estimation for both correlation and regression analyses. Chapter 16, multiple and curvilinear regression, now includes a more general approach to multiple regression and the use of general linear models for anova. The section on path analysis has been extended to include structural equation modeling. The section on the selection of independent variables has been shortened but with new material added on the use of the Akaike's information criterion and related coefficients. Chapter 17 has been extensively edited to include a more standard presentation of the log-likelihood ratio test, a corrected Kolmogorov–Smirnov test, a revised discussion of log-odds ratios and logistic regression, and a new section on effect sizes, power, and estimated sample sizes. Finally, Chapter 18 contains a major new section on meta-analysis and a revised section on the Mantel test. An Appendix B has been added to give an overview of matrix algebra that is now needed for an understanding of several methods described in Chapter 16.

Recent statistical literature has presented many improvements to standard statistical tests. In mentioning some of these and omitting others, we have tried to strike a balance between adding every suggestion in the literature and ignoring them all. The danger in preparing a new edition of a text such as this is adding too many minute changes, making it difficult for the reader to see the main outlines of the methods and to decide among the various alternatives offered. Too many methods are confusing to the novice. Inevitably, however, the book has become more diverse, and the number of topics has increased.

The tabular guides to types of problems encountered frequently in biological research, found in the endpapers, have been updated to include the new methods introduced in the Fourth Edition.

Accompanying the publication of the Fourth Edition is a new version (4) of the BIOMstat programs. These programs are based on the Fourth Edition and feature the major methods listed in the new edition. The programs have been made easier to use and more powerful. Written by F. James Rohlf, the programs are distributed by Exeter Software, 47 Route 25A, Suite 2, Setauket, New York 11733.

The Fourth Edition continues to reflect our many years of experience teaching biometry to classes of graduate students in various departments in the life, health, and earth sciences at the University of Kansas, the University of California, Santa

Barbara, Stony Brook University and various universities abroad. In these courses we have emphasized "doing." For the last thirty years or so our students have been taught to use computer programs. By a series of assignments they are then taught how to solve biological problems statistically, both by thinking through the nature of the problem and by carrying out the computations. Restricting homework exercises to those that are simple to do (hence not time consuming) was an approach we rejected even during the era of the mechanical calculator. With the present general availability of computers, there is no further reason to shield students from real-life data. Students have to be taught the solution of real biological problems involving complicated and copious data, similar to those that would be obtained in an actual research project.

The fear expressed by some teachers that students would not learn statistics well if they were permitted to use computer programs has not been realized in our experience. A careful monitoring of achievement levels before and after the introduction of computers in the teaching of our course revealed no appreciable change in students' performance.

1 Introduction

Dear Reader:

If you are starting your study of this book without looking at the preface, we respectfully but firmly urge you to turn back and read it. We realize that prefaces are often uninspiring, but the preface in this book contains important information for the user. There we explain our reasons for originally writing and now revising this book and describe the audience we address. More important, however, we explain our philosophy of teaching statistics and how we expect the arrangement of this book to meet our objectives. Unless the system of the book is understood, we fear that you are not likely to receive full benefit from reading it. Thank you.

The Authors

This introductory chapter is concerned with setting the stage for your study of biometry. First we define the field itself (Section 1.1). Then we cast a necessarily brief glance at its historical development (Section 1.2). Section 1.3 concludes the chapter with a discussion of the attitudes someone trained in statistics brings to biological research. The application of statistics to biology has revolutionized not only the methodology of research but also the very interpretation of the phenomena under study. The effects this philosophical attitude has had and continues to have on biology are outlined briefly. We hope that the readerss of this book will absorb and appropriate some of these points of view.

1.1 Some Definitions

Scientific etiquette demands that a field be defined before its study is begun. The Greek roots of "biometry" are *bios* ("life") and *metron* ("measure"); hence, **biometry** is literally the measurement of life. For the purposes of this book, biometry is conceived in a very broad sense. We define it as *the application of statistical methods to the solution of biological problems*. As seen in Section 1.2, the original meaning of biometry was much narrower and implied a special field related to the study of evolution and natural selection; however, the wider definition is customary now. Biometry is also called *biological statistics* or simply *biostatistics*.

Our definition tells us that biometry is the application of statistics to biological problems, but the definition leaves us somewhat up in the air because *statistics* has not been defined. Statistics is a science that by name at least is well known even to the layperson, who quite often unjustly abuses the term. The number of definitions you can find is limited only by the number of books you wish to consult. In its modern sense, we might define **statistics** as *the scientific study of data describing natural variation*. All parts of this definition are important and deserve emphasis.

Scientific study: We are concerned with the commonly accepted criteria of validity of scientific evidence. Objectivity in presenting and evaluating data and the general ethical code of scientific methodology must constantly be in evidence if the old canard that "figures never lie, only statisticians do" is not to be revived.

Data: Statistics generally deals with populations or groups of individuals; hence, it deals with *quantities* of information, not with a single *datum*. Thus, the measurement of a single animal or the response from a single biochemical test will generally not be of interest; unless a sample of animals is measured or several such tests are performed, statistics ordinarily can play no role. (We are old-fashioned enough to insist that *data* be used only in its plural sense, despite the latest dictionaries.) Unless the data of a study can be quantified in one way or another, they are not amenable to statistical analysis. The data can be measurements (the length or width of a structure or the amount of a chemical in a body fluid) or counts (the number of bristles or teeth, or frequencies of different qualitative states such as mutant eye colors). Types of variables will be discussed in greater detail in Chapter 2.

Natural variation: We use this term in a wide sense to include all those events that happen in animate and inanimate nature not under the direct control of the investigator, plus those that are evoked by the scientist and are partly under his or her control, as in an experiment. Different biologists concern themselves with different levels of variation, and other kinds of scientists with yet different levels—but all scientists agree that variation in the chirping frequencies of crickets, the number of peas in a pod, and the age at maturity in a chicken is natural. The heartbeat of rats in response to adrenalin or the mutation rate in maize after irradiation may still be considered natural, even though the researcher has manipulated the phenomenon in an experiment. The average biologist, however, would not consider variation in the number of high-definition television sets bought by persons in different states in a given year to be natural, although sociologists or human ecologists might, thus deeming it worthy of study. The qualification "variation in nature" is included in the definition of statistics largely to make certain that the phenomena studied are not arbitrary ones that are entirely under the will and control of the researcher, such as the number of animals employed by the experimenter.

The word *statistics* is also used in another, related sense: as the plural of the noun *statistic,* which refers to any one of many computed or estimated statistical quantities, such as the mean, the standard deviation, or the correlation coefficient. Each one of these is a statistic. (By extension, the term *statistic* has also been used to indicate figures or pieces of data—that is, the individuals or items of data to be defined. This is the way the word is used in the familiar warning to the careless driver not to become a statistic. This is not, however, a recommended use of the word.)

1.2 The Development of Biometry

Neither the nature of this book nor the space at our disposal permit us to indulge in an extended historical review. The origin of modern statistics can be traced back to the 17th century, when it derived from two sources. The first of these related to political science and developed as a quantitative description of the various aspects of the affairs of a government or state (hence the term *statistics*). This subject also became known as *political arithmetic*. Taxes and insurance caused people to become interested in problems of censuses, longevity, and mortality. Such considerations assumed increasing importance, especially in England, as the country prospered during the development of its empire. John Graunt (1620–1674) and William Petty (1623–1687) were early students of vital statistics, and others followed in their footsteps.

The second root of modern statistics originated during the 17th century as well: The mathematical theory of probability was engendered by the interest in games of chance among the leisure classes of the time. Important contributions to this theory were made by French mathematicians Blaise Pascal (1623–1662) and Pierre de Fermat (1601–1665). Swiss mathematician Jacques Bernoulli (1654–1705) laid the foundation of modern probability theory in *Ars Conjectandi,* published posthumously. Abraham de Moivre (1667–1754), a Frenchman living in England, was the first to combine the statistics of his day with probability theory in working out annuity values. De Moivre also was the first to approximate the important normal distribution through the expansion of the binomial.

Subsequent stimulus for the development of statistics came from the science of astronomy, in which many individual observations had to be digested into a coherent theory. Many of the famous astronomers and mathematicians of the 18th century, such as Pierre-Simon de Laplace (1749–1827) in France and Carl Friedrich Gauss (1777–1855) in Germany, were among the leaders in this field. Gauss's lasting contribution to statistics is the development of the method of least squares, which will be encountered later in this book in a variety of forms. Gauss already realized the importance of his ideas to any situation in which errors in observations occur.

Perhaps the earliest important figure in biometric thought was Adolphe Quetelet (1796–1874), a Belgian astronomer and mathematician, who combined the theory and practical methods of statistics and applied them to problems of biology, medicine, and sociology. He introduced the concept of the "average man" and was the first to recognize the significance of the constancy of large numbers—an idea of prime importance in modern statistical work. He also developed the notion of statistical variation and distribution.

Much progress was made in the theory of statistics by mathematicians in the 19th century, but this development is of less interest to us than the work of Francis Galton (1822–1911), a cousin of Charles Darwin. Galton has been called the father of biometry and eugenics (a branch of genetics), two subjects that he studied interrelatedly. The publication of Darwin's work and the inadequacy of Darwin's genetic theories stimulated Galton to try to solve the problems of heredity. Galton's major contribution to biology is his application of statistical methodology to the analysis of biological variation, such as the analysis of variability and his study of regression and correlation in biological measurements. His hope of unraveling the laws of genetics through these procedures was in vain. He started with the most difficult material and with the wrong assumptions. His methodology, however, has become the foundation for the application of statistics to biology.

A contemporary of Galton's, Florence Nightingale (1820–1910), was the first distinguished female statistician. In addition to being the founder of modern nursing, for which she is universally famous, she was an excellent mathematician who pioneered the compilation and graphic presentation of vital and medical statistics.

Karl Pearson (1857–1936) at University College, London, became interested in the application of statistical methods to biology, particularly in the demonstration of natural selection, through the influence of W. F. R. Weldon (1860–1906), a zoologist at the same institution. Weldon, incidentally, is credited with coining the term *biometry* for the type of studies he pursued. Pearson continued in the tradition of Galton and laid the foundation for much of descriptive and correlational statistics.

The dominant figure in statistics and biometry in the 20th century was Ronald A. Fisher (1890–1962). His many contributions to statistical theory will become obvious even to the casual reader of this book. Other important contributors to statistics in the 20th century were Bradley Efron (b. 1938), who developed resampling methods and the bootstrap (see Chapter 7); Harold Hotelling (1895–1973), whose name is associated with multivariate analysis of variance and principal components analysis (see Chapter 16); Jerzy Neymann (1894–1981), who developed a new theory of hypothesis testing and confidence limits together with Egon Pearson (1895–1980), Karl Pearson's son, who is also well known for the statistical tables he co-edited with H. O. Hartley; John Tukey (1915–2000), who specialized in data analysis and robust statistical techniques; John von Neumann (1903–1957), who developed the theory of games and made numerous contributions to computer science; and Abraham Wald (1902–1950), who laid the foundations for econometrics, sequential analysis, and statistical decision theory. All of these statisticians lived and worked in the United States, except for Egon Pearson, who was British, although three of the six American statisticians were born abroad and got their early training there. We will acquaint you with the contributions of others of the last century's eminent statisticians as we discuss the topics they developed.

Currently, statistics is a broad and extremely active field with applications that touch almost every science and even the humanities. New applications for statistics are constantly being found, and no one can predict from what branch of statistics new applications to biology will be made. The prospective student of biological statistics need not be overwhelmed by the formidable size of the field but should be aware that statistics is a constantly changing and growing science; in this respect, it is quite different from a subject such as elementary algebra or trigonometry. Methods of teaching these subjects may have changed over the last 50 years, but not the basic principles and theorems. Some of the methods presented in this book were published only recently in the technical literature. Time-honored methods such as tests for normality and statistics of skewness and kurtosis, after a period of disuse, are again commonly employed because high-speed computers make their calculation feasible. Other techniques, such as the chi-square test and the *t*-test, may become obsolete for a variety of reasons. The entire field of data analysis has become possible only recently because of the development of inexpensive, high-speed computation.

Anyone would be rash to try to predict the scope and practice of biometry 20 years from now. Whatever its nature, however, biometry is more than likely to occupy an even more commanding position in biology over the rest of this century than it does now.

1.3 The Statistical Frame of Mind

The ever-increasing importance and application of statistics to biological data is evident even on cursory inspection. Figure 1.1 shows the results of a survey of 12 decennial volumes of *The American Naturalist;* because of its wide coverage, this journal is a general indicator of trends in biological research. Papers in the decennial volumes were cataloged into those containing no numerical results, those containing numbers but no statistical computations, those containing simple statistics, and those having a major emphasis on mathematics or statistics. The dramatic increase of the percentage of papers involving quantitative approaches is shown clearly in the graph, in spite of the fluctuation to be expected in a sample of this sort. By 2000, 97% of the papers published during that year employed some statistical tests or were more heavily statistical or mathematical. This trend can be observed in most biological journals.

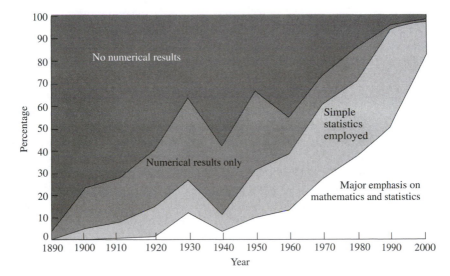

FIGURE 1.1 Proportions of articles involving numerical and statistical work in decennial issues of *The American Naturalist.*

Why has there been such a marked increase in the use of statistics in biology? It is apparently a result of the realization that in biology the interplay of causal and response variables obeys laws that are not in the classic mold of 19th-century physical science. In that century, biologists such as Robert Mayer and Hermann von Helmholtz, among others, tried to demonstrate that biological processes were nothing but physicochemical phenomena; in doing so, they helped create the impression that the experimental methods and natural philosophy that had led to such dramatic progress in the physical sciences should be imitated fully in biology. Regrettably, opposition to this point of view was confounded with the vitalistic movement, which led to unproductive theorizing.

Thus, many biologists even to this day have retained the tradition of strictly mechanistic and deterministic concepts of thinking, whereas physicists, as their science became more refined and came to deal with ever more "elementary" particles, began to resort to statistical approaches. In biology, most phenomena are affected by many causal factors that are uncontrollable in their variation and often unidentifiable. Statistics is needed to measure such variable phenomena with a predictable error and to ascertain the reality of minute but important differences. Whether biological phenomena are in fact fundamentally deterministic and only the variety of causal variables and our inability to control them make these phenomena appear probabilistic, or whether biological processes are truly probabilistic, as postulated in quantum mechanics for elementary particles, is a deep philosophical question beyond the scope of this book and the competence of the authors. The fact remains that most biological phenomena can be discussed only within a probabilistic framework.

A misunderstanding of these principles and relationships has given rise to the attitude of some biologists that if differences induced by an experiment or observed in nature are not clear on plain inspection (and therefore not in need of statistical analysis), they are not worth investigating. This attitude is also related to the search for all-or-none responses, common in fields such as molecular biology. There are still a few legitimate fields of inquiry in which, from the nature of the phenomena studied, statistical investigation is unnecessary. When all animals given injections of a pathogenic organism contract the disease, while none of the controls fall ill, statistical testing is hardly needed. Much more frequent, however, are investigations, such as those determining the relation between smoking and heart disease, where the variability of the outcomes necessitates statistical analysis.

We should stress that statistical thinking is not really different from ordinary disciplined scientific thinking in which we try to quantify our observations. In statistics, we express our degree of belief or disbelief as a probability rather than as a vague, general statement. For example, biologists often make statements such as "species A is larger than species B" or "females are more often found sitting on tree M rather than on tree N." Such statements can and should be expressed more precisely in quantitative form.

The human mind is a remarkable statistical machine, absorbing many facts from the outside world, digesting them, and regurgitating them in simple summary form.

From our experience, we know that certain events occur frequently, others rarely. "Someone smoking a cigarette" is (regrettably) still a frequently observed event; "someone slipping on a banana peel" is rare. We know from experience that Japanese are, on the average, shorter than Englishmen, and that Egyptians have, on average, darker skin than Swedes. We associate thunder with lightning almost always, flies with garbage cans frequently in the summer, but snow with the southern Californian desert extremely rarely. All such knowledge comes to us as a result of our lifetime experience, both directly and through that of others by direct communication or through reading. All these facts have been processed by that remarkable computer, the human brain, which furnishes an abstract. This summary is constantly under revision, and, though occasionally faulty and biased, it is on the whole astonishingly sound; it is our knowledge of the moment.

Although statistics arose to satisfy the needs of scientific research, the development of its methodology in turn has affected the sciences in which statistics is applied. Thus, in a positive feedback loop, statistics, created to serve the needs of natural science, has itself affected the philosophy of the biological sciences. Consider the following two examples. First, analysis of variance has had a tremendous effect in influencing the types of experiments researchers carry out. For example, the whole field of quantitative genetics, one of whose problems is the separation of environmental from genetic effects, depends on the analysis of variance for its realization, and many of the concepts of quantitative genetics have been built directly around the analysis of variance. Second, the techniques of multivariate analysis have given rise to numerous studies designed especially to exploit their ability to analyze many variables simultaneously.

2 Data in Biology

In Section 2.1, we explain the statistical meaning of *sample* and *population,* terms used throughout this book. Then we come to the types of observations obtained from biological research material, with which we shall perform the computations in the rest of this book (Section 2.2). In obtaining data, we shall run into the problem of the degree of accuracy necessary for recording the data. This problem and the procedure for rounding off figures are discussed in Section 2.3, after which we will be ready to consider in Section 2.4 certain kinds of derived data, such as ratios and indexes, frequently used in biological science, which present peculiar problems with respect to their accuracy and distribution. Knowing how to arrange data as frequency distributions is important because such arrangements permit us to get an overall impression of the shape of the variation present in a sample. Frequency distributions, as well as the presentation of numerical data, are discussed in the last section (2.5) of this chapter.

2.1 Samples and Populations

We shall now define several important terms necessary for an understanding of biological data. The data in a biometric study are generally based on **individual observations,** which are *observations or measurements taken on the smallest sampling unit.* These smallest sampling units frequently, but not necessarily, are also individuals in the ordinary biological sense. If we measure weight in 100 rats, then the weight of each rat is an individual observation; the hundred rat weights together represent the **sample of observations,** defined as *a collection of individual observations selected by a specified procedure.* In this instance, one individual observation is based on one individual in a biological sense—that is, one rat. However, if we had studied weight in a single rat over a period of time, the sample of individual observations would be all the weights recorded on one rat at successive times. In a study of temperature in ant colonies, where each colony is a basic sampling unit, each temperature reading for one colony is an individual observation, and the sample of observations is the temperatures for all the colonies considered. An estimate of the amount of DNA expressed as a weight per sperm cell is an individual observation, and the corresponding sample of observations is the estimates of DNA content of all other sperm cells studied in one individual mammal. *Item* is a synonym for individual observation.

Up to now, we have carefully avoided specifying the particular variable being studied because "individual observation" and "sample of observations" as we just used them define only the structure but not the nature of the data in a study. *The actual property measured by the individual observations* is the **variable,** or *character.* The more common term employed in general statistics is *variable.* In evolutionary and systematic biology, however, *character* is frequently used synonymously. More than one variable can be measured on each smallest sampling unit. Thus, in a group of 25 mice we might measure the blood pH and the erythrocyte count in a specified volume of blood. The mouse (a biological individual) would be the smallest sampling unit; blood pH and cell count would be the two variables studied. In this example, the pH readings and cell counts are individual observations, and two samples of 25 observations on pH and erythrocyte count would result. Alternatively, we may call this example a **bivariate sample** of 25 observations, each referring to a pH reading paired with an erythrocyte count.

Next we define **population.** The biological definition of this term is well known: It refers to all the individuals of a given species (perhaps of a given life history stage or sex) found in a circumscribed area at a given time. In statistics, population always means the *totality of individual observations about which inferences are to be made, existing anywhere in the world or at least within a definitely specified sampling area limited in space and time.* If you take five humans and study the number of leucocytes in 1 mL of their peripheral blood and you are prepared to draw conclusions about all humans from this sample of five, then the population from which the sample has been drawn represents the leucocyte counts of all humankind—that is, all extant members of the species *Homo sapiens.* If, on the other hand, you restrict yourself to a more narrowly specified sample, such as five male Chinese, aged 20, and you are restricting your conclusions to this particular group, then the population from which you are sampling will be leucocyte numbers of all Chinese males of age 20. The population in this statistical sense is sometimes referred to as the *universe.*

A population may refer to variables of a concrete collection of objects or creatures—such as the tail lengths of all the white mice in the world, the leucocyte counts of all the Chinese men in the world of age 20, or the amount of DNA in all the hamster sperm cells in existence—or it may refer to the outcomes of experiments—such as all the heartbeat frequencies produced in guinea pigs by injections of adrenalin. In the first three cases, the population is finite. Although in practice it would be impossible to collect, count, and examine all white mice, all Chinese men of age 20, or all hamster sperm cells in the world, these populations are finite. Certain smaller populations, such as all the whooping cranes in North America or all the pocket gophers in a given colony, may lie within reach of a total census. By contrast, an experiment can be repeated an infinite number of times (at least in theory). An experiment such as the administration of adrenalin to guinea pigs could be repeated as long as the experimenter could obtain material and his or her health, patience, and funds held out. The sample of experiments performed is a sample from an infinite number that *could* be performed. Some of the statistical methods to be developed later make a distinction between sampling from finite and from infinite populations. However, although populations are theoretically finite in most applications in biology, they are generally so much larger than samples drawn from them that they can be considered as de facto infinitely sized populations.

2.2 Variables in Biology

Enumerating all the possible kinds of variables that can be studied in biological research would be a hopeless task. Each discipline has its own set of variables, which may include morphological measurements, concentrations of chemicals in body fluids, rates of certain biological processes, frequencies of certain events, physical readings of optical or electronic machinery used in biological research, and many more. We assume that persons reading this book already have special interests and have become acquainted with the methodologies of research in their areas of interest so that the variables commonly studied in their fields are at least somewhat familiar. In any case, the problems for measurement and enumeration must suggest themselves to the researcher; in general, statistics will not contribute to the discovery and definition of such variables.

Some exception must be made to this statement. Once a variable has been chosen, statistical analysis may demonstrate it to be unreliable. If several variables are studied, certain elaborate procedures of multivariate analysis can assign weights to them, indicating their relative value for a given procedure. For example, in taxonomy and various other applications, the method of discriminant functions can identify the combination of a series of variables that best distinguishes between two groups (see Section 16.11). Other multivariate techniques, such as multivariate analysis of variance, principal components analysis, or factor analysis, can specify characters that best represent or summarize certain patterns of variation (Krzanowski, 1988; Jackson, 2003). As a general rule, however, and particularly within the framework of this book, choosing a variable as well as defining the problem to be solved is primarily the responsibility of the biological researcher.

A more precise definition of variable than the one given earlier is desirable. It is *a property with respect to which individuals in a sample differ in some ascertainable way.* If the property does not differ within a sample at hand or at least among the samples being studied, it cannot be of statistical interest. Being entirely uniform, such a property would also not be a variable from the etymological point of view and should not be so called. Length, height, weight, number of teeth, amount of vitamin C, and genotypes are examples of variables in ordinary, genetically and phenotypically diverse groups of organisms. Warm-bloodedness in a group of mammals is not a variable because they are all alike in this regard, although body temperature of individual mammals would, of course, be a variable. Also, if we had a heterogeneous group of animals, of which some were homeothermic and others were not, then body temperature regulation (with its two states or forms of expression, "warm-blooded" and "cold-blooded") would be a variable. The term *variate* is sometimes used as a synonym for variable, but it is defined as a variable that has a specified sampling distribution.

We can classify variables as follows:

Qualitative variables
Quantitative variables

Continuous variables
Discontinuous or meristic variables

Qualitative variables are not numerical but must be expressed qualitatively. Some examples are color (black or white), wingedness (winged or nonwinged insects), gender (male or female), leaf shape (lanceolate, ovate, palmate, etc.), and survival (dead or alive). Such variables are also known as *attributes,* or as *categorical variables.* When such attributes are combined with frequencies, they can be treated statistically. Of 80 mice, for instance, we may state that 4 are black and the rest gray. When attributes are combined with frequencies into tables suitable for statistical analysis, they are referred to as **enumeration data.** The enumeration data on color in mice just mentioned can be arranged as follows:

Color	Frequency
Black	4
Gray	76
Total number of mice	80

In some cases, if desired, qualitative variables can be changed into quantitative variables (discussed immediately below). Thus, colors can be changed into wavelengths or color chart values, which are continuous variables. Certain other attributes can be ranked or ordered as ranked variables. For example, three descriptors referring to a structure as "poorly developed," "well developed," and "hypertrophied" could conveniently be coded 1, 2, and 3. Importantly, these values imply the rank order of development but not the relative magnitudes of these attribute states.

Quantitative variables are numerical. There are two types of such variables: continuous and discontinuous. **Continuous variables** can take on a continuous range of values and can assume an infinite number of values between any two fixed points. For example, between the two length measurements 1.5 cm and 1.6 cm, an infinite number of lengths could be measured if one were so inclined and had a measuring instrument with sufficiently precise calibration. Any given reading of a continuous variable, such as a length of 1.57 cm, is an approximation to the exact reading, which in practice cannot be known. For purposes of computation, however, these approximations are usually sufficient and, as will be seen below, may be made even more approximate by rounding. Many of the variables studied in biology are continuous. Examples are length, area, volume, weight, angle, temperature, period of time, percentage, and rate.

Discontinuous or **meristic variables** are those whose values are limited to discrete states with no intermediate values possible. The number of segments in a certain insect appendage, for instance, may be 4, 5, or 6 but never $5\frac{1}{2}$ or 4.3. Examples of discontinuous variables are number of a certain structure (such as segments, bristles, teeth, or glands), number of offspring, number of colonies of microorganisms or animals, or number of plants in a given quadrat.

A word of caution: Not all variables restricted to integral numerical values are meristic. An example will illustrate this point. If an animal behaviorist were to code the reactions of animals in a series of experiments as (1) very aggressive, (2) aggressive, (3) neutral, (4) submissive, and (5) very submissive, we might be tempted to believe that these five different states of the variable were meristic because they assume integral values. However, they are clearly only arbitrary points (class marks, see Section 2.5) along a continuum of aggressiveness; the only reason that no values such as 1.5 occur is that the experimenter did not wish to subdivide the behavior classes too finely, either for convenience or because of an inability to determine more than five subdivisions of this spectrum of behavior with precision. Thus, this variable is clearly continuous rather than meristic, as it might have appeared at first sight.

Variables can also be described in terms of the scales of measurement. These distinctions are important because different statistical methods are needed for different types of variables.

Nominal scale
Ordinal scale
Interval scale
Ratio scale

Variables that are recorded on a **nominal scale** are those that are in discrete states without any implied ordering. Qualitative variables are examples of a nominal scale. Variables recorded on an **ordinal scale** are also in discrete states, but these states can be ordered or ranked. Examples include speed (fastest, next fastest, etc. but without specifying an actual measurement of the speed), parity (first born, second born, etc. in a litter), and performance on a test (best, second best, etc.). Special methods for dealing with ranked variables have been developed, and several are described in this book. By expressing a variable as a series of ranks, such as 1, 2, 3, 4, 5, we do not imply that the difference in magnitude between, say, ranks 1 and 2 is identical to or even proportional to the difference between 2 and 3. Such an assumption is made, however, for continuous variables, discussed above. Variables measured on an **interval scale** are recorded using an arbitrary scale. Examples are temperature (in centigrade or Fahrenheit) and IQ tests. These are always continuous variables. In contrast, variables measured on a **ratio scale** are recorded along a scale that has a meaningful zero. Examples are age, number of chromosomes, temperature in degrees Kelvin, and the length of a structure.

Many different terms are used to identify a single reading, score, or observation of a given variable in a sample. In this text, we identify variables by capital italic letters, the most common symbol being Y. Thus, Y may stand for tail length of mice. A single observation of a given length measurement, Y_i, is the measurement of tail length of the ith mouse, and Y_4 is the measurement of tail length of the fourth mouse in our sample.

2.3 Accuracy and Precision of Data

Scientists are generally aware of the importance of accuracy and precision in their work, and it is self-evident that accuracy must extend to the numerical results of their work and to the processing of these data. Because of the great diversity of approaches employed in different disciplines of biology, it seems futile to attempt to furnish specific rules for obtaining data accurately. We should make certain, however, that whatever the method of obtaining data, it is consistent so that given the same observational or experimental setup, different numerical readings of the same structure or event would be identical or within acceptable and predictable limits of each other.

Accuracy and precision are used synonymously in everyday speech, but in statistics we distinguish between them. **Accuracy** is *the closeness of a measured or computed value to its true value;* **precision** is *the closeness of repeated measurements of the same quantity* to each other. A biased but sensitive scale might yield inaccurate but precise weight readings. By chance, an insensitive scale might result in an accurate reading, but the reading would be imprecise because another weighing of the same object would be unlikely to yield an equally accurate weight. Unless there is bias in a measuring instrument, precision will lead to accuracy. We need therefore to be mainly concerned with the former.

Precise observations are usually but not necessarily whole numbers. Thus, when we count four eggs in a nest, there is no doubt about the exact number of eggs in the nest if we have counted correctly; it is four, not three or five, and clearly it could not be four plus or minus a fractional part. Meristic variables are generally measured as exact numbers. Seemingly, continuous variables derived from meristic ones can, under certain conditions, also be exact numbers. For instance, ratios of exact numbers are themselves also exact. In a colony of animals, if there are 18 females and 12 males, the ratio of females to males is 1.5, an observation for a continuous variable but also an exact number.

Most continuous variables, however, are approximate: The exact value of the single measurement is unknown and probably unknowable. The last digit of the measurement stated should imply precision—that is, the limits on the measurement scale between which we believe the true measurement to lie. Thus, a length measurement of 12.3 mm implies that the true length of the structure lies somewhere between 12.25 and 12.35 mm. Exactly where between these **implied limits** the real length is we do not know. Some might object to defining the limits of the number as 12.25 to 12.35 mm because the adjacent measurement of 12.2 would imply limits of 12.15 to 12.25. Where then, they would ask, would a true measurement of 12.25 fall? Would it not equally likely fall in either of the two classes 12.2 and 12.3, clearly an unsatisfactory state of affairs? For this reason, some authors define the implied limits of 12.3 as 12.25 to 12.34999. . . . Such an argument is correct, but when we record a number as either 12.2 or 12.3, we imply that the decision whether to put it into the higher or lower class has already been made. This decision was not made arbitrarily but presumably based on the best available measurement. If the scale of measurement is so precise that a value of 12.25 would clearly have been recognized, then the measurement should have been recorded originally to four significant figures. Implied limits therefore always carry one figure beyond the last significant one measured by the observer.

Hence it follows that if we record the measurement as 12.32, we are implying that the true value lies between 12.315 and 12.325. If this is what we mean, then there is no need to add another decimal figure to our original measurements. If we add another figure, we imply an increase in precision. We see therefore that accuracy and precision in numbers is not an absolute concept but is relative. Assuming there is no bias, a number becomes increasingly accurate as we are able to write more significant figures for it (that is, increase its precision). To illustrate the relativity of accuracy, consider the following three numbers:

Implied Limits	
193	192.5–193.5
192.8	192.75–192.85
192.76	192.755–192.765

Imagine these numbers to be the recorded measurements of the same structure. If we assume that we know the true length of the given structure to be 192.758 units, then the three measurements increased in accuracy from the top down, as the interval between their implied limits decreases.

Observations of meristic variables, though usually exact, may be recorded approximately when large numbers are involved. Thus, when counts are reported to the nearest thousand, a count of 36,000 insects in a cubic meter of soil implies that the true number lies between 35,500 and 36,500 insects.

To how many significant figures should we record measurements? If we array the sample by order of magnitude from the smallest individual to the largest one, an easy rule to remember is that *the number of unit steps from the smallest to the largest measurement in an array should usually be between 30 and 300.* For example, if we are measuring a series of shells to the nearest millimeter and the largest is 8 mm and the smallest 4 mm wide, there are only four unit steps between the largest and the smallest measurement. Hence we should have measured our shells to one more significant decimal place. Then the two extreme measurements might have been 8.2 mm and 4.1 mm, with 41 unit steps between them (counting the last significant digit as the unit); this would have been an adequate number of unit steps. The reason for such a role is that an error of 1 in the last significant digit of a reading of 4 mm would constitute an inadmissible error of 25%, but an error of 1 in the last digit of 4.1 is less than 2.5%. By contrast, if we had measured the height of the tallest of a series of plants as 173.2 cm and that of the shortest of these plants as 26.6 cm, the difference between these limits would comprise 1,466 unit steps (of 0.1 cm each), which are far too many. We should therefore have recorded the heights to the nearest centimeter, as follows: 173 cm for the tallest and 27 cm for the shortest, yielding 146 unit steps between.

Using the rule stated above, we will record two or three digits for most measurements. On occasion, however, more digits will be necessary—when the leading digits are constant, as in the following example. In a representative series of pH readings measured with great precision, the highest value might be 7.456 and the lowest value 7.434. The first two digits of such measurements are constant for pH readings of the material under study. It is therefore necessary to use three decimal places to obtain an adequate amount of variation for analysis. The perceptive reader may have noticed that the number of unit steps from lowest to highest observations in the pH example was only 22, or less than 30 given as the lower limit for the desired number of steps. This rule, as are all such rules of thumb, is to be taken with a grain of statistical salt. These rules are general guidelines but not laws whose mild infraction would immediately invalidate all further work. With some experience, an appropriate adherence to these rules becomes second nature to the biologist engaging in statistical analysis.

The last digit of an approximate number should always be significant; that is, it should imply a range for the true measurement of from half a unit step below to half a unit step above the recorded score, as illustrated earlier. This rule applies to all digits, zero included. Zeros should therefore not be written at the end of approximate numbers to the right of the decimal point unless they are meant to be significant digits. The measurement 7.80 implies the limits 7.795 to 7.805. If 7.75 to 7.85 is meant to be implied, then the measurement should be recorded as 7.8.

When the number of significant digits is reduced, we carry out the process of **rounding** numbers. The rules for rounding are very simple. A digit to be rounded is not changed if it is followed by a digit less than 5. If the digit to be rounded is followed by a digit greater than 5 or by 5 followed by other nonzero digits, it is increased by one. When the digit to be rounded is followed by a 5 standing alone or followed by zeros, it is unchanged if it is even but increased by one if it is odd. The reason for this last rule is that when such numbers are summed in a long series, we should have as many digits raised as are being lowered on the average; these changes would therefore balance out. Practice these rules by rounding the following numbers to the indicated number of significant digits:

Number	Significant Digits Desired	Answer
26.58	2	27
133.7137	5	133.71
0.03725	3	0.0372
0.03715	3	0.0372
18,316	2	18,000
17.3476	3	17.3

Most pocket calculators or larger computers round their displays using a different rule: They increase the preceding digit when the following digit is a 5 standing alone or with trailing zeros. However, because most of the machines that can be used for statistics also retain 8 or 16 significant figures internally, the accumulation of rounding errors is minimized. Incidentally, if two calculators give answers with slight differences in the final (least significant) digits, suspect a different number of significant digits in memory as a cause of the disagreement.

2.4 Derived Variables

Most variables in biometric work are observations recorded as direct measurements or counts of biological material or as readings that are the output of various types of instruments. However, there is an important class of variables in biological research, which we may call the **derived** (or computed) **variables,** that are generally based on two or more independently measured variables whose relations are expressed in a certain way. We are referring to ratios, percentages, indices, rates, and the like.

A **ratio** expresses as a single value the relation between two variables. In its simplest form, a ratio is expressed, for example, as 64:24, which may represent the number of wild-type versus mutant individuals, the number of males versus females, a count of parasitized individuals versus those not parasitized, and so on. These examples imply ratios based on counts; a ratio based on a continuous variable might be similarly expressed as 1.2:1.8, which could represent the ratio of width to length in a sclerite of an insect or the ratio between the concentrations of two minerals contained in water or soil. Ratios may also be expressed as fractions; thus, the two ratios above could be expressed as $\frac{64}{24}$ and $\frac{1.2}{1.8}$. For computational purposes, however, it is most useful to express the ratio as a quotient. The two ratios cited here would therefore be 2.666 . . . and 0.666 . . . , respectively. These are pure numbers, not expressed in measurement units of any kind. It is this form for ratios that we will consider further. *Percentages* are also a type of ratio. Ratios and percentages are basic quantities in much biological research. They are widely used and generally familiar.

Not all derived variables are ratios or percentages. The term **index** is used in a general sense for derived variables, although some would limit it to the ratio of one anatomic variable divided by a larger, so-called standard variable. For instance, in a study of the cranial dimensions of cats, Haltenorth (1937) divided all measurements by the basal length of the skull. Thus, each measurement was a proportion of the basal length of the skull of the cat being measured. Another example of an index in this sense is the well-known cephalic index in physical anthropology. In quantitative genetics, an index is constructed by combining measurements of economically important traits. Individuals are then selected on the basis of this index (see Falconer and MacKay, 1996, for more information). An index could be a weighted average of two measurements, such as $1/3[(2 \times$ length of $A) +$ length of $B]$ or a more complicated function of the values of several traits. An index may refer to the summation of a series of numerically scored properties. Thus, if an animal is given six behavioral tests in which its score can range from 0 to 4, an index of its behavior might be the sum of the scores of the six tests. Similar indices have been described for determining the degree of hybridity in organisms (the hybrid indices of various authors).

Rates are important in many experimental fields of biology. The amount of a substance liberated per unit weight or volume of biological material, weight gain per unit time, reproductive rates per unit population size and time (birthrates), and death rates would fall in this category. Many counts are really ratios or rates—the number of pulse beats observed in one minute or the number of birds of a given species found in some quadrat. In general, counts are ratios if the unit over which they are counted is not natural, as, for example, an arbitrary time or space interval.

As we shall see, there are some serious drawbacks to the use of ratios and percentages in statistical work. In spite of these disadvantages, the use of these variables is deeply ingrained in scientific thought processes and is not likely to be abandoned.

Furthermore, ratios may be the only meaningful way to interpret and understand certain types of biological problems. If the biological process being investigated operates on the ratio of the variables studied, then one must examine this ratio to understand the process. Thus, in their classic study, Sinnott and Hammond (1935) found that inheritance of the shapes of the squash *Cucurbita pepo* could be interpreted by a simple index based on a length–width ratio, but not in terms of the length and width dimensions themselves. Similarly, the evolution of shape in many burrowing animals is a function of the cross-sectional profile of the animal rather than of a single dimension. Selection affecting body proportions will be found to exist in the evolution of almost any organism when properly investigated.

The disadvantages of using ratios are several. First is their relative inaccuracy. Let us return to the ratio $\frac{1.2}{1.8}$ and recall from the previous section that a measurement of 1.2 indicates a true range of measurement of the variable from 1.15 to 1.25; similarly, a measurement of 1.8 implies a range from 1.75 to 1.85. We realize, therefore, that the true ratio may vary anywhere from $\frac{1.15}{1.85}$ to $\frac{1.25}{1.75}$, or 0.622 and 0.714, respectively. We note a possible maximal error of 4.2% if 1.2 were an original measurement: $(1.25 - 1.2)/1.2$. The corresponding maximal error for the ratio is 7.0%: $(0.714 - 0.667)/0.667$. Furthermore, the best estimate of a ratio is not usually the midpoint between its possible ranges. Thus, in our example, the midpoint between the implied limits is 0.668 and the true ratio is 0.666 . . . a difference that here is only slight but that would be greater if the denominator were smaller. In many cases, therefore, ratios are not as accurate as measurements obtained directly. This liability can be overcome by empirical determinations of the variability of ratios through measures of their variance (described later in this book).

A second drawback to ratios and percentages is that they may not be approximately normally distributed (see Chapter 6), as required by many statistical tests. This difficulty frequently can be overcome by transformation of the variable (see Chapter 13). Another disadvantage of ratios is that they do not provide information on *how* the two variables whose ratio is being taken are related. Often more may be learned by studying the variables singly first and then examining their relation to each other (bivariate and multivariate analysis).

Finally, when ratios involve enumeration data or meristic variables, occasionally they give rise to curious distributions. As an example, we cite the percentages obtained from an experiment performed many years ago by Sokal. In this experiment, 10 *Drosophila* eggs were put in a vial and the positions of the pupae noted after pupation. Some pupae pupated at the margin or the wall of the vials; these were called *peripheral*. Others, which pupated away from the wall, were called *central*. Ideally, there should have been 10 pupae—if all the eggs had hatched and there had been no larval mortality whatsoever. In fact, however, because of natural mortality and because of errors in preparing the vials, 10 pupae were not always found. The logical minimum number of survivors on which a result could be reported was one pupa. The proportion of peripheral pupae was then calculated by dividing the number of such pupae in a vial by the total number of pupae found in that vial.

Although ratios thus obtained are continuous variables in appearance, they are not so, in fact, because certain values can never be obtained. For instance, by limiting the maximum number of pupae to 10, we cannot obtain percentages of peripheral pupation between 0% and 10% or between 90% and 100%. Also, we see that some percentages are given by several ratios, others by only one: 33% is obtained by $\frac{3}{9}$ or $\frac{2}{6}$ or $\frac{1}{3}$, but 57% can be obtained only by having 4 out of 7 pupae peripheral. It follows that if we make the simplest assumption—that there is an equal probability for any given experimental outcome—some values are more likely to be encountered. In using ratios of meristic variables or of counts, such discontinuities and peculiarities of distribution must be taken into account. The problem in this instance was solved by transforming the data to probits (see Section 14.10). For a more detailed discussion of the problems encountered with ratios of discrete variables, see Johnston et al. (1995), who also furnish recommendations for overcoming the statistical artifacts.

2.5 Frequency Distributions

If we were to sample a population of birth weights of infants, we could represent each measurement by a point along an axis denoting magnitude of birth weights. This example is illustrated in Figure 2.1a for a sample of 25 birth weights. Note that because of the limited precision of the data, we have to place some of these points on top of other points along the axis in order to record them all correctly (this type of display is called a *dot plot*). If we sample repeatedly from the population and obtain 100 birth weights, the number of multiple points at numerous axis locations increases (Figure 2.1b). As we continue sampling additional hundreds and thousands of birth weights (Figure 2.1c, 2.1d), plots of individual points become difficult to absorb, and they are conveniently replaced by histogram bars whose heights are proportional to the number of points at a given axis location. The assemblage of bars increases in height and will assume a fairly definite shape. The curve tracing the outline of the mound of points approximates the distribution of the variable. Remember that a continuous variable such as birth weight can assume an infinity of values between any two points on the abscissa. The refinement of our measurements determines the number of recorded divisions between any two points along the axis.

The distribution of a variable is of considerable biological interest. An asymmetrical distribution drawn out in one direction indicates possible selection for or against organisms falling in one of the tails of the distribution or that the scale of measurement chosen may be causing a distortion of the distribution. A bimodal (two-peaked) distribution of measurements in a sample of immature insects suggests that the population is dimorphic; different species or races may have become intermingled in our sample, or the dimorphism could have arisen from the presence of both sexes or of different instars. There are several characteristic shapes of frequency distributions, the most common of which is the symmetrical bell-shaped distribution (approximated by the graph in Figure 2.1d), the normal frequency distribution discussed in Chapter 6. There are also skewed distributions (drawn out more at one tail than at

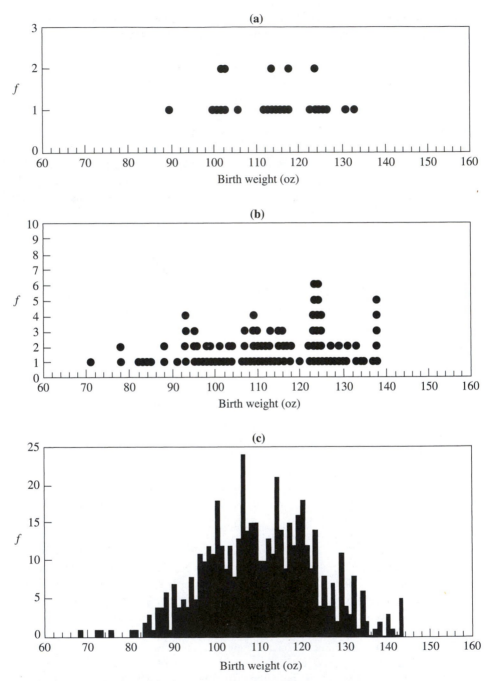

FIGURE 2.1 Sampling from a population of birth weights of infants (a continuous variable). **(a)** A sample of 25. **(b)** A sample of 100. **(c)** A sample of 500. **(d)** A sample of 2,000. (continued)

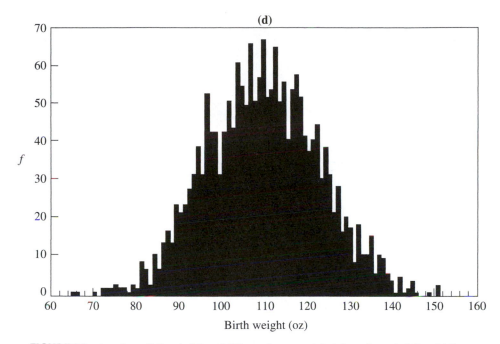

FIGURE 2.1 (continued) Panels **(a)** and **(b)** are shown as dot plots and panels **(c)** and **(d)** as histograms. The width of the classes is 1 oz.

the other), ∟-shaped distributions as in Figure 2.2, U-shaped distributions, and others that impart significant information about certain types of relationships. We will have more to say about the implications of various types of distributions later.

After data have been obtained in a given study, they must be arranged in a form suitable for computation and interpretation. We may assume that observations are randomly ordered initially or are in the order in which the measurements were taken. Researchers should be aware that the sequence of observations in a sample may contain valuable information that is accessible with techniques discussed in Chapter 18. Records of original observations and their sequence should not be discarded; they may be important in tracing and solving problems of sampling or of measurement.

A simple arrangement would be an *array* of the data by order of magnitude. Thus, for example, the observations 7, 6, 5, 7, 8, 9, 6, 7, 4, 6, 7 could be arrayed in order of increasing magnitude as follows: 4, 5, 6, 6, 6, 7, 7, 7, 7, 8, 9. Where there are some observations of the same value, such as the 6s and 7s in this fictitious example, a time-saving device might immediately have occurred to you—namely, to list a frequency for each of the recurring observations: 4, 5, 6(3×), 7(4×), 8, 9. Such shorthand notation is one way to represent a **frequency distribution,** which is simply *an arrangement of the classes of observations with the frequencies of each*

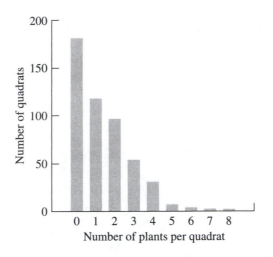

FIGURE 2.2   Bar diagram. Frequency of the sedge *Carex flacca* in 500 quadrats. Data from Table 2.2.

class indicated. Conventionally, a frequency distribution is stated in tabular form, as follows for the above example:

Variable Y	Frequency f
4	1
5	1
6	3
7	4
8	1
9	1

This case is an example of a *quantitative* frequency distribution because Y is clearly a measurement variable. However, arrays and frequency distributions need not be limited to such variables. We can make *qualitative* frequency distributions of attributes. In these, the various classes are listed in some logical or arbitrary order. For example, in genetics we might have a qualitative frequency distribution as follows:

	f
A–	86
aa	32

This distribution tells us that there are two classes of individuals, those identified by the A– phenotype, of which 86 were found, and the homozygote recessive *aa*, of which there were 32 in the sample. In ecology, it is quite common to have a species list of inhabitants of a sampled ecological area. The arrangement of such tables is

| TABLE 2.1 | A Qualitative Frequency Distribution

Number of individuals of Heteroptera tabulated by family in total samples from a summer foliage insect community.

Family	Observed frequency f
Alydidae	2
Anthocoridae	37
Coreidae	2
Lygaeidae	318
Miridae	373
Nabidae	3
Neididae	5
Pentatomidae	25
Piesmidae	1
Reduvidae	3
Rhopalidae	2
Saldidae	1
Thyreocoridae	10
Tingidae	69
Total Heteroptera	851

SOURCE: Data from Whittaker (1952).

usually alphabetical. An example of an ecological list constituting a qualitative frequency distribution of families is shown in Table 2.1. In this instance, the families of insects are listed alphabetically; in other cases, the sequence may be by convention, as for the families of flowering plants in botany.

A quantitative frequency distribution based on a meristic variable is shown in Table 2.2. This example comes from plant ecology. The number of plants per quadrat sampled are listed at the left in the variable column; the observed frequency is shown at the right. We have already shown a graph of this distribution in Figure 2.2.

Quantitative frequency distributions based on a continuous variable are the most commonly employed frequency distributions, and you should become thoroughly familiar with them. An example is shown in Box 2.1 (pages 26–27). It is based on 25 femur lengths of stem mothers (one of the life history stages) of a species of aphids. The 25 readings (measured in coded micrometer units) are shown at the top of the box in the order (reading down columns from left to right) in which they were obtained as measurements. The observations increase in magnitude from 3.3 to 4.7 by unit steps of 0.1. Most statistical computer programs will prepare a frequency distribution from a set of input data. The program will request the desired class interval (also called the *bin width*). Initially, we specified 0.1, which means that observations as much as 0.1 or more apart will end up in different classes or bin. The resulting frequency distribution (the leftmost one shown) summarizes the distribution of the 25 femur lengths without losing any information in the original data. The sum of the frequencies is indicated by Σf or n.

TABLE 2.2 A Quantitative Frequency Distribution Based on a Meristic Variable

Number of plants of the sedge *Carex flacca* found in 500 quadrats.

No. of plants per quadrat Y	Observed frequency f
0	181
1	118
2	97
3	54
4	32
5	9
6	5
7	3
8	1
Total	500

SOURCE: Data from Archibald (1950).

Let us look at what we have achieved in summarizing our data. The original 25 observations are now represented by only 15 classes. We find that observations 3.6, 3.8, and 4.3 have the highest frequencies. However, we also note several classes, such as 3.4 or 3.7, that are not represented by a single aphid. This gives the frequency distribution a drawn-out and scattered appearance. The reason is that we have only 25 aphids, too few to put into a frequency distribution with 15 classes. To obtain a more cohesive and smooth-looking distribution, we have to condense our data into fewer classes. This process, known as **grouping the observations into classes** of frequency distributions, is illustrated in Box 2.1 (pages 26–27) and described in the following paragraphs. In the statistical literature in other disciplines, such grouping is also called *binning,* each class being a *bin.*

What we are doing when we group individual observations into classes of wider range is only an extension of the same process that took place when we obtained the initial measurement. Thus, as we saw in Section 2.3, when we measure an aphid and record its femur length as 3.3 units, we imply that the true measurement lies along the interval from 3.25 to 3.35 units, but that we were unable to measure to the second decimal place. In recording the measurement initially as 3.3 units, we estimated that it fell within this range. Had we estimated that it exceeded the value of 3.35, for example, then we would have given it the next higher score, 3.4. Therefore, all the measurements between 3.25 and 3.35 were, in fact, grouped into the class identified by the **class mark** or class midpoint 3.3. The width of our **class interval** (or *bin width*) was 0.1 units. If we now make wider class intervals, we are doing nothing but extending the range within which measurements are placed into one class.

Referring to Box 2.1 will clarify this process. We group the data twice to emphasize the flexibility of the process. In the first example of grouping, the class interval has been doubled in width to 0.2 units. If we start at the lower end, the implied class limits will now be from 3.25 to 3.45, the limits for the next class from 3.45 to 3.65, and so forth.

Next we examine the class marks. This task was simple in the frequency distribution shown at the left side of Box 2.1, in which the original measurements had been used as class marks. However, now we are using a class interval that is twice as wide, and the class marks are calculated by taking the midpoint of the new class intervals. For the first class, we take the midpoint between 3.25 and 3.45, which is 3.35. Note that the class mark has one more decimal place than do the original measurements. Do not believe that we have suddenly achieved greater precision. Whenever we designate a class interval whose last *significant* digit is even (0.2 in this case), the class mark will carry one more decimal place than the original measurements do. On the right side of the table in Box 2.1, the data are grouped once again, using a class interval of 0.3. Because its last significant digit is odd, the class mark now shows as many decimal places as the original observations do, the midpoint between 3.25 and 3.55 being 3.4. The wider the class interval, the fewer classes result. In the three frequency distributions, the number of classes or bins is 15, 8, and 5, respectively. It should be obvious that the wider the class intervals, the more compact, but also the less precise, the frequency distribution becomes. However, looking at the frequency distribution of aphid femur lengths in Box 2.1, we notice that the initial chaotic structure is simplified by grouping. When we group the frequency distribution into five classes with a class interval of 0.3 units, it becomes notably bimodal (that is, it possesses two peaks of frequencies).

In setting up frequency distributions, from 12 to 20 classes should be established. This rule need not be adhered to slavishly, but should be employed with some of the common sense that comes from experience in handling statistical data. The number of classes depends largely on the size of the sample studied. Samples of less than 40 to 50 should rarely be given as many as 12 classes because that would provide too few frequencies per class. On the other hand, samples of several thousand may profitably be grouped into more than 20 classes. If the aphid data of Box 2.1 need to be grouped, they should probably not be grouped into more than 6 classes.

When the sample size is small or the class intervals are wide, the general impression one obtains of a frequency distribution may depend in part on exactly how the class limits are defined. Shifting the class limits may remove the appearance of bimodality or change the apparent location of the mode.

If the original data provide us with fewer classes than we think we should have, then nothing can be done if the variable is meristic because this is the nature of the data in question. With a continuous variable, however, a scarcity of classes indicates that we probably have not made our measurements with sufficient precision. If we had followed the rule on the number of significant figures stated in Section 2.3, this could not have happened.

Whenever there are more than the desired number of classes, grouping should be undertaken. When the data are meristic, the implied limits of continuous variables are meaningless. Yet with many meristic variables, such as a bristle number varying from 13 to 81, it would probably be wise to group them into classes, each containing

BOX 2.1   Preparation of Frequency Distribution and Grouping into Fewer Classes with Wider Class Intervals

Twenty-five femur lengths of stem mothers of the aphid *Pemphigus populitransversus*. Measurements are in mm $\times 10^{-1}$.

Original measurements

3.8	3.6	4.3	3.5	4.3
3.3	4.3	3.9	4.3	3.8
3.9	4.4	3.8	4.7	3.6
4.1	4.4	4.5	3.6	3.8
4.4	4.1	3.6	4.2	3.9

Original frequency distribution

Implied limits	Y	f
3.25–3.35	3.3	1
3.35–3.45	3.4	0
3.45–3.55	3.5	1
3.55–3.65	3.6	4
3.65–3.75	3.7	0
3.75–3.85	3.8	4
3.85–3.95	3.9	3
3.95–4.05	4.0	0

Grouping into 8 classes of interval 0.2

Implied limits	Class mark	f
3.25–3.45	3.35	1
3.45–3.65	3.55	5
3.65–3.85	3.75	4
3.85–4.05	3.95	3

Grouping into 5 classes of interval 0.3

Implied limits	Class mark	f
3.25–3.55	3.4	2
3.55–3.85	3.7	8
3.85–4.15	4.0	5

4.05–4.15	4.1	2	4.05–4.25	4.15	3	4.15–4.45	4.3	8
4.15–4.25	4.2	1						
4.25–4.35	4.3	4	4.25–4.45	4.35	7			
4.35–4.45	4.4	3				4.45–4.75	4.6	2
4.45–4.55	4.5	1	4.45–4.65	4.55	1			
4.55–4.65	4.6	0						
4.65–4.75	4.7	$\dfrac{1}{25}$	4.65–4.85	4.75	$\dfrac{1}{25}$			$\overline{25}$
Σf		25						

Histogram of the original frequency distribution shown above and of the grouped distribution with 5 classes. Line below abscissa shows class marks for the grouped frequency distribution. Shaded bars represent original frequency distribution, hollow bars represent grouped distribution.

Y (femur length, in units of 0.1 mm)

f (frequency)

several counts. This grouping can best be done by using an odd number as a class interval so that the class mark representing the data is a whole rather than a fractional number. Thus, if we were to group the bristle numbers 13, 14, 15, and 16 into one class, the class mark would have to be 14.5, a meaningless value in terms of bristle number. It would therefore be better to use a class width of 3 or 5 bristle numbers.

Grouping data into frequency distributions was necessary when computations were done by pencil and paper or with mechanical calculators. Nowadays, even thousands of observations can be processed efficiently by computer without prior grouping. However, frequency distributions are still extremely useful as a tool for data analysis, especially in an age when it is all too easy for a researcher to obtain a numerical result from a computer program without ever really examining the data for outliers (extreme values) or for other ways in which the sample may not conform to the assumptions of the statistical methods. For this reason, most modern statistical computer programs furnish some graphic output of the frequency distribution of the observations.

An alternative to setting up a frequency distribution as shown in Box 2.1 is the so-called **stem-and-leaf display** suggested by Tukey (1977). The advantage of this technique is that it not only results in a frequency distribution of the observations of a sample but also presents them in a form that makes a ranked (ordered) array very easy to construct. It also effectively creates a list of these values and is easy to check. This technique is therefore useful in computing the median of a sample (see Section 4.3) and in computing various nonparametric statistics that require ordered arrays of the sample observations (see Sections 13.11 and 13.12). Many computer programs for basic statistics provide the option of furnishing a stem-and-leaf display of the data. Below we show you what these programs accomplish.

In Box 15.7, we feature measurements of total length recorded for a sample of 15 aphid stem mothers. The unordered measurements are reproduced here: 8.7, 8.5, 9.4, 10.0, 6.3, 7.8, 11.9, 6.5, 6.6, 10.6, 10.2, 7.2, 8.6, 11.1, 11.6. To prepare a stem-and-leaf display, we write down the leading digit or digits of the observations in the sample to the left of a vertical line (the "stem") as shown below; we then put the next digit of the first observation (a "leaf") at that level of the stem corresponding to its leading digit(s):

Step 1	Step 2 . . .	Step 7 . . .	Completed Array (Step 15)	Ordered Array
6 |	6 |	6 | 3	6 | 356	6 | 356
7 |	7 |	7 | 8	7 | 82	7 | 28
8 | 7	8 | 75	8 | 75	8 | 756	8 | 567
9 |	9 |	9 | 4	9 | 4	9 | 4
10 |	10 |	10 | 0	10 | 062	10 | 026
11 |	11 |	11 | 9	11 | 916	11 | 169

The first observation in our sample is 8.7. We therefore place a 7 next to the 8. The next observation is 8.5. It is entered by finding the stem level for the leading digit 8 and recording a 5 next to the 7 that is already there. Similarly for the third observation, 9.4, we record a 4 next to the 9, and so on until all 15 observations have been entered (as "leaves") in sequence along the appropriate leading digits of the stem. Finally, we order the leaves from smallest (0) to largest (9).

The ordered array is equivalent to a frequency distribution and has the appearance of a histogram or bar diagram (see below), but it also efficiently orders the observations. Thus, from the ordered array, it becomes obvious that the appropriate ordering of the 15 observations is: 6.3, 6.5, 6.6, 7.2, 7.8, 8.5, 8.6, 8.7, 9.4, 10.0, 10.2, 10.6, 11.1, 11.6, 11.9. The median, the observation having an equal number of observations on either side, can easily be read off the stem-and-leaf display. It is 8.7.

In biometric work, we frequently wish to compare two samples to see if they differ. In such cases, we may employ **back-to-back stem-and-leaf displays,** illustrated here:

Sample A		Sample B
94	10	05778
988642	11	16
88866531	12	013

This example is taken from Box 13.7 and describes a morphological measurement obtained from two samples of chiggers. By setting up the stem in such a way that it may serve for both samples, we can easily compare the frequencies. Even though these data furnish only three classes for the stem, we can readily see that the samples differ in their distributions. Sample A has by far the higher readings.

When the shape of a frequency distribution is of particular interest, we may often wish to present the distribution in graphic form when discussing the results. This is generally done with frequency diagrams, of which there are two common types. For a distribution of meristic data, we use a **bar diagram,** as shown in Figure 2.2 for the sedge data of Table 2.2. The abscissa represents the variable (in our case, the number of plants per quadrat), and the ordinate represents the frequencies. What is important about such a diagram is that the bars do not touch each other, which indicates that the variable is not continuous.

By contrast, continuous variables, such as the frequency distribution of the femur lengths of aphid stem mothers, are graphed as a **histogram,** in which the width of each bar along the abscissa represents a class interval of the frequency distribution and the bars touch each other to show that the actual limits of the classes are contiguous. The midpoint of the bar corresponds to the class mark. At the bottom of Box 2.1 are histograms of the frequency distribution of the aphid data, ungrouped and grouped. The height of the bars represents the frequency of each class. Occasionally, the class intervals of a grouped continuous frequency distribution are unequal. For instance, in a frequency distribution of ages, we might have more detail on the different stages of young individuals and less accurate identification of the ages of old individuals. In such cases, the class intervals for the older age groups would be wider,

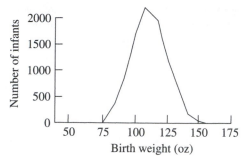

FIGURE 2.3 Frequency polygon. Birth weights of 9465 male infants. Chinese third-class patients in Singapore, 1950 and 1951. (Data from Millis and Seng, 1954.)

those for the younger age groups, narrower. In representations of such data, the bars of the histogram are drawn with different widths.

Figure 2.3 shows another graphic mode of representing a frequency distribution of a continuous variable—birth weight in infants. This is a *frequency polygon,* in which the heights of the class marks in a histogram are connected by straight lines. As we shall see later, the shapes of frequency distributions as seen in these various graphs can reveal much about the biological situations affecting a given variable.

A method nowadays called *dot plots* graphically displays the distribution of observations in a relatively small sample by a technique that dates back for at least a century. We show two examples in Figures 2.1a and 2.1b. Several commercially available basic statistics programs produce dot plots of datasets. However, these programs treat such displays as though they were substitutes for histograms. The dots for any one class interval are piled on top of each other as though they were indicators of the height of the corresponding bar in a histogram (our Figures 2.1a and 2.1b also are constructed in this manner). Wilkinson (1999) points out that that the correct method for graphing dot plots is to place individual observations exactly along the axis describing the variable, making allowance for fully or partially overlapping dots (tied or numerically close observations). He provides algorithms for creating such displays.

EXERCISES 2

2.1 Differentiate between the following pairs of terms and give an example of each. (a) Statistical and biological populations. (b) Observation and individual. (c) Accuracy and precision (repeatability). (d) Class interval and class mark. (e) Bar diagram and histogram. (f) Abscissa and ordinate.

2.2 Round the following numbers to three significant figures: 106.55, 0.06819, 3.0495, 7815.01, 2.9149, and 20.1500. What are the implied limits before and after rounding? Round these same numbers to one decimal place.

2.3 Given 200 measurements ranging from 1.32 to 2.95 mm, how would you group them into a frequency distribution? Give class limits as well as class marks.

2.4 Group the following 40 measurements of interorbital width of a sample of domestic pigeons into a frequency distribution and draw its histogram (data from Olson and Miller, 1958). Measurements are in millimeters.

12.2	12.9	11.8	11.9	11.6	11.1	12.3	12.2	11.8	11.8
10.7	11.5	11.3	11.2	11.6	11.9	13.3	11.2	10.5	11.1
12.1	11.9	10.4	10.7	10.8	11.0	11.9	10.2	10.9	11.6
10.8	11.6	10.4	10.7	12.0	12.4	11.7	11.8	11.3	11.1

2.5 How precisely should you measure the wing length of a species of mosquitoes in a study of geographic variation if the smallest specimen has a length of about 2.8 mm and the largest a length of about 3.5 mm?

2.6 Transform the 40 measurements in Exercise 2.4 into common logarithms and make a frequency distribution of these transformed observations. Comment on the resulting change in the pattern of the frequency distribution from that found before.

2.7 In Exercise 4.3, we feature 120 percentages of butterfat from Ayrshire cows. Make a frequency distribution of these values using a stem-and-leaf display. Prepare an ordered array of these observations from the display. Save the display for use later in Exercise 4.3(b).

3

Computers and
Data Analysis

Because the emphasis of this book is largely on practical statistics, it is important to discuss the various techniques available for carrying out statistical computations. We discuss computer hardware in Section 3.1 and software in Section 3.2. In Section 3.3, we focus on the important question of which computational devices and types of software are most appropriate in terms of efficiency and economy.

Without mechanical computational aids, we would be reduced to carrying out statistical computations by the so-called pencil-and-paper methods once heavily emphasized in textbooks. They used to contain extensive sections dealing with clever ways to make computation by hand feasible. Such methods are now very inefficient uses of time and energy. Even handheld calculators are now used much less than they once were, although more advanced models are capable of performing many of the simpler analyses presented in this text. Calculators are now considered a less convenient option because they are not able to save data sets for reuse in other analyses. With desktop and laptop microcomputers in every laboratory, computations by hand or with calculators are mostly just to verify the reasonableness of results produced by computer or to express the results in a different form.

The benefit from these developments is that researchers can concentrate on the more-important problems of proper experimental design and interpreting the statistical results rather than spending tedious hours on just the computations. In each edition, we reduced and then finally eliminated explanations of all the special methods for hand or calculator computation. This also made the material in this text easier to learn because these special computational formulas tended to obscure their meaning. In the current edition, we have continued that trend and have included methods that would be impractical without using computers and for which describing the computational details would not help in interpreting the results.

An obvious benefit of the use of computers is that more powerful and complex analyses are now feasible and larger datasets can be easily analyzed. Some of these methods are able to give more exact probability values for standard tests of significance. Other methods (see Chapters 7 and 18) enable us to make statistical tests for nonstandard statistics or for situations in which the usual distributional assumptions are not likely to be true.

3.1 Computers

Since the publication of the first edition of *Biometry*, the revolution in the types of equipment available for performing statistical computations has continued. The once standard mechanical desk calculators used for routine statistical calculations in biology have been discarded or relegated to museums. They were replaced first by a wide variety of electronic calculators (ranging from pocket to desktop models). Calculators range from devices that can only add, subtract, multiply, and divide to scientific calculators with many built-in mathematical and statistical functions. Their principal limitation for use in statistics is that they usually can store only limited amounts of data. This means that data must be reentered if an alternative analysis is to be performed. Reentry is tedious and greatly increases the chance for error. Although some calculators are programmable, they usually use a language unique to the manufacturer of the device. This means that considerable effort may be required to repeat an analysis on a new device. The capabilities of high-end calculators have actually been reduced since the 1970s. At one time, small microcomputers were marketed as "calculators" to seem more user friendly and to avoid institutional rules that restricted the purchase of computers because their widespread use threatened the importance of the large and expensive centralized computer centers (this prediction was indeed borne out).

Computer hardware can be classified into three main categories: the central processing unit (CPU), which performs the calculations and controls the other components; the main memory, usually random access memory (RAM), which stores both data and instructions; and peripheral devices, which handle the input and output of information and the mass storage of data and libraries of programs. Different computers vary considerably in the capabilities of these three components. Currently, standard microcomputers usually have 2×10^9 to 4×10^9 eight-bit numbers or characters (bytes, 10^9 bytes is a Gigabyte) of main memory, 500×10^9 or more bytes of disk storage, a CPU chip (that may contain two to four processing cores that act as separate processors) that can perform about 10^9 average floating-point operations per second (FLOPS, the average number of arithmetic operations per second on numbers with a decimal place). Larger computers range up to supercomputers that now can have several hundred thousand processing cores each with its own main memory (each one may not be very large) with a total processing power of 10^{15} FLOPS (with 10^{18} FLOPS expected within a few years). Such computational power and storage capacity was difficult to imagine when the first edition of this book was published. As in the past editions of this text, we expect that the continued exponential increase in computer performance and memory capacity will render the quantities quoted above obsolete by the time this book is in print and to seem somewhat ridiculous by the time this book is next revised.

The ever-increasing speed of computation is important for many applications and biology. Although computers perform no operations that could not, in principle, be done by hand, tasks that current microcomputers can perform in a fraction of a second might take many thousands of years for humans to complete. In fact, it is now possible to solve problems on supercomputers that one would scarcely even have contemplated before computers were developed.

Although many biologists find that they never use more than a fraction of the storage capacity of even entry-level computers, research in some areas of biology may be limited by available technology. For example, in human microbiome studies, sequence data for both the human host and all of the different microbes inhabiting different parts of the human body at different times need to be stored in order to analyze their possible effect on human health. However, the statistical methods described in this book do not require such high-performance computers. They can be performed quite easily on modern microcomputers.

Generally speaking, the largest and fastest computers are the most efficient and economical for carrying out a specified amount of computation, but other factors such as convenience influence the decision of which computing hardware should actually be used. Availability is, of course, very important. Because of their cost, very-large-scale computers are located only at major universities or laboratories, and their use is shared among hundreds or thousands of users. Often the most important factor is the availability of appropriate software, which is discussed in the next section.

3.2 Software

In the early days of computing, users usually had to develop their own software by expressing the desired computations in a computer language specific to the type of computer being used. If the program was written using just numerical codes for the operations and specific memory locations, then it is called a *first-generation computer language* (1GL).

In second-generation computer languages (2GL), one was able to use mnemonic codes and symbolic memory addresses rather than only numerical codes. These codes were easier to remember, but programming was still a tedious task because the operations had to be specified at a very low level (load a number from memory into a register, add another number to it, store the result back into memory, etc.). The development of the second-generation languages made it practical to develop code for common computations that could be reused in other programs. There was still the limitation that the programs would work on only one type of computer. This promoted loyalty to the existing computer manufacturer because a major effort was often required to switch to computers from a different company. This also limited the ability of researchers to share software unless their colleagues happened to also use the same type of computer.

The next step was the development of general languages for specifying the desired computation that were largely independent of the details of the hardware of any particular type of computer. These are third-generation computer languages (3GL). This was an important step because it meant that the programs written in such a language could be used on any computer for which a program (called a **compiler**) was available that could translate it into the machine language for that computer. This was not perfect because there were still differing results because of differences in the precision of the computations and because there were different "dialects" of the languages on different computer systems. However, it was then possible to write software that was portable across different computers if one was sufficiently careful to use only the most standard features of the language.

Some of the best-known 3GLs used for scientific computations are C, C++, FORTRAN, Java, Python, and Pascal. There are many others. FORTRAN is one of the oldest computer languages but it has evolved and is still very popular for numerical calculations. It is also used for some types of large-scale computations on many supercomputers.

There are also fourth-generation computer languages (4GL) for special areas of application. These usually act as **interpreters**—each instruction is executed rather than translating the entire program into machine language before execution begins. Although the resulting computations are slower than for a compiled program, the speed of modern computers is usually sufficient for many types of scientific applications. These languages are usually designed for specific types of computations, which makes them very convenient because many of the most common operations and data structures required for such applications are built into the language. This also means that these programs contain many fewer commands than a 3GL program and are thus quicker to prepare and simpler to debug. Some popular examples used for statistical computations are Matlab, SAS IML, S+, and R. These include commands for statistical, matrix, and graphical operations. This means that a user can work at a high level to create software appropriate for a certain specialized task. It does, however, still require a researcher to be computer literate and have a good understanding of the tool that is being used. In recent years, R has become especially popular not only because it is free but also because many researchers have contributed the code that they have used for their analyses (including functions that can be used in other analyses).

The alternative is to use specialized user-friendly programs that have already been set up to perform certain analyses. These do not provide the flexibility that one has when one writes one's own software, but they offer the convenience of ready-to-use software and the simplicity of being designed for just a certain range of applications. Most of the methods presented in this book are relatively standard statistical methods that have been included in many statistical packages—too many to attempt to list here. One example is the BIOMstat software that was developed by one of us (FJR) that runs under MS Windows to carry out the computations for most of the topics covered in this book. Version 4 of this program is keyed to this fourth edition of *Biometry*. It is available from Exeter Software at the address given in the Preface of this book (in "Notes on the Fourth Edition"). There are also large comprehensive statistical packages that combine many types of analysis. In addition to selecting methods by using menus, the user can also specify the computations using its own built-in command language (an example of a 4GL) that allows the user to operate the program by either typing commands or submitting a file of commands that can be run in batch mode. The use of batch mode is especially efficient for processing large datasets because one does not have to wait until one operation is completed before the next command is given. This type of software requires more effort to learn to use, but once learned many types of analyses can be performed. Some well-known examples are BMDP, SAS, SPSS, and STATA.

Spreadsheet programs such as MS Excel (and a number of others) do not easily fit in the classification given above. Their very different mode for computation is often useful. They simulate a large sheet of paper that has been divided into cells. Each cell can contain text, a numeric value, or a formula for some computation using the other cells in the spreadsheet. The formulas can produce a numeric value or a graph. Probably the most important feature is that when a user changes a cell, any and all results that depend on the value in that cell are immediately recomputed and displayed. This feature makes it very easy to experiment with data and see the results of alternative analyses. Spreadsheets also provide a convenient environment for setting up arrays of data for input to other software. Computations can be carried out interactively without any programming, or one can create macros for repetitive computations or for computations that one plans to repeat with other data. One can also take advantage of libraries of statistical functions. Perhaps the main limitation of spreadsheets is the difficulty in debugging a complex spreadsheet—especially one that contains macros.

Web-based software is a relatively new development that seems likely to become more popular. Such programs are run by using a browser to visit a Web site. The Web page displays a form in which a user can specify various options and either enter data directly or specify a data file to be loaded. Clicking a button runs the program, which can display the results as another Web page. An advantage of this approach is that it does not depend on the operating system of the user's computer—but there can be compatibility problems with different browsers.

Not all software, however, is of the same quality. There can be different degrees of rounding error, and the software may sometimes produce erroneous results because of bugs in the program. This seems inevitable as software becomes ever more complex. It is a good idea to check a software package by using standard datasets (such as the examples in the boxes in this text) before running one's own data. It is also useful to read reviews of software before investing a lot of effort in converting data files to compatible formats as well as learning how to use the new software.

3.3 Efficiency and Economy in Data Processing

From the foregoing description, it might seem to the reader that a computer would always be preferred over a calculator. This is usually the case, but simpler calculations can often be done more quickly on a calculator. (One may be done with a simple calculation before most computers have finished booting and loading software—although that time is now very short on some computers.) On the other hand, even simple computations often need to be checked or repeated in a slightly different way—for example, different transformations can be tried or possible outliers can be removed. It is inefficient to have to reenter the data more than once. It is also important to have a record of exactly what operations were performed. On a calculator, one does not know for sure if one pressed the correct sequence of buttons—one can only verify that a repetition of the calculations gives the same result.

The difficult decision is not whether to use a computer but which computer and software to use. Software for most of the computations described in this book is available on most computers, whether they use Microsoft Windows, Apple's Mac operating system (OS), Linux, or another OS. With the availability of emulators and virtual machine software, it is even possible to run Windows programs on a Mac, Linux, or other UNIX-based systems. Although it violates Apple's license agreements, it has also been shown that the Mac OS can be run on a Windows computer. Thus, the choice of hardware and OS is usually determined by such factors as cost, the type of computer one is already familiar with, and what support is readily available. The computations described in this book can be performed on standard laptop or desktop computers. They do not require large-scale, high-performance computing systems (unless they are, perhaps, just one step in a large-scale simulation).

Although communication speeds are increasing rapidly, many operations seem more convenient if performed locally. This is one reason why powerful microcomputers and personal workstations are popular. Other reasons for the popularity of small, powerful computers are perhaps more psychological. The user generally feels freer to experiment because the computer is always available and there are no usage charges (nor does one face the formality of applying in advance for an allocation of time on the computer). Because modern microcomputers have sufficient power for most statistical analyses, the most important factor in choosing a system should be the availability of software to perform the desired tasks. An enormous quantity of software for personal computers has been developed in recent years.

One danger inherent in computer processing of data is that the user may simply obtain the numerical results of a statistical test without observing problems in the data (perhaps without even seeing the data if they are collected automatically by various data-acquisition devices). It is important to take advantage of any options available to display the data that could lead to interesting new insights into their nature, to the rejection of some outlying observations, or to suggestions that the data do not conform to the assumptions of a particular statistical test. Most statistical packages are capable of providing such graphics, but they may not be displayed by default. We strongly urge research workers to make use of such options. Another danger is that it is too easy to blindly use only whatever tests are provided with a particular program without understanding their meanings and assumptions (Searle, 1989). The availability of computers relieves the tedium of computation, but not the necessity to understand the methods being employed.

4 Descriptive Statistics

An early and fundamental stage in any science is the descriptive stage. Until the facts can be described accurately, analysis of their causes is premature. The question "What?" must come before "How?" The application of statistics to biology has followed these general trends. Before Francis Galton could begin to think about the relations between the heights of fathers and those of their sons, he had to have adequate tools for measuring and describing heights in a population. Similarly, unless we know something about the usual distribution of the sugar content of blood in a population of guinea pigs, as well as its fluctuations from day to day and within days, we cannot ascertain the effect of a given dose of a drug on this variable.

In a sizable sample, obtaining knowledge of the material by contemplating all the individual observations would be tedious and difficult to communicate to others. We need some form of summary to deal with the data in manageable form, as well as to share our findings with others in scientific talks and publications. A histogram or bar diagram of the frequency distribution is one type of summary. For most purposes, however, a numerical summary is needed to describe the specific properties of the observed frequency distribution concisely and accurately. Quantities providing such a summary are called **descriptive statistics.** This chapter will introduce you to some of them and show how they are computed.

Two kinds of descriptive statistics will be discussed in this chapter: statistics of location and statistics of dispersion. **Statistics of location** describe the position of a sample along a given dimension representing a variable. For example, we might like to know whether the sampled observations measuring the length of certain animals usually lie in the vicinity of 2 cm or 20 cm. A statistic of location must yield a typical or representative value for the sample of observations. However, such a statistic (sometimes also known as a *measure of central tendency*) does not describe the amount of variation in a sample. To this end, we need to define and study **statistics of dispersion.** These also do not describe the shape of a frequency distribution. This distribution may be humped or U-shaped; it may contain two humps, or it may be markedly asymmetrical. Quantitative measures of such aspects of frequency distributions are also required.

The arithmetic mean described in Section 4.1 is undoubtedly the most important single statistic of location, but others (the geometric mean, the harmonic mean, the median, and the mode) are introduced in Sections 4.2, 4.3, and 4.4. Section 4.5 is our first encounter with the distinction between sample statistics and population parameters. A simple statistic of dispersion, the range, is briefly noted in Section 4.6; and in Section 4.7, we describe the standard deviation, the most common statistic for describing dispersion. Section 4.8 contains a description of methods of coding data that are sometimes used to simplify the computations. The coefficient of variation (a statistic that permits us to compare the relative amount of dispersion in different samples) is explained in the last section (4.9).

The techniques that will be at your disposal after you have mastered this chapter are simple, but they are indispensable tools for any further work in biometry.

IMPORTANT NOTE: We will encounter the use of logarithms in this chapter for the first time. To avoid confusion here and in subsequent chapters, common logarithms have been abbreviated consistently as log, and natural logarithms as ln. Thus, $\log x$ means $\log_{10} x$, and $\ln x$ means $\log_e x$.

4.1 The Arithmetic Mean

The most common statistic of location should be familiar. It is the **arithmetic mean,** commonly called the *mean* or *average.* One calculates the mean by summing all the individual observations or items of a sample and dividing this sum by the number of items in the sample. Let us apply this procedure to the 25 aphid femur lengths of Box 2.1. We sum the values in the 5×5 table of observations at the head of Box 2.1, reading off the data row by row. This yields 25 numbers, the first three of which are 3.8, 3.6, and 4.3; the last two are 4.2 and 3.9, all in mm $\times 10^{-1}$. The sum of all 25 observations equals 100.1, and the mean of the sample is $100.1 \div 25 = 4.004$ mm $\times 10^{-1}$.

Calculating a mean presents an opportunity for learning statistical symbolism. We have already seen that an individual observation is symbolized by Y_i, which stands for the ith observation in the sample (see Section 2.2). Twenty-five observations could be written symbolically as follows:

$$Y_1, Y_2, Y_3, \ldots, Y_{24}, Y_{25}$$

We shall define n, the **sample size,** as the number of items in a sample. In this particular instance, the sample size n is 25. In general, for any sample, we can symbolize the array from the first to the nth item as follows:

$$Y_1, Y_2, \ldots, Y_n$$

When we wish to sum items, we use the following notation:

$$\sum_{i=1}^{i=n} Y_i = Y_1 + Y_2 + \cdots + Y_n$$

The capital Greek sigma, Σ, means sum the items indicated. The expression below Σ (the subscript) gives the first item to be included in the sum; the expression above Σ (the superscript) gives the last item. Thus, in this example, $i = 1$ tells us to begin summing with the first item, Y_1; and $i = n$ tells us to end with the nth item, Y_n. The subscript and superscript are necessary to indicate how many items should be summed. The "$i =$" in the superscript is usually omitted as superfluous. For instance, if we had wished to sum only the first three items, we could have written $\sum_{i=1}^{3} Y_i$. On the other hand, had we wished to sum all of them except the first one, we could have written $\sum_{i=2}^{n} Y_i$.

We call Σ an operator symbol. It indicates what to do with the variables that follow it. Summation signs with subscripts and superscripts give an explicit mathematical formulation of the summation operation. However, with some exceptions (which will appear in later chapters), it is convenient, for the sake of simplicity, to omit subscripts and superscripts. Such notation generally adds to the apparent complexity of the formula and, when unnecessary, distracts from the important relations expressed by it. Below are increasing simplifications of the complete summation notation shown at the extreme left:

$$\sum_{i=1}^{i=n} Y_i = \sum_{i=1}^{n} Y_i = \sum_{i} Y_i = \sum^{n} Y = \sum Y$$

The third of the symbols might be interpreted as: Sum the Y_i's over all available values of i. This is a frequently used notation, although we shall not employ it in this book. The next notation, with n as a superscript, tells us to sum n items of Y; note that the i subscript of the Y has been dropped as unnecessary. Finally, the simplest notation is shown at the right. It says simply, sum the Y's. We will use this form most frequently; the summation will be understood to be over n items (all the items in the sample) unless subscripts or superscripts specifically tell us otherwise.

We will use the symbol $\overline{Y}$ for the arithmetic mean of the variable Y. Some textbooks use $\bar{y}$. Many other textbooks use the symbol X for a variable and, consequently, use $\overline{X}$ or $\bar{x}$ for the mean. We have adopted Y, for reasons that will be explained in Chapter 14. Statistical symbolism is, regrettably, still far from uniform, although laudable progress toward uniformity has been made in recent years. From time to time, we shall point out common alternative symbolisms to those presented in the text to prepare you for encounters with the statistical writings of others.

Our newly acquired knowledge of symbols enables us to write the formula for the arithmetic mean:

$$\overline{Y} = \frac{\sum Y}{n} \tag{4.1}$$

This formula tells us to sum all (n) items and divide the sum by n. Examples of actual computations of the arithmetic mean are shown in Box 4.1 for unordered data and in Box 4.2 for data grouped in a frequency distribution.

The mean of a sample represents the center of the observations in the sample. If you were to draw a histogram of a frequency distribution on a sheet of cardboard, cut out the histogram and lay it flat against a blackboard, and support it with a pencil beneath, chances are that it would be out of balance, toppling left or right. If you moved the supporting pencil point to a position about which the histogram would exactly balance, this point of balance would be the arithmetic mean. In fact, this would be an empirical but imprecise method of finding the arithmetic mean of a frequency distribution.

It is often necessary to average means or other statistics that may differ in their reliabilities because they are based on different sample sizes or for other reasons.

BOX 4.1 Calculation of $\overline{Y}$ and s from Unordered Data

Based on 25 aphid femur lengths, unordered, as shown at the head of Box 2.1.

Computation

$n = 25; \Sigma Y = 100.1; \overline{Y} = 4.004$

1. $\Sigma y^2 = \Sigma (Y - \overline{Y})^2 = 3.2096$

2. $s^2 = \dfrac{\Sigma y^2}{n - 1} = \dfrac{3.2096}{24} = 0.1337$

3. $s = \sqrt{0.1337} = 0.3657$

BOX 4.2 Calculation of $\overline{Y}$, s, and V from a Frequency Distribution

Birth weights of 9465 male Chinese in ounces

(1) Class mark Y	(2) f	(3) Coded class mark Y_c
59.5	2	0
67.5	6	1
75.5	39	2
83.5	385	3
91.5	888	4
99.5	1729	5
107.5	2240	6
115.5	2007	7
123.5	1233	8
131.5	641	9
139.5	201	10
147.5	74	11
155.5	14	12
163.5	5	13
171.5	1	14
	9465 = n	

SOURCE: Millis and Seng (1954).

Box 4.2 (continued)

Coding and Decoding

To code:

$$Y_c = \frac{Y - 59.5}{8}$$

To decode $\bar{Y}_c$:

$$\bar{Y} = 8\bar{Y}_c + 59.5$$
$$= 50.4 + 59.5$$
$$= 109.9 \text{ oz}$$

To decode s_c:

$$s = 8s_c = 13.594 \text{ oz}$$

Computation

Without coding	With coding

Without coding

$\Sigma fY = 1,040,199.5$

$\bar{Y} = 109.8996$

$\Sigma fy^2 = \Sigma f(Y - \bar{Y})^2$
$\quad = 1,748,956.7983$

$s^2 = \dfrac{\Sigma fy^2}{n-1} = 184.8010$

$s = 13.5942$

$V = \dfrac{s}{\bar{Y}} \times 100 = \dfrac{13.594}{109.9} \times 100 = 12.370\%$

With coding

$\Sigma fY_c = 59,629$

$\bar{Y}_c = 6.300$

$\Sigma fy_c^2 = \Sigma f(Y_c - \bar{Y}_c)^2$
$\quad = 27,327.450$

$s_c^2 = \dfrac{\Sigma fy_c^2}{n-1} = 2.888$

$s_c = 1.6993$

In such cases, a **weighted average** is used. A general formula for calculating the weighted average of a set of values Y_i is as follows:

$$\bar{Y}_w = \frac{\sum\limits_{}^{n} w_i Y_i}{\sum\limits_{}^{n} w_i} \tag{4.2}$$

where n observations, each weighted by factor w_i, are being averaged. The values of Y_i in such cases are unlikely to represent single observations. They are more likely to be sample means, $\bar{Y}_i$, or other statistics of different reliabilities. Thus, if the following three means are to be averaged:

$\bar{Y}_i$	n_i
3.85	12
5.21	25
4.70	8

their weighted average using sample sizes n_i as weights w_i will be

$$\overline{Y}_w = \frac{(12)(3.85) + (25)(5.21) + (8)(4.70)}{12 + 25 + 8} = 4.76$$

In this example, computation of the weighted mean is exactly equivalent to add-ing up all the original measurements and dividing the sum by the total number of the measurements. Thus, the largest sample, with 25 observations, will influence the weighted average in proportion to the size of the sample. With data such as the percentage pupation at a given site, discussed in the last paragraph of Section 2.4, we might wish to weight each percentage by its sample size. Clearly, a value of 33% is more reliable when based on nine pupae than it is when based on three pupae.

Occasionally, it is desirable to employ an **unweighted average,** which for the three means would be $(3.85 + 5.21 + 4.70)/3 = 4.59$. This type of average would be appropriate if we considered each sample mean equally reliable and wished to represent the entire range of variation of the variable without weighting each mean by the size of its sample.

4.2 Other Means

Two other types of means are used occasionally in biometric work. The **geometric mean** is the nth root of the product of the n observations in a sample:

$$GM_Y = \sqrt[n]{Y_1 Y_2 Y_3 \cdots Y_n} \tag{4.3}$$

This formula can be condensed with the help of the product operator symbol: capital pi, Π. Just as Σ symbolizes *summation* of the items that follow it, so does Π symbol-ize the *multiplication* of the items that follow it. Subscripts and superscripts of the product operator have exactly the same meaning as in the summation case. Thus, Expression (4.3) for the geometric mean can be rewritten more compactly as follows:

$$GM_Y = \sqrt[n]{\prod_{i=1}^{n} Y_i} \tag{4.3a}$$

For any but the very smallest samples, computing the geometric mean by Expression (4.3 or 4.3a) is impractical because the resulting product becomes very big; thus, this computation is carried out by transforming the observations into logarithms.

We will see in Chapters 13 and 14 that variables are sometimes transformed into their logarithms or reciprocals. If we calculate the mean of such a transformed vari-able and then change the mean back into the original scale, this mean will not be the same as if we had computed the arithmetic mean of the original variable.

$$GM_Y = \text{antilog} \frac{1}{n} \sum \log Y \tag{4.4}$$

which indicates that the geometric mean, GM_Y, is the antilogarithm of the mean of the logarithms of variable Y. Recall that adding logarithms is equivalent to multiply-ing their antilogarithms.

The geometric mean is employed for growth rates measured at equal time intervals. It is the average rate of growth per unit of time a structure (or a savings account) must maintain to achieve the observed growth over the total amount of time of a study.

The **harmonic mean** is the reciprocal of the arithmetic mean of reciprocals. If we symbolize it by H_Y, the formula for the harmonic mean can be written in concise form (without subscripts and superscripts) as

$$\frac{1}{H_Y} = \frac{1}{n}\sum\frac{1}{Y} \tag{4.5}$$

The harmonic mean is employed for rates stated as reciprocals, as in x meters per hour. It is the correct average to use when computing the average speed of a process.

You may wish to convince yourself that the geometric mean and the harmonic mean of the 25 aphid femur lengths of Box 2.1, whose arithmetic mean was computed in Section 4.1 and found to be 4.0040 mm $\times 10^{-1}$, are 3.9879 and 3.9716 mm $\times 10^{-1}$, respectively. Unless the individual items do not vary, the geometric mean is always less than the arithmetic mean, and the harmonic mean is always less than the geometric mean.

Some beginners in statistics have difficulty accepting the fact that measures of location or central tendency other than the arithmetic mean are permissible or even desirable. They feel that the arithmetic mean is the "logical" average and that any other mean would distort the data. This attitude raises the question of the proper scale of measurement for representing data; this scale is not always the linear scale familiar to everyone but is sometimes, by preference, a logarithmic or reciprocal scale. If you have doubts about this question, they may be allayed in Chapter 13, where we discuss the reasons for transforming variables.

4.3 The Median

The **median** (M), a statistic of location often useful in biological research, is defined as the value of the variable (in an ordered array) that has an equal number of items on either side of it. Thus, the median divides a frequency distribution into two halves. In the following ordered sample of five measurements,

$$14, 15, 16, 19, 23$$

$M = 16$ because the third observation has an equal number of observations on either side of it. We can visualize the median easily if we think of an array from largest to smallest—for example, a row of men lined up by their heights. The median individual has an equal number of men on his right and left sides. His height will be the median height of the sample. This quantity is evaluated easily from a sample array with an odd number of individuals. The median in such a case is the value of the $[(n/2) + 1]$th observation. When the number in the sample is even, the median is conventionally calculated as the midpoint between two observations—the $(n/2)$th and the $[(n/2) + 1]$th item. Thus, for the sample of four ordered measurements,

$$14, 15, 16, 19$$

the median would be any point between the second and third items. By convention, it is taken as the midpoint, or 15.5. Stem-and-leaf displays (Section 2.5) are helpful in locating the median.

When observations are grouped in classes, we interpolate rather than just take the midpoint of the class containing the median. The aphid femur length data of Box 2.1 are such an example. In such cases, we compute the median of a frequency distribution as the value of the $(n/2)$th item. In the 25 femur lengths, this formula yields the 12.5th item. The measurement value associated with this point halfway between the 12th and the 13th observations is calculated as follows. Below we show once more the frequency distribution of the 25 femur lengths grouped into 15 classes of width 0.1. Following the frequency column f, there is a cumulative frequency column, labeled F, that is produced by successive additions of the frequencies of each class to the sum of the frequencies of the prior classes. The first 1 in column F is simply a copy of the 1 to its left; the second 1 is the sum of 0 and 1; the third frequency 2 is the sum of 1 and 1; the fourth, 6, is the sum of 4 and 2; and so forth. The cumulative frequencies represent the total number of items below the upper class limit of their class. Thus, there are 16 observations below 4.25, the upper limit of the 4.2 class. The final value in the cumulative frequency column is the sum of all the frequencies in the sample.

The value of the 12.5th observation is located in the femur length class of 3.9, somewhere between the implied limits of 3.85 and 3.95. This class contains 3 observations, and item 12.5 is the $12.5 - 10 = 2.5$th item in it. Assuming equal distributions of items in the class, we consider item 12.5 to be $2.5 \div 3 = 0.8333$ of the total class interval, or 83.33% of the distance from the lower class limit to the upper class limit. Because each class interval is 0.1 mm, the median item is $0.8333 \times 0.1 = 0.0833$ above the lower class limit (3.85); that is, the median femur length is located at $3.85 + 0.0833 = 3.9333$ mm $\times 10^{-1}$.

Y	f	F
3.3	1	1
3.4	0	1
3.5	1	2
3.6	4	6
3.7	0	6
3.8	4	10
3.9	3	13
4.0	0	13
4.1	2	15
4.2	1	16
4.3	4	20
4.4	3	23
4.5	1	24
4.6	0	24
4.7	1	25

The median is just one of a family of statistics dividing a frequency distribution into equal proportions. It divides the distribution into two halves. **Quartiles,** on the other hand, cut the distribution at the 25%, 50%, and 75% points—that is, at points dividing the distribution into first, second, third, and fourth quarters of their total frequencies. The second quartile is, of course, the median. (There are also quintiles, deciles, and percentiles, dividing the distribution into 5, 10, and 100 equal portions, respectively.) A general term for such quantities is **quantiles.**

Medians are most often used for distributions that do not conform to the standard probability models and that therefore require nonparametric methods (see Chapter 13). Sometimes the median is considered a more representative measure of location than is the arithmetic mean. Such instances almost always involve asymmetric distributions. An often-quoted example from economics is a suitable measure of location for the "typical" salary of an employee of a corporation. The very high salaries of the few senior executives shift the arithmetic mean toward a completely unrepresentative value. The median, on the other hand, is little affected by a few high salaries.

In biology, an example in which the application of a median is preferred over the arithmetic mean is in populations showing skewed distribution, such as weights. Thus, the median weight of American males 50 years old may be a more meaningful statistic than the average weight. The median is also useful in cases where it may be difficult or impossible to obtain and measure all the items of a sample necessary to calculate a mean. Consider the following example: An animal behaviorist is studying the time it takes for a sample of animals to perform a certain behavioral step. The variable measured is the time from the beginning of the experiment until each individual has performed. What the behaviorist wants to obtain is an average time of performance. Such an average time, however, could be calculated only after records have been obtained on all the individuals. It may take a long time for the slowest animals to complete their performance, longer than the observer wishes to spend looking at them. Moreover, some of them may never respond appropriately, making the computation of a mean impossible. Therefore, a convenient statistic of location to describe these animals may be the median time of performance, or a related statistic, such as the 75th or 90th percentile. Thus, as long as the observer knows the total sample size, measurements for the right-hand tail of the distribution are not needed. Similar examples are the responses to a drug or poison in a group of individuals (the median lethal or effective dose, LD_{50} or ED_{50}) or the median time for a mutation to appear in a number of genetic lines.

4.4 The Mode

The **mode** refers to the most "fashionable" value of the variable in a frequency distribution, or the value represented by the greatest number of individuals. On a frequency distribution, the mode is the value of the variable at which the curve peaks. In grouped frequency distributions, the mode as a point does not have much meaning. Identifying the modal class is usually sufficient. In biology, the mode does not have many applications.

Distributions with two peaks (equal or unequal in height) are called **bimodal;** those with more than two peaks are **multimodal.** In those rare distributions that are U-shaped, we refer to the low point at the middle of the distribution as an **antimode.**

In evaluating the relative merits of the arithmetic mean, the median, and the mode, we must keep a number of considerations in mind. The mean is generally preferred in statistics because it has a smaller standard error than do other statistics of location (see Section 7.2), is easier to work with mathematically, and has an additional desirable property (see Section 7.1): It tends to be distributed normally even if the original data are not. A disadvantage of the mean is that it is markedly affected by outlying observations, whereas the median and mode are not. The mean is generally more sensitive to changes in the shape of a frequency distribution, so if a statistic reflecting such changes is desired, the mean may be preferred.

In unimodal, symmetrical distribution, the mean, the median, and the mode are all identical. A prime example of this is the well-known normal distribution discussed in Chapter 6. In a typical asymmetrical distribution such as the one in Figure 4.1, the

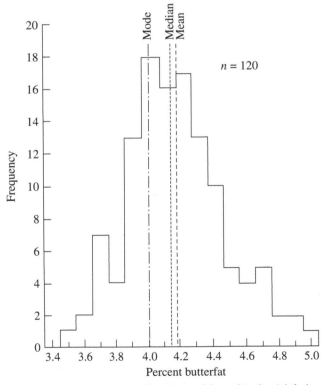

FIGURE 4.1   An asymmetrical frequency distribution (skewed to the right) showing location of the mean, median, and mode. Percent butterfat in 120 samples of milk (from a Canadian cattle breeder's record book).

relative positions of the mode, median, and mean are generally these: The mean is closest to the drawn-out tail of the distribution, the mode is farthest away from the tail, and the median is in between these. An easy way to remember this sequence is to recall that they occur in alphabetical order from the longer tail of the distribution.

4.5 Sample Statistics and Parameters

Up to now, we have calculated statistics from samples without giving too much thought to what these statistics represent. When correctly calculated, any statistic of location, such as a mean, median, or mode, is always an absolutely true measure for the sample on which it is based. Thus, the true mean of the 25 aphid femur lengths in Section 4.1 is 4.004 mm $\times$ 10^{-1} *for this particular sample.* Similar considerations will hold for the measures of dispersion, such as the standard deviation, discussed later in this chapter. However, rarely in biology (or in science in general, for that matter) are we interested in measures of location and dispersion only as descriptive summaries of the samples we have studied. Almost always, we are interested in the *populations* from which the samples were taken. We therefore would like to know, for example, not the mean of the particular 25 aphid femur lengths, but the true femur length of the aphid stem mother population from which the 25 measurements were sampled. When studying dispersion, we generally wish to learn the true standard deviations of the populations, not those of the samples. These population statistics, however, are unknown and (generally speaking) are unknowable. Who would be able to collect all the stem mothers of this particular aphid population and measure them? Thus, we must use **sample statistics** as estimators of *population statistics, or* **parameters.**

It is conventional in statistics to use Greek letters for population parameters and Roman letters for sample statistics. Thus, the sample mean $\overline{Y}$ estimates μ, the parametric mean of the population. Similarly, a sample variance (see Section 4.7), symbolized by s^2, estimates a parametric variance, symbolized by σ^2. Such estimators should be **unbiased.** By this we mean that samples (regardless of the sample size), taken from a population, should give sample statistics which, when averaged, will give the parametric value. An estimator that does not do so is called **biased.** The sample mean $\overline{Y}$ is an unbiased estimator of the parametric mean μ.

4.6 The Range

We now turn to measures of dispersion or spread. Figure 4.2 demonstrates that distributions that look radically different may possess the identical arithmetic mean. Other ways of characterizing distributions must therefore be found.

A simple measure of dispersion is the **range,** the difference between the largest and the smallest items in a sample. The range of the 25 aphid femur lengths from Section 4.1 and Box 4.1 is

$$\text{Range} = 4.7 - 3.3 = 1.4 \, \text{mm} \times 10^{-1}$$

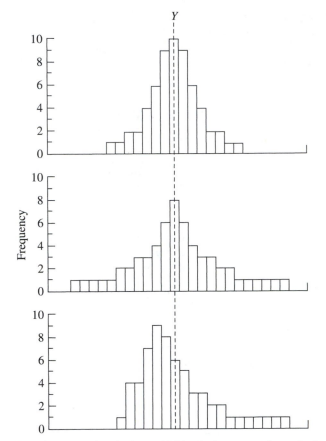

FIGURE 4.2   Three frequency distributions with identical means and sample sizes but different dispersion patterns.

and the range of the birth weights in Box 4.2 is

$$\text{Range} = 171.5 - 59.5 = 112.0 \text{ oz}$$

Because the range is a measure of the span of the observations along the scale of the variable, it is expressed in the same units as the original measurements. The range is clearly affected by even a single outlying value; for this reason, it is only a rough estimate of the dispersion of all the items in the sample. The range is also affected by sample size: The greater the sample, the wider the range of observations, on the average, because more deviant observations are likely to be observed in larger samples.

4.7 The Standard Deviation

A desirable measure of dispersion would take all items of a distribution into consideration, weighting each item by its distance from the center of the distribution. We will now try to construct such a statistic. Table 4.1 is based on the 25 femur lengths of the aphid stem mothers from Box 2.1, arranged as an array. To conserve space, we only show the first four and the last two observations, separating the two tails of the distribution by ellipses. If readers wish to check the sums on the full array, they can easily set up the data on a spreadsheet and repeat the computations. The first column, containing the 25 observations Y_i, is needed for computing the mean of the *sample*, shown beneath the table. The mean femur length is 4.004 mm $\times$ 10^{-1}.

In the second column of Table 4.1, we compute the difference between each observation and the mean as

$$y = Y - \bar{Y}$$

Each individual deviation, or **deviate,** is by convention computed as the individual observation minus the mean, $Y - \bar{Y}$, rather than the reverse, $\bar{Y} - Y$. Deviates are symbolized by lowercase letters corresponding to the capital letters of the variables.

It might seem reasonable to summarize the amount of variation by calculating an average deviation based on the sum of all the deviates divided by the number of

TABLE 4.1 Deviations from the Mean

Femur lengths of aphids (from Box 2.1). Only the first four and the last two observations are shown here.

	Y	$y = Y - \bar{Y}$	$|y|$	y^2
	3.3	−0.704	0.704	0.4956
	3.5	−0.504	0.504	0.2540
	3.6	−0.404	0.404	0.1632
	3.6	−0.404	0.404	0.1632
	⋮	⋮	⋮	⋮
	4.5	0.496	0.496	0.2460
	4.7	0.696	0.696	0.4844
SUM	100.1	≈ 0	7.904	3.2096
$\Sigma \div n$	4.004	≈ 0	0.31616	0.12838

$$\bar{Y} = \frac{\Sigma Y}{n} = \frac{100.1}{25} = 4.004 \qquad AD = \frac{\Sigma |y|}{n} = \frac{7.904}{25} = 0.31616 \qquad \Sigma y^2 = 3.2096$$

$$\text{Variance} = \frac{\Sigma y^2}{n} = \frac{3.2096}{25} = 0.12838 \qquad \text{Standard deviation} = \sqrt{0.12838} = 0.3583$$

$$\text{Bias-corrected Variance} = s^2 = \frac{\Sigma y^2}{n - 1} = \frac{3.2096}{24} = 0.13373$$

$$\text{Bias-corrected Standard deviation} = s = \sqrt{0.13373} = 0.3657$$

deviates in the sample. However, when we sum our deviates, we note that, except for a minuscule quantity caused by rounding error, they add to zero, as shown at the bottom of the second column. This is always true of the sum of deviations from the arithmetic mean. Consequently, an average based on the sum of deviations would also always equal zero. Study Proof A.1 in Appendix A, which demonstrates that the sum of deviations around the mean of a sample is always equal to zero.

What we are really interested in, however, are the magnitudes of the deviations—not their signs. To this end, we compute these differences as the absolute values (also known as values with sign neglected) of the deviates. Such values are positive whether their actual signs are positive or negative. An absolute quantity is symbolized by placement between two vertical lines, as shown at the head of the third column in Table 4.1. The sum of these deviates, 7.904, is shown at the foot of the column. The *mean deviation,* or **average deviation** (abbreviated *AD*), is then computed as shown at the bottom of Table 4.1. Its value is 0.31616 in the original units of measurement, mm $\times$ 10^{-1}. This statistic, once very popular, has now been almost entirely abandoned in favor of the standard deviation, discussed below.

An alternative way of measuring deviations from the mean is in terms of squared distances from the mean. The sums of such squared deviations have desirable statistical properties, which we will discuss later. The fourth column in Table 4.1 lists the squared deviates, all of which are, of course, positive. The sum of these squared deviates is 3.2096. This sum is a very important quantity in statistics, known for short as the **sum of squares** and identified symbolically as Σy^2. Another common symbol for the sum of squares is *SS*.

The next step, shown in the second line under Table 4.1, is to obtain the average of the *n* squared deviations. The resulting quantity is known as the **variance** or the *mean square* (a term that is short for "mean squared deviation from the mean"):

$$\text{Variance} = \frac{\Sigma y^2}{n} = \frac{3.2096}{25} = 0.12838$$

The variance is a measure of fundamental importance in statistics, and we will employ it throughout this book. At the moment, we need only remember that because of the squaring of the deviations, the variance is expressed in squared units. To undo the effect of the squaring, we now take the positive square root of the variance and obtain the **standard deviation:**

$$\text{Standard deviation} = +\sqrt{\frac{\Sigma y^2}{n}} = 0.3583$$

A standard deviation is expressed in the original units of measurement because it is a square root of the squared units of the variance.

The formula you have just learned is not the one generally employed for computing a standard deviation. A correction for bias must still be applied. This is shown in the last two lines of Table 4.1.

The sample variance as computed above will on the average underestimate the magnitude of the population variance σ^2. To overcome this bias, mathematical statisticians have shown that when sums of squares are divided by $n - 1$ rather than by n, the resulting sample variances are unbiased estimators of the population variance. For this reason, variances are usually computed by dividing the sum of squares by $n - 1$.

$$s^2 = \frac{\sum y^2}{n - 1} \tag{4.6}$$

This leads to the following formula for the sample standard deviation:

$$s = +\sqrt{\frac{\sum y^2}{n - 1}} \tag{4.7}$$

In the case of the aphid stem mother data (see Table 4.1), the standard deviation would thus be computed as

$$s = \sqrt{\frac{3.2096}{24}} = 0.3657$$

This value is only slightly larger than our previous estimate of 0.3583. Of course, the greater the sample size, the less difference there will be between division by n and division by $n - 1$. Regardless of sample size, however, it is good practice to divide a sum of squares by $n - 1$ when computing a variance or standard deviation. It should be assumed that the symbol s^2 refers to a variance obtained by division of the sum of squares by the **degrees of freedom,** as the quantity $n - 1$ is generally called. The only time when division of the sum of squares by n is appropriate is when the sample at hand is considered to represent an entire population. In the rare cases in which the investigator possesses data on the entire population, division by n is justified because then the investigator is evaluating rather than estimating a parameter. Thus, for example, the variance of the wing lengths of all adult whooping cranes would be a parametric value; similarly, if the IQs of all winners of the Nobel Prize in physics had been measured, their variance would be a parameter because it is based on the entire population.

Readers who seek greater than usual accuracy when estimating the standard deviation should note that Gurland and Tripathi (1971) have pointed out that Expression (4.7) consistently underestimates σ. They suggest computing $s \times C_n$, where C_n is a correction factor tabled in Statistical Table **II** for sample sizes up to $n = 30$ (for larger sample sizes, a simple approximate formula is furnished). This factor, which is always greater than 1.0, closely approaches 1.0 as sample size n increases much beyond 100 and can be ignored.

Recapitulating, there are three steps necessary for computing the standard deviation: (1) finding $\sum y^2$, the sum of squares; (2) dividing by $n - 1$ to estimate the variance; and (3) taking the square root of the variance to obtain the standard deviation. The procedure used to compute the sum of squares can be expressed by the following formula:

$$\sum y^2 = \sum (Y - \bar{Y})^2 \tag{4.8}$$

When the data are unordered, the computation proceeds as in Box 4.1, which is based on the unordered aphid femur length data shown at the head of Box 2.1.

Occasionally, original data are already in the form of a frequency distribution, or the person computing the statistics may want to avoid manual entry of large numbers of individual observations, in which case setting up a frequency distribution simplifies the input. An example is shown in Box 4.2. Computer programs that compute the basic statistics $\overline{Y}$, s^2, s, and others that we have not yet discussed are furnished in many available programs. The BIOMstat program accepts raw, unordered observations as input, as well as data in the form of a frequency distribution.

An approximate method for estimating statistics is useful when checking the results of calculations because it enables the detection of gross errors in computation. A simple method for estimating the mean is to average the largest and smallest observations to obtain the **midrange.** For the aphid stem mother data of Box 2.1, this value is $(4.7 + 3.3)/2 = 4.0$, which happens to fall almost exactly on the computed sample mean (but, of course, this will not be true of other datasets). Standard deviations can be estimated from ranges by appropriate division of the range:

For samples of . . .	. . . divide the range by
10	3
30	4
100	5
500	6
1000	$6\frac{1}{2}$

The range of the aphid data is 1.4. When this value is divided by 4, we get an estimate of the standard deviation of 0.35, which compares not too badly with the calculated value of 0.3657 in Box 4.1.

A more accurate procedure for estimating the standard deviation is to use Statistical Table **I**, which furnishes the mean range for different sample sizes of a normal distribution (see Chapter 6) with a variance of one. When we divide the range of a sample by a mean range from Statistical Table **I**, we obtain an estimate of the standard deviation of the population from which the sample was taken. Thus, for the aphid data we look up $n = 25$ in Statistical Table **I** and obtain 3.931. We estimate $s = 1.4/3.931 = 0.356$, a value closer to the sample standard deviation than that obtained by the rougher method discussed above (which, however, is based on the same principle and assumptions).

4.8 Coding Data Before Computation

Coding the original data is a far less important subject currently because efficient computational devices are readily available; in earlier days, however, coding was essential for carrying out most computations. By **coding** we mean the addition or subtraction of a constant number to the original data, the multiplication or division of these data by a constant so as to obtain simpler values (fewer digits), or both. Data may need to be coded because they were originally expressed in too many digits or are very large numbers that may cause difficulties and errors during data handling. Coding can therefore simplify computation appreciably, and for certain techniques it can reduce rounding error (Bradley and Srivastava, 1979). The types of coding shown here are linear transformations of the variables. Persons using statistics should know the effects of such transformations on means, standard deviations, and any other statistics they intend to employ.

Additive coding is the addition or subtraction of a constant (because subtraction is addition of a negative number). Similarly, **multiplicative coding** is the multiplication or division by a constant (because division is multiplication by the reciprocal of the divisor). **Combination coding** is the application of both additive and multiplicative coding to the same set of data. In Section A.2 of Appendix A, we examine the consequences of the three types of coding for computing means, variances, and standard deviations.

For the case of means, the formula for combination coding and decoding is the most generally applicable one. If the coded variable is $Y_c = D(Y + C)$, then

$$\overline{Y} = \frac{\overline{Y}_c}{D} - C$$

where C is an additive code and D is a multiplicative code. Additive codes have no effect, however, on the sums of squares, variances, or standard deviations. The mathematical proof is given in Appendix A.2, but this can be seen intuitively because an additive code has no effect on the distance of an item from its mean. For example, the distance from an item of 15 to its mean of 10 would be 5. If we were to code the observations by subtracting a constant of 10, the item would now be 5 and the mean zero, but the difference between them would still be 5. Multiplicative coding, on the other hand, does have an effect on sums of squares, variances, and standard deviations. The standard deviations have to be divided by the multiplicative code, just as had to be done for the mean; the sums of squares or variances have to be divided by the square of the multiplicative codes because they are squared terms, and the multiplicative factor became squared during the operations. In combination coding, the additive code can be ignored.

An example of coding and decoding data is shown in Box 4.2, where we coded each class mark by subtracting 59.5, the lowest class mark of the array. The resulting class marks are the values 0, 8, 16, 24, 32, and so on. Dividing these values by 8 changes them to 0, 1, 2, 3, 4, and so on, which is the desired format, shown in column (3). Details of the computation are in Box 4.2.

4.9 The Coefficient of Variation

Having obtained the standard deviation as a measure of the amount of variation in the data, you may legitimately ask, "What can I do with it?" So far, the only use that

we might have for the standard deviation is as an estimate of the amount of variation in a population. Thus, we may wish to compare the magnitudes of the standard
deviations of similar populations to see whether population A is more or less variable
than population B. When populations differ appreciably in their means, however, the
direct comparison of their variances or standard deviations is less useful because
larger organisms usually vary more than smaller ones. For instance, the standard
deviation of the tail lengths of elephants is obviously much greater than the entire
tail length of a mouse. To compare the relative amounts of variation in populations
having different means, one can use the **coefficient of variation,** symbolized by V
(or occasionally CV). This coefficient is simply the standard deviation expressed as a
percentage of the mean. Its formula is

$$V = \frac{s \times 100}{\overline{Y}} \tag{4.9}$$

For example, the coefficient of variation of the birth weights in Box 4.2 is 12.37%,
as shown at the bottom of that box. The coefficient of variation is independent of the
unit of measurement and is expressed as a percentage. The coefficient of variation
as computed above is a biased estimator of the population V. The following estimate
V^* is corrected for bias:

$$V^* = \left(1 + \frac{1}{4n}\right)V \tag{4.10}$$

In small samples, this correction can make an appreciable difference. Note that when
using Expression (4.9), the standard deviation used to compute V should not be corrected using C_n because this would result in an overcorrection. Note also that the
correction factor approximates C_n, the correction factor used in Section 4.7.

Coefficients of variation are used extensively when comparing the variation of
two populations independently of the magnitude of their means. Whether the birth
weights of the Chinese children (see Box 4.2) are more or less variable than the
femur lengths of the aphid stem mothers (see Box 2.1) is probably of little interest, but we can calculate the latter as $0.3656 \times 100/4.004 = 9.13\%$, which would
suggest that the birth weights are more variable. More commonly, we might wish
to test whether a given biological sample is more variable for one character than
for another. For example, for a sample of rats, is body weight more variable than
blood sugar content? Another frequent type of comparison, especially in systematics, is among different populations for the same character. If, for instance, we had
measured wing length in samples of birds from several localities, we might wish
to know whether any one of these populations is more variable than the others. An
answer to this question can be obtained by examining the coefficients of variation
of wing length in these samples.

Computing a coefficient of variation is meaningful only for variables that can
be measured on a *ratio scale,* which is an interval scale possessing a true zero value,
such as length, height, weight, and temperature expressed in degrees Kelvin, but not
in degrees Fahrenheit or Celsius. Workers should be cautious about indiscriminate

use of coefficients of variation, especially in cases where the ratio scale requirement is not met. Employing the coefficient of variation in a comparison between two variables or two populations assumes that the variable in the second sample is proportional to that in the first. Thus, we could write $Y_1 = kY_2$, where k is a constant of proportionality. If two variables are related in this manner, their coefficients of variation should be identical. This should be obvious if we remember that k is a multiplicative code. Thus, $\overline{Y}_1 = k\overline{Y}_2$ and $s_1 = ks_2$, and consequently

$$V_1 = \frac{100s_1}{\overline{Y}_1} = \frac{100ks_2}{k\overline{Y}_2} = \frac{100s_2}{\overline{Y}_2} = V_2$$

If the variables are transformed to logarithms (see Section 13.7), the relationship between them can be written as $\ln Y_1 = \ln k + \ln Y_2$, and because $\ln k$ is a constant, the variances of $\ln Y_1$ and $\ln Y_2$ are identical. This relation can lead to a test of equality of coefficients of variation (see Lewontin, 1966; Sokal and Braumann, 1980). This relationship is most likely to be satisfied for variables that are lognormally distributed (see Chapter 6).

At one time, systematists put great stock in coefficients of variation and even based some classification decisions on the magnitude of these coefficients. However, there is little, if any, foundation for such actions. More extensive discussion of the coefficient of variation, especially as it relates to systematics, can be found in Simpson et al. (1960), Lewontin (1966), Lande (1977), and Sokal and Braumann (1980).

EXERCISES 4

4.1 Find the mean, standard deviation, and coefficient of variation for the pigeon data given in Exercise 2.4. Group the data into 10 classes, recompute $\overline{Y}$ and s, and compare them with the results obtained from the ungrouped data. Compute the median for the grouped data.

ANSWER: For ungrouped data, $\overline{Y} = 11.48$ and $s = 0.69178$.

4.2 Find $\overline{Y}$, s, V, and the median for the following data (milligrams of glycine per milligram of creatinine in the urine of 37 chimpanzees; from Gartler et al., 1956).

.008	.025	.011	.100
.018	.036	.060	.155
.056	.043	.070	.370
.055	.100	.050	.019
.135	.120	.080	.100
.052	.110	.110	.100
.077	.100	.110	.116
.026	.350	.120	
.440	.100	.133	
.300	.300	.100	

4.3   The following are percentages of butterfat from 120 registered 3-year-old Ayrshire cows selected at random from a Canadian stock record book.

4.32	4.25	4.82	4.17	4.24	4.28	3.91	3.97	4.29	4.03	4.71	4.20
4.00	4.42	3.96	4.51	3.96	4.09	3.66	3.86	4.48	4.15	4.10	4.36
3.89	4.29	4.38	4.18	4.02	4.27	4.16	4.24	3.74	4.38	3.77	4.05
4.42	4.49	4.40	4.05	4.20	4.05	4.06	3.56	3.87	3.97	4.08	3.94
4.10	4.32	3.66	3.89	4.00	4.67	4.70	4.58	4.33	4.11	3.97	3.99
3.81	4.24	3.97	4.17	4.33	5.00	4.20	3.82	4.16	4.60	4.41	3.70
3.88	4.38	4.31	4.33	4.81	3.72	3.70	4.06	4.23	3.99	3.83	3.89
4.67	4.00	4.24	4.07	3.74	4.46	4.30	3.58	3.93	4.88	4.20	4.28
3.89	3.98	4.60	3.86	4.38	4.58	4.14	4.66	3.97	4.22	3.47	3.92
4.91	3.95	4.38	4.12	4.52	4.35	3.91	4.10	4.09	4.09	4.34	4.09

(a)   Calculate $\overline{Y}$, s, and V directly from the data.

(b)   Group the data in a frequency distribution and again calculate $\overline{Y}$, s, and V. Compare the results with those of (a). How have the results been affected by grouping? Also calculate the median.

ANSWER: For ungrouped data, $\overline{Y} = 4.16608$, $s = 0.30238$, $V = 7.25815$.

4.4   What effect would adding the constant 5.2 to all observations have on the numerical values of the following statistics: $\overline{Y}$, s, V, average deviation, median, mode, and range? What would be the effect of adding 5.2 and then multiplying the sums by 8.0? Would it make any difference in the above statistics if we multiplied by 8.0 first and then added 5.2?

4.5   Estimate μ and σ using the midrange and the range (see Section 4.9) for the data in Exercises 4.1, 4.2, and 4.3. How well do these estimates agree with the estimates given by $\overline{Y}$ and s?

ANSWER: Estimates of μ and σ for Exercise 4.2 are 0.224 and 0.1014, respectively.

4.6   Show that the equation for variance can also be written as

$$s^2 = \frac{\sum Y^2 - n\overline{Y}^2}{n - 1} \quad \text{or} \quad \frac{\sum Y^2 - \dfrac{\left(\sum Y\right)^2}{n}}{n - 1}$$

4.7   Apply the C_n correction to the estimated standard deviation of the data in Exercise 2.4. Also compute the unbiased estimate of the coefficient of variation, V. Comment on the importance of these corrections.

ANSWER: $s/C_{40} = 0.69621$, $V* = 6.06365$.

4.8   Calculate the median birth weight in ounces of the sample of 9465 Chinese males presented in Box 4.2. (*Hint:* You can obtain the correct result using either the original or the coded class marks.)

5 Introduction to Probability Distributions: Binomial and Poisson

Section 2.5 was our first discussion of frequency distributions. For example, Table 2.2 shows a distribution for a meristic, or discrete (discontinuous) variable, the number of sedge plants per quadrat. Examples of distributions for continuous variables are the femur lengths of aphid stem mothers in Box 2.1 or the human birth weights in Box 4.2. Each of these distributions informs us about the absolute frequency of any given class and permits computation of the relative frequencies of any class of variable. Thus, most of the quadrats contained either no sedges or just one or two plants. In the 139.5-oz class of birth weights, we find only 201 out of the 9465 babies recorded; that is, approximately only 2.1% of the infants are in that birth-weight class.

Of course, these frequency distributions are only samples from given populations. The birth weights represent a population of male Chinese infants from a given geographical area. If we knew our sample to be representative of that population, however, we could make all sorts of predictions based on the frequency distribution of the sample. For instance, we could say that approximately 2.1% of male Chinese babies born in this population weigh between 135.5 and 143.5 oz at birth. Similarly, we might say that the probability of the weight at birth of any one baby in this population being in the 139.5-oz birth class is quite low. If each of the 9465 weights were mixed up in a hat, and we pulled one out, the probability that we would pull out one of the 201 in the 139.5-oz class would be very low indeed—only 2.1%. It would be much more probable that we would sample an infant of 107.5 or 115.5 oz, because the infants in these classes are represented by the frequencies 2240 and 2007, respectively.

Finally, if we were to sample from an unknown population of babies and find that the very first individual sampled had a birth weight of 170 oz, we would probably reject any hypothesis that the unknown population was the same as that sampled in Box 4.2. We would arrive at this conclusion because, in the distribution in Box 4.2, only one out of almost 10,000 infants had a birth weight that high. Although it is possible to have sampled from the population of male Chinese babies and obtained a birth weight of 170 oz, the probability that the first individual sampled would have such a value is very low indeed. It is much more reasonable to suppose that the unknown population from which we are sampling is different in mean and possibly in variance from the one in Box 4.2.

We have used this empirical frequency distribution to make certain predictions (with what frequency a given event will occur) or make judgments and decisions (whether it is likely that an infant of a given birth weight belongs to this population). In many cases in biology, however, we will make such predictions not from empirical distributions, but on the basis of theoretical considerations that in our judgment are pertinent. We may feel that the data should be distributed in a certain way because of basic assumptions about the nature of the forces acting on the example at hand. If our observed data do not sufficiently conform to the values expected on the basis of these assumptions, we will have serious doubts about our assumptions. This is a common use of frequency distributions in biology. The assumptions being tested generally lead to a theoretical frequency distribution, known also as a **probability distribution.**

A probability distribution may be a simple two-valued distribution such as the 3:1 ratio in a Mendelian cross, or it may be a more complicated function that is intended to predict the number of plants in a quadrat. If we find that the observed data do not fit the expectations on the basis of theory, we are often led to the discovery of some biological mechanism causing this deviation from expectation. The phenomena of linkage in genetics, of preferential mating between different phenotypes in animal behavior, of congregation of animals at certain favored places or, conversely, their territorial dispersion are cases in point. We will thus make use of probability theory to test our assumptions about the laws of occurrence of certain biological phenomena. Probability theory underlies the entire structure of statistics, a fact that, because of the nonmathematical orientation of this book, may not be entirely obvious.

In Section 5.1, we present an elementary discussion of probability, limited to what is necessary for comprehension of the sections that follow. The binomial frequency distribution—which not only is important in certain types of studies, such as genetics, but also is fundamental to an understanding of the kinds of probability distributions discussed in this book—is covered in Section 5.2. The Poisson distribution, which follows in Section 5.3, is widely applicable in biology, especially for tests of randomness of the occurrence of certain events. Both the binomial and Poisson distributions are discrete probability distributions. Some other discrete distributions—the hypergeometric, multinomial, negative binomial, and logarithmic distributions—are mentioned briefly in Section 5.4. The entire chapter therefore deals with discrete probability distributions. The most common continuous probability distribution is the normal frequency distribution, discussed in Chapter 6.

5.1 Probability, Random Sampling, and Hypothesis Testing

We will start this discussion with an example that is not biometrical or biological in the strict sense. We have often found it pedagogically effective to introduce new concepts through situations thoroughly familiar to the student, even if the example is not relevant to the general subject matter of biometry.

Let us imagine ourselves at Matchless University, a state institution somewhere between the Appalachians and the Rockies. Its enrollment figures yield the following breakdown of the student body: 70% of the students are American undergraduates (AU), 26% are American graduate students (AG), and the remaining 4% are from abroad. Of these, 1% are foreign undergraduates (FU) and 3% are foreign graduate students (FG). In much of our work, we use proportions rather than percentages as a useful convention. Thus, the enrollment consists of 0.70 AUs, 0.26 AGs, 0.01 FUs, and 0.03 FGs. The total student body, corresponding to 100%, is represented by the figure 1.0.

If we sample 100 students at random, we expect that, on the average, 3 will be foreign graduate students. The actual outcome might vary. There might not be a single FG student among the 100 sampled or there may be quite a few more than 3, and the estimate of the proportion of FGs may therefore range from 0 to greater than 0.03. If we increase our sample size to 500 or 1000, it is less likely that the ratio will fluctuate widely around 0.03. The larger the sample, the closer to 0.03 the ratio of FG students sampled to total students sampled will be. In fact, the **probability** of sampling a foreign student can be defined as the limit reached by the ratio of foreign students to the total number of students sampled as sample size increases. Thus, we may formally summarize by stating that the probability of a student at Matchless University being a foreign graduate student is $P[FG] = 0.03$. Similarly, the probability of sampling a foreign undergraduate is $P[FU] = 0.01$, that of sampling an American undergraduate is $P[AU] = 0.70$, and that for American graduate students, $P[AG] = 0.26$.

Now imagine the following experiment. We try to sample a student at random from the student body at Matchless University. This task is not as easy as might be imagined. If we wanted to do the operation physically, we would have to set up a collection or trapping station somewhere on campus. And to make certain that the sample is truly random with respect to the entire student population, we would have to know the ecology of students on campus very thoroughly so that we could locate our trap at a station where each student had an equal probability of passing. Few, if any, such places can be found in a university. The student union facilities are likely to be frequented more by independent and foreign students, less so by those living in organized houses and dormitories. Fewer foreign and graduate students might be found along fraternity row. Clearly, we would not wish to place our trap near the International Club or International House because our probability of sampling a foreign student would be greatly enhanced. In front of the bursar's window we might sample students paying tuition. The time of sampling is equally important in both seasonal and diurnal cycles. There seems no easy solution in sight.

Those of you who are interested in sampling organisms from nature will already have perceived parallel problems in your work. If we were to sample only students wearing turbans or saris, their probability of being foreign students would be almost 1. We could no longer speak of a random sample. In the familiar ecosystem of the university, these violations of proper sampling procedure are obvious to all of us, but they are not nearly so obvious in real biological instances that are less well known.

How should we proceed to obtain a random sample of leaves from a tree, of insects from a field, or of mutations in a culture? In sampling at random we are attempting to permit the frequencies of various events occurring in nature to be reproduced unaltered in our records; that is, we hope that on the average the frequencies of these events in our sample will be the same as they are in the natural situation. Another way of stating this goal is that in a random sample we want every individual in the population being sampled to have an equal probability of being included in the sample.

We might go about obtaining a random sample by using records representing the student body, such as the student directory, selecting a page from it at random and a name at random from the page. Or we could assign an arbitrary number to each student, write each on a chip or disk, put these in a large container, stir well, and then pull out a number.

Imagine now that we sample a single student by the best random procedure we can devise. What are the possible outcomes? Clearly, the student could be an AU, AG, FU, or FG. This *set* of four possible outcomes exhausts the possibilities of the experiment. This set, which we can represent as {AU, AG, FU, FG}, is called the **sample space.** Any single experiment would result in only one of the four possible outcomes (elements) in the set. Such an element in a sample space is called a **simple event.** It is distinguished from an **event,** which is any subset of the sample space. Thus, in the sample space defined above, {AU}, {AG}, {FU}, or {FG} are each simple events. The following sampling results are some of the possible events: {AU, AG, FU}, {AU, AG, FG}, {AG, FG}, {AU, FG}. The meaning of these events, in order, implies being an American, an undergraduate, or both; being an American, a graduate student, or both; being a graduate student; being an American undergraduate or a foreign graduate student. By the definition of event, simple events, as well as the entire sample space, are also events.

Given the sampling space described above, the event **A** = {AU, AG} encompasses all possible outcomes in the space yielding an American student. Similarly, the event **B** = {AG, FG} summarizes the possibilities for obtaining a graduate student. The **intersection** of events **A** and **B,** written **A** ∩ **B,** contains only the simple events that occur in both **A** and **B.** Clearly, only AG qualifies, as can be seen here:

$$\mathbf{A} = \{AU, AG\}$$

$$\mathbf{B} = \{AG, FG\}$$

Thus, **A** ∩ **B** is the event in the sample space that gives rise to the sampling of an American graduate student. When the intersection of two events is empty, as in **B** ∩ **C,** where **C** = {AU, FU}, the events **B** and **C** are mutually exclusive. These two events have no common elements in the sampling space.

We may also define events that are **unions** of two other events in the sample space. Thus, **A** ∪ **B** indicates that **A** or **B** or both **A** and **B** occur. As defined above, **A** ∪ **B** describes all students who are Americans, graduate students, or are both (American graduate students).

This discussion of events makes the following relations almost self-evident. Probabilities are necessarily bounded by 0 and 1. Therefore,

$$0 \leq P[\mathbf{A}] \leq 1 \tag{5.1}$$

The probability of an entire sample space is

$$P[\mathbf{S}] = 1 \tag{5.2}$$

where $\mathbf{S}$ is the sample space of all possible events. Also, for any event $\mathbf{A}$, the probability of it not occurring is $1 - P[\mathbf{A}]$. This is known as the *complement rule*:

$$P[\mathbf{A}^C] = 1 - P[\mathbf{A}] \tag{5.3}$$

where $\mathbf{A}^C$ stands for all events that are not $\mathbf{A}$.

Why are we concerned with defining sample spaces and events? Because these concepts lead us to useful definitions and operations regarding the probability of various outcomes. If we can assign a number $0 \le p \le 1$ to each simple event in a sample space such that the sum of these p's over all simple events in the space equals unity, then the space becomes a (finite) **probability space.** In our example, the following numbers were associated with the appropriate simple events in the sample space:

$$\{AU, AG, FU, FG\}$$

$$\{0.70, 0.26, 0.01, 0.03\}$$

Given this probability space, we are now able to make statements regarding the probability of given events. For example, what is the probability of a student sampled at random being an American graduate student? Clearly, it is $P[\{AG\}] = 0.26$. What is the probability that a student is either American or a graduate student? The general formula for the union of the probabilities of events $\mathbf{A}$ and $\mathbf{B}$ is

$$P[\mathbf{A} \cup \mathbf{B}] = P[\mathbf{A}] + P[\mathbf{B}] - P[\mathbf{A} \cap \mathbf{B}] \tag{5.4}$$

In terms of the events defined earlier, we must subtract $P[\mathbf{A} \cap \mathbf{B}] = P[\{AG\}]$ because if we did not do so it would be included twice, once in $P[\mathbf{A}]$ and once in $P[\mathbf{B}]$, and this could lead to the absurd result of a probability greater than 1.

Now let us assume that we have sampled our single student from the student body of Matchless University. He turns out to be a foreign graduate student. What can we conclude? By chance alone, this result has a probability of 0.03, or would happen 3% of the time—that is, not very frequently. The assumption that we have sampled at random should probably be rejected because if we accept the hypothesis of random sampling, then the outcome of the experiment is improbable. Please note that we said *improbable,* not *impossible.* We could have chanced on an FG as the first one to be sampled, but it is not very likely. The probability is 0.97 that a single student sampled would be a non-FG. If we could be certain that our sampling method was random (as when drawing student numbers out of a container), we would of course have to reach the conclusion that an improbable event has occurred. The decisions of this paragraph are all based on our definite knowledge that the proportion of students at Matchless University is as specified by the probability space. If we were uncertain about this, the outcome of our sampling experiment would lead us to assume a higher proportion of foreign graduate students.

Next, we will sample two students rather than just one. What are the possible outcomes of this sampling experiment? The new sample space can best be depicted

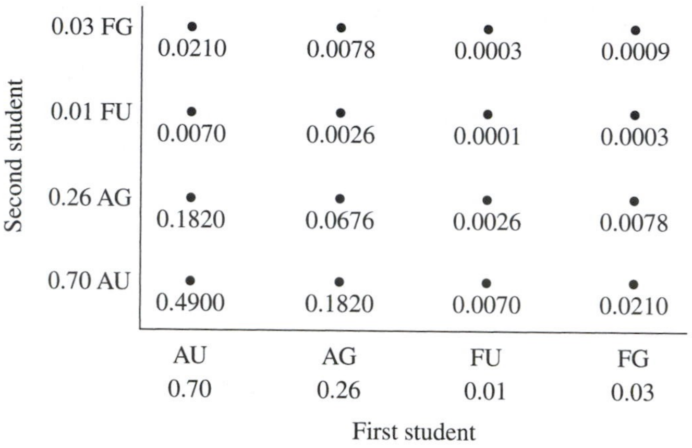

FIGURE 5.1 Sample space for sampling two students from Matchless University.

by a diagram that shows the set of the 16 possible simple events as points in a lattice (Figure 5.1). The possible combinations, ignoring which student was sampled first, are {AU, AU}, {AU, AG}, {AU, FU}, {AU, FG}, {AG, AG}, {AG, FU}, {AG, FG}, {FU, FU}, {FU, FG}, and {FG, FG}.

What would be the expected probabilities of these new outcomes? Now the nature of the sampling procedure becomes quite important. We may sample with or without **replacement;** that is, we may return the first student sampled to the population or may keep him out of the pool of individuals to be sampled. If we do not replace the first individual sampled, the probability of sampling a foreign graduate student will no longer be exactly 0.03. We can visualize this easily. Assume that Matchless University has 10,000 students. Because 3% are foreign graduate students, there must be 300 FG students at the university. After obtaining a foreign graduate student in the first sample, this number is reduced to 299 out of 9999 students. Consequently, the probability of sampling an FG student now becomes 299/9999 = 0.0299, a slightly lower probability than the value of 0.03 for sampling the first FG student. If, on the other hand, we return the original foreign student to the student population and make certain that the population is thoroughly randomized before being sampled again (that is, give him a chance to lose himself among the campus crowd or, in drawing student numbers out of a container, mix up the disks with the numbers on them), the probability of sampling a second FG student is the same as before—0.03. In fact, if we continue to replace the sampled individuals in the original population, we can sample from it as though it were an infinitely sized population.

Biological populations are, of course, finite, but they are frequently so large that for purposes of sampling experiments we can consider them effectively infinite, whether we replace sampled individuals or not. After all, even in this relatively small population of 10,000 students, the probability of sampling a second foreign graduate student (without replacement) is only minutely different from 0.03. For the rest of this section, we will assume that sampling is with replacement, so the probability of obtaining a foreign student will not change.

There is a second potential source of difficulty in this design. We have to assume not only that the probability of sampling a second foreign student is equal to that of the first but also that these events are **independent.** By independence of events, we mean that the probability of one event is not affected by whether or not another event has occurred. In the case of the students, having sampled one foreign student, is it more or less likely that the second student we sampled is also a foreign student? Independence of the events may depend on where we sample the students or on the method of sampling. If we sampled students on campus, it is quite likely that the events are not independent; that is, having sampled one foreign student, the probability that the second student we sample is foreign increases because foreign students tend to congregate. Thus, at Matchless University the probability that a student walking with a foreign graduate student is also an FG would be greater than 0.03.

Events **D** and **E** in a sample space are defined as independent whenever

$$P[D \cap E] = P[D]P[E] \tag{5.5}$$

The probability values assigned to the 16 points in the sample space of Figure 5.1 have been computed to satisfy this condition. Thus, letting $P[D]$ equal the probability that the first student sampled is an AU—that is, $P[\{AU_1AU_2, AU_1AG_2, AU_1FU_2, AU_1FG_2\}]$—and $P[E]$ equal the probability that the second student is an FG—that is, $P[\{AU_1FG_2, AG_1FG_2, FU_1FG_2, FG_1FG_2\}]$—we note that the intersection **D** $\cap$ **E** is $\{AU_1FG_2\}$. This event has a value of 0.0210 in the probability space of Figure 5.1. We find that this value is the product of $P[\{AU\}]P[\{FG\}] = 0.70 \times 0.03 = 0.0210$. These mutually independent relations have been deliberately imposed on all points in the probability space. Therefore, if the sampling probabilities for the second student are independent of the type of student sampled first, we can compute the probabilities of the outcomes simply as the product of the independent probabilities. Thus, the probability of obtaining two FG students is $P[\{FG\}]P[\{FG\}] = 0.03 \times 0.03 = 0.0009$.

The probability of obtaining one AU and one FG student in the sample might seem to be the product 0.70×0.03. However, it is twice that probability. It is easy to see why. There is only one way of obtaining two FG students—namely, by sampling first one FG and then again another FG. Similarly, there is only one way to sample two AU students. However, sampling one of each type of student can be done by first sampling an AU followed by an FG or by first sampling an FG followed by an AU. Thus, the probability is $2P[\{AU\}]P[\{FG\}] = 2 \times 0.70 \times 0.03 = 0.0420$.

If we conducted such an experiment and obtained a sample of two FG students, we would be led to the following conclusions: Because only 0.0009 of the samples (9/100th of 1%, or 9 out of 10,000 cases) would be expected to consist of two foreign graduate students, obtaining such a result by chance alone is quite improbable. Given that $P[\{FG\}] = 0.03$ is true, we would suspect that sampling was not random or that the events were not independent (or that both assumptions—random sampling and independence of events—were incorrect).

Random sampling is sometimes confused with randomness in nature. The former is the faithful representation in the sample of the distribution of the events in nature; the latter is the *independence* of the events in nature. Random sampling generally is or should be under the control of the experimenter and is related to the strategy of good sampling. Randomness in nature generally describes an innate property of the objects being sampled and thus is of greater biological interest. The confusion between random sampling and independence of events arises because lack of either can yield observed frequencies of events differing from expectation. We have already seen how lack of independence in samples of foreign students can be interpreted from both points of view in our example from Matchless University.

Expression (5.5) can be used to test whether two events are independent. It is well known that foreign students at American universities are predominantly graduate students, quite unlike the situation for American students. Can we demonstrate this for Matchless University? The proportion of foreign students at the university $P[F] = P[FU] + P[FG] = 0.04$; for graduate students, it is $P[G] = P[AG] + P[FG] = 0.29$. If the properties of being a foreign student and of being a graduate student were independent, then $P[FG]$ would equal $P[F]P[G]$. In fact, it does not, because 0.04×0.29 is only 0.0116, not 0.03, the actual proportion of foreign graduate students. Thus, nationality and academic rank are not independent. The probability of being a graduate student differs, depending on the nationality: Among American students, the proportion of graduate students is $0.26/0.96 = 0.27$, but among foreign students it is $0.03/0.04 = 0.75$.

This dependence leads us directly to the notion of **conditional probability.** We speak of the conditional probability of event **A**, given event **B**, symbolized as $P[\mathbf{A}|\mathbf{B}]$. This quantity can be computed as follows:

$$P[\mathbf{A}|\mathbf{B}] = \frac{P[\mathbf{A} \cap \mathbf{B}]}{P[\mathbf{B}]} \tag{5.6}$$

Let us apply this formula to our Matchless University example. What is the conditional probability of being a graduate student, given that the student is foreign? We calculate $P[G|F] = P[G \cap F]/P[F] = 0.03/0.04 = 0.75$. Thus, 75% of the foreign students are graduate students. Similarly, the conditional probability of being a graduate student, given that the student is American, is $0.26/0.96 = 0.2708$, a much lower value. Only 27.08% of the American students are graduate students.

Here is another example that requires Expression (5.6) for conditional probability: Let **C** be the event that a person has a specific cancer and $P[\mathbf{C}]$ denote the probability that a member of a particular population has that cancer. Hence, $P[\mathbf{C}^c] = 1 - P[\mathbf{C}]$, the probability of not having the cancer. In epidemiology, $P[\mathbf{C}]$ is known as the **prevalence** of the disease. Let **T** be the event that a certain diagnostic test for the given cancer is positive—that is, that it indicates that a person has the cancer. On receiving a positive test result for a patient, what the physician would like to know is $P[\mathbf{C}|\mathbf{T}]$, the probability that the patient has the cancer, given that the results were positive. Applying Expression (5.6) yields

$$P[\mathbf{C}|\mathbf{T}] = \frac{P[\mathbf{C} \cap \mathbf{T}]}{P[\mathbf{T}]} \tag{5.7}$$

However, the information that one collects to determine the reliability of such tests is not usually in the form required by the above equation. Results are collected from tests on individuals known to have cancer and on those known to be cancer free. Then the proportion of positive tests results among the known cancer cases—$P[T|C]$—is determined. Epidemiologists call this proportion the **sensitivity** of the test. This value ideally is close to 1. The proportion of negative test results for cancer-free persons—$P[T^C|C^C]$—is determined next. This value is known as the **specificity** of the test. Again, this value should be as close to 1 as possible. It follows that $P[T|C^C]$, the probability of a positive result in cancer-free patients equals $1 - P[T^C|C^C]$, the one-complement of the specificity.

From Expression (5.7), by interchanging **C** and **T** we can show that $P[C \cap T] = P[T|C] \, P[C]$. The probability of an event such as **T** can be written as

$$P[T] = P[T|C] \, P[C] + P[T|C^C] \, P[C^C]$$

which is the sum of the probabilities of having a positive test among those who have cancer and among those who do not have cancer—each weighted by the frequencies of the two populations. Substituting these two results into Expression (5.7) yields

$$P[C|T] = \frac{P[T|C]P[C]}{P[T|C]P[C] + P[T|C^C]P[C^C]} \tag{5.8}$$

This expression is known as **Bayes' theorem** and can be generalized to allow for an event **C** having more than just two states (the denominator is summed over all events C_i rather than just **C** and its complement). This famous formula, published posthumously, was derived by the 18th-century English clergyman Thomas Bayes, and has led to much controversy over the interpretation of the quantity $P[C|T]$.

Earlier we defined "probability" as the proportion of times that an event occurs in a large number of trials. In the current example, we have only a single patient, who either does or does not have cancer. The patient does not have cancer some proportion of the time. Thus, the meaning of $P[C|T]$ in this case is the degree of one's belief, or the likelihood that this patient has cancer. This *Bayesian* approach to statistical inference contrasts with the *frequentist* approach, which is the one commonly used in most applications of statistics in the life and social sciences; it is the approach we follow in this text because we wish our readers to be knowledgeable about the statistics they will encounter in the literature and that they will themselves employ for their work. In the frequentist approach, we base our inferences entirely on repeated sampling from the population, or employ a statistical estimate for what the outcome of such a sampling experiment might yield. Nevertheless, there is much to be said for a Bayesian approach because in everyday life, as well as in our scientific work, we condition our interpretation of results by prior knowledge of the distribution of our experimental or observational material. In various advanced applications, Bayesian approaches have been used successfully. Felsenstein (2004) furnishes a detailed discussion of the application of Bayesian methods to phylogeny estimation. For a primer on Bayesian methods, see Berry (1995). Kotz and Stroup (1983) give a good introduction to the idea that probability refers to uncertainty of knowledge rather than of events.

Consider the following example, in which Bayes' theorem was applied to a diagnostic test. The figures are based on Watson and Tang (1980). The sensitivity of the radioimmunoassay for prostatic acid phosphatase (RIA-PAP) as a test for prostate cancer is 0.70. Its specificity is 0.94. The prevalence of prostate cancer in the white male population is 35 per 100,000, or 0.00035. Applying these values to Expression (5.8), we find

$$P[C|T] = \frac{P[T|C]P[C]}{P[T|C]P[C] + P[T|C^C]P[C^C]}$$

$$= \frac{0.70 \times 0.00035}{(0.70 \times 0.00035) + [(1 - 0.94)(1 - 0.00035)]} = 0.0041$$

The rather surprising result is that the likelihood that a white male who tests positive for the RIA-PAP test actually has prostate cancer is only 0.41%. This probability is known in epidemiology as the **positive predictive value.** Even if the test had been much more sensitive—say, 0.95 rather than 0.70—the positive predictive value would have been low: 0.55 percent. Only for a perfect test (i.e., sensitivity and specificity both = 1) would a positive test imply with certainty that the patient had prostate cancer.

The paradoxically low positive predictive value is a consequence of its dependence on the prevalence of the disease. Only if the prevalence of prostate cancer were 7895 per 100,000 would there be a 50:50 chance that a patient with a positive test result has cancer. This is more than 127 times the highest prevalence ever reported from a population in the United States. Watson and Tang (1980) use these findings (erroneously reported as 1440 per 100,000) and further analyses to make the point that using the RIA-PAP test as a routine screening procedure for prostate cancer is not worthwhile.

Readers interested in extending their knowledge of probability should refer to general texts such as Dekking et al. (2005), Denny and Gaines (2002), Galambo (1984), or Kotz and Stroup (1983).

5.2 The Binomial Distribution

In the following discussion, we will simplify our sample space to consist of only two elements, foreign and American students, represented by {F, A}, and ignore whether they are undergraduates or graduates. Let us symbolize the probability space by $\{p, q\}$, where $p = P[F]$, the probability of being a foreign student, and $q = P[A]$, the probability of being an American student. As before, we can compute the probability space of samples of two students as follows:

$$\{FF, FA, AA\}$$

$$\{p^2, 2pq, q^2\}$$

If we were to sample three students independently, the probability space of the sample would be

$$\{FFF, FFA, FAA, AAA\}$$

$$\{p^3, 3p^2q, 3pq^2, q^3\}$$

Samples of three foreign or three American students can be obtained in only one way, and their probabilities are p^3 and q^3, respectively. In samples of three, however, there are three ways of obtaining two students of one kind and one student of the other. As before, if A stands for American and F stands for foreign, then the sampling sequence could be AFF, FAF, and FFA for two foreign students and one American. Thus, the probability of this outcome will be $3p^2q$. Similarly, the probability for two Americans and one foreign student is $3pq^2$.

A convenient way to summarize these results is by the binomial expansion, which is applicable to samples of any size from populations in which objects occur independently in only two classes—students who may be foreign or American, individuals who may be dead or alive, male or female, black or white, rough or smooth, and so forth. This summary is accomplished by expanding the binomial term $(p + q)^k$, where k equals sample size, p equals the probability of occurrence of the first class, and q equals the probability of occurrence of the second class. By definition, $p + q = 1$; hence, q is a function of p: $q = 1 - p$. We will expand the expression for samples of k from 1 to 3:

For samples of 1,

$$(p + q)^1 = p + q$$

For samples of 2,

$$(p + q)^2 = p^2 + 2pq + q^2$$

For samples of 3,

$$(p + q)^3 = p^3 + 3p^2q + 3pq^2 + q^3$$

These expressions yield the same outcomes discussed previously. The coefficients (the numbers before the powers of p and q) express the number of ways a particular outcome is obtained.

A general formula that gives both the powers of p and q, as well as the binomial coefficients, is

$$\binom{k}{Y}p^Yq^{k-Y} = \frac{k!}{Y!(k - Y)!}p^Y(1 - p)^{k-Y} \tag{5.9}$$

In this formula, k, p, and q retain their earlier meaning, while Y stands for the number or count of "successes," the items that interest us and whose probability of occurrence is symbolized by p. In our example, Y designates the number of foreign students. The expression $\binom{k}{Y}$ stands for the number of combinations that can be formed from k items taken Y at a time. This expression can be evaluated as $k!/[Y!(k - Y)!]$, where ! means *factorial*. In mathematics, k factorial is the product of all the integers from 1 up to and including k. Thus: $5! = 1 \times 2 \times 3 \times 4 \times 5 = 120$. By convention,

$0! = 1$. Note that in working out fractions containing factorials, a factorial always cancels against a higher factorial. Thus, $5!/3! = (5 \times 4 \times 3!)/3! = 5 \times 4$. For example, the binomial coefficient for the expected frequency of samples of 5 students containing 2 foreign students is $\binom{5}{2} = 5!/2!3! = (5 \times 4)/2 = 10$.

Now let us turn to a biological example. Suppose we have a population of insects, exactly 40% of which are infected with a given virus X. If we take samples of $k = 5$ insects each and examine each insect separately for the presence of virus, what distribution of samples could we expect if the probability of infection of each insect in a sample was independent from that of other insects in the sample? In this case, $p = 0.4$, the proportion infected, and $q = 0.6$, the proportion not infected. The population is assumed to be so large that the question of whether sampling is with or without replacement is irrelevant for practical purposes. The expected frequencies would be the expansion of the binomial:

$$(p + q)^k = (0.4 + 0.6)^5$$

With the aid of Expression (5.9), this expansion is

$$p^5 + 5p^4q + 10p^3q^2 + 10p^2q^3 + 5pq^4 + q^4$$

or

$$(0.4)^5 + 5(0.4)^4(0.6) + 10(0.4)^3(0.6)^2 + 10(0.4)^2(0.6)^3 + 5(0.4)(0.6)^4 + (0.6)^5$$

representing the expected proportions of samples of five infected insects, four infected and one noninfected insects, three infected and two noninfected insects, and so on.

By now you have probably realized that the terms of the binomial expansion yield a type of frequency distribution for these different outcomes. Associated with each outcome, such as "five infected insects," is a probability of occurrence—in this case, $(0.4)^5 = 0.01024$. This is a theoretical frequency distribution, or **probability distribution,** of events that can occur in two classes. It describes the expected distribution of outcomes in random samples of five insects, 40% of which are infected. The probability distribution described here is known as the **binomial distribution;** the binomial expansion yields the expected frequencies of the classes of the binomial distribution.

A convenient layout for presentation and computation of a binomial distribution is shown in Table 5.1, based on Expression (5.9). In the first column, which lists the number of infected insects per sample, note that we have revised the order of the terms to indicate a progression from $Y = 0$ successes (infected insects) to $Y = k$ successes. The second column features the binomial coefficient as given by the combinatorial portion of Expression (5.9). Column (3) shows increasing powers of p from p^0 to p^5, and column (4) shows decreasing powers of q from q^5 to q^0. The **relative expected frequencies,** which are the probabilities of the various outcomes, are shown in column (5). We label such expected frequencies $\hat{f}_{rel}$. They are the product of columns (2), (3), and (4), and their sum is equal to 1.0 because the events in column

TABLE 5.1 Expected Frequencies of Infected Insects in Samples of Five Insects Sampled from an Infinitely Large Population with an Assumed Infection Rate of 40%

(1) Number of infected insects per sample Y	(2) Binomial coefficients $\binom{k}{Y}$	(3) Powers of $p = 0.4$	(4) Powers of $q = 0.6$	(5) Relative expected frequencies $\hat{f}_{rel}$	(6) Absolute expected frequencies $\hat{f}$	(7) Observed frequencies f
0	1	1.00000	0.07776	0.07776	188.4	202
1	5	0.40000	0.12960	0.25920	628.0	643
2	10	0.16000	0.21600	0.34560	837.4	817
3	10	0.06400	0.36000	0.23040	558.3	535
4	5	0.02560	0.60000	0.07680	186.1	197
5	1	0.01024	1.00000	0.01024	24.8	29
		$\Sigma \hat{f}$ or $\Sigma f (= n)$		1.00000	2423.0	2423
		Mean		2.00000	2.00004	1.98721
		Standard deviation		1.09545	1.09543	1.11934

(1) exhaust the possible outcomes. We see from column (5) that only about 1% of the samples are expected to consist of 5 infected insects, and 25.9% are expected to contain 1 infected and 4 noninfected insects. We will now investigate whether these predictions hold in a sampling experiment.

One can simulate a model for the infected insects by employing a sequence of random numbers generated by computer or from a table of random numbers such as Statistical Table **FF**. These are randomly chosen one-digit numbers in which each digit 0 through 9 has an equal probability of appearing. Such numbers are obtained by pseudorandom number-generating algorithms in computer programs or, alternatively, from (pseudo) random number keys on some pocket calculators. Since there is an equal probability for any particular digit to appear, you can let any four digits (say, 0, 1, 2, 3) stand for the infected insects and the remaining digits (4, 5, 6, 7, 8, 9) stand for the noninfected insects. The probability that any one digit selected from the table represents an infected insect (that is, a 0, 1, 2, or 3) is therefore 0.4, since these are four of the ten possible digits. Also, successive digits are assumed to be independent of the values of previous digits. Thus, the assumptions of the binomial distribution should be met in this experiment. If you use a table of random numbers, enter the table at an arbitrary point (not always at the beginning!) and look at successive groups of five digits, noting in each group how many of the digits are 0, 1, 2, or 3. Take as many groups of five as you can find time to do, but no fewer than 100 groups. Note, however, the existence of some programs that specialize in simulating sampling experiments. With the help of such programs, much larger numbers of samples can be obtained.

Column (7) in Table 5.1 shows the results of such an experiment performed many years ago by a biometry class (it would now be much easier to use computer software). A total of 2423 samples of five numbers were obtained by the class, and the distribution of the four digits simulating the percentage of infection is shown in this column. The observed frequencies are labeled f. To calculate the expected frequencies for this example, we multiplied the relative expected frequencies, $\hat{f}_{rel}$, of column (5) by $n = 2423$, the number of samples taken. These calculations resulted in **absolute expected frequencies, $\hat{f}$**, shown in column (6). When we compare the observed frequencies in column (7) with the expected frequencies in column (6), we note general agreement between the two columns of figures. The two distributions are illustrated in Figure 5.2. If the observed frequencies did not fit expected frequencies, we might believe that the lack of fit resulted from chance alone. Alternatively, we might be led to reject one or more of the following hypotheses: (1) that the true proportion of digits 0, 1, 2, and 3 is 0.4 (rejection of this hypothesis would normally not be reasonable, for we may rely on the fact that the proportion of digits 0, 1, 2, and 3 in a sequence of random numbers *is* 0.4 or very close to it); (2) that sampling was random; and (3) that events are independent.

These statements can be reinterpreted in terms of the original infection model with which we started this discussion. If, this had been a real sampling experiment of insects, instead of a sampling experiment of digits by a biometry class, we would conclude that the insects had indeed been randomly sampled and that we had no evidence to reject the hypothesis that the proportion of infected insects was 40%. If the observed frequencies had not fit the expected frequencies, then the lack of fit might be attributed to chance or to the conclusion that the true proportion of infection is not 0.4, or we would have to reject one or both of the following assumptions: (1) that sampling was at random, and (2) that the occurrence of infected insects in these samples was independent.

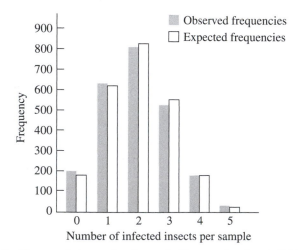

FIGURE 5.2   Bar diagram of observed and expected frequencies given in Table 5.1.

Expected binomial frequencies are readily computed by programs in numerous statistical computer program packages. The BIOMstat program includes such an option. When no such program is available, and samples are of small size, as in the infected insects where $k = 5$, expected binomial frequencies can be conveniently computed with a setup as in Table 5.1. For larger k, an alternative efficient method for calculating expected binomial frequencies is to use the following recursion formula:

$$\hat{f}_i = \hat{f}_{i-1} Q \left(\frac{k - i + 2}{i - 1} \right) \text{ for } i = 2, 3, \ldots k+1 \qquad (5.10)$$

where the subscripts i indicate the row numbers, *not* Y, the number of infected insects in Table 5.1, $Q = p/q$, and $\hat{f}_1 = q^k$. This formula yields the relative expected frequencies. If you desire absolute expected frequencies, multiply $\hat{f}_1$ by n (number of samples in the study). These computations can be easily carried out on a spreadsheet.

The sampling experiment described above was designed to yield random samples and independent events. How could we simulate a sampling procedure in which the occurrences of the digits 0, 1, 2, and 3 were not independent? We could, for example, instruct the sampler to sample as indicated previously, but every time he found a 3, to search through the succeeding digits until he found another one of the four digits standing for infected individuals and to incorporate this in the sample. Thus, once a 3 was found, the probability would be 1.0 that another one of the indicated digits would be included in the sample. After repeated samples, this procedure would result in higher frequencies of classes of two or more indicated digits and in lower frequencies than expected (on the basis of the binomial distribution) of classes of one event. Many such sampling schemes could be devised. It should be clear that the probability of the second event occurring would be different from and dependent on that of the first.

How would we interpret a large departure of the observed frequencies from expected frequencies in another example? We have not yet learned techniques for testing whether observed frequencies differ from those expected by more than can be attributed to chance alone. This topic will be taken up in Chapter 17. Assume that such a test has been carried out and that it has shown that our observed frequencies are more different from the expected frequencies than one would usually anticipate because of chance. Two main types of departure from expectation are likely: (1) **clumping** and (2) **repulsion,** shown in fictitious examples in Table 5.2. In real examples, we would have no a priori notions about the magnitude of p, the probability of one of the two possible outcomes. In such cases, it is customary to obtain p from the observed sample and to calculate the expected frequencies using the sample p. This would mean that the hypothesis that p is a given value cannot be tested because, by design, the expected frequencies will have the same p value as the observed frequencies. Therefore, the hypotheses tested are whether the samples are random and the events independent.

The clumped frequencies in Table 5.2 have an excess of observations at the tails of the frequency distribution and, consequently, a shortage of observations at the center. Such a distribution is also called **contagious.** Remember that the total number of

TABLE 5.2 Artificial Distributions to Illustrate Clumping and Repulsion

Expected frequencies from Table 5.1.

(1) Number of infected insects per sample Y	(2) Absolute expected frequencies $\hat{f}$	(3) Clumped (contagious) frequencies f	(4) Deviation from expectation	(5) Repulsed frequencies f	(6) Deviation from expectation
0	188.4	225	+	143	−
1	628.0	703	+	618	−
2	837.4	663	−	943	+
3	558.3	558	0	548	−
4	186.1	227	+	157	−
5	24.8	47	+	14	−
Σf or n	2423.0	2423		2423	
Mean	2.00004	2.00000		2.00000	
Standard deviation	1.09543	1.20074		1.01435	

items must be the same in both observed and expected frequencies in order to make them comparable. In the repulsed frequency distribution, there are more observations than expected at the center of the distribution and fewer at the tails. These discrepancies are most obvious in columns (4) and (6) of Table 5.2, where the deviations of observed from expected frequencies are shown as plus or minus signs. (These two types of distributions are also called *overdispersed* and *underdispersed*, but there has been some confusion in the literature about the meaning of these terms, so we will not use them here.)

What do these phenomena imply? In the clumped frequencies, more samples were entirely infected (or largely infected) and, similarly, more samples were entirely noninfected (or largely noninfected) than you would expect if probabilities of infection were independent. This result could be the result of poor sampling design. If, for example, the investigator, in collecting samples of five insects, always tended to pick out like ones—that is, infected ones or noninfected ones—then such a result would likely appear. If the sampling design is sound, however, the results become more interesting. Clumping would then mean that the samples of five are in some way related—that is, if one insect is infected, others in the same sample are more likely to be infected. This relation could be true if the insects came from adjacent locations in a situation in which neighbors are easily infected. Or the insects could be siblings exposed simultaneously to a source of infection. Or the infection could spread among members of a sample between the time the insects are sampled and the time they are examined.

The opposite phenomenon, repulsion, is more difficult to interpret biologically. There are fewer homogeneous groups and more mixed groups in such a distribution, which implies a compensatory phenomenon: If some of the insects in a sample are infected, the others in the sample are less likely to be. If the infected insects in the sample could transmit immunity to their associates in the sample, such a situation could arise logically, but it is biologically improbable. A more reasonable interpretation of such a finding is that for each sampling unit, a limited number of pathogens are available and that once several insects have become infected, the others go free of infection simply because there are no further infectious agents. This situation is unlikely in microbial infections, but in situations in which a limited number of parasites enter the body of the host, repulsion is more reasonable.

From the expected and observed frequencies in Table 5.1, we may calculate the mean and standard deviation of the number of infected insects per sample. These values are given at the bottom of columns (5), (6), and (7) in Table 5.1. We note that the means and standard deviations in columns (5) and (6) are almost identical and differ only trivially because of rounding errors. Column (7), however, differs, being based on 2423 random samples of 5 insects from a population whose parameters are the same as those of the expected frequency distribution in columns (5) or (6). The mean is slightly smaller, and the standard deviation is slightly greater than in the expected frequencies. If we wish to know the mean and standard deviation of expected binomial frequency distributions, we need not go through the computations shown in Table 5.1. The mean and standard deviation of a binomial frequency distribution are, respectively,

$$\mu = kp \qquad \sigma = kpq$$

Substituting the values $k = 5$, $p = 0.4$, and $q = 0.6$ from the preceding example, we obtain $\mu = 2.0$ and $\sigma = 1.09545$, which are identical to the values computed from column (5) in Table 5.1. Note that we use the Greek parametric notation here because μ and σ are parameters of an expected frequency distribution, not sample statistics, as are the mean and standard deviation in column (7). The proportions p and q are parametric values also and, strictly speaking, should be distinguished from sample proportions. In fact, in later chapters we resort to π and $1 - \pi$ for parametric proportions. Note that this standard notation is possibly confusing, but from the context it should be clear that we do not mean the constant π, the ratio of the circumference to the diameter of a circle. To reduce confusion, we will usually subscript π when we mean the parametric proportion. Here, however, we prefer to keep our notation simple and just use p and q.

It is interesting to look at the standard deviations of the clumped and repulsed frequency distributions of Table 5.2. We note that the clumped distribution has a standard deviation greater than expected, and that of the repulsed one is less than expected. Comparison of sample standard deviations with their expected values is a useful measure of dispersion in such instances. If we wish to express our variable as a proportion rather than as a count—that is, to indicate mean incidence of infection in the insects as 0.4, rather than as 2 per sample of 5—we can use other formulas for the mean and standard deviation in a binomial distribution:

$$\mu = p \qquad \sigma = \sqrt{pq/k}$$

We will now use the binomial distribution to solve a biological problem. On the basis of our knowledge of the cytology and biology of species A, we expect the sex ratio among its offspring to be 1:1. The study of a litter in nature reveals that of 17 offspring, 3 were males and 14 were females. What conclusions can we draw from this evidence? Assuming that p_δ (the probability of being a male offspring) = 0.5 and that this probability is independent among the members of the sample, the pertinent probability distribution is the binomial for sample size $k = 17$. Expanding the binomial to the power 17 is a nontrivial task, which, as we shall see, fortunately need not be done in its entirety.

The setup of this example is shown in Table 5.3. For the purposes of our problem, we need not proceed beyond the term for 4 males and 13 females. Calculating the relative expected frequencies in column (3), we note that the probability of 3 males and 14 females is 0.005,188,40, a very small value. If we add to this value all "worse" outcomes—that is, all outcomes that are even more unlikely than 14 females and 3 males on the assumption of a 1:1 hypothesis—we obtain a probability of 0.006,363,42, still a very small value. In statistics, one often needs to calculate the probability of observing a deviation as large or larger than a given value.

On the basis of these findings, one or more of the following assumptions is unlikely: (1) that the true sex ratio in species A is 1:1, (2) that we have sampled at random in the sense of obtaining an unbiased sample, or (3) that the sexes of the offspring are independent of one another. Lack of independence of events may mean that, although the average sex ratio is 1:1, the individual sibships, or litters, are largely unisexual—that is, the offspring from a given mating tend to be all (or largely) males or all (or largely) females. To confirm this hypothesis, we would need

TABLE 5.3 Some Expected Frequencies of Males and Females for Samples of 17 Offspring on the Assumption That the Sex Ratio Is 1:1

$$[p_\delta = 0.5, p_\female = 0.5, (p_\delta + p_\female)^k = (0.5 + 0.5)^{17}]$$

(1)	(2)	(3)
		Relative
		expected
Y	**k − Y**	**frequencies**
♂♂	♀♀	$\hat{f}_{rel}$
0	17	0.000,007,63 �️
1	16	0.000,129,71
2	15	0.001,037,68 ⎬ 0.006,363,42
3	14	0.005,188,40 ⎦
4	13	0.018,157,91

to have more samples and then examine the distribution of samples for clumping, which would indicate a tendency for unisexual sibships.

We must be very precise about the questions we ask of our data. There are really two questions we can ask about the sex ratio: (1) Are the sexes unequal in frequency so that females appear more often than males? (2) Are the sexes unequal in frequency? We may be concerned with only the first of these questions because we know from past experience that the males are never more frequent than females in this particular group of organisms. In such a case, the reasoning followed above is appropriate. However, if we know very little about this group of organisms and if our question is simply whether the sexes among the offspring are unequal in frequency, then we have to consider both tails of the binomial frequency distribution; departures from the 1:1 ratio could occur in either direction. We should then consider not only the probabilities of samples with 3 males and 14 females (and all worse cases) but also the probability of samples of 14 males and 3 females (and all worse cases in that direction). Because this probability distribution is symmetrical (because $p_{\delta} = q_{\varphi} = 0.5$), we simply double the cumulative probability of 0.006,363,42 obtained previously, which results in 0.012,726,84. This new value is still very small, making it quite unlikely that the true sex ratio is 1:1.

This is your first experience with one of the most important applications of statistics—hypothesis testing. A formal introduction to this field will be deferred until Section 7.8. We simply point out here that the two approaches just described are known as **one-tailed tests** and **two-tailed tests,** respectively. Students sometimes have difficulty knowing which of the two tests to apply. In future examples, we will try to explain why a one-tailed or a two-tailed test is being used.

We have said that a tendency for unisexual sibships would result in a clumped distribution of observed frequencies. An actual case of this phenomenon in nature is a classic in the literature, the sex-ratio data obtained by Geissler (1889) from hospital records in Saxony. Table 5.4 shows the sex ratios of 6115 sibships of 12 children, each from the more extensive study by Geissler. All columns of the table should by now be familiar. To keep you on your toes and to conform to the layout of the original publication, the meaning of p and q have been reversed from that in the earlier sex-ratio example. Now p_{φ} is the proportion of females and q_{δ} that of males. In a binomial, which of the two outcomes is p and which is q is simply a matter of convenience.

The expected frequencies in this example were not calculated on the basis of a 1:1 hypothesis because it is known that the sex ratio at birth is not 1:1 in human populations. Because the sex ratio varies in different human populations, the best estimate of it for the population in Saxony was obtained simply by using the mean proportion of males in these data. This value can be obtained by calculating the average number of females per sibship ($\bar{Y} = 5.76942$) for the 6115 sibships and converting this into a proportion. This value is 0.480,785. Consequently, the proportion of males = 0.519,215. In the deviations of the observed frequencies from the absolute expected frequencies shown in column (6) of Table 5.4, we notice considerable clumping. There are many more instances of families with all female or all male children (or nearly so) than independent probabilities would indicate. The genetic basis for this

TABLE 5.4 Sex Ratios in 6115 Sibships of 12 in Saxony

(1)	(2)	(3) Relative expected frequencies	(4) Absolute expected frequencies	(5) Observed frequencies	(6) Deviation from expectation
Y ♀♀	$k - Y$ ♂♂	$\hat{f}_{rel}$	$\hat{f}$	f	$f - \hat{f}$
0	12	0.000,384	2.3	7	+
1	11	0.004,264	26.1	45	+
2	10	0.021,725	132.8	181	+
3	9	0.067,041	410.0	478	+
4	8	0.139,703	854.3	829	−
5	7	0.206,973	1265.6	1112	−
6	6	0.223,590	1367.3	1343	−
7	5	0.177,459	1085.2	1033	−
8	4	0.102,708	628.1	670	+
9	3	0.042,280	258.5	286	+
10	2	0.011,743	71.8	104	+
11	1	0.001,975	12.1	24	+
12	0	0.000,153	0.9	3	+
Total		0.999,998	6115.0	6115	

$$\bar{Y} = 5.76942 \qquad s^2 = 3.48985$$

is not clear, but it is evident that there are some families that "run to girls" and similarly others that "run to boys." Other evidence of clumping is the fact that s^2 is much larger than we would expect on the basis of the binomial distribution [$\sigma^2 = kpq = 12(0.480785)0.519215 = 2.99557$].

There is a distinct contrast between the data in Table 5.1 and those in Table 5.4. In the insect infection data of Table 5.1, we had a hypothetical proportion of infection based on outside knowledge. In the sex-ratio data of Table 5.4, we had no such knowledge; we used an *empirical value of* p *obtained from the data,* rather than a *hypothetical value external to the data.* The importance of this distinction will become apparent later. In the sex-ratio data of Table 5.3, as in much work in Mendelian genetics, a hypothetical value of p is used.

5.3 The Poisson Distribution

In the typical application of the binomial, we had relatively small samples (2 students, 5 insects, 17 offspring, 12 siblings), in which two alternative states occurred at varying frequencies (American and foreign, infected and noninfected, male and female). Quite frequently, however, we study cases in which sample size k is very large, and one of the events (represented by probability q) is much more frequent than the other (represented by p). Suppose you had to expand the expression $(0.001 + 0.999)^{1000}$. We have seen that the expansion of the binomial $(p + q)^k$ is quite tiresome when k is large. In such cases, we are generally interested in one tail of the distribution only. This is the tail represented by the terms

$$p^0 q^k, \quad \binom{k}{1} p^1 q^{k-1}, \quad \binom{k}{2} p^2 q^{k-2}, \quad \binom{k}{3} p^3 q^{k-3}, \dots$$

The first term represents no rare events and k frequent events in a sample of k events; the second term represents 1 rare event and $k-1$ frequent events; the third term, 2 rare events and $k-2$ frequent events, and so forth. The expressions of the form $\binom{k}{i}$ are the binomial coefficients, discussed in the previous section. Although the desired tail of the curve could be computed by the methods of Section 5.2, as long as sufficient decimal accuracy is maintained, it is far easier in such cases to compute another distribution, the Poisson distribution, which closely approximates the desired results. As a rough guide, the Poisson may be used to approximate the binomial when the probability of the rare event $p < 0.1$ and the product of sample size and probability $kp < 5$.

The **Poisson distribution** is also a discrete frequency distribution of the number of times a rare event occurs. In contrast to the binomial distribution, however, the number of times that an event does not occur is infinitely large. For our purposes, a Poisson variable will be studied in either a spatial or temporal sample. Examples of the former are the number of moss plants in a sampling quadrat on a hillside or the number of parasites on an individual host; examples of the latter are the number of mutations occurring in a genetic strain in the time interval of one month or the reported cases of influenza in one town during one week.

The Poisson variable, Y, will be the number of (rare) events per sample. It can assume discrete (integer) values from 0 on up. To be distributed in Poisson fashion, the variable must have two properties:

1. Its mean must be small relative to the maximum possible number of events per sampling unit. Thus, the event should be "rare." This requires that the sampling unit should be of sufficient size. For example, a quadrat in which moss plants are counted must be large enough that a substantial number of moss plants could occur there physically if the biological conditions favored the development of moss plants in the quadrat. A quadrat of 1 cm^2 would be far too small for mosses to be distributed in Poisson fashion. Similarly, a time span of 1 minute would be unrealistic for reporting new influenza cases in a town, but within 1 week a great many such cases could occur.
2. An occurrence of the event must be independent of prior occurrences within the sampling unit. Thus, the presence of one moss plant in a quadrat must not enhance or diminish the probability of other moss plants developing in the quadrat. Similarly, the fact that one influenza case has been reported must not affect the probability of subsequent influenza cases being reported. Events that meet these conditions (rare and random events) should be distributed in Poisson fashion.

The purpose of fitting a Poisson distribution to numbers of rare events in nature is to test whether the events occur independently of each other. If they do, they will follow the Poisson distribution. If the occurrence of one event enhances the probability of a second such event, we obtain a clumped or contagious distribution. If the occurrence of one event impedes that of a second such event in the sampling unit, we obtain a repulsed, spatially or temporally uniform distribution. The Poisson distribution can be used as a test for randomness or independence of distribution not only spatially but also temporally, as the examples that follow show.

This distribution, named after the French mathematician Siméon-Denis Poisson, who described it in 1837, is an infinite series whose terms add to one (as must be true for any probability distribution). The series can be represented as

$$\frac{1}{e^\mu}, \quad \frac{\mu}{1!e^\mu}, \quad \frac{\mu^2}{2!e^\mu}, \quad \frac{\mu^3}{3!e^\mu}, \quad \frac{\mu^4}{4!e^\mu}, \quad \cdots, \quad \frac{\mu^r}{r!e^\mu}, \quad \cdots \qquad (5.11)$$

which are the relative expected frequencies corresponding to the following counts of the rare event Y:

$$0, 1, 2, 3, 4, \ldots, r, \ldots$$

Thus, the first term represents the relative expected frequency of samples containing no rare event; the second term, one rare event; the third term, two rare events; and so on. The denominator of each term contains e^μ, where e is the base of the natural, or Napierian, logarithm, a constant whose value, accurate to five decimal places, is 2.71828. We recognize μ as the parametric mean of the distribution; it is a constant for any given problem. The exclamation mark after the coefficient in the denominator means "factorial," as explained in the previous section.

One way to learn more about the Poisson distribution is to apply it to an actual case. Table 5.5 is a well-known result from the early statistical literature that tests the distribution of yeast cells in 400 squares of a hemacytometer, a counting chamber used in counting blood cells and other microscopic structures suspended in liquid. Column (1) lists the number of yeast cells observed in each hemacytometer square, and column (2) gives the observed frequency—the number of squares containing a given number of yeast cells. Note that 75 squares contained no yeast cells but that most squares held either 1 or 2 cells. Only 17 squares contained 5 or more yeast cells.

Why would we expect this frequency distribution to be distributed in Poisson fashion? We have here a relatively rare event, the frequency of yeast cells per hemacytometer square, the mean of which has been determined as 1.8; that is, on the average there are 1.8 cells per square. Relative to the amount of space in each square and the number of cells that could have come to rest in any one square, the number found is low indeed. We might also expect that the occurrence of individual yeast cells in a square is independent of the occurrence of other yeast cells. This is a commonly encountered class of results tested against a Poisson distribution.

The mean of the rare event is the only quantity that we need to know to calculate the relative expected frequencies of a Poisson distribution. Because we do not know the parametric mean of the yeast cells in this problem, we employ an estimate (the sample mean) and calculate expected frequencies of a Poisson distribution whose μ equals the sample mean of the observed frequency distribution of Table 5.5. It is

TABLE 5.5 Yeast Cells in 400 Squares of a Hemacytometer

$\overline{Y}$ = 1.8 cells per square; n = 400 squares sampled.

(1) Number of cells per square Y	(2) Observed frequencies f	(3) Absolute expected frequencies $\hat{f}$	(4) Deviation from expectation $f - \hat{f}$
0	75	66.1	+
1	103	119.0	−
2	121	107.1	+
3	54	64.3	−
4	30	28.9	+
5	13 ⎫	10.4 ⎫	+ ⎫
6	2 ⎪	3.1 ⎪	− ⎪
7	1 ⎬ 17	0.8 ⎬ 14.5	+ ⎬ +
8	0 ⎪	0.2 ⎪	− ⎪
9+	1 ⎭	0.0 ⎭	+ ⎭
	400	399.9	

convenient for the purpose of computation to rewrite Expression (5.11) as the following recursion formula:

$$\hat{f}_i = \hat{f}_{i-1}\left(\frac{\overline{Y}}{i}\right) \qquad \text{for } i = 1, 2, \ldots, \text{ where } \hat{f}_0 = e^{-\overline{Y}} \qquad (5.12)$$

Note that the parametric mean μ has been replaced by the sample mean $\overline{Y}$. Each term developed by this recursion formula is mathematically exactly the same as its corresponding term in Expression (5.11). It is important to make no computational error because, in such a chain multiplication, the correctness of each term depends on the accuracy of the term before it. Expression (5.12) yields relative expected frequencies. If, as is more usual, absolute expected frequencies are desired, simply set the first term $\hat{f}_0$ to $n/e^{\overline{Y}}$, where n is the number of samples, and then proceed with the computation as before. We list the expected frequencies so obtained in column (3) of Table 5.5.

What have we learned from this computation? When we compare the observed frequencies with the expected frequencies, we notice quite a good fit of our observed frequencies to a Poisson distribution of mean 1.8. A statistical test for goodness of fit will be given in Chapter 17. No clear pattern of deviations from expectation is shown. We cannot test a hypothesis about the mean because the mean of the expected distribution was taken from the sample mean of the observed variates. As in the binomial distribution, clumping or aggregation would mean that the probability that a second yeast cell will be found in a square is not independent of the presence of the first one:

It is higher. This dependence would result in a clumping of the items in the classes at the tails of the distribution, so there would be some squares with many yeast cells.

The biological interpretation of the dispersion pattern varies with the problem. In this example, the yeast cells seem to be randomly distributed in the counting chamber, indicating thorough mixing of the suspension. Unless the proper suspension fluid is used, however, red blood cells often will stick together because of an electrical charge. This so-called rouleaux effect would be indicated by clumping of the observed frequencies.

Note that in Table 5.5, as in the subsequent tables giving examples of the application of the Poisson distribution, we group the low frequencies at one tail of the curve, uniting them by means of a bracket. This grouping tends to simplify the patterns of distribution, but the main reason for it is related to the tests for goodness of fit (of observed to expected frequencies), which are discussed in Section 17.2. For these tests, no expected frequency $\hat{f}$ should be less than 5.

Before we turn to other examples, we need to learn a few more facts about the Poisson distribution. You probably noticed that in computing expected frequencies we needed to know only one parameter—the mean of the distribution. By comparison, in the binomial we needed two parameters, p and k. Thus, the mean completely defines the shape of a given Poisson distribution. It follows, then, that the variance is a function of the mean. In a Poisson distribution, we have a simple relationship between the two: $\mu = \sigma^2$, the variance being equal to the mean. The variance of the number of yeast cells per square based on the observed frequencies in Table 5.5 equals 1.965, not much larger than the mean of 1.80, indicating again that the yeast cells are distributed approximately in Poisson fashion—that is, randomly. This relationship between variance and mean suggests a rapid test of whether an observed frequency distribution is distributed in Poisson fashion even without fitting expected frequencies to the data. We simply compute a **coefficient of dispersion:**

$$CD = \frac{s^2}{\bar{Y}}$$

This value will be near 1 in distributions that are essentially Poisson, >1 in clumped samples, and <1 in cases of repulsion. In the yeast cell example, $CD = 1.092$. Expected Poisson frequencies are calculated by programs in numerous statistical computer program packages. Program BIOMstat includes such an option. Lacking such programs, it is easy to arrange the flow of the recursive Expression (5.12) on a spreadsheet.

Figure 5.3 will give you an idea of the shapes of the Poisson distributions of different means. It shows frequency polygons (the lines connecting the midpoints of a bar diagram) for five Poisson distributions. For the low value of $\mu = 0.1$, the frequency polygon is extremely ∟-shaped, but with an increase in the value of μ, the distributions become humped and eventually nearly symmetrical.

We conclude our study of the Poisson distribution by considering several examples from diverse fields. The first example, which is analogous to that of the yeast cells, is from an ecological study of mosses of the species *Hypnum schreberi*

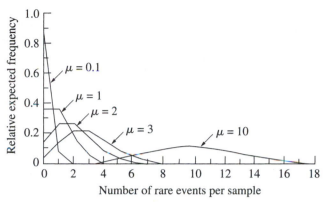

FIGURE 5.3 Frequency polygons of the Poisson distributions for various values of the mean.

invading mica residues of china clay (Table 5.6). These residues occur on deposited "dams" (often 5000 yd² in area), on which the ecologists laid out 126 quadrats. In each quadrat, they counted the number of moss shoots. Expected frequencies are again computed, using the mean number of moss shoots, $\overline{Y}$ = 0.4841, as an estimate of μ. There are many more quadrats than expected at the two tails of the distribution than at its center. Thus, although we would expect only approximately 78 quadrats without a moss plant, we find 100. Similarly, although there are 11 quadrats containing 3 or more moss shoots, the Poisson distribution predicted only 1.7 quadrats. By way of compensation, the central classes are shy of expectation. Instead of the

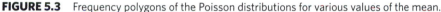

TABLE 5.6	Number of Moss Shoots (*Hypnum schreberi*) Per Quadrat on China Clay Residues (Mica)		
(1) **Number of moss shoots per quadrat** Y	**(2)** **Observed frequencies** f	**(3)** **Absolute expected frequencies** $\hat{f}$	**(4)** **Deviation from expectation** $f - \hat{f}$
0	100	77.7	+
1	9	37.6	−
2	6 ⎫	9.1 ⎫	− ⎫
3	8 ⎪	1.5 ⎪	+ ⎪
4	1 ⎬ 17	0.2 ⎬ 10.8	+ ⎬ +
5	0 ⎪	0.0 ⎪	0 ⎪
6+	2 ⎭	0.0 ⎭	+ ⎭
Total	126	126.1	

$$\overline{Y} = 0.4841 \qquad s^2 = 1.308 \qquad CD = 2.702$$

SOURCE: Data from Barnes and Stanbury (1951).

nearly 38 expected quadrats with one moss plant each, only 9 were found, and there is a slight deficiency also in the 2-mosses-per-quadrat class. This case is another illustration of clumping, which was encountered first in the binomial distribution. The sample variance $s^2 = 1.308$, much larger than $\overline{Y} = 0.4841$, yields a coefficient of dispersion $CD = 2.702$.

Searching for a biological explanation of the clumping, the investigators found that the protonemata, or spores, of the moss were carried in by water and deposited at random but that each protonema gave rise to a number of upright shoots, so counts of the latter indicated a clumped distribution. In fact, when the clumps instead of individual shoots were used as variates, the investigators found that the clumps followed a Poisson distribution—that is, they were randomly distributed. There are problems in directly applying this approach because the size of the sampling units can influence the results. This would happen, for example, if the plants produced substances that prevented other plants from growing very close to them (repulsed distribution), but on a larger scale the plants were clumped in favorable regions (clumped distribution). Organisms moving under their own power, when distributed in clumped fashion, may have responded to various social instincts or may have accumulated in clumps as a result of, or in response to, environmental forces.

The second example tests the randomness of the distribution of weed seeds in samples of grass seed (Table 5.7). Here the total number of seeds in a quarter-ounce sample could be counted, so we could estimate k (which is several thousand), and q, which represents the large proportion of grass seed, as contrasted with p, the small proportion of weed seed. The data are therefore structured as in a binomial

TABLE 5.7 *Potentilla* (Weed) Seeds in 98 Quarter-Ounce Samples of Grass Seeds (*Phleum pratense*)

(1) Number of weed seeds per sample of seeds Y	(2) Observed frequencies f	(3) Poisson expected frequencies $\hat{f}$	(4) Deviation from expectation $f - \hat{f}$
0	37	31.3	+
1	32	35.7	−
2	16	20.4	−
3	9 ⎫	7.8 ⎫	+ ⎫
4	2 ⎪	2.2 ⎪	− ⎪
5	0 ⎬ 13	0.5 ⎬ 10.6	− ⎬ +
6	1 ⎪	0.1 ⎪	+ ⎪
7+	1 ⎭	0.0 ⎭	+ ⎭
Total	98	98.0	

$$\overline{Y} = 1.1429 \qquad s^2 = 1.711 \qquad CD = 1.497$$

SOURCE: Modified from Leggatt (1935).

distribution with alternative states: weed seed and grass seed. For purposes of computation, however, only the number of weed seeds must be considered. This is an example of a case mentioned at the beginning of this section, a binomial in which the frequency of one outcome is very much smaller than that of the other and the sample size is large. We can use the Poisson distribution as a useful approximation of the expected binomial frequencies for the tail of the distribution. We use the average number of weed seeds per sample of seeds as our estimate of the mean and calculate expected Poisson frequencies for this mean. Although the pattern of the deviations and the coefficient of dispersion again indicate clumping, this tendency is not pronounced, and the methods of Chapter 17 would indicate insufficient evidence for the conclusion that the distribution is not a Poisson. We conclude that the weed seeds are randomly distributed through the samples. If clumping had been found, it might have been because the weed seeds stuck together, possibly because of interlocking hairs, sticky envelopes, or the like.

Perhaps you need to be convinced that the binomial under these conditions does approach the Poisson distribution. The mathematical proof is too involved for this text, but we can give an empirical example (Table 5.8). Here the expected binomial frequencies for the expression $(0.001 + 0.999)^{1000}$ are given in column (2); in column (3) these frequencies are approximated by a Poisson distribution with mean equal to 1 because the expected value for p is 0.001, which is one event for a sample size of $k = 1000$. The two columns of expected frequencies are extremely similar.

TABLE 5.8 Expected Frequencies Compared for Binomial and Poisson Distributions

(1) Y	(2) Expected binomial frequencies $p = 0.001$ $k = 1000$ $\hat{f}_{rel}$	(3) Expected frequencies approximate by Poisson $\mu = 1$ $\hat{f}_{rel}$
0	0.367,695	0.367,879
1	0.368,063	0.367,879
2	0.184,032	0.183,940
3	0.061,283	0.061,313
4	0.015,290	0.015,328
5	0.003,049	0.003,066
6	0.000,506	0.000,511
7	0.000,072	0.000,073
8	0.000,009	0.000,009
9	0.000,001	0.000,001
Total	1.000,000	0.999,999

TABLE 5.9 Azuki Bean Weevils (*Callosobruchus chinensis*) Emerging from 112 Azuki Beans (*Phaseolus radiatus*)

(1) Number of weevils emerging per bean Y	(2) Observed frequencies f	(3) Poisson expected frequencies $\hat{f}$	(4) Deviation from expectation $f - \hat{f}$
0	61	70.4	$-$
1	50	32.7	$+$
2	1 ⎫	7.6 ⎫	$-$ ⎫
3	0 ⎬ 1	1.2 ⎬ 8.9	$-$ ⎬ $-$
4+	0 ⎭	0.1 ⎭	$-$ ⎭
Total	112	112.0	

$$\bar{Y} = 0.4643 \qquad s^2 = 0.269 \qquad CD = 0.579$$

SOURCE: Data from Utida (1943).

The next distribution is extracted from an experimental study of populations of the azuki bean weevil (Table 5.9). Larvae of these weevils enter the beans, feed and pupate inside them, and then emerge through a hole. Thus, the number of emergence holes per bean is a good measure of the number of adults that have emerged. The rare event in this case is the weevil present in the bean. We note that the distribution is strongly repulsed, a far rarer occurrence than a clumped distribution. There are many more beans containing one weevil than the Poisson distribution would predict. A statistical finding of this sort leads us to investigate the biology of the phenomenon. In this case, it was found that the adult female weevils tended to deposit their eggs evenly rather than randomly over the available beans. This prevented too many eggs being placed on any one bean and precluded heavy competition among the developing larvae. A contributing factor was competition between larvae feeding on the same bean, generally resulting in all but one being killed or driven out. Thus, it is easy to understand how these biological phenomena would give rise to a repulsed distribution.

We will end this discussion with a classic, tragicomic application of the Poisson distribution. Table 5.10 is a frequency distribution of men killed by being kicked by a horse in 10 Prussian army corps over 20 years. The basic sampling unit is temporal in this case: one army corps per year. We are not certain exactly how many men are involved, but most likely a considerable number. The 0.610 men killed per army corps per year is the rare event. If we knew the number of men in each army corps, we could calculate the probability of not being killed by a horse in one year. This calculation would give us a binomial that could be approximated by the Poisson distribution. Knowing that the sample size (the number of men in an army corps) is large, however, we need not concern ourselves with values of p and k and can consider the

| TABLE 5.10 | Men Killed by Being Kicked by a Horse in 10 Prussian Army Corps in the Course of 20 Years |

(1) Number of men killed per year per army corps	(2) Observed frequencies f	(3) Poisson expected frequencies $\hat{f}$	(4) Deviation from expectation $f - \hat{f}$
0	109	108.7	+
1	65	66.3	−
2	22	20.2	+
3	3 ⎫	4.1 ⎫	− ⎫
4	1 ⎬ 4	0.6 ⎬ 4.8	+ ⎬ −
5+	0 ⎭	0.1 ⎭	− ⎭
Total	200	200.0	

$$\overline{Y} = 0.610 \qquad s^2 = 0.611 \qquad CD = 1.002$$

SOURCE: Data from Bortkiewicz (1898).

example simply from the Poisson model, using the observed mean number of men killed per army corps per year as an estimate of μ.

This example is an almost perfect fit of expected frequencies to observed ones. What would clumping have meant in such an example? An unusual number of deaths in a certain army corps in a given year could have been the result of poor discipline in the particular corps or of a particularly vicious horse that killed several men before the corps got rid of it. Repulsion might mean that one death per year per corps occurred more frequently than expected. This might have been so if men in each corps were careless each year until someone had been killed, after which time they all became more careful for a while.

Occasionally, samples of items that could be distributed in Poisson fashion are taken in such a way that the zero class is missing. For example, an entomologist studying the number of mites per leaf of a tree may not sample leaves at random from the tree but collect only leaves containing mites, excluding those without mites. For these so-called **truncated Poisson distributions,** special methods can be used to estimate parameters and to calculate expected frequencies. Cohen (1960) and Zelterman (1988) give tables and graphs for rapid estimation.

5.4 Other Discrete Probability Distributions

The binomial and Poisson distributions are both examples of discrete probability distributions. For the moss example of the previous section (Table 5.5), this means that there is either no moss plant per quadrat or one moss plant, two plants, three plants, and so forth, but no values in between. Other discrete distributions are known in probability theory, and in recent years many have been suggested as suitable for one application or another. As a beginner, you need not concern yourself with these, but it might be well for you to be familiar at least by name with the four that follow.

The **hypergeometric distribution** is the distribution equivalent to the binomial case but sampled from a *finite population without replacement*. In Section 5.1, we estimated the probability of finding a second foreign student in Matchless University. We pointed out that if there are 10,000 students at Matchless University, 400 of whom are foreign, then the probability of sampling one foreigner is indeed 0.04. Once a foreign student has been sampled, however, the probability of sampling another foreign student is no longer 0.04 (even when independent) but is 399/9999, or 0.0399. This probability would be 0.04 only *with replacement*—that is, if we returned the first foreign student sampled to the university population before we sampled again.

The binomial distribution is entirely correct only in cases of sampling with replacement or with infinite population size (which amounts to the same thing). For practical purposes, when small samples are taken from large populations, as in the case of Matchless University, these populations can be considered infinite. But if you have a population of 100 animals, 4% of which carry a mutation, then a sample of one mutant reduces the population to 3 out of 99, or from 4% to 3.03%. Thus, repeated samples of 5 from this population would follow not the binomial distribution but a different distribution—the hypergeometric distribution. The individual terms of the hypergeometric distribution are given by the expression

$$\frac{\binom{pN}{r}\binom{qN}{k-r}}{\binom{N}{k}} \tag{5.13}$$

which gives the probability of sampling r items of the type represented by probability p out of a sample of k items from a population of size N. The mean and variance of a hypergeometric distribution are kp and $kpq(N-k)/(N-1)$.

Note that the mean is the same as that of the binomial distribution, and the variance is that of the binomial multiplied by $(N-k)/(N-1)$. When N is very large as compared with k, this term is approximately 1, as expected because in such cases the hypergeometric distribution approximates the binomial. The expected frequencies for any sizable distribution are tedious to compute. In such cases, computer software using sufficient precision arithmetic is required.

In biology, sampling of small samples from a finite distribution occurs in certain problems of evolutionary genetics. Another application is in mark–recapture studies in which a certain proportion of a population is caught, marked, released, and subsequently recaptured, leading to estimates of the number of the entire population.

The **multinomial distribution** is a discrete probability distribution that extends the binomial distribution to cases where an attribute has more than two classes. We feature it in some detail in Section 17.4, where it is needed for the *G*-test of independence in a two-way frequency table.

We have used the binomial and Poisson distributions to examine whether given data are random or whether they show marked departure from random expectations—either clumping or repulsion. In some cases, it may seem unreasonable to assume random occurrences. In social organisms, for example, aggregation is a given. Statisticians and biologists have developed models for such cases, which lead to the so-called contagious distributions. Although we cannot discuss these in detail in this volume, we mention two here briefly.

The **negative binomial distribution** has been used probably more frequently than any other contagious distribution. The theoretical conditions that would give rise to a negative binomial distribution are discussed by Bliss and Calhoun (1954), who also give methods of calculating expected frequencies. The account in Bliss and Fisher (1953) is somewhat more rigorous. Krebs (1989) describes methods for estimation and significance testing and provides computer software for the calculations.

The **logarithmic distribution** (or logarithmic series) has been used extensively in studying the distribution of taxonomic units in faunal samples. This distribution has been employed frequently by C. B. Williams (see Williams, 1964). Johnson and Kotz (1969) give general information on many types of discrete distributions.

For more detail on these two distributions as well as other contagious distributions in ecology, consult the following references. The difficulty of fitting these varies with the distribution. Greig-Smith (1964) gives a nontechnical account of the application of these distributions to ecology and provides references that may lead the interested reader deeper into the subject. More quantitative accounts are given by Pielou (1977) and Krebs (1989). Many ecological examples of the application of contagious distributions are furnished in Williams (1964). Evans et al. (1992) is a convenient handbook describing many types of distributions of interest to biologists.

EXERCISES 5

5.1   In humans, the sex ratio of newborn infants is about $100 ♀ ♀ : 105 ♂ ♂$. If we were to take 10,000 random samples of 6 newborn infants each from the total population of such infants for 1 year, what would be the expected frequency of groups of 6 males, 5 males, 4 males, and so on?

   ANSWER: For 4 males, $\hat{f} = 2456.5$.

5.2   Show algebraically why the computational method of Expression 5.10 works.

5.3   The two columns below give fertility of eggs of the CP strain of *Drosophila melanogaster* raised in 100 vials of 10 eggs each (data from R. R. Sokal). Find the expected frequencies, assuming that the mortality for each egg in a vial is independent. Use the observed mean. Calculate the expected variance and compare it with the observed variance. Interpret the results, knowing that the eggs of each vial are siblings and that the different vials contain descendants from different parent pairs.

   ANSWER: $\sigma^2 = 2.417$, $s^2 = 6.628$.

Number of Eggs Hatched Y	Number of Vials f
0	1
1	3
2	8
3	10
4	6
5	15
6	14
7	12
8	13
9	9
10	9

5.4 Calculate Poisson expected frequencies for the frequency distribution given in Table 2.2 (number of plants of the sedge *Carex flacca* found in 500 quadrats).

5.5 The Army Medical Corps is concerned about intestinal disease X. From previous experience, the corps knows that soldiers suffering from the disease invariably harbor the pathogenic organism in their feces and that, for all practical purposes, every diseased stool specimen contains these organisms. The organisms are never abundant, however, and thus only 20% of all slides prepared by the standard procedure contain some of them (we assume that if an organism is present on a slide, it will be seen). How many slides per stool specimen should the laboratory technicians prepare and examine to ensure that, if a specimen is positive, it will be erroneously diagnosed negative in less than 1% of the cases (on the average)? On the basis of your answer, would you recommend that the corps attempt to improve its diagnostic methods?

ANSWER: 21 slides.

5.6 A cross is made in a genetic experiment in drosophila in which it is expected that one-fourth of the progeny will have white eyes and one-half will have the trait called *singed bristles*. Assume that the two gene loci segregate independently. (a) What proportion of the progeny should exhibit both traits simultaneously? (b) If four flies are sampled at random, what is the probability that all will have white eyes? (c) What is the probability that none of the four flies will have either white eyes or singed bristles? (d) If two flies are sampled, what is the probability that at least one of the flies will have white eyes or singed bristles or both traits?

5.7 Readers who have had a semester or two of calculus may wish to try to prove that Expression (5.9) tends to Expression (5.11) as k becomes indefinitely large (and p becomes infinitesimal, so that $\mu = kp$ remains constant).

$$Hint: \left(1 - \frac{x}{n}\right)^n \rightarrow e^{-x} \text{ as } n \rightarrow \infty$$

5.8 Summarize and compare the assumptions and parameters on which the binomial and Poisson distributions are based.

5.9 The two columns below present an accident record of 647 women working in a munitions plant during a five-week period reported by Greenwood and Yule (1920). The sampling unit is one woman during this period. The model assumes that the accidents are not fatal or very serious and thus do not remove the individual from further exposure.

Compute the expected frequency distribution for number of accidents per woman. Do the observed frequencies fit these expectations? Can you come up with an explanation for your findings? Calculate and interpret the coefficient of dispersion.

ANSWER: $CD = 1.488$.

Number of Accidents per Woman Y	Observed Frequencies f
0	447
1	132
2	42
3	21 ⎫
4	3 ⎬ 26
5+	2 ⎭
Total	647

5.10 Population geneticists work with a variable called *gene frequency*, which is the proportion of an allele found in the gametes of a population. Thus, if a genetic locus has two alternative alleles, A and a, p is the gene frequency or proportion of A, and $q = 1 - p$ is the gene frequency of a. In an insect population, the gene frequency of A is 0.6. If mating is random in the population, what will be the frequencies (that is, the proportions) of the three possible genotypes: AA, Aa, and aa?

Hint: Random mating can be considered as sampling two gametes at random from the gametic pool. The formula that will solve this problem describes the most fundamental relation in population genetics, known as the Hardy–Weinberg Law.

5.11 In Exercise 5.10, allele A is dominant over allele a, so genotypes AA and Aa are phenotypically indistinguishable. When a geneticist examines a single dominant specimen from the population, what is the probability that the specimen is a heterozygote?

ANSWER: $P[Aa|D] = 0.5714$.

5.12 The following data were collected on the numbers of mites (*Arrenurus* sp.) found on a sample of Chironomid flies (*Calopsectra akrina*). Give at least two possible explanations for the pattern of deviations from a Poisson distribution.

Number of Mites per Fly Y	Observed Frequencies f
0	442
1	91
2	29
3	14
4	4
5	6
6	2
7	0
8+	1

6 The Normal Probability Distribution

The theoretical frequency distributions in Chapter 5 were all discrete. Their variables assumed values that changed in integral steps (meristic variables). Thus, the number of infected insects per sample was 0 or 1 or 2 but could not assume a value between these. Similarly, the number of yeast cells per hemacytometer square is a meristic variable and requires a discrete probability function to describe it. Most variables encountered in biology, however, are continuous (such as the infant birth weights or the aphid femur lengths used as examples in Chapters 2 and 4). This chapter deals more extensively with the distributions of continuous variables.

Section 6.1 introduces frequency distributions of continuous variables. In Section 6.2, we examine the properties of the most common of such distributions, the normal probability distribution, and we present a model for generating this distribution in Section 6.3. A few applications of the normal distribution are illustrated in Section 6.4. The technique of fitting a normal distribution to observed data is described in Section 6.5, and Section 6.6 discusses some of the reasons for departure from normality in observed frequency distributions and describes techniques for measuring the nature and amount of such departures. Graphic techniques for identifying departures from normality and for estimating the mean and the standard deviation in approximately normal distributions are given in Section 6.7. Section 6.8 comments briefly on other types of theoretical frequency distributions of continuous variables.

6.1 Frequency Distributions of Continuous Variables

For continuous variables, the theoretical probability distribution, or **probability density function,** can be represented by a continuous curve (Figure 6.1a). The height of the curve gives the density for a given value of the variable. By **density,** we mean the relative concentration of observations along the Y-axis (as indicated in Figure 2.1). Note that the Y-axis is the abscissa (horizontal axis) in the figure, and the frequency f, or the density, is the ordinate (vertical axis). To compare the theoretical with the observed frequency distribution, the two must be divided into corresponding classes, as shown by the vertical lines in Figure 6.1b. Probability density functions are defined so that the expected frequency of observations between two class limits (vertical lines) is represented by the area between these limits under the curve. The total area under the curve is therefore equal to the sum of the expected frequencies (1.0 or n, depending on whether relative or absolute expected frequencies were calculated).

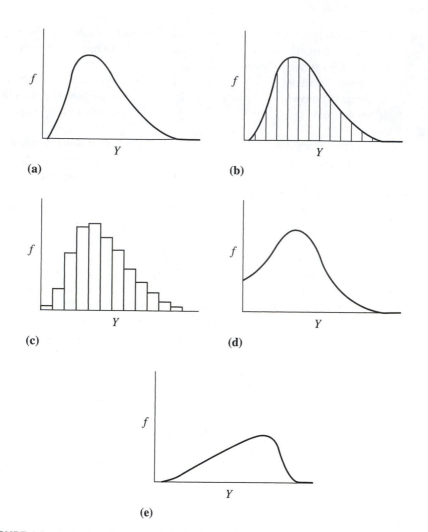

FIGURE 6.1   Examples of some probability distributions of continuous variables. (See text for explanation.)

When forming a frequency distribution of observations of a continuous variable, the choice of class limits is arbitrary because all values of a variable are theoretically possible. In a continuous distribution, one cannot evaluate the probability of the variable being exactly equal to a given value such as 3 or 3.5. One can only estimate the frequency of observations falling between two limits because the area of the curve corresponding to any point along the curve is infinitesimal.

The usual way to graph continuous distributions is in the form of a histogram—that is, to make each of the segments in Figure 6.1b into rectangles of the same width, as shown in Figure 6.1c. As long as each class interval is the same width, the height of each rectangle is directly proportional to its area, and the frequency of each class may be read directly off the ordinate.

Thus, in order to calculate expected frequencies for a continuous distribution, we must calculate the proportions of the area under the curve between the class limits. In Sections 6.2, 6.4, and 6.5, we will see how this is done for normal frequency distributions.

Continuous frequency distributions may start and terminate at finite points along the Y-axis, as in Figure 6.1a, or one or both ends of the curve may extend indefinitely, as in Figures 6.1d and 6.1e. The idea of an area under a curve, when one or both ends go to infinity, may trouble those of you not acquainted with the calculus. Fortunately, however, this is not a great conceptual stumbling block because, in all the cases we will encounter, the tail of the curve approaches the Y-axis rapidly enough that the portion of the area beyond a certain point is, for all practical purposes, zero, and the frequencies it represents are infinitesimal.

We may fit continuous frequency distributions to some sets of meristic data, such as the number of teeth in an organism. In such cases, we have reason to believe that underlying biological variables causing the differences in numbers of the structure are continuous, even though expressed as a discrete variable.

We will now discuss the most important probability density function in statistics, the normal frequency distribution.

6.2 Properties of the Normal Distribution

Formally, the **normal probability density function** can be represented by the expression

$$Z = \frac{1}{\sigma \sqrt{2\pi}} e^{-\frac{1}{2}[(Y-\mu)/\sigma]^2} \tag{6.1}$$

where Z indicates the height of the ordinate of the curve, which represents the density of the items. Z is the dependent variable in the expression because it is a function of the variable Y. There are two constants in the equation: π, well known to be approximately 3.14159, making $1/\sqrt{2\pi}$ equal 0.39894; and e, the base of the Napierian, or natural, logarithms, whose value approximates 2.71828. A normal probability density function has two parameters: the parametric mean μ and the parametric standard deviation σ, which determine the location and shape of the distribution, respectively.

Thus, there is not just one normal distribution, as one might naively believe when encountering the same bell-shaped image in elementary textbooks. Rather, the number of such curves is infinite because these parameters can assume an infinite number of values. This variation is illustrated by the three normal curves in Figure 6.2, each of which represents the same total frequencies. Curves A and B, which differ in location but not in shape, represent populations with different means. Curves B and C represent populations that have identical means but different standard deviations. Because the standard deviation of curve C is only half that of curve B, curve C is much narrower.

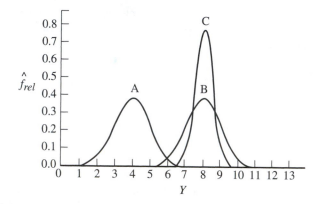

FIGURE 6.2   Changes in the two parameters of the normal distribution affect the shape and location of the normal probability density function. **(a)** $\mu = 4, \sigma = 1$; **(b)** $\mu = 8, \sigma = 1$; **(c)** $\mu = 8, \sigma = 0.5$.

In theory, a normal frequency distribution extends from negative infinity to positive infinity along the axis of the variable Y, which here is shown as the abscissa. A normally distributed variable thus can assume any value, however large or small, although values farther from the mean than ± 3 standard deviations are quite rare because their relative expected frequencies are very small. This tendency can be seen from Expression (6.1). When Y is very large or very small, the term $\frac{1}{2}[(Y - \mu)/\sigma]^2$ necessarily becomes very large. Hence, e raised to the negative power of that term is very small and Z is therefore very small. The ordinate Z for a normal distribution is tabulated in many books as a function of distance from the mean in standard deviation units (e.g., Pearson and Hartley, 1958, Table 1). It can also be obtained using the NORMDIST function in the spreadsheet program Excel.

The curve is symmetrical around the mean. Therefore, the mean, median, and mode of the normal distribution are all at the same point. The following percentages of items in a normal frequency distribution lie within the indicated limits.

$$\mu \pm \sigma \text{ contains 68.27\% of the items}$$

$$\mu \pm 2\sigma \text{ contains 95.45\% of the items}$$

$$\mu \pm 3\sigma \text{ contains 99.73\% of the items}$$

Conversely,

$$50\% \text{ of the items fall between } \mu \pm 0.674\sigma$$

$$95\% \text{ of the items fall between } \mu \pm 1.960\sigma$$

$$99\% \text{ of the items fall between } \mu \pm 2.576\sigma$$

These relations are shown in Figure 6.3. How were these percentages calculated? The direct calculation of any portion of the area under the normal curve requires an

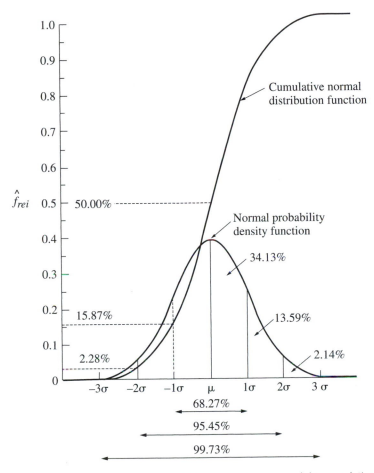

FIGURE 6.3   Areas under the normal probability density function and the cumulative normal distribution function.

integration of the function shown as Expression (6.1). For Figure 6.3, this integral has been evaluated and is presented in an alternative form: the **normal distribution function** (the theoretical *cumulative* distribution function of the normal probability density function). This function gives the total frequency from negative infinity up to any point along the abscissa. We can therefore directly look up the probability that an observation will be less than a specified value of Y. For example, Figure 6.3 shows that the total frequency up to the mean is 50.00% and that the frequency up to a point 1 standard deviation below the mean is 15.87%. These frequencies are found, graphically, by raising a vertical line from a point, such as $-\sigma$, until it intersects the cumulative distribution curve, and then reading the frequency off the ordinate. The probability of an observation falling between two arbitrary points can be found by subtracting the probability of an observation falling below the lower point from the probability of an observation falling below the upper point. For example, we can see from Figure 6.3 that the probability of an observation falling between the mean and a point 1 standard deviation below the mean is $0.5000 - 0.1587 = 0.3413$.

The normal distribution function is tabulated in Statistical Table **A,** which gives areas of the normal curve for the normal frequency distribution with the following parameters: mean $= 0$ and standard deviation $= 1$. For convenience in later calculations, 0.5 has been subtracted from all the entries. This table therefore lists the proportion of the area between the mean and any point a given number of standard deviations above it. Thus, for example, the area between the mean and the point 0.50 standard deviation above the mean is 0.1915 of the total area of the curve. Similarly, the area between the mean and the point 2.64 standard deviations above the mean is 0.4959 of the curve. A point 4.0 standard deviations from the mean includes 0.499968 of the area between it and the mean. Because the normal distribution extends from negative to positive infinity, however, one needs to go an infinite distance from the mean to reach fully half of the area under the curve. Section 6.4 will explain how to use the table of areas of the normal curve.

We carried out two sampling experiments to give you a feel for the distribution of items randomly sampled from a normal distribution, as well as one skewed positively (strongly drawn out to the right; see Section 6.6). The first dataset is based on a sample ($n = 100$) of wing lengths of houseflies adapted from Sokal and Hunter (1955). The observations closely approximated a normal frequency distribution. For convenience, we sampled from a normal distribution, employing the observed sample mean (45.5 mm $\times 10^{-1}$) and standard deviation (3.90) as its parameters. The second dataset consists of a sample of 100 total annual milk yields taken from a large population of Jersey cows in Canadian government records. It deviates strongly from normality, but we found it to be consistent with a lognormal distribution (see Section 6.8). We therefore employed a lognormal distribution with the observed sample mean (66.61 lbs $\times 10^2$) and standard deviation (11.1570) as its parameters. We used these parameters to sample 1000 items from each of the two populations.

This can be done using a spreadsheet program such as Excel, but more extensive sampling experiments can be performed using programs such as MATLAB, Mathematica, or R. Such experiments are easy to carry out using specialized software which can draw samples from various theoretical distributions and can calculate means, variances, and so on for each sample.

For each variable, we calculated the expression $(Y_i - \mu)/\sigma$ for each observation Y_i, where μ and σ are the values we adopted above for the true (or parametric) means and standard deviations of the two distributions. Thus, for a housefly wing length observation of 41, we evaluate

$$\frac{41 - 45.5}{3.90} = -1.1538$$

This result means that this wing length is 1.1538 standard deviations below the true mean of the population. The deviation from the mean measured in standard deviation units is called a *standardized deviate,* or **standard deviate.** The arguments of

Statistical Table **A,** expressing distance from the mean in units of σ, are called **standard normal deviates.** We grouped all 1000 standard deviates of the wing lengths in a frequency distribution with class intervals of $\frac{1}{2}\sigma$; then we did the same for the milk yields. Frequency distributions of the standard deviates of both variables together with the expected frequencies under the hypothesis of normality are shown in Table 6.1. Note the marked differences in distribution between the housefly wing lengths and the milk yields.

TABLE 6.1 Observed and Expected Frequencies of 1000 Standard Deviates Sampled from Populations with the Same Parameters as the Housefly Wing Lengths ($\mu = 45.5$, $\sigma^2 = 15.21$, $\sigma = 3.90$) and Milk-Yield Data ($\mu = 66.61$, $\sigma^2 = 124.4779$, $\sigma = 11.1597$)

Column 1. Class marks in standard deviates units. Column 2. Expected frequencies $\hat{f}$ for a normal distribution of the classes implied by the class marks. Columns 3 and 5. Distribution into these classes of the 1000 sampled observations of the housefly wing lengths (subscripted HF) and the milk yields (subscripted MY), respectively. Columns 4 and 6. Deviations of observed from expected frequencies for these two datasets.

(1) Class Mark	(2) $\hat{f}$	(3) f_{HF}	(4) $f_{HF} - \hat{f}$	(5) f_{MY}	(6) $f_{MY} - \hat{f}$
−5.0000	0.0	0	−0.0	0	−0.0
−4.5000	0.0	0	−0.0	0	−0.0
−4.0000	0.1	0	−0.1	0	−0.1
−3.5000	0.5	0	−0.5	0	−0.5
−3.0000	2.4	5	2.6	0	−2.4
−2.5000	9.2	9	−0.2	4	−5.2
−2.0000	27.8	32	4.2	22	−5.8
−1.5000	65.6	62	−3.6	68	2.4
−1.0000	121.0	104	−17.0	133	12.0
−0.5000	174.7	173	−1.7	185	10.3
0.0000	197.4	196	−1.4	214	16.6
0.5000	174.7	182	7.3	161	−13.7
1.0000	121.0	112	−9.0	95	−26.0
1.5000	65.6	75	9.4	57	−8.6
2.0000	27.8	33	5.2	29	1.2
2.5000	9.2	12	2.8	22	12.8
3.0000	2.4	5	2.6	6	3.6
3.5000	0.5	0	−0.5	2	1.5
4.0000	0.1	0	−0.1	0	−0.1
4.5000	0.0	0	−0.0	2	2.0
5.0000	0.0	0	−0.0	0	−0.0

6.3 A Model for the Normal Distribution

When we study continuous frequency distributions in nature, we find that a large proportion of them approximate the normal distribution. Why? There are several ways of generating the normal frequency distribution from models based on elementary assumptions. Many of these require more mathematics than we expect of our readers. We therefore use a largely intuitive approach that we have found to be of heuristic value.

Let us consider a binomial distribution of the familiar form $(p + q)^k$, in which k becomes indefinitely large. What type of biological situation could give rise to such a binomial distribution? An example is one in which many factors cooperate additively to produce a biological result. The following hypothetical case is perhaps not too far removed from reality. The intensity of skin pigmentation in an animal is the sum of many factors, some genetic and others environmental. To simplify this example, let us assume that every factor can occur in two states only: present or absent. When the factor is present, it contributes one unit of pigmentation to skin color, but it contributes nothing to pigmentation when it is absent. Each factor, regardless of its nature or origin, has the same effect, and the effects are additive; if three out of five possible factors are present in an individual, the pigmentation intensity would be three units (the sum of three contributions of one unit each). Let us also assume that each factor has an equal probability of being present or absent in a given individual. Thus, $p_F = 0.5$, the probability of the factor being present, while $q_f = 0.5$, the probability of the factor being absent.

With only one factor ($k = 1$), expansion of the binomial $(p_F + q_f)^1$ would yield the following two pigmentation classes among the animals:

F	f	pigmentation classes
0.5	0.5	expected frequency
1	0	pigmentation intensity

Half the animals would have intensity 1, the other half 0. With $k = 2$ factors present in the population (the factors are assumed to occur independently of each other), the distribution of pigmentation intensities would be represented by the expansion of the binomial $(p_F + p_f)^2$:

FF	Ff	ff	pigmentation classes
0.25	0.50	0.25	expected frequency
2	1	0	pigmentation intensity

One-fourth of the individuals would have pigmentation intensity 2, one-half 1, and the remaining one-fourth 0.

The expected frequencies for 4, 10, and 50 factors are graphed as histograms in Figure 6.4 (rather than as bar diagrams, as they should be drawn). They approach the

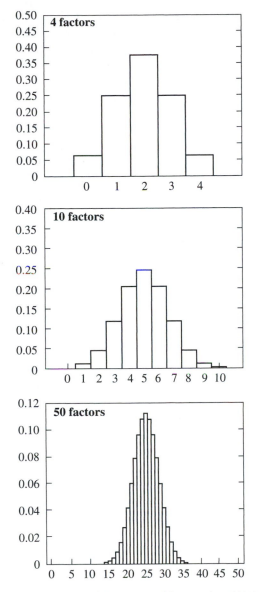

FIGURE 6.4 Histograms showing relative expected frequencies of Y, the pigmentation intensity for k = 4, 10, and 50 factors. These frequencies are based on the expansion of the binomial $(0.5 + 0.5)^k$.

familiar bell-shaped outline of the normal frequency distribution (see Figures 6.2 and 6.3). For $k = 50$, our histogram is so close to a normal frequency distribution that it is difficult to show the difference between them on a graph the size of this page. Of course, simply stating that the expected frequencies of the binomial approach the shape of a normal probability distribution is not acceptable mathematical proof. The mathematically inclined reader will find a relatively simple proof of this proposition in Section 2.7 of Ross (1985). Other readers must take it on faith that the model postulated above, the result of many factors acting independently and additively, will approach normality. At the beginning of this procedure, however, we made some severely limiting assumptions for the sake of simplicity. Let us now return to these and see if we can remove them.

1. The two states of the factors occur in equal frequency—that is, $p = q$. When $p \neq q$, the distribution also approaches normality as k approaches infinity. This is difficult to see intuitively because when $p \neq q$, the histogram is asymmetrical at first. It can be shown mathematically, however, that when k, p, and q are such that $kpq \geq 3$, the normal distribution will be closely approximated.

2. Factors occur in two states only—present or absent. In a more realistic situation, factors would occur in several states, one state making a large contribution, a second state making a smaller contribution, and so forth. It can also be shown, however, that the multinomial $(p + q + r + \cdots + z)^k$ approaches the normal frequency distribution as k approaches infinity.

3. Each factor has the same effect. In reality, different factors may be present in different frequencies and also may have different quantitative effects. As long as these are additive and independent, normality is still approached when k approaches infinity.

Lifting these restrictions makes the assumptions leading to a normal distribution compatible with innumerable biological situations. It is therefore not surprising that many biological variables are approximately normally distributed.

Let us summarize the conditions that tend to produce normal frequency distributions: (1) if there are many factors that are single or composite; (2) if these factors are independent in occurrence; (3) if the factors are independent in effect—that is, if their effects are additive; and (4) if they make equal contributions to the variance. The fourth condition we are not yet in a position to discuss and we mention it here only for completeness; we shall discuss it in Chapter 8.

6.4 Applications of the Normal Distribution

The normal frequency distribution is the most widely used distribution in statistics, and we will have recourse to it in a variety of situations. For the moment, we may subdivide its applications as follows:

1. We sometimes have to know whether a given sample is distributed normally before we can apply a certain test to it. To test a sample for normal distribution, we have to calculate expected frequencies for a normal curve of the same mean and standard deviation. The next section will show you how to do this, and Section 6.7 provides shortcuts.

2. Knowing whether a sample is distributed normally may confirm or reject certain underlying hypotheses about the nature of the factors affecting the phenomenon studied. Refer to the conditions for normality in a frequency distribution (see Section 6.3). Thus, if we find a given variable to be distributed normally, we have no reason for rejecting the hypothesis that the causal factors affecting the variable are additive and independent and of equal variance. On the other hand, departures from normality may indicate that certain forces, such as selection, are affecting the variable under study. Skewness of milk-yield data may reflect the fact that these were records of selected cows and that substandard milk cows were not included in the records. Bimodality (see Section 4.4) may indicate a mixture of observations from two populations. Even though measurement variables from a sample are distributed normally, areas and volumes from such samples will not be distributed normally. This is because products and nonlinear transformations of normally distributed variables will not themselves be distributed normally. In many cases, transformations of nonnormal variables (see Chapter 13) change the distribution of the transformed variable to normality.

3. If we assume a given distribution to be normal, we may make predictions and tests of given hypotheses based on this assumption. The following is an example of such an application.

Recall the birth weights of male Chinese children, last illustrated in Box 4.2. The mean of this sample of 9465 birth weights is 109.9 oz, and its standard deviation is 13.593 oz. Sampling at random from the birth records of this population, what is the chance of obtaining a birth weight of 151 oz or heavier? Such a birth weight is considerably above the mean of our sample (the difference is $151 - 109.9 = 41.1$ oz). With a difference in ounces, however, we cannot consult the table of areas of the normal curve. We must **standardize** the difference—that is, divide it by the standard deviation to convert it into a standard deviate. When we divide the difference by the standard deviation, we obtain $41.1/13.593 = 3.02$. Thus, a birth weight of 151 oz is 3.02 standard deviation units greater than the mean.

Assuming that the birth weights are distributed normally, we may consult the table of areas of the normal curve (Statistical Table **A**), where we find a value of 0.4987 for 3.02 standard deviations. Alternatively, we could use the NORMSDIST function of Excel (or equivalent functions in other software) to obtain the cumulative area under the curve from its left tail up to 3.02 standard deviations above the mean. The function will return 0.998736. Subtracting 0.5 for the area to the left of the mean, we obtain the same value as in Statistical Table **A**. Function NORMDIST will yield the same number if you give it the values of the single observation, the mean, and the standard deviation of the sample. The number 0.4987 means that 49.87% of the area of the curve lies between the mean and a point 3.02 standard deviations from it.

Conversely, 0.0013 or 0.13% of the area lies beyond 3.02 standard deviation units above the mean. Thus, assuming that we have a normal distribution of birth weights and that $\sigma = 13.593$, only 0.13%, or 13 out of 10,000 of the infants, would have a birth weight of 151 oz *or* a weight *greater than* the mean. That a single sampled item from that population would deviate by so much from the mean is quite improbable. If a random sample of one such weight was obtained from the records of an unspecified population, we might therefore be justified in doubting that the observation came from a normally distributed population of human birth weights with a mean and standard deviation as specified above.

The probability in this example was calculated from one tail of the distribution. We found the probability that an individual is 3.02 standard deviations *greater* than the mean. If we have no prior knowledge that the individual will be either heavier or lighter than the mean but are merely concerned with how different it is from the population mean, the following question is appropriate: Assuming that the individual belongs to the population—and that the latter is normally distributed—what is the probability of observing a birth weight of an individual as deviant or more deviant from the mean in either direction? This probability, which must be computed by using both tails of the distribution, can be obtained simply by doubling the previous probability because the normal curve is symmetrical. Thus, $2 \times 0.0013 = 0.0026$. This value, too, is so small that we would conclude that a birth weight as deviant as 41.1 oz from its mean is unlikely to have come from the population represented by our sample of male Chinese children—*if it was normally distributed.*

However, the *if* in the preceding clause above should be written in bold. Our assumption has been that the birth weights are distributed normally. Inspection of the frequency distribution in Box 4.2, however, shows that the distribution is asymmetrical, tapering to the right. Although there are eight classes above the mean class, there are only six classes below it. In view of this asymmetry, conclusions about one tail of the distribution would not necessarily pertain to the second tail. Conclusions based on the assumption of normality might be in error if the exact probability were critical for a given test. Our statistical conclusions are only as valid as our assumptions about the data.

6.5 Fitting a Normal Distribution to Observed Data

Graphic fitting of a normal curve to an observed frequency distribution in the form of a histogram is done only rarely (mostly by writers of statistics textbooks or as an aid in constructing hanging histograms—see Section 6.7). Ordinates are computed by modifying Expression (6.1) to conform to a frequency distribution:

$$Z = \frac{ni}{s\sqrt{2\pi}} e^{-\frac{1}{2}[(Y - \bar{Y})/s]^2} \tag{6.2}$$

In this expression, n is the sample size and i the class interval of the frequency distribution. (We show the normal curve superimposed on the Chinese birth-weight data later in Figure 6.7**a**.)

More common is the technique of calculating expected frequencies for a normal distribution of the same mean and standard deviation as the observed sample. This is the same approach we used for calculating expected binomial or Poisson frequencies (see Chapter 5). We shall apply the procedure to the frequency distribution of birth weights of male Chinese children in Box 4.2. The computation of expected normal frequencies is featured in numerous statistical packages such as BIOMstat. However, if you have some knowledge of spreadsheet calculations, you can easily perform your own computations. We shall follow this approach for two reasons. Almost everyone has access to a spreadsheet program, and setting up the instructions for the calculation of expected normal frequencies will demystify the procedure for the novice.

The first two columns of the spreadsheet will contain the class marks and frequencies from Box 4.2. Next, we compute the mean and standard deviation of the frequency distribution, as well as the sum of the frequencies, and store these quantities in three cells beneath the last row needed for the frequency distribution. Column 3 is the lower class limit of each class; it is obtained by subtracting half the class interval from the class mark. In our case, the class interval is 8 oz; therefore, we subtract 4 from each class mark. Column 4 is the area under the normal curve from its left tail (negative infinity) to the lower limit of the class in a given row of the spreadsheet. It is obtained in Excel by means of the NORMDIST function, which requires four arguments—the lower limit of the class (in column 3), the mean and the standard deviation in their respective cells, and a code (the numeral 1) to indicate that the cumulative area is desired. Column 5 is the relative expected frequency of each class. It is obtained by subtracting the area under the curve to the left of the lower class limit of a given class from the area to the left of the lower class limit of the next class. These quantities are found in successive rows of column 4. Finally, column 6 is the absolute expected frequency of each class. It is simply the relative expected frequency in column 5 multiplied by the sum of the frequencies taken from the cell where it has been stored. To enable you to check your results, we feature columns 1 through 6 for the first two rows of the spreadsheet we constructed.

59.5	2	55.5	0.0000	0.0003	2.7
67.5	6	63.5	0.0003	0.0020	19.4

If the intermediate answers are not required, the equations for columns 3, 4, and 5 can be combined to yield column 6 directly because the latter is usually the column of principal interest. In Table 6.2, we show observed and expected normal frequencies for the Chinese birth weights. It is customary to add a final column listing the signs of the departures from expectation. These can be written down by inspection or can be computed automatically on the spreadsheet by appropriate IF statements.

Note that classes above the mean generally show an excess, and those below the mean show a shortage of observed frequencies with respect to expectation. This is what you would expect in a distribution skewed to the right (see Section 6.6). As with the binomial and Poisson distributions, we have not yet learned a computational technique for testing the departures from expectation (covered in Chapter 17). Graphic comparisons of observed and expected frequencies are discussed in Section 6.7.

TABLE 6.2 Observed and Expected Normal Frequencies

Birth weights of male Chinese in ounces (from Box 4.2):
$n = 9465$; $\bar{Y} = 109.8996$; $s = 13.5942$

Wt oz	f	$\hat{f}$	$f - \hat{f}$
59.5	2	2.7	−
67.5	6	19.4	−
75.5	39	97.5	−
83.5	385	350.5	+
91.5	888	899.6	−
99.5	1729	1648.4	+
107.5	2240	2157.6	+
115.5	2007	2017.1	−
123.5	1233	1347.0	−
131.5	641	642.4	−
139.5	201	218.8	−
147.5	74	53.2	+
155.5	14	9.2	+
163.5	5	1.1	+
171.5	1	0.1	+
179.5	0		
	$\Sigma f = 9465$		

These figures were obtained by spreadsheet calculations described in Section 6.5.

6.6 Skewness and Kurtosis

The previous section showed how to calculate expected frequencies for a normal distribution. In many cases, an observed frequency distribution departs obviously from normality; thus, statistics that measure the nature and amount of departure are useful. We will emphasize two types of departure from normality. One is **skewness,** which is another name for asymmetry; skewness means that one tail of the curve is drawn out more than the other. In such curves, the mean and the median do not coincide. Curves are called skewed to the right or left, depending upon whether the right or left tails are drawn out.

The other type of departure from normality, **kurtosis,** is a more complicated change in distribution. If a symmetrical distribution is considered to have a center, two shoulders, and two tails, kurtosis describes the proportions found in the center and in the tails in relation to those in the shoulders. A **leptokurtic curve** has more items near the center and at the tails, with fewer items in the shoulders relative to a normal distribution with the same mean and variance. A **platykurtic curve** has fewer items at the center and at the tails than does the normal curve, but it has more items in the shoulders. A bimodal distribution is an extreme platykurtic distribution.

Both skewness and kurtosis are vague concepts (as, incidentally, are location and dispersion, or scale) with many ways of being evaluated. For a review of approaches to kurtosis, see Balanda and MacGillivray (1988). In this book, we use the conventional sample statistics for measuring skewness and kurtosis, called g_1 and g_2, to represent population parameters γ_1 and γ_2. Their computation is tedious and should always be done by computer. Program BIOMstat computes g_1 and g_2 when evaluating basic statistics for both ungrouped and grouped data.

The formulas for g_1 and g_2 involve **moment statistics.** A central moment in statistics, as in physics, is $(1/n)\Sigma^n(Y - \overline{Y})^r$, the average of the deviations of all items from the mean, each raised to the power r. The first central moment, $(1/n)\Sigma(Y - \overline{Y})$, always equals zero, as we saw in Section 4.6. The second moment, $(1/n)\Sigma(Y - \overline{Y})^2$, is the variance. The statistic g_1 is the third central moment divided by the cube of the standard deviation, $(1/ns^3)\Sigma(Y - \overline{Y})^3$, and g_2 is 3 less than the fourth central moment divided by the fourth power of the standard deviation, $(1/ns^4)\Sigma(Y - \overline{Y})^4 - 3$. The computation of these two statistics is outlined in Box 6.1 for the birth weights of Chinese males. Just as the sample variance had to be corrected for bias by dividing $\Sigma(Y - \overline{Y})$ by $n - 1$ rather than by n, so sample g_1 and g_2 values need to be corrected to allow for similar bias. These corrections are complicated and are included in Box 6.1. These statistics can also be computed for ungrouped data, in which case the formulas in Box 6.1 would simply drop the f's.

In a normal frequency distribution, both γ_1 and γ_2 are zero. A negative g_1 indicates skewness to the left, a positive g_1 skewness to the right. A negative g_2 indicates platykurtosis, while a positive g_2 shows leptokurtosis. Thus, a repulsed distribution would have a positive g_2 but a clumped distribution should have a negative g_2. The absolute magnitudes of g_1 and g_2 do not mean much; these statistics have to be tested to see whether they differ from zero by more than what one would expect just due to chance by methods yet to be learned (see Section 7.7 and Box 7.2). It appears from the results in Box 6.1 that the Chinese birth weights are skewed to the right (weights generally are skewed in this manner) and that a very slight tendency to leptokurtosis is also present.

When only small samples are taken from highly skewed populations, $|g_1|$ will tend to underestimate $|\gamma_1|$ because in a sample of size n it can be shown that $|g_1| \leq \sqrt{n}$ (Wallis et al., 1974). Thus, in a random sample of 5 milk yields from Table 6.1, we found $g_1 = 1.944$, which is close to its upper bound, $\sqrt{5} = 2.236$. By contrast, the parametric γ_1 of the 100 milk yields is actually 0.9408. Similar problems exist with estimates of γ_2 in small samples from highly kurtotic populations. The bounds on g_2 are

$$\frac{-2(n - 1)}{n - 3} \leq g_2 \leq n$$

Thus, large samples must be obtained to estimate adequately the parameters of strongly skewed or kurtotic populations.

BOX 6.1 | Computation of g_1 and g_2 from a Frequency Distribution

These computations are based on the birth weights of male Chinese in ounces, arranged as in the first two columns of Box 4.2.

Basic sums

$$n = \Sigma f = 9465 \qquad\qquad \Sigma Y = \Sigma fY = 1{,}040{,}199.5$$

$$\bar{Y} = \frac{1}{n}\Sigma fY = 109.8996 \quad \Sigma y^2 = \Sigma f(Y - \bar{Y})^2 = 1{,}748{,}956.7983$$

$$\Sigma y^3 = \Sigma f(Y - \bar{Y})^3 = 4{,}501{,}097.3779$$

$$\Sigma y^4 = \Sigma f(Y - \bar{Y})^4 = 998{,}109{,}887.1459$$

Computation

1. $s^2 = \dfrac{\Sigma y^2}{n-1} = \dfrac{1{,}748{,}956.7983}{9464} = 184.8010$

$\quad s = 13.5942$

2. $g_1 = \dfrac{n\,\Sigma y^3}{(n-1)(n-2)s^3} = \dfrac{9465(4{,}501{,}097.3779)}{(9464)(9463)(13.5942)^3} = 0.18936$

3. $g_2 = \dfrac{(n+1)n\,\Sigma y^4}{(n-1)(n-2)(n-3)s^4} - \dfrac{3(n-1)^2}{(n-2)(n-3)}$

$\quad = \dfrac{(9466)(9465)(998{,}109{,}887.1459)}{(9464)(9463)(9462)(13.5942)^4} - \dfrac{3(9464)^2}{(9463)(9462)}$

$\quad = 0.08913$

4. If the variable Y has been coded, g_1 and g_2 do not require decoding because they are scale free. In a normal frequency distribution, γ_1 and γ_2 are both zero. A negative g_1 indicates skewness to the left, a positive g_1 skewness to the right. A negative g_2 indicates platykurtosis and a positive g_2 shows leptokurtosis.

Compare these results with the graphic analysis in Box 6.2.

6.7 Graphic Methods

Many computer programs furnish graphs of the data as by-products of their computations of statistics describing the frequency distributions. This feature permits visual inspection of the data and is often useful for suggesting the way in which a sample deviates from expectation (such as bimodality or the presence of outliers). To take advantage of such features, the user needs to know how such graphs are constructed.

We begin by describing graphs based on cumulative frequency distributions. We first met cumulative frequencies in Section 4.3 when computing the median. In Figure 6.3, we saw that a normal frequency distribution graphed in cumulative fashion describes an S-shaped curve, called a *sigmoid curve.* If we drop perpendiculars to the abscissa from the cumulative normal curve, at the level of a given proportion we obtain the corresponding quantile (see Section 4.3). In Figure 6.3, the quantile is scaled as the number of standard deviations above or below the mean equivalent to a given proportion. For a normal distribution, these cumulative proportions, transformed into standard deviation scale, are called **normal quantiles.** We show such a transformation in Figure 6.5. Note that there are no 0% or 100% points on the ordinate. These values are not possible because the normal frequency distribution extends from negative to positive infinity, so however long we made our line, we would never reach the limiting values of 0% and 100%.

We now construct a graph whose abscissa is in the measurement scale Y of some observed variable and whose ordinate is in normal quantile units. Assume that we plot a normal frequency distribution above the abscissa. If we measure off suitably spaced quantiles along the abscissa, transform them into normal quantiles, and plot these points with respect to the measurement and normal quantile scales, the points

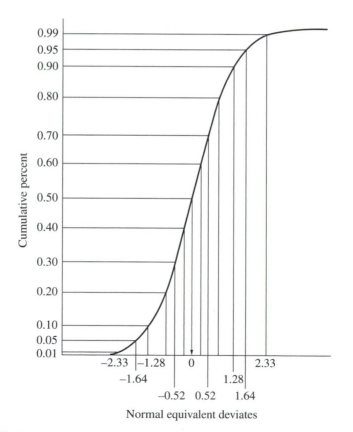

FIGURE 6.5 Transformation of cumulative percentages into normal quantiles.

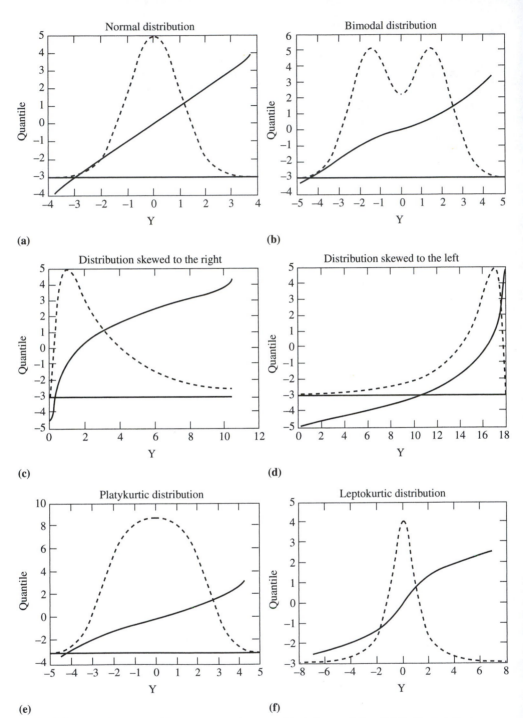

FIGURE 6.6 Examples of frequency distributions with their cumulative distributions plotted as normal quantile plots. (See Box 6.2 for explanation.) The ordinates, in normal quantile scale, provide the scale for the normal quantile plots, not for the frequencies of the frequency distributions shown.

should fall along a straight line because the normal quantile transformation is based on the area under a cumulative normal curve (Figure 6.6a). Therefore, if we take an observed sample with an unknown frequency distribution, plot it in this manner, and find that normal quantiles against variables fall more or less along a straight line, we may conclude that the sample is consistent with a normal distribution. Such graphs have become known as *quantile–quantile,* or *Q–Q, plots* or as **normal quantile plots,** the term we prefer. Nowadays such plots are usually produced by computer programs. Box 6.2, part I, explains what these programs present, and Figure 6.6 illustrates how to interpret departures from normality in normal quantile plots.

Normal quantile plots work best for fairly large ($n > 50$) samples (see Box 6.2, parts I and II). In smaller samples, a difference of one item per class would make a substantial difference in the cumulative percentage in the tails. For small samples ($n < 50$), the method of *ranked normal deviates,* or **rankits,** is preferable. With this method, instead of quantiles we use the ranks of each observation in the sample; and

---

BOX 6.2 Graphic Test for Normality of Samples by Means of Normal Quantile Plots

I. *Routine Computer-Processed Samples*

The following steps, performed in a computer program, will result in a normal quantile plot of the data. These steps can also be adapted for spreadsheet computation.

1. Array observations in order of increasing magnitude.

2. For each observation, compute the quantity $p = (i - 1/2)/n$, where i is the rank order of the ith observation in the array, and n is the sample size. The correction of 1/2 prevents the last observation from yielding $p = 1.0$, for which the normal quantile would be positive infinity. For tied values of i, calculate an average p.

3. For each value of p, evaluate the corresponding normal quantile.

4. The normal quantiles are plotted against the original observations. Examine the scatterplot visually for linearity. Some programs also plot the straight line expected when the sample is perfectly normally distributed to serve as a benchmark against which the observed scatterplot can be judged. Alternatively, a straight line is fitted to the points by eye, preferably using a transparent plastic ruler, which permits all the points to be seen as the line is drawn. In drawing the line, most weight should be given to the points between cumulative frequencies of 25% to 75% because a difference of a single item may make appreciable changes in the percentages at the tails. Some programs plot normal quantiles along the abscissa and the variable Y along the ordinate. We prefer the more common arrangement shown in Figure 6.6 and the figures in this box.

Figure A shows 1400 housefly wing lengths randomly sampled from Table 6.1 as a normal quantile plot. Because these are approximately normal data, it is no surprise that the scatterplot forms a nearly straight line.

Box 6.2 (continued)

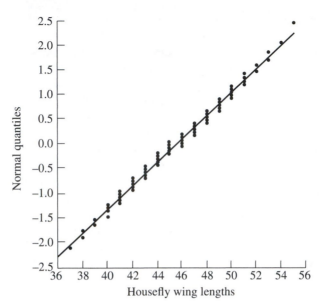

FIGURE A   Normal quantile plot of 1400 randomly sampled housefly wing lengths from Table 6.1.

II. *When Data Are Already Grouped into Frequency Distributions and Sample Size* n
Is Large

The normal quantile plot can be applied to the frequency distribution as follows. We employ the by now thoroughly familiar Chinese birth weights.

Birth weights of male Chinese in ounces, from Box 4.2.

(1) Class mark Y	(2) Upper class limit	(3) f	(4) Cumulative frequencies F	(5) $p = (F - \frac{1}{2})/n$
59.5	63.5	2	2	0.0002
67.5	71.5	6	8	0.0008
75.5	79.5	39	47	0.0049
83.5	87.5	385	432	0.0456
91.5	95.5	888	1320	0.1394
99.5	103.5	1729	3049	0.3221
107.5	111.5	2240	5289	0.5587
115.5	119.5	2007	7296	0.7708
123.5	127.5	1233	8529	0.9011
131.5	135.5	641	9170	0.9688
139.5	143.5	201	9371	0.9900
147.5	151.5	74	9445	0.9978
155.5	159.5	14	9459	0.9993
163.5	167.5	5	9464	0.9998
171.5	175.5	1	9465	0.9999
		9465		

Box 6.2 (continued)

Computation

1. Prepare a frequency distribution as shown in columns (1), (2), (3), and (4).

2. In column (5), express the cumulative frequencies as percentile *p*-values using the formula at the head of the column.

3. Graph the normal quantiles, corresponding to these *p*-values against the upper class limit of each class (Figure B). Notice that the upper frequencies deviate to the right of the straight line. This is typical of data that are skewed to the right (see Figure 6.6**d**).

4. If a computer is not handy at any given point in time, the graph in Figure B (and those in Figures A and C in this box) can also be used for rapid estimation of the mean and standard deviation of a sample. The mean is approximated by a graphic estimation of the median. The more normal the distribution is, the closer the mean

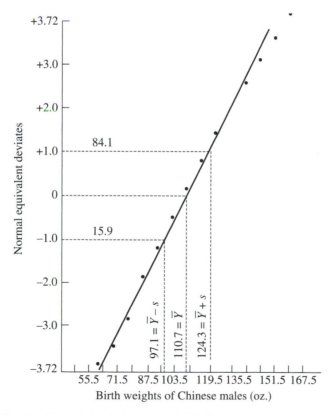

FIGURE B Graphic analysis of a frequency distribution.

Box 6.2 (continued)

will be to the median. The median is estimated by dropping a perpendicular from the intersection of the 0 point on the ordinate and the cumulative frequency curve to the abscissa (see Figure B). The estimate of the mean of 110.7 oz is quite close to the computed mean of 109.9 oz. The standard deviation is estimated by dropping similar perpendiculars from the intersections of the -1 and the $+1$ points with the cumulative curve, respectively. These points enclose the portion of a normal curve represented by $\mu \pm \sigma$. By measuring the difference between these perpendiculars and dividing this value by 2, we obtain an estimate of 1 standard deviation. In this instance, the estimate is $s = 13.6$ oz, since the difference is 27.2 oz divided by 2. This is a close approximation to the computed value of 13.59 oz.

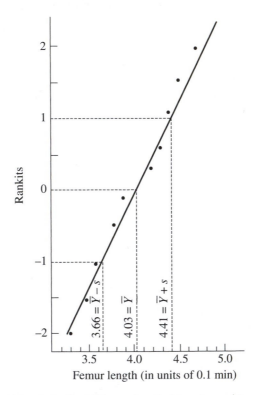

FIGURE C   Graphic analysis of aphid femur length data using rankits.

Box 6.2 (continued)

III. *Small Samples (n ≤ 50)*

Femur lengths of aphid stem mothers, from Box 2.1: $n = 25$.

1. Enter the sample arrayed in increasing order of magnitude in column (1). In column (2), put corresponding rankits from Statistical Table **H**. The table gives only the rankits for the half of each distribution greater than the median for any sample size. The other half is the same but is negative in sign. All samples containing an odd number of observations (such as this one) have 0 as the median value. The rankits for this example are looked up under sample size $n = 25$.

2. A special problem illustrated in this example is the case of ties, or observations of identical magnitude. In such a case, we sum the rankit values for the corresponding ranks and find their mean. Thus, the -1.00 occupying lines 3 to 6 in column (3) is the average of rankits -1.26, -1.07, -0.91, and -0.76 in column (2), which are the third, fourth, fifth, and sixth rankits for a sample of 25, respectively. Care must be taken when ties include the median because then we must sum positive as well as negative rankits. For example, four rankits of values -0.24, -0.14, -0.05, and $+0.05$, respectively, would give an average rankit of -0.095.

3. The rankit values (ordinate) are plotted against the variable (abscissa) (see Figure C), and a straight line is fitted by eye. The values do not lie exactly on a straight line and there is some suggestion of bimodality in the data, as is also shown in the figure at the bottom of Box 2.1.

4. Mean and standard deviation can be computed as shown in part II.

instead of normal quantiles, we plot values from a table of rankits, the average positions in standard deviation units of the ranked items in a normally distributed sample of n items. We could obtain these values empirically by sampling from a standard normal distribution, one whose mean $\mu = 0$ and whose variance (and standard deviation) is $\sigma^2 = \sigma = 1$. If we sample five observations from such a distribution, order them from lowest to highest observations, and record their scores, we might get a sample as follows: -1.8, -0.6, -0.1, $+0.9$, $+2.1$. If we repeat this process many times, we can calculate the average score of the first item in the ordered sample, of the second, and so forth. Statistical Table **H** shows these average standardized scores, or rankits. We find the rankits for a sample of 5 to be -1.163, -0.495, 0.0, 0.495, 1.163.

These rankits have a variety of uses; one is for testing normality of a frequency distribution, as we have shown in Box 6.2. When rankits are plotted against a ranked array of a normally distributed variable, the points will again lie in a straight line. An illustration of the application of rankits to testing for normality and for obtaining estimates of μ and σ is shown in Box 6.2, part III. Figure 6.6 may again be used to interpret the meaning of the resulting curves. Not too much reliance should be placed on single samples, but repeated trends in different samples are likely to be meaningful.

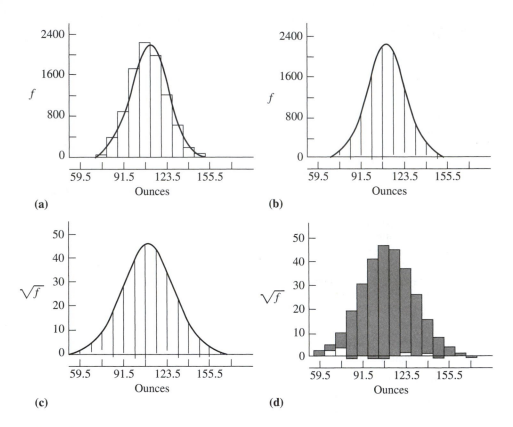

FIGURE 6.7 **(a)** Histogram of the observed frequency distribution of birth weights of male Chinese from Box 6.1 with the expected normal curve superimposed. **(b)** The data in part (a) displayed as a hanging histogram. Frequency bars, narrowed so that they are represented by a single line at the midpoint of each class, are suspended from the expected normal curve. Lines that do not reach the abscissa indicate deficiencies from expectations. Lines extending below the abscissa indicate observed frequencies in excess of expectation. **(c)** The data in part (b) shown as a hanging rootogram. Observed and expected frequencies are given as square roots of the actual values. Departures from expectation in the tails of the distribution are thus accentuated. **(d)** Comparison of the observed and expected frequencies for the birth weight data in part (a). The histogram of the skyline indicates the expected frequencies; that of the "inverted skyline" indicates the observed frequencies. Both frequencies are given in square roots of actual values. Where the inverted skyline does not reach the abscissa, there are fewer observed than expected frequencies. Wherever it reaches below the abscissa, there is an excess of observed frequencies over expected frequencies.

Often we want to compare observed frequency distributions with their expectations without resorting to cumulative frequency distributions. An example is the superimposition of a normal curve on the histogram of an observed frequency distribution, as discussed in Section 6.5. In Figure 6.7**a**, we show the frequency distribution of birth weights of male Chinese from Box 6.2 with the ordinates of the normal curve superimposed. An excess of observed frequencies appears at the right tail because of the skewness of the distribution.

Many people find it difficult to compare the heights of bars against the arch of a curve. For this reason, John Tukey (see Wainer, 1974, or Velleman and Hoaglin, 1981) suggested that the bars of the histograms be suspended from the curve. Their departures from expectation are then obvious against the straight abscissa of the graph. Such a **hanging histogram** is shown in Figure 6.7**b**. The departure from normality is now much clearer.

Because important departures frequently occur in the tails of a curve, it has been suggested that square roots of expected frequencies should be compared with the square roots of observed frequencies. Such a **hanging rootogram** is shown in Figure 6.7**c**. Note the accentuation of the departure from normality. Finally, one can also use this technique for comparing expected with observed histograms. Figure 6.7**d** shows the same data plotted in this manner. Square roots of frequencies are again shown. The excess of observed over expected frequencies in the right tail of the distribution is quite evident. Hazelton (2003) proposed another graphic technique that also emphasizes any deviations in the tails of the distribution but does not require the data to be grouped.

6.8 Other Continuous Distributions

Statisticians have described numerous continuous probability functions and studied their properties. Some observed nonnormal distributions can be made normal by transformation to a suitable scale (see Chapter 13). A common example is the lognormal distribution, which becomes normally distributed when the observations Y_i are transformed to $Y_i' = \ln Y_i$. Quite frequently, morphological variables are lognormally distributed. Even when a variable is normally distributed, various statistics obtained from it (such as the variance) are not normally distributed. We will study several of these distributions and their applications in subsequent chapters.

EXERCISES 6

6.1 Perform the following operations on the data of Exercise 2.4. (a) If you have not already done so, make a frequency distribution from the data and graph the results in the form of a histogram. (b) Compute the expected frequencies for each class based on a normal distribution with $\mu = \overline{Y}$ and $\sigma = s$. (c) Graph the expected frequencies in the form of a histogram and compare them with the observed frequencies. (d) Comment on the degree of agreement between observed and expected frequencies.

6.2 Carry out the operations in Exercise 6.1 on the transformed data generated in Exercise 2.6.

6.3   Assume that the petal length of a population of plants of species X is normally distributed with a mean of $\mu = 3.2$ cm and a standard deviation of $\sigma = 0.8$ cm. What proportion of the population would be expected to have a petal length (a) greater than 4.5 cm?
(b) greater than 1.78 cm? (c) between 2.9 and 3.6 cm?

ANSWER: (a) $= 0.0521$ (b) $= 0.9621$ (c) $= 0.3376$

6.4   Perform a graphic analysis on the data of Exercises 2.4 and 2.6; examine for normality
and estimate the mean and standard deviation. Use a table of rankits (Statistical Table
H).

6.5   Compute g_1 and g_2 for the data in Exercises 2.4 and 2.6. Do your results agree with what
you would expect on the basis of the graphic analyses performed in Exercise 6.4?

ANSWER: For Exercise 2.4, $g_1 = 0.2981$ and $g_2 = 0.0512$.

6.6   Perform a graphic analysis (normal quantile plot) of the butterfat data in Exercise 4.3. In
addition, plot the data with the abscissa in logarithmic units. Compare the results of the
two analyses. Draw hanging rootograms for these two distributions.

6.7   The following data are total lengths (in centimeters) of a collection of bass from a southern lake. Make a histogram of the data, compute g_1 and g_2, and perform the graphic tests.
Try to diagnose the departure from normality, if any.

29.9	40.2	37.8	19.7	30.0	29.7	19.4	39.2	24.7	20.4
19.1	34.7	33.5	18.3	19.4	27.3	38.2	16.2	36.8	33.1
41.4	13.6	32.2	24.3	19.1	37.4	23.8	33.3	31.6	20.1
17.2	13.3	37.7	12.6	39.6	24.6	18.6	18.0	33.7	38.2

ANSWER: Suggestion of bimodality, $= g_1 -0.03691$, $g_2 = 1.42854$.

6.8   Perform a graphic analysis on the following measurements. Are they consistent with
what one would expect in sampling from a normal distribution?

11.44	12.88	11.06	7.02	10.25	6.26	7.92	12.53	6.74
15.81	9.46	21.27	9.72	6.37	5.40	3.21	6.50	3.40
5.60	14.20	6.60	10.42	8.18	11.09	8.74		

6.9   Compute a normal quantile plot for the milk-yield data of Table 6.1. What do we learn
from this graph?

7

Hypothesis Testing and Interval Estimation

You are now ready for two very important steps toward learning statistics. In this chapter, we provide methods for answering two fundamental statistical questions that all biologists must answer repeatedly in the course of their work: (1) How probable is it that the differences between observed results and those expected on the basis of a hypothesis result from chance alone? (2) How reliable are one's estimates of a parameter? The first question involves the topic of hypothesis testing. The second question is answered by using intervals as estimates rather than a single value (a point estimate). These intervals are called *confidence intervals*. Both subjects belong to the field of statistical inference. The subject matter in this chapter is fundamental to an understanding of all the subsequent chapters. We therefore urge you to study the material until it has been thoroughly mastered.

In Section 7.1, we introduce the theory of hypothesis testing by means of randomization procedures. This section is fundamental to almost all of the chapters that follow. The next three sections contain background material required for an understanding of further developments in hypothesis testing. Section 7.2 describes the form of the distribution of means and their variance. In Section 7.3, we examine the distributions and variances of statistics other than the mean. Because one usually deals with samples whose true standard deviation is unknown and has to be estimated from the sample, we introduce, in Section 7.4, the *t*-distribution, which is applied in such cases. In Section 7.5, we continue with our discussion of hypothesis testing, this time employing continuous and normally distributed data. We conclude our presentation of the theory of hypothesis testing in Section 7.6 by presenting the concept of the power of a test.

In the next section, 7.7, we finally apply our newly acquired knowledge of hypothesis testing to some simple cases. In Section 7.8, another important distribution (chi-square) is introduced and then applied in Section 7.9 to hypothesis testing for variances. The next major topic in this chapter covers statistics that measure the reliability of an estimate. Confidence limits provide bounds to our estimates of population parameters. We develop the idea of a confidence interval in Section 7.10 and show its application to samples where the true standard deviation is known. However, one usually deals with samples whose true standard deviation is unknown and has to be estimated from the sample. We show how to do this in Section 7.11 using the *t*-distribution. In the next section of this chapter, 7.12, we learn to set confidence limits for the variance both by means of the chi-square distribution as well as by tabulated coefficients yielding the shortest unbiased confidence intervals. Finally, the last topic, in Section 7.13, presents the jackknife and the bootstrap, two applications of randomization tests that are useful for various estimation and significance-testing problems.

7.1 Introduction to Hypothesis Testing: Randomization Approaches

The most frequent application of statistical methods in biological research is to test a statistical hypothesis using data collected from an experiment or survey. Statistical hypotheses are assertions such as a sample was drawn from a population with a specified mean or, more commonly, two or more samples were drawn from populations with the same mean. There will also be one or more competing alternative hypotheses (often simply that the first hypothesis is false). Statistical tests are used to allow an investigator to decide between competing hypotheses, taking sampling error into account. Statistics is important in biology because results of experiments are usually not clear-cut. Statistical methods are needed to help investigators decide objectively whether the results of an experiment are consistent with the hypothesis being tested. The decision is about whether there are sufficient data to be able to reject it and accept an alternative one. The nature of the tests varies with the data and the hypothesis, but the same general philosophy of hypothesis testing is common to all tests. Study the material in this section very carefully because it is fundamental to an understanding of every subsequent chapter in this book.

Let us start with a simple example that nevertheless corresponds to one of the most commonly analyzed designs in biological and biomedical research. It is a comparison of the means of two independent samples. The dataset is from a study of skulls of ancient Egyptian males at five different historical periods (original data by Thomson and Randall-Maciver, 1905; more modern sources of the same data are Manly, 1997, and Hand et al., 1994). Among the measurements taken in this study is the maximum breadth of the skull, abbreviated MB and measured in millimeters. We chose data from two periods, 3300 B.C. (late predynastic) and 150 A.D. (Roman). For each period, 30 skulls had been measured, but for our illustration here, smaller samples are used initially. For the first sample, we randomly sampled (without replacement) 8 observations from the 3300 B.C. sample. For the second sample, we randomly selected 10 observations from the 150 A.D. sample. The actual observations in these subsamples as well as their means are shown in the following table.

Subsample 1, $n = 8$	Subsample 2, $n = 10$
131	147
126	130
131	137
132	129
135	135
137	128
130	138
132	136
	145
	134
$\bar{Y}_1 = 131.75$	$\bar{Y}_2 = 135.90$

We have estimated the means of groups 1 and 2 from the means of the subsamples of 8 and 10 observations. We would like to know whether the difference between these means is larger than what one would expect by chance if one were actually sampling from a single population. From the data we have available, we estimate the difference between their means as

$$D_{2,1} = \overline{Y}_2 - \overline{Y}_1 = 135.90 - 131.75 = 4.15 \text{ mm}$$

We introduce hypothesis testing by means of a randomization experiment. We begin with a brief digression on **randomization tests**. Such tests involve three steps:

1. Enumerate all possible outcomes, assuming the samples had been sampled from a single population. You will soon learn that such an assumption corresponds to what is called a *null hypothesis.*

2. Order these outcomes in terms of deviation from expectation under this assumption.

3. Count them to determine the proportion of outcomes that are as or more deviant than the observed difference. This proportion is interpreted as the probability of obtaining an observation as deviant as, or more deviant than, the single observed outcome.

If this probability is less than some predetermined "significance level" (usually 0.05), then the null hypothesis that the samples were drawn from populations with the same means is rejected. Although some randomization tests involve heavy computation, the general availability of computers has made them feasible, so they are applied frequently. For comprehensive treatments of the subject, see Manly (1997) and Good (2000).

We can compute the probability of a randomization test in at least three ways, although the terminological distinctions among them are not always made. If the structure of the problem is not too complicated, we can sometimes compute the number of possible outcomes using probability theory. We can then perform an **exact probability test** and base our decision on the proportion of possible outcomes that are as deviant as or more deviant than the observed sample in one or both tails. If this proportion were less than, say, 5 percent, then we would reject the hypothesis that our observed outcome is simply a chance event. You are already familiar with one such example. In Section 5.2, we worked out the probability of sibships with 14 females and 3 males on the assumption that all simple events were equally likely (because the assumed sex ratio was 50:50). In Section 17.4, we shall use 2×2 tests of independence to evaluate the probability of a given outcome (and of more deviant outcomes), assuming independence of two criteria of classification and the equal likelihood of all possible distributions of the n observations into the four cells, given various assumptions regarding marginal totals.

In the example featured in this section, it would be difficult to work out analytically the probability of a mean difference as large as or larger than the one observed. We therefore turned to the second way of carrying out a randomization test. This is the **complete enumeration** of all possible outcomes and the location of the observed outcome in the resulting distribution of outcomes. This is also an exact test, but we do not arrive at it by probability theory and algebra; instead, we use "brute force" enumeration of all possible outcomes. Another example will be presented in Chapter 18, using an example from numerical taxonomy.

Finally, the third way of obtaining a probability for a randomization test is as an estimate from a **sampled randomization test,** also known as a **Monte Carlo test.** It is used whenever the number of possible sampling outcomes is so large that total enumeration is impractical. We take random samples, compute our statistic, empirically find a distribution of this statistic, and estimate the probability of obtaining our original observed statistic or one more deviant. We shall briefly encounter such a test in this section, and others are treated in greater detail in Sections 8.3, 13.10, and 17.4.

Armed with an introductory notion of randomization tests, we now return to the main theme of this section—hypothesis testing—and to the example of the Egyptian skulls. Do the MB means differ between these two archaeological periods or is the observed difference what we might expect by chance when drawing two samples from the same population of skulls? It is this latter hypothesis that the investigator needs to dismiss in order to decide whether or not there is a difference in the MB means between the two periods. We call this hypothesis H_0, the **null hypothesis,** because it states that there is no real difference between the true, parametric value of $\delta_{2,1}$ in the populations from which we sampled and the hypothesized value of $\delta_{2,1} = 0$.

We now describe a procedure for testing the null hypothesis in this example. If there were no true difference between the means of the two populations from which we sampled, then the difference of 4.15 obtained by the observer could result (not infrequently) from a random partitioning of the data. There are $\binom{18}{8} = 43{,}758$ ways of dividing the 18 observations into two subsamples—one of 8 observations and the other of 10. For each of these we could compute the difference between the means in order to determine what proportion of these samples yield differences equal to or greater than what we obtained for our actual data. Earlier during our careers, it would have been a prohibitive task to investigate that many partitions. However, nowadays, this task can be carried out in a very few seconds on a personal computer. As we have learned above, such a procedure is termed an *exact randomization test* of the null hypothesis by complete enumeration.

The results of this computation are summarized in Figure 7.1, which shows the distribution of the 43,758 $D_{2,1}$-values in histogram form. The distribution appears to be symmetrical and close to a normal distribution, but we will not make use of that information here. An arrow points to the location of the observed difference of 4.15. Two dashed lines mark off 2.5% of the area in the lower and upper tails of the distribution. In Chapter 6, we learned that it is conventional to include all "worse" outcomes when stating a probability for continuous data—that is, all those that deviate even more from the outcome expected on the hypothesis $\delta_{2,1} = 0$. Including all worse outcomes from among the 43,758 partitions, 2685 ($= 0.0614$) are ≥ 4.15 and 2840 ($= 0.0649$) are ≤ -4.15. Thus, the overall proportion deviant by as much as 4.15 or more from the expectation $\delta_{2,1} = 0$ is 0.1263.

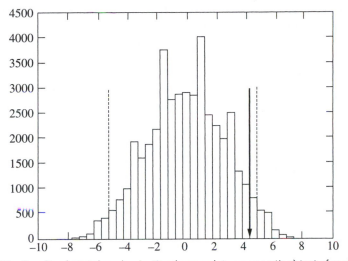

FIGURE 7.1 Results of a total randomization (= complete enumeration) test of ancient Egyptian skull data. The histogram depicts the distribution of the 43,758 different partitions of 8 and 10 measurements of maximum skull breadth from samples dated 3300 B.C. and 150 A.D., respectively. The statistic recorded is the difference between the mean of subsample 2 less the mean of subsample 1. The 95% critical values are shown by dashed lines at −5.075 and 4.825. The arrow shows the position of the observed difference (4.15) in this distribution.

If the null hypothesis H_0: $\delta_{2,1} = 0$ is true, then 12.6%, approximately $\frac{1}{8}$ of the samples will be as deviant or more deviant than this one in either direction *by chance alone*. This corresponds to a probability of 0.126 of getting a deviation at least as large as ±4.15 even though there is no true difference. This is fairly frequent and more than twice the conventional cutoff of 5%. Based on these findings alone, we would probably decide that the data are consistent with the null hypothesis of no difference, so we would, at least provisionally, accept the null hypothesis. Such results are often described by saying that no "statistically significant" difference was found with respect to the variable that was studied (in our case, the maximal breadth of skull). Although standard, such terminology is unfortunate. A statistical test simply allows one to make a decision as to whether or not the observed data are consistent within sampling error of what one expects under the null hypothesis. It does not imply that the observed difference is of such a magnitude as to be a biologically important difference. Likewise, when no statistically significant difference is found, this does not mean that the null hypothesis is necessarily true. It could just mean that the sample size was too small to allow us to confidently assert that there is a real difference. Whenever we carry out the test of a null hypothesis, we may make one of two decisions: (1) The null hypothesis is true (that is, $\delta_{2,1} = 0$) or (2) such a deviant sample is too improbable an event to justify acceptance of the null hypothesis—that is, that the hypothesis H_0: $\delta_{2,1} = 0$ is not true.

Because statistical tests are so important, a more formal description follows. Either of the decisions described above may be correct, depending on the truth of the matter. If the $\delta_{2,1} = 0$ hypothesis is correct, then the first decision (to accept the null hypothesis) will be correct. If we decide to reject the hypothesis under these circumstances, we commit an error. *The rejection of a true null hypothesis* is called a **type I error.** On the other hand, if the true difference of means between the two sets of samples is other than zero, the first decision (to accept the $\delta_{2,1} = 0$ hypothesis) is an error, a so-called **type II error,** which is *the acceptance of a false null hypothesis.* Finally, if the $\delta_{2,1} = 0$ hypothesis is not true and we do decide to reject it, then again we make the correct decision. Thus, there are two kinds of correct decisions—accepting a true null hypothesis and rejecting a false null hypothesis—and two kinds of errors—rejecting a true null hypothesis (type I) and accepting a false null hypothesis (type II). The relationships between hypotheses and decisions can be summarized as follows:

		Null hypothesis	
		Accepted	Rejected
Null hypothesis	True	Correct decision	Type I error
	False	Type II error	Correct decision

Before we carry out a test, we have to decide how frequently we are willing to allow ourselves to make a type I error (rejecting a true null hypothesis). The choice of level is a personal preference and not a value obtained by analyzing data. There will always be some samples that by chance are very deviant. The most deviant of these are likely to mislead us into believing the null hypothesis is false. If we permit 5% of samples to lead us into making a type I error, then we will expect on average to falsely reject 5 out of 100 samples when there is actually no difference, deciding that these samples were not drawn from the given population. In the distribution under study, this means that we would reject all sampled $D_{2,1}$-values equal to or greater than 4.825 or equal to or less than -5.075. These are the cutoff values of 2.5% for both tails of the distribution. Together, they cut off exactly 5% of the area. These cutoff values are called *critical values,* and they define the values of $D_{2,1}$ that lead to the acceptance or rejection of the null hypothesis. If we decide to allow an approximate 1% type I error rate, then we reject the hypothesis $\delta_{2,1} = 0$ for all samples with $D_{2,1}$-values ≥ 5.950 or ≤ -6.200. Thus, the smaller the type I error rates that we are prepared to accept, the wider the region becomes in which we would accept H_0 and thus the more deviant a sample has to be for us to reject the null hypothesis.

Your natural inclination might be to allow as little error as possible and thus use an extremely small type I error rate, such as 0.1% or even 0.01%, accepting the null hypothesis unless the sample is extremely deviant. The difficulty with such an approach is that although you are minimizing the chance of making an error of the first kind, you will be increasing the chance of making an error of the second kind (type II), accepting the null hypothesis when, in fact, it is not true and an alternative hypothesis H_1 is true. We will show later in the discussion how this comes about.

Let us learn some more terminology. The probability of a type I error is usually given the symbol α (lowercase Greek alpha). When expressed as a percentage, it is known as the **significance level.** Thus, a type I error probability of $\alpha = 0.05$ corresponds to a significance level of 5% for a given test. This is standard terminology, but it is not ideal because it seems to imply that the amount of difference that would lead to a "significant" result is in some sense important. When we cut off areas proportional to α (the type I error) on a frequency distribution, the portion of the abscissa under the area that has been cut off is called the **rejection region,** or *critical region,* of a test, and the portion of the abscissa that would lead to acceptance of the null hypothesis is called the **acceptance region.** In Figure 7.1, the histogram showing the observed distribution of $D_{2,1}$-values for MB of the Egyptian skulls, given H_0, the dashed lines separate approximate 5% rejection regions (2.5% in each tail) from the 95% acceptance region.

On the basis of a complete enumeration of partitions into 8 and 10 observations, we concluded that we would not reject the null hypothesis of no difference between the means of the two samples. This was because a difference as large as or larger than the observed difference of 4.15 mm could be obtained fairly frequently by chance. Would our results have changed had we employed larger sample sizes? Using the original data with sample sizes of 30 observations would require more than 10^{17} partitions, which would be impractical using ordinary computer facilities. A complete enumeration of such a large number of partitions is also not necessary. We can obtain a very good estimate of the probability of obtaining a difference equal to or larger than the observed difference by computing the differences obtained for a large number of partitions sampled at random from the population of all possible partitions. As we have seen, this is called a *Monte Carlo randomization test,* or a sampled randomization test, which is distinct from an exact randomization test based on a complete enumeration.

Employing a Monte Carlo randomization test allows us to analyze much larger problems. Using the original Egyptian skull data ($n = 30$ for each sample), differences between the group means were computed for a random sample of 50,000 partitions. The results are shown in Figure 7.2. In this distribution, the left tail 2.5% cutoff is at -2.733 and the right tail 2.5% is at 2.733. The observed difference between the two means is 3.80, which falls well into the rejection region of the distribution. Thus, a difference such as that observed is quite unlikely to be obtained by chance if there was actually no true difference between the mean. The proportion of random partitions that yielded differences equal to or larger than the observed difference irrespective of its sign was 0.0047.

The conclusions just cited are based on testing against an alternative hypothesis, $H_1\colon \mu_1 \neq \mu_2$, that ignores which of the μ's is larger. This is a two-tailed test. In this example, however, it might be reasonable to assume that μ_2 could only be larger than μ_1 because it is known from more extensive data that there is a trend toward an

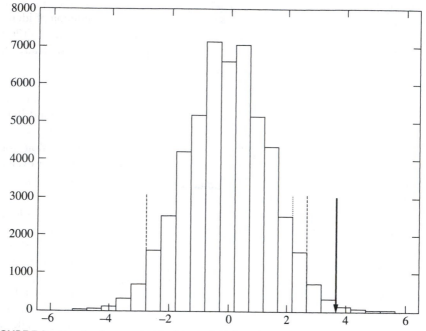

FIGURE 7.2   Results of a sampled (= Monte Carlo) randomization test of ancient Egyptian skull data. The histogram depicts the distribution of 50,000 random partitions of the full dataset, 30 measurements each, of maximum skull breadth from samples dated 3300 B.C. and 150 A.D. The statistic recorded is the difference between the mean of subsample 2 less the mean of subsample 1. The observed difference (3.80) is shown by the arrow. The 95% critical values (−2.733 and 2.733) are shown with dashed lines. The one-tailed upper 5% critical value (2.333) is shown as a dotted line.

increase in skull breadth with the more recent archaeological period. In such a case, a more appropriate alternative hypothesis is H_1: $\mu_2 > \mu_1$. This leads to a one-tailed test in which there is only a single cutoff in the upper tail of the distribution. We find that the upper 5% of the null distribution starts at 2.333. Hence, our observed difference of 3.80 is far less likely than 5%. Indeed, the one-tailed probability of the observed difference is $P = 0.0028$. By restricting the alternative hypothesis, we have lowered the P-value of the test.

Randomization tests are distribution free—that is, they are not dependent on an assumption that the observations are drawn from any particular statistical distribution such as the normal distribution. However, that does not mean that there are no assumptions necessary for the validity of the test. The assumptions made for the above test were twofold: (1) Observations were exchangeable and (2) both samples had identical distributions. The first of these assumptions means that if the null hypothesis were true, then all the observations would have been sampled from the same population, and any one observation can exchange places with any other observation in the same sample or the other one. If there is any reason why this assumption is invalid, the permutation test becomes unreliable. The second assumption of identity of distributions is necessary for exactness and efficiency. However, as Good (2000) shows in a review of the literature, mild violations of this assumption still lead to approximately correct solutions.

So far, we have not yet said much about type II error. Randomization tests are not ideal material with which to delve into this topic. We therefore present a discussion of type II error below using a different statistical approach and a different biological example.

We do not need to perform the computationally intensive randomization test if we are willing to make some assumptions about the distribution and can evaluate the probability of a particular outcome directly from probability theory. Recall the sex-ratio example considered in Section 5.2. We observed a litter of 3 males and 14 females and wanted to know whether a null hypothesis of a 1:1 sex ratio was tenable. The exact probabilities of the 18 possible outcomes in sex ratio for a litter of 17 offspring can be obtained from the binomial frequency distribution (see Tables 5.3 and 7.1). We found that the probability of obtaining a deviation as great as or greater than 3.14 in both tails of the distribution sums to 0.012,725,8. Although possible, such a large deviation is infrequent, so we are led to the conclusion that it is unlikely that our litter came from a population producing equal numbers of male and female offspring. We show a bar diagram of the expected distribution of sex ratios under the null hypothesis in Figure 7.3**a**.

Now let us take a closer look at type II errors, the accepting of a null hypothesis when it is false. If you try to evaluate the probability, β, of a type II error, you immediately run into a problem. If the null hypothesis H_0 is false, then some other hypothesis H_1 must be true. But unless you can specify H_1, you are not able to calculate a type II error. Suppose in our sex-ratio case, based on our knowledge of the genetics of this species, we believe that there are only two reasonable possibilities—(1) our initial hypothesis H_0: $p_\female = q_\male$, or (2) an alternative hypothesis H_1: $p_\female = 2q_\male$, which states that the sex ratio is 2:1 in favor of females, so $p_\female = \frac{2}{3}$ and $q_\male = \frac{1}{3}$. We now have to calculate expected frequencies for the binomial distribution $(p_\female + q_\male)^k = (\frac{2}{3} + \frac{1}{3})^{17}$ to find the probabilities of the outcomes assuming that the second hypothesis is true. These frequencies are shown graphically in Figure 7.3**b** and are tabulated and compared with expected frequencies of the earlier distribution in Table 7.1.

Suppose we had decided on a type I error of $\alpha \approx 0.01$, as shown in Figure 7.3**a**. If we allow a 1% type I error rate, then we would accept the H_0 for all samples of 17 having 13 or fewer animals of one sex. Approximately 99% of all samples will fall into this category. However, what if H_0 is not true and H_1 is true? Clearly, from the population implied by hypothesis H_1, we could also obtain outcomes in which one sex was represented 13 or fewer times in samples of 17. If we calculate the proportion of the curve based on hypothesis H_1 that overlaps the acceptance region of the distribution based on hypothesis H_0, we find that $\beta = 0.8695$ of the distribution based on H_1 overlaps the acceptance region of H_0 (see Figure 7.3**b**). Thus, if H_1 is really true (and H_0 correspondingly false), we would erroneously accept the null hypothesis 86.95% of the time.

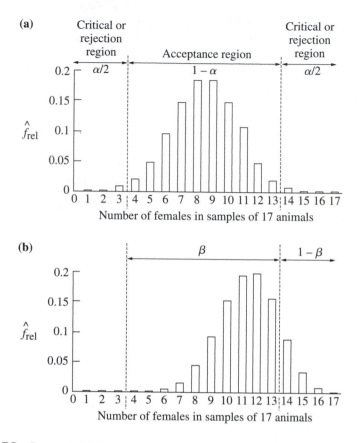

FIGURE 7.3 Expected distributions of outcomes when sampling 17 animals from two hypothetical populations. **(a)** $H_0: p_♀ = q_♂ = \frac{1}{2}$. **(b)** $H_1: p_♀ = \frac{2}{3}$ and $q_♂ = \frac{1}{3}$. Dashed lines separate rejection, or critical, regions from the acceptance region of the distribution of part (a). Type I error α equals approximately 0.01.

This percentage corresponds to the expected proportion of samples from H_1 that fall within the limits of the acceptance regions of H_0. This proportion is called β, the type II error expressed as a proportion. In this example, β is quite large. If H_0 were true, then only 1% of the samples would lead us to falsely reject H_0. However, if H_1 were true, then 87% of the samples would lead us to falsely reject H_1. This unsatisfactory situation is due to the relatively small sample size, $k = 17$. The probability of a type II error also depends on the amount of difference between the distributions implied by the two hypotheses. As H_1 approaches H_0 (as in $H_1: p_♀ = 0.55$, for example), the two distributions would overlap more and more and the magnitude of β would increase, making discrimination between the hypotheses even less likely. Conversely, if H_1 states that $p_♀ = 0.9$, the distributions would be much farther apart and type II error β would be reduced. Clearly, then, the magnitude of β depends, among other

TABLE 7.1 Relative Expected Frequencies for Samples of 17 Animals Under Two Hypotheses

Binomial Distribution

(1)	(2)	(3)	(4)
		$H_0: p_♀ = q_♂ = \frac{1}{2}$	$H_1: p_♀ = 2q_♂ = \frac{2}{3}$
♀ ♀	♂ ♂	$\hat{f}_{rel}$	$\hat{f}_{rel}$
17	0	0.000,007,6	0.001,015,0
16	1	0.000,129,7	0.008,627,2
15	2	0.001,037,6	0.034,508,6
14	3	0.005,188,0	0.086,271,5
13	4	0.018,158,0	0.150,975,2
12	5	0.047,210,7	0.196,267,7
11	6	0.094,421,4	0.196,267,7
10	7	0.148,376,5	0.154,210,4
9	8	0.185,470,6	0.096,381,5
8	9	0.185,470,6	0.048,190,7
7	10	0.148,376,5	0.019,276,3
6	11	0.094,421,4	0.006,133,4
5	12	0.047,210,7	0.001,533,3
4	13	0.018,158,0	0.000,294,9
3	14	0.005,188,0	0.000,042,1
2	15	0.001,037,6	0.000,004,2
1	16	0.000,129,7	0.000,000,3
0	17	0.000,007,6	0.000,000,0
Total		1.000,000,2	1.000,000,0

things, on the parameters of the alternative hypothesis H_1 and cannot be specified without knowledge of these parameters.

When the alternative hypothesis is fixed, as in the previous example ($H_1: p_♀ = 2q_♂$), the magnitude of the type I error α we are prepared to tolerate will determine the magnitude of the type II error β. The smaller the rejection region α in the distribution under H_0, the greater the acceptance region $1 - \alpha$ will be in this distribution. The greater $1 - \alpha$, however, the greater its overlap will be with the distribution representing H_1, and hence the greater β. Convince yourself of this in Figure 7.3. By moving the dashed lines outward, we are reducing the critical regions representing type I error α in diagram (a). But as the dashed lines move outward, more of the distribution based on H_1 in diagram (b) will lie in the acceptance region of the null hypothesis. Thus, by decreasing α we are increasing β and in a sense defeating our own purposes.

In most applications, scientists want to keep both of these errors small because they do not wish to reject a null hypothesis when it is true, nor do they wish to accept it when another hypothesis is correct. We will see what steps can be taken to decrease

β while holding α constant at a preset level. Note, however, that there are special applications, often nonscientific, in which one type of error is less serious than the other, and our strategy of testing or method of procedure would obviously take this into account. Thus, if you were a manufacturer producing a certain item according to specifications (which correspond to the null hypothesis in this case), you would wish to maximize your profits and to reject as few as possible, which is equivalent to making α small. That is why in industrial statistics, α is known as **producers' risk.** You might not be as concerned with samples that come from a population specified by the alternative hypothesis H_1 but that appear to conform with H_0 because they fall within its acceptance range. Such products could conceivably be marketed as conforming to specification H_0.

On the other hand, as a consumer you would not mind so much a large value of α, representing a large proportion of rejects in the manufacturing process. You would, however, be greatly concerned about keeping β as small as possible because you would not wish to accept items as conforming to H_0, which in reality were samples from the population specified by H_1 that might be of inferior quality. For this reason, β is known in industrial statistics as **consumers' risk.**

Let us summarize what we have learned up to this point. When we have to carry out a statistical test, we first specify a null hypothesis H_0 and set a type I error probability of α. In the case of the sex ratios, we defined H_0: $p_{\female} = q_{\male}$ and $\alpha \approx 0.01$. Having done this, we take a sample and see whether the sample statistic is within the acceptance region of the null hypothesis. Our sample turned out to be 14 females and 3 males. Because this sample statistic falls in the rejection region, we reject the null hypothesis and conclude that this sample came from a population in which $p_{\female} \neq q_{\male}$.

If we can specify an alternative hypothesis precisely, we can calculate the probability of type II error. For example, if H_1: $p_{\female} = 2q_{\male}$, then $\beta = 0.8695$. In certain special situations in which the alternative hypotheses can be clearly specified, as in genetics and in this example, one could then test the previous alternative hypothesis, changing it into the null hypothesis. Thus, we might now wish to test whether the true sex ratio is $2\female\female:1\male$. From Figure 7.3b it is obvious that the probability of 14 males and 3 females is very small and can be ignored. The probability of obtaining 14 or more females under the new null hypothesis is $1 - \beta$ of the old alternative hypothesis, as illustrated in Figure 7.3b. Hence, this probability would be 0.1305, and, if we use $\alpha = 0.05$, we cannot reject the new null hypothesis. Note that in this somewhat unusual case of having two precisely specified competing hypotheses, one would not arbitrarily designate one as the null hypothesis and the other as the alternative because, as we have just seen, the decision to accept one hypothesis over the other would depend on that arbitrary choice. In practice, one would compute the expected frequencies as shown in Table 7.1 and then select the hypothesis that has the larger expected frequency for the observed data. In examining Table 7.1, one can see that the hypothesis that $p_{\female} = q_{\male}$ would be accepted if there were 9 or fewer females (with a probability of 0.314529 of falsely rejecting this hypothesis) and rejected if there were more than 10 females (with a probability of 0.171,856,7 of falsely rejecting the hypothesis that $p_{\female} = 2q_{\male}$).

The type I error probabilities can be varied at will by the investigator. Most computer programs now compute probabilities for the outcomes under the null hypothesis. In earlier years, the choices were limited, however, because cumulative probabilities of the appropriate distributions had not been tabulated for many tests. One had to use published probability levels, which commonly were 0.05, 0.01, and 0.001. In such cases, when a null hypothesis has been rejected at a specified level of α, we would often say that the sample is **significantly different** from the parametric or hypothetical population at probability $P \leq \alpha$. Nowadays, when the program yields a computed P-value, it should be reported as is, rather than expressed as an inequality against one of the conventional critical values of 0.5, 0.01, and 0.001. Generally, values of α greater than 0.05 are not considered to correspond to a **statistically significant** difference. A type I error probability of 5% ($P = 0.05$) corresponds to a frequency of one type I error in 20 trials, and a level of 1% ($P = 0.01$) to one error in 100 trials. Type I error probabilities less than 1% ($P \leq 0.01$) nearly always lead an investigator to reject a null hypothesis; those between 5% and 1% may also lead to the rejection of H_0 at the discretion of the investigator. Statistical significance has a special technical meaning (H_0 rejected at $P \leq \alpha$). It does not mean that the difference is necessarily of such a magnitude as to be biologically important. We shall use the adjective *significant* only in the statistical sense in this book. We discourage its use in scientific papers and reports, unless its technical meaning is clearly implied. For general descriptive purposes, synonyms such as *important, meaningful, marked,* and *noticeable* can serve to underscore large differences and effects.

A brief remark on null hypotheses represented by asymmetrical probability distributions is in order here. Suppose our null hypothesis in the sex-ratio case had been $H_0\colon p_\female = \frac{2}{3}$, as discussed above. The distribution of samples of 17 offspring from such a population is shown in Figure 7.3**b**. Because this distribution is clearly asymmetrical, the critical regions have to be defined independently. Before we can present methods for normally distributed variables, we have to furnish you with some essential background information in the next three sections.

7.2 Distribution and Variance of Means

We commence our study of the distribution and variance of means with a sampling experiment. In a previous sampling experiment (Section 6.1), 1000 samples of 5 housefly wing lengths and similar means of milk yields were collected. For each of these two variables, we also obtained the mean of samples of 35 observations. Here again we will make a frequency distribution of these means.

Figure 7.4**a** shows a histogram of the 1000 means of samples of 5 housefly wing lengths. The mean and standard deviation of the 1000 means are given. Normal quantile plots for these data are superimposed in Figure 7.4**a**. Note that the distribution appears quite normal, as does that of the means based on 1000 samples of 35 wing lengths (Figure 7.4**b**). This illustrates an important theorem: *The means of samples from a normally distributed population are themselves normally distributed regardless of sample size n.* Thus, we note that the means of samples from the normally

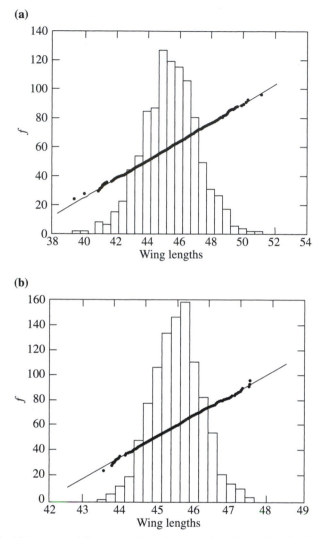

(a)

(b)

FIGURE 7.4   Histograms of the means of random samples of housefly wing lengths with normal quantile plots superimposed. The scales for the normal quantile plots are not shown. **(a)** 1000 random samples of size $n = 5$. $\overline{Y} = 45.462$, $s = 1.682$, $\sigma_{\overline{Y}} = 1.744$. **(b)** 1000 random samples of size $n = 35$. $\overline{Y} = 45.536$, $s = 0.666$, $\sigma_{\overline{Y}} = 0.659$.

distributed housefly wing lengths are distributed normally whether they are based on 5 or 35 individual readings.

Both distributions of means of the heavily skewed milk yields (Figures 7.5a and 7.5b) appear to be close to normal distributions. The means based on 5 milk yields, however, do not agree with the normal distribution nearly as well as do the means based on 35 observations. This illustrates another theorem of fundamental importance

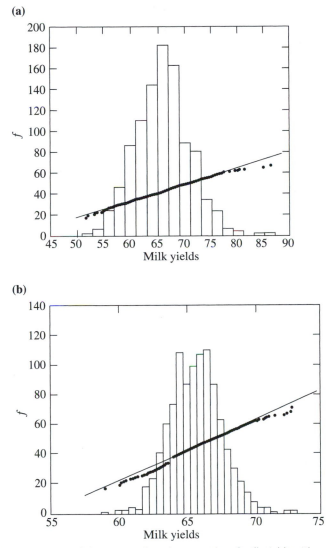

FIGURE 7.5 Histograms of the means of random samples of milk yields with normal quantile plots superimposed. Layout as in Figure 7.4. **(a)** 1000 random samples of size $n = 5$. $\overline{Y} = 65.832$, $s = 4.862$, $\sigma_{\overline{Y}} = 4.991$. **(b)** 1000 random samples of size $n = 35$. $\overline{Y} = 65.6928$, $s = 1.9077$, $\sigma_{\overline{Y}} = 1.886$.

in statistics: *As sample size increases, the means of samples drawn from a population of any distribution will approach the normal distribution.* This theorem, when rigorously stated (about sampling from populations with finite variances), is known as the **central limit theorem** and is illustrated by the near normality of the milk yield means based on 35 observations (Figure 7.5b) as compared with the considerable skewness of the means based on 5 milk yields (Figure 7.5a). The importance of this theorem is that if n is large enough, it permits us to use the normal distribution to make statistical inferences about means of populations even though the observations are not at all normally distributed. The necessary size of n depends on the distribution (skewed populations require larger samples).

Another important fact is that the range of the wing length means is considerably less than that of the original observations, which ranged from 36 to 55. Thus, in Figure 7.4a, the classes of the wing length means in samples of 5 range from 39.5 to 51.0 with class intervals of 0.5. Means in Figure 7.4b are based on samples of 35 and range from 43.5 to 47.5 with class intervals of 0.25. The classes of the milk yield means range from 52 to 86 in samples of 5 (see Figure 7.5a, where the class interval is 2.0) and from 60 to 72 in samples of 35 (see Figure 7.5b, class interval = 0.5), but the individual milk yields range from 51 to 98. Not only do means show less scatter than the observations on which they are based (an easily understood phenomenon if you give some thought to it) but also the range of the distribution of the means diminishes as the sample size on which the means are based increases.

The differences in ranges are reflected in differences in the standard deviation of these distributions. If we calculate the standard deviations of the means in the four distributions in Figures 7.4a through 7.5b, we obtain the following values:

	Observed Standard Deviation of Distribution Means	
	$n = 5$	$n = 35$
Wing lengths	1.682	0.666
Milk yields	4.862	1.908

Note that the standard deviations of the sample means based on 35 observations are considerably less than those based on 5. This is also intuitively obvious. Means based on large samples should be close to the parametric mean and will not vary as much as will means based on small samples. The variance of means is therefore partly a function of the sample size on which the means are based. It is also a function of the variance of the observations in the samples. Thus, in the table above, the means of milk yields have a much greater standard deviation than means of wing lengths based on comparable sample size simply because the standard deviation of the individual milk yields (11.1575) is greater than that of individual wing lengths (3.90).

That the variance of means decreases as sample size n increases can also be seen in Figures 7.4a through 7.5b. The slopes of the normal quantile plots in Figures 7.4b and 7.5b (samples of 35 each) are steeper than those in Figures 7.4a and 7.5a (samples of 5 each), respectively, indicating less scatter (hence less variance) in the means based on the larger samples. The increased steepness may not be obvious on casual inspection because both ordinate and abscissa have been rescaled in Figures 7.4b and 7.5b over those of Figures 7.4a and 7.5a, respectively.

The expected value of the variance of sample means can be calculated. By *expected value*, we mean the average value to be obtained by infinitely repeated sampling. Thus, if we were to take samples of *a* means of *n* observations repeatedly and were to calculate the variance of these *a* means each time, the average of these variances would be the expected value. Usually, we obtain the expected value of the variance of sample means by a formula, the derivation of which is shown below. We defer to Section 15.3 (and Appendix A.13), the proof that the variance of the sum of *independent* variables $Y_1 + Y_2 + \cdots \ldots + Y_n$ equals $\sigma_1^2 + \sigma_2^2 + \cdots + \sigma_n^2 = \sum^n \sigma_i^2$. If these variables are also *identically distributed*, which they would be if they were observations from the same statistical population, this sum simplifies to $n\sigma^2$. However, this is the variance of $\sum Y$, and we wish to have the variance of $\overline{Y}$. Because $\overline{Y} = \sum Y/n$, we follow Section 4.8 and Appendix A.2 and apply a squared multiplicative code $(1/n)^2$ to the variance of the sum of the observations with the result that

$$\sigma_{\overline{Y}}^2 = \frac{n\sigma^2}{n^2} = \frac{\sigma^2}{n} \tag{7.1}$$

Consequently, the expected standard deviation of means is

$$\sigma_{\overline{Y}} = \frac{\sigma}{\sqrt{n}} \tag{7.2}$$

This formula makes it clear that the standard deviation of means is a function of the standard deviation of observations, as well as of the sample size of means. The greater the sample size, the smaller the standard deviation of means. In fact, as sample size increases to a very large number, the standard deviation of means becomes vanishingly small. This makes good sense. Large sample sizes, averaging many observations, should yield estimates of means less variable than those based on a few items.

When working with samples from a population, we do not, of course, know the parametric standard deviation σ of the population but can only obtain a sample estimate, s, of it. Also, we would be unlikely to have numerous samples of size n from which to compute the standard deviation of means directly. Therefore, we usually have to estimate the standard deviation of means from a single sample by using Expression (7.2), substituting s for σ:

$$s_{\overline{Y}} = \frac{s}{\sqrt{n}} \tag{7.3}$$

Thus, we obtain, from the standard deviation of a single sample, an estimate of the standard deviation of means we would expect were we to obtain a collection of means based on equal-sized samples of n observations from the same population. As we shall see, this estimate of the standard deviation of a mean is a very important and frequently used statistic.

Table 7.2 illustrates some estimates of the standard deviations of means that might be obtained from random samples of the two populations that we have been discussing.

TABLE 7.2 Means, Standard Deviations, and Standard Deviations of Means (Standard Errors) of 5 Random Samples of 5 and 35 Housefly Wing Lengths and Jersey Cow Milk Yields

Parametric values for the statistics are given in the sixth line of each category.

	(1) $\bar{Y}$	(2) s	(3) $s_{\bar{Y}}$
	Wing lengths		
	45.8	1.095	0.490
	45.6	3.209	1.435
$n = 5$	43.6	4.827	2.159
	44.8	4.764	2.131
	46.8	1.095	0.490
$\mu = 45.5$	$\sigma = 3.90$	$\sigma_{\bar{Y}} = 1.744$	
	45.37	3.812	0.644
	45.00	3.850	0.651
$n = 35$	45.74	3.576	0.604
	45.29	4.198	0.710
	45.91	3.958	0.669
$\mu = 45.5$	$\sigma = 3.90$	$\sigma_{\bar{Y}} = 0.659$	
	Milk yields		
	66.0	6.205	2.775
	61.6	4.278	1.913
$n = 5$	67.6	16.072	7.188
	65.0	14.195	6.348
	62.2	5.215	2.332
$\mu = 66.61$	$\sigma = 11.160$	$\sigma_{\bar{Y}} = 4.991$	
	65.429	11.003	1.860
	64.971	11.221	1.897
$n = 35$	66.543	9.978	1.687
	64.400	9.001	1.521
	68.914	12.415	2.099
$\mu = 66.61$	$\sigma = 11.160$	$\sigma_{\bar{Y}} = 1.886$	

Sampled from Table 6.1.

The means of 5 samples of wing lengths based on 5 individuals ranged from 43.6 to 46.8, their standard deviations from 1.095 to 4.827, and the estimate of standard deviation of the means from 0.490 to 2.159. Ranges for the other categories of samples in Table 7.2 similarly include the parametric values of these statistics. The estimates of the standard deviations of the means of the milk yields cluster around the expected value because they are not dependent on normality of the individual observations. However, in a particular sample in which, by chance, the sample standard deviation is a poor estimate of the population standard deviation (as in the second sample of five milk yields), the estimate of the standard deviation of means is equally wide of the mark.

We should emphasize one difference between the standard deviation of observations and the standard deviation of sample means: If we estimate a population standard deviation through the standard deviation of a sample, the magnitude of the estimate will not change as we increase the sample size. We may expect that the estimate will improve and approach the true standard deviation of the population; however, its order of magnitude will be the same, whether the sample is based on 3, 30, or 3000 individuals, as Table 7.2 clearly shows. The values of s are closer to σ in the samples based on $n = 35$ than in samples of $n = 5$. Yet the general magnitude is the same in both instances. The standard deviation of means, however, decreases as sample size increases, as is obvious from Expression (7.3). Thus, means based on 3000 observations will have a standard deviation only 1/10th that of means based on 30 measurements, as illustrated here:

$$\frac{s}{\sqrt{3000}} = \frac{s}{\sqrt{30} \times \sqrt{100}} = \frac{s}{\sqrt{30} \times 10}$$

Because $s = \sqrt{s^2}$ is a biased estimate of σ, $s_{\bar{Y}}$ will also be biased. If an unbiased estimate is desired, $s_{\bar{Y}}$ can be corrected by the procedure described in Section 4.7.

7.3 Distribution and Variance of Other Statistics

Just as we obtained a mean and a standard deviation from each sample of the wing lengths and milk yields, so we could also have obtained other statistics from each sample, such as variance, median, coefficient of variation, or g_1. After repeated sampling and computation, we would have frequency distributions for these statistics and would be able to compute their standard deviations just as we did for the frequency distribution of means. In many cases, the statistics are distributed normally, as was true for the means. In other cases, the statistics will be distributed normally only if they are based on samples from a normally distributed population, or if they are based on large samples, or if both these conditions hold. Figure 7.6 shows a frequency distribution of the variances from the 1000 samples of 5 housefly wing lengths (see Figure 7.4a). Notice that the distribution is strongly skewed to the right, which is characteristic of the distribution of variances of small samples.

Standard deviations of statistics are generally known as **standard errors.** Beginners sometimes get confused by an imagined distinction between standard deviations and standard errors. The standard error of a statistic such as the mean (or g_1) is the standard deviation of a distribution of means (or g_1's) for samples of a given sample size n. Thus, the terms *standard error* and *standard deviation* are used synonymously with the following exception: It is not customary to use standard error as a synonym for standard deviation of observations in a sample or population. Standard error or standard deviation has to be qualified by referring to a given statistic, such as the standard deviation of g_1, which is the same as the standard error g_1. Used without

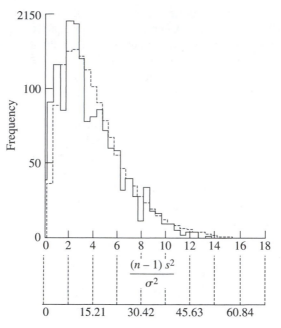

FIGURE 7.6   Histograms of variances (solid line) based on 1000 samples of five housefly wing lengths from Table 6.1 and of expected $\chi^2_{[4]}$ distribution (dotted line). Abscissa is given in terms of both s^2 and $(n-1)s^2/\sigma^2$.

any qualification, "standard error" conventionally implies the standard error of the mean. "Standard deviation" used without qualification generally means standard deviation of observations in a sample or population. Having means, standard deviations, standard errors, and coefficients of variation in a table signifies that arithmetic means, standard deviations of measurements in samples, standard deviations of their means (= standard errors of means), and coefficients of variation are displayed. The following summary of terms may be helpful:

$$\text{Standard deviation} = s = \sqrt{\sum y^2/(n-1)}$$

Standard deviation of a statistic St = standard error of a statistic St = s_{St}

Standard error = standard error of a mean = standard deviation of a mean = $s_{\bar{Y}}$

Standard errors are usually not obtained from a frequency distribution by repeated sampling but are estimated from only a single sample and represent the expected standard deviation of the statistic if a large number of such samples had been obtained. Remember that we estimated the standard error of a distribution of means from a single sample in this manner in the previous section.

Box 7.1 lists the standard errors of common statistics. Column (1) lists the statistic whose standard error is described; column (2) shows the formula for the estimated

BOX 7.1 Standard Errors for Common Statistics

(1) Statistic	(2) Estimate of standard error	(3) df	(4) Comments on applicability
1 $\bar{Y}$	$s_{\bar{Y}} = \dfrac{s}{\sqrt{n}} = \dfrac{s_Y}{\sqrt{n}} = \sqrt{\dfrac{s^2}{n}}$	$n-1$	True for any population with finite variance
2 Median	$s_{\text{med}} = (1.2533)s_{\bar{Y}}$	$n-1$	Large samples from normal populations
3 Average deviation (AD)	$s_{\text{AD}} \approx (0.602{,}810{,}3)\dfrac{s}{\sqrt{n}}$	$n-1$	For large n (>100)
4 s	$s_s = \dfrac{s}{n}\sqrt{2(n-1)\left[\dfrac{\pi}{2} + \sqrt{n(n-1)} - n - \arcsin\left\{\dfrac{1}{(n-1)}\right\}\right]}$ $= (0.707{,}106{,}8)\dfrac{s}{\sqrt{n}}$	$n-1$	Samples from normal populations ($n > 15$)
5 V	$s_V \approx \dfrac{V}{\sqrt{2n}}\sqrt{1 + 2\left(\dfrac{V}{100}\right)^2}$ $s_V \approx \dfrac{V}{\sqrt{2n}}$	$n-1$	Samples from normal populations
5* $V*$	$s_{V*} = \left(1 + \dfrac{1}{4n}\right)s_V$	$n-1$	Used when $V < 15$
6 g_1	$s_{g_1} = \sqrt{\dfrac{6n(n-1)}{(n-2)(n+1)(n+3)}} \approx \sqrt{\dfrac{6}{n}}$	8	Samples from normal populations. The approximate formula is for large n (>150)
7 g_2	$s_{g_2} = \sqrt{\dfrac{24n(n-1)^2}{(n-3)(n-2)(n+3)(n+5)}} \approx \sqrt{\dfrac{24}{n}}$	8	Samples from normal populations. The approximate formula is for large n (>150)

standard error; column (3) gives the degrees of freedom on which the standard error is based (their use is explained in Section 7.5); column (4) provides comments on the range of application of the standard error. The uses of these standard errors will be illustrated in subsequent sections.

7.4 The t-Distribution

The deviations $\overline{Y} - \mu$ of sample means from the parametric mean of a normal distribution are themselves distributed normally. If these deviations are divided by the parametric standard deviation, $(\overline{Y} - \mu)/\sigma_{\overline{Y}}$, they are still distributed normally, with $\mu = 0$ and $\sigma = 1$. In fact, they form a standard normal distribution (see Section 6.7). Subtracting the constant μ from every $\overline{Y}$ is simply an additive code (see Section 4.8) and will not change the form of the distribution of sample means, which is normal (see Section 7.2). Dividing each deviation by the constant $\sigma_{\overline{Y}}$ reduces the variance to unity, but it does so proportionally for the entire distribution, so its shape is not altered and a previously normal distribution remains so.

If, on the other hand, we had calculated the variance s_i^2 of each of the samples and calculated the deviation for each mean $\overline{Y}_i$ as $(\overline{Y}_i - \mu)/s_{\overline{Y}_i}$, where $s_{\overline{Y}_i}$ stands for the estimate of the standard error of the mean of the ith sample, we would have found the distribution of the deviations to be leptokurtic (more items near the center and at the tails and fewer items at the shoulders relative to that found in a normal distribution; see Section 6.6). This distribution is illustrated in Figure 7.7, which shows the ratio $(\overline{Y}_i - \mu)/s_{\overline{Y}_i}$ for the 1000 samples of 5 housefly wing lengths of Figure 7.4a. The new distribution is wider than the corresponding normal distribution because the denominator is the sample standard error rather than the parametric standard error and thus will sometimes be smaller, at other times greater, than expected. This increased variation will be reflected in the greater variance of the ratio $(\overline{Y}_i - \mu)/s_{\overline{Y}_i}$. The expected distribution of this ratio, called the **t-distribution,** is also known as *Student's distribution* after W. S. Gossett, who first described it publishing under the pseudonym "Student." The t-distribution is a function with a complicated mathematical formula that need not be presented here.

Note that the sample standard deviation s that enters into the denominator of the t-ratio must *not* be corrected for bias by multiplying by C_n as shown in Section 4.7. The t-distribution itself allows for the bias inherent in the estimate of s.

Like the normal distribution, the t-distribution is symmetric and extends from negative to positive infinity. It differs from the normal distribution, however, in that it assumes different shapes, depending on the number of degrees of freedom. By "degrees of freedom," we mean the quantity $n - 1$, where n is the sample size for the variance. Remember that the quantity $n - 1$ is the divisor in obtaining an unbiased estimate of the variance from a sum of squares (see Section 4.7). The number of degrees of freedom pertinent to a given t-distribution is the same as the number of degrees of freedom of the standard deviation in the ratio $(\overline{Y} - \mu)/s_{\overline{Y}}$. Degrees of freedom (abbreviated df or symbolized by the Greek letter ν) can range from 1 to infinity. A t-distribution for $df = 1$ deviates the most from the normal distribution. As

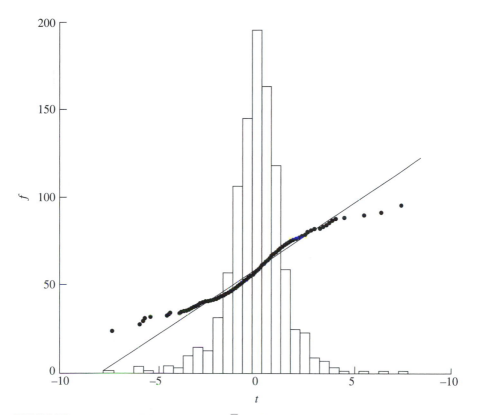

FIGURE 7.7 Histogram of the quantity $t_s = (\bar{Y} - \mu)/s_{\bar{Y}}$ with a normal quantile plot superimposed for 1000 samples of 5 housefly wing lengths. The scale for the normal quantile plot is not shown.

the number of degrees of freedom increases, the *t*-distribution approaches the shape of the standard normal distribution ($\mu = 0$, $\sigma = 1$) ever more closely, and in a graph the size of this page, a *t*-distribution of $df = 30$ is essentially indistinguishable from a normal distribution. At $df = \infty$, the *t*-distribution *is* the normal distribution. Thus, we can think of the *t*-distribution as the general case and the normal distribution as the special case of the *t*-distribution in which $df = \infty$. Figure 7.8 shows *t*-distributions for 1 and 2 degrees of freedom compared with a normal frequency distribution.

We were able to employ a single table for the areas of the normal curve by coding the argument in standard deviation units (see Section 6.4). However, because the *t*-distributions differ in shape for differing degrees of freedom, a separate table of areas under the *t*-distribution is needed for each value of *df*. This would make for a very cumbersome and elaborate set of tables. Cumulative *t*-distributions for selected degrees of freedom are given in Table 9 of Pearson and Hartley (1958). Nowadays, such tables are rarely required because computer subroutines have been developed that furnish the cumulative area of Student's distribution for any degree of freedom

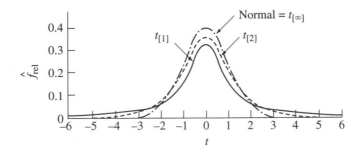

FIGURE 7.8   Frequency curves of t-distributions for 1 and 2 degrees of freedom compared with the normal distribution.

and any value of t. Such subroutines are built into most programs for statistical tests that require the t-distribution. If you have access to such a program or to the Pearson and Hartley table, try looking up a given value of t, say, $t = 1.0$, for $df = 5$. The cumulative area includes 0.81839 of the area of the t-distribution for $df = 5$. The corresponding value for the normal distribution, from a table of the areas of the normal curve, is 0.84134. The latter figure can also be found in the table of the cumulative t-distribution under $df = \infty$. Because a greater proportion of the area is drawn out into the tails, less of the area is found between one standard deviation above and below the mean in the t-distribution than in the normal distribution.

When neither a program nor the Pearson and Hartley table are available, data analysts must resort to conventional t-tables. These, however, are arranged differently. Statistical Table **B** shows degrees of freedom and probability as arguments and the corresponding values of t as functions. The probabilities indicate the percent of the area in both tails of the curve (to the right and left of the mean) beyond the indicated value of t. Thus, looking up the *critical value* of t at probability $P = 0.05$ and $df = 5$, we find $t = 2.571$ in Statistical Table **B**. Because this is a two-tailed table, the probability of 0.05 means that 0.025 of the area in each tail will be beyond a t-value of ± 2.571. Recall that the corresponding value for infinite degrees of freedom (for the normal curve) is ± 1.960. Only those probabilities generally used are shown in Statistical Table **B**. A fairly conventional symbolism is $t_{\alpha[\nu]}$, meaning the tabled t-value for ν degrees of freedom and proportion α in both tails ($\alpha/2$ in each tail), which is equivalent to the t-value for the cumulative probability of $1 - \alpha/2$. Try looking up some of these values to become familiar with the table. For example, convince yourself that $t_{.05[7]}$, $t_{.01[3]}$, $t_{.02[10]}$, and $t_{.05[\infty]}$ correspond to 2.365, 5.841, 2.764, and 1.960, respectively.

We will use the t-distribution in the next section's continuation of hypothesis testing, as well as later on in this chapter and elsewhere in this book.

7.5 More on Hypothesis Testing: Normally Distributed Data

We review now what we have learned in Section 7.1 by means of another example, this time involving a continuous frequency distribution—the normally distributed housefly wing lengths from Table 6.1—of parametric mean $\mu = 45.5$ and variance $\sigma^2 = 15.21$. Means based on five observations sampled from these are also distributed normally (see Figures 7.4**a, b**). Assume that someone presents you with a single sample of five housefly wing lengths and you wish to test whether it could have come from the specified population. Your null hypothesis will be $H_0: \mu = 45.5$ or $H_0: \mu = \mu_0$, where μ is the true mean of the population from which you sampled and μ_0 stands for the hypothetical parametric mean of 45.5. Assume for the moment that we have no evidence that the variance of our sample is very much greater or smaller than the parametric variance of the housefly wing lengths. (If it were, it would be unreasonable to assume that our sample comes from the specified population. There is an important test of the assumption about the sample variance, which we will discuss later.) The curve at the center of Figure 7.9 represents the expected distribution of means of samples of five housefly wing lengths from the specified population. Acceptance and rejection regions for the null hypothesis at a type I error $\alpha = 0.05$ are delimited along the abscissa. The critical values that delimit the rejection regions are computed as follows (remember that $t_{[\infty]}$ is equivalent to the normal distribution):

$$L_1 = \mu_0 - t_{.05[\infty]}\,\sigma_{\bar{Y}} = 45.5 - (1.96)(1.744) = 42.08$$

and

$$L_2 = \mu_0 + t_{.05[\infty]}\,\sigma_{\bar{Y}} = 45.5 + (1.96)(1.744) = 48.92$$

Thus, we would consider it improbable for means less than 42.08 or greater than 48.92 to have been sampled from this population. For such sample means, we would therefore reject the null hypothesis. We compute critical values for a two-tailed test because we have no a priori assumption about the possible alternatives to our null hypothesis. If we could assume that the true mean of the population from which the sample was taken could be only equal to or greater than 45.5, then we would use a one-tailed test.

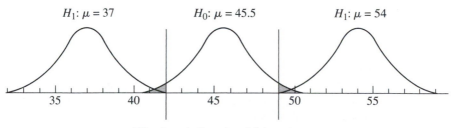

$H_1: \mu = 37$ $H_0: \mu = 45.5$ $H_1: \mu = 54$

35   40   45   50   55

Wing length (in units of 0.1 mm)

FIGURE 7.9 Expected distribution of means of samples of 5 housefly wing lengths from normal populations specified by μ as shown above curves and $\sigma_{\bar{Y}} = 1.744$. Center curve represents null hypothesis, $H_0: \mu = 45.5$, curves at sides represent alternative hypotheses, $\mu = 37$ or $\mu = 54$. Vertical lines delimit 5% rejection regions for the null hypothesis (2.5% in each tail, shaded).

Now let us examine alternative hypotheses. One alternative hypothesis might be that the true mean of the population from which our sample stems is 54.0 (but with the same variance as before). We can express this assumption as H_1: $\mu = 54.0$ or H_1: $\mu = \mu_1$, where μ_1 stands for the alternative parametric mean 54.0. From the table of the areas of the normal curve (Statistical Table A) and our knowledge of the variance of the means, we can calculate the proportion of the distribution implied by H_1 that would overlap the acceptance region implied by H_0. We find that 54.0 is 5.08 measurement units from 48.92, the upper boundary of the acceptance region of H_0. This corresponds to $5.08/1.744 = 2.91\sigma_{\bar{Y}}$ units. From Statistical Table A, we find that 0.0018 of the area will lie beyond 2.91 σ at one tail of the curve. Thus, under this alternative hypothesis, 0.0018 of the distribution of H_1 will overlap the acceptance region of H_0. This is β, the type II error under this alternative hypothesis. Actually, this is not entirely correct. Because the left tail of the H_1 distribution goes all the way to negative infinity, it will leave the acceptance region and cross over into the left-hand rejection region of H_0. However, this represents only an infinitesimal amount of the area of H_1 (the lower critical boundary of H_0, 42.08, is $6.83\sigma_{\bar{Y}}$ units from $\mu_1 = 54.0$; the area is less than 1×10^{-9}) and can be ignored.

Our alternative hypothesis H_1 specified that μ_1 is 8.5 units greater than μ_0. As we said, however, we may have no a priori reason to believe that the true mean of our sample is either greater or less than μ. Therefore, we may simply assume that the true mean is 8.5 measurement units away from 45.5. In such a case, we must similarly calculate β for the alternative hypothesis: $\mu_1 = \mu_0 - 8.5$. Thus, the alternative hypothesis becomes H_1: $\mu = 54.0$ or 37.0, or H_1: $\mu = \mu_1$, where μ_1 represents either 54.0 or 37.0, the alternative parametric means. Because the distributions are symmetrical, β is the same for both alternative hypotheses. Type II error for hypothesis H_1 is therefore 0.0018, regardless of which of the two alternative hypotheses is correct. If H_1 is really true, then 18 out of 10,000 samples would lead to an incorrect acceptance of H_0, a very low proportion of error. These relations are shown in Figure 7.9.

You may rightly ask what reason we have to believe that the alternative parametric value for the mean is 8.5 measurement units to either side of $\mu_0 = 45.5$. Any justification for such a belief would be quite unusual. As a matter of fact, the true mean may just as well be 7.5 or 6.0 or any number of units to either side of μ_0. If we draw curves for H_1: $\mu = \mu_0 \pm 7.5$, we find that β has increased considerably because the curves for H_0 and H_1 are now closer together. Thus, the magnitude of β depends on how far the alternative parametric mean is from the parametric mean of the null hypothesis. As the alternative mean approaches the parametric mean, β increases to a maximum value of $1 - \alpha$, which is the area of the acceptance region under the null hypothesis. At this maximum, the two distributions would be superimposed on each other. Figure 7.10 illustrates the increase in β as μ_1 approaches μ, starting with the test illustrated in Figure 7.9. To simplify the graph, the alternative distributions are shown for one tail only. Thus, we see clearly that β is not a fixed value but depends on how the alternative hypothesis is specified.

Let us look briefly at a one-tailed test. The null hypothesis is H_0: $\mu_0 = 45.5$ as before. The alternative hypothesis, however, assumes that we have reason to believe

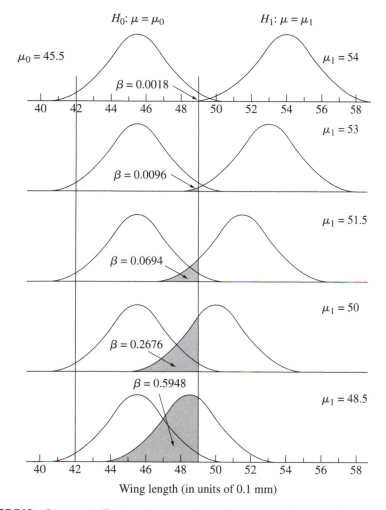

FIGURE 7.10 Diagram to illustrate increases in type II error, β, as alternative hypothesis, H_1, approaches null hypothesis, H_0—that is, μ_1 approaches μ. Shading represents β. Vertical lines mark off 5% critical regions (2.5% in each tail) for the null hypothesis. To simplify the graph, the alternative distributions are shown for one tail only. Data identical to those in Figure 7.9.

that the parametric mean of the population from which our sample has been taken cannot be less than $\mu_0 = 45.5$. If it is different from that value, then it can be only greater than 45.5. We might have two grounds for such a hypothesis. First, we might have a biological reason for such a belief. For example, our benchmark flies might have come from a dwarf population, making wing lengths from any other population necessarily larger. A second reason might be that we are interested in only one direction of difference. For example, we may be testing the effect of a chemical in the larval food intended to increase the size of the sample of flies. Therefore, we would

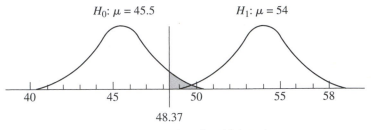

$H_0: \mu = 45.5$ $H_1: \mu = 54$

40   45   50   55   58

48.37

Wing length (in units of 0.1 mm)

FIGURE 7.11   One-tailed test for the distribution of Figure 7.9. The vertical line now cuts off 5% rejection region from *one* tail of the distribution (shaded area).

expect that $\mu_1 \geq \mu_0$, and we are not interested in testing for any μ_1 that is less than μ_0 because such an effect is the opposite of what we anticipate. Similarly, if we are investigating the effect of a certain drug as a cure for cancer, we might wish to compare the untreated population that has a mean fatality rate θ (from cancer) with the treated population, whose rate is θ_1. Our alternative hypotheses would be $H_1: \theta_1 < \theta$; that is, we would not be interested in any θ_1 that is greater than θ because if our drug increases mortality from cancer, it certainly is not much of a prospect for a cure.

When such a one-tailed test is performed, the rejection region along the abscissa is only under one tail of the curve representing the null hypothesis. Thus, for our housefly data (distribution of means of sample size $n = 5$) the rejection region will be in one tail of the curve only, and for a 5% type I error it will appear as shown in Figure 7.11. We compute the critical boundary as $45.5 + (1.645)(1.744) = 48.37$. (The 1.645 is $t_{.10[\infty]}$, which corresponds to the 5% value for a one-tailed test.) Compare this rejection region, which rejects the null hypothesis for all sample means greater than 48.37, with the two rejection regions in Figure 7.10, which reject the null hypothesis for means lower than 42.08 and greater than 48.92. In Figure 7.11, the alternative hypothesis is considered for one tail of the distribution only.

7.6 Power of a Test

An important concept in connection with hypothesis testing is the **power** of a test. The power of a test is $1 - \beta$, the complement of β, and is the probability of rejecting the null hypothesis when it is false and the alternative hypothesis is correct. For any given test, we would like to have the quantity $1 - \beta$ be as large as possible and the quantity β as small as possible. Because we generally cannot specify a given alternative hypothesis, we have to describe β or $1 - \beta$ for a continuum of alternative values. When $1 - \beta$ is graphed in this manner, the result is called a **power curve** for the test under consideration. Figure 7.12 shows the power curve for the housefly wing length example just discussed. This figure can be compared with Figure 7.10, from which it is directly derived.

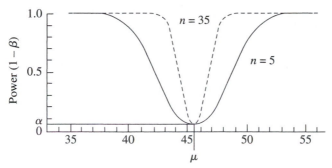

Wing length (in units of 0.1 mm)

FIGURE 7.12   Power curves for testing H_0: μ = 45.5. H_1: $\mu \neq$ 45.5 for n = 5 (as in Figures 7.9 and 7.10) and for n = 35.

Figure 7.10 emphasizes the type II error β, and Figure 7.12 graphs the complement of this value, $1 - \beta$. Note that the power of the test falls off sharply as the alternative hypothesis approaches the null hypothesis. Common sense confirms these conclusions: We can make clear and firm decisions about whether our sample comes from a population of mean 45.5 or 60.0. The power is essentially 1. But if the alternative hypothesis is that μ_1 = 45.6, differing only by 0.1 from the value assumed under the null hypothesis, then deciding which hypothesis is true is difficult and the power will be very low.

To improve the power of a given test (that is, decrease β) while keeping α constant for a stated null hypothesis, we must increase sample size. If, instead of sampling 5 wing lengths we had sampled 35, the distribution of means would be much narrower. Thus, rejection regions for the identical type I error would now commence at 44.21 and 46.79. Although the acceptance and rejection regions would remain the same proportionately, the acceptance region would become much narrower in absolute value. Previously, we could not, with confidence, reject the null hypothesis for a sample mean of 48.0. Now, when based on 35 individuals, a mean as deviant as 48.0 would occur only 15 times out of 100,000, and the hypothesis would, therefore, be rejected. What has happened to type II error? Because the distribution curves are not as wide as before, they overlap less. If the alternative hypothesis H_1: μ = 54.0 or 37.0 is true, the probability that the null hypothesis could be accepted by mistake (type II error) is infinitesimally small. If we let μ_1 approach μ_0, β will increase, of course, but it will always be smaller than the corresponding value for sample size n = 5. This comparison is shown in Figure 7.12, where the power for the test with n = 35 is much higher than that for n = 5. If we were to increase our sample size to 100 or 1000, the power would be increased still further. Thus, we reach an important conclusion: If a given test is not sensitive enough, we can increase its sensitivity (i.e., power) by increasing sample size.

There is yet another way of increasing the power of a test. If we cannot increase sample size, we may increase the power by changing the nature of the test. Different statistical techniques testing roughly the same hypothesis may differ substantially in both the magnitude and the slopes of their power curves. Tests that maintain higher power levels over substantial ranges of alternative hypotheses are clearly to be preferred. The popularity of the nonparametric tests described in several chapters of this book has grown not only because of their computational simplicity but also in many cases because their power curves are less affected by failure of assumptions than are those of the parametric methods. Nonparametric tests, however, have lower overall power than do parametric ones, when all the assumptions of the parametric test are met.

While on the subject of different tests, we should discuss briefly the **size** of a test. When we follow the protocols of a given statistical test and decide on a specified type I error α, the correctness of that value depends on how sensitive the test is to departures from its assumptions in the sample analyzed. The intended type I error rate may be only nominal, and the actual error rate—termed, somewhat infelicitously, the *size* of the test—may be greater or smaller. In the former case, we speak of a **liberal test** (i.e., we reject the null hypothesis more often than we should); the latter case represents a **conservative test** (i.e., we reject the null hypothesis less often).

In Figure 7.11, the alternative hypothesis is considered for one tail of the distribution only. If we were to draw a power curve for this test, it would not be symmetrical because the null hypothesis would always be accepted for any sample with a mean less than or equal to 45.5.

7.7 Tests of Simple Hypotheses Using the Normal and *t*-Distributions

We will now apply our newly acquired knowledge of hypothesis testing to some simple examples involving the normal and *t*-distributions.

The first test is fairly uncommon. Given a single observation Y_1 or a sample mean, we wish to know whether it differs significantly from a parametric mean μ— that is, whether it is likely to have been sampled from a population with parameter μ for the variable in question. Such a test assumes that we know the parametric mean of a population, which is not the usual case. It is more likely that μ represents a generally accepted standard or one promulgated by a government control or inspection agency. Examples might be the mean percentage of butterfat in a type of milk or the mean number of microorganisms permitted in a grade of milk or in a vaccine.

If we know the variance of the standard or parametric population, and if the variable is distributed normally, the test is straightforward. Either

$$\frac{Y_1 - \mu_0}{\sigma} \tag{7.4}$$

or

$$\frac{\bar{Y} - \mu_0}{\sigma_{\bar{Y}}} \tag{7.5}$$

would be distributed normally. As an example, we will test whether a single housefly with wing length of 43 units belongs to the population we are familiar with ($\mu = 45.5$, $\sigma = 3.90$). We compute Expression (7.4):

$$\frac{43 - 45.5}{3.90} = \frac{-2.5}{3.90} = -0.641$$

We ignore the negative sign of the numerator because this is a two-tailed test, so it is immaterial in which direction the item deviates from the mean. This is a two-tailed test because the question posed is whether the sample could belong to the population (H_0: $\mu = 45.5$). The alternative hypothesis is that the sample belongs to a population whose mean is either greater than or less than 45.5 (H_1: $\mu \neq 45.5$). If a computer program furnishing the exact probability of a deviation of 0.641 standard deviations in a normal distribution is unavailable to yield the correct value (0.5215), we can employ linear interpolation in Statistical Table **A**. We find that 0.2392 of the observations under the normal curve lie between μ_0 and 0.641 standard deviations, and hence 0.2608 of the observations lie beyond 0.641 standard deviations on one side of the mean; thus, 0.5216 (a close approximation to the earlier computed value) of the observations will be 0.641 or more standard deviations from the mean on both sides of the distribution. Therefore, by any of the conventional type I probability levels (5% or 1%), we would accept the null hypothesis and conclude that the observed housefly wing length of 43 units is consistent with sampling from the specified population. If we were satisfied with a less precise answer, we would simply use the t-table (Statistical Table **B**) and look up the value using the row corresponding to $df = \infty$. We would note that $t_{0.9[\infty]} = 0.126$ and $t_{0.5[\infty]} = 0.674$. Therefore, the probability of such a deviation is between 0.50 and 0.90 but closer to 0.50, as we found more accurately from the table of the areas of the normal curve.

Similarly, a sample of 100 milk yields that gave a mean of 70.05 could be tested to determine whether it belonged to the population of Table 6.1 ($\mu = 66.61$, $\sigma = 11.1597$). Expression (7.5) in this case yields $(70.05 - 66.61)/(11.1597/\sqrt{100}) = 3.08$, which indicates that the sample mean is 3.08 standard deviations above the parametric mean. Only 0.0011 of the area of the normal curve (from Statistical Table **A**) extends beyond 3.08 standard deviations, and even when doubled for the two-tailed test appropriate in this case (0.0022 or 0.22% of the area of the curve), the percentage is much smaller than the conventional type I error probabilities. We are led to reject the null hypothesis in this case. The mean based on the sample of 100 observations does not seem likely to belong to the parametric population we know.

Now let us proceed to a test that is similar, except that we do not know the parametric standard deviation. Government regulations prescribe that the standard dosage in a certain biological preparation should be 600 activity units per cubic centimeter. We prepare 10 samples of this preparation and test each for potency. We find that the mean number of activity units per sample is 592.5 units per cubic centimeter and that the standard deviation of the samples is 11.2. Does our sample conform to the government standard? Stated more precisely, our null hypothesis is H_0: $\mu = \mu_0$, where μ equals the true mean of the population from which our samples were taken, and μ_0 is the standard prescribed by the government agency. The alternative hypothesis is that the dosage is not equal to 600, or H_1: $\mu \neq \mu_0$. We calculate the probability of the deviation $\overline{Y} - \mu_0$ expressed in standard deviation units. The appropriate standard deviation is that of means (the standard error of the mean), *not* the standard deviation of observations, because the deviation is that of a sample mean around a parametric mean. We therefore calculate $s_{\overline{Y}} = s/\sqrt{n} = 11.2/\sqrt{10} = 3.542$. We next test the deviation $(\overline{Y} - \mu_0)/s_{\overline{Y}}$. We saw in Section 7.4 that a deviation divided by an estimated standard deviation will be distributed according to the t-distribution with $n - 1$ degrees of freedom. We therefore write

$$t_s = \frac{\overline{Y} - \mu_0}{s_{\overline{Y}}} \tag{7.6}$$

This expression indicates that we would expect this deviation to follow a t-distribution. Note that in Expression (7.6), we wrote t_s. In most textbooks, you will find this ratio simply identified as t, but in fact the t-distribution is a parametric and theoretical distribution that generally is only approached, but never equaled, by observed, sampled data. This may seem a minor distinction, but readers should be quite clear that in any hypothesis testing of samples we are only *assuming* that the distributions of the tested variables follow certain theoretical probability distributions. To conform with general statistical practice, the t-distribution should really have a Greek letter (such as τ), with t serving as the sample statistic. Because this would violate long-standing practice, however, we prefer to use the subscript s to indicate the sample value.

The test is very simple. We calculate Expression (7.6),

$$t_s = \frac{592.5 - 600}{3.542} = -2.12 \qquad df = n - 1 = 9$$

and compare it with the expected values for t at 9 degrees of freedom. We either obtain the probability associated with a t-value of 2.12 in a t-distribution for 9 degrees of freedom by a computer program (we obtained $P = 0.0633$, using an argument of $t_s = 2.1174$), or we have to turn to Statistical Table **B**. Because the t-distribution is symmetrical, we ignore the sign of t_s and always look up its positive value in Statistical Table **B**. The two values on either side of $t_s = 2.12$ are $t_{.05[9]} = 2.26$ and $t_{.1[9]} = 1.83$. These are t-values for two-tailed tests, appropriate in this instance because the alternative hypothesis is that $\mu \neq 600$; that is, μ can be smaller or greater. The probability of our value of t_s appears to be between 5% and 10%; if the null hypothesis is true, the probability of obtaining a deviation as great or greater than 7.5 is somewhere between 0.05 and 0.10. By conventional type I error rates, this probability is insufficient for declaring the sample mean to be different from the standard. We thus accept the null hypothesis. In conventional language, we would report the results of the statistical analysis as follows: "The sample mean is not statistically different from the accepted standard." Such a statement in a scientific report should always be backed up by a probability value—in this case, the computed value $P = 0.0633$. When the result is based on a table of the t-distribution such as Statistical Table **B,** the proper presentation is $0.10 > P > 0.05$, which means that, assuming the null hypothesis is true, the probability of such a deviation is between 0.05 and 0.10. Another way of saying this is that the value of t_s is *not statistically significant* (frequently abbreviated *ns*).

Asterisks often appear after the computed value of a statistical test—for example, $t_s = 2.86**$. The symbols generally represent the following probability ranges:

$$* = 0.05 \geq P > 0.01 \qquad ** = 0.01 \geq P > 0.001 \qquad *** = 0.001 \geq P$$

However, because some authors occasionally imply other ranges by these asterisks, the meaning of the symbols should be specified in each scientific report.

It might be argued that in the case of the biological preparation, the concern of the tester should not be whether the sample mean differs from a standard but whether it is *below* the standard. This may be one of those biological preparations in which an excess of the active component is of no harm but a shortage would make the preparation ineffective at the conventional dosage. Then the test becomes one-tailed, performed in exactly the same manner except that the critical values of t for a one-tailed test are at half the probabilities of the two-tailed test. Thus, 2.26, the former 0.05-value, becomes $t_{.025[9]}$, and 1.83, the former 0.10-value, becomes $t_{.05[9]}$, making our observed t_s-value of 2.12 "significant at the 5% level" or, more precisely stated, significant at $0.05 > P > 0.025$. If we are prepared to accept a 5% chance of a type I error, we would consider the preparation to be below the standard.

You may be surprised that the same example, employing the same data and statistical tests, can lead to two different conclusions, and you may begin to wonder whether some of the things you hear about statistics and statisticians are not, after all, correct. The explanation lies in the fact that the two results are answers to different questions. If we test whether our sample mean is different from the standard in either direction, we must conclude that it is not different enough for us to reject the null hypothesis. If, on the other hand, we exclude from consideration the fact that the true sample mean μ could be greater than the established standard μ_0, the difference we found leads us to now reject the null hypothesis of no difference. It is obvious from this example that in any statistical test one must clearly state whether a one-tailed or a two-tailed test has been performed, if the nature of the example is such that there could be any doubt about the matter. We should also point out that such a difference in the outcome of the results is not necessarily typical and occurred only because the outcome in this case is in a borderline area. Had the difference between sample and standard been 10.5 activity units, the null hypothesis would have been rejected by both the one-tailed and the two-tailed tests.

The promulgation of a standard mean is generally insufficient for establishing a rigid standard for a product. If the variance among the samples is very large, it may not be possible to reject the null hypothesis of no difference between the mean of the population from which the samples are drawn and the population corresponding to the standard. This point is important and should be quite clear to you. Remember that the standard error can be increased in two ways—by lowering sample size or by increasing the standard deviation of the replicates. Both of these are undesirable

---

BOX 7.2 | **Testing a Deviation of a Normally Distributed Statistic from a Parameter**

Computation

1. Compute t_s as the following ratio

$$t_s = \frac{St - St_p}{s_{St}}$$

where St is a sample statistic, St_p is the parametric value against which the sample statistic is to be tested, and s_{St} is its estimated standard error, obtained from Box 7.1.

2. The pertinent hypotheses are

$$H_0: St = St_p \qquad H_1: St \neq St_p$$

for a two-tailed test, and

$$H_0: St = St_p \qquad H_1: St > St_p$$

or

$$H_0: St = St_p \qquad H_1: St < St_p$$

for a one-tailed test.

3. In the two-tailed test look up the critical value of $t_{\alpha[\nu]}$ in Statistical Table **B**, where α is the type 1 error agreed upon and ν is the degrees of freedom pertinent to the standard error employed (see Box 7.1). In the one-tailed test look up the critical value of $t_{2\alpha[\nu]}$ for a significance level of α.

4. Accept or reject the appropriate hypothesis in step **2** on the basis of the t_s value in step **1** compared with critical values of t in step **3**.

Example of computation

Test the null hypothesis that the g_1 value from Box 6.1 ($g_1 = 0.18936$, $n = 9465$) is equal to zero. Note from Box 7.1 that the standard deviation of g_1 is $\sqrt{6/n}$ for samples of $n > 150$.

1. $t_s = \dfrac{(g_1 - \gamma_1)}{s_{g_1}}$

$$= \frac{0.18936 - 0}{\sqrt{6/9465}} = 7.52$$

2. The null hypothesis is that the distribution is not skewed—that is, that $\gamma_1 = 0$. The pertinent test is clearly two-tailed, since g_1 can be either negative (skewed to the left) or positive (skewed to the right). We wish to test whether there is *any* skewness; that is, the alternative hypothesis is that $\gamma_1 \neq 0$. Thus

$$H_0: \gamma_1 = 0 \qquad H_1: \gamma_1 \neq 0$$

Box 7.2 (continued)

3. The appropriate number of degrees of freedom employed with s_{g_1} is ∞ (see Box 7.1). We therefore employ critical values of t with degrees of freedom $\nu = \infty$:

$$t_{.05[\infty]} = 1.960 \qquad t_{.01[\infty]} = 2.576 \qquad t_{.001[\infty]} = 3.291$$

4. $t_s = 7.52^{***}$. This expression is a terse symbolism for stating that if we assume that the null hypothesis is true then the probability of the observed g_1 is < 0.001 and that we have therefore rejected the null hypothesis. The actual probability is much less, $P = 5.9 \times 10^{-14}$. We accept the alternative hypothesis and conclude that $\gamma_1 \neq 0$. Because g_1 is positive, this indicates that the birth weights are skewed to the right. Thus the distribution of birth weights is asymmetrical, drawn out at the right tail.

---

aspects of any experimental setup. Yet, in the relatively rare cases where the purpose of the investigator is to show *no* difference rather than to show a significant difference, this result can be obtained trivially simply by having a large enough standard error because of small sample size or poor technique.

One of us once had to review a research article, the main point of which was that the unpleasant side effects of drug B (newly developed) were no different from those of drug A (which was well established). The writers of the article used the appropriate statistical test but used it on only very few individuals. They apparently failed to understand that by using few enough individuals you could never refute the null hypothesis (that side effects of drug B equal side effects of drug A). One needs to calculate beforehand the sample size necessary to establish differences of a given magnitude at a certain type I error level (see Section 12.4).

The test described earlier for the biological preparation leads us to a general test for any statistic—that is, for the deviation of any statistic from a hypothesized value—which is illustrated in Box 7.2. Such a test applies whenever the statistics are expected to be distributed normally. When the standard error is estimated from the sample, the t-distribution is used. However, because the normal distribution is just a special case, $t_{[\infty]}$, of the t-distribution, most statisticians uniformly apply the t-distribution with the appropriate degrees of freedom from 1 to infinity.

In Box 7.2, as an example, we are testing whether the deviation of the observed value of g_1 differs sufficiently from the expected value of γ_1 for a normal distribution, which is zero, to decide that the sample was not drawn from a normally distributed population. Tests for coefficients, such as g_1, g_2, regression coefficients, and correlation coefficients, are tested in this manner by using the appropriate standard errors. Values of g_1 and g_2 based on small samples are only approximately normally distributed. For critical tests of such values, consult D'Agostino and Tietjen (1973) and D'Agostino and Pearson (1973).

In Box 7.2, the conclusions are stated in statistical shorthand (the conventional way in which such results are reported in the scientific literature) and in terms of the actual detailed implications of the test, which you should understand thoroughly before reading further.

7.8 The Chi-Square Distribution

Another continuous distribution of great importance in statistics is the χ^2-distribution (read *chi-square distribution*). We need to learn it now in connection with the distribution and confidence limits of variances.

The **chi-square distribution** is a probability density function for a variable whose values range from zero to positive infinity. Thus, unlike the normal distribution or t, the function approaches the χ^2-axis asymptotically only at the right-hand tail of the curve, not at both tails. The function describing the χ^2-distribution is complicated and will not be given here. As in the case of t, there is not merely one χ^2-distribution, but one distribution for each number of degrees of freedom. Therefore, χ^2 is a function of ν, the number of degrees of freedom. Figure 7.13 shows probability density functions for the χ^2-distributions for 1, 2, 3, and 6 degrees of freedom. Notice that the curves are strongly skewed to the right, L-shaped at first (sometimes called a *reverse J-shaped distribution*), but more or less approaching symmetry for higher degrees of freedom.

We can generate a χ^2-distribution from a normal distribution. Recall that we standardize a variable Y_i by subjecting it to the operation $(Y_i - \mu)/\sigma$ (see Section 6.2). Let us symbolize a standardized variable as $Y'_i = (Y_i - \mu)/\sigma$. Now imagine repeated samples of n observations Y_i from a normal population with mean μ and standard deviation σ. For each sample, we transform every reading Y_i to Y'_i, as defined above. The quantities $X^2 = \sum^n Y'^2_i$ computed for each sample will be

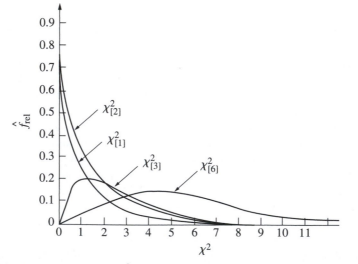

FIGURE 7.13   Frequency curves of χ^2-distributions for 1, 2, 3, and 6 degrees of freedom.

distributed as a χ^2-distribution with n degrees of freedom. Using the definition of Y'_i, we can rewrite X^2 as

$$X^2 = \sum^n \frac{(Y_i - \mu)^2}{\sigma^2} = \frac{1}{\sigma^2} \sum^n (Y_i - \mu)^2 \tag{7.7}$$

When we change the parametric mean μ to a sample mean, this expression becomes

$$\frac{1}{\sigma^2} \sum^n (Y_i - \bar{Y})^2 \tag{7.8}$$

which is simply the sum of squares of the variable divided by a constant, the parametric variance. Another common way of stating this expression is

$$\frac{(n - 1)s^2}{\sigma^2} \tag{7.9}$$

Here we have replaced the numerator of Expression (7.8) with $n - 1$ times the sample variance, which is, of course, the sum of squares.

If we were to sample repeatedly n observations from a normally distributed population, Expression (7.9) computed for each sample would yield a χ^2-distribution with $n - 1$ degrees of freedom. Notice that, although we have samples of n observations, we have lost a degree of freedom because we are now employing a sample mean rather than the parametric mean. The earlier Figure 7.6, a sample distribution of variances, has a second scale along the abscissa, which is the first scale multiplied by the constant $(n - 1)/\sigma^2$. This scale converts the sample variances s^2 of the first scale into Expression (7.9). Because the second scale is proportional to s^2, the distribution of the sample variances serves to illustrate a sample distribution approximating χ^2. The distribution is strongly skewed to the right, as would be expected in a χ^2-distribution.

Figure 7.6 also shows the absolute expected frequencies of a χ^2-distribution for 4 degrees of freedom. The observed and expected frequencies do not agree well. There are more observations near the low values of χ^2, probably because the housefly wing length data of Table 6.1 are not quite normal. They do not extend from $-\infty$ to $+\infty$ but only from -2.44σ to $+2.44\sigma$. Means sampled from these data apparently were approximately normally distributed, but these ratios seem more sensitive to a departure from normality in the original population. To bring the population closer to normality, we would have had to furnish a base population of $n = 1000$ or more, rather than the mere 100 wing lengths of Table 6.1.

The expected frequencies are obtained from tables (for example, Table 7 in Pearson and Hartley, 1958) or by direct computation of the cumulative frequencies of the χ^2-distribution by one of the numerous available functions for doing so. Values of χ^2 and ν (degrees of freedom) serve as arguments, and the portion of the area of the curve to the right of a given value of χ^2 and for the given df is the function. Extensive tables of this sort are too cumbersome and also do not furnish the exact probability levels customarily required. Therefore, conventional χ^2-tables, such as

Statistical Table **D**, give these common probabilities and degrees of freedom as arguments and list the χ^2 corresponding to the probability and the *df* as the functions. Each chi-square in Statistical Table **D** is the value of χ^2 beyond which the area under the χ^2-distribution for ν degrees of freedom represents the indicated probability. Just as we used subscripts to indicate the cumulative proportion of the area, as well as the degrees of freedom represented by a given value of t, we subscript χ^2 as follows: $\chi^2_{\alpha[\nu]}$ indicates the χ^2-value, to the right of which is found proportion α of the area under a χ^2-distribution for ν degrees of freedom.

Let us learn how to use Statistical Table **D**. Looking at the distribution of $\chi^2_{[2]}$, we note that 90% of all values of $\chi^2_{[2]}$ would be to the right of 0.211, but only 5% of all values of $\chi^2_{[2]}$ would be greater than 5.991. Mathematical statisticians have shown that the expected value of $\chi^2_{[\nu]}$ (the mean of a χ^2-distribution) equals its degrees of freedom ν. Thus, the expected value of a $\chi^2_{[5]}$-distribution is 5. When we examine 50% values (the medians) in the χ^2-table, we notice that they are generally lower than the expected values (the means). Thus, for $\chi^2_{[5]}$ the 50% point is 4.351, which illustrates the asymmetry of the χ^2-distribution: The mean is to the right of the median. In the next section of this chapter, as well as in its last section, we shall show you some applications of the χ^2-distribution. Its most extensive use, however, will occur in Chapter 17.

7.9 Testing the Hypothesis H_0: $\sigma^2 = \sigma_0^2$

The method of Box 7.2 can be used only if the statistic is distributed normally. In the case of the variance, this is not so. As we saw in the last section, sums of squares divided by σ^2 follow the χ^2-distribution. Therefore, testing the null hypothesis that a sample variance does not differ from a parametric variance requires that we employ the χ^2-distribution.

Let us use the biological preparation of the last section as an example. We were told that the standard deviation based on 10 samples was 11.2. Therefore, the sample variance must have been 125.44. Suppose the government postulates that the variance of samples from the preparation should be no greater than 100.0. Is our sample variance sufficiently above 100.0 to allow us to reject the null hypothesis that $\sigma^2 = 100$? Remembering from Expression (7.9) that $(n - 1)s^2/\sigma^2$ is distributed as $\chi^2_{[n-1]}$, we proceed as follows. We first calculate

$$X^2 = (n - 1)s^2/\sigma^2$$

$$= (9)125.44/100 = 11.290$$

Note that we call this quantity X^2 rather than χ^2, again to emphasize that we are obtaining a sample statistic which we shall compare to the parametric distribution. Following the general outline of Box 7.2, we next establish our null and alternative hypotheses, which are H_0: $\sigma^2 \leq \sigma_0^2$ and H_1: $\sigma^2 > \sigma_0^2$, where σ_0^2 is the postulated variance. As stated, this is a one-tailed test. The critical value of χ^2 is found next as $\chi^2_{\alpha[\nu]}$, where α indicates the type I error and ν the pertinent degrees of freedom. Quantity α represents the proportion of the χ^2-distribution to the right of the given value as described in Section 7.8, and you see now why we used the symbol α for that portion of the area; it corresponds with the type I error. For 9 degrees of freedom, we find in Statistical Table **D** that

$$\chi^2_{.05[9]} = 16.919 \qquad \chi^2_{.1[9]} = 14.684 \qquad \chi^2_{.5[9]} = 8.343$$

Note that the probability of getting a χ^2 as large as 11.290 is therefore higher than 0.10 but less than 0.50, assuming that the null hypothesis is true. Thus, X^2 does not exceed the 5% level; we do not reject the null hypothesis and conclude that the variance of the 10 samples of the biological preparation may be no greater than the standard permitted by the government. If we had decided to test whether the variance is different from the standard, permitting it to deviate in either direction, then the hypotheses for this two-tailed test would have been H_0: $\sigma^2 = \sigma_0^2$ and H_1: $\sigma^2 \neq \sigma_0^2$, and a 5% type I error would have yielded the following critical values for the two-tailed test:

$$\chi^2_{.975[9]} = 2.700 \qquad \chi^2_{.025[9]} = 19.023$$

The values represent chi-squares at points cutting off 2.5% rejection regions at each tail of the χ^2-distribution. A value of $\chi^2 < 2.700$ or >19.023 would have been evidence that the sample variance did not belong to this population. Our value of $\chi^2 = 11.290$ would have led again to an acceptance of the null hypothesis.

In the next chapter, we will introduce another statistical test for the hypotheses about variances of this section. This test is the mathematically equivalent F-test, which is a more general test, allowing us to test the hypothesis that two sample variances come from populations with equal variances.

> Investigators who wish to test the null hypothesis H_0: $\sigma^2 = \sigma_0^2$ may wish to know how large a sample they should use to test this hypothesis with a specified power. An example of how to determine the appropriate sample size is featured in Zar (2010, Section 8.7).

7.10 Introduction to Interval Estimation (Confidence Limits)

The sample statistics we have been obtaining, such as mean, standard deviation, or g_1, are estimates of the population parameters μ, σ, or γ_1, respectively. So far, we have not discussed the reliability of these estimates. One matter we may wish to know is whether the sample statistics are *unbiased estimators* of the population parameters, as discussed in Section 4.7.

But knowing, for example, that $\overline{Y}$ is an unbiased estimate of μ is not enough. We would like to find out how reliable a measure of μ it is. Of course, what we really wish to know are the actual values of μ, σ^2, and other parameters. However, this is an impossible task, unless we include all the individuals of the population in our study. Thus, the true values of the parameters almost always remain unknown, and we commonly indicate the reliability of a sample statistic by using an interval rather than a point estimate of the parameter. Estimates such as $\overline{Y}$ are called *point estimates* because they furnish a particular value as an estimate of the population parameter, μ. Although it may be close, it is unlikely that the estimate will be exactly correct. An alternative is to compute an interval around our sample estimate that is sufficiently wide that it is likely to include the unknown value μ.

To begin our discussion of this topic, let us start with the unusual case of a normally distributed population whose parametric mean and standard deviation are known to be μ and σ, respectively. Recall from Section 7.2 that $\sigma_{\bar{Y}} = \sigma/\sqrt{n}$. From Section 6.2, we know that the region from $1.96\sigma_{\bar{Y}}$ below μ to $1.96\sigma_{\bar{Y}}$ above μ includes 95% of the sample means of size n. Another way of expressing this is by using the quantity $(\bar{Y} - \mu)/\sigma_{\bar{Y}}$. This ratio is the standardized deviation of a sample mean from its parametric value. Because $\bar{Y}$ is normally distributed, this ratio is also normally distributed; thus we expect 95% of such standard deviates to fall between -1.96 and $+1.96$. We can express this statement symbolically as follows:

$$P\left\{-1.96 \leq \frac{\bar{Y} - \mu}{\sigma_{\bar{Y}}} \leq 1.96\right\} = 0.95$$

This expression means that the probability, P, that the sample means $\bar{Y}$ will differ by no more than 1.96 standard errors, $\sigma/\sqrt{n}$, from the parametric mean μ is 0.95. The expression in brackets is an inequality, all terms of which can be multiplied by $\sigma/\sqrt{n}$ to yield

$$\{-1.96\sigma_{\bar{Y}} \leq \bar{Y} - \mu \leq 1.96\sigma_{\bar{Y}}\}$$

We can rewrite this expression as

$$\{-1.96\sigma_{\bar{Y}} \leq \mu - \bar{Y} \leq 1.96\sigma_{\bar{Y}}\}$$

because $-a \leq b \leq a$ implies $a \geq -b \geq -a$, and this can be written as $-a \leq -b \leq a$. Finally, we can transfer $-\bar{Y}$ across the inequality signs, just as it could be transferred across the equal sign in an equation to yield the final desired expression:

$$P\{\bar{Y} - 1.96\sigma_{\bar{Y}} \leq \mu \leq \bar{Y} + 1.96\sigma_{\bar{Y}}\} = 0.95 \tag{7.10}$$

Thus, the probability, P, is 0.95 that the interval between the limits $L_1 = \bar{Y} - 1.96\sigma_{\bar{Y}}$ and $L_2 = \bar{Y} + 1.96\sigma_{\bar{Y}}$ will contain the parametric mean μ. The interval from L_1 to L_2 is called a **confidence interval.** The two limits, L_1 and L_2, are called the lower and upper 95% **confidence limits** of the mean.

One has to be careful of the way Equation (7.10) is interpreted. If we repeatedly obtained samples of size n from the population and constructed these limits for each, we could expect 95% of the intervals between these limits to contain the true mean, and only 5% of the intervals would miss μ. It is not correct to say that there is a 95% chance that the true mean is between the confidence limits computed for a particular sample. Another way of interpreting a confidence interval that links it to the concepts of hypothesis testing is as the set of null hypotheses H_0 for which a statistical test would result in a nonsignificant P-value (e.g., $\alpha \geq 0.05$). The values bounding this set are the confidence limits of the mean. We can see that these different ways of defining the confidence limits are mathematically equivalent by expressing the third definition as an equation.

$$P\left\{-1.96 \le \frac{\overline{Y} - \mu_0}{\sigma_{\overline{Y}}} \ge 1.96\right\} = 0.05$$

In this expression, μ_0 represents the variable set of null hypotheses H_0. By comparing this expression with the first display in this section, it is easily seen that the two expressions are complements of each other, the former display defining the acceptance region, the latter the rejection region of the null hypothesis $H_0: \mu = \mu_0$.

If you were not satisfied using confidence intervals that are expected to contain the true mean only 95% of the time, you might employ 2.576 as a coefficient in place of 1.960. You may remember that 99% of the area of the normal curve lies between $\overline{Y} \pm 2.576\sigma_{\overline{Y}}$ (see Section 6.2). Thus, to calculate 99% confidence limits, compute the two quantities $L_1 = \overline{Y} - 2.576\sigma_{\overline{Y}}$ and $L_2 = \overline{Y} + 2.576\sigma_{\overline{Y}}$ as lower and upper confidence limits, respectively. In this case, on average 99% of the confidence intervals obtained in repeated sampling would be expected to contain the true mean. The new confidence interval is wider than the 95% interval (because a wider interval has a greater chance of containing the true mean).

If you were still not satisfied with 99% likelihood that the confidence limits enclose the mean, you could increase it, multiplying the standard error of the mean by 3.291 in order to obtain 99.9% confidence limits. This value, or 3.891 for 99.99% limits, is obtained by employing an inverse function from a computer program for the area of the normal curve. Alternatively, such values can be found by inverse interpolation in an extensive table of areas of the normal curve (for example, Table 1 in Pearson and Hartley, 1976) or directly in a table of the inverse of the normal probability distribution (for example, Table 1.2 in Owen, 1962). The new coefficients would widen the interval further. But as your confidence increases, your statement becomes vaguer and vaguer, as the confidence interval widens. Consider the following example. We obtain a sample of 35 housefly wing lengths from the population of Table 6.1 with known mean ($\mu = 45.5$) and standard deviation ($\sigma = 3.90$). Let us assume that the sample mean is 44.8. We can expect the standard deviation of means based on samples of 35 observations to be $\sigma_{\overline{Y}} = \sigma/\sqrt{n} = 3.90/\sqrt{35} = 0.6592$. We compute 95% confidence limits as follows:

Lower limit is $L_1 = 44.8 - (1.960)(0.6592) = 43.51$

Upper limit is $L_2 = 44.8 + (1.960)(0.6592) = 46.09$

Remember that this is an unusual case in which we happen to know the true mean of the population ($\mu = 45.5$), and hence we know that the confidence limits enclose the mean in our example. We expect 95% of such confidence intervals obtained in repeated sampling to include the parametric mean.

For each of the samples of 35 housefly wing lengths described in the sampling experiment in Section 7.2, 95% confidence limits were computed for the parametric mean for each sample. The standard errors of the means were based on the parametric standard deviations of these populations (housefly wing lengths: $\sigma = 3.90$; milk yields: $\sigma = 11.1597$). The number of correct confidence intervals were recorded—that is, the number of confidence intervals that contained the parametric mean of the population.

In Figure 7.14**a**, we show the first 100 of these confidence intervals plotted parallel to the ordinate. Of these, 95 (95.0%) cross the parametric mean of the population. In the next 100 (not shown), 99 cross the mean line, yielding an overall count of 194 (97.0%) confidence intervals including the true mean. This should give you a concrete idea of the reliability of confidence limits.

To reduce the width of the confidence interval, we have to reduce the standard error of the mean, which, because $\sigma_{\bar{Y}} = \sigma/\sqrt{n}$, can be done by reducing either the standard deviation of the observations or by increasing the sample size. The first of these alternatives is frequently not available. If we are sampling from a population in nature, ordinarily we have no way of reducing its standard deviation, except by making more precise measurements. In many experimental procedures, however, we may be able to reduce the variance of the data. For example, if we are studying heart weight in rats and find that its variance is rather large, we might be able to reduce this variance by taking rats of only one age group, in which the variation of heart weight would be considerably less. Thus, by controlling one of the variables of the experiment, the variance of the response variable, heart weight, is reduced. Similarly, by keeping temperature or other environmental variables constant in a procedure, we can frequently reduce the variance of our response variable and hence obtain more precise estimates of population parameters.

A more common way to reduce the standard error is by increasing sample size. It is obvious from Expression (7.2) that as n increases, the standard error decreases; hence, as n approaches infinity, the standard error and the lengths of confidence intervals approach zero. This relationship ties in with what we have already learned: In samples whose size approaches infinity, the variation of sample means around the parametric mean becomes infinitesimal.

We must guard against a common mistake in expressing the meaning of the confidence limits of a statistic. When we have set lower and upper limits (L_1 and L_2, respectively) to a statistic, we imply that the probability of this interval covering the mean is 0.95 or, expressed in another way, that on the average, 95 out of 100 confidence intervals similarly obtained would cover the mean. We *cannot state* that there is a probability of 0.95 that the true mean is contained within any particular observed confidence limits, although this may seem to be saying the same thing. The last statement is incorrect because the true mean is a parameter; hence, it is a fixed value and is therefore either inside or outside of the interval. It cannot be inside a particular interval 95% of the time. It is important, therefore, to learn the correct statement and meaning of confidence limits.

Up to now, we have considered only means based on normally distributed samples with known parametric standard deviations. We can, however, extend the methods we have learned to samples from populations with unknown standard deviations but where the statistic being tested is known to be normally distributed and the sample sizes are large—say, $n \geq 100$. In such cases, we use the sample standard deviation for computing the standard error of the statistic.

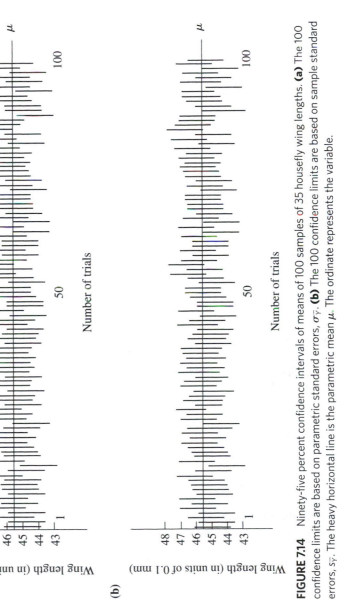

FIGURE 7.14 Ninety-five percent confidence intervals of means of 100 samples of 35 housefly wing lengths. **(a)** The 100 confidence limits are based on parametric standard errors, $\sigma_{\bar{Y}}$. **(b)** The 100 confidence limits are based on sample standard errors, $s_{\bar{Y}}$. The heavy horizontal line is the parametric mean μ. The ordinate represents the variable.

When the samples are small ($n < 100$) and we lack knowledge of the parametric standard deviation, however, we must consider the reliability of our sample standard deviation, which requires making use of the t-distribution, first encountered in Section 7.4. We illustrate this procedure in the next section.

7.11 Confidence Limits Using Sample Standard Deviations

The t-distribution enables us to set confidence limits to the means of samples from a normal frequency distribution whose parametric standard deviation is unknown. The limits are computed as $L_1 = \bar{Y} - t_{\alpha[n-1]}s_{\bar{Y}}$ and $L_2 = \bar{Y} + t_{\alpha[n-1]}s_{\bar{Y}}$ for confidence limits of probability $P = 1 - \alpha$. Thus, for 95% confidence limits, we use values of $t_{.05[n-1]}$. We can rewrite Expression (7.10) as

$$P\{\bar{Y} - t_{\alpha[n-1]}s_{\bar{Y}} \leq \mu \leq \bar{Y} + t_{\alpha[n-1]}s_{\bar{Y}}\} = 1 - \alpha \tag{7.11}$$

Box 7.3 shows an example of the application of this expression. The following sampling experiment illustrates the appropriateness of the t-distribution for setting confidence limits to means with unknown σ.

The computations and procedures of the sampling experiment described in Section 7.10 were repeated, but using standard errors of the means based on the observed standard deviations computed for each sample and the appropriate t-value. Figure 7.14b shows 95% confidence limits of the first 100 sampled means of 35 housefly wing lengths, computed with t and $s_{\bar{Y}}$ rather than with the normal curve and $\sigma_{\bar{Y}}$ as in the previous sampling experiment (see Figure 7.14a). Note that 94 (94.0%) of these first 100 confidence intervals cross the line corresponding to the parametric mean. In the next 100 (not shown), 97 cross the mean line, yielding an overall count of 191 (95.5%) confidence intervals that include the true mean.

We can use the same technique for setting confidence limits to any statistic, as long as the statistic follows the normal distribution. This technique applies in an approximate way to all the statistics of Box 7.1. Thus, for example, we may set confidence limits to the coefficient of variation of the aphid femur lengths of Boxes 2.1 and 4.1. These are computed as

$$P\{V^* - t_{\alpha[n-1]}s_{V^*} \leq V_p \leq V^* + t_{\alpha[n-1]}s_{V^*}\} = 1 - \alpha$$

where V_p stands for the parametric value of the coefficient of variation. Because the standard error of the coefficient of variation approximately equals $s_V = V/\sqrt{2n}$ and $s_{V^*} = (1 + \frac{1}{4})s_V$, we proceed as follows:

$$V = \frac{100s}{\bar{Y}} = \frac{100(0.3656)}{4.004} = 9.1309$$

$$V^* = \left(1 + \frac{1}{4n}\right)V = \left(1 + \frac{1}{4(25)}\right)9.1309 = 9.2222$$

$$s_V = \frac{9.1309}{\sqrt{2 \times 25}} = \frac{9.1309}{7.0711} = 1.2913$$

$$s_{V*} = \left(1 + \frac{1}{4n}\right)s_V = 1.01(1.2913) = 1.3042$$

$$L_1 = V* - t_{.05[24]}s_{V*}$$

$$= 9.2222 - 2.064(1.3042)$$

$$= 6.5303$$

$$L_2 = V* + t_{.05[24]}s_{V*}$$

$$= 9.2222 + 2.064(1.3042)$$

$$= 11.9141$$

BOX 7.3 Confidence Limits for μ

Aphid stem mother femur lengths (from Boxes 2.1 and 4.1): $\overline{Y} = 4.004$; $s = 0.366$; $n = 25$.

Values for $t_{\alpha[n-1]}$ from a two-tailed t-table (Statistical Table **B**), where $1 - \alpha$ is the proportion expressing confidence and $n - 1$ is the degrees of freedom:

$$t_{.05[24]} = 2.064 \qquad t_{.01[24]} = 2.797$$

The 95% confidence limits for the population mean, μ, are given by the equations

$$L_1 \text{ (lower limit)} = \overline{Y} - t_{.05[n-1]}\frac{s}{\sqrt{n}}$$

$$= 4.004 - \left(2.064\frac{0.366}{\sqrt{25}}\right) = 3.853$$

$$L_2 \text{ (upper limit)} = \overline{Y} + t_{.05[n-1]}\frac{s}{\sqrt{n}}$$

$$= 4.155$$

The 99% confidence limits are

$$L_1 = \overline{Y} - t_{.05[24]}\frac{s}{\sqrt{n}}$$

$$= 4.004 - \left(2.797\frac{0.366}{\sqrt{25}}\right) = 3.799$$

$$L_2 = \overline{Y} + t_{.01[24]}\frac{s}{\sqrt{n}}$$

$$= 4.209$$

The distribution of the coefficient of variation is quite sensitive to departures from normality in the distribution of Y, so this procedure should not be used when Y is markedly nonnormally distributed.

When sample size is very large or when σ is known, the distribution of the statistic V is effectively normal. Rather than turn to the table of areas of the normal curve, however, we usually employ $t_{.05[\infty]}$, the t-distribution with infinite degrees of freedom.

Although confidence limits are useful indicators of the reliability of a sample statistic, they are not commonly given in scientific publications; instead, the statistic $\pm$ its standard error is cited. Thus, you will frequently see column headings such as "Mean $\pm$ SE," indicating that the reader is free to use the standard error to set confidence limits if so inclined. It should be obvious to you from studying the t-distribution that you cannot set confidence limits to a statistic without knowing the sample size on which it is based because n is necessary to compute the correct degrees of freedom. Thus, one should not cite means and standard errors without also stating sample size n.

You must also be careful in interpreting such tables of results. Often they are cited as "Mean and Standard Deviation." The meaning of standard deviation is ambiguous here. Is it the standard deviation of the observations or the standard deviation of their mean? Desirable headings for $\overline{Y} \pm s_{\overline{Y}}$ are "Means and Their Standard Errors" or "Means and Their Standard Deviations." If the heading says "Means and Standard Deviations," one would ordinarily assume s, not $s_{\overline{Y}}$. Also, if for any reason you wish to cite confidence limits, do not give them as "Statistic $\pm$ ½ (Confidence Interval)" because readers would confuse this with the conventional "Mean $\pm$ Standard Error"; state limits as L_1 and L_2, respectively.

It is important to state a statistic and its standard error to a sufficient number of decimal places. The following rule of thumb helps. *Divide the standard error by three, then note the decimal place of the first nonzero digit of the quotient; give the statistic significant to that decimal place and provide one further decimal for the standard error.* This rule is quite simple, as an example will illustrate. If the mean and standard error of a sample are computed as 2.354 ± 0.363, we divide 0.363 by 3, which yields 0.121. Therefore, the mean should be reported precisely to one decimal place, and the standard error should be reported precisely to two decimal places. Thus, we report this result as 2.4 ± 0.36. If, on the other hand, the same mean had a standard error of 0.243, dividing this standard error by 3 would have yielded 0.081, and the first nonzero digit would have been in the second decimal place. Thus, the mean should have been reported as 2.35 ± 0.243.

Graphic representations of the means of samples often indicate confidence limits by a simple method (Figure 7.15). A common alternative is to use wide bars with error bars superimposed on them. We show two versions of such graphs in Figure 7.16. This format is appropriate for frequency data, as in the example we show. Representing means of continuous variables and their standard errors in this way, however, tends to mislead the reader into assuming the bars are frequencies (as in Figures 2.2 or 5.2). There are two ways of indicating the error bar or "whisker." The one shown on the left in Figure 7.16 marks off one standard error above the mean; the one

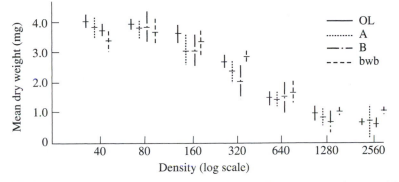

FIGURE 7.15 Weights of four strains of houseflies in pure culture at varying densities. Means and their 95% confidence limits are shown. The abscissa, density in logarithmic scale, refers to number of eggs per 36 g of medium. The four strains are indicated by different patterns of lines, as shown at upper right. (Data from Sullivan and Sokal, 1965.)

on the right marks off two standard errors, one above, the other below the mean. We prefer the right-hand version because most people find it difficult to mentally reflect the error bar below the mean. This recommendation, however, requires that the bar be white, or at least not solid black, so that the whisker beneath the mean is visible. You can see that there is no perfect solution for this scheme.

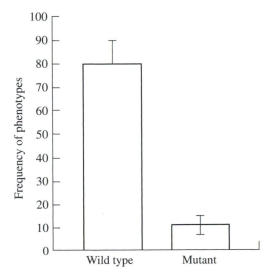

FIGURE 7.16 Examples of error bars showing one standard error above the mean (left) or one standard error above and below the mean (right) for the observed frequencies given in Table 17.1. We assume that these frequencies (counts) are Poisson distributed and their standard errors (= standard deviations) are the square roots of the frequencies.

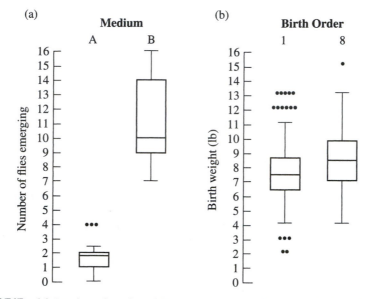

FIGURE 7.17   **(a)** Boxplots of number of flies emerging from different media (Table 13.3). **(b)** Boxplots of birth weight data from Exercise 9.3.

A more elaborate scheme, which has become quite common, is the method of **boxplots** (Tukey, 1977; McGill et al., 1978). A boxplot generally accompanies computer output of basic statistics programs. Such plots for two datasets are shown in Figure 7.17. The sample is represented as a box whose top and bottom are drawn at the lower and upper quartiles. The box is divided at the median. The length of the box is known as the **interquartile range**. A vertical line is drawn from the top of the box to the largest observation within 1.5 interquartile ranges of the top. A comparable line is drawn from the bottom to the smallest observation within 1.5 interquartile ranges of the bottom. All observations beyond these limits are plotted individually. Observations more than 3 interquartile ranges away from the box are given special prominence.

The great utility of boxplots is that they furnish measures of location (the median line), dispersion (the length of the box and the distance between the upper and lower whiskers), skewness (asymmetry of the upper and lower portions of the box, asymmetry of the upper and lower whiskers, or inequalities in the number of extreme individual observations plotted), and long-drawn-out tails (distance between the ends of the whiskers in relation to the length of the box). Boxplots are useful when two or more samples must be compared, as in both Figures 7.17a and 7.17b. In Figure 7.17a, we note very clearly that more flies emerged from medium B than from medium A. There is no overlap between the two distributions. In Figure 7.17b, we note that the birth weights of firstborns appear to be lighter than those of birth order 8. The lengths of the whiskers and the larger number of observations in the upper tails of the distributions show that the data are skewed to the right. This is to be expected in birth weights. For further elaborations of boxplots, see Benjamini (1988).

7.12 Confidence Limits for Variances

We saw earlier (Section 7.8) that the quantity $(n - 1)s^2/\sigma^2$ is distributed as χ^2 with $n - 1$ degrees of freedom. We take advantage of this fact when we set confidence limits to variances.

We can make the following statement about the quantity $(n - 1)s^2/\sigma^2$:

$$P\left\{\chi^2_{(1-(\alpha/2))[n-1]} \leq \frac{(n - 1)s^2}{\sigma^2} \leq \chi^2_{(\alpha/2)[n-1]}\right\} = 1 - \alpha$$

This expression is similar to those in Section 7.10 and implies that the probability, P, that a sample value of $(n - 1)s^2/\sigma^2$ would lie between the indicated critical values of $\chi^2_{[n-1]}$ is $1 - \alpha$. Simple algebraic manipulation of the quantities in the inequality within brackets yields

$$P\left\{\frac{(n - 1)s^2}{\chi^2_{(\alpha/2)[n-1]}} \leq \sigma^2 \leq \frac{(n - 1)s^2}{\chi^2_{(1-(\alpha/2))[n-1]}}\right\} = 1 - \alpha \tag{7.12}$$

Because $(n - 1)s^2 = \sum y^2$, we can simplify Expression (7.12) to

$$P\left\{\frac{\sum y^2}{\chi^2_{(\alpha/2)[n-1]}} \leq \sigma^2 \leq \frac{\sum y^2}{\chi^2_{(1-(\alpha/2))[n-1]}}\right\} = 1 - \alpha \tag{7.13}$$

This expression still looks formidable, but it means simply that if we divide the sum of squares $\sum y^2$ by the two values of $\chi^2_{[n-1]}$ bounding $1 - \alpha$ of the area of the $\chi^2_{[n-1]}$-distribution, the two quotients will enclose the true value of the variance σ^2 with a probability of $P = 1 - \alpha$.

A numerical example will make this clear. Suppose we have a sample of 5 housefly wing lengths with a sample variance of $s^2 = 13.52$. If we wish to set 95% confidence limits to the parametric variance, we evaluate Expression (7.12) for the sample variance s^2. We first calculate the sum of squares for this sample: $4 \times 13.52 = 54.08$. Then we look up the values for $\chi^2_{.025[4]}$ and $\chi^2_{.975[4]}$. Because 95% confidence limits are required, α in this case is equal to 0.05. These χ^2-values span between them 95% of the area under the χ^2-curve. They correspond to 11.143 and 0.484, respectively, and the limits in Expression (7.13) then become

$$L_1 = 54.08/11.143 = 4.85 \text{ and } L_2 = 54.08/0.484 = 111.74$$

This confidence interval is very wide, but we must not forget that the sample variance is, after all, based on only 5 individuals. Note also that the interval is asymmetrical around 13.52, the sample variance, in contrast to the confidence intervals encountered earlier, which were symmetrical around the sample statistic.

This method is called the **equal tails method** because an equal amount of probability is placed in each tail (for example, 2.5%). It can be shown that in view of the

---

BOX 7.4 Confidence Limits for σ^2. Method of Shortest Unbiased Confidence Intervals

Aphid stem mother femur lengths (from Boxes 2.1 and 4.1): $n = 25$; $s^2 = 0.1337$.

The factors from Statistical Table **O** for $\nu = n - 1 = 24$ df and confidence coefficient $1 - \alpha = 0.95$ are

$$f_1 = 0.5943 \qquad f_2 = 1.876$$

and for a confidence coefficient of 0.99 they are

$$f_1 = 0.5139 \qquad f_2 = 2.351$$

The 95% confidence limits for the population variance, σ^2, are given by the equations

$$L_1 = \text{(lower limit)} = f_1 s^2 = 0.5943(0.1337) = 0.07946$$

$$L_2 = \text{(upper limit)} = f_2 s^2 = 1.876(0.1337) = 0.25082$$

The 99% confidence limits are

$$L_1 = f_1 s^2 = 0.5139(0.1337) = 0.06871$$

$$L_2 = f_2 s^2 = 2.351(0.1337) = 0.31433$$

---

skewness of the distribution of variances, this method does not yield the shortest possible confidence intervals. One may wish the confidence interval to be "shortest" in the sense that the ratio L_2/L_1 be as small as possible. Box 7.4 shows how to obtain these **shortest unbiased confidence intervals** for σ^2 using Statistical Table **O**, based on the method of Tate and Klett (1959). With the help of Statistical Table **O**, which gives $(n - 1)/\chi^2_{p[n-1]}$, where p is an adjusted value of $\alpha/2$ or $1 - \alpha/2$ designed to yield the shortest unbiased confidence intervals, the computation is very simple and much faster than calculating confidence limits by means of χ^2.

Similar shortest unbiased confidence limits can be found for binomial parameters π and for the mean λ of a Poisson variable. Box 7.5 shows the simple computation by means of, respectively, Statistical Tables **P** and **N**, based on methods described in Crow (1956) and Crow and Gardner (1959).

7.13 The Jackknife and the Bootstrap

This chapter concludes with two general-purpose techniques, based on the randomization procedures learned in Section 7.1. These techniques are useful for analyzing either a novel statistic for which the mathematical distribution has not been fully worked out or a more ordinary statistic for which the distributional assumptions for a test may not be true. The **jackknife** technique was developed by John W. Tukey as a rough-and-ready statistical tool. It is a procedure that allows us to reduce the bias in

---

BOX 7.5 Confidence Limits for π, the Binomial Parameter, and λ, the Mean of the Poisson Distribution. Method of Shortest Unbiased Confidence Intervals

Binomial Parameter, π

In Section 5.2, we described a sibship consisting of 17 offspring, 3 males and 14 females. The proportion of males is $3/17 = 0.1765$. If we look up the 95% shortest unbiased confidence limits for $k = 17$ and $Y = 3$ in Statistical Table **P**, we find $L_1 = 0.0499$ and $L_2 = 0.4165$. The limits are broad, indicating uncertainty in the location of the true proportion (although we can be confident in excluding 0.5).

Poisson Parameter, λ

In a bacterial culture, six mutations of a certain type were found in one generation. What are the 99% confidence limits of the mean? According to Statistical Table **N**, 99% limits are $L_1 = 1.786$ and $L_2 = 15.813$. Such intervals have a 99% chance of including the true mean number of mutations.

---

our estimate of the population value for a statistic and provides a standard error for the statistic. The standard error allows us to make t-tests and to compute confidence intervals for the statistic (assuming the statistic is at least approximately normally distributed). Note that we are assuming normality for the jackknifed statistic, not for the individual observations themselves.

The idea is to split the observed data into groups, usually of size 1—that is, in a sample of n observations, there will be n "groups" of one observation each. Then one computes the desired statistic, each time ignoring the data in a different one of the groups of observations. Only when n is exceedingly large or the statistic very complex would we make an effort to reduce the computational load by working with fewer groups. The average of these estimates is used to provide a less biased estimate of the statistic, and the variability among these values is used to estimate the standard error of the statistic. An outline of the procedure is given in the first part of Box 7.6. For statistics that are bounded in range, such as a variance ($s^2 \geq 0$), the jackknife procedure often works better if the statistics are first transformed to have a distribution closer to the normal. For example, the jackknife procedure is usually applied to the logarithm of a variance.

An example of the application of the jackknife procedure is shown in Box 7.6. It provides a less biased estimate of the coefficient of variation, V, and its standard error, s_V. This allows us to set confidence limits to our estimate of V_P, the parametric value of the coefficient of variation. The usual estimate of V is biased to yield an underestimate in small samples. The sample consists of n observations Y_i. To jackknife V, n different coefficients of variation are then computed with each observation ignored in turn. Thus, V_{-1} is the coefficient of variation based on all $(n - 1)$

BOX 7.6   Tukey's Jackknife Method

Computation

1. Compute the desired sample statistic, St, based on the complete sample (of size n).
2. Compute the corresponding statistics St_{-i} based on the sample data with each of the observations i ignored in turn. If the sample size is very large, one may compute the statistics corresponding to the deletion of groups of k observations and then substitute the number of such groups for n in the following equations.
3. Compute the so-called pseudovalues, ϕ_i as follows:

$$\phi_i = nSt - (n-1)St_{-i}$$

4. The jackknifed estimate of the statistic is then simply

$$\widehat{St} = \frac{\sum \phi_i}{n} = \overline{\phi}$$

We may now assign approximate confidence limits to the statistic and perform approximate t-tests. Usually, $s_{\widehat{St}}$ is considered to have $n-1$ degrees of freedom. If the statistic is such that the ϕ_i can take on only a distinct values, however, the degrees of freedom are taken as $a-1$. For example, if one were to jackknife the range, there would be only three different pseudovalues possible, generated by dropping the smallest, the largest, or any intermediate value. In such a case, $a = 3$, and the degrees of freedom are 2.

Example

Estimate V and s_V by the jackknife method and set 95% confidence limits to V_P, the parametric coefficient of variation, using the aphid femur lengths from Box 2.1. For convenience, we have ordered (but *not grouped*) by magnitude the $n = 25$ observations in the table at the head of Box 2.1.

1. The observed V based on $n = 25$ is 9.1333. The observed standard deviation of the coefficient of variation, s_V, following step **5** in Box 7.1 (simplified formula $V/\sqrt{2n}$), is 1.2916.
2. The coefficients of variation, V_{-i}, are then computed. For example, V_{21} is computed using only observations 2 through 25, V_{-2} is obtained from observations 1 plus 3 through 25, and so forth. The V_{-i}-values are shown in the following table, along with their pseudovalues, $\Phi_i = nV - (n-1) V_{-i}$. For example, $\Phi_1 = 25(9.1333) - 24(8.4843) = 24.7078$. (*Note:* These figures do not exactly match those given for the same data in Section 7.11 because the present calculations were carried out with more decimal places.)

Box 7.6 (continued)

i	V_{-i}	ϕ_i	i	V_{-i}	ϕ_i
1	8.4843	24.7078	14	9.3250	4.5303
2	8.8902	14.9660	15	9.3250	4.5303
3	9.0412	11.3422	16	9.2903	5.3642
4	9.0412	11.3422	17	9.2245	6.9433
5	9.0412	11.3422	18	9.2245	6.9433
6	9.0412	11.3422	19	9.2245	6.9433
7	9.2468	6.4072	20	9.2245	6.9433
8	9.2468	6.4072	21	9.1268	9.2885
9	9.2468	6.4072	22	9.1268	9.2885
10	9.2468	6.4072	23	9.1268	9.2885
11	9.3032	5.0538	24	8.9959	12.4294
12	9.3032	5.0538	25	8.6275	21.2726
13	9.3032	5.0538		$\Sigma\phi_i$	229.5983
				$\overline{\phi}$	9.1839

3. The jackknife estimate $\hat{V}$ is then simply $\overline{\phi}$, which is 9.1839, and is slightly larger than our original estimate.

4. The estimated standard error of $\hat{V}$ is $\sqrt{s_\phi^2/n} = 1.0081$ with $n - 1 = 24$ degrees of freedom.

5. The final step is to set confidence limits. We set 95% confidence limits to $\hat{V}$:

$$L_1 = \hat{V} - t_{.05[24]}s_{\hat{V}}$$
$$= 9.1839 - 2.0639(1.0081) = 7.1033$$
$$L_2 = \hat{V} + t_{.05[24]}s_{\hat{V}}$$
$$= 9.1839 + 2.0639(1.0081) = 11.2645$$

For these data, the jackknife estimate of the coefficient of variation is somewhat larger than the conventional estimate, V, which is known to be an underestimate for small samples. The confidence intervals are also shorter because the estimated standard error from the jackknife is smaller.

---

observations except Y_1. These coefficients of variation are converted to "pseudovalues," ϕ_i, as defined in step **3** of Box 7.6. The average of the ϕ-values is our jackknifed estimate of V. In this particular example, the jackknife estimate of V_P is slightly larger than the simple estimate based on all the data. This is likely to be a better estimate because the usual estimate is known to be biased downward in small samples. The confidence limits are somewhat shorter than those that would be obtained using the methods described in Section 7.11 because the estimated standard error is smaller.

Another application of this technique is in testing functions of sample variances. For example, we can jackknife the statistic $\delta = \ln(s_1/s_2)$ to test for the equality of two variances. The jackknife procedure correctly estimates the variability of δ for all underlying distributions, whereas the F-test, Bartlett's test, and related tests that are discussed in subsequent chapters can be very inaccurate when the distributions are not normal (Miller, 1968).

The jackknife does not always work. For example, the distribution of the jackknifed largest sample observation can be degenerate or nonnormal. The jackknife is also not useful for correcting for the effects of outliers because they seem to affect the jackknifed estimate as much as they affect the original estimates. In such cases, the jackknifed estimate of the standard error is a severe underestimate. By examining the pseudovalues, however, one can ascertain the effect of deleting variates. If the deletion of a particular observation yields a pseudovalue very different from the rest, that observation should be tested to see if it is an outlier. The quantity $\bar{\phi} - \phi_i$, called the **sample influence function** by Devlin et al. (1975), is a convenient measure of the influence that a given observation has on the statistic being studied. Observations with large influence values are potentially outliers.

Miller (1974) reviews the statistical properties of the jackknife procedure and provides an extensive bibliography. Manly (1977) gives an example of the calculation of approximate variances for estimates of population parameters using the jackknife procedure. Other applications are described in Efron (1987) and Efron and Gong (1983).

The **bootstrap** is a related technique for obtaining standard errors and confidence limits of various statistics. The basic idea of the bootstrap is quite simple. We take the sample for which we hope to estimate a parameter π using a sample statistic p. This statistic can be as simple as the arithmetic mean (for which, of course, analytical formulas are readily available, as we saw earlier in this chapter, making the bootstrap unnecessary) or as complicated as the statistics already discussed in this section for the jackknife method. We compute p for the observed sample. If the sample is of size n, we then carry out a random resampling procedure by repeatedly sampling n items *with replacement* using the n observations in the sample as a parent population. For each of these samples, we estimate the desired parametric statistic π. Because we are sampling with replacement, most samples will contain two or more replicates of the same observation in the observed sample and consequently will lack some of the observations from the observed sample.

It has been shown that the mean of estimated statistics from these bootstrapped samples estimates the true value of the statistic in the population and that the standard deviation of such an estimate approximates the standard error of the statistic as if we had repeatedly sampled from the unknown population without replacement. This is an extremely important result because it permits us to calculate standard errors for all but a few refractory statistics (such as the sample median).

We can apply the bootstrap to the data we jackknifed to calculate the coefficient of variation and its standard error in Box 7.6. We will not illustrate this example in a box, but we describe it here. We took 10,000 samples of size 25, sampled with replacement from the observed 25 femur lengths of Box 2.1. We compute the coefficient of variation for each of these samples. Next, we compute the mean of these 10,000 V-values, as well as their standard deviation, which is the standard error of our estimate. One method for setting confidence limits makes use of the fact that the estimates should approximately follow the normal distribution because it is the mean of a large number of samples (see Section 7.2). Another method of setting confidence limits is called the percentile method. For it one examines the distribution of bootstrap estimates and uses the values that cut off in each tail as the upper and lower confidence limits. This method works well unless, as in the present example, the distribution is skewed. In such cases the bias-corrected or the accelerated bootstrap methods should be used although the estimates still may not be very satisfactory unless the original sample size is large. See Dixon (2001) for details. Table 7.3 gives a comparison of asymptotic, jackknife, and bootstrap estimates of V_p, its standard error, as well as 95% confidence limits, for the 25 femur lengths. Note that it has been shown that the jackknife and the bootstrap estimates approach each other asymptotically for large samples (Efron and Gong, 1983). In this example, the estimates are rather different—probably because the sample is relatively small ($n = 25$) and the distribution of the statistic is not symmetrical.

The bootstrap, although it can be computationally very intensive, is a simple method for obtaining standard errors and confidence limits of complicated statistics. Considered to be an improvement over the jackknife procedure, the bootstrap has been used extensively in biostatistical applications. For example, it has become quite popular in numerical taxonomic and phylogenetic research, where it was introduced by Felsenstein (1985). In phylogenetic estimation, the branching sequence of t species is estimated from a set of n characters, which vary in their states among species.

TABLE 7.3 Comparison of Asymptotic, Jackknife, and Bootstrap Estimates, Standard Errors, and Confidence Limits

Statistic	Method of Estimation		
	Asymptotic	**Jackknife**	**Bootstrap**
$\overline{Y}$	4.0040		
s	0.3657		
V	9.1333	9.1839	8.8973
s_V	1.2916	1.0081	0.9439
L_1	6.4674	7.1033	7.0409
L_2	11.7991	11.2645	10.7437

Data from Box 2.1. The percentile method was used to construct the bootstrap confidence limits.

The estimates are based, for example, on constructing the shortest tree, in terms of the amount of implied evolutionary change for a given data set. As a result of constructing such a tree, the systematist may conclude that species A and B are closer to each other than they are to other species—that is, that they form a monophyletic taxon.

How reliable is this taxon? Felsenstein suggested sampling n characters with replacement from the original data set to create new data sets from which new minimum-length trees are constructed. The result would be a number, say, $m = 100$ trees. If 95% or more of the trees contain the taxon {A, B}, then this branch of the tree is considered well substantiated. If, by contrast, only 30% of the bootstrapped trees show that set, little reliance can be placed on the taxonomic structure of that portion of the tree. Other applications are described in Chernick (2007), Davison and Hinkley (1997), Dixon (2001, 2002), Efron (1987), Efron and Gong (1983), and Efron and Tibshirani (1994).

The BIOMstat computer program carries out jackknife and bootstrap computations for selected statistics.

EXERCISES 7

7.1 Differentiate between type I and type II errors. What do we mean by the power of a statistical test?

7.2 Because it is possible to test a statistical hypothesis with a sample of any size, why are larger sample sizes preferred?

7.3 The 95% confidence limits for μ as obtained in a given sample were 4.91 and 5.67 g. Is it correct to say that 95 times out of 100 the population mean, μ, falls inside the interval from 4.91 to 6.57 g? If not, what would the correct statement be?

7.4 Set 95% confidence limits to the means in Table 7.1. Are these limits all correct? (In other words, do they contain μ?)

7.5 Set 99% confidence limits to the mean, median, coefficient of variation, variance, and g_2 for the birth weight data given in Section 4.3 and Boxes 4.2 and 6.1.

ANSWER: The lower limits are 109.540, 109.060, 12.136, 178.698, and −0.0405, respectively.

7.6 In Section 5.3, the coefficient of dispersion was given as an index of whether or not data agree with a Poisson distribution. Because the mean, μ, in a true Poisson distribution equals the parametric variance, σ^2, the coefficient of dispersion is analogous to Expression (7.9). Using the weed seed data from Table 5.7, test the hypothesis that the true variance is equal to the sample mean—in other words, that we have sampled from a Poisson distribution (in which the coefficient of dispersion should equal unity).

7.7 In a study measuring bill length of the dusky flycatcher, Johnson (1966) found that the bill length for the males had a mean of 8.14 ± 0.021 and a coefficient of variation of 4.67%. On the basis of this information, infer how many specimens must have been used.

ANSWER: $n = 328$.

7.8 In a study of mating calls in the tree toad *Hyla ewingi*, Littlejohn (1965) found the note duration of the call in a sample of 39 observations from Tasmania to have a mean of 189 ms and a standard deviation of 32 ms. Assign 95% confidence intervals to the mean and to the variance.

7.9   Using the method described in Exercise 7.6, test the agreement of the observed distributions to Poisson distributions by testing the hypothesis that the true coefficient of dispersion equals unity for the data of Tables 5.5, 5.6, 5.9, and 5.10. Note that in these examples, the chi-square table is not adequate, so approximate critical values must be computed using the method given with Statistical Table **D**. Section 8.3 presents an alternative test that avoids this problem.

   ANSWER: For the data in Table 5.6, $(n - 1) \times CD = 1308.30$, $\chi^2_{.025[588]} \approx 656.61$.

7.10   In direct klinokinetic behavior relating to temperature, animals turn more often in the warm end of a gradient and less often in the cold end, but the direction of turning is random. In a computer simulation of such behavior, the following results were found. The mean position along a temperature gradient was found to be -1.352. The standard deviation was 12.267, and n equaled 500 individuals. The gradient was marked off in units: Zero corresponded to the middle of the gradient, the starting point of the animals; minus corresponded to the cold end; and plus corresponded to the warmer end. Test the hypothesis that direct klinokinetic behavior did not result in a tendency toward aggregation in either the warm or the cold end; that is, test the hypothesis that μ, the mean position along the gradient, was zero.

7.11   Distinguish between power and size of a test.

7.12   Box 9.4 shows the effects of five treatments on growth in plant tissue cultures. For each sample, prepare a boxplot. Line up the boxplots to permit comparison among the treatments.

7.13   We observe 10 weed seeds in a 2-ounce sample of grass seeds. Estimate the 95% confidence limits of the number of weed seeds *per ounce* of grass seed.

   ANSWER: Lower limit, 1.971 to upper limit, 11.799. *Hint:* Use Statistical Table **N**.

7.14   Prepare a boxplot for the birth weight data of Section 4.3 and Box 4.2. What do you learn from inspecting this plot?

7.15   Compute the jackknife estimate of the biased estimate of the variance, $\Sigma y^2/n$, for the following data:

$$41\ 46\ 54\ 44\ 42$$

Compare your result with the usual unbiased estimate s^2. Which observation has the greatest influence on your sample estimate?

7.16   Compute the jackknife estimate of the coefficient of variation and its standard deviation for the data of Exercise 4.2.

   ANSWER: $V = 92.1462$, $s_V = 57.5$.

7.17   Compute the bootstrap estimate of the mean and its confidence limits for the data of Exercise 4.2. Compare the confidence limits with those computed in the usual way. Use at least 1000 resamplings.

8 Introduction to Analysis of Variance

This chapter focuses on the analysis of variance, developed by R. A. Fisher, which is fundamental to much of the application of statistics in biology and especially to experimental design. Analysis of variance is used to test for differences among sample means and differences among linear combinations of means. Its name is derived from the fact that variances are used to measure the differences among means. A simple application of the analysis of variance is to test whether two or more sample means could have been obtained from populations with the same parametric mean. Where only two samples are involved, the *t*-test has traditionally been used to test differences between means. The analysis of variance, however, provides a more general test, which permits testing two samples as well as many. We are therefore introducing it at this early stage in order to equip the reader with a powerful weapon for his or her statistical arsenal. We will discuss the *t*-test for two samples as a special case in Section 9.3.

A knowledge of analysis of variance is indispensable to any modern biologist, and after you have mastered it, you will undoubtedly use it many times to test scientific hypotheses. The analysis of variance, however, is more than a technique for statistical analysis. Once it is understood, analysis of variance is a tool that can provide an insight into the nature of variation of natural events—into Nature, in short—which is possibly of even greater value than the knowledge of the method as such. Like other models in science, however, analysis of variance may create constructions of nature in the mind of the scientist that give rise to misleading or unproductive conclusions. See Lewontin (1974) for an in-depth discussion of this issue.

In Section 8.1, we approach this subject via familiar ground, the sampling experiment of the housefly wing lengths. From these samples, we will obtain two independent estimates of the population variance. We digress in Section 8.2 to introduce yet another continuous distribution, the *F*-distribution, needed for the significance test in analysis of variance. Section 8.3 is another digression, in which we show how the *F*-distribution can be used to test whether two samples come from populations with the same variance. We are then ready for Section 8.4, in which we examine the effects of subjecting the samples to different treatments. Section 8.5 describes the partitioning of sums of squares and their degrees of freedom, the actual *analysis* of variance. Sections 8.6 and 8.7 take up in a more formal way the two scientific models for which the analysis of variance is appropriate, the so-called fixed treatment effects model (Model I) and the variance component model (Model II).

177

Except for Section 8.3, the entire chapter is largely theoretical. We will postpone the practical details of computation until Chapter 9. A thorough understanding of the material in Chapter 8 is necessary, however, for working out the examples of analysis of variance in Chapter 9.

One final comment: The late Professor J. W. Tukey of Princeton University has contributed many a well-turned phrase or term (in addition to many important theoretical concepts) to the field of statistics. The abbreviation *anova* for "analysis of variance" is one of them, and it quickly became standard. We will use *anova* interchangeably with *analysis of variance* throughout the text.

8.1   Variances of Samples and Their Means

In this section, we discuss analysis of variance in terms of the familiar housefly wing lengths (see Section 6.2 and Table 6.1). We obtained seven random samples of 5 wing lengths, which we have reproduced in Table 8.1. The seven samples of 5, here called groups, are listed vertically in the upper half of the table. Before we explain Table 8.1 further, we must become familiar with some additional terminology and symbolism for dealing with this kind of problem. We call our samples **groups;** they are sometimes called *classes* or other names that we will learn later. In any analysis-of-variance problem, we will have two or more such samples or groups, and the symbol a will represent the number of groups. Thus, in this example $a = 7$. Each group or sample is based on n observations as before; in Table 8.1, $n = 5$. The total number of observations in the table is a times n, which in this case equals 7×5, or 35.

In an anova, summation signs can no longer be as simple as heretofore. We can sum either the items of one group only or the items of the entire table. We therefore have to use superscripts with the summation symbol. In line with our policy of using the simplest possible notation whenever this is not likely to lead to misunderstanding, we will use $\Sigma^n Y$ to indicate the sum of the items of a group and $\Sigma^{an} Y$ to indicate the sum of all the items in the table. The mean of each group, symbolized by $\overline{Y}_i$, is shown in the row underneath the horizontal dividing line and is computed simply as $\Sigma^n Y_i / n$. The bottom row in Table 8.1 contains $\Sigma^n y_i^2$, the sum of squares of $\overline{Y}$, separately for each group.

From each of these sums of squares we can obtain an estimate of the population variance of housefly wing length. Thus, in the first group $\Sigma^n y^2 = 29.2$. Therefore, our estimate of the population variance is

$$s^2 = \frac{\overset{n}{\sum} y^2}{n-1} = \frac{29.2}{4} = 7.3$$

which is a rather low estimate compared to what we know the true variance to be. Because we have a sum of squares for each group, we could obtain an estimate of the population variance from each of these. It stands to reason, however, that we would get a better estimate if we averaged these separate variance estimates in some way.

TABLE 8.1 Seven Samples (Groups) of 5 Wing Lengths of Randomly Selected Houseflies

Parametric mean, $\mu = 45.5$; variance, $\sigma^2 = 15.21$.

	a groups ($a = 7$)							Computation of sum of squares of means	Computation of total sum of squares
	1	**2**	**3**	**4**	**5**	**6**	**7**		
n individuals per group ($n = 5$)	41	48	40	40	49	40	41		
	44	49	50	39	41	48	46		
	48	49	44	46	50	51	54		
	43	49	48	46	39	47	44		
	42	45	50	41	42	51	42		
$\bar{Y}_i$	43.6	48.0	46.4	42.4	44.2	47.4	45.4	$\bar{\bar{Y}} = 45.34$	$\bar{\bar{Y}} = 45.34$
$\sum\limits^{n} y_i^2$	29.2	12.0	75.2	45.2	98.8	81.2	107.2	$\sum\limits^{a}(\bar{Y} - \bar{\bar{Y}})^2 = 25.417$	$\sum\limits^{an} y^2 = 575.886$

Data from Table 6.1.

We can obtain this value by computing the weighted average of the variances using Expression (4.2) in Section 4.1. Actually, in this instance, a simple average would suffice because each estimate of the variance is based on samples of the same size. We prefer to give the general formula, however, which works equally well for this case and for instances of unequal sample sizes, where the weighted average formula is necessary.

In this case, each sample variance s_i^2 is weighted by its degrees of freedom, $w_i = n_i - 1$, resulting in a sum of squares ($\Sigma^n y_i^2$) because $(n_i - 1)s_i^2 = \Sigma y_i^2$. Thus, the numerator of Expression (4.2) is the sum of the sums of squares. The denominator is $\Sigma^a(n_i - 1) = 7 \times 4 = 28$, the sum of the degrees of freedom of each group. The average variance is therefore

$$s^2 = \frac{29.2 + 12.0 + 75.2 + 45.2 + 98.8 + 81.2 + 107.2}{28} = \frac{448.8}{28} = 16.029$$

This quantity is an estimate of 15.21, the parametric variance of housefly wing lengths. We call this estimate, based on 7 independent estimates of variances of groups, the *average variance within groups,* or simply **variance within groups.** Note that we use the expression *within* groups, although in previous chapters we used the term variance *of* groups. The reason is that up to now the variance estimates used for computing the average variance have all come from sums of squares measuring the variation within one column. As we will see, one can also compute variances among groups, cutting across group boundaries.

To obtain a second estimate of the population variance, we treat the seven group means, $\bar{Y}$, as though they were a sample of seven observations. The resulting statistics are shown in Table 8.1, at the bottom of the column headed "Computation of sum of squares of means." There are seven means in this example; in the general case, there will be a means. First, we compute $\bar{\bar{Y}}$, the grand mean of the group means, as $\bar{\bar{Y}} = \Sigma^a\bar{Y}/a$. The sum of the seven means is $\Sigma^a\bar{Y} = 317.4$ and the grand mean is $\bar{\bar{Y}} = 45.34$, a fairly close approximation to the parametric mean $\mu = 45.5$. The sum of squares represents the deviations of the group means from the grand mean, $\Sigma^a(\bar{Y} - \bar{\bar{Y}})^2$, which, when evaluated, is 25.417.

From the sum of squares of the means we obtain a **variance among the means** in the conventional way, as follows: $\Sigma^a(\bar{Y} - \bar{\bar{Y}})^2/(a - 1)$. We divide by $a - 1$ rather than $n - 1$ because the sum of squares was based on a items (means). Thus, variance of the means $s_{\bar{Y}}^2 = 25.417/6 = 4.2362$. We learned in Expression (7.1) that when randomly sampling from a single population,

$$\sigma_{\bar{Y}}^2 = \frac{\sigma^2}{n}$$

and hence

$$\sigma^2 = n\sigma_{\bar{Y}}^2$$

Thus, we can estimate a variance of items by multiplying the variance of means by the sample size on which the means are based (assuming we have sampled at random from a common population). When we do this for our present example, we obtain s^2 = 5 × 4.2362 = 21.181. This value is another estimate of the parametric variance 15.21. It is not as close to the true value as the previous estimate based on the average variance within groups, but this is to be expected because it is based on only 7 "observations." To distinguish it from the variance of means from which it was computed, as well as from the variance within groups with which it will be compared, we call this variance the **variance among groups;** its value, n times the variance of means, is an independent estimate of the parametric variance σ^2 of the housefly wing lengths. It may not be clear at this stage why the two estimates of σ^2 that we have obtained, the variance within groups and the variance among groups, are independent. We ask you to take it on faith that they are.

Let us review what we have done so far by expressing in a more formal way the data table for an anova and the computations carried out up to now. Table 8.2 is a generalized table for data such as the samples of housefly wing lengths. Do not be put off by the unfamiliar notation! Each individual wing length is represented by Y, subscripted to indicate its position in the data table. The wing length of the jth fly from the ith sample or group is given by Y_{ij}. Thus, you will notice that the first subscript changes with each column representing a group in the table and that the second subscript changes with each row representing an individual item. (Those of you familiar with matrix algebra may wonder why the subscript notation here is the reverse of the one in that field. It is simply a well-established usage in anova, and we prefer not to change it.) Using this notation, we can compute the variance of sample 1 as

$$\frac{1}{n-1}\sum_{j=1}^{j=n}(Y_{1j}-\bar{Y}_1)^2$$

The variance within groups, which is the average variance of the samples, is computed as

$$\frac{1}{a(n-1)}\sum_{i=1}^{i=a}\sum_{j=1}^{j=n}(Y_{ij}-\bar{Y}_i)^2$$

Note the double summation. It means that we first set $i = 1$ (i being the index of the outer Σ) when we start with the first group. We sum the squared deviations of all items from the mean of the first group, changing index j of the inner Σ from 1 to n in the process. We then return to the outer summation, set $i = 2$, and sum the squared deviations for group 2 from $j = 1$ to $j = n$. This process is continued until i, the index of the outer Σ, is set to a. In other words, we sum all the squared deviations within one group first and add this sum to similar sums from all the other groups. The variance among groups is computed as

$$\frac{n}{a-1}\sum_{i=1}^{i=a}(\bar{Y}_i-\bar{\bar{Y}})^2$$

TABLE 8.2 Data Arranged for Simple Analysis of Variance, Single Classification, Completely Randomized

		\multicolumn{7}{c}{a groups}					
		1	2	3	$\cdots$	i	$\cdots$	a
	1	Y_{11}	Y_{21}	Y_{31}	$\cdots$	Y_{i1}	$\cdots$	Y_{a1}
	2	Y_{12}	Y_{22}	Y_{32}	$\cdots$	Y_{i2}	$\cdots$	Y_{a2}
	3	Y_{13}	Y_{23}	Y_{33}	$\cdots$	Y_{i3}	$\cdots$	Y_{a3}
n items	$\vdots$	$\vdots$	$\vdots$	$\vdots$		$\vdots$		$\vdots$
	j	Y_{1j}	Y_{2j}	Y_{3j}	$\cdots$	Y_{ij}	$\cdots$	Y_{aj}
	$\vdots$	$\vdots$	$\vdots$	$\vdots$		$\vdots$		$\vdots$
	n	Y_{1n}	Y_{2n}	Y_{3n}	$\cdots$	Y_{in}	$\cdots$	Y_{an}
Means	$\bar{Y}$	$\bar{Y}_1$	$\bar{Y}_2$	$\bar{Y}_3$	$\cdots$	$\bar{Y}_i$	$\cdots$	$\bar{Y}_a$

Now that we have two independent estimates of the population variance, what shall we do with them? We might wish to find out whether they do, in fact, estimate the same parameter. To test this hypothesis, we need a statistical test that will evaluate the probability that the two sample variances are from the same population. Such a test employs the F-distribution, which we discuss in the next section.

8.2 The F-Distribution

Let us devise yet another sampling experiment. Assume that you are sampling at random from a normally distributed population, such as the housefly wing lengths with mean μ and variance σ^2. The sampling procedure consists of first sampling n_1 items and calculating their variance s_1^2, followed by sampling n_2 items and calculating their variance s_2^2. The degrees of freedom of the two variances are $\nu_1 = n_1 - 1$ and $\nu_2 = n_2 - 1$. Sample sizes n_1 and n_2 may or may not be equal to each other but are fixed for any one sampling experiment. Thus, for example, we might always sample 8 wing lengths for the first sample (n_1) and 6 wing lengths for the second sample (n_2). After each pair of values (s_1^2 and s_2^2) has been obtained, we calculate

$$F_s = \frac{s_1^2}{s_2^2}$$

This ratio will be close to 1 because these variances are estimates of the same quantity. Its value will depend on the relative magnitudes of variances s_1^2 and s_2^2. If we repeatedly take samples of sizes n_1 and n_2, calculating the ratios F_s of their variances, the average of these ratios will approach $(n_2 - 1)/(n_2 - 3)$, which is close to 1.0 when n_2 is large. Statisticians have worked out the expected distribution of this statistic, which is called the **F-distribution** in honor of R. A. Fisher. This is another distribution described by a complicated mathematical function that need not concern us here.

Unlike the t- and χ^2-distributions, the shape of the F-distribution is determined by *two* values for degrees of freedom, ν_1 and ν_2. Thus, for every possible combination of values ν_1, ν_2, each ν ranging from 1 to infinity, there exists a separate F-distribution. Remember that, like the t-distribution and the χ^2-distribution, the F-distribution is a theoretical probability distribution. Variance ratios based on sample variances, s_1^2/s_2^2, are sample statistics that may or may not follow the F-distribution. We have, therefore, distinguished the sample variance ratio by calling it F_s, conforming to our convention of separate symbols for sample statistics as distinct from probability distributions (such as t_s and X^2 for t and χ^2, respectively).

We have discussed the generation of an F-distribution by repeatedly taking two samples from the same normal distribution. We could also have generated the F-distribution by sampling from two separate normal distributions differing in their mean but identical in their parametric variances—that is, with $\mu_1 \neq \mu_2$ but $\sigma_1^2 = \sigma_2^2$. Thus, we obtain an F-distribution whether the samples come from the same normal population or from different ones, as long as their variances are identical.

Figure 8.1 shows several representative F-distributions. For very low degrees of freedom, the distribution is L-shaped, but it becomes humped and strongly skewed to the right as both degrees of freedom increase. P-values are frequently furnished by computer routines built into analysis of variance programs. They require the upper and lower degrees of freedom for the variance ratio as well as its observed value and return the appropriate one-tailed P-value. Absent such programs, workers are

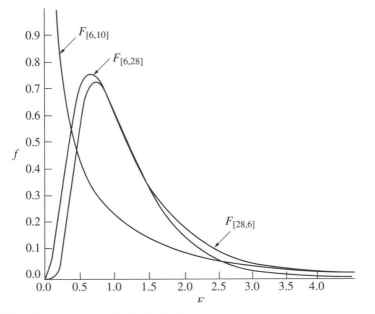

FIGURE 8.1 Three representative F-distributions.

required to look up the critical F-value in a statistical table. Our Statistical Table **F** shows the cumulative probability distribution of F for several selected probability values. The values in the table represent $F_{\alpha[\nu_1,\nu_2]}$, where α is the proportion of the F-distribution to the right of the given F-value (in one tail) and ν_1, ν_2 are the degrees of freedom pertaining to the numerator and the denominator of the variance ratio, respectively. The table is arranged with ν_1, the degrees of freedom pertaining to the upper (numerator) variance, across the top and ν_2, the degrees of freedom pertaining to the lower (denominator) variance, along the left margin.

At each intersection of degree of freedom values, we list several values of F for decreasing magnitudes of α. The first value is $\alpha = 0.75$. Note that for $\alpha = 0.5$, F is not $(n_2 - 1)/(n_2 - 3)$ because F for $\alpha = 0.5$ is the median and not the mean of the asymmetrical F-distribution. In tests of hypotheses, we are usually interested in small values of α, so the right tail of the curve is especially emphasized in most F-tables. For example, an F-distribution with $\nu_1 = 6$, $\nu_2 = 28$ is 2.45 at $\alpha = 0.05$. Thus, 0.05 of the area under the curve lies to the right of $F = 2.45$, as Figure 8.2 shows. Only 0.01 of the area under the curve lies to the right of $F = 3.53$. Thus, if we have a null hypothesis $H_0\colon \sigma_1^2 = \sigma_2^2$, with the alternative hypothesis $H_1\colon \sigma_1^2 > \sigma_2^2$, we would use a one-tailed F-test as illustrated by Figure 8.2.

We can now test the two variances obtained in the sampling experiment of Section 8.1 and Table 8.1. The variance among groups based on 7 means was 21.180, and the variance within 7 groups of 5 individuals was 16.029. Our null hypothesis

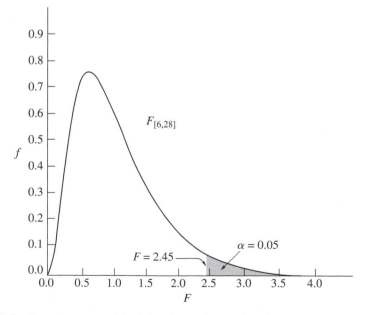

FIGURE 8.2 Frequency curve of the F-distribution for 6 and 28 degrees of freedom, respectively. A one-tailed 5% rejection region is marked off at $F = 2.45$.

is that the two variances estimate the same parametric variance; the alternative hypothesis in an anova is always that the parametric variance estimated by the variance among groups is greater than that estimated by the variance within groups. The reason for this restrictive alternative hypothesis, which leads to a one-tailed test, will be explained in Section 8.4.

We calculate the variance ratio: $F_s = s_1^2/s_2^2 = 21.181/16.029 = 1.32$. Before we can evaluate a probability based on an F-distribution, we have to know the appropriate degrees of freedom for this variance ratio. We will learn simple formulas for degrees of freedom in an anova later, but at the moment let us reason it out for ourselves. The upper variance (among groups) was based on the variance of $a = 7$ means; hence, it should have $a - 1 = 6$ degrees of freedom. The lower variance was based on an average of $a = 7$ variances, each of them based on $n = 5$ individuals yielding 4 degrees of freedom per variance; $a(n - 1) = 7 \times 4 = 28$ degrees of freedom. Thus, the upper variance has 6, the lower variance 28 degrees of freedom. If we check Statistical Table **F** for $\nu_1 = 6$, $\nu_2 = 28$, we find that $F_{.05[6,28]} = 2.45$. For $F = 1.32$, corresponding to the F_s-value actually obtained, α is between 0.5 and 0.25. Thus, we may expect between half and one-fourth of all variance ratios of samples based on 6 and 28 degrees of freedom, respectively, to have F_s-values greater than 1.32.

When we furnished the F_s-value and the degrees of freedom to a computer program (BIOMstat), we obtained a P-value of 0.2811. From either result, we have no evidence to reject the null hypothesis and therefore conclude that the two sample variances estimate the same parametric variance. This conclusion corresponds, of course, to what we knew anyway from our sampling experiment. Because the seven samples were taken from the same population, the estimate using the variance of their means is expected to yield another estimate of the parametric variance of housefly wing length.

Whenever the alternative hypothesis is more general—that is, the two parametric variances are unequal (rather than the restrictive hypothesis $H_1: \sigma_1^2 > \sigma_2^2$)—the sample variance s_1^2 could be smaller as well as greater than s_2^2. This leads to a two-tailed test, and in such cases a 5% type I error means that rejection regions of $2\frac{1}{2}\%$ will occur at each tail of the curve. Because the F-distribution is asymmetrical, these rejection regions will not look identical, as they do in the t-distribution. Figure 8.3 shows the 5% two-tailed rejection regions for the same F-distribution illustrated in Figure 8.2. Since usually only the right half of the F-distribution is tabulated (Statistical Table **F** shows only $\alpha = 0.75$ in the left half of the curve), when faced with a two-tailed test we deliberately divide the greater variance by the lesser variance; for a type I error of α, we look up the F-value for $\alpha/2$ in the F-table. Computational details of this method are illustrated in Section 8.3.

It is sometimes necessary to obtain F-values for $\alpha > 0.5$ (that is, in the left half of the F-distribution). Because these values are rarely tabulated, as we have just learned, they can be obtained by using the following simple relationship:

$$F_{\alpha[\nu_1, \nu_2]} = \frac{1}{F_{(1-\alpha)[\nu_2, \nu_1]}} \tag{8.1}$$

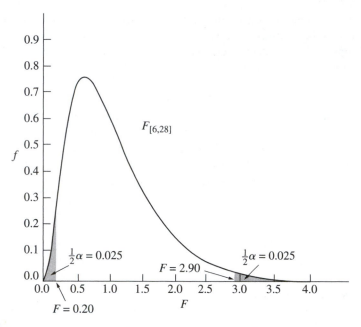

FIGURE 8.3   Frequency curve of the F-distribution for 6 and 28 degrees of freedom, respectively. A two-tailed 5% rejection region ($2\frac{1}{2}$% in each tail) is marked off at $F = 0.20$ and $F = 2.90$, respectively.

Note that $F_{.05[5,24]} = 2.62$. If we wish to obtain $F_{.95[5,24]}$ (the F-value to the right of which lies 95% of the area of the F-distribution with 5 and 24 degrees of freedom, respectively), we first have to find $F_{.05[24,5]} = 4.53$. Then $F_{.95[5,24]}$ is the reciprocal of 4.53, which equals 0.221. Thus, 95% of an F-distribution with 5 and 24 degrees of freedom lies to the right of 0.221.

There is an important relationship between the F-distribution and the χ^2-distribution. You may remember that the ratio $X^2 = \Sigma y^2/\sigma^2$ was distributed as a χ^2 with $n - 1$ degrees of freedom. If you divide the numerator of this expression by $n - 1$, you obtain the ratio $F_s = s^2/\sigma^2$, which is a variance ratio with an expected distribution of $F_{[n-1,\infty]}$. The upper degrees of freedom are $n - 1$ (the degrees of freedom of the sum of squares or sample variance). The lower degrees of freedom are infinite because only on the basis of an infinite number of items can we obtain the true, parametric variance of a population. Therefore, by dividing a value of X^2 by $n - 1$ degrees of freedom, we obtain an F_s-value with $n - 1$ and ∞ df, respectively. In general, $\chi^2_{[\nu]}/\nu = F_{[\nu,\infty]}$. We can convince ourselves of this fact by inspecting the F- and χ^2-tables. From the χ^2-table (Statistical Table D) we find that $\chi^2_{.05[10]} = 18.307$. Dividing this value by 10 df, we obtain 1.8307. From the F-table (Statistical Table F) we find, for $\nu_1 = 10$, $\nu_2 = \infty$, that $F_{.05[10,\infty]} = 1.83$. Thus, the two test statistics are closely related and, lacking a χ^2-table, we could make do with an F-table alone, using the values of $\nu F_{[\nu,\infty]}$ in place of $\chi^2_{[\nu]}$.

Before we return to analysis of variance, we will quickly apply our new knowledge of the F-distribution to testing a hypothesis about two sample variances.

8.3 The Hypothesis H_0: $\sigma_1^2 = \sigma_2^2$

Box 8.1 illustrates a test of the null hypothesis that two normal populations represented by two samples have the same variance. As we will see later, some other tests leading to a decision about whether the means of two samples come from the same population assume that the population variances are equal. The test illustrated in Box 8.1 is useful for examining the validity of this assumption. However, this test is of interest in its own right. It is often necessary to test whether two samples could have come from populations with the same variance. In genetics, we may need to know whether an offspring generation is more variable for a character than the parent generation. In systematics, we might like to find out whether two local populations are equally variable. In experimental biology, we may wish to demonstrate under which of two experimental setups the readings will be more variable. In general, the less variable setup would be preferred; if both setups were equally variable, the experimenter would pursue the one that is simpler or less costly to undertake.

BOX 8.1 | **Testing the Difference Between Two Variances**

Survival in days of the cockroach *Blattella vaga* when kept without food or water.

Females	$n_1 = 10$	$\overline{Y}_1 = 8.5$ days	$s_1^2 = 3.6$
Males	$n_2 = 10$	$\overline{Y}_2 = 4.8$ days	$s_2^2 = 0.9$
	H_0: $\sigma_1^2 = \sigma_2^2$		H_1: $\sigma_1^2 \neq \sigma_2^2$

Source: Data modified from Willis and Lewis (1957).

The alternative hypothesis is that the two variances are unequal. We have no reason to suppose that one sex should be more variable than the other. In view of the alternative hypothesis, this is a two-tailed test. Because only the right tail of the F-distribution is tabulated extensively in Statistical Table **F** and in most other F-tables, we calculate F_s as the ratio of the greater variance over the lesser one:

$$F_s = \frac{s_1^2}{s_2^2} = \frac{3.6}{0.9} = 4.00$$

Because the test is two-tailed, we look up the critical value $F_{\alpha/2[\nu_1, \nu_2]}$, where α is the type I error accepted, $\nu_1 = n_1 - 1$ and $\nu_2 = n_2 - 1$, the degrees of freedom for the upper and lower variance, respectively. Whether we look up $F_{\alpha/2[\nu_1, \nu_2]}$ or $F_{\alpha/2[\nu_2, \nu_1]}$ depends on whether sample 1 or sample 2 has the greater variance and was placed in the numerator.

Box 8.1 (continued)

From Statistical Table **F**, we find $F_{.025[9,9]} = 4.03$, $F_{.05[9,9]} = 3.18$, and $F_{.10[9,9]} = 2.44$. Because this is a two-tailed test, we double these probabilities. Thus, the F-value of 4.03 represents a probability of $\alpha = 0.05$ because the right-hand tail area of $\alpha = 0.025$ is matched by a similar left-hand area to the left of $F_{.975[9,9]} = 1/F_{.025[9,9]} = 0.248$. Therefore, assuming the null hypothesis is true, the probability of observing an F-value greater than 4.00 and smaller than $1/4.00 = 0.25$ is $0.10 > P > 0.05$. Alternatively, we can employ a computer program; using BIOMstat, we obtain $P = 0.0255$ for $F = 4.00$, $df = 9, 9$. We double this value for a one-tailed test, which yields $P = 0.051$. Strictly speaking, the two sample variances are not significantly different—we do not reject the null hypothesis that the two sexes were equally variable in their duration of survival. The outcome is close enough to the 5% significance level, however, to make us suspicious that the variances were, in fact, different. It would be desirable to repeat this experiment with larger sample sizes in the hope that more decisive results would emerge.

The test employed is usually a two-tailed test because the alternative hypothesis is generally H_1: $\sigma_1^2 \neq \sigma_2^2$.

There are instances when it is not legitimate to employ the F-distribution to test the hypothesis H_0: $\sigma_1^2 = \sigma_2^2$. The following example is a case in point. In an experiment studying the effects of selection, 25 larval ticks were exposed to cold shock treatment. Only 9 larvae survived. Measurements of the scutum (in micrometers) were made for both dead and surviving larvae. The results are as follows:

Surviving larvae	$n_1 = 9$,	$\bar{Y}_1 = 210.63$,	$s_1^2 = 16.4025$
211.3	218.5	211.2	205.1
211.9	204.9	211.4	211.9
209.5			

Killed larvae	$n_2 = 16$,	$\bar{Y}_2 = 214.43$,	$s_2^2 = 51.0036$
219.2	211.1	210.4	219.5
205.1	222.8	210.1	218.4
213.4	210.2	213.1	204.6
206.7	212.7	224.4	229.2

The investigator hypothesized that variation in scutum length reflects general genetic variability and that only ticks from a restricted portion of the genetic spectrum would survive. This hypothesis would be supported if the variance of scutum length of the sample of ticks surviving the cold shock is lower than that for the ticks that were killed.

How can we test this hypothesis? We can calculate a variance, s_2^2, of the scutum length of the 16 nonsurviving ticks and divide it by the variance, s_1^2, based on the 9 survivors. In the observed data, $s_1^2 = 16.4025$ and $s_2^2 = 51.0036$, so $s_2^2/s_1^2 = 3.1095$. This value suggests that the hypothesis is justified. However, this ratio does not follow the F-distribution. The two variances are obtained not from independent, normally distributed samples but from a (nonrandom) sorting of individuals within a circumscribed universe, the 25 experimental organisms. Also, on inspection, the scutum lengths do not appear to be distributed normally. The investigator in this example, who was unaware of any appropriate statistical technique for such a test, proceeded with a randomization test. There are $\binom{n_1 + n_2}{n_1} = \binom{25}{9} = 25!/(9! \, 16!) = 2{,}042{,}975$

ways of taking 9 ticks out of the sample of 25. For each of these, we can compute the two variances and their ratio, s_2^2/s_1^2, and then determine whether the ratio observed in the sample is sufficiently deviant in terms of the total distribution of the sampled ratios to merit rejection of the null hypothesis of random assortment. Such an undertaking, however, would involve a very large number of computations. The investigator therefore carried out a sampled randomization test (Monte Carlo randomization) and obtained 500 random partitions of the data into samples of 9 and 16 from the population of the 25 scutum lengths and computed s_1^2, s_2^2, and s_2^2/s_1^2 for each case. Figure 8.4 illustrates the distribution of ratios obtained by running the problem on a computer. The ratio 3.1095 obviously is not in the upper 5% tail of the distribution of 500 empirically obtained ratios. Thus, we cannot support our hypothesis with the evidence from this experiment. Of the randomized variance ratios, 8% are equal to or greater than the observed value 3.1095. Based on a sample of 500, the 99% confidence limits

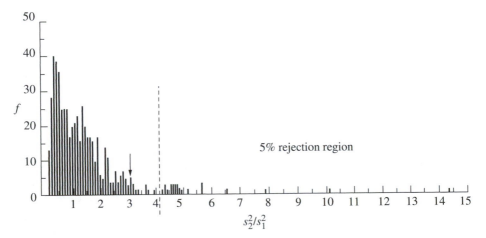

FIGURE 8.4 Distribution of ratios s_2^2/s_1^2 obtained from 500 random partitions of the 25 scutum lengths into sets of 9 and 16 each. The one-tailed 5% rejection region is marked off by a broken line. The arrow identifies the observed ratio being tested.

of this percentage are 5.23 and 11.60 (see Statistical Table **P**). Thus, it is improbable that the true probability based on the entire population of 2,042,975 partitions is less than 5%.

The BIOMstat computer program carries out randomization tests for selected statistics.

8.4   Heterogeneity Among Sample Means

We will now modify the data of Table 8.1 (see Section 8.1). Suppose these seven groups of houseflies did not represent random samples from the same population but resulted instead from the following experiment: Each sample was reared in a separate culture jar, and the medium in each of the culture jars was prepared in a different way. Some had more water added, others more sugar, yet others more solid matter. Let us assume that sample 7 represents the standard medium against which we propose to compare the other samples. The changes in the medium affect the sizes of the flies that emerge from it; the sizes of the flies, in turn, affect the wing lengths we have been measuring.

We shall assume the following effects resulting from treatment of the medium:

Medium 1—decreases average wing length of a sample by 5 units
    2—decreases average wing length of a sample by 2 units
    3—does not change average wing length of a sample
    4—increases average wing length of a sample by 1 unit
    5—increases average wing length of a sample by 1 unit
    6—increases average wing length of a sample by 5 units
    7—(control) does not change average wing length of a sample

We will symbolize the effect of treatment i as α_i. (Please note that this use of α is not related to its use as a symbol for type I error. There are, regrettably, more symbols needed in statistics than there are letters in the Greek and Roman alphabets.) Thus, α_i assumes the following values for these treatment effects:

$$\alpha_1 = -5 \qquad \alpha_4 = 1$$
$$\alpha_2 = -2 \qquad \alpha_5 = 1$$
$$\alpha_3 = 0 \qquad \alpha_6 = 5$$
$$\alpha_7 = 0$$

Note that $\Sigma^a \alpha_i = 0$; that is, the sum of the effects cancels out. This cancellation is a convenient property that is generally postulated, but it is unnecessary for our argument. We can now modify Table 8.1 by adding the appropriate values of α_i to each sample. In sample 1, the value of α_1 is -5, so the first wing length, 41, now becomes 36; the second wing length, 44, becomes 39; and so on. For the second sample, α_2 is -2, changing the first wing length from 48 to 46. Where α_i is 0, the wing lengths do not change; where α_i is positive, they are increased by the magnitude indicated. The changed values can be inspected in Table 8.3, arranged identically to Table 8.1.

TABLE 8.3 Data of Table 8.1 with Fixed-Treatment Effects α_i or Random Effects A_i Added to Each Sample

	a groups (a = 7)							Computation of sum of squares of means	Computation of total sum of squares
	1	2	3	4	5	6	7		
α_i or A_i	−5	−2	0	+1	+1	+5	0		
n individuals per group ($n = 5$)	36	46	40	41	50	45	41		
	39	47	50	40	42	53	46		
	43	47	44	47	51	56	54		
	38	47	48	47	40	52	44		
	37	43	50	42	43	56	42		
$\bar{Y}_i$	38.6	46.0	46.4	43.4	45.2	52.4	45.4	$\bar{\bar{Y}} = 45.34$	$\bar{\bar{Y}} = 45.34$
$\sum\limits^{n} y_i^2$	29.2	12.0	75.2	45.2	98.8	81.2	107.2	$\sum\limits^{a}(\bar{Y} - \bar{\bar{Y}})^2 = 100.617$	$\sum\limits^{an} y^2 = 951.886$

We now repeat our previous computations. We calculate the sum of squares of the first sample and find it to be 29.2. If you compare this value with the sum of squares of the first sample in Table 8.1, you find the two values to be identical. Similarly, all other values of $\Sigma^n y_i^2$, the sum of squares of each group, are identical to their previous values. Why is this so? The effect of adding α_i to each group is simply that of an additive code because α_i is constant for any one group. Appendix A.2 shows that additive codes do not affect sums of squares or variances. Therefore, not only is each separate sum of squares the same as before but also the average variance within groups is still 16.029.

Now let us compute the variance of the means. It is $100.617/6 = 16.770$, a value much higher than the variance of means found before, 4.236. When we multiply by $n = 5$ to get an estimate of σ^2, we obtain the variance of groups, which now is 83.848—no longer even close to an estimate of σ^2. We then repeat the F-test with the new variances and find that $F_s = 83.848/16.029 = 5.23$, which is much greater than the critical value of $F_{.05[6,28]} = 2.45$. In fact, the observed value of F_s is greater than $F_{.005[6,28]} = 4.02$. Clearly, the upper variance, representing the variance among groups, has become significantly larger. The two variances are most unlikely to represent the same parametric variance.

What has happened? Table 8.4 explains by representing Table 8.3 symbolically in the manner that Table 8.2 represented Table 8.1. Note that each group has a constant α_i added and that this constant changes the means of these groups by α_i. In Section 8.1, we computed the variance within groups as

$$\frac{1}{a(n-1)}\sum_{i=1}^{i=a}\sum_{j=1}^{j=n}(Y_{ij}-\bar{Y}_i)^2$$

If we try to repeat this computation, our formula becomes more complicated because to each Y_{ij} and each $\bar{Y}_i$ there has now been added α_i. We therefore write

$$\frac{1}{a(n-1)}\sum_{i=1}^{i=a}\sum_{j=1}^{j=n}[(Y_{ij}+\alpha_i)-(\bar{Y}_i+\alpha_i)]^2$$

When we open the parentheses inside the square brackets, the second α_i changes sign and the α_i's cancel out, leaving the expression exactly as before, substantiating our earlier observation that the variance within groups does not change, despite the treatment effects.

The variance of means was previously calculated by the formula

$$\frac{1}{a-1}\sum_{i=1}^{i=a}(\bar{Y}_i-\bar{\bar{Y}})^2$$

From Table 8.4, however, we see that the new grand mean equals

$$\frac{1}{a}\sum_{i=1}^{i=a}(\bar{Y}_i+\alpha_i)=\frac{1}{a}\sum_{i=1}^{i=a}\bar{Y}_i+\frac{1}{a}\sum_{i=1}^{i=a}\alpha_i=\bar{\bar{Y}}+\bar{\alpha}$$

TABLE 8.4 Data of Table 8.3 Arranged in the Manner of Table 8.2

		1	2	*a* groups 3	$\cdots$	*i*	$\cdots$	*a*
	1	$Y_{11} + \alpha_1$	$Y_{21} + \alpha_2$	$Y_{31} + \alpha_3$	$\cdots$	$Y_{i1} + \alpha_i$	$\cdots$	$Y_{a1} + \alpha_a$
	2	$Y_{12} + \alpha_1$	$Y_{22} + \alpha_2$	$Y_{32} + \alpha_3$	$\cdots$	$Y_{i2} + \alpha_i$	$\cdots$	$Y_{a2} + \alpha_a$
	3	$Y_{13} + \alpha_1$	$Y_{23} + \alpha_2$	$Y_{33} + \alpha_3$	$\cdots$	$Y_{i3} + \alpha_i$	$\cdots$	$Y_{a3} + \alpha_a$
n items	j	$Y_{1j} + \alpha_1$	$Y_{2j} + \alpha_2$	$Y_{3j} + \alpha_3$	$\cdots$	$Y_{ij} + \alpha_i$	$\cdots$	$Y_{aj} + \alpha_a$
	n	$Y_{1n} + \alpha_1$	$Y_{2n} + \alpha_2$	$Y_{3n} + \alpha_3$	$\cdots$	$Y_{in} + \alpha_i$	$\cdots$	$Y_{an} + \alpha_a$
Means		$\bar{Y}_1 + \alpha_1$	$\bar{Y}_2 + \alpha_2$	$\bar{Y}_3 + \alpha_3$	$\cdots$	$\bar{Y}_i + \alpha_i$	$\cdots$	$\bar{Y}_a + \alpha_a$

When we substitute the new values for the group means and the grand mean, the formula appears as

$$\frac{1}{a-1}\sum_{i=1}^{i=a}[(\bar{Y}_i + \alpha_i) - (\bar{\bar{Y}} + \bar{\alpha})]^2$$

which, in turn, yields

$$\frac{1}{a-1}\sum_{i=1}^{i=a}[(\bar{Y}_i - \bar{\bar{Y}}) + (\alpha_i - \bar{\alpha})]^2$$

Squaring the expression in the square brackets, we obtain the terms

$$\frac{1}{a-1}\sum_{i=1}^{i=a}(\bar{Y}_i - \bar{\bar{Y}})^2 + \frac{1}{a-1}\sum_{i=1}^{i=a}(\alpha_i - \bar{\alpha})^2 + \frac{2}{a-1}\sum_{i=1}^{i=a}(\bar{Y}_i - \bar{\bar{Y}})(\alpha_i - \bar{\alpha})$$

The first of these terms we recognize immediately as the previous variance of the means, $s_{\bar{Y}}^2$. The second is a new quantity but is familiar by general appearance as a variance or at least a quantity akin to a variance. The third expression is a new quantity, a so-called covariance, that we have not yet encountered. We will not be concerned with it at this stage except to say that in cases such as this one, where the magnitude of treatment effects α_i is assumed to be independent of the $\bar{Y}_i$ to which they are added, the expected value of this quantity is zero; hence, it would not contribute to the new variance of means. The assumed independence of the treatment effects from their sample means is also the reason why the variance estimates among groups and within groups are independent, as stated earlier in Section 8.1.

The independence of the treatment effects and the sample means is an important concept. If we had not applied different treatments to the medium jars but simply had treated all jars as controls, we would still have obtained differences among the wing length means. These are the differences found in Table 8.1 with random sampling from the same population. By chance, some of these means are greater, some are smaller. In planning the experiment, we had no way of predicting which sample means would be small and which would be large. Therefore, in planning our treatments we had no way of matching up large treatment effect, such as that of medium 6, with the mean that by chance would be the greatest, as that in sample 2. Also, the smallest sample mean (sample 4) is not associated with the smallest treatment effect. Only if the magnitude of the treatment effects were deliberately correlated with the sample means (this would be difficult to do in the experiment designed here) would the third term in the expression, the covariance, have an expected value other than zero.

We have seen that the first quantity is simply the variance of means as described before. The second quantity clearly is added as a result of the treatment effects and is analogous to a variance but cannot be called such because it is not based on a random variable; rather, it is based on deliberately chosen treatments largely under our control. By changing the magnitude and nature of the treatments, we can more or less alter this variance-like quantity at will. We shall, therefore, call it the **added component due to treatment effects.** Because the α_i's are coded so that $\bar{\alpha} = 0$, we can rewrite the middle term as

$$\frac{1}{a-1}\sum_{i=1}^{i=a}(\alpha_i - \bar{\alpha})^2 = \frac{1}{a-1}\sum_{i=1}^{i=a}\alpha_i^2 = \frac{1}{a-1}\sum^{a}\alpha^2$$

In analysis of variance, we multiply the variance of the means by n to estimate the parametric variance of the items. As you know, we call the quantity obtained in this way the *variance of groups*. When we do this for the case in which treatment effects are present, we obtain

$$n\left(s_{\bar{Y}}^2 + \frac{1}{a-1}\sum^{a}\alpha^2\right) = s^2 + \frac{n}{a-1}\sum^{a}\alpha^2$$

Thus, we see that the estimate of the parametric variance of the population is increased by the quantity

$$\frac{n}{a-1}\sum^{a}\alpha^2$$

which is n times the added component due to treatment effects. We found the variance ratio F_s to be greater than could be expected if the null hypothesis were true. It is now obvious why. We were testing the variance ratio expecting to find F approximately equal to $\sigma^2/\sigma^2 = 1$. In fact, however, the expected ratio is

$$\frac{\sigma^2 + \dfrac{n}{a-1}\sum^{a}\alpha^2}{\sigma^2}$$

It is clear from this formula, deliberately displayed in this lopsided manner, that the F-test is sensitive to the presence of the added component due to treatment effects.

At this point, you have an additional insight into the analysis of variance. The method permits us to test whether there are added treatment effects—that is, to test whether a group of means can be considered random samples from the same population or whether we have sufficient evidence to conclude that the treatments that have affected each group separately have resulted in shifting these means sufficiently so that they can no longer be considered samples from the same population. If the latter is true, an added component due to treatment effects will be present and may be detected by an F-test.

In such a study, we are generally not interested in the magnitude of

$$\frac{n}{a-1}\sum^{a}\alpha^{2}$$

but we are interested in the magnitude of the separate values of α_i. In our example, these are the effects of different formulations of the medium on wing length. If instead of housefly wing length we were measuring blood pressure in samples of rats, and if the different groups had been subjected to different drugs or different doses of the same drug, the quantities α_i would represent the effects of drugs on the blood pressure, the issue of interest to the investigator. We may also be interested in studying differences of the type $\alpha_1 - \alpha_2$, leading us to test hypotheses about the differences between the effects of any two types of medium or any two drugs. But we are a little ahead of our story.

Analysis of variance that involves treatment effects of the type just studied is called a **Model I anova,** or a *fixed effects model.* In Section 8.6, Model I will be defined precisely. There is another model, called a **Model II anova**, in which the added effects for each group are not fixed treatments but are *random effects*—that is, we have not deliberately planned or fixed the treatment for any one group, and the effects on each group are random and only partly under our control.

Suppose that the seven samples of houseflies in Table 8.3 represented the offspring of seven randomly selected females from a population reared on a uniform medium. There would be genetic differences among these females that would be reflected in their seven broods. The exact nature of these differences is unclear and unpredictable. Before measuring them, we have no way of knowing whether brood 1 would have longer wings than brood 2, nor have we any way of controlling this experiment so that brood 1 would grow longer wings. As far as we can ascertain, the genetic factors for wing length are distributed in an unknown manner in the population of houseflies (we might hope that they are distributed normally), and our sample of seven is a random sample of these factors.

Another example of a Model II anova is that instead of making up our seven cultures from a single batch of medium, we have prepared seven batches separately, one right after the other, and are now analyzing the variation among the batches. We would not be interested in the exact differences from batch to batch. Even if we measured these, we would not be in a position to interpret them. Not having deliberately varied batch 3, we have no idea why it should produce longer wings than batch 2, for example. We would, however, be interested in the magnitude of the variance of the added effects. Thus, if we used seven jars of medium derived from one batch, we could expect the variance of the jar means to be $\sigma^2/5$ because there were 5 flies per jar. When based on different batches of medium, however, the variance could be expected to be greater because all the imponderable accidents of formulation and environmental differences during medium preparation that make one batch of medium different from another would come into play. Interest would focus on the added variance component arising from differences among batches. Similarly, in the first example, we would be interested in the added variance component arising from genetic differences among the females.

We will now take a quick look at the algebraic formulation of the Model II anova. In Table 8.3, the first row at the head of the data columns shows not only α_i but also A_i, which is the symbol we will use for a random group effect. We use a capital letter to indicate that the effect is a variable. As with the treatment effects, we will consider the sum of the random effects to be zero:

$$\frac{1}{a}\sum_{}^{a} A_i = \bar{A} = 0$$

The algebra of calculating the two estimates of the population variance is the same as in Model I, except that in place of α_i we imagine A_i substituted in Table 8.4. The estimate of the variance among means now represents the quantity

$$\frac{1}{a-1}\sum_{i=1}^{i=a}(\bar{Y}_i - \bar{\bar{Y}})^2 + \frac{1}{a-1}\sum_{i=1}^{i=a}(A_i - \bar{A})^2 + \frac{2}{a-1}\sum_{i=1}^{i=a}(\bar{Y}_i - \bar{\bar{Y}})(A_i - \bar{A})$$

The first term is the variance of means $s_{\bar{Y}}^2$ as before, and the last term is the covariance between the group means and the random effects A_i, the expected value of which is zero (as before) because the random effects are independent of the magnitude of the means. The middle term is a true variance, since A_i is a random variable. We symbolize it by s_A^2 and call it the **added variance component among groups**. In the previous two examples, this term would represent the added variance component among females or among medium batches. The existence of this added variance component is tested for by the F-test. If the groups are random samples, we may expect F to approximate $\sigma^2/\sigma^2 = 1$, but with an added variance component the expected ratio is

$$\frac{\sigma^2 + n\sigma_A^2}{\sigma^2}$$

Note that σ_A^2, the parametric value of s_A^2, is multiplied by n because we have to multiply the variance of means by n to obtain an independent estimate of the variance of the population.

In a Model II anova, we are not interested in the magnitude of any A_i or in differences such as $A_1 - A_2$, but we are interested in the magnitude of σ_A^2 and its relative magnitude with respect to σ^2, which is generally expressed as the percentage $100s_A^2/(s^2 + s_A^2)$. Because the variance among groups estimates $\sigma^2 + n\sigma_A^2$, we can calculate s_A^2 as

$\frac{1}{n}$ (variance among groups $-$ variance within groups)

$$= \frac{1}{n}[(s^2 + ns_A^2) - s^2] = \frac{1}{n}(ns_A^2) = s_A^2$$

If we interpret the example in Table 8.3 as a Model II anova,

$$s_A^2 = \tfrac{1}{5}(83.848 - 16.029) = 13.56$$

This added variance component among groups is

$$\frac{100 \times 13.56}{(16.029 + 13.56)} = 45.83\%$$

of the sum of the variances among and within groups.

Model II anovas will be discussed formally in Section 8.7; the methods of estimating variance components are treated in detail in the next chapter.

8.5 Partitioning the Total Sum of Squares and Degrees of Freedom

You may be discouraged to find more algebraic formulas ahead in this section, but these describe only simple properties and, given a little good will, can be mastered by anyone. We present these quantities to round out our presentation of basic analysis of variance.

Until now, we have ignored one other variance that can be computed from the data in Table 8.1. If we remove the classification into groups, we can consider the housefly data to be a single sample of $an = 35$ wing lengths and calculate the mean and variance of these items in the conventional manner. The quantities necessary for this computation are shown in the last column at the right in Tables 8.1 and 8.3, headed "Computation of total sum of squares." We obtain a mean of $\overline{Y} = 45.34$ for the sample in Table 8.1, which is, of course, the same as the quantity $\overline{\overline{Y}}$ computed previously from the seven group means. The sum of squares of the 35 items is 575.886, which gives a variance of 16.938 when divided by 34 degrees of freedom. Repeating these computations for the data in Table 8.3, we obtain $\overline{Y} = 45.34$ (the same as in Table 8.1 because $\Sigma^a \alpha_i = 0$) and $s^2 = 27.997$, which is considerably greater than the corresponding variance from Table 8.1. The total variance computed from all an items is another estimate of σ^2. It is a good estimate in the first case, but in the second sample (Table 8.3), where added components from treatment effects or added variance components are present, it is a poor estimate of the population variance.

However, the purpose of calculating the total variance in an anova is not to use it as yet another estimate of σ^2 but to introduce an important mathematical relationship between it and the other variances. This relationship is most obvious when we arrange our results in a conventional *analysis of variance table*, or **anova table** (Table 8.5). Such a table is divided into four columns. The first identifies the source of variation as among groups, within groups, and total (groups amalgamated to form a single

TABLE 8.5 Anova Table for Data in Table 8.1

	(1) Source of variation	(2) df	(3) Sums of squares, SS	(4) Mean squares, MS
$\bar{Y} - \bar{\bar{Y}}$	Among groups	6	127.086	21.181
$Y - \bar{Y}$	Within groups	28	448.800	16.029
$Y - \bar{\bar{Y}}$	Total	34	575.886	16.938

sample). The column headed *df* gives the degrees of freedom by which the sums of squares pertinent to that source of variation must be divided to yield the variances. The degrees of freedom for variation among groups is $a - 1$, for variation within groups $a(n - 1)$, and for total variation $an - 1$. The next two columns show sums of squares and variances, respectively. Notice that the sums of squares entered in the anova table are the sum of squares among groups, the sum of squares within groups, and the sum of squares of the total sample of *an* items. In anovas, variances are not called such but are generally called **mean squares** because, in a Model I anova, they do not estimate a population variance. These quantities are not true *mean* squares because the sums of squares are divided by the degrees of freedom rather than by sample size. The sums of squares and mean squares are frequently abbreviated *SS* and *MS*, respectively.

The sums of squares and mean squares in Table 8.5 are the same as those obtained previously, except for minute rounding errors. Note, however, an important property of the sums of squares: They are obtained independently of each other, but when we add the *SS* among groups to the *SS* within groups, we obtain the total *SS*. The sums of squares are additive! Another way of saying this is that we can decompose the total sum of squares into a portion due to variation among groups and another portion due to variation within groups. In Appendix A.3, we demonstrate why the sums of squares within and among groups add up to the total sum of squares. Observe that the degrees of freedom are also additive and that the total of 34 *df* can be decomposed into 6 *df* among groups and 28 *df* within groups. Thus, if we know any two of the sums of squares (and their appropriate degrees of freedom), we can compute the third and complete our analysis of variance. Note that the mean squares, however, are not additive, which should be obvious, because generally $(a + b)/(c + d) \neq a/c + b/d$.

Before we continue, let us review the meaning of the three mean squares in the anova. The total *MS* is a statistic of dispersion of the 35 (*an*) items around their mean, the grand mean (in this example, 45.34). It describes the variance in the entire sample resulting from all and sundry causes and estimates σ^2 when there are no added treatment effects or variance components among groups. The within-group *MS*, also known as the *individual* or *intragroup* or **error mean square**, gives the average dispersion of the 5 (*n*) items in each group around the group means. If the *a* groups are random samples from a common homogeneous population, the within-group *MS*

TABLE 8.6 Anova Table for Data in Table 8.3

	(1) Source of variation	(2) df	(3) Sums of squares, SS	(4) Mean squares, MS
$Y - \overline{Y}$	Among groups	6	503.086	83.848
$\overline{Y} - \overline{\overline{Y}}$	Within groups	28	448.800	16.029
$Y - \overline{\overline{Y}}$	Total	34	951.886	27.997

should estimate σ^2. The MS among groups is based on the variance of group means, which describes the dispersion of the 7 (a) group means around the grand mean. If the groups are random samples from a homogeneous population, the expected variance of their mean will be σ^2/n. Therefore, in order to have the error variance term σ^2 in the among-groups MS be comparable to that in the within-groups MS, we multiply the variance of means by n to obtain the variance among groups. If there are no added treatment effects or variance components, the MS among groups is an estimate of σ^2. Otherwise, it is an estimate of

$$\sigma^2 + \frac{n}{a-1}\sum^{a}\alpha^2 \quad \text{or} \quad \sigma^2 + n\sigma_A^2$$

depending on whether the anova is Model I or II.

The additivity relations we have just learned are independent of the presence of added treatment or random effects. We could show this algebraically, but Table 8.6 makes this clear simply by summarizing the anova of Table 8.3, in which α_i or A_i is added to each sample. The additivity relation still holds, although the values for group SS and total SS are different from those in Table 8.5.

Another way of looking at the partitioning of the variation is to study the deviation from means in a particular case. Referring to Table 8.1, we can look at the wing length of the first individual in the seventh group, which is 41. Its deviation from its group mean is

$$Y_{71} - \overline{Y}_7 = 41 - 45.4 = -4.4$$

The deviation of the group mean from the grand mean is

$$\overline{Y}_7 - \overline{\overline{Y}} = 45.4 - 45.34 = 0.06$$

and the deviation of the individual wing length from the grand mean is

$$Y_{71} - \overline{\overline{Y}} = 41 - 45.34 = -4.34$$

Note that these deviations are additive. The deviation of the item from the group mean and that of the group mean from the grand mean add to the total deviation of the item from the grand mean. These deviations are stated algebraically as

$(Y - \bar{Y}) + (\bar{Y} - \bar{\bar{Y}}) = (Y - \bar{\bar{Y}})$. Squaring and summing these deviations for an items will result in

$$\sum^{a}\sum^{n}(Y - \bar{Y})^2 + n\sum^{a}(\bar{Y} - \bar{\bar{Y}})^2 = \sum^{an}(Y - \bar{\bar{Y}})^2$$

Before squaring, the deviations were in the relationship $a + b = c$. After squaring, we would expect them to be in the form $a^2 + b^2 + 2ab = c^2$. What happened to the cross-product term corresponding to $2ab$? This is

$$2\sum^{an}(Y - \bar{Y})(\bar{Y} - \bar{\bar{Y}}) = 2\sum^{a}\left[(\bar{Y} - \bar{\bar{Y}})\sum^{n}(Y - \bar{Y})\right]$$

which is a covariance-type term that is always zero because $\Sigma^n(Y - \bar{Y}) = 0$ for each of the a groups (see proof in Appendix A.1).

For pedagogical reasons, we identify the deviations represented by each level of variation at the left margins of the anova tables (see Tables 8.5 and 8.6). Note that the deviations add up correctly: The deviation among groups plus the deviation within groups equals the total deviation of items in the analysis of variance, $(\bar{Y} - \bar{\bar{Y}}) + (Y - \bar{Y}) = (Y - \bar{\bar{Y}})$.

8.6   Model I Anova

We have already mentioned the two different models in analysis of variance. An important point is that the basic setup of data, as well as the computation and significance test, is the same for both models in most cases. The purposes of analysis of variance differ for the two models, as do some of the supplementary tests and computations following the initial test of the null hypothesis. The two models were first defined by Eisenhart (1947). Although analysis of variance dates back to the 1930s, the distinction between the models had not been made in earlier work.

Let us now try to resolve the variation found in an analysis of variance case. This process not only will lead us to a more formal interpretation of anova but also will give us a deeper understanding of the nature of variation itself. For this discussion, we will refer to the housefly wing lengths of Table 8.3. We ask the following question: What makes any given housefly wing length assume the value it does? The third wing length of the first sample of flies is 43 units. How can we explain such a reading?

If we knew nothing else about this individual housefly, our best guess as to its wing length would be the parametric mean of the population, which we know to be $\mu = 45.5$. However, we have additional information about this fly. It is a member of group 1, which has undergone a treatment shifting the mean of the group downward by 5 units. Therefore, $\alpha_1 = -5$, and we would expect our individual Y_{13} (the third individual of group 1) to measure $45.5 - 5 = 40.5$ units. In fact, however, it is 43 units, which is 2.5 units above this latest expectation. To what can we ascribe this deviation? Clearly, this is individual variation of the flies within a group as demonstrated by the variance of individuals in the population ($\sigma^2 = 15.21$). All the genetic and environmental effects that make one housefly different from another housefly come into play to produce this variance.

By means of carefully designed experiments, we might learn something about what causes this variance and attribute it to certain specific genetic or environmental factors. We might also be able to eliminate some of the variance. For instance, by using only full sibs (brothers and sisters) in any one culture jar, we would decrease the genetic variation in individuals, and undoubtedly the variance within groups would be smaller. However, it is hopeless to try to eliminate all variance completely. Even if we could remove all genetic variance, there would still be environmental variance, and even in the most improbable case in which we could remove both, there would remain measurement error; thus, we would never obtain exactly the same reading, even on the same individual fly. The within-groups *MS* always remains as a residual, greater or smaller from experiment to experiment—part of the nature of things. This is why the within-groups variance is also called the *error variance* or *error mean square*. It is not an error in the sense of someone having made a mistake but rather in the sense of providing you with a measure of the variation you have to contend with when trying to estimate differences among the groups. The error variance is composed of deviations for each individual, symbolized by ϵ_{ij}, the random component of the *j*th individual observation in the *i*th group. In our case, $\epsilon_{13} = 2.5$ because the actual observed value is 2.5 units above its expectation of 40.5.

We will now state this relationship more formally. In a Model I analysis of variance, we assume that the differences among group means, if any, result from the fixed treatment effects applied by the experimenter. The purpose of the analysis of variance is to estimate the true differences among the group means. Any single observation can be decomposed as follows:

$$Y_{ij} = \mu + \alpha_i + \epsilon_{ij} \tag{8.2}$$

where $i = 1, \ldots, a, j = 1, \ldots, n$, ϵ_{ij} represents an independent, normally distributed variable with mean $\epsilon_{ij} = 0$ and variance $\sigma_\epsilon^2 = \sigma^2$. Therefore, a given reading is composed of the grand mean μ of the population, a fixed deviation α_i of the mean of group *i* from the grand mean μ, and a random deviation ϵ_{ij} of the *j*th individual of group *i* from its expectation, which is ($\mu + \alpha_i$). Remember that both α_i and ϵ_{ij} can be either positive or negative. The expected value (mean) of the ϵ_{ij}'s is zero, and their variance is the parametric variance of the population, σ^2. For all the assumptions of the analysis of variance to hold, the distribution of ϵ_{ij} must be normal.

In a Model I anova, we test for differences of the type $\alpha_1 - \alpha_2$ among the group means by testing for the presence of an added component due to treatments. If we find that such a component is present, we reject the null hypothesis that the groups come from the same population and accept the alternative hypothesis that at least some of the group means are different from each other, which indicates that at least some of the α_i's are unequal in magnitude. Next, we generally wish to test which α_i's are different from each other. This is done by significance tests, with alternative hypotheses such as $H_1: \alpha_1 > \alpha_2$ or $H_1: \frac{1}{2}(\alpha_1 + \alpha_2) > \alpha_3$. In words, these test whether the mean of group 1 is greater than the mean of group 2, or whether the mean of group 3 is smaller than the average of the means of groups 1 and 2.

Some examples of Model I anovas in biological disciplines follow. An experiment in which we try the effects of different drugs on batches of animals results in a Model I anova. We are interested in the results of the treatments and the differences between them. The treatments are fixed and determined by the experimenter. The same is true when we test the effects of different doses of a given factor—for example, a chemical or the amount of light to which a plant has been exposed, or temperatures at which culture bottles of insects have been reared. The treatment does not have to be entirely understood and manipulated by the experimenter; as long as it is fixed and repeatable, Model I will apply. Comparing the birth weights of the Chinese children in a hospital in Singapore with weights of Chinese children born in a hospital in China would also be a Model I anova. The treatment effects then would be "China versus Singapore," which sums up a whole series of different factors, genetic and environmental—some known to us, but most of them not understood. However, this is a definite treatment that we can describe and also repeat; namely, we can, if we wish, again sample birth weights of infants in Singapore as well as in China.

Another example of a Model I anova is the study of body weights for several age groups of animals. The treatments would be the ages, which are fixed. If we find ourselves able to reject the null hypothesis of no difference in weight among the ages, we might follow this up with the question of whether there is a difference from age 2 to age 3 or only from age 1 to age 2. To a very large extent, Model I anovas are the result of an experiment and of deliberate manipulation of factors by the experimenter. However, the study of differences such as the comparison of birth weights from two countries, although not an experiment proper, also falls into this category.

The null hypothesis in a Model I anova can be written as H_0: $\alpha_1 = \alpha_2 = \cdots = \alpha_a$. What this means in words is that the different levels of the treatment do not differ in their effect on variable Y. This does not say, however, that the treatment variable itself has no effect on the observed variable, only that differences in the level or quality of the treatment do not affect the response. Thus, we emphasize contrasts of the type $(\alpha_1 - \alpha_2)$ among the groups. Accepting the null hypothesis implies that $(\alpha_1 - \alpha_2) = 0$ and all other contrasts among treatment levels are equally zero, but the factor itself may well have an effect on the variable. Thus, different oxygen concentrations may have no effect on the heartbeat of samples of animals, yet the presence of oxygen is essential—in its absence, the heartbeat would drop to zero. What Model I anova tests is the *differential* effects of the treatment.

Let us review once more the idea of decomposing an individual reading, using the birth weights as an example. An individual birth weight of an infant in Singapore can be thought of as consisting of the parametric mean μ of the birth weights of all Chinese children plus the treatment effects of being in Singapore α_1, plus a random component ϵ_{1j}, which represents the genetic and environmental differences making this individual infant deviate from its population mean.

8.7   Model II Anova

The structure of variation in a Model II anova is similar to that in Model I:

$$Y_{ij} = \mu + A_i + \epsilon_{ij} \qquad (8.3)$$

where $i = 1, \ldots, a, j = 1, \ldots, n$, ϵ_{ij} represents an independent, normally distributed variable with mean $\bar{\epsilon}_{ij} = 0$ and variance $\sigma_\epsilon^2 = \sigma^2$, and A_i represents a normally distributed variable, independent of all ϵ's, with mean $\bar{A} = 0$ and variance σ_A^2. The main distinction is that in place of fixed treatment effects α_i, we now consider random effects A_i, which differ from group to group. Because the effects are random, it is futile to estimate the magnitude of these random effects for any one group, or the differences from group to group, but we can estimate their variance, the added variance component among groups σ_A^2. We test for its presence and estimate its magnitude s_A^2, as well as its percentage contribution to the variation in a Model II analysis of variance.

The following examples illustrate applications of Model II anova. Suppose we wish to determine the DNA content of rat liver cells. We take five rats and make three preparations from each of the five livers obtained. The assay readings will be for $a = 5$ groups with $n = 3$ readings per group. The five rats presumably are sampled at random from the colony available to the experimenter. They must be different in various ways, genetically and environmentally, but we have no definite information about the nature of these differences. Thus, if we learn that rat 2 has slightly more DNA in its liver cells than rat 3, we can do little with this information because we are unlikely to have any basis for following up this problem.

We will, however, be interested in estimating the variance of the three replicates within any one liver and the variance among the five rats; that is, we want to know if there is variance σ_A^2 among rats in addition to the variance σ^2 expected on the basis of the three replicates. The variance among the three preparations presumably arises only from differences in technique and possibly from differences in DNA content in different parts of the liver (unlikely in a homogenate). Added variance among rats, if it existed, might result from differences in ploidy or related phenomena. The relative amounts of variation among rats and "within" rats (= among preparations) would guide us in designing further studies of this sort. If there were little variance among the preparations and relatively more variation among the rats, we would need fewer preparations and more rats. On the other hand, if the variance among rats were proportionally smaller, we would use fewer rats and more preparations per rat. We will discuss the efficient planning of experiments in Section 10.4.

In a study of the amount of variation in skin pigment in human populations, we might wish to look at different families within a homogeneous ethnic group and brothers and sisters within each family. The variance within families would be the error mean square, and we would test for an added variance component among families. We would expect an added variance component σ_A^2 because there are genetic differences among families that determine amount of skin pigmentation. We would be especially interested in the relative proportions of the two variances σ^2 and σ_A^2

because they would provide us with important genetic information. From our knowledge of genetic theory we would expect the variance among families to be greater than the variance among brothers and sisters within a family.

These examples illustrate the two types of problems involving Model II anovas that are most likely to arise in biological work. One type is concerned with the general problem of the design of an experiment and the magnitude of the experimental error at different levels of replication, such as error among replicates within rat livers and, among rats, error among batches or experiments, and so forth. The other relates to variation among and within families, among and within females, among and within populations, and so forth, being concerned with the general problem of the relation between genetic and phenotypic variation.

It is sometimes difficult to decide whether a given study is a Model I or Model II anova. The distinction depends not on the nature of the factor differentiating the groups but on whether levels of that factor can be considered a random sample of more such levels or are fixed treatments whose differences the investigator wishes to contrast. Thus, the factor "years" might simply be a kind of replication in which it would be a random factor. On the other hand, one might be interested in specific years because of events that occurred in those years, which would make years a fixed factor.

In the next chapter, we will practice computing analyses of variance using several different examples. These examples will provide an opportunity to distinguish cases of Model I and Model II anovas and to learn the proper way of analyzing each model and of presenting the results of such analyses.

EXERCISES 8

8.1 In a study comparing the chemical composition of the urine of chimpanzees and gorillas (Gartler et al., 1956), the following results were obtained. For 37 chimpanzees, the variance for the amount of glutamic acid in milligrams per milligram of creatinine was 0.01082 (see Exercise 4.2). A similar study based on six gorillas yielded a variance of 0.12442. Test whether there is a difference between the variability in chimpanzees and gorillas.

   ANSWER: $F_s = 11.499$, $F_{.025[5,36]} = 2.94$, $P = 1.1 \times 10^{-6}$.

8.2 Show that it is possible to represent the value of an individual observation as follows: $Y_{ij} = (\bar{\bar{Y}}) + (\bar{Y}_i - \bar{\bar{Y}}) + (Y_{ij} - \bar{Y}_i)$. What do each of the terms in parentheses estimate in a Model I anova and in a Model II anova?

8.3 The following data are from an experiment by Sewall Wright, in which he crossed Polish and Flemish giant rabbits and obtained 27 F_1 rabbits. These were inbred, and 112 F_2 rabbits were obtained. We have extracted the following data on femur length of these rabbits

	n	$\bar{Y}$	s
F_1	27	83.39	1.65
F_2	112	80.50	3.81

Is there a greater amount of variability in femur lengths among the F_2 than among the F_1 rabbits? What well-known genetic phenomenon is illustrated by these data?

ANSWER: Yes; segregation and recombination.

8.4 For the data in Table 8.3, make tables to represent partitioning of the value of each observation into its three components: $\overline{\overline{Y}}$, $(\overline{Y}_i - \overline{\overline{Y}})$, and $(Y_{ij} - \overline{Y}_i)$. The first table will consist of 35 values, all equal to the grand mean. In the second table, all entries in a given column will be equal to the difference between the mean of that column and the grand mean. The last table will consist of the deviations of each individual observation from its column mean. These tables represent estimates of the individual components of Expression (8.2). Compute the mean and sum of squares for each table.

9 Single-Classification Analysis of Variance

We are now ready to study applications of the analysis of variance. This chapter deals with the simplest kind of anova, **single-classification analysis of variance**, in which the groups of samples are classified (grouped) by only a single criterion. Both interpretations of the seven samples of housefly wing lengths studied in the last chapter—different medium formulations (Model I) and progenies of different females (Model II)—would represent a single criterion for classification. Other examples are different temperatures at which groups of animals were raised or different soils in which samples of plants have been grown.

Section 9.1 restates the basic computational formulas for analysis of variance, based on the topics covered in the previous chapter. Section 9.2 first gives an example of the general case with unequal sample sizes because all groups in the anova need not necessarily have the same sample size. We will illustrate this case using a Model II anova. Some computations unique to a Model II anova—estimating variance components and their confidence intervals—are also shown. Because the basic computations for the analysis of variance are the same in either model, it is not necessary to repeat the illustration with a Model I anova. Section 9.2 also features a single-classification analysis of variance for groups with equal sample sizes, using an example of a Model I anova. The computational simplification for equal sample sizes, important in earlier years, has now become of less consequence when most users of these methods employ computer programs. Formulas become especially simple for the two-sample case, as we will see in Section 9.3. In this case, a t-test can be applied as well, and we shall show that it is mathematically equivalent to an analysis of variance F-test when there are only two groups. The two-group case includes instances in which one of the groups is represented by a single specimen.

When the null hypothesis of a Model I analysis of variance has been rejected, leading to the conclusion that the means are not from the same population, we wish to test the means in various ways to discover which pairs of means are different from each other and whether the means can be divided into groups that are different from each other. These so-called multiple comparisons tests are covered in Sections 9.4 and 9.5. Section 9.4 deals with the necessary background material, including orthogonal comparisons which are independent pre-planned tests of the group means, per-comparison and familywide error rates, and the Dunn-Šidák and Bonferroni methods of evaluating such error rates. Section 9.5 discusses special methods for multiple comparisons tests in a variety of designs.

In contrast to some previous chapters that dealt largely with theory, much of the discussion in this chapter uses boxes to illustrate practical applications and discuss computational details; general implications and additional comments are found in the text.

9.1 Computational Formulas

We saw in Section 8.5 that the total sum of squares and degrees of freedom can be partitioned additively into those of variation among groups and those of variation within groups. For the analysis of variance proper, we need only the sum of squares among groups and the sum of squares within groups, not the total sum of squares. Not all computer programs furnish all three sums of squares. When only two are given, the third can be obtained easily from the other two by addition or subtraction, depending on which two are initially available. The three sums of squares are computed as follows:

$$SS_{among} = \sum_{i=1}^{a} n_i (\bar{Y}_i - \bar{\bar{Y}})^2$$

$$SS_{within} = \sum_{i=1}^{a} \sum_{j=1}^{n_i} (Y_{ij} - \bar{Y}_i)^2$$

$$SS_{total} = \sum_{i=1}^{a} \sum_{j=1}^{n_i} (Y_{ij} - \bar{\bar{Y}})^2$$

The formula for the sum of squares among groups (SS_{among}) is based on a general model for single-classification anova allowing for unequal sample sizes and all groups in the analysis being tested. Note that $\bar{\bar{Y}}$ is a weighted mean, $\bar{Y}_w$. The weighting allows each group mean, $\bar{Y}_i$, to have its own, different sample size n_i to serve as the weights, w_i as in Expression (4.2).

9.2 General Case: Unequal and Equal *n*

Our first computational example of a single-classification analysis of variance has unequal sample sizes. We will develop general formulas for carrying out such an analysis. Later in this section, we will see that an anova with equal sample sizes is only a special case that is slightly simpler to compute. The example, shown in Box 9.1, is a Model II anova. Up to and including the *F*-test for significance, the computations are the same whether the anova is Model I or Model II. We will identify the stage in the computations at which the operations change depending on the model.

The data in Box 9.1 are a series of morphological measurements of the width of the scutum (dorsal shield) of samples of tick larvae obtained from four different host individuals of the cottontail rabbit. These four hosts were obtained at random from

BOX 9.1	Single-Classification Model II Anova (Unequal Sample Sizes)

Width (in μm) of scutum (dorsal shield) of larvae of the tick *Haemaphysalis leporis-palustris* in samples from 4 cottontail rabbits. This is a Model II anova.

	Hosts ($a = 4$)				
	1	**2**	**3**	**4**	$\overset{a}{\Sigma}$
	380	350	354	376	
	376	356	360	344	
	360	358	362	342	
	368	376	352	372	
	372	338	366	374	
	366	342	372	360	
	374	366	362		
	382	350	344		
		344	342		
		364	358		
			351		
			348		
			348		
$\overset{n_i}{\underset{}{\Sigma}} Y$	2978	3544	4619	2168	13,309
$\bar{Y}_i$	372.25	354.40	355.31	361.33	359.70
n_i	8	10	13	6	37
$\overset{n_i}{\underset{}{\Sigma}} y_i^2$	379.50	1278.40	954.77	1165.33	3778.00

SOURCE: Data from P. A. Thomas (unpublished results).

Preliminary computations

1. Grand mean $= \bar{\bar{Y}} = \dfrac{1}{\overset{a}{\underset{}{\Sigma}} n_i} \overset{a}{\underset{i=1}{\Sigma}} \overset{n_i}{\underset{j=1}{\Sigma}} Y_{ij} = \dfrac{1}{37} (2978 + 3544 + \ldots + 2168)$

$= 359.70$

IMPORTANT NOTE: For simplicity we will abbreviate our notation in the rest of this box as follows:

$$\sum = \overset{a}{\underset{i=1}{\sum}} = \Sigma^a \qquad \overset{n}{\sum} = \overset{n_i}{\underset{j=1}{\sum}} \qquad Y = Y_{ij}$$

Box 9.1 (continued)

2. SS_{among}, sum of squares among groups $= \Sigma^a n_i (\bar{Y}_i - \bar{\bar{Y}})^2$

$$= 8(372.25 - 359.70)^2$$
$$+ 10(354.40 - 359.70)^2 + \ldots$$
$$+ 6(361.33 - 359.70)^2 = 1807.73$$

3. SS_{within}, sum of squares within groups, $= \displaystyle\sum^a \sum^n y^2 = \sum^a \sum^n (Y - \bar{Y}_i)^2 = 3778.00$

4. $SS_{total} = SS_{among} + SS_{within} =$ quantity 2 + quantity 3 $= 1807.73 + 3778.00$
$$= 5585.73$$

The anova table of formulas is constructed as follows. (Boldface numbers represent quantities calculated in the corresponding computational steps.)

Source of variation		df	SS	MS	F_s	Expected MS
$\bar{Y} - \bar{\bar{Y}}$	Among groups	$a - 1$	**2**	$\dfrac{\mathbf{2}}{a-1}$	$\dfrac{MS_{among}}{MS_{within}}$	$\sigma^2 + n_0 \sigma_A^2$
$Y - \bar{Y}$	Within groups	$\displaystyle\sum^a n_i - a$	**3**	$\dfrac{\mathbf{3}}{\displaystyle\sum^a n_i - a}$		σ^2
$Y - \bar{\bar{Y}}$	Total	$\displaystyle\sum^a n_i - 1$	**4**			

Substituting the computed values, we obtain the following completed anova table.

Source of variation		df	SS	MS	F_s	P
$\bar{Y} - \bar{\bar{Y}}$	Among groups (among hosts)	3	1807.73	602.58	5.26	0.0045
$Y - \bar{Y}$	Within groups (error; among larvae on a host)	33	3778.00	114.48		
$Y - \bar{\bar{Y}}$	Total	36	5585.73			

Conclusion

Because only 45 out of 10,000 random samples would be as diverse or more so by chance alone, we reject the null hypothesis that the added variance component among groups σ_A^2 is zero. There is an added variance component among hosts for width of scutum in larval ticks.

one locality. We know nothing about their origins or their genetic constitution. They represent a random sample of the population of host individuals from the given locality. We would not be in a position to interpret differences between larvae from any given host because we know nothing of the origins of the individual rabbits. Population biologists are nevertheless interested in such analyses because they provide an answer to the following question: Are the variances of means of larval characters among hosts greater than expected on the basis of variances of the characters within hosts? We can calculate the average variance of width of larval scutum on a host. This will be our "error" term in the analysis of variance. We then test the observed mean square among groups to see if it contains an added component of variance. What would such an added component of variance represent? The mean square within host individuals (that is, of larvae on any one host) represents genetic differences among larvae and differences in environmental experiences of these larvae. Added variance among hosts demonstrates differentiation among the larvae possibly because of differences among the hosts affecting the larvae or to genetic differences among the larvae (assuming each host carries a family of ticks, or at least a population whose individuals are more related to each other than they are to tick larvae on other host individuals). The emphasis in this example is on the magnitudes of the variances. In view of the random choice of hosts, this is a clear case of a Model II anova.

Note that in order to compute quantity **2**, we feature the means of the four groups in Box 9.1. However, because this is a Model II anova in which the added group effects A_i are assumed to be random, the means hold no interest for us except as an aid to computation. We are not interested in the individual means or possible differences among them.

The computation is illustrated in Box 9.1. After quantities **2** through **4** have been evaluated, they are entered into an anova table, as shown in the box. General formulas for such a table are shown first, followed by a table filled in with actual values for the specific example. We note 3 degrees of freedom (*df*) among groups, because there are four host individuals, and 33 *df* within groups (the sum of 7, 9, 12, and 5 degrees of freedom for groups 1 through 4, respectively). Note that the mean square among groups (MS_{among}) is considerably greater than the error mean square (MS_{within}), engendering suspicion that an added variance component is present. If MS_{among} is equal to or less than MS_{within}, we do not bother going on with the analysis, for we would not have evidence for the presence of an added variance component. You may wonder how MS_{among} can be less than MS_{within}. Remember that these two values are independent estimates. If there is no added variance component among groups, the estimate of the variance among groups is as likely to be less than as it is to be greater than the variance within groups.

Expressions for the expected values of the mean squares are also shown in the first anova table of Box 9.1. These are the expressions you learned in the previous chapter for a Model II anova, with one exception: *n* has been replaced by n_0. Because sample size n_i differs among groups in this example, no single value of *n* is appropriate in the formula. We therefore use an average *n*; this, however, is not simply

$\bar{n}$, the arithmetic mean of the n_i's. The "classic" anova method is to use the following special average

$$n_0 = \frac{1}{a-1}\left(\sum^a n_i - \frac{\sum^a n_i^2}{\sum^a n_i}\right) \tag{9.1}$$

which is an average usually close to, but always less than $\bar{n}$, unless sample sizes are equal, in which case $n_0 = \bar{n} = n$. In Box 9.2, we shall employ a later developed method that uses the harmonic mean of the sample sizes, H_n, along with a mean square among groups, $\widetilde{MS}_{among}$, that treats the group means as if they were all based on this sample size. This procedure produces estimates, $\widetilde{s}_A^2$, of the variance component that are, on average, closer to their parametric values (which we verified by simulations). It also enables the construction of more accurate confidence intervals.

The probability of obtaining an F_s-value equal to or larger than our observed value can be determined in three ways. The first is by consulting a table of critical values of the F-distribution. The critical 5%, 1%, and 0.1% values of F for Box 9.1 are 2.89, 4.44, and 6.89, respectively. You should confirm them for yourself in Statistical Table **F**. Note that the argument $\nu_2 = 33$ is not given. You must therefore interpolate between arguments representing 30 and 40 degrees of freedom, respectively. The values were approximated using harmonic interpolation (see section on interpolation in the Introduction to *Statistical Tables*). It is frequently not necessary, however, to carry out such an interpolation. One can simply look up the "conservative" value of F, which is the next tabled F with fewer degrees of freedom—in our case, $F_{\alpha[3,30]}$, which is 2.92 for $\alpha = 0.05$ and 4.51 for $\alpha = 0.01$. The observed value of F_s is 5.26, considerably greater than the interpolated as well as the conservative value of $F_{.01}$. Second, many of the statistical computer programs employed nowadays enable one to obtain probabilities without having to look up values in tables. When we applied the statistical calculator of the BIOMstat program to our F_s-value of 5.26, we obtained $P = 0.0045$. Finally, many anova computer programs evaluate the probability corresponding to a given variance ratio internally and show it as part of the output. We feature it in the last column of the anova table as $P = 0.0045$. By any and all of these three approaches, we reject the null hypothesis (H_0: $\sigma_A^2 = 0$) that there is no added variance component among groups and that the two mean squares estimate the same variance, allowing a type I error of less than 1% or, more specifically, 0.45%. We accept instead the alternative hypothesis of the existence of an added variance component σ_A^2.

What is the biological meaning of this conclusion? The ticks on different host individuals differ more from each other than do individual ticks on any one host either because of some modifying influence of individual hosts on the ticks (e.g., bio-chemical differences in blood, differences in the skin, differences in the environment of the host individual—all of which are rather unlikely in this case) or because of genetic differences among the ticks. The ticks on each host may represent a sibship

---

BOX 9.2 | Estimation of Variance Components and Their Confidence Limits in a Single-Classification Model II Anova with Unequal Sample Sizes: The MLS Method

Data from Box 9.1, *but note* that the among-groups and expected among-groups mean squares are not the same as in that box. For explanation, see step **1**, below.

Anova Table

Source of variation	df	MS	F'_s	Expected MS
Among groups (among hosts)	3	576.2263	5.0332	$\sigma^2 + H_n\sigma_A^2$
Within groups (error, among larvae on a host)	33	114.4849		σ^2

Computations

1. Thomas and Hultquist (1978) proposed estimating the variance component using an unweighted sums of squares to compute the mean square among groups. We first compute the harmonic mean of the sample sizes and the unweighted mean of the group means.

$$H_n = \frac{a}{\sum^a \frac{1}{n_i}} = \frac{4}{0.4686} = 8.5363, \text{ the harmonic mean of the sample sizes.}$$

$$\overline{\overline{Y}}' = \frac{\sum^a \overline{Y}_i}{a} = \frac{372.25 + 354.40 + 355.31 + 361.33}{4}$$

$$= 360.8228, \text{ the unweighted mean of the group means.}$$

Then compute the unweighted mean square among groups

$$\widetilde{MS}_{among} = \frac{H_n \sum^a (\overline{Y}_i - \overline{\overline{Y}}')^2}{a - 1} = \frac{8.5363 \times 202.5103}{4 - 1} = 576.2263$$

The variance component, σ_A^2, is then estimated as

$$\widetilde{s}_A^2 = \frac{\widetilde{MS}_{among} - MS_{within}}{H_n} = \frac{576.2263 - 114.4849}{8.5363} = 54.0918$$

Ting et al.'s (1990) modified large sample (MLS) method to construct approximate confidence intervals for a variance component is somewhat complex. Thomas and Hultquist (1978) had already shown that the approximation used by Ting et al.'s method works well unless the parametric intraclass correlation coefficient, $\rho_I = \sigma_A^2/(\sigma^2 + \sigma_A^2)$, presented at the bottom of this box, is less than 0.2 and the sample sizes are very unequal.

Box 9.2 (continued)

2. For degrees of freedom $\nu_1 = df_{among} = 3$ and $\nu_2 = df_{within} = 33$, $1 - \alpha =$ desired level of confidence $= 0.95$ (for 95% confidence intervals), look up the following tabulated χ^2 and F-values in Statistical Tables **D** and **F** or compute them by BIOMstat.

$$\chi^2_{\alpha/2[\nu_1]} = \chi^2_{0.025[3]} = 9.3484, \quad \chi^2_{\alpha/2[\nu_2]} = \chi^2_{0.025[33]} = 50.7251$$

$$\chi^2_{1-\alpha/2[\nu_1]} = \chi^2_{0.975[3]} = 0.2158, \quad \chi^2_{1-\alpha/2[\nu_2]} = \chi^2_{0.975[33]} = 19.0467$$

$$F_1 = F_{\alpha/2[\nu_1,\nu_2]} = F_{0.025[3,33]} = 3.5429, \quad F_2 = F_{1-\alpha/2[\nu_1,\nu_2]} = F_{0.975[3,33]} = 0.0711$$

3. Using these values, calculate the following auxiliary quantities:

$$G_1 = 1 - \frac{\nu_1}{\chi^2_{\alpha/2[\nu_1]}} = 1 - \frac{3}{9.3484} = 0.6791, \quad G_2 = 1 - \frac{\nu_2}{\chi^2_{\alpha/2[\nu_2]}} = 1 - \frac{33}{50.7251} = 0.3494$$

$$H_1 = \frac{\nu_1}{\chi^2_{1-\alpha/2[\nu_1]}} - 1 = \frac{3}{0.2158} - 1 = 12.9021, \quad H_2 = \frac{\nu_2}{\chi^2_{1-\alpha/2[\nu_2]}} - 1 = \frac{33}{19.0467} - 1 = 0.7326$$

$$G_{12} = \frac{(F_1 - 1)^2 - G_1^2 F_1^2 - H_2^2}{F_1} = \frac{(3.5429 - 1)^2 - 0.6791^2 \times 3.5429^2 - 12.9021^2}{3.5429} = 0.0398$$

$$H_{12} = \frac{(1 - F_2)^2 - H_1^2 F_2^2 - G_2^2}{F_2} = \frac{(1 - 0.0711)^2 - 12.9021^2 \times 0.0711^2 - 0.3494^2}{0.0711} = -1.4173$$

4. The $100(1 - \alpha)\%$ confidence limits for σ_A^2 can then be computed as follows:

$$L_1 = \tilde{s}_A^2 - \frac{\sqrt{(G_1 \widetilde{MS}_{among})^2 + (H_2 MS_{within})^2 + (G_{12} \widetilde{MS}_{among} MS_{within})}}{H_n} = 54.0918$$

$$- \frac{\sqrt{(0.6791 \times 576.2263)^2 + (0.7326 \times 114.4849)^2 + (0.0398 \times 576.2263 \times 114.4849)}}{8.5363}$$

$$= 6.8271$$

$$L_2 = \tilde{s}_A^2 + \frac{\sqrt{(H_1 \widetilde{MS}_{among})^2 + (G_2 MS_{within})^2 + (H_{12} \widetilde{MS}_{among} MS_{within})}}{H_n} = 54.0918$$

$$+ \frac{\sqrt{(12.9021 \times 576.2263)^2 + (0.3494 \times 114.4849)^2 + (-1.4173 \times 576.2263 \times 114.4849)}}{8.5363}$$

$$= 924.3012$$

The lower limit is greater than zero, which is consistent with the fact that the null hypothesis of no added variance component was rejected. Note that the limits are asymmetrical around the estimate of 54.0918. The interval is very wide, implying that the estimate is quite unreliable. This is partly because of the small sample sizes, especially the number of groups.

Box 9.2 (continued)

The intraclass correlation coefficient, r_I, is a measure of the relative amount of variation among groups. It can be estimated as

$$r_I = \frac{\widetilde{s}_A^2}{s^2 + \widetilde{s}_A^2} = \frac{54.0918}{114.4849 + 54.0918} = 0.3209.$$

The percentage of variation among groups (hosts) is $100r_I = 32.1\%$, and the percent variation within groups (among larvae on a host) is $100(1 - r_I) = 67.9\%$.

Sahai and Ojeda (2005) give the following method to set approximate $100(1 - \alpha)\%$ confidence limits to the intraclass correlation coefficient when sample sizes are unequal. For $\alpha = 0.05$, the limits are

$$L_1 = \frac{F_s' - F_{\alpha/2[\nu_1,\nu_2]}}{F_s' + (H_n - 1)F_{\alpha/2[\nu_1,\nu_2]}} = \frac{5.0332 - 3.5429}{5.0332 + (8.5363 - 1)3.5429} = 0.0470 \text{ and}$$

$$L_2 = \frac{F_s' - F_{1-\alpha/2[\nu_1,\nu_2]}}{F_s' + (H_n - 1)F_{1-\alpha/2[\nu_1,\nu_2]}} = \frac{5.0332 - 0.0711}{5.0332 + (8.5363 - 1)0.0711} = 0.8910,$$

where $F_s' = \dfrac{MS'_{\text{among}}}{MS_{\text{within}}}$ is the F_s-ratio based on the unweighted sums of squares among groups.

These intervals should be sufficiently accurate unless the sample sizes are very unequal and $\rho_I < 0.2$, in which case the intervals tend to be too short.

(the descendants of a single pair of parents), and thus the differences of the ticks among host individuals could represent genetic differences among families; or selection may have acted differently on the tick populations on each host; or the hosts may have migrated to the collection locality from different geographic areas in which the ticks differ in width of scutum. Of these possibilities, genetic differences among sibships seem most reasonable in view of the biology of the organism.

The computations up to this point would have been identical in a Model I anova. (If this had been Model I, the conclusion would have been that there is a treatment effect rather than an added variance component.) Now, however, we must complete the computations appropriate to a Model II anova. These may include estimating the added variance component, computing its confidence limits, and calculating percentage variation at the two levels. Computer programs often include the computations for variance components, regardless of whether the anova is Model I or Model II, rather than have the user specify which model is intended. The user must then decide on the correct model and should ignore variance components when these are inappropriate. The BIOMstat computer programs for single-classification and nested anovas operate in this way.

The steps for obtaining estimates of the variance components and their confidence intervals for the example in Box 9.1 are given in Box 9.2. The expressions for the expected mean squares in the anova table indicate how the variance component among groups σ_A^2 and the error variance σ^2 can be obtained. Of course, because the values we obtain are sample estimates, they are written as s_A^2 and s^2. The unequal sample sizes in this example introduce complications. A special mean square is used that is based on a sum of squares among group means that ignores the differences in sample sizes among the different means and instead uses a harmonic mean of the sample sizes. There has been a considerable amount of work done in recent years developing good approximations for the confidence intervals of σ_A^2 (no exact methods exist even for equal sample sizes). The modified large sample method of Ting et al. (1990) employed in Box 9.2 was recommended by Li and Li (2007)—except in situations where sample sizes are extremely unequal and the intraclass correlation (see below) is thought to be small. In other cases, methods based on generalized confidence intervals are recommended, but they require much more extensive computations and are not shown here. Note that the harmonic mean sample size is used here as the average sample size. We are frequently interested not so much in the absolute values of these variance components as in their relative magnitudes. For this purpose, we sum these two values and express each as a percentage of this sum. Thus, $s^2 + s_A^2 = 114.4849 + 54.0918 = 168.5767$; s^2 and s_A^2 are 67.9% and 32.1% of this sum, respectively.

Relatively more variation occurs within groups (larvae on a host) than among groups (hosts). The proportion of the variation among groups is also known as r_I, the **coefficient of intraclass correlation.** A high value of this coefficient means that most of the variation in the sample is among groups. A value of unity would indicate that all of the variance in the sample is among groups—that is, there is no variance within groups. In such a case, all the observations within each group would be identical. This coefficient is a measure of the similarity, or *correlation*, among the individuals within a group relative to the amount of differences found among the groups. This meaning is not the sense in which we use correlation nowadays (see Chapter 15), and the percentage variation among and within groups seems a more expressive way to represent this statistic.

The formula we furnish for the intraclass correlation coefficient originated in the early 20th century and is well established. In recent years, proposals for other formulations of the relation between the different levels of variation have been made. Shrout and Fleiss (1979) distinguish between six kinds of r_I, depending on the design and model of the anova. The coefficient is frequently used in the social sciences, where researchers use it to quantify the ratings of judges and the consistency of responses of experimental subjects. Among biologists, r_I is more often employed by workers in quantitative genetics (for certain experimental designs, the intraclass correlation coefficient corresponds to an estimate of heritability), morphometrics, and related fields, and we believe that the standard coefficient given in Box 9.2 should suffice. For an assessment of several intraclass correlation coefficients, see McGraw and Wong (1996). For the special case of unordered pairs, Ernst, Guerra, and Schucany (1996) suggest a method of graphic representation of r_I. A method for constructing approximate confidence intervals for ρ_I is also given in Box 9.2.

In Box 9.3, we describe the computation of confidence intervals for variance components of another anova with *equal sample sizes*. The method used here is the same as the modified large sample method shown in Box 9.2 except that the un-weighted sums of squares among groups and the harmonic mean of the sample sizes are replaced with the usual sums of squares among groups and the fixed sample size n. The example in Box 9.3 expresses fertility as the number of larvae hatched out of 10 eggs laid by females of *Drosophila melanogaster*. From each of the 7 females in the study, which had been chosen randomly from a certain genetic strain, 3 replicate egg samples (of 10 eggs each) were tested for hatchability. We therefore have a basic data table (not shown) of 7 groups (females) and 3 replicates (hatch counts) within groups. The anova table is shown in Box 9.3. Note that the F-value is not large enough to allow us to reject the null hypothesis that the variance component is zero. However, we will set confidence limits to σ_A^2 in order to provide an interval estimate for this statistic. The lower confidence limit is negative. This result was to be expected because we could not reject the null hypothesis that the variance component among groups is zero. Confidence limits for the intraclass correlation are also given.

BOX 9.3 **Estimation of Variance Components in a Single-Classification Model II ANOVA Using the MLS Method: Equal Sample Sizes**

Fertility (number of larvae hatched out of 10 eggs deposited) for the offspring of 7 females of *Drosophila melanogaster*, PP strain ($a = 7$ females). There were $n = 3$ replicates for each female.

Anova table

Source of variation	df	MS	Expected MS	F_s	P
Among groups (among females)	6	9.111	$\sigma^2 + n\sigma_A^2$	2.20	0.1052
Within groups (replicates)	14	4.142	σ^2		
Total	20				

SOURCE: Data from R. R. Sokal (unpublished results)

When sample sizes are equal, the standard MS_{among} of Box 9.1 and the mean square using the unweighted sum of squares of Box 9.2 yield the identical estimates of the added variance component due to differences among the females, σ_A^2:

$$s_A^2 = \frac{MS_{among} - MS_{within}}{n}$$

$$= \frac{9.111 - 4.142}{3} = 1.6563$$

BOX 9.3 (continued)

The computation of approximate confidence limits for the variance component by the modified large sample (MLS) procedure of Ting et al. (1990) for equal n is almost identical to that of for unequal sample sizes already described in Box 9.2. You can employ the same formulas presented in that box with three minor changes. The average sample size changes from H_n, the harmonic mean of the sample sizes to n; the added variance component is now computed by the conventional method as shown above rather than by the unweighted sum of squares method of Box 9.2; and the MS_{among} is the unmodified mean square of the standard anova, 9.111 in the present example, rather than the unweighted mean square among groups, MS'_{among}, of Box 9.2. Once these substitutions are made, the formulas of Box 9.2, supplied with the data from this box yield estimated confidence limits of -1.0259 and 13.2824. When, as in this case, the lower limit is negative, it is set equal to zero. Thus, $L_1 = 0$. This result is expected when the null hypothesis of no added variance component is accepted.

The intraclass correlation coefficient, r_I, and its confidence limits are computed as in Box 9.2, after replacing s_A^2 by s_A^2, F'_s by F_s (the F-ratio of the conventional anova, which, to 4 decimal places, is 2.1997) and H_n by n. After doing so, we obtain $r_I = 0.2857$, which is a measure of the relative amount of variation *among* groups. The 95% confidence limits computed by the method of Sahai and Ojeda (2004) were found to be -0.1415 and 0.7802. Because r_I estimates the proportion of the added variance component among groups, negative estimates and negative lower confidence limits are set equal to zero. Thus, the 95% confidence interval is from 0 to 0.78. Rather large samples are required in order to obtain narrow confidence intervals.

In recent years, there has been considerable progress on the development of methods for the analysis of variance components and various functions of variance components for single classification and more complex designs. There are now a number of specialized books on the subject. Burdick and Graybill (1992) covers confidence intervals. Burdick et al. (2005) is quite accessible and includes both theory and numerical examples of practical applications. The two volumes by Sahai and Ojeda (2004, 2005) give a comprehensive discussion of Model II anova methods.

When sample sizes in a single-classification analysis of variance are equal, the computational formulas from Box 9.1 are somewhat simplified. In the old days, when computation was slow and troublesome, these simplifications made a substantial difference. Nowadays, this is no longer necessary, and we therefore feature the next example, a Model I single-classification anova with equal sample sizes in Table 9.1 without illustrating the computations. It should be understood that steps **1** through **4** in Box 9.1 would lead to the results shown. We have introduced a Model II anova with unequal n in Box 9.1, and we are about to introduce you to a Model I anova with equal n in Table 9.1. Do not link sample size equality with model in your mind. All four combinations of model (I *versus* II) and sample size (equal *versus* unequal n) are

| TABLE 9.1 | Single-Classification Model I Anova (Equal Sample Sizes)

The effect of different sugars on length of pea sections grown in tissue culture with auxin present. Measured in ocular units (each 0.114 mm), transformed by taking their reciprocals, multiplying by 100, and then rounding to two decimal places to give values in a convenient range. Reasons for such a transformation are given in Chapter 13.

			Treatments ($a = 5$)		
				1% Glucose	
		2%	2%	+	2%
Observation		Glucose	Fructose	1% Fructose	Sucrose
(i.e., Replications)	Control	Added	Added	Added	Added
1	1.33	1.75	1.72	1.72	1.61
2	1.49	1.72	1.64	1.69	1.52
3	1.43	1.67	1.79	1.72	1.54
4	1.33	1.69	1.72	1.64	1.59
5	1.54	1.61	1.75	1.75	1.56
6	1.41	1.67	1.79	1.79	1.61
7	1.49	1.67	1.64	1.72	1.54
8	1.49	1.75	1.67	1.75	1.54
9	1.32	1.69	1.75	1.75	1.61
10	1.47	1.64	1.72	1.69	1.49
$\sum\limits^{n} Y$	14.300	16.860	17.190	17.220	15.610
$\overline{Y}$	1.430	1.686	1.719	1.722	1.561

SOURCE: Data from W. Purves (unpublished results).

Anova Table: Formulas

	Source of Variation	df	SS	MS	F_s	Expected MS
$\overline{Y} - \overline{\overline{Y}}$	Among groups	$a - 1$	SS_{among}	$\dfrac{SS_{among}}{a - 1}$	$\dfrac{MS_{among}}{MS_{within}}$	$\sigma^2 + \dfrac{n}{a - 1}\sum\limits^{a}\alpha^2$
$Y - \overline{Y}$	Within groups	$a(n - 1)$	SS_{within}	$\dfrac{SS_{within}}{a(n - 1)}$		σ^2
$Y - \overline{\overline{Y}}$	Total	$an - 1$	SS_{total}			

Completed Anova Table

	Source of Variation	df	SS	MS	F_s	P
$\overline{Y} - \overline{\overline{Y}}$	Among groups	4	0.6408	0.16019	53.88	10^{-16}
$Y - \overline{Y}$	Within groups	45	0.1338	0.00297		
$Y - \overline{\overline{Y}}$	Total	49	0.7746			

Conclusions

Because P is such an exceedingly small value, we reject the null hypothesis of no added treatment effects with near certainty. We conclude that there is a large effect of the sugars tested on growth of the pea sections.

See Sections 9.4, 9.5, Boxes 9.6, and 9.7 for the completion of a Model I analysis of variance—i.e., to determine which means are different from each other.

possible, and you will encounter them in this and subsequent chapters in the tables, boxes, and exercises. As we have stated before, the computation up to and including the first test of significance is identical for both models.

The data in Table 9.1, which are from an experiment in plant physiology, record the length in coded units of pea sections grown in tissue culture with auxin present. The purpose of the experiment was to test the effects of various sugars on growth as measured by length. Four experimental groups, representing three different sugars and one mixture of sugars, were used, plus one control without sugar. Ten observations (replicates) were made for each treatment. (The term *treatment* already implies a Model I anova.) The five groups obviously do not represent random samples from all possible experimental conditions but were deliberately designed to test the effects of certain sugars on the growth rate. We are interested in the effect of the sugars on length, so our null hypothesis is that there is no added component resulting from treatment effects among the five groups; that is, the population means are all assumed to be equal. The anova table is similar to that for the general case (see Box 9.1), except that the expressions for numbers of degrees of freedom are simpler.

The combination of degrees of freedom needed here are not available in Statistical Table **F**. However, comparison of the observed variance ratio $F_s = 53.88$ with $F_{.001[4,40]} = 5.70$, the conservative critical value in Statistical Table **F**, would convince you to reject the null hypothesis. The probability that the five groups differ as much as they do by chance is almost infinitesimally small. Furthermore, most computer programs, BIOMstat included, will furnish a computed P-value. Table 9.1 gives the value 10^{-16}. Clearly, the sugars produce an added treatment effect, increasing the means of the groups receiving sugars, that corresponds to reducing the lengths of the pea sections when the variable is uncoded.

At this stage, we are not in a position to say whether each treatment is different from every other treatment, or whether the sugars are different from the control but not different from each other. Such tests are necessary to complete a Model I analysis, but we defer their discussion until Sections 9.4 and 9.5.

9.3 Special Case: Two Groups

A frequent test in biological research is to evaluate the probability of the null hypothesis of no difference between two sample means. Such a test, colloquially but imprecisely termed a test of the *difference between two means*, can be done easily by means of an *analysis of variance for two groups*. Such designs are computed in the same manner as those for $a > 2$ groups shown in Box 9.1 and Table 9.1. The layout of the data and the results are found in Table 9.2, which illustrates a Model I anova, the more frequent model in the two-sample case. If the assumptions of a Model II anova are met, however, it too can be carried out for two groups.

The data in Table 9.2 stem from an experiment with water fleas, *Daphnia longispina,* in which the onset of reproductive maturity is measured as the average age (in days) at the beginning of reproduction. Because each observation in the table is an average, a possible flaw in the analysis is that these averages may not be based on

TABLE 9.2 Single-Classification Model I Anova with Two Groups (Equal Sample Sizes)

Average age (in days) at beginning of reproduction in *Daphnia longispina* (each observation is a mean based on approximately similar numbers of females). Two series derived from different genetic crosses and containing 7 clones each are compared; $n = 7$ clones per series. This is a Model I anova.

	Series ($a = 2$)	
	I	**II**
	7.2	8.8
	7.1	7.5
	9.1	7.7
	7.2	7.6
	7.3	7.4
	7.2	6.7
	7.5	7.2
$\sum^{n} Y$	52.6	52.9
$\bar{Y}$	7.5143	7.5571
s^2	0.5047	0.4095

SOURCE: Data by Ordway from Banta (1939).

Anova Table

	Source of Variation	*df*	*SS*	*MS*	F_s	*P*
$\bar{Y} - \bar{\bar{Y}}$	Between groups (series)	1	0.00641	0.00641	0.0140	0.9078
$Y - \bar{Y}$	Within groups (error; clones within series)	12	5.48571	0.45714		
$Y - \bar{\bar{Y}}$	Total	13	5.49212			

Conclusions

Because the *P*-value corresponding to the observed F_s-value is $\gg 0.05$, the null hypothesis is accepted. The means of the two series are not different; that is, the two series do not differ in average age at beginning of reproduction.

equal sample sizes. We are not given this information, however, and therefore must proceed on the assumption that each reading in the table is an equally reliable observation. The two series represent different genetic crosses, and the seven replicates in each series are clones derived from the same genetic cross. This example is clearly a Model I because the question to be answered is whether average age at the beginning of reproduction in series I differs from that in series II. Inspection of the data shows that the mean age at beginning of reproduction is very similar for the two series. We would be surprised therefore to find that we could reject the null hypothesis, but we will carry out a test anyway. As you realize by now, one cannot determine the probability of a difference from its absolute magnitude. The probability depends as well on the magnitude of the error mean square, representing the variance within series.

The computations are analogous to those of Box 9.1. With equal sample sizes and only two groups, we can compute quantity 2, SS_{among}, as $n/2$ times the squared difference between the means. There is only 1 degree of freedom between the two groups. The P-value associated with F_s is so high that we readily accept the null hypothesis. Inspection of the mean squares in the anova shows that MS_{among} is much smaller than MS_{within}; therefore, the value of F_s is far below unity, so we cannot estimate an added component resulting from treatment effects between the series. In cases where $MS_{among} \leq MS_{within}$, we do not usually bother to calculate F_s because we could not possibly reject the null hypothesis.

This example is unusual in that the mean square between groups is very much lower than that within groups. You may wonder whether these two could really be estimates of the same parametric variance, as they should be according to theory. Let us test whether the two variances are equal. Our null hypothesis is $H_0: \sigma^2_{among} = \sigma^2_{within}$, and the alternative hypothesis is $H_1: \sigma^2_{among} \neq \sigma^2_{within}$. In view of the alternative hypothesis, this is a two-tailed test analogous to that in Box 8.1. We calculate $F_s = MS_{within}/MS_{among} = 0.45714/0.00643 = 71.09$. Because $F_{.025[12,1]} = 977$, we accept H_0 and conclude that the two mean squares are estimates of the same parametric variance. We should not be surprised that there is such a great discrepancy between the variance estimates. After all, the mean square between groups is based on only 1 degree of freedom.

Some statisticians would object to the procedure just described on the grounds that, after a one-tailed test has been made with the alternative hypothesis ($H_1: \sigma^2_{among} > \sigma^2_{within}$) and the null hypothesis has been accepted, it is not valid to retest the exact same difference with another alternative hypothesis ($H_1: \sigma^2_{among} \neq \sigma^2_{within}$). The net effect of such a procedure is that of a two-tailed test, with an expected error rate 1.5 times larger than intended.

A mean square among groups much lower than the error mean square would mean either that an unusual event has occurred by chance (type I error) or that the group means are less variable than they should be on the basis of the variation of their items. A low mean square can occur whenever items are not sampled at random but are sampled with a type of compensatory bias. A grocer filling baskets of strawberries may not wish to fill them at random but may fill them so that the smaller, less desirable strawberries are at the bottom of each basket and the larger, more attractive ones are on top on view to the customer. Filling the strawberry baskets in this way would result in considerable within-basket variance of strawberry size but remarkable constancy in means among baskets (if the grocer has done a thorough job of biased filling of baskets) so that the mean square among groups (baskets) would be remarkably low. In theory at least, MS_{among} would be zero if the grocer is able to fill each basket with an identical assortment of strawberries.

In scientific work, similar poor sampling may result from unrecognized biases that cause the person doing the sampling to compensate unconsciously for obvious deviations from what he or she believes to be the norm. When biased sampling can be ruled out, the underdispersion of sample means may result from a compensation effect in the natural phenomenon under study. Whenever resources for a process are limited, some individuals may obtain a large part and others may have to be satisfied with only a small part. Thus, if there is a limited space in which to live, several large individuals may occupy most of it, leaving space for only a few small ones. This phenomenon is related to the idea of repulsion encountered in the binomial distribution (see Section 5.2). Situations such as these indicate that the basic assumptions of an anova are not met (see Chapter 13), and ordinarily one would not continue the anova.

The reader should not confuse two distinct sets of null and alternative hypotheses that can be considered in connection with an example such as the two strains of water fleas. We have just discussed the one-sidedness of the alternative hypothesis $H_1: \sigma^2_{among} > \sigma^2_{within}$. We might refer to this hypothesis and its accompanying null hypothesis as **operational hypotheses** because they are related to the set of necessary computational operations; that is, to carry out an analysis of variance we have to calculate a mean square among groups and test whether it is greater than the mean square within groups. Thus, we use an operationally one-tailed hypothesis, and it is for this reason that the F-table (Statistical Table **F**) is presented as a one-tailed table. These operational hypotheses, however, should be distinguished from **conceptual hypotheses**, which are concerned with the nature of the original experiment or analysis undertaken.

In the water flea example, we had no a priori reason for the mean of series I to be greater than or less than that of series II, so we applied a two-sided alternative hypothesis $H_1: \mu_I \neq \mu_{II}$. Had there been reason to formulate the conceptual alternative hypothesis in a one-sided manner, say, $H_1: \mu_I > \mu_{II}$, then the critical value of F to be looked up (in Statistical Table **F** or on a statistical calculator) should have been for probability 2α, where the desired type I error is α. Keep in mind the distinction between a conceptual and an operational alternative hypothesis; we will refer to other conceptually one-sided tests from time to time. Note that for analyses of variance with more than a single degree of freedom in the numerator of an F-statistic ($a > 2$ groups), a one-sided conceptual alternative hypothesis cannot be defined.

Another method of solving a Model I two-sample analysis of variance is a ***t*-test of the differences between two means**. This t-test is the traditional method of solving such a problem and may already be familiar to you from previous acquaintance with statistical work. It has no real advantage in ease of either computation or understanding, and as you will see below, it is mathematically equivalent to the anova shown in Table 9.2. The t-test is presented here mainly for completeness. It would seem too much of a break with tradition not to have the t-test in a biometry text.

In Section 7.4, we learned about the t-distribution and saw that a t-distribution of $n - 1$ df could be obtained from a distribution of the term $(\overline{Y}_i - \mu)/s_{\overline{Y}_i}$, where $s_{\overline{Y}_i}$ has

$n - 1$ degrees of freedom and $\overline{Y}$ is normally distributed. The numerator of this term represents a deviation of a sample mean from a parametric mean, and the denominator represents a standard error for such a deviation. The expression

$$t_s = \frac{(\overline{Y}_1 - \overline{Y}_2) - (\mu_1 - \mu_2)}{\sqrt{\left[\dfrac{(n_1 - 1)s_1^2 + (n_2 - 1)s_2^2}{n_1 + n_2 - 2} \right]\left(\dfrac{n_1 + n_2}{n_1 n_2} \right)}} \tag{9.2}$$

is also distributed as t. Expression (9.2) looks complicated, but it really has the same structure as the simpler term for t. The numerator is a deviation, this time not between a single sample mean and the parametric mean but between a difference between two sample means, $\overline{Y}_1$ and $\overline{Y}_2$, and the true difference between the means of the populations represented by these means. In a test of this sort, our null hypothesis is that the two samples come from the same population; that is, they must have the same parametric mean. Thus, the difference $\mu_1 - \mu_2$ is assumed to be zero. We therefore test the deviation of the difference $\overline{Y}_1 - \overline{Y}_2$ from zero. The denominator of Expression (9.2) is a standard error, the standard error of the difference between two means, $s_{\overline{Y}_1 - \overline{Y}_2}$. The left portion of the denominator (in square brackets) is a weighted average of the variances of the two samples, s_1^2 and s_2^2, computed in the manner of Section 8.1. The right term of the standard error can be rewritten as $(1/n_1) + (1/n_2)$, which is the factor by which the average variance within groups must be multiplied to convert it into a variance of the difference of means. The analogy to multiplying a sample variance s^2 by $1/n$ to transform it into a variance of a mean $s_{\overline{Y}}^2$ should be obvious.

The test as outlined here assumes equal variances in the two samples. This assumption has been made with all the analyses of variance we have carried out thus far, although we have not stressed this. If the variances of the two samples are very different from each other, other methods have to be applied. These methods are discussed in Section 13.3, where we also present the tests for checking whether several variances are equal. With only two variances, equality may be tested by the procedure in Box 8.1.

Two simplifications of the denominator of Expression (9.2) useful to persons using pocket calculators are shown in Appendix A.4. The first is for equal sample sizes in a two-sample test, the second applies when the sample sizes are unequal but rather large and hence the differences between n_i and $n_i - 1$ are relatively trivial. In the first case, $n_1 = n_2 = n$; in the second case, $(n_1 - 1) \approx n_1$ and $(n_2 - 1) \approx n_2$. The pertinent degrees of freedom for Expression (9.2) and for the second simplification are $n_1 + n_2 - 2$ and for the first simplification df is $2(n - 1)$. When the calculation is done by computer, there is, of course, no reason to use these simplifications of the standard errors.

The test of significance for differences between means using the t-test is shown in Box 9.4. This is a two-tailed test because our alternative hypothesis is $H_1: \mu_1 \neq \mu_2$. The results of this test are identical to those of the anova in Table 9.2: the two means are not significantly different. We mentioned earlier that these two results are, in fact, mathematically equivalent. We can show this most simply by squaring the obtained value for t_s, which should be identical to the F_s-value of the corresponding analysis of variance. In Box 9.4 $t_s = -0.1184$, so $t_s^2 = 0.0140$. This value equals the F_s obtained in the anova of Table 9.2. Why? We will show two reasons for this. We learned that $t_{[\nu]} = (\overline{Y} - \mu/s_{\overline{Y}})$, where ν is the degrees of freedom of the variance of the mean $s_{\overline{Y}}^2$; therefore $t_{[\nu]}^2 = (\overline{Y} - \mu)^2/s_{\overline{Y}}^2$. However, this expression can be regarded as a variance ratio. The denominator is clearly a variance with ν degrees of freedom. The numerator, also a variance, is a single deviation squared, which represents a sum of squares possessing 1 rather than zero degrees of freedom (because it is a deviation from the true mean ν rather than from a sample mean). A sum of squares based on 1 degree of freedom is also a variance. Thus, t^2 is a variance ratio; specifically, $t_{[\nu]}^2 = F_{[1,\nu]}$, as we have already seen. In Appendix A.5, we demonstrate algebraically that the t_s^2 obtained in Box 9.4 and the F_s-value obtained in Table 9.2 are identical quantities.

We can also demonstrate the relationship between t and F numerically from the Statistical Tables. In Statistical Table **F**, we find that $F_{.05[1,10]} = 4.96$. This value should be equal to $t_{.05[10]}^2$. The square root of 4.96 is 2.227 and when we look up $t_{.05[10]}$ in Statistical Table **B**, we find it to be 2.228, which is within acceptable rounding error of 2.227. Because t approaches the normal distribution as its degrees of freedom

---

BOX 9.4 | **A *t*-Test of the Hypothesis That Two Sample Means Come from a Population with Equal μ; Also Confidence Limits for the Difference Between Two Means**

Average age at beginning of reproduction in *Daphnia*. Data from Table 9.2.

The use of this test assumes that the variances in the populations from which the two samples were taken are identical. If in doubt about this hypothesis, test by the method of Box 8.1; if the hypothesis of homogeneity of variances of the two samples is rejected, do not employ this test but carry out the test shown in Section 13.3.

The appropriate formula for t_s is Expression (9.2), but because this example has equal sample sizes, we can use the first simplification mentioned in the text. This yields

$$t_s = \frac{(\overline{Y}_1 - \overline{Y}_2) - (\mu_1 - \mu_2)}{\sqrt{\dfrac{1}{n}(s_1^2 + s_2^2)}}$$

Because we are testing the null hypothesis that $\mu_1 - \mu_2 = 0$, we replace $\mu_1 - \mu_2$ in this equation by zero. Thus

$$t_s = \frac{7.5143 - 7.5571}{\sqrt{(0.5047 + 0.4095)/7}} = -0.1184$$

The degrees of freedom for this example are $2(n - 1) = 2 \times 6 = 12$. The critical value (from Table **B**) of $t_{.05[12]} = 2.179$. Because the absolute value of our observed t_s is less than the critical *t*-value, the means are found to be not significantly different, which is the same result as was obtained by the anova of Table 9.2.

BOX 9.4 (continued)

A comparison of the results of the present t-test with those of the F-test of Table 9.2 shows close correspondence. We have learned that $F_{[1,\nu_2]} = t^2_{[\nu_2]}$. Because $t_s = -0.1184$, $t^2_s = 0.0140$, which is equal to the F_s-value from Table 9.2.

Confidence limits of the difference between two means

$$L_1 = (\overline{Y}_1 - \overline{Y}_2) - t_{\alpha[\nu]}s_{\overline{Y}_1 - \overline{Y}_2}$$

$$L_2 = (\overline{Y}_1 - \overline{Y}_2) + t_{\alpha[\nu]}s_{\overline{Y}_1 - \overline{Y}_2}$$

In this case $\overline{Y}_1 - \overline{Y}_2 = -0.0428$, $t_{.05[12]} = 2.179$, and $s_{\overline{Y}_1 - \overline{Y}_2} = 0.3614$ as computed earlier as the denominator of the t-test. Therefore

$$L_1 = -0.0428 - (2.179)(0.3614) = -0.8303$$

$$L_2 = 20.0428 + (2.179)(0.3614) = 0.7447$$

The 95% confidence limits contain the zero point (no difference), as was to be expected, because the difference $\overline{Y}_1 - \overline{Y}_2$ was found to be not significant.

approach infinity, $F_{\alpha[1,\nu]}$ approaches the distribution of the square of the normal deviate as $\nu \to \infty$.

The relation between t^2 and F leads to another relation. We have just learned that when $\nu_1 = 1$, $F_{[\nu_1,\nu_2]} = t^2_{[\nu_2]}$. We know from Section 8.2 that $\chi^2_{[\nu_1]}/\nu_1 = F_{[\nu_1,\infty]}$. Therefore, when $\nu_1 = 1$ and $\nu_2 = \infty$, $\chi^2_{[1]} = F_{[1,\infty]} = t^2_{[\infty]}$. This relation can be demonstrated from Statistical Tables **D**, **F**, and **B**:

$$\chi^2_{.05[1]} = 3.841$$

$$F_{.05[1,\infty]} = 3.841$$

$$t_{.05[\infty]} = 1.960 \qquad t^2_{.05[\infty]} = 3.8416$$

The interrelationships between the t- and F-distributions (as well as the normal and χ^2-distributions) are evident in a diagrammatic view of an F-table, as shown in Figure 9.1.

The t-test for differences between two means is useful when we wish to estimate confidence intervals for such a difference. Box 9.4 shows how to calculate 95% confidence intervals for the difference between the series means in the *Daphnia* example. The appropriate standard error and degrees of freedom are the ones chosen for t_s. We are not surprised to find that the confidence limits of the difference in this case enclose the value of zero, ranging from -0.8303 to $+0.7447$. This condition must be true when a difference is found to be not significantly different from zero. Thus, we cannot exclude zero as the true value of the difference between the means of the two series. In fact, this confidence interval contains all parameter values for the difference between two means, for which hypotheses would not be rejected.

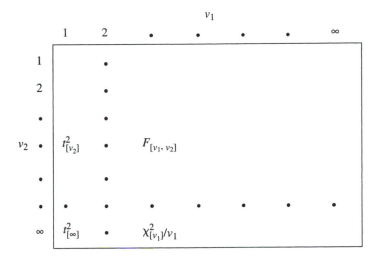

FIGURE 9.1 Diagrammatic representation on an F-table. The leftmost column is $F_{[1,\nu_2]}$, which equals $t^2_{[\nu_2]}$. The bottom row is $F_{[\nu,\infty]}$, which equals $\chi^2_{[\nu_1]}/\nu_1$. The entry in the lower left-hand corner is the intersection of $F_{[1,\nu_2]}$ and $F_{[\nu,\infty]}$, which is $F_{[1,\infty]}$ and equals $t^2_{[\infty]}$.

Had the alternative hypothesis been one-sided conceptually, as for example, $H_1: \mu_I > \mu_{II}$, then we should have employed the t-table using probability 2α for an intended type I error of α.

A special case of testing the difference between two groups occurs when one of the groups consists of only a single observation. The test is whether a single specimen sampled at random could belong to a given population. If the parameters and distribution of the population were known, the test would present no new features. Take the case of a normal distribution with mean μ and standard deviation σ. If we sample a single specimen and measure it for the variable in question, recording it as Y_1, we can test whether it belongs to the population by considering $(Y_1 - \mu)/\sigma$ as a standard normal deviate (see Section 6.2) and computing the probability of such a deviate using the table of areas of the normal curve (Statistical Table A).

If the population is known to us only through a sample, however, the way to deal with this problem is to consider it a two-sample case similar to the one discussed earlier in this section. The new aspect here is that one of the samples is represented by only a single observation, so it does not contribute to the degrees of freedom or to the estimate of the variance within groups. You can substitute in Expression (9.2) as follows: Y_1 replaces $\overline{Y}_1$ and 1 replaces n_1; $s_1 = 0$. The other symbols—$\overline{Y}_2$, n_2, and s_2—refer to the sample. We show an example of such a test below.

The problem analyzed is whether the wing length of a single mosquito collected in Kansas could belong to a population with the same mean wing length as a sample from Minnesota (unpublished data by F. J. Rohlf). This example is a two-tailed test because our alternative hypothesis is that the mean of the Kansas population from which the single specimen comes could be greater or less than that of the Minnesota population (H_1: $\mu_1 \neq \mu_2$), which is known from a sample of $n = 5$ specimens. The wing lengths (in millimeters) are as follows: 4.02, 3.88, 3.34, 3.87, and 3.18. Mean wing length $\overline{Y}_2$ is 3.658, and the sample standard deviation s_2 is 0.3725. The single observation Y_1 from Kansas is 3.02, and the degrees of freedom are as before: $n_1 + n_2 - 2$, which reduces to $n_2 - 1 = 4$. We can substitute the appropriate statistics in Expression (9.2) or we can simplify that expression algebraically to yield

$$t_s = \frac{\overline{Y}_1 - \overline{Y}_2 - (\mu_1 - \mu_2)}{s_2\sqrt{\dfrac{n_2 + 1}{n_2}}} \tag{9.3}$$

and substitute the above statistics. When we do that, we obtain $t_s = -0.64/0.40805 = -1.5684$. The BIOMstat statistical calculator yields a P-value of 0.1919. Alternatively, we could have consulted Statistical Table **B**. We would find that for 4 degrees of freedom, $t_s = -1.5684$ lies between α-values of 0.1 and 0.2, but quite close to the latter value. Thus, approximately 19% of wing lengths sampled from the Minnesota population would be as far from the Minnesota mean in either direction as is the Kansas specimen. We therefore do not have sufficient evidence to reject the null hypothesis of no difference.

Note that such a test has few, if any, direct taxonomic implications. It does not prove or disprove the assertion that the Kansas specimen is in the same species, subspecies, or in any other taxon with the Minnesota population. The Kansas specimen could well be in another genus and still not be different in wing length. Conversely, although there may be a statistical difference in wing length between the Kansas population and the Minnesota population for a variety of reasons, it may not be wise to recognize this specimen in any formal sense as a taxonomically distinct unit. In conventional taxonomy, criteria for specific distinction are usually not statistical but biological (Mayr and Ashlock, 1991). Criteria for subspecies may involve tests for spatial homogeneity with respect to many characters rather than simple tests of significance.

> Occasionally, the comparison of single individuals with sample means is of individuals that are very deviant from their sample mean. This situation involves considerations of statistical theory that are beyond the scope of this text. Chapter 13 describes methods that minimize the effects of outliers. If you wish to carry out such tests for outliers, consult Barnett and Lewis (1994), Dixon and Massey (1969), or Gumbel (1954), who have treated this problem in depth.

9.4 Comparisons Among Means in a Model I Anova: Essential Background

As we have seen in Section 9.2, the initial significance test is the same for both anova models. We completed a Model II anova by estimating the added variance components. In contrast, a Model I anova of more than two groups is usually completed by examining the data in greater detail and testing which means are different from which other means or which groups of means are different from other such groups or from single means. Two-sample cases, such as the example of Table 9.2, examining the average age of water fleas in two series, are of no interest here. Their mean square between groups possesses only one degree of freedom and admits of only one test—that is, do the two means differ? This is indeed the test that the basic anova carried out.

When the number of groups, a, is greater than two, however, further tests are possible and usually are of interest. For example, although we conclude that the means of the lengths of pea sections in Table 9.1 are not all equal, we do not know which ones differ from which other ones. This leads us to the subject of tests among pairs and groups of means. For example, we might test the control against the average of the four experimental treatments representing added sugars. The question to be tested would be: Does the addition of sugars have an effect on length of pea sections? We might also test for differences among the sugar treatments. A reasonable test might be the average of the pure sugars (glucose, fructose, and sucrose) versus the mixed sugar treatment (1% glucose, 1% fructose).

Before we discuss the general subject of multiple comparisons, we need to dispose of a special case for which the complications described below do not apply. If the sum of the degrees of freedom of the comparisons does not exceed the $a - 1$ degrees of freedom among groups, and if the set of planned comparisons exhibits a property known as *orthogonality* (explained later in this section), we should use the method of **orthogonal comparisons**. There are two ways to compute orthogonal comparisons: We can subdivide the treatment sum of squares and the treatment degrees of freedom into separate comparisons, carrying these out as part of the analysis of variance, using F-tests as tests of significance. Alternatively, we can carry out t-tests for comparisons between two groups, but these are mathematically equivalent to the F-tests (as shown in Section 9.3). We prefer the F-tests because of their simplicity and the elegance of their decomposition of the sum of squares for groups.

The general procedure for making a planned comparison is simple and relates to the method for obtaining the sum of squares for any set of groups (see Section 9.1). To compare k groups of any size n_i, compute $SS = \Sigma^k n_i (\bar{Y}_i - \bar{\bar{Y}})^2$, where the summation is over the k groups, $\bar{Y}_i$ is the mean of each group, and $\bar{\bar{Y}} = \Sigma n_i \bar{Y}_i / \Sigma n_i$ is the weighted average over the k groups. The following example will illustrate the computations.

Table 9.3 lists the means and sample sizes of the experiment with the pea sections from Table 9.1. We now wish to test whether the controls are different from the four treatments representing addition of sugars. There will thus be two groups, the control group and the "sugars" group. The mean of the sugars group is 1.672, and its sample size is 40. We therefore compute

$$SS_{(\text{control vs. sugars})} = \sum n_i (\bar{Y}_i - \bar{\bar{Y}})^2$$
$$= 10(1.430 - 1.624)^2 + 40(1.672 - 1.624)^2$$
$$= 0.4685$$

TABLE 9.3   Means and Sample Sizes from the Data in Table 9.1

Length of pea sections (in ocular units, transformed as described in Table 9.1) grown in tissue culture. Coefficients of linear comparisons are also given for two comparisons.

	Control	2% glucose added	2% fructose added	1% glucose + 1% fructose added	2% sucrose added	Σ
$\overline{Y}$	1.430	1.686	1.719	1.722	1.561	$(1.624 = \overline{\overline{Y}})$
n	10	10	10	10	10	50
Comparisons						
Control vs. treatments (c_{i1})	+4	−1	−1	−1	−1	
Mixed vs. pure sugars (c_{i2})	0	−1	−1	+3	−1	

In this case, $\overline{\overline{Y}}$ is the same as for the anova because it involves all the groups of the study. The result is a sum of squares for the comparison between these two groups. Because a comparison between two groups has only 1 degree of freedom, the sum of squares is also a mean square. This mean square is tested over the error mean square of the anova to give the following variance ratio test:

$$F_s = \frac{MS_{\text{(control vs. sugars)}}}{MS_{\text{within}}} = \frac{0.4685}{0.002973} = 157.59$$

The probability corresponding to this F-value is extremely small ($<10^{-16}$), showing that there is strong evidence that the additions of sugars increased the coded ocular units, which corresponds to retarding the growth of the pea sections.

Next we test whether the mixture of sugars is different from the pure sugars. We first calculate the mean of the pure sugars:

$$\overline{Y}_{\text{pure sugars}} = \frac{1.686 + 1.719 + 1.561}{3} = 1.655$$

We use a simple average because the sample sizes are equal. Then we evaluate the new grand mean (of all sugars):

$$\overline{\overline{Y}}_{\text{sugars}} = \frac{1.686 + 1.719 + 1.722 + 1.561}{4} = 1.672$$

Thus,

$$SS_{\text{(mixed sugar vs. pure sugars)}} = 10(1.722 - 1.672)^2 + 30(1.655 - 1.672)^2 = 0.03333$$

Here, the $\overline{\overline{Y}}$ is different because it is based on the sugars only. The appropriate test statistic is

$$F_s = \frac{MS_{\text{(mixed sugars vs. pure sugars)}}}{MS_{\text{within}}} = \frac{0.03333}{0.002973} = 11.21$$

This F-value yields a P-value of 0.0017, which suggests that this result is very improbable, given the null hypothesis. We consequently reject the null hypothesis. It is likely that mixed sugars have higher coded means than pure sugars—that is, pea-section growth is retarded more by mixed than by pure sugars.

Next, we test whether the disaccharide sugar (sucrose) differs in effect from the monosaccharides (glucose and fructose). We compute the monosaccharide mean as

$$\overline{Y}_{\text{(monosaccharides)}} = \frac{1.686 + 1.719}{2} = 1.7025$$

The grand mean $\overline{\overline{Y}}$ is based on all pure sugars, and we have calculated it earlier as 1.655. The sum of squares for this contrast is

$$SS_{\text{(monosaccharides vs. disaccharide)}} = 20(1.7025 - 1.655)^2 + 10(1.561 - 1.655)^2$$
$$= 0.1335$$

This yields

$$F_s = \frac{MS_{\text{(monosaccharides vs. disaccharide)}}}{MS_{\text{within}}} = \frac{0.1335}{0.002973} = 44.90$$

This F_s-value indicates a very low probability ($P < 10^{-7}$) that the null hypothesis is correct. Consequently, we reject H_0 and conclude that sucrose (a disaccharide) has a lower coded mean than glucose and fructose (both monosaccharides); the latter inhibit pea-section growth more than sucrose.

Finally, we test glucose versus fructose. All the quantities we need are already available.

$$SS_{\text{(glucose vs. fructose)}} = 10(1.686 - 1.7025)^2 + 10(1.719 - 1.7025)^2$$
$$= 0.005445$$

$$F_s = \frac{MS_{\text{(glucose vs. fructose)}}}{MS_{\text{within}}} = \frac{0.005445}{0.002973} = 1.83$$

This variance ratio yields $P = 0.1829$. We therefore accept the null hypothesis that the two monosaccharides do not differ in their effects on the pea-section lengths.

We conclude that the addition of the three sugars retards growth in the pea sections and that mixed sugars retard growth more than pure sugars, probably because the mixed sugars are monosaccharides whereas the pure sugars include the disaccharide sucrose, which inhibits growth far less than the former. We did not find differences in growth inhibition between glucose and fructose.

Our orthogonal comparisons could have been different, however, depending entirely on our choice of hypotheses to test. Thus, we could have planned to test control versus sugars, disaccharides (sucrose) versus monosaccharides (glucose, fructose, glucose + fructose), mixed versus pure monosaccharides, and finally glucose versus fructose.

The pattern and number of planned comparison tests are determined by one's hypotheses about the data. There are certain restrictions, however, if the comparisons are to be orthogonal. As stated earlier, we cannot use up more than the $a - 1$ degrees of freedom among the groups, and we should structure the tests in such a way that each comparison tests an independent relationship among the means (as was done in the sugars example). For example, we would prefer not to test whether means one, two, and three differed if we had already found that mean one differed from mean three because significance of the latter implies heterogeneity of the former. The following discussion provides rules for establishing the orthogonality of the planned comparisons.

At the bottom of Table 9.3 are two rows of **coefficients of linear comparisons** labeled c_{ij}. The subscript i refers to the group $(1, 2, \ldots, i, \ldots, a)$ in the anova, and j identifies the particular linear comparison $(1, 2, \ldots, j, \ldots, a - 1)$. How are these coefficients determined? For convenience, they are usually expressed as integers, although they may be fractional. For any one comparison, the coefficients must sum to zero when weighted by the sample sizes of the groups; $\sum_{i=1}^{a} n_i c_{ij} = 0$. The purpose of the first row of coefficients is to test whether the control is different from the mean of the four treatment groups. Note that to indicate the contrast, the four treatment groups are given a minus sign and the control group is given a plus sign. To obtain the mean of the four treatment groups, we have to compute the following: $\frac{1}{4}(\overline{Y}_2 + \overline{Y}_3 + \overline{Y}_4 + \overline{Y}_5)$. Thus, each treatment mean is weighted by a coefficient of $\frac{1}{4}$, as contrasted with the coefficient 1 for the control. To avoid fractions, we multiply the coefficients by 4.

An easy way to obtain the coefficients to be assigned to each sum in a linear comparison is to assign it the integer representing the *number of samples in the contrasted set*. In this example, there is a control set and a sugar set. The control set has one sample; the sugar set has four samples. Therefore, we assign a coefficient of 4 to the control set and a coefficient of 1 to each member of the sugar set. The signs of the two sets must be unlike, but computationally it makes no difference whether the plus is assigned to the 1's or to the 4.

To reinforce what we have just learned, let us work out the second row of coefficients in Table 9.3. We now wish to contrast mixed sugars and pure sugars—that is, to test the mean of the three pure sugars against the single mixed sugar, leaving out the control group altogether. The control group does not enter in this comparison, so we assign it a value $c_{1,2} = 0$. The comparison is between $\frac{1}{3}(\overline{Y}_2 + \overline{Y}_3 + \overline{Y}_5)$ and $\overline{Y}_4$. By the rule given here, the single mixed sugar is assigned a coefficient of 3 and the pure sugars each a coefficient of 1 because the comparison is between a set of three pure sugars and a set of one mixed sugar. Again, we assign negative signs to the 1's and a plus sign to the 3, but we could just as well switch the signs around.

A table of coefficients permits an easy test for orthogonality of comparisons. On the basis of such a table, two comparisons are orthogonal if and only if the sum of the sample sizes times the products of their coefficients equals zero, or, expressed as a formula, $\Sigma_{i=1}^{a} n_i c_{ij} c_{ik} = 0$. The subscripts j and k refer to two separate comparisons and i to the various groups. Let us test the two comparisons in Table 9.3, which we have just completed. Because n_i is constant ($= 10$) for all groups in this example, this factor can be ignored.

$$(4 \times 0) + (-1 \times -1) + (-1 \times -1) + (-1 \times 3) + (-1 \times -1) = 0$$

This calculation shows that these two comparisons are orthogonal. The minor complications of specifying the coefficients when sample sizes are unequal are presented in Section 14.9.

A property of an orthogonal set of comparisons is that their sums of squares add to the sum of squares among groups. The four sums of squares we have obtained, each based on 1 df, together add to the sum of squares among treatments of the original analysis of variance based on 4 degrees of freedom:

$SS_{\text{(control vs. sugars)}}$	$= 0.468512$	1
$SS_{\text{(mixed vs. pure sugars)}}$	$= 0.033333$	1
$SS_{\text{(monosaccharides vs. disaccharide)}}$	$= 0.133482$	1
$SS_{\text{(glucose vs. fructose)}}$	$= \underline{0.005445}$	$\underline{1}$
$SS_{\text{(among treatments)}}$	$= 0.640772$	4

This property illustrates once again the elegance of analysis of variance. The treatment sums of squares can be decomposed into separate parts that are sums of squares in their own right, with degrees of freedom pertaining to them. We can present all of these results as an anova table as shown in Table 9.4.

As an example of a *non*orthogonal set of comparisons, we replace $SS_{\text{(mixed sugar vs. pure sugars)}}$ with the SS for the average of glucose and fructose versus glucose and fructose mixed. What would be the coefficients of linear comparisons for this test?

TABLE 9.4 Anova Table from Table 9.1 with Treatment Sum of Squares Decomposed into Orthogonal Comparisons

Source of Variation	SS	df	MS	F_s	P
Treatments	0.640772	4	0.160193	53.88462	10^{-16}
Control vs. sugars	0.468512	1	0.468512	157.59486	$<10^{-16}$
Mixed vs. pure sugars	0.033333	1	0.033333	11.212438	0.001650
Monosaccharides vs. disaccharides	0.133482	1	0.133482	44.89959	2.83×10^{-8}
Glucose vs. fructose	0.005445	1	0.005445	1.83154	0.182705
Within	0.133780	45	0.0029729		
Total	0.774552	49			

Because control and sucrose are not involved in the comparison, they are assigned values of zero. The average of the two pure sugars is represented by 1's for the sugars and the mixed sugar by a 2 to balance the number of samples in the contrasted set. Thus, the coefficients are 0, 1, 1, −2, 0. Although this comparison is orthogonal to the first of our tests (control versus treatments), it is not orthogonal to the test of mixed versus pure sugars. To confirm these statements, we may compute $(4 \times 0) + (−1 \times 1) + (−1 \times 1) + (−1 \times −2) + (−1 \times 0) = 0$ and $(0 \times 0) + (−1 \times 1) + (−1 \times 1) + (+3 \times −2) + (−1 \times 0) = −8$. The second sum of products is not equal to zero, confirming the nonorthogonality of mixed versus pure sugars to the average of glucose and fructose versus mixed sugars in the same set.

To calculate the sum of squares for the comparison, we first compute the mean of glucose and fructose: $(1.686 + 1.719)/2 = 1.7025$. Then we obtain the grand mean of glucose, fructose, and mixed sugars: $(1.686 + 1.719 + 1.722)/3 = 1.709$. Finally, SS for the average of glucose and fructose versus glucose and fructose mixed is calculated as follows:

$$20(1.7025 − 1.709)^2 + 10(1.722 − 1.709)^2 = 0.002535$$

With 0.002535 replacing 0.033333 (see Table 9.4), the sums of squares obviously will no longer add to 0.640772.

Now let us apply these techniques to another example. Box 9.5 shows egg-laying records from 25 females of three genetic lines of *Drosophila*, two selected for resistance and susceptibility to DDT, the third a nonselected control line. The data show that average fecundity is considerably higher in the nonselected line. Because we are interested in differences among the lines, each of which was treated in a specific manner, this is clearly a Model I anova. The computations are the same as for Table 9.1, which also was a Model I single-classification anova with equal sample sizes. The steps are shown less fully because we assume that you are by now familiar with the basic computations of the analysis of variance. The anova table shows that the observed difference among lines is very unlikely to be obtained by chance (4 out of 10,000 times) if there actually were no differences among the means. The P-value indicates only one chance out of 10,000 that the observed difference could be obtained if the null hypothesis of no differences in fecundity among the lines were correct.

The investigator planned to test two hypotheses: Did the selected lines (RS and SS) differ in fecundity from the nonselected line (NS)? Second, did the line selected for resistance differ in fecundity from that selected for susceptibility? These hypotheses form a set of orthogonal comparisons and are tested in the manner you just learned. The results show that the selected lines have a significantly different (lower) fecundity from that of the nonselected line and that the two selected lines do not differ from each other in fecundity. These data are tabulated in Box 9.5 in a new anova table, which includes both one-degree-of-freedom tests. Again, you can see that the sums of squares are additive, together yielding the sum of squares among lines. Note also that this anova table is given in more abbreviated form than earlier ones to show you how such tables might be published in the literature. Neither the F's nor the total sum of squares is given in this table. As you will see later, sometimes even the sums of squares are not stated, leaving only the minimally essential information: the degrees of freedom and the mean squares.

Let us formalize the kinds of comparisons that we might carry out by a simple, generally applicable symbolism. We relabel the five treatments of Table 9.1 with the numbers 1 to 5, starting with the control treatment. Then our intended tests might be (mean of) 1 versus (mean of) 2, 3, 4, and 5, or (mean of) 2, 3, and 5 versus 4, as described above. However, another set of comparisons might be control versus each of the treatments, yielding 1 versus 2, 1 versus 3, 1 versus 4, and 1 versus 5. Alternatively, we might want to test each treatment mean against every other mean, yielding 1 versus 2, 1 versus 3, . . . , 4 versus 5, making altogether $a(a - 1)/2 = 5 \times 4/2 = 10$ comparisons. A recent term for each of the sets of comparisons is to call it a *family* of hypotheses. Each family will contain $k \geq 2$ comparisons. The fewer comparisons there are in a family, the more powerful the tests will be. It is up to the investigator to formulate the hypothesis tests in such a way that all the important questions are

---

BOX 9.5 Orthogonal Comparisons of Means and Groups of Means in a Single-Classification Model I Anova

Per diem fecundity (number of eggs laid per female per day for the first 14 days of life) for 25 females from each of 3 lines of *Drosophila melanogaster*. The RS and SS lines were selected for resistance and for susceptibility to DDT, respectively; and the NS line is a nonselected control strain; $n = 25$ females per line.

Lines ($a = 3$)					
Resistant (RS)		**Susceptible (SS)**		**Nonselected (NS)**	
12.8	22.4	38.4	23.1	35.4	22.6
21.6	27.5	32.9	29.4	27.4	40.4
14.8	20.3	48.5	16.0	19.3	34.4
23.1	38.7	20.9	20.1	41.8	30.4
34.6	26.4	11.6	23.3	20.3	14.9
19.7	23.7	22.3	22.9	37.6	51.8
22.6	26.1	30.2	22.5	36.9	33.8
29.6	29.5	33.4	15.1	37.3	37.9
16.4	38.6	26.7	31.0	28.2	29.5
20.3	44.4	39.0	16.9	23.4	42.4
29.3	23.2	12.8	16.1	33.7	36.6
14.9	23.6	14.6	10.8	29.2	47.4
27.3		12.2		41.7	
$\sum^{n} Y$	631.4		590.7		834.3
$\bar{Y}$	25.256		23.628		33.372

SOURCE: Data from R. R. Sokal (unpublished results).

Box 9.5 (continued)

I. *Anova*

1. $\overline{\overline{Y}} = \dfrac{1}{an} \sum\limits^{a} \sum\limits^{n} Y = \dfrac{1}{75}(631.4 + 590.7 + 834.3) = 27.419$

2. $SS_{\text{among}} = n \sum\limits^{a}(\overline{Y}_i - \overline{\overline{Y}})^2$

$= 25((25.256 - 27.419)^2 + (23.628 - 27.419)^2 + (33.372 - 27.419)^2)$

$= 1362.21$

3. $SS_{\text{within}} = \sum\limits^{a}\sum\limits^{n}(Y - \overline{Y}_i)^2 = 5659.02$

4. $SS_{\text{total}} = SS_{\text{among}} + SS_{\text{within}} = 7021.23$

Anova table

Source of variation		df	SS	MS	F_s	P
$\overline{Y} - \overline{\overline{Y}}$	Among groups (among genetic lines)	2	1362.21	681.11	8.67	0.0004
$Y - \overline{Y}$	Within groups	72	5659.02	78.60		
$Y - \overline{\overline{Y}}$	Total	74	7021.23			

II. *Planned tests among the means*

There were two interesting a-priori hypotheses. Their rationale is discussed in the text.

1. $H_0: \mu_{(\text{RS and SS})} = \mu_{\text{NS}}$ $H_1: \mu_{(\text{RS and SS})} \neq \mu_{\text{NS}}$

The mean for RS and SS is $(25.256 + 23.628)/2 = 24.442$. The grand mean is quantity **1** from part **I.**

$SS_{(\text{RS}+\text{SS})\text{ vs. NS}} = 50(24.442 - 27.419)^2 + 25(33.372 - 27.419)^2 = 1329.08$

Because there is only one *df* for this comparison, $SS = MS$.
The significance test for this hypothesis is

$$F_s = \dfrac{MS_{(\text{RS}+\text{SS})\text{ vs. NS}}}{MS_{\text{within}}} = \dfrac{1329.08}{78.60} = 16.91 \quad P = 0.001$$

We can reject the null hypothesis of no difference in fecundity between the selected and non-selected strains with high confidence.

2. $H_0: \mu_{\text{RS}} = \mu_{\text{SS}}$ $H_1: \mu_{\text{RS}} \neq \mu_{\text{SS}}$

Because in this case, the two groups to be compared have equal sample sizes, instead of testing them with the equation

$$n_{\text{RS}}(\overline{Y}_{\text{RS}} - \overline{\overline{Y}})^2 + n_{\text{SS}}(\overline{Y}_{\text{SS}} - \overline{\overline{Y}})^2$$

Box 9.5 (continued)

we may use a similar equation for the sum of squares.

$$SS_{(RS \text{ vs. } SS)} = \frac{n}{2} (\bar{Y}_{RS} - \bar{Y}_{NS})^2 = \frac{25}{2} (25.256 - 23.628)^2 - 33.13$$

There is one df for this comparison, and again $SS = MS$.
The significance test for this hypothesis is

$$F_s = \frac{MS_{RS \text{ vs. } SS}}{MS_{within}} = \frac{33.13}{78.60} = 0.4215 \quad P = 0.5183$$

which is not rejected.

Anova table

Source of variation	df	SS	MS	P
Among groups (among genetic lines)	2	1362.21	681.11	0.0004
Selected vs. nonselected lines	1	1329.08	1329.08	0.0001
RS vs. SS	1	33.13	33.13	0.5183
Within groups (error; females within genetic lines)	72	5659.02	78.60	

answered with a minimum number of tests. The arithmetic involved in such computations is simple to carry out and is based on familiar formulas. However, unlike the orthogonal comparisons discussed above, there are complications with the probability distributions with which these hypothesis tests are associated.

Let us assume we have sampled from an approximately normal population of heights of men. We have computed their mean and standard deviation. If we sample two men at a time from this population, we can predict the difference between them on the basis of ordinary statistical theory. Some men will be very similar, others relatively different. Their differences will be distributed normally, with a mean of 0 and an expected variance of $2\sigma^2$, for reasons explained in Section 15.3. Thus, a large difference between a randomly sampled pair of men will have to be a sufficient number of standard deviations greater than zero in order for us to reject our null hypothesis that the two men come from the specified population.

If, on the other hand, we were to look at the heights of the people before sampling them and then take pairs that seem to be very different, we would repeatedly obtain differences between pairs of men that are several standard deviations apart. Such differences would be outliers in the expected frequency distribution of differences, and time and again we would reject our null hypothesis when, in fact, it is true. The men would have been sampled from the same population. However, they would not have been sampled at random but would have been inspected before being sampled so that the probability distribution on which our hypothesis rested would no longer be valid. Obviously, the tails in a large sample from a normal distribution will be anywhere from 5 to 7 standard deviations apart, and if we deliberately take individuals from each tail and compare them, they will appear to be highly significantly different from each other, even though they belong to the same population.

When we select means differing greatly from each other in an analysis of variance, we are doing exactly the same thing as taking a tall and a short man from the frequency distribution of heights. The key issue to be considered is whether the individual tests that make up a family of hypothesis tests are independent of each other or whether these tests lack mutual independence. If the outcome of one test depends on the outcome of another test, then these two tests cannot be independent.

Returning to the example of the heights of men, when we compare the height of the tallest with the height of the shortest man in our sample, we are not really carrying out a single test only. To find the tallest and shortest man in our study, we have effectively compared the height of each man with that of every other man in the sample. So we have implicitly made $a(a - 1)/2$ comparisons instead of a single one. Although we did not compute the results of all these comparisons, each of the heights would be used repeatedly. The actual test that we propose to do—that is, shortest versus tallest man—is therefore dependent on the outcomes of these other implicitly performed tests. Similarly, when we undertake to test various treatments against a control, as in the example on the pea sections, we customarily use only a single control mean. If that particular mean is unusually large or small because of chance sampling, this would affect the outcome of all the tests of treatments against the control.

In the hypothesis tests considered earlier in this book, we have always tested a single hypothesis only. The type I error α for these single tests will now be called the **per-comparison error rate,** or, alternatively, the *comparisonwise error rate*. As you recall, it is the probability that a true null hypothesis will be rejected erroneously. Such a decision is usually made when an observed value lies at or beyond the critical value corresponding to a predetermined α (for a conventional hypothesis test) or when the confidence interval for the observed estimates does not include zero (for a hypothesis test by means of confidence intervals). When we have an entire family of hypotheses to test, we employ a conservative approach in which one lowers the type I error for each comparison so that the probability of making any type I error at all in the entire series of tests does not exceed α. This probability is called the **familywide or familywise error rate.** (The first of the adjectives is preferable because it employs a more meaningful suffix when compared with the meaningless suffix *-wise*. An earlier term for this error rate was the *experimentwise error rate,* but this is not sufficiently inclusive because not every model I analysis of variance is based on an experiment.) Assuming that an investigator plans to make k independent significance tests, each using a critical value corresponding to an adjusted per-comparison error rate level of α', then the probability of making no type I errors in any of the k tests is $(1 - \alpha')^k$. The familywide error rate (i.e., the probability of making at least one type I error) is then

$$\alpha = 1 - (1 - \alpha')^k \tag{9.4}$$

If we were to set the level of α' equal to 0.05 per comparison, then, for $k = 10$ tests, the familywide error rate would rise to 14.13%. For $k = 20$ and 30 comparisons, the corresponding figures are 64.15% and 78.54%. Clearly, such values are unacceptably high. In consequence, we have to lower the per-comparison error rate to less than 0.05, to make the familywide error rate α equal to 0.05. To achieve a specified familywide error rate, α, Expression (9.4) can be rearranged to yield an adjusted per-comparison error rate of

$$\alpha' = 1 - (1 - \alpha)^{1/k} \tag{9.5}$$

This method has been called the **Dunn-Šidák method** by Ury (1976), who showed that employing α' for each comparison results in a conservative test (familywide error rate $\leq \alpha$) when the members of each family of significance tests do not deviate too much from independence. (It is exact when the tests are independent.)

If the k significance tests are not independent of each other, we can employ an alternative method for limiting the overall familywide error rate by testing each comparison using an adjusted per-comparison error rate of

$$\alpha'' = \alpha/k \tag{9.6}$$

This is called the **Bonferroni method.** It has been shown to be slightly more conservative (its familywide error rate is slightly smaller) than the Dunn-Šidák method. The calculation of α'' (Bonferroni method) is slightly simpler than that of α' (Dunn-Šidák method). Unless the number of comparisons, k, yields a tabled probability value of α'', consulting a standard t- or F-table will be tedious, requiring interpolation. In such cases, it is best to turn to a statistical calculator for the specified P-value and degrees of freedom, unless the Bonferroni critical values are automatically generated by the computer program being used. The principal consideration for deciding which of these two methods to use should be your judgment about the validity of the assumption of independence of the tests. It is more conservative to use the Bonferroni test. Its P-values will always be equal to or greater than those of the Dunn-Šidák method. Hence, fewer null hypotheses will be rejected when Bonferroni is employed. Note also that the Bonferroni method, the Dunn-Šidák method, and other similar approaches are frequently referred to as Bonferroni techniques, using the term in a more inclusive sense.

Applying these approaches to the pea-section data from Table 9.1, let us assume that the investigator has good reason to test five contrasts between the treatments specified in the first tabular display in Box 9.6. The 5 degrees of freedom for these tests are greater than the available $a - 1 = 4\ df$ among treatments. Hence, these tests cannot all be independent of each other. We therefore chose the Bonferroni method to adjust the per-comparison error rate for each contrast. For the five tests, the adjusted error rate is

$$\alpha'' = \alpha/k = 0.05/5 = 0.01000$$

for a familywide error rate of $\alpha = 0.05$. Thus, the critical value for the F_s-ratios of these comparisons is adjusted to $F_{.01[1,45]}$, which can be looked up in an ordinary

F-table such as Statistical Table **F**. However, if we had wanted to test six rather than five contrasts, the adjusted per-comparison error rate would have been $\alpha'' = 0.05/6 = 0.00833$. Such nonstandard values of α are not found in any F-tables. The critical value of $F_{.00833[1,45]}$ for an adjusted per-comparison error rate can be obtained from a statistical calculator program such as is furnished in BIOMstat, which yields 7.6178.

Alternatively, we can circumvent this problem for critical values of $F_{\alpha'[1,\nu_2]}$ by making use of the relationship $F_{\alpha'[1,\nu_2]} = t^2_{\alpha'[\nu_2]}$. These values can be found by turning to Statistical Table **C**, which contains critical values of $t_{\alpha'[\nu_2]}$ for familywide error rate α and degrees of freedom ν_2 for k comparisons. For critical values of $F_{\alpha'[\nu_1,\nu_2]}$, where $\nu_1 \geq 2$, there are no exact tables available. One must either interpolate in the F-table between the bracketing values of α (unless the α' or α'' values are beyond the range of the table) or use a computer program to compute the desired critical value.

Suppose that we had done a larger study with more, say, 10, treatments, yielding $a - 1 = 9$ degrees of freedom. Then, if all we wanted was to test five contrasts, *and* we had reason to know or assume that the five contrasts were independent, we would be led to choose the Dunn-Šidák method. This would adjust each per-comparison error rate as

$$\alpha' = 1 - (1 - 0.05)^{1/5} = 0.01021$$

for a familywide error rate of $\alpha = 0.05$. The F-ratio from each test would therefore be compared with the critical value $F_{.01021[1,45]}$. A numerical value for this quantity, 7.1900, is obtained in the same manner as discussed above for the Bonferroni approach.

The nature of the five contrasts in these data is such that adjusting the test results for multiple comparisons makes little difference. The three tests that are significant by conventional criteria maintain their significance when adjusted for Bonferroni or for Dunn-Šidák. This may not be true in other cases. Tests that are seemingly significant by conventional tests may become nonsignificant when the familywide error rate is used. In any case, the appropriate choice between these two methods is for Bonferroni, because we cannot assume that the individual tests are independent.

Although it is easy enough to calculate the adjusted per-comparison error rates α' and α'' for any given familywide error rate and then compare the observed F-values with the critical F-values based of the adjusted error rates, the results of such comparisons can only be expressed as inequalities, such as $P < 0.01$ or the equivalent asterisk symbolism. Moreover, readers of the results may become confused on whether the reported probabilities are per-comparison or familywide error rates. Statisticians have therefore developed **adjusted P-values,** which are the lowest familywide error rates at which a given null hypothesis could be rejected. The adjusted P-values for Bonferroni are computed as $P_{adj} = kP_i$, where P_i equals the probability associated with the ith test. (When $kP_i > 1$, set $P_{adj} = 1$.) For Dunn-Šidák, $P_{adj} = 1 - (1 - P_i)k$.

In Box 9.6, we show these values for the family of five tests investigated in the pea-section data. It is easy to see that by conventional criteria for significance tests,

BOX 9.6 | Simultaneous and Sequential Bonferroni Tests: k Comparisons by Dunn-Šidák and Bonferroni Methods

We decided to test five single degree of freedom contrasts on the pea-section data of Table 9.1. These tests cannot all be mutually orthogonal because they sum to 5 df, whereas the mean square among the 5 treatments possesses only 4 df. We list the tests in arbitrary order. All tests in the table below have been carried out previously in Section 9.4. We furnish verbal as well as symbolic descriptions of the contrasts, F_s-values for each test, and their associated probabilities, obtained from BIOMstat. Readers without this program can obtain these P-values from similar programs for the F- or t-distribution or from extended t-tables for single degrees of freedom with detailed interpolation.

Contrasts	F_s	Probability P_i
Sugars vs. control (G, F, G+F, S)(C)	157.59	2.2204×10^{-16}
Mixed vs. pure sugars (G+F)(G, F, S)	11.21	0.001652
Glucose vs. fructose	1.83	0.1829
Mixed vs. average of G and F (G+F)(G, F)	0.8527	0.3607
Monosaccharides vs. disaccharides (G, F)(S)	44.90	2.8349×10^{-8}

To carry out the desired $k = 5$ multiple contrasts tests with a familywide error rate $\alpha = 0.05$, we compute an adjusted per-comparison error rate of $\alpha'' = \alpha/k = 0.05/5 = 0.01000$ by the Bonferroni method using Expression (9.7) and a similarly adjusted rate of $\alpha' = 1 - (1 - 0.05)^{1/5} = 0.01021$ by the Dunn-Šidák method from Expression (9.6). We then examine the conventional P-values in the light of these adjusted error rates. We find that the null hypothesis can be firmly rejected for three contrasts—sugars versus control, mixed versus pure sugars, and monosaccharides versus disaccharides— using either method. For the other two contrasts, glucose versus fructose and mixed versus average of these two sugars, we cannot reject the null hypothesis of no difference by conventional standards.

An alternative approach is to compute adjusted P-values for each of the two methods. For Bonferroni, these are obtained as $P_{adj} = kP_i$, where P_i equals the conventional probability associated with the ith test. (When $kP_i > 1$, set $P_{adj} = 1$.) For Dunn-Šidák adjusted P-values, $P_{adj} = 1 - (1 - P_i)^k$. These adjusted values are shown below. In this and the following table, the names of the contrasts have been abbreviated.

Contrast	Simultaneous P	Bonferroni Adj. P	Dunn-Šidák Adj. P
Sug-Contr	2.2204×10^{-16}	1.1102×10^{-15}	$<10^{-16}$
Mix-Pure	0.0016520	0.0082600	0.0082328
Gluc-Fruct	0.1829000	0.9145000	0.6357697
Mix-Ave	0.3607000	1.0000000	0.8932117
Mono-Di	2.8349×10^{-8}	1.4175×10^{-7}	1.4174×10^{-7}

Necessarily, the adjusted P-values corroborate the findings obtained by using critical P-values as shown earlier. However, in this example, the corrections for multiple testing

Box 9.6 (continued)

do not alter the results obtained by inspection of the conventional *P*-values. This may not be true in other cases. Comparisons in which we are able to reject the null hypothesis using simultaneous tests may no longer be rejected when the more conservative familywide error rate is used. Note that the adjusted probabilities are always slightly less for Dunn-Šidák than for Bonferroni, showing the latter to be more conservative.

Sequential Bonferroni and Dunn-Šidák Tests

These tests increase the power of the simultaneous Bonferroni procedures described above.

1. Order the contrasts by their simultaneous probability values from smallest to largest. The re-ordered contrasts and their simultaneous (conventional) *P*-values are shown in the first two columns of the table below.

Contrast	Simultaneous P	Bonferroni Adj. P	Dunn-Šidák Adj. P	Bonferroni Adj. Per-Comp. Error Rates	Dunn-Šidák Adj. Per-Comp. Error Rates
Sug-Contr	2.2204×10^{-16}	1.1102×10^{-15}	$<10^{-16}$	0.01000	0.01020
Mono-Di	2.8349×10^{-8}	1.1340×10^{-7}	1.1340×10^{-7}	0.01250	0.01274
Mix-Pure	0.0016520	0.0049560	0.0049478	0.01667	0.01695
Gluc-Fruct	0.1829000	0.3658000	0.3323476	0.02500	0.02532
Mix-Ave	0.3607000	0.3607000	0.3607000	0.05000	0.05000

2. Compare the smallest simultaneous probability P_1 to adjusted per comparison error rates α/k for a sequential Bonferroni test or $1 - (1 - \alpha)^{1/k}$ for a sequential Dunn-Šidák test. For a 5% familywide error rate, this comparison yields $0.05/5 = 0.01$ for Bonferroni and $1 - (1 - 0.05)^{1/5} = 0.0102$ for Dunn-Šidák. If P_1 is greater than this threshold value, accept the null hypothesis for all tests. If $P_1 \le$ the critical value, as in our example, reject the null hypothesis and compare the second smallest probability, P_2, to $\alpha/(k - 1) = 0.0125$ for Bonferroni and to $1 - (1 - \alpha)^{1/(k-1)} = 0.0127$ for Dunn-Šidák. If P_2 is greater than either of these critical values, then accept the null hypothesis for the remaining tests by the respective test. If $P_2 \le$ either of these critical values, as in our example, reject the null hypothesis that this contrast is equal to zero and continue testing the successively larger probabilities in the same manner, proceeding to $(k - 2), (k - 3) \ldots (1)$.

Simultaneous Confidence Intervals of Means

We shall set such confidence limits and intervals for the five treatment means of the pea-section data. In this case, $k = a = 5$. We need the following quantities: the (pooled) standard error of each treatment mean, computed as $s_{\bar{Y}_i} = \sqrt{MS_{within}/n_i} = \sqrt{0.002973/10} = 0.017242$; the *t*-values for $t_{\alpha''[45]}$ and $t_{\alpha'[45]}$ adjusted for Bonferroni

Box 9.6 (continued)

and Dunn-Šidák probabilities, as shown earlier in this box. The 45 degrees of freedom are the (pooled) within groups degrees of freedom pertaining to MS_{within}. For $\alpha = 0.05$ and $k = 5$, these values are $t_{.01[45]} = 2.6895$ and $t_{.010206[45]} = 2.6816$ for Bonferroni and Dunn-Šidák, respectively. They were obtained from BIOMstat. Readers without this program can obtain these t-values from similar programs for the t-distribution or from Table **C** in the Statistical Tables by interpolation. For comparative purposes, we also record the conventional t-value, $t_{\alpha[45]} = t_{.05[45]} = 2.0141$. Each confidence limit, ordinary or simultaneous, is computed as $\bar{Y}_i \pm t_{\alpha[45]} s_{\bar{Y}_i}$. Thus, for the lower limit L_1 of the control mean using Bonferroni-adjusted t-values, we obtain $L_1(\text{Bon}) = 1.430 - (2.6895)$ $(0.017242) = 1.3836$.

Treatment	Control	Glucose	Fructose	Glucose + Fructose	Sucrose
Mean	1.430	1.686	1.719	1.722	1.561
L_1 (t)	1.395	1.651	1.684	1.687	1.526
L_2 (t)	1.465	1.721	1.754	1.757	1.596
CI (t)	0.069	0.069	0.069	0.069	0.069
L_1 (Bon)	1.384	1.640	1.673	1.676	1.515
L_2 (Bon)	1.476	1.732	1.765	1.768	1.607
CI (Bon)	0.093	0.093	0.093	0.093	0.093
L_1 (D-S)	1.384	1.640	1.673	1.676	1.514
L_2 (D-S)	1.476	1.732	1.765	1.768	1.606
CI (D-S)	0.092	0.092	0.092	0.092	0.092

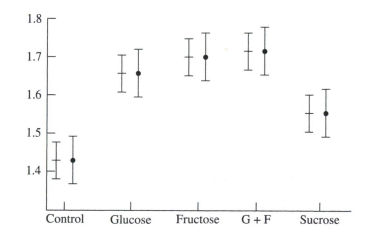

For each treatment, the conventional confidence interval is given at the left and its simultaneous confidence interval on the right. As expected, the simultaneous confidence intervals are broader in order to ensure that there is a 95% chance that they will all simultaneously include the true means for each treatment.

the first, second, and last tests the null hypothesis is rejected, whereas for the third and fourth contrasts is it accepted. We need not compare the adjusted P-values with any critical values. Furthermore, using and displaying adjusted P-values is more in tune with the current practice of presenting specific probabilities for each statistical test rather than mere inequalities.

The approaches we have just described have the desired property of controlling the probability of making any type I error at all in a family of k significance tests. As a consequence, each individual test is rather conservative. The power of the tests is thus low if more than one null hypothesis is false. Holm (1979) developed a sequential test that increases the power of the individual tests when the overall null hypothesis is rejected. Although the original proposal was based on the Bonferroni method, it can be applied to the Dunn-Šidák method as well. Both applications are shown in Box 9.6. The individual tests can be evaluated against critical values or expressed as adjusted P-values, as before. Applying this procedure, we decide that the null hypothesis can be rejected for the first three tests but not for the last two. Note, too, that for these data, the results do not differ from what we have already found. Nevertheless, the multiple comparison procedures are preferable because they are the correct tests when the tests exhibit nonorthogonality (which must be the case whenever more tests are made than there are degrees of freedom among groups).

Furthermore, the use of a simultaneous Bonferroni procedure for all tests is less powerful than the corresponding sequential method. Although in this dataset the results obtained by Holm's procedure lead to the same conclusions as before, his procedure may make a difference when the original results are borderline. Once again, we note that the Bonferroni results yield slightly higher P-values than those from Dunn-Šidák and hence are more conservative.

There is yet another statistical calculation in which multiple comparisons play a part. When we analyze a Model I anova, we may wish to calculate confidence limits for the means. For an entire family of confidence intervals, we again wish to maintain a familywide type I error rate of α. That means that we expect that 95% of the time all k confidence intervals should include the true parametric means for each group. We show how to do this in Box 9.6 for the pea-section data. Note that, as expected, the Bonferroni confidence intervals (0.93 transformed ocular units) are wider than those for the unadjusted confidence intervals (0.69 units). The Dunn-Šidák confidence intervals are only slightly less conservative than the Bonferroni intervals, but the difference is trivial (0.92 versus 0.93).

Earlier in this section, we learned how to analyze a Model I anova with more than two groups for which we wish to test the relations of subsets of groups. We can now furnish preliminary advice that might save you considerable time in analyzing such data. We saw that carrying out these tests on a comparison-wise basis, using per-comparison error rates, will yield the smallest probabilities and the shortest confidence intervals. However, these results would not be legitimate because we would not have allowed for the familywide error. At the other extreme are the Bonferroni tests, which furnish larger P-values and the widest confidence intervals. The Dunn-Šidák results are very similar but invariably yield slightly smaller probabilities and somewhat narrower confidence intervals than Bonferroni's formula.

We can use the two sets of extremes as bracketing values for the outcome of our analysis. If your data are not significant by the per-comparison tests—that is, the typical t- and F-tests—they cannot be significant by any more conservative test such as the Bonferroni method, and no further tests are necessary. By contrast, if all your tests reject the null hypothesis by Bonferroni tests, then any tests carried out by other, less conservative methods such as Dunn-Šidák or any of the tests enumerated in the next section will necessarily reject the null hypothesis as well, and again there is no need for further testing. It is only when the results are not as clear-cut as the extreme results we have just mentioned that we have to resort to other techniques presented in Section 9.5.

When considering multiple comparison techniques, we need to be aware of two desirable properties that these techniques ideally should possess (Gabriel and Sokal, 1969). The first is known as *coherence,* or *transitivity*. It states that if the null hypothesis of homogeneity is accepted for a set of means, then it should not be rejected for any subset thereof. Thus, if we computed samples for all groups in the analysis and accept the overall null hypothesis of homogeneity—meaning that all samples come from the same population—we should not then discover some subset of these means for which we have to reject that null hypothesis. Fortunately, almost all multiple comparison tests are coherent. The complementary property is *consonance*, which implies that if a set of means is heterogeneous, then at least one pair of means in the set must also be heterogeneous. Most multiple comparison procedures are *dissonant*—they may be heterogeneous overall without at least any one pair of means being heterogeneous as well.

In addition to Types I and II, per-comparison, and familywide error rates, readers may encounter mention of two other error rates. A directional error (also known as a type III error), is defined as the probability of misclassifying the sign of an effect. This can occur when μ_1, the parametric mean of a sample of observations, is located to one side of μ_0, the parameter defining the null hypothesis, whereas the sample mean by chance is located on the opposite side of μ_0. A directional familywide error rate is the probability that the sign of any tested effect is misclassified. For further details, see Westfall et al. (1999).

The second of these errors is the false discovery rate, defined by Benjamini and Hochberg (1995) as the expected proportion of falsely rejected null hypotheses among the set of rejected hypotheses in the family of hypotheses being tested. Obviously, one would like to make the false discovery rate as small as possible because one would like to minimize cases of mistaken rejections of the null hypothesis. It is easy to demonstrate that the false discovery rate is equal to or smaller than the familywide error rate, making it easier to control. These other error rates are often used in genetic studies in which many thousands of tests may be performed.

In the next section, we turn to a discussion of the various methods of multiple comparison that have been proposed for cases that fall between the two bracketing extreme outcomes discussed above.

9.5 Comparisons Among Means: Special Methods

The development of proper methods for testing hypotheses suggested by inspection of the results of analysis of variance has occupied the attention of statisticians since the 1950s. Several different approaches have been proposed, and a complete consensus has not yet been achieved on which method to employ when. There have been so many diverse approaches because the problem of testing all of the "interesting" differences—those differences among means that loom large in the observed sample—is subject to a variety of interpretations. The problem is the fact that these tests depend on other tests in complicated ways so that a general method cannot be developed. We furnish below methods for important special cases such as comparisons among all pairs of means or comparisons of all means against a particular mean (e.g., the control).

Toward the end of the previous section, we saw that it was profitable to examine the outcome of two sets of tests bracketing the extreme possible outcomes of multiple comparisons tests. These were testing contrasts as though they were independent (orthogonal) or subjecting them to Bonferroni tests (possibly of the sequential kind). In the former case, if we are not able to reject any of the null hypotheses of no difference, then no further testing of this dataset seems useful; in the latter case, if all contrasts are significant for this very conservative test, again, no further testing seems necessary, and we can declare all contrasts significant at the familywide error rate chosen. In other words, we reject the null hypotheses of all individual contrasts at the familywide error rate α for that many contrasts. However, what are we to do when the results fall somewhere in between these extreme outcomes? This section discusses such cases. Fortunately, the commonly used pairwise methods follow the same general outline, which we shall present below, leaving a brief mention of some alternative approaches for the end of the section.

Let us assume we have a family of independent orthogonal comparisons such as we encountered in the previous section. Although we tested these by F-tests there, we pointed out that they could equally well be tested by mathematically equivalent t-tests, which is what we will do here in order to construct a generally useful model for one-degree-of-freedom tests. The two groups in the contrast are represented by their means. For unequal sample sizes, the formula for such a test is

$$t_s = \frac{(\bar{Y}_1 - \bar{Y}_2) - (\mu_1 - \mu_2)}{\sqrt{\left[\left(\dfrac{1}{n_1} + \dfrac{1}{n_2}\right)MS_{\text{within}}\right]}} \tag{9.7}$$

For equal sample sizes, it simplifies to

$$t_s = \frac{(\bar{Y}_1 - \bar{Y}_2) - (\mu_1 - \mu_2)}{\sqrt{\dfrac{2}{n}MS_{\text{within}}}} \tag{9.8}$$

with the same number of degrees of freedom as possessed by the error mean square.

If we assume that $\mu_1 - \mu_2 = 0$ and set t_s equal to $t_{\alpha[\nu]}$, Expression (9.8) for equal sample sizes can be rearranged to read

$$|\bar{Y}_1 - \bar{Y}_2| = t_{\alpha[\nu]}\sqrt{\frac{2}{n}MS_{\text{within}}} = LSD \tag{9.9}$$

This quantity is called the **least significant difference (LSD)**. It should be obvious that the quantity LSD is the smallest difference between two means that will cause one to reject the null hypothesis of no difference. Any two means that differ by at least this amount, when put into Expression (9.7), will yield a t_s-value that will just yield a probability of α. The computation of an LSD is convenient because it is a single value that permits the investigator to look at any pair of means in a study and tell whether they are sufficiently different to allow one to reject the null hypothesis of no difference. In the *Drosophila* example of Box 9.5, we can compute

$$LSD = t_{.01[72]}\sqrt{\frac{2}{n}MS_{\text{within}}} = 2.646 \times 2.507 = 6.634$$

Thus, any two means that we planned to compare and that differ by 6.634 or more are significantly different from each other at $P \leq 0.01$. The LSD indicates that the nonselected line differs from either selected line in fecundity but that the two selected lines are not different from each other. However, as we have pointed out already, *such an approach is valid for independent comparisons only.*

To generalize the procedure for nonindependent comparisons, we rewrite the formula for the LSD as

$$MSD = (\text{critical value}) \times SE \tag{9.10}$$

where MSD stands for **minimum significant difference**, the *critical value* refers to the statistical distribution appropriate to the given test, and SE is the appropriate standard error. Most of the tests discussed below will be based on this general formula, but the critical values and the standard errors will vary from test to test.

We have organized our presentation of the multiple comparisons methods into two tables and one box. Table 9.5 is a guide to the 12 approaches featured in this book arranged by their designs. Note that the number k of possible comparisons increases from the top of the table downward. In its last column, we indicate location of a more detailed discussion for each specific method. For the sake of completeness, we have also included in Table 9.5 methods discussed in the preceding section. For the seven

TABLE 9.5 Designs for Multiple Comparisons Tests in Analysis of Variance.

Type of Multiple Comparisons	Number k of Comparisons	Method (and Additional Properties)	Described in
Orthogonal comparisons	$k \le (a - 1)$	Orthogonal decomposition of MS_{among}	Section 9.4
		Independent tests by LSD method	Section 9.5 & Table 9.6, row 1
All treatments vs. a control	$k = (a - 1)$	Dunnett's test	Section 9.5 & Table 9.6, row 2
Tests corresponding to confidence intervals	$k < a(a - 1)/2$	Dunn-Šidák method (independent tests)[1]	Section 9.4
		Bonferroni method (dependent tests)[1]	Section 9.4
All pairwise comparisons	$k = k*$ $= a(a - 1)/2$	GT2-method (independent tests, unequal n)	Section 9.5 & Table 9.6, row 5
		T-method (dependent tests, equal n)	Section 9.5 & Table 9.6, row 3
		Welsch method (dependent tests, equal n)	Section 9.5 & Table 9.6, row 4
		T'-method (dependent tests, nearly equal n)	Section 9.5 & Table 9.6, row 6
		Tukey-Kramer method (independent tests, unequal n)	Section 9.5 & Table 9.6, row 7
More than all pairwise comparisons	$k > a(a - 1)/2$	Scheffé method (all possible contrasts)	Section 9.5 & Box 9.8
	$k \gg a(a - 1)/2$	SS-STP (all possible sets of comparisons)	Section 9.5 & Box 9.9

[1]Sequential Dunn-Šidák or Bonferroni cannot be used to compute confidence intervals for mean differences, but power can be improved dramatically while retaining familywide error control.

tests that can be carried out using Expression (9.10), Table 9.6 furnishes locations of computational layouts, formulas for the critical values with their parameters, and the corresponding standard errors. We shall take up the multiple comparisons methods in the order in which they are presented in Table 9.6.

The first row of Table 9.6 is the least significant difference (*LSD*) that would apply to fully orthogonal single degree of freedom tests. These are based on Expression (9.9). An application of such a test to actual data, the *Drosophila* fecundity data of Box 9.5, has been shown above, immediately following that expression.

In numerous situations, especially those arising from the results of experiments, investigators are not interested in testing the differences among *all* pairs of treatment means. Rather, researchers would like to compare each treatment mean against a single control or standard mean. **Dunnett's test**, listed in the second row in Table 9.6, is such a test. It is based on *Dunnett's 1- and 2-sided range distribution* for equal and unequal sample sizes, respectively. We can use the pea-section data of Table 9.1 as an example, if we decide that the only question we wish to ask of these data is whether the effects of the four sugar treatments on growth of pea sections differ from that of the controls.

We present an outline of Dunnett's test in part I of Box 9.7. There will be four $(a - 1)$ separate tests of the five (a) means, one for each comparison of a sugar treatment with the control. To carry out Dunnett's test, we compute a column of the differences between the control mean and the four treatment means. For convenience, these differences are listed in an ascending order. In our example, the first in the column of four differences is that between control and sucrose, which equals -0.130, the smallest absolute difference.

If we have no idea in which direction the treatment effect will depart from the control effect, the two-sided test is appropriate. To carry out the Dunnett's test, we need to obtain the critical value $Q^{2S}_{a-1, \nu}$ by computer or by using Table 9.6. The computations are laid out in Box 9.7, part I. The resulting MSD^{2S} for the two-sided test is 0.061716. This quantity is subtracted from and added to each of the four mean differences to yield the lower and upper confidence limits for each difference. Thus, the first treatment mean difference from the control, -0.130, yields a lower limit L_1 of -0.192 and an upper limit L_2 of -0.068. Note that the confidence intervals of all four mean differences exclude zero (the upper and lower confidence limits are all negative), implying that the hypothesis of no difference can be rejected and that all sugar treatments have higher means than the control. Remember that the variables have been transformed so that enhancements of the coded variable by the sugars actually represent inhibition of growth by the sugars.

When there is only one direction of change that the investigator is interested in, we carry out one-sided tests, which are performed similarly. Such a situation might arise in the pea-section data if previous research or theoretical considerations had convinced the data analyst that sugars always raised the mean transformed variable. We now need to obtain the one-sided critical value $Q^{1S}_{a-1, \nu}$ by computer or by using Table 9.6. The computations are shown in Box 9.7, part I, steps **4** and **5**. The formula for the *MSD* is the same as before, except that $Q^{2S}_{a-1, \nu}$ is replaced by $Q^{1S}_{a-1, \nu}$. The resulting MSD^{1S} for the one-sided tests is 0.054181. This quantity is added to each of the four mean differences to yield the upper confidence limits for each difference. The first such limit turns out to be -0.076. Because none of the upper limits is positive, the results for the one-sided tests are the same as those for the two-sided tests. Limiting our consideration to one-sided tests has increased the power of the test noticeably, as is evident from the smaller *MSD* for the one-sided tests.

Rather than consider a new example for a Dunnett's test on an anova with unequal sample sizes n, we modify the familiar pea-section data to resemble a situation

TABLE 9.6 Multiple Comparisons Tests in Anova Employing an $MSD = $ (Critical Value) $\times$ SE Formulation: Computational Layouts, Critical Values, Their Parameters, and Appropriate Standard Errors

Method	Computational Layout	Critical Value	Sources and Parameters of Critical Values	SE
Independent tests by LSD method	Shown in text in Section 9.5	$t_{\alpha[\nu]}$	Table **B** $k \leq (a-1)$ $\nu = a(n-1)$	$SE1 = \sqrt{\dfrac{2}{n}MS_{within}}$
Dunnett's test (choices involve 1-sided vs. 2-sided tests, and equal or unequal sample sizes)	See Dunnett's test for all treatments vs. a control Box 9.7, part I	$Q^{1S\ or\ 2S}_{a-1,\nu}$	Tables **OO** and **PP** $k = (a-1)$ $\nu = a(n-1)$ or $\nu = \displaystyle\sum^{a}(n_i - 1)$	$SE1 = \sqrt{\dfrac{2}{n}MS_{within}}$ or $SE3 = \sqrt{\left(\dfrac{1}{n_i}+\dfrac{1}{n_j}\right)MS_{within}}$
T-method	See Ordered means and comparison limits, Box 9.7, part II	$Q_{\alpha[k,\nu]}$	Table **J** $k = a$ $\nu = a(n-1)$	$SE2 = \sqrt{\dfrac{1}{n}MS_{within}}$
Welsch method	See Ordered means tested sequentially, Box 9.7, part III	$Q_{\alpha(l)[k,\nu]}$	Table **K** $k = a$ $j = 2, 3, \ldots, k$ $\alpha(j) = \alpha j/k$ $\nu = a(n-1)$	$SE2 = \sqrt{\dfrac{1}{n}MS_{within}}$

Method				
GT2-method	See All pairs matrix layout, Box 9.7, part IV	$m_{\alpha[k^*,\nu]}$	Table **M** $k=a$ $k^*=a(a-1)/2$ $\nu=\sum^a (n_i-1)$	$SE3 = \sqrt{\left(\dfrac{1}{n_i}+\dfrac{1}{n_j}\right)MS_{within}}$
T'-method	See All pairs matrix layout, Box 9.7, part IV	$Q'_{\alpha[k,\nu]}$	Table **L** $k=a$ $\nu=\sum^a(n_i-1)$	$SE4 = \sqrt{\dfrac{1}{\min(n_i,n_j)}MS_{within}}$
Tukey-Kramer method	See All pairs matrix layout, Box 9.7, part IV	$Q_{\alpha[k,\nu]}$	Table **J** $k=a$ $\nu=\sum^a(n_i-1)$	$SE5 = \sqrt{\dfrac{1}{2}\left(\dfrac{1}{n_i}+\dfrac{1}{n_j}\right)MS_{within}}$

BOX 9.7 Computational Layouts for Multiple Comparison Methods Featured in Table 9.6

Part I. Dunnett's Test for All Treatments vs. a Control

Coded pea-section lengths from Table 9.1. Names of the control and sugar treatments have been abbreviated.

Difference	(1) $\bar{Y}_C - \bar{Y}_i$	(2) $l_i = (1) - MSD^{2S}$	(3) $u_i = (1) + MSD^{2S}$	(4) $l_i = (1) - MSD^{1S}$	(5) $u_i = (1) + MSD^{1S}$
Contr-Sug	−0.130	−0.192	−0.068*	−∞	−0.076*
Contr-Gluc	−0.256	−0.318	−0.194*	−∞	−0.202*
Contr-Fruct	−0.289	−0.351	−0.227*	−∞	−0.235*
Contr-(Gluc+Fruct)	−0.292	−0.354	−0.230*	−∞	−0.238*

1. Calculate the difference $\bar{Y}_C - \bar{Y}_i$ between the control mean and the $a - 1 = 4$ treatment means and array the differences in ascending or descending order. In column **(1)**, we have ordered them from least to most different.

2. For the two-sided test, evaluate the critical value $Q_{a-1,\nu}^{2S}$ as per Table 9.6. This is found in Statistical Table **PP**; some program packages compute the critical value directly. We look up the arguments $k = a - 1 = 4$ and $\nu = a(n - 1) = 45$, and, using harmonic interpolation, we find that the 95% critical value is 2.531. We multiply this critical value by the standard error $SE1$ shown in Table 9.6. MS_{within}, the mean square error within, is 0.002973 (see Table 9.1). $MSD^{2S} = Q_{a-1,\nu}^{2S} \times SE1 = 2.531 \times 0.024384 = 0.061716$.

3. In columns **(2)** and **(3)** we calculate l_i and u_i, the lower and upper confidence limits for the mean differences as $(1) \pm MSD^{2S}$. Because all four mean differences are negative and their 95% confidence intervals cover only negative values, we can conclude that we should reject the null hypothesis of no difference from the control mean for all sugar treatments.

4. For a one-sided test, we have to evaluate the one-sided critical value $Q_{a-1,\nu}^{1S}$. Its parameter values are found in Table 9.6, and it can be looked up in Statistical Table **OO** or obtained by computer program. We obtained 2.222 using the BIOMstat software. MSD^{1S} is obtained by a formula analogous to the one used in step **2**. $MSD^{1S} = Q_{a-1,\nu}^{1S} \times SE1 = 2.222 \times 0.024384 = 0.054181$.

Box 9.7 (continued)

5. In this one-sided test, we expect the treatments to have higher means than the control. To reject the null hypothesis $H_0: \mu_i \geq \mu_C$, we need cases in which $\bar{Y}_C - \bar{Y}_i$ is a positive number. For this reason, we are not really interested in the lower limit l_i and set the entire column (4) to $-\infty$. In column (5), we calculate u_i, the upper confidence limits for the mean differences as (1) + MSD^{lS}. All four upper limits are still negative, leading us to reject the null hypothesis for the four sugar treatments.

   For an example of Dunnett's test when sample sizes are unequal, we have modified the coded pea-section lengths from Table 9.1, as follows. Ten observations were randomly deleted from the data, resulting in a data table with unequal sample sizes. A table of the new means and sample sizes is given below, followed by an anova table for these data. The numbers in theses two tables furnish all the information required to carry out Dunnett's test of treatments against a control, when sample sizes are unequal.

New Means and Sample Sizes

Treatments	C	G	F	G+F	S
$\bar{Y}_i$	1.446	1.686	1.716	1.730	1.567
n_i	8	10	9	6	7

New Anova Table

Source	df	MS	F_s	P
Treatments	4	0.11375	38.2103	2.587×10^{-12}
Within	35	0.00298		

6. The computations for two-sided and one-sided Dunnett's tests of unbalanced anovas (those with unequal sample sizes) are carried out similarly to the procedures described above for equal sample sizes with only two changes. First, the critical values must be directly computed. We obtained $Q_{a-1,\nu}^{2S} = 2.53265$ and $Q_{a-1,\nu}^{1S} = 2.24251$ using the BIOMstat software. Second, the standard error changes to $SE3$ as shown in Table 9.6. This standard error is a function of n_i and n_j, the sizes of the samples being compared. As a result, a separate standard error has to be calculated for each comparison; consequently, there will be a different MSD for each comparison as well.

Box 9.7 (continued)

Dunnett's Test with Unequal Sample Sizes

Difference	(1) $\bar{Y}_C - \bar{Y}_i$	(2) n_i, n_j	(3) MSD_{ij}^{2S}	(4) $l_i = (1) - (3)$	(5) $u_i = (1) + (3)$	(6) MSD_{ij}^{1S}	(7) $l_i = (1) - (6)$	(8) $u_i = (1) + (6)$
C-S	-0.121	8, 7	0.071554	-0.192	-0.049	0.063356	-∞	-0.077
C-G	-0.240	8, 10	0.065580	-0.305	-0.174	0.058067	-∞	-0.202
C-F	-0.269	8, 9	0.067180	-0.336	-0.202	0.059484	-∞	-0.235
C-(G+F)	-0.284	8, 6	0.074666	-0.358	-0.209	0.066112	-∞	-0.238

Part II. Ordered Means and Comparison Limits (the T-Method)

Coded pea-section lengths from Table 9.1. Names of the control and sugar treatments have been abbreviated.

1. The formula for the *MSD* is $Q_{\alpha[k,\nu]} \times SE2$. The critical value $Q_{\alpha[k,\nu]}$ is the studentized range found in Table **J** for $k = a = 5$ and $\nu = a$ $(n - 1) = 45$. Harmonic interpolation in Table **J** yields $Q_{.05[5,45]} = 4.018$. Standard error $SE2 = (MS_{within}/n)^{\frac{1}{2}} = (0.00297/10)^{\frac{1}{2}} = 0.01723$. Therefore, $MSD = 0.06923$. We can reject the null hypothesis of no difference for any pair of means $\bar{Y}_i$ and $\bar{Y}_j$ at the familywide error rate α if and only if the difference between the sample means equals or exceeds the critical difference *MSD*. An efficient arrangement for testing is to array the means by their magnitudes as shown in the table below. It is obvious that once we have established that the absolute difference between $\bar{Y}_1$ and $\bar{Y}_2$ ($= 0.131$) is greater than the $MSD = 0.06923$, all other means will be even more different from mean $\bar{Y}_1$; hence, the null hypotheses will be rejected at $P \leq 0.05$. Conversely, if a given range of means is less than the *MSD*, as is the case for $|\bar{Y}_3 - \bar{Y}_5|$, there is no point testing any range within this span such as the means of treatments 3 and 4. This simplifies testing considerably, and we arrive at the following conclusions: The control mean is less than that of any of the sugars, and the mean of the disaccharide sucrose is less than the means of the other three (monosaccharide) sugar treatments, which do not differ from each other.

2. A convenient method for systematically testing all $a(a - 1)/2$ pairs of means is to compute upper and lower comparison limits for each mean such that the null hypothesis for two means can be rejected with confidence (i.e., the two means differ) if and only if their limits do not overlap. The limits (Gabriel, 1978) are simply

Box 9.7 (continued)

$$l_i = \bar{Y}_i - \tfrac{1}{2} MSD = \bar{Y}_i - 0.03462$$

$$u_i = \bar{Y}_i + \tfrac{1}{2} MSD = \bar{Y}_i + 0.03462$$

Because the *MSD* above was calculated for $\alpha = 0.05$, these comparisons limits have a familywide error rate of 0.05. The ordered treatment means and their limits are given in the table below.

Treatment Rank Number	Sugar	$\bar{Y}_i$	l_i	u_i
1	C	1.430	1.395	1.465
2	S	1.561	1.526	1.596
3	G	1.686	1.651	1.721
4	F	1.719	1.684	1.754
5	G + F	1.722	1.687	1.757

Treatment means whose comparison intervals overlap (for example, glucose, fructose, and glucose + fructose) are not different at the family-wide error rate of $\alpha = 0.05$. Those whose intervals do not overlap are different. We can display the results conveniently by plotting the treatment means and their comparison intervals, as shown in the figure below. Clearly, control and sucrose means differ from each other and from all other means at the 5% familywide error rate. The glucose, fructose, and glucose + fructose means do not differ from each other.

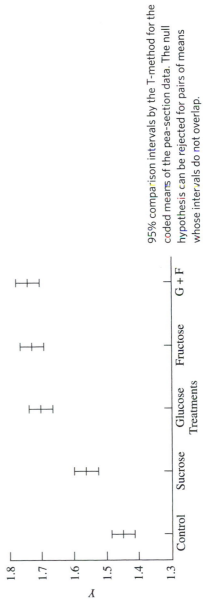

95% comparison intervals by the T-method for the coded means of the pea-section data. The null hypothesis can be rejected for pairs of means whose intervals do not overlap.

Box 9.7 (continued)

Part III. Ordered Means Tested Sequentially (the Welsch Method)

Coded pea-section lengths from Table 9.1. Names of the control and sugar treatments have been abbreviated.

1. The Welsch step-up procedure employs minimum significant ranges (MSR_j) in a manner analogous to minimum significant differences (MSD). We compute $MSR_j = Q_{\alpha_j[k,\nu]} \times SE2$. The value of $SE2$ is calculated as for the T-method in step 1 of part II, where we find $SE2 = 0.01723$. The standard error is based on $a(n-1) = 45$ degrees of freedom (the df of MS_{within}).

2. Next evaluate $Q_{\alpha_j[k,\nu]}$ for subranges of $j = 2, 3, \ldots, k$ means. The critical value $Q_{\alpha_j[k,\nu]}$ is the studentized range tabled for nonstandard probabilities $\alpha_j = \alpha j/k$ (except for $j = k-1$, where $\alpha_{k-1} = \alpha$). Its value can be found in Statistical Table **K**. Harmonic interpolation is necessary for $\nu = 45$. For a familywide error rate $\alpha = 0.05$, $\nu = 45$ degrees of freedom, and k set to equal $a = 5$, the α_j's, $Q_{\alpha_j[k,\nu]}$'s, and MSR_j's are the following:

$j =$	2	3	4	5
α_j	0.02	0.03	0.05	0.05
$Q_{\alpha_j[k,\nu]}$	3.41	3.77	3.77	4.07
MSR_j	0.0589	0.0650	0.0651	0.0701

The group means are then arrayed in order of magnitude:

C	S	G	F	G+F
$\bar{Y}_1$	$\bar{Y}_2$	$\bar{Y}_3$	$\bar{Y}_4$	$\bar{Y}_5$
1.430	1.561	1.686	1.719	1.722

The sequence of steps for testing is as follows:

1. Compare ranges of adjacent pairs of means with MSR for $j = 2$, which equals 0.0589.

Box 9.7 (continued)

$$\bar{Y}_5 - \bar{Y}_4 = 1.722 - 1.719 = 0.003 < 0.0589$$

$$\bar{Y}_4 - \bar{Y}_3 = 1.719 - 1.686 = 0.033 < 0.0589$$

$$\bar{Y}_3 - \bar{Y}_2 = 1.686 - 1.561 = 0.125 > 0.0589$$

$$\bar{Y}_2 - \bar{Y}_1 = 1.561 - 1.430 = 0.131 > 0.0589$$

We do not test any wider range that includes the significant ranges $(\bar{Y}_3 - \bar{Y}_2)$ and $(\bar{Y}_2 - \bar{Y}_1)$ because any range including a significant subrange is declared significant by this test.

2. Compare the range of three means with MSR for $j = 3$, which equals 0.0650. For example, $\bar{Y}_5 - \bar{Y}_3 = 1.722 - 1.686 = 0.036 < 0.0650$, so this group is considered homogeneous.

Our conclusions can be summarized as $\mu_1 < \mu_2 < \mu_3 = \mu_4 = \mu_5$. Conclusions are often shown diagrammatically by underlining the sets of means for which the null hypothesis cannot be rejected.

C	S	G	F	G+F
$\bar{Y}_1$	$\bar{Y}_2$	$\bar{Y}_3$	$\bar{Y}_4$	$\bar{Y}_5$
1.430	1.561	1.686	1.719	1.722

Part IV. All Pairs Matrix Layout (for GT2, T', and Tukey-Kramer Tests)

All three tests can be carried out using the same general design. The critical values and standard errors of the three methods differ. Note that the formulas for the standard errors involved—$SE3$, $SE4$, and $SE5$—all contain the terms n_i and n_j. Therefore, standard errors will not only vary with the test employed but also, unlike the balanced anovas in parts I–III, each comparison of two means will potentially involve a standard error with a numerical value unique to that pair of means. This property will result in a more complex layout than heretofore in this box.

Box 9.7 (continued)

Head Widths (in Millimeters) of a Species of Tiger Beetle (*Cicindela circumpicta*) Collected from 8 Localities

Localities (ordered by magnitude of mean)

	Barnard, Kansas	Roswell, New Mexico	Nebraska	Talmo, Kansas	Okeene, Oklahoma	Kackley, Kansas	Mayfield, Oklahoma	Stafford Co., Kansas
$\bar{Y}$	3.5123	3.6940	3.7223	3.7229	3.7376	3.7479	3.7843	3.8451
n_i	20	20	20	14	10	15	11	20
s_i^2	0.0357	0.0445	0.0135	0.0284	0.0204	0.0184	0.0338	0.0157

SOURCE: Selected data from H. L. Willis (unpublished results).

MS_{within} is 0.02645 with $\nu = 122$ df. See text for mode of computation. We find it convenient to arrange the layout to carry out the simultaneous computation of all three tests. Readers may wish to employ the results of only one test, basing their choice on the properties of their specific datasets. However, we prefer to carry out all three tests for reasons given in the text.

The basic computations are fundamentally similar to those described earlier in this box; that is, they are a calculation of an *MSD* as the product of a critical value with a standard error, followed by comparisons of the *MSD* with the observed differences of means. The columns in the partially reproduced table below list the steps necessary to obtain the desired results. Column (**1**) indicates the numerical values of *j* and *k*, the pairs of means being compared. Columns (**2**) and (**3**) furnish sample sizes for means *j* and *k*, respectively. Columns (**4**) to (**6**) are the *MSD*s for the three methods, which are computed as the critical value times the correct standard error, *SE3* through *SE5*, respectively. The critical values are named in Table 9.6 with an indication of their sources and the standard errors. Columns (**7**) and (**8**) furnish the actual numerical values of the means *j* and *k*, and column (**9**) is the absolute difference between them. Finally, columns (**10**) through (**12**) receive an asterisk whenever the absolute difference between a pair of means is greater than the corresponding *MSD*.

Box 9.7 (continued)

Note that the results given below indicate that mean 1 differs from all other means, but that these other means do not differ from each other.

(1)	(2)	(3)	(4)	(5)	(6)	(7)	(8)	(9)	(10)	(11)	(12)
$i{:}j$	n_i	n_j	$MSD(GT2)$	$MSD(T')$	$MSD(T\text{-}K)$	$\bar{Y}_i$	$\bar{Y}_j$	$|\bar{Y}_i - \bar{Y}_j|$	$GT2$	T'	$T\text{-}K$
1:2	20	20	0.163647	0.158621	0.158621	3.5123	3.6940	0.1817	*	*	*
1:3	20	20	0.163647	0.158621	0.158621	3.5123	3.7223	0.2100	*	*	*
1:4	20	14	0.180330	0.189589	0.174792	3.5123	3.7229	0.2106	*	*	*
1:5	20	10	0.200426	0.224325	0.194271	3.5123	3.7376	0.2253	*	*	*
1:6	20	15	0.176759	0.183160	0.171331	3.5123	3.7479	0.2356	*	*	*
1:7	20	11	0.194257	0.213885	0.188292	3.5123	3.7843	0.2720	*	*	*
1:8	20	20	0.163647	0.158621	0.158621	3.5123	3.8451	0.3328	*	*	*
2:3	20	20	0.163647	0.158621	0.158621	3.6940	3.7223	0.0283			
2:4	20	14	0.180330	0.189589	0.174792	3.6940	3.7229	0.0289			
2:5	20	10	0.200426	0.224325	0.194271	3.6940	3.7376	0.0436			
⋮											
6:7	15	11	0.205425	0.213885	0.199116	3.7479	3.7843	0.0364			
6:8	15	20	0.176759	0.183160	0.171331	3.7479	3.8451	0.0972			
7:8	11	20	0.194257	0.213885	0.188292	3.7843	3.8451	0.0608			

that occurs frequently in experimental research. Even though investigators may design a study with an equal number of replications per group or sample, they may not achieve that goal for a variety of natural or accidental reasons. We have pretended that 10 of the 50 dishes prepared at the beginning of the experiment were lost for one reason or another, leaving us with only 40 replicates. Ten readings were randomly removed from the study, resulting in sample sizes of 8, 10, 9, 6, and 7 for groups C, G, F, G + F, and S, respectively. The means and sample sizes of these modified data are shown in Box 9.7, part I. The average within-groups variance is almost identical to the previous one, 0.00298 as compared to 0.00297 for the full dataset. Owing to the smaller sample sizes, one expects a reduction in the power of the tests, and thus we are not surprised that the F_s-value decreased from 53.8846 to 38.2103. Nevertheless, the null hypothesis of homogeneity of the sugar means is still rejected very strongly.

The next quantity required is the critical value C_α for Dunnett's test of means with unequal n. Programs that feature the test for unequal n should furnish the critical value. Because different anovas will vary in their distributions of sample sizes, it is difficult to create a useful table to look up such values. We have obtained estimates for the one-sided and two-sided critical values using the BIOMstat software.

Because the standard error changes for every mean compared to the control, there is a different *MSD* for each mean, which must be employed for estimating confidence intervals. In other respects, the computation is the same as in the equal n case already discussed. Inspection of the layout for this test in Box 9.7, part I, together with the discussion of this case in step **6,** should enable the reader to reproduce the results obtained by us. The latter agree with the results obtained from the complete dataset. Clearly, all sugars have higher transformed means than the control whether tested using a one-sided or two-sided null hypothesis.

If we wish to test all pairwise comparisons for a balanced design, the most common method is the **T-method**, listed in Row 3 of Table 9.6. It is named after the late John Tukey, a leading statistician of the second half of the 20th century. It is applied to balanced datasets (equal sample sizes n). The *MSD* (minimum significant difference) is computed as $Q_{\alpha[k,\nu]} \times SE2$, where $Q_{\alpha[k,\nu]}$ is the studentized range from Statistical Table **J**. The *MSD* is 0.06923. The null hypothesis of no difference can be rejected for all observed differences equal to or greater than this quantity, with a familywide Type I error rate of α. An efficient way of calculating the results of the test is furnished in step **1** of part II of Box 9.7. An alternative approach, yielding mathematically identical results, is by calculating and graphing comparison limits as shown in step **2** of part II. The final result by either approach is that, at a 5% familywide error rate, control and sucrose differ from each other and from all other means. These monosaccharide sugar means do not appear to differ from each other.

Row 4 of Table 9.6 lists an alternative technique for testing all differences between pairs of means in a balanced dataset, the **Welsch step-up procedure** (Welsch, 1977). It employs a stepwise approach in which the means are arrayed in order of magnitude. Then pairs of adjacent means are tested. Only if the null hypothesis is not rejected for a given pair does one proceed to test larger subsets of adjacent means of size j that include the pair. We, therefore, assume coherence or transitivity in our data (see Section 9.4). The procedure differs from others discussed in this section in that the error rate for each test is adjusted to reflect the size of the subset being tested. The adjustment is computed as $\alpha_j = \alpha(j/k)$, where α_j is the type I error for testing a range of j adjacent means.

The test is illustrated in part III of Box 9.7, where we first evaluate adjusted critical values before testing the array of means. The Welsch procedure has greater power than the T- and GT2-methods discussed in this section. This method does not, however, permit the construction of simultaneous confidence intervals to differences or of graphical tests by comparison limits. Because the T- and GT2-methods are very simple and fast, many workers prefer them to the Welsch test, even though they are less powerful. The BIOMstat computer program carries out the Welsch procedure.

We now turn to an example with unequal sample sizes. This consists of measurements of head widths (in millimeters) of a species of tiger beetle at eight localities in the Midwest. These are shown in Box 9.7, part IV. These data, in a format quite common in scientific publications, are not ready for multiple comparisons analysis without a preliminary analysis of variance to determine whether there are overall differences among the group means and, also, to obtain the variance within groups, MS_{within}, which is needed for computing the standard errors. Although the original observations are not available, we do have the variances for each sample, so we can calculate

$$MS_{\text{within}} = \frac{\sum\limits^{a} (n_i - 1)s_s^2}{\sum\limits^{a} (n_i - 1)} =$$

$$\frac{(20 - 1)0.0357 + (20 - 1)0.0445 + \ldots + (20 - 1)0.0157}{19 + 19 + \ldots + 10 + 19} = 0.02645$$

Three techniques for testing all differences between pairs of means have been widely recommended: the **GT2-method** (Hochberg, 1974), the **T'-method** (Spjøtvoll and Stoline, 1973), and the **Tukey-Kramer procedure** (Dunnett, 1980). The three differ in the distributions that they employ for their critical values and in their standard errors. The GT2-method employs the **studentized** maximum modulus distribution m as a critical value, the T'-method is based on the studentized **augmented range** distribution Q', and the Tukey-Kramer procedure employs the studentized range distribution Q.

The **studentized maximum modulus** is the maximum over k comparisons of the quantity $|\overline{Y}_i - \mu|/s_{\overline{Y}}$. The $\overline{Y}_i$ are assumed to be normally distributed, with a mean of μ and standard deviation $\sigma_{\overline{Y}}$. Critical values of these distributions are given in Statistical Tables **M** and **L**, respectively. The procedures are explained in detail in Box 9.7, part IV. For GT2 in the tiger beetle example, $a = 8$, hence $k^* = 8 \times 7/2 = 28$, and $\nu = 122$, the degrees of freedom pertaining to the MS_{within}. For a 5% familywide error rate, we use the tabled value $m_{.05[28,120]} = 3.183$ as a conservative approximation of the desired value of $m_{.05[28,122]}$. When the required degrees of freedom are not as close to a tabled value of ν as in this case, it is recommended that the average between the linearly and harmonically interpolated values of m be employed. Note that the standard error of difference between any pair of means includes the pooled variance within groups but with divisors using the sample sizes for both means. For this reason, there is a separate standard error for each pair of means. For example, to calculate the SE_{14}, we evaluate $[(1/n_1 + 1/n_4) MS_{within}]^{\frac{1}{2}} = [(1/20 + 1/14) 0.026451]^{\frac{1}{2}} = 0.056674$. Then $MSD_{14} = m_{.05[28,120]} SE_{14} = 3.183(0.056674) = 0.1804$.

We now examine the absolute difference $|\overline{Y}_1 - \overline{Y}_4| = 0.2106$ to see whether it is $\geq MSD_{14}$. Because $0.2106 > 0.1804$, we reject the null hypothesis of no difference between the means for samples 1 and 4 and conclude that these two means differ at a familywide error rate of 0.05. It is convenient to arrange the MSD values in tabular form for each and every pair of means and pair these with the absolute differences between the respective pair of means. When we do this, we conclude that sample 1 from Barnard, Kansas, differs from each of the other seven samples, but these others do not differ from each other.

Alternatively, we may employ a simple approximate method proposed by Gabriel (1978), which is to compute the lower and upper comparison limits (l, u) for each sample mean such that the null hypothesis of no difference between two means is rejected if and only if their intervals do not overlap. For the ith mean, the limits are simply

$$l_i = \overline{Y}_i - \sqrt{\frac{1}{2}} \times m_{\alpha[k^*,\nu]} \, s_{\overline{Y}_i}$$

$$u_i = \overline{Y}_i + \sqrt{\frac{1}{2}} \times m_{\alpha[k^*,\nu]} \, s_{\overline{Y}_i}$$

We compute $s_{\overline{Y}_i}$ as $[MS_{within}/n_i]^{\frac{1}{2}}$, basing the standard error on the pooled mean square within groups. For the first mean, these limits are

$$l_1 = 3.5123 - \sqrt{\frac{1}{2}} (3.183)(0.02645/20)^{\frac{1}{2}} = 3.4304$$

$$u_1 = 3.5123 + \sqrt{\frac{1}{2}} (3.183)(0.02645/20)^{\frac{1}{2}} = 3.5942$$

For sample mean 4, the limits, similarly computed, are

$$l_4 = 3.6251$$

$$u_4 = 3.8207$$

We note that the comparison intervals of the two means do not overlap, which indicates that the two means differ. After we have computed comparison intervals for all eight sample means, it is convenient to graph them as shown in Figure 9.2. We note that only the limits for locality 1 do not overlap those of the other localities, all of

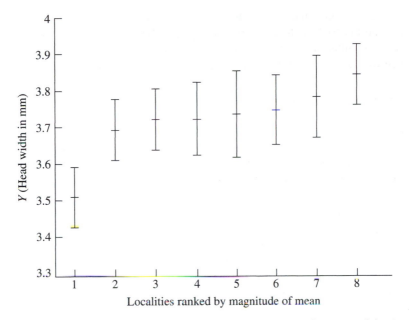

FIGURE 9.2 The 95% comparison intervals by the GT2-method for the means of the tiger beetle data. Means whose intervals do not overlap are significantly different.

which overlap mutually. Thus, we conclude that the mean for locality 1 is different from the others, which do not differ from one another.

The T′-method is carried out using an algorithm analogous to that of the GT2-procedure. Only the critical values and the standard errors differ. We interpolate in Statistical Table **L** to obtain the critical value $Q'_{.05[8,122]}$ as 4.3617. The standard error $SE4$ for comparing means 3 and 4 with sample sizes 20 and 14, respectively, evaluates to $(0.02645/14)^{\frac{1}{2}} = 0.04347$. The MSD for comparing means 3 and 4 is 0.189589, whereas the absolute difference between these means is only 0.0006. Clearly, we cannot reject the null hypothesis of no difference between these two means. The results of all pairwise T′-tests agree with those of the GT2-tests: the sample from Barnard, Kansas, differs from all other samples, which do not differ among themselves. The T′-method can also be done approximately by comparison limits constructed by the following formulas:

$$l_i = \overline{Y}_i - \tfrac{1}{2}Q'_{\alpha[k,\nu]}s_{\overline{Y}}$$

$$u_i = \overline{Y}_i + \tfrac{1}{2}Q'_{\alpha[k,\nu]}s_{\overline{Y}}$$

The **Tukey-Kramer procedure,** a third, older method for unequal sample sizes, was revived by Dunnett (1980) following a simulation study. It employs the student-ized range but uses average sample sizes in the standard errors. Dunnett's results suggest that the T′- and GT2-methods are too conservative, whereas the Tukey-Kramer method yields results that, while still conservative, are much closer to the intended significance level α. The Tukey-Kramer method is also given in Box 9.7, part IV. The algorithm is similar to that used above. The critical value $Q_{0.05[8,122]} = 4.3617$, and the *MSD* for means 3 and 4 is 0.174792. The results agree with those found using the GT2 and T'-methods.

When in doubt about which method to use, carry out all three procedures and employ the one that yields the smallest *MSD*. Choice of the method that gives the best result (i.e., the smallest *MSD*) is valid because the decision is not based on a consideration of sample statistics but only on the sample sizes. When the sample sizes are nearly equal, the T'-method is expected to yield the smallest *MSD*. The BIOMstat computer program carries out all three methods of multiple comparisons simultaneously.

When we presented the Bonferroni techniques earlier, we showed that a family of simultaneous confidence intervals to the group means in an anova has to take the familywide error rate into consideration. This issue comes up in data suitable for the GT2-method as well. To calculate simultaneous confidence intervals for all eight means in the tiger beetle example, we need to evaluate $m_{\alpha[k^*, \nu]}$ for $k^* = a$ because we only need to set $a = 8$ confidence intervals. The confidence limits are

$$L_{1(i)} = \overline{Y}_i - m_{\alpha[a,\nu]}\, s_{\overline{Y}_i} \text{ and } L_{2(i)} = \overline{Y}_i + m_{\alpha[a,\nu]}\, s_{\overline{Y}_i}$$

The value of $m_{\alpha[a,\nu]}$ can be looked up in Statistical Table **M** for $k^* = 8$ and $\nu = 120$, which is a close approximation to the correct value of 122. Because there is no tabled value for $k^* = 8$, we have to obtain it by the process of double interpolation explained in the introduction to the table. We obtain $m_{.05[8,120]} = 2.772$, and the confidence limits to the first mean of the tiger beetle data are 3.4115 to 3.6131. We show the results for all eight means in Figure 9.3. They appear similar to the comparison limits from Figure 9.2, but the two kinds of intervals are interpreted differently. The **simultaneous** or **multiple confidence intervals** have a probability of 95% that they all enclose their true parametric group means. By contrast, the **comparison intervals** are chosen such that nonoverlap between two intervals indicates (approximate) rejection of the null hypothesis of no difference between the two group means, again with a familywide error rate of 5% for $a(a-1)/2$ comparisons.

In addition to testing pairs of means, the investigator may wish to test certain *contrasts among means* as suggested by the results of the experiment. To carry out such tests, we employ the coefficients of linear comparisons introduced in the previous section. In Box 9.8, we feature three approaches—the Scheffé method, the T-method, and the GT2-method. The Scheffé method is most powerful when a and ν are both small and most coefficients of linear comparisons c_i are nonzero. With equal or nearly equal sample sizes, the T-method will be the best method, while in other cases the GT2-method should be employed. Borderline cases may again require trying all three methods to find the most sensitive one. The computations are not difficult and are laid out in detail in Box 9.8.

The **Scheffé method**, as carried out in Box 9.8, step **2a,** uses the same critical SS as the **sum of squares simultaneous test procedure** (*SS-STP*) described next.

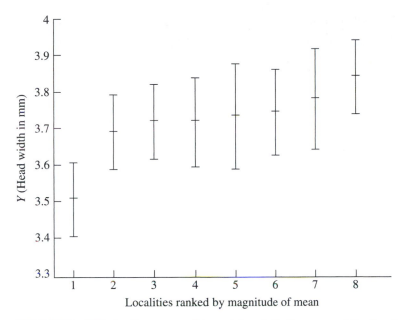

FIGURE 9.3 95% simultaneous confidence intervals for the means of the tiger beetle data using the GT2-method.

More than just pairs of means and contrasts between averages based on two groups of means may be of interest to an investigator examining the results of an analysis of variance. Do certain partitions of the set of means show more variation than one would expect by chance? To return to the pea-section example, it seems that the effects of the control differs from sucrose, which in turn differs from the three other sugar treatments; but these three seem homogeneous—that is, they do not differ.

Many hypotheses similar to the one just stated could be tested. *SS-STP* is a conservative method that allows one to test *all possible sets of comparisons* among means with a familywide error rate α. It is an extension of the Scheffé method, the relation between the Scheffé test and *SS-STP* being analogous to that between the *t*-test and the *F*-test. "All possible sets of comparisons" subsumes all pairs of means, all possible contrasts between groups of means, and all possible partitions of the a means—which is $2^a - a - 1$, a large number of tests indeed.

The critical value for the *SS-STP* method can be developed as follows. In a single-classification anova, a null hypothesis is rejected if

$$\frac{MS_{\text{among}}}{MS_{\text{within}}} \geq F_{\alpha[a-1,a(n-1)]} \tag{9.11}$$

Because $MS_{\text{among}}/MS_{\text{within}} = SS_{\text{among}}/[(a-1)MS_{\text{within}}]$, we can rewrite Expression (9.11) as

$$SS_{\text{among}} \geq (a-1)MS_{\text{within}}F_{\alpha[a-1,a(n-1)]} \tag{9.12}$$

BOX 9.8 **Simultaneous Confidence Limits and Tests of Contrasts Among Means Using the Scheffé, T-, and GT2-Methods**

Length of pea sections grown in tissue culture. Data from Table 9.1.

1. Set up the desired contrasts. For example, in the pea-section data the mean for sucrose is considerably higher than the average of the other sugars. The contrast would be as follows. (For convenience the coefficients c_i have been scaled so that their sum in each direction equals unity, but the coefficients could have been written 0, 1, 1, 1, -3.)

	Treatments ($a = 5$)				
	Control	2% Glucose Added	2% Fructose Added	1% Glucose + 1% Fructose Added	2% Sucrose Added
$\bar{Y}$	1.430	1.686	1.719	1.722	1.561
n_i	10	10	10	10	10
c_i	0	$\dfrac{1}{3}$	$\dfrac{1}{3}$	$\dfrac{1}{3}$	-1

$$\sum n_i c_i = 0$$

The contrast between the means is C, computed as follows:

$$C = \sum c_i \bar{Y}_i = 0 + \tfrac{1}{3}1.686 + \tfrac{1}{3}1.719 + \tfrac{1}{3}1.722 - 1(1.561) = 0.148$$

2. For simultaneous confidence limits and tests of significance with familywide error, the most efficient method depends on a, the number of means, ν, the degrees of freedom of MS_{within}, and the particular contrast being tested. Generally the Scheffé procedure will be the most powerful (yield the shortest confidence intervals) for small values of a and ν and contrasts with nonzero c_i for most of the means. The T-method will be best if the sample sizes are equal or nearly equal. In other cases, the GT2-method (or its Gabriel approximation) should be used. To be certain, one has selected the optimal method (with the shortest interval); all three methods should be tried in critical studies.

2a. Scheffé method

The simultaneous $100(1 - \alpha)\%$ confidence limits for a contrast $C = \sum c_i \bar{Y}_i$ are

$$C \pm [(a - 1) F_{\alpha[a-1, \nu_2]}]^{1/2} s_c$$

where

$$s_c = \left(MS_{within} \sum \frac{c_i^2}{n_i} \right)^{1/2}$$

In this example,

$$s_c = \left\{ 0.002973 \left(0 + \frac{(\tfrac{1}{3})^2}{10} + \frac{(\tfrac{1}{3})^2}{10} + \frac{(\tfrac{1}{3})^2}{10} + \frac{(-1)^2}{10} \right) \right\}^{1/2}$$

$$= [0.002973(0.13333)]^{1/2} = 0.01991$$

Box 9.8 (continued)

Also, $F_{.05[4, 45]} = 2.579$, using program BIOMstat.

Then the limits to $C = 0.148$ are

$$0.148 \pm [(5 - 1)2.579]^{1/2}0.01991 = 0.148 \pm 0.063948$$

Because the limits (0.08405, 0.21195) exclude zero, C is different from zero.

2b. *The T-method*

The simultaneous $100(1 - \alpha)\%$ confidence limits for a contrast are

$$C \pm Q_{\alpha[a,\nu]}s_c$$

Note that the critical value is Q, the studentized range from Statistical Table **J**. The standard error is computed as follows: Let c_i^+ and c_i^- denote those c_i's that are positive and negative, respectively. Then

$$s_c = [MS_{\text{within}}]^{1/2} \max \left\{ \sum \frac{c_i^+}{\sqrt{n_i}}, -\sum \frac{c_i^-}{\sqrt{n_i}} \right\}$$

In our example

$$s_c = (0.002973)^{1/2} \max \left\{ \frac{\frac{1}{3}}{\sqrt{10}} + \frac{\frac{1}{3}}{\sqrt{10}} + \frac{\frac{1}{3}}{\sqrt{10}}, -\frac{(-1)}{\sqrt{10}} \right\}$$

$$= 0.05453 \max \{0.31623, 0.31623\} = 0.017243$$

After obtaining $Q_{\alpha[a,\nu]}$ by interpolating in Table **J**, the 95% limits are computed as

$$0.148 \pm 4.018(0.017243) = 0.148 \pm 0.06928$$

These limits (0.07872 to 0.21728) are slightly wider than those found using the Scheffé method. When a and ν are not small, one expects the Scheffé limits to be wider and hence to provide a less powerful test.

2c. *The GT2-method*

The simultaneous $100(1 - \alpha)\%$ confidence limits for a contrast, C, are

$$C \pm m_{\alpha[k^*,\nu_2]}s_c$$

where the standard error s_c is computed as follows. (As in the T-method let c_i^+ and c_i^- denote the positive and negative coefficients, respectively.)

$$s_c = MS_{\text{within}}^{1/2} \frac{\sum c_i^+ |c_j^-| \left(\frac{1}{n_i} + \frac{1}{n_j} \right)^{1/2}}{\frac{1}{2} \sum |c_i|}$$

The summation in the numerator is over all possible combinations of a positive c_i with a negative c_j. The summation in the denominator is over all c_i,

$$s_c = \sqrt{0.002973} \frac{\frac{1}{3}(1)(\frac{1}{10} + \frac{1}{10})^{1/2} + \frac{1}{3}(1)(\frac{1}{10} + \frac{1}{10})^{1/2} + \frac{1}{3}(1)(\frac{1}{10} + \frac{1}{10})^{1/2}}{\frac{1}{2}(0 + \frac{1}{3} + \frac{1}{3} + \frac{1}{3} + |-1|)}$$

$$= 0.054525 \frac{(\frac{1}{10} + \frac{1}{10})^{1/2}}{\frac{1}{2}(2)} = 0.024384$$

Box 9.8 (continued)

After $m_{\alpha[k^*,\nu]}$ is obtained from Table **M** as 2.935, the 95% limits are computed as

$$0.148 \pm 2.935(0.024384) = 0.148 \pm 0.071567$$

These limits (0.07643, 0.21957) are even wider than the limits obtained by the T-method, which is to be expected because the sample sizes are equal.

3. Gabriel (1978) proposed a simple approximation to the T or GT2 confidence limits for a contrast making use of the comparison limits given in Box 9.7. The limits are

$$l = \sum c_i^+ l_i - \sum |c_i^-| u_i$$

and

$$u = \sum c_i^+ u_i - \sum |c_i^-| l_i$$

where c_i^+ and c_i^- are the positive and negative coefficients of linear comparison, respectively, and l_i and u_i are the lower and upper comparison limits as shown in Box 9.7, parts II and IV. In our example, because the sample sizes are equal, we apply the T-method and make use of the comparison limits from Box 9.7, part II. We obtain

$$l = [\tfrac{1}{3}(1.651) + \tfrac{1}{3}(1.684) + \tfrac{1}{3}(1.687)] - |-1|1.595 = 1.674 - 1.595 = 0.0790$$

$$u = [\tfrac{1}{3}(1.721) + \tfrac{1}{3}(1.687) + \tfrac{1}{3}(1.757)] - |-1|1.525 = 1.722 - 1.525 = 0.1967$$

The approximate 95% limits are close to the exact ones obtained by the T-method in Step **2b.** If sample sizes are unequal, these approximate limits require much less effort when many contrasts are to be tested.

For example, in Table 9.1, where the null hypothesis is rejected, $SS_{among} = 0.6408$. Substituting into Expression (9.11), we obtain

$$0.6408 > (5 - 1)(0.00297)(2.579) = 0.03064$$

It is therefore possible to compute a critical SS value for a test of significance of an anova.

Another way to perform a significance test is to see whether SS_{among} is greater than this critical SS. It is of interest to investigate why the SS_{among} is as large as it is and to test the contributions made to this SS by differences among the sample means. This was discussed in the previous section, where separate sums of squares were computed based on comparisons among means planned before the data had been examined. A comparison was called significant if its F_s-ratio was greater than $F_{\alpha[k-1,a(n-1)]}$, where k is the number of means being compared. We can now also state this in terms of sums of squares. An SS is significant if it is greater than $(k - 1)MS_{within}F_{\alpha[k-1,a(n-1)]}$.

Following a procedure applied in other *unplanned* comparisons, we set $k = a$ in the last formula, no matter how many means we compare. Thus, the critical value of the SS will be larger than in the previous method, making it more difficult to reject the null hypothesis for a sample SS. We do this to allow for the fact that we chose to test those differences between group means that appear to contribute substantially to the overall SS. Two applications of the SS-STP test are illustrated in Box 9.9.

The diagrammatic representation of the results of a multiple comparisons test will not always be as simple as those in Box 9.7. Often there are no clear boundaries (gaps) between sets of means not significantly different from each other, and such sets overlap. For example, in the following data, which represent means of the character thorax width for the aphid *Pemphigus populitransversus* studied in 23 localities in eastern North America, the means are arrayed by order of magnitude in coded units; the lines beside them represent nonsignificant sets of means. Sokal and Rinkel (1963) employed this symbolism at the margin of a geographic distribution map to provide an easy visual test of significance of any pair of means. When there are too

---

BOX 9.9 **The Sum of Squares Simultaneous Test Procedure (SS-STP)**

Applied to the pea-section data of Table 9.1.
Compute a critical sum of squares from Expression (9.12).

$$SS_{\text{CRIT}} = (a-1)MS_{\text{within}}F_{\alpha[a-1,a(n-1)]} = (5-1) \times 0.00297 \times 2.579 = 0.03064$$

We are now able to test any differences between means and groups of means that attract our attention. For example, are the three sugar treatments other than sucrose different from each other? To test this, we compute the SS among these three means by the usual formula:

$$SS = n\sum(\bar{Y}_i - \bar{\bar{Y}})^2 = 10[(1.686 - 1.709)^2 + (1.719 - 1.709)^2 + (1.722 - 1.709)^2]$$

$$= 0.00798$$

Note that the grand mean $\bar{\bar{Y}} = 1.709$ is computed only from the three means involved in this contrast, not from all five means in the pea-section data. These means are not considered different because this SS is less than the critical SS (0.03064) calculated earlier.

For a second example, the sucrose mean looks suspiciously different from the means of the other sugars. To test this, we compute

$$SS = 10(1.561 - 1.672)^2 + 30(1.709 - 1.672)^2 = 0.1643$$

Note that the grand mean, 1.672, is calculated from the four means involved in the contrast, and that 1.709 is the mean of the three sugars other than sucrose. The resulting SS is greater than SS_{CRIT}. We conclude, therefore, that the mean for sucrose is less than that of the other sugars. We may continue in this fashion, testing all the differences that look suspicious or even testing all possible sets of means, considering them 2, 3, 4, and 5 at a time. This approach may require a computer if there are more than 5 means to be compared because very many tests could be made. Computer program BIOMstat carries out such tests.

many overlapping lines, they are difficult to keep apart by eye and a different technique (illustrated below on the upper right) is preferable (Sokal and Thomas, 1965).

25	25—39
37	37—57
39	39—58
39	45—62
45	55—72
45	
47	
47	
47	
49	
50	
50	
50	
54	
55	
55	
57	
58	
58	
59	
62	
62	
72	

Any pair of means enclosed by the range of any one line is not significantly different. Thus, means 47 and 50 are not different from each other because they are enclosed in the range 39 to 58, but means 39 and 62 are different from each other because there is no range that encloses them both.

Although all the procedures described here lead to the same conclusions in the pea-section data (see Table 9.1 and Boxes 9.7 and 9.8), such agreement need not, in general, occur. When the tiger beetle data in Box 9.7, part IV were analyzed by the *SS-STP*, we found the following pattern:

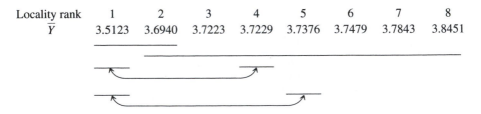

Locality rank	1	2	3	4	5	6	7	8
$\bar{Y}$	3.5123	3.6940	3.7223	3.7229	3.7376	3.7479	3.7843	3.8451

This is not consistent with the results indicated in Box 9.7, part IV, even though all these methods are testing the same null hypothesis: equality of means. Because sample sizes are unequal, means in nonsignificant sets are not necessarily adjacent, as indicated by the double-headed arrows. A reason for this difference is that *SS* and simultaneous confidence interval tests differ in their sensitivity to various alternative hypotheses.

EXERCISES 9

9.1 The following is an example with easy numbers to help you become familiar with the analysis of variance. A plant ecologist wishes to test the hypothesis that the height of plant species X depends on the type of soil it grows in. He measures the height of 3 plants in each of 4 plots representing different soil types, all 4 plots being contained in an area 2 miles square. His results are tabulated below. (Height is given in centimeters.) Does your analysis support this hypothesis?

ANSWER: $F_s = 6.950$, $F_{.05[3,8]} = 4.07$, $P = 0.012830$.

	Plots			
Observation Number	1	2	3	4
1	15	25	17	10
2	9	21	23	13
3	14	19	20	16

9.2 The following are measurements (in coded micrometer units) of the thorax length of the aphid *Pemphigus populitransversus*. The aphids were collected in 28 galls on the cottonwood, *Populus deltoides*. Four alate (winged) aphids were randomly selected from each gall and measured. The alate aphids of each gall are mostly isogenic (identical twins), being descended parthenogenetically from one stem mother. Thus, any variance within galls should only be the result of environment. Variance among different galls may be caused by differences in genotype and also environmental differences among galls. If this character, thorax length, is affected by genetic variation, then intergall variance must be present. The converse is not necessarily true; intergall variance need not indicate genetic variation—it could as well be the result of environmental differences (data from Sokal, 1952).

Gall No.	Thorax length				Gall No.	Thorax length			
1.	6.1	6.0	5.7	6.0	15.	6.3	6.5	6.1	6.3
2.	6.2	5.1	6.1	5.3	16.	5.9	6.1	6.1	6.0
3.	6.2	6.2	5.3	6.3	17.	5.8	6.0	5.9	5.7
4.	5.1	6.0	5.8	5.9	18.	6.5	6.3	6.5	7.0
5.	4.4	4.9	4.7	4.8	19.	5.9	5.2	5.7	5.7
6.	5.7	5.1	5.8	5.5	20.	5.2	5.3	5.4	5.3
7.	6.3	6.6	6.4	6.3	21.	5.4	5.5	5.2	6.3
8.	4.5	4.5	4.0	3.7	22.	4.3	4.7	4.5	4.4
9.	6.3	6.2	5.9	6.2	23.	6.0	5.8	5.7	5.9
10.	5.4	5.3	5.0	5.3	24.	5.5	6.1	5.5	6.1
11.	5.9	5.8	6.3	5.7	25.	4.0	4.2	4.3	4.4
12.	5.9	5.9	5.5	5.5	26.	5.8	5.6	5.6	6.1
13.	5.8	5.9	5.4	5.5	27.	4.3	4.0	4.4	4.6
14.	5.6	6.4	6.4	6.1	28.	6.1	6.0	5.6	6.5

Analyze the variance of thorax length. Is there significant intergall variance? Give estimates of the added component of intergall variance and estimate confidence intervals for it. What percentage of the variance is controlled by intragall and what percentage by intergall factors? Discuss your results.

9.3 Millis and Seng (1954) published a study on the relation of birth order to the birth weights of infants. The data below on first-born and eighth-born infants are extracted from a table of birth weights of male infants of Chinese third-class patients at the Kandang Kerbau Maternity Hospital in Singapore in 1950 and 1951.

Birth weight (lb : oz)	Birth order 1	Birth order 8
3:0–3:7	—	—
3:8–3:15	2	—
4:0–4:7	3	—
4:8–4:15	7	4
5:0–5:7	111	5
5:8–5:15	267	19
6:0–6:7	457	52
6:8–6:15	485	55
7:0–7:7	363	61
7:8–7:15	162	48
8:0–8:7	64	39
8:8–8:15	6	19
9:0–9:7	5	4
9:8–9:15	—	—
10:0–10:7	—	1
10:8–10:15	—	—
	1932	307

Which birth order appears to be accompanied by heavier infants? Can one reject the null hypothesis of no difference? Can you conclude that birth order causes differences in birth weight? Reanalyze using the t-test, and verify that $t_s^2 = F_s$.

ANSWER: $t_s = 11.016$.

9.4 The following cytochrome oxidase assessments of male *Periplaneta* roaches in cubic millimeters per 10 minutes per milligram were taken from a larger study by Brown and Brown (1956):

	n	$\overline{Y}$	$s_{\overline{Y}}$
24 hours after methoxychlor injection	5	24.8	0.9
Control	3	19.7	1.4

Can one reject the null hypothesis of no difference?

9.5 The following data are measurements of five random samples of domestic pigeons collected during the months of January, February, and March in Chicago in 1955. The variable is the length from the anterior end of the narial opening to the tip of the bony beak and is recorded in millimeters. Data from Olson and Miller (1958).

	Samples				
	1	2	3	4	5
	5.4	5.2	5.5	5.1	5.1
	5.3	5.1	4.7	4.6	5.5
	5.2	4.7	4.8	5.4	5.9
	4.5	5.0	4.9	5.5	6.1
	5.0	5.9	5.9	5.2	5.2
	5.4	5.3	5.2	5.0	5.0
	3.8	6.0	4.8	4.8	5.9
	5.9	5.2	4.9	5.1	5.0
	5.4	6.6	6.4	4.4	4.9
	5.1	5.6	5.1	6.5	5.3
	5.4	5.1	5.1	4.8	5.3
	4.1	5.7	4.5	4.9	5.1
	5.2	5.1	5.3	6.0	4.9
	4.8	4.7	4.8	4.8	5.8
	4.6	6.5	5.3	5.7	5.0
	5.7	5.1	5.4	5.5	5.6
	5.9	5.4	4.9	5.8	6.1
	5.8	5.8	4.7	5.6	5.1
	5.0	5.8	4.8	5.5	4.8
	5.0	5.9	5.0	5.0	4.9

Are the five samples homogeneous?

ANSWER: $F_s = 1.977$, $df = 4, 95$.

9.6 P. E. Hunter (1959, detailed data unpublished) selected two strains of *Drosophila mela-nogaster*, one for short larval period (SL) and one for long larval periods (LL). A nonse-lected control strain (CS) was also maintained. At generation 42, the following data were obtained for the larval period (measured in hours). Analyze and interpret.

	Strain		
	SL	CS	LL
n_i	80	69	33
$\sum\limits^{n_i} Y$	8070	7291	3640

$$\sum^{3}\sum^{n_i} Y^2 = 1,994,650$$

Note that part of the computation has already been performed for you. Perform planned tests among the three means (short versus long larval periods and each against the control). Are these three tests independent? Estimate 95% confidence intervals for the differences of means for which these comparisons are made.

ANSWER: $MS_{(SL \text{ vs. } LL)} = 2076.6697$.

9.7 The following data were taken from a study of the systematics of honeybees (DuPraw, 1965). Of the 15 variables reported in the original study, the means and sample sizes for wing length variable *A-O* (in mm × 83) are given below for 15 localities. The pooled standard deviation within localities is 9.284. Perform an analysis of variance and compare all pairs of means to determine which are significantly different using a familywide error rate of 5%. Graph the comparison limits for the means. Interpret your analysis. You may wish to compare your results with those in the original article based on all 15 measurements.

				Locality				
	1	2	3	5	6	7	8	
n	20	20	20	16	12	11	16	
$\overline{Y}$	379.9	384.1	376.8	362.9	374.1	360.4	341.8	

				Locality				
	9	10	11	13	14	15	16	17
n	20	20	20	10	10	10	10	10
$\overline{Y}$	371.0	353.3	354.2	330.7	342.6	378.6	368.5	368.8

Locality codes:
 1. Europe northwest of the Alps
 2. Europe southeast of the Alps
 3. Italy
 5. Cyprus
 6. Caucasus
 7. Eastern Mediterranean
 8. Egypt
 9. North Africa
10. Central Africa
11. Cape of Good Hope
13. India and Pakistan
14. Japan
15. Mt. Kilimanjaro and vicinity
16. Yugoslavia
17. Yellow "oasis bee" from Sahara

9.8 The following data were taken from a study of blood protein variations in deer (Cowan and Johnston, 1962). The variable is the mobility of serum protein fraction *II* expressed as 10^{-5} cm^2/V·s.

Samples	$\overline{Y}$	$s_{\overline{Y}}$
1 Sitka	2.8	0.07
2 California blacktail	2.5	0.05
3 Vancouver Island blacktail	2.9	0.05
4 Mule deer	2.5	0.05
5 Whitetail	2.8	0.07

Sample size $n = 12$ for each mean. Perform an analysis of variance and a multiple-comparisons test using a variety of procedures. Compare the results of the various multiple-comparisons tests. Which method is preferred in this example?

ANSWER: $MS_{within} = 0.0416$, maximal nonsignificant sets (at $P = 0.05$) are samples (1, 3, 5) and (2, 4) by the *SS-STP* procedure. The corresponding sets for the T-method (the preferred method) are (1, 5), (2, 4), and (3).

9.9 In a study of the effect of density on dry weight of hybrid houseflies, Bhalla and Sokal (1964) obtained the following results. Density is given as number of individuals per 36 grams of medium; the means $\bar{Y}$ are stated in milligrams and are based on 5 replicates.

Density	$\bar{Y}$	$s_{\bar{Y}}$
40	3.23	0.146
80	2.63	0.153
160	2.24	0.141
320	1.57	0.047
640	1.40	0.201
1280	0.73	0.031
2560	0.74	0.065

Carry out an analysis of variance. Earlier experiments had demonstrated the effect of density on percent emergence of these flies to be the following: density 40—suboptimal; densities 80 to 320—optimal; density 640—some stress; densities 1280 and 2560—severe stress. Design an orthogonal set of comparisons to test whether differences in stress on these houseflies affect their dry weights.

9.10 An investigator testing the effect of two heavy metal ions on the growth of rainbow trout established the following treatment classes:

1. Control
2. Copper, low concentration
3. Copper, high concentration
4. Zinc, low concentration
5. Copper and zinc, both at low concentrations
6. Copper and zinc, both at high concentrations

Give an example of a set of orthogonal single-degree-of-freedom comparisons and a set of nonorthogonal comparisons.

ANSWER: One such set of orthogonal comparisons is:

Control	Cu low	Cu high	Zn low	Cu + Zn low	Cu + Zn high
−5	1	1	1	1	1
0	2	2	2	−3	−3
0	1	1	−2	0	0
0	0	0	0	1	+1
0	1	−1	0	0	0

10 Nested Analysis of Variance

In this chapter, we discuss an extension of single-classification anova to handle cases where each class or group is divided into two or more randomly chosen subgroups. Section 10.1 explains the reasons for employing such a design. Section 10.2 shows the computation for subsamples of equal size. Some complications ensue when subsamples are of unequal sizes; these are treated in Section 10.3.

10.1 Nested Anova: Design

The simple design of a single-classification anova (see Chapter 9) is frequently insufficient to represent the complexity of a given experiment and to extract all the relevant information from it. This chapter deals with cases in which each major grouping or class is divided into randomly chosen subgroups. Thus, for each of the seven batches of medium in the example from Section 8.4, we might prepare five jars, and a sample of houseflies could be reared from each of these jars. Each group would represent a type of medium, each subgroup a jar for one medium, and the observations in a subgroup would be measurements of individual flies from each jar. Chapter 11 deals with designs in which each group is classified by two or more criteria rather than being subdivided.

What other information do we obtain from such a design? In an experiment with a single jar for each different medium formulation, we have no way of knowing how much of the observed differences among the flies in different jars are due to different medium formulations or just the effect of being in different jars. Environmental and other unintentional differences among the jars could result in morphological differences among the flies, which would have occurred even if the same formulation had been used in all of the jars. The only way to separate these two effects is to have two or more jars for each medium formulation. If we do not find any differences among the jars *within* medium formulations—beyond what we should expect on the basis of the amount of variation observed within a jar—we can ascribe differences among flies to differences among the formulations of their media. Even if we find a large amount of added variance among the jars within a formulation, however, we can still test whether the variation among the media is greater than that to be expected on the basis of the observed variation among the jars that were treated alike. If the tested mean square is larger than we would expect by chance, we can conclude that differences among medium formulations are present above and beyond differences expected among jars.

We call the analysis of such a design a **nested analysis of variance** because the subordinate classification is nested within the higher level of classification. Thus, in our example, the jars are subordinate to the medium formulations. The design is also often called a *hierarchic analysis of variance*. A crucial requirement of this type of analysis is that groups representing a subordinate level of classification be randomly chosen. Therefore, the replicate jars within one medium formulation must not be deliberately picked to represent certain microenvironments. The subordinate level of a nested anova is always Model II. The highest level of classification in a nested anova may be Model I or Model II. If it is Model II, we speak of a **pure Model II nested anova**. If the highest level is Model I, we called it a **mixed-model nested anova**.

There are two general classes of applications of nested anova. The first is similar to the hypothetical example just discussed. This application commonly is used to ascertain the magnitude of error at various stages of an experiment or an industrial process. In the hypothetical housefly example, we wish to know whether there is substantial variation among jars as well as among flies within jars, and we need to know the magnitude of this variation to enable us to test the null hypothesis of no added component due to treatments (= different formulations of medium). Nested anovas are not limited to the two levels discussed so far. We can divide the subgroups into subsubgroups and even further, as long as these are chosen randomly.

Thus, we might design an experiment in which we test the effect of five drugs on the concentration of pigment in the skin of an animal. The five drugs and one control (six groups) are the major classification and clearly are fixed treatment effects (Model I). For each drug, we might use five randomly selected rats. These would provide a measure of variance in pigment among rats within a drug class or of "rats within drugs," as one would say in statistical jargon. From each rat we might take three small skin samples at random from a defined area on its ventral side. This presents us with a new subordinate level of variation—skin samples within rats. Each skin sample is macerated and divided into two lots, which are hydrolyzed separately. This level is hydrolysates within skin samples. Finally, the concentration of pigment might be read as an optical density, and two replicate readings might be made of each hydrolysate. The error variance would be the variance of the replicate readings per hydrolysate, but we would also have estimates of the variance between hydrolysates within one skin sample, among skin samples within one rat, and among rats within one drug.

These variance estimates are important in designing similar experiments because they indicate at which level of the experiment most of our sampling efforts should be concentrated. The most variable aspect of our experiment will need the greatest replication or will need better experimental control. Thus, if we find that the two hydrolysates show the greatest proportion of variance, our method of hydrolysis clearly is not standardized enough and should be improved, or if this is impossible, we should divide each macerated skin sample into more lots to provide several more hydrolysates. We discuss this topic in greater detail in Section 12.7. The preceding example of drugs, rats, skin samples, hydrolysates and optical density is clearly a mixed-model nested anova, with drugs representing the fixed treatment effect and all other levels representing the random effects.

The second major use of a nested analysis of variance is in cases that are usually pure Model II. Such cases frequently are from the field of quantitative genetics, where we wish to know the magnitude of the variance attributable to various levels of variation in a study. Systematists interested in discovering the sources of variation in natural populations also use this type of approach. To illustrate such an application, we propose an experiment in which separate litters by the same mother have been sired by different fathers. If there are 10 females (dams), each with 5 litters sired by 5 different males (sires), the variance can be subdivided into that among dams, that among sires within the same dam, and that among offspring within the same sire. Genetic theory (using the statistical consequences of Mendelian genetics) can predict the relative magnitudes of such components on various assumptions. One of the aims of the field of quantitative genetics has been to verify such predictions and analyze departures from them. A similar approach can be used in geographic variation studies in which variances among samples within one locality and among localities can be estimated and may lead to important conclusions about the population distribution pattern of an organism. For further information on the application of analysis of variance to quantitative genetics, refer to the very readable account by Falconer and Mackay (1996) and the more extended treatment by Weir (1990).

Before we carry out an actual nested anova, we should examine the linear model on which it is based. In Expressions (8.2) and (8.3), we gave equations for decomposing observations in a single-classification anova for Model I and Model II, respectively. For a two-level nested anova (Model II) this equation becomes

$$Y_{ijk} = \mu + A_i + B_{ij} + \epsilon_{ijk} \tag{10.1}$$

where Y_{ijk} is the kth observation in the jth subgroup of the ith group, μ is the parametric mean of the population, A_i is the random contribution for the ith group of the higher level A, B_{ij} is the random contribution for the jth subgroup (level B) of the ith group, and ϵ_{ijk} is the error term of the kth item in the jth subgroup of the ith group. We assume that A_i, B_{ij}, and ϵ_{ijk} are distributed normally, with means of zero and variances of σ_A^2, $\sigma_{B \subset A}^2$, and σ^2, respectively. We use the symbol $\sigma_{B \subset A}^2$, rather than σ_B^2, to indicate that the variance is of level B *within* level A. We shall use $A, B, C, \ldots$, as subscripts for levels of variance components. In a particular application, researchers frequently use letters that help them remember the meaning of variance components. For example, σ_G^2 might signify genetic variance, σ_D^2 variance among dogs, and so on. In a textbook, however, a standard system is necessary.

When the model is mixed (that is, the highest level has fixed treatment effects), we decompose an observation as follows:

$$Y_{ijk} = \mu + \alpha_i + B_{ij} + \epsilon_{ijk} \tag{10.2}$$

This expression is the same as Expression (10.1) except for the use of α_i for the fixed treatment effect, rather than A_i for a random effect in the earlier model. Expressions (10.1) and (10.2) can be expanded to further levels, which are identified by C, D, and so forth, with the subscripts extended to l, m, and so on.

We will now carry out the computations for a nested anova.

10.2 Nested Anova: Computation

To illustrate the computation of a nested analysis of variance, we have chosen an example that is a pure Model II anova as described in Section 10.1. This anova estimates the percentage of phenotypic variation at two sampling levels. The data on rearing field-caught mosquito pupae (*Aedes intrudens*) in the laboratory were obtained by Rohlf. Twelve female pupae were brought into the laboratory and divided at random into three rearing cages, into which four pupae each were placed. The purpose of dividing them among different cages was to see whether the differences among cages would add variance to the overall variation among females that is due to their presumable genetic and ecological differences. Length of the left wing was measured on each female, and the measurement was repeated once, the mosquitoes being presented to the measurer in a random order so as not to bias the results. The measurements are shown at the top of Box 10.1. The structure of the experiment and of the analysis, therefore, is three cages each with four females, each measured twice.

Because this is a pure Model II nested anova, the means of particular groups (cages) and subgroups (females) are not of interest. However, they are furnished in Box 10.1 because they are needed for the computation. Rather than employing the more precise notation $\overline{Y}_{ij}$ and $\overline{Y}_i$ for subgroup and group means, respectively, we use the simpler $\overline{Y}_B$ and $\overline{Y}_A$ to indicate levels B and A. The sum of squares of groups (cages) is computed as in the single-classification anova of Chapter 9. In step **1**, we compute the grand mean of all the observations in the usual way. In step **2**, the deviations of the $a = 3$ group means $\overline{Y}_A$ from the grand mean $\overline{\overline{Y}}$ are squared, summed, and then multiplied by their sample sizes (nb) because we need sums of squares of groups rather than of means. Because there are a group means, the degrees of freedom pertaining to the sum of squares of groups is $a - 1$ which in this case equals 2.

Next, in step **3**, we compute the *sum of squares of subgroups within groups*. The deviations between subgroup means $\overline{Y}_B$ and group means $\overline{Y}_A$ are squared and summed for the b subgroups within a group and over the a groups. To turn this SS of means into one of groups, we multiply by n, the sample size of each subgroup. The degrees of freedom in each subgroup are $b - 1$, where b is the number of subgroups per group. Because there are a groups, the df for subgroups within groups is $a(b - 1)$, which for the example in Box 10.1 is 9. Finally, in step **4**, we compute the error sum of squares, the variation within subgroups. This value is the sum of the squared deviations of the items Y from the subgroup means $\overline{Y}_B$, the summation extending over all subgroups and all groups. Because there are n items per subgroup, there are $n - 1$ degrees of freedom for each of the ab subgroups in the experiment. Thus, the error df is computed as $ab(n - 1)$, which here is 12.

We now arrange the sums of squares in an anova table, as shown in Box 10.1. Mean squares are obtained as before by dividing the sums of squares by their respective degrees of freedom. Note that the total SS is merely the sum of the other three sums of squares and that the degrees of freedom are similarly additive. The total SS is the quantity you would obtain if you calculated the sum of squares of the

BOX 10.1 Two-Level Nested Anova: Equal Sample Sizes and Pure Model II

Two independent measurements of the left wings of each of 4 female mosquitoes (*Aedes intrudens*) reared in each of 3 cages; $b = 4$ females. The data are given in micrometers. This is a Model II anova.

	Groups of females ($a = 3$)											
	Cage 1				**Cage 2**				**Cage 3**			
	1	2	3	4	1	2	3	4	1	2	3	4
Measurements 1	58.5	77.8	84.0	70.1	69.8	56.0	50.7	63.8	56.6	77.8	69.9	62.1
($n = 2$) 2	59.5	80.9	83.6	68.3	69.8	54.5	49.3	65.8	57.5	79.2	69.2	64.5
Subgroup sums $\sum\limits^{n} Y$	118.0	158.7	167.6	138.4	139.6	110.5	100.0	129.6	114.1	157.0	139.1	126.6
Subgroup means $\bar{Y}_B$	59.0	79.35	83.8	69.2	69.8	55.25	50.0	64.8	57.05	78.5	69.55	63.3
Group means $\bar{Y}_A$	72.8375				59.9625				67.1000			

SOURCE: Unpublished results from F. J. Rohlf.

Preliminary Computations

1. Grand mean $\bar{\bar{Y}} = \dfrac{1}{abn} \sum\limits^{a}\sum\limits^{b}\sum\limits^{n} Y = 66.6333$

2. SS_{among} (among groups) $= nb \sum\limits^{a} (\bar{Y}_A - \bar{\bar{Y}})^2 = 665.6758$

3. SS_{subgr} (subgroups within groups) $= n \sum\limits^{a}\sum\limits^{b} (\bar{Y}_B - \bar{Y}_A)^2 = 1720.6775$

4. SS_{within} (within subgroups; error SS) $= \sum\limits^{a}\sum\limits^{b}\sum\limits^{n} (Y - \bar{Y}_B)^2 = 15.6200$

Box 10.1 (continued)

Anova table: formulas

	Source of variation	df	SS	MS	F_s	Expected MS
$\bar{Y}_A - \bar{\bar{Y}}$	Among groups	$a - 1$	2	$\dfrac{2}{a-1}$	$\dfrac{MS_{among}}{MS_{subgr}}$	$\sigma^2 + no\sigma^2_{B \subset A} + nbo\sigma^2_A$
$\bar{Y}_B - \bar{Y}_A$	Among subgroups within groups	$a(b-1)$	3	$\dfrac{3}{a(b-1)}$	$\dfrac{MS_{subgr}}{MS_{within}}$	$\sigma^2 + no\sigma^2_{B \subset A}$
$Y - \bar{Y}_B$	Within subgroups	$ab(n-1)$	4	$\dfrac{4}{ab(n-1)}$		σ^2
$Y - \bar{\bar{Y}}$	Total	$abn - 1$	$2 + 3 + 4$			

Completed anova

	Source of variation	df	SS	MS	F_s	P
$\bar{Y}_A - \bar{\bar{Y}}$	Among groups (among cages)	2	665.6758	332.8379	1.741	0.2295
$\bar{Y}_B - \bar{Y}_A$	Among subgroups within groups (among females within cages)	9	1720.6775	191.1864	146.88	1.3467×10^{-7}
$Y - \bar{Y}_B$	Within subgroups (error; between measurements on each female)	12	15.6200	1.3017		
$Y - \bar{\bar{Y}}$	Total	23	2401.9733			

Conclusions

There is little evidence ($P = 0.2295$) for a variance component among cages, but there is strong evidence ($P = 1.3467 \times 10^{-7}$) for an added variance component among females for wing length in these mosquitoes. See Section 12.1 for ways to measure the size of the effects.

Box 10.1 (continued)

From the components of the expected mean squares shown in the anova table presenting general formulas, we can derive the standard anova procedure for estimating the variance components:

Within subgroups (error; between measurements on each female) $= s^2 = 1.3017$

Among subgroups within groups (among females within cages) $= s^2_{B \subset A} = \dfrac{MS_{subgr} - MS_{within}}{n} = 94.9424$

Among groups (among cages) $= s^2_A = \dfrac{MS_{among} - MS_{subgr}}{nb} = 17.7064$

For these data, the estimated variance components are all positive. If one or more were negative, then the REML estimates of Sahai and Ojeda (2004) should be used.

$$s^2_A = \frac{1}{bn} \left\{ MS_{among} - \max \left(MS_{subgr}, \frac{SS_{within} + SS_{subgr}}{ab(n-1) + a(b-1)} \right) \right\}$$

$$s^2_{B \subset A} = \frac{1}{n} \left\{ \min \left(MS_{subgr}, \frac{SS_{subgr} + SS_{among}}{a(b-1) + (a-1)} \right) - MS_{within} \right\}$$

$$s^2 = \min \left(MS_{within}, \frac{SS_{within} + SS_{subgr}}{ab(n-1) + a(b-1)}, \frac{SS_{within} + SS_{subgr} + SS_{among}}{ab(n-1) + a(b-1) + (a-1)} \right)$$

If any of these estimates are negative, then they should be set to zero.

Box 10.1 (continued)

Confidence Intervals for σ_A^2 and σ_{BCA}^2 Using the Procedure of Ting et al. (1990)

First look up or compute the following chi-square and F-values that are needed for the computations of the confidence limits.

$$F_1 = F_{\alpha/2[\nu_{among},\nu_{subgr}]} = 5.7147 \qquad X_1 = \chi^2_{\alpha/2[\nu_{among}]} = 7.3778$$

$$F_2 = F_{1-\alpha/2[\nu_{among},\nu_{subgr}]} = 0.0254 \qquad X_2 = \chi^2_{\alpha/2[\nu_{subgr}]} = 19.0228$$

$$F_3 = F_{\alpha/2[\nu_{subgr},\nu_{within}]} = 3.4358 \qquad X_3 = \chi^2_{\alpha/2[\nu_{within}]} = 23.3367$$

$$\qquad\qquad\qquad X_4 = \chi^2_{1-\alpha/2[\nu_{among}]} = 0.0506$$

$$F_4 = F_{1-\alpha/2[\nu_{subgr},\nu_{within}]} = 0.2585 \qquad X_5 = \chi^2_{1-\alpha/2[\nu_{subgr}]} = 2.7004$$

$$\qquad\qquad\qquad X_6 = \chi^2_{1-\alpha/2[\nu_{within}]} = 4.4038$$

Then compute the following quantities.

$$G_1 = 1 - \frac{\nu_{among}}{X_1} = 1 - \frac{2}{7.3778} = 0.7289, \quad G_2 = 1 - \frac{\nu_{subgr}}{X_2} = 1 - \frac{9}{19.0228} = 0.5269,$$

$$G_3 = 1 - \frac{\nu_{within}}{X_3} = 1 - \frac{12}{23.3367} = 0.4858, \quad H_1 = \frac{\nu_{among}}{X_4} - 1 = \frac{2}{0.0506} - 1 = 38.4979,$$

$$H_2 = \frac{\nu_{subgr}}{X_5} - 1 = \frac{9}{2.7004} - 1 = 2.3329, \quad H_3 = \frac{\nu_{within}}{X_6} - 1 = \frac{12}{4.4038} - 1 = 1.7249$$

$$G_{12} = \frac{(F_1 - 1)^2 - G_1^2 F_1^2 - H_2^2}{F_1} = \frac{(5.7147 - 1)^2 - 0.7289^2 \times 5.7147^2 - 2.3329^2}{5.7147} = -0.0989$$

$$G_{23} = \frac{(F_3 - 1)^2 - G_2^2 F_3^2 - H_3^2}{F_3} = \frac{(3.4358 - 1)^2 - 0.5269^2 \times 3.4358^2 - 1.7249^2}{3.4358} = -0.0929$$

Box 10.1 (continued)

$$H_{12} = \frac{(1-F_2)^2 - H_1^2 F_2^2 - G_2^2}{F_2} = \frac{(1-0.0254)^2 - 38.4979^2 \times 0.0254^2 - 0.5269^2}{0.0254} = -11.1507$$

$$H_{23} = \frac{(1-F_4)^2 - H_2^2 F_4^2 - G_3^2}{F_4} = \frac{(1-0.2585)^2 - 2.3329^2 \times 0.2585^2 - 0.4858^2}{0.2585} = -0.1930$$

$$L_A = G_1^2 MS_{among}^2 + H_2^2 MS_{subgr}^2 + G_{12} MS_{among} MS_{subgr}$$
$$= 0.7289^2 \times 332.8379^2 + 2.3329^2 \times 191.1864^2 + (-0.0989) \times 332.8379 \times 191.1864 = 2.5149 \times 10^5$$

$$U_A = H_1^2 MS_{among}^2 + G_2^2 MS_{subgr}^2 + H_{12} MS_{among} MS_{subgr}$$
$$= 38.4979^2 \times 332.8379^2 + 0.5269^2 \times 191.1864^2 + (-11.1507) \times 332.8379 \times 191.1864 = 1.6349 \times 10^8$$

$$L_B = G_2^2 MS_{subgr}^2 + H_3^2 MS_{within}^2 + G_{23} MS_{subgr} MS_{within}$$
$$= 0.5269^2 \times 191.1864^2 + 1.7249^2 \times 1.3017^2 + (-0.0929) \times 191.1864 \times 1.3017 = 1.0129 \times 10^4$$

$$U_B = H_2^2 MS_{subgr}^2 + G_3^2 MS_{within}^2 + H_{23} MS_{subgr} MS_{within}$$
$$= 2.3329^2 \times 191.1864^2 + 0.4858^2 \times 1.3017^2 + (-0.1930) \times 191.1864 \times 1.3017 = 1.9888 \times 10^5$$

The confidence limits for σ_A^2 then are

$$L_1 = \frac{MS_{among} - MS_{subgr} - \sqrt{L_A}}{bn} = \frac{332.8379 - 191.1864 - \sqrt{2.5149 \times 10^5}}{4 \times 2} = -44.9793$$

$$L_2 = \frac{MS_{among} - MS_{subgr} + \sqrt{U_A}}{bn} = \frac{332.8379 - 191.1864 + \sqrt{1.6349 \times 10^8}}{4 \times 2} = 1616.0$$

Box 10.1 (continued)

And for σ_{BCA}^2

$$L_1 = \frac{MS_{subgr} - MS_w - \sqrt{L_B}}{n} = \frac{191.1864 - 1.3017 - \sqrt{1.0129 \times 10^4}}{2} = 44.6208$$

$$L_2 = \frac{MS_{subgr} - MS_w + \sqrt{U_B}}{n} = \frac{191.1864 - 1.3017 + \sqrt{1.9888 \times 10^5}}{2} = 317.9205$$

Because we are frequently interested only in their relative magnitudes, the variance components are often expressed as proportions or as percentages of the sum of their variance components:

$$s^2 + s_{BCA}^2 + s_A^2 = 113.9505$$

s^2 represents $\dfrac{100 \times 1.3017}{113.9505} = 1.14\%$

s_{BCA}^2 represents $\dfrac{100 \times 94.9424}{113.9505} = 83.32\%$

s_A^2 represents $\dfrac{100 \times 17.7064}{113.9505} = 15.54\%$

Confidence intervals for these as proportions are shown in the following. Multiply them by 100 to obtain confidence limits as percentages. First look up or compute the additional chi-square and F-values that are needed for the computations of the confidence limits.

$$F_5 = F_{\alpha/2[\nu_{among}, \nu_{within}]} = 5.0959$$

$$F_6 = F_{1-\alpha/2[\nu_{among}, \nu_{within}]} = 0.0254$$

$$F_7 = F_{\alpha/2[\nu_{subgr}, \nu_{among}]} = 39.3869$$

$$F_8 = F_{1-\alpha/2[\nu_{subgr}, \nu_{among}]} = 0.1750$$

$$X_7 = X_{\alpha/2[\nu_{among} + \nu_{subgr}]}^2 = 21.9200$$

Box 10.1 (continued)

For the intraclass correlation, $\rho_A = \sigma_A^2/(\sigma^2 + \sigma_{BCA}^2 + \sigma_A^2)$, the 95% confidence limits can be computed directly.

$$L_1 = \frac{MS_{among} - F_1 MS_{subgr}}{MS_{among} + (b-1)F_1 MS_{subgr} + b(n-1)F_5 MS_{within}^2}$$

$$= \frac{332.8379 - 5.7147 \times 191.1864}{332.8379 + (4-1) \times 5.7147 \times 191.1864 + 4 \times (2-1) \times 5.0959 \times 1.3017^2} = -0.2089 = 0$$

$$L_2 = \frac{MS_{among} - F_2 MS_{subgr}}{MS_{among} + (b-1)F_2 MS_{subgr} + b(n-1)F_6 MS_{within}^2}$$

$$= \frac{332.8379 - 0.0254 \times 191.1864}{332.8379 + (4-1) \times 0.0254 \times 191.1864 + 4 \times (2-1) \times 0.0254 \times 1.3017^2} = 0.9438$$

The lower limit is negative and is thus replaced by zero. The confidence interval is very broad because of the small sample sizes.

For $\rho_{BCA} = \sigma_{BCA}^2/(\sigma^2 + \sigma_{BCA}^2 + \sigma_A^2)$, one must first compute the following two quantities:

$$L_B = \frac{MS_{subgr}^2 - \dfrac{X_2}{\nu_{subgr}} MS_{subgr} MS_{within} - \left(F_3 - \dfrac{X_2}{\nu_{subgr}}\right) F_3 MS_{within}^2}{F_7 MS_{among} MS_{subgr} + \dfrac{(bn-1)X_2}{\nu_{subgr}} MS_{subgr} MS_{within}}$$

$$= \frac{191.1864^2 - \dfrac{19.0228}{9} 191.1864 \times 1.3017 - \left(3.4358 - \dfrac{19.0228}{9}\right) 3.4358 \times 1.3017^2}{39.3869 \times 332.8379 \times 191.1864 + \dfrac{(4 \times 2 - 1) \times 19.0228}{9} 191.1864 \times 1.3017} = 0.0143$$

Box 10.1 (continued)

$$U_B = \frac{MS_{subgr}^2 - \frac{X_5}{\nu_{subgr}} MS_{subgr} MS_{within} - \left(F_4 - \frac{X_5}{\nu_{subgr}}\right) F_4 MS_{within}^2}{F_8 MS_{among} MS_{subgr} + \frac{(bn-1)X_5}{\nu_{subgr}} MS_{subgr} MS_{within}}$$

$$= \frac{191.1864^2 - \frac{2.7004}{9} 191.1864 \times 1.3017 - \left(0.2585 - \frac{2.7004}{9}\right) 0.2585 \times 1.3017^2}{0.1750 \times 332.8379 \times 191.1864 + \frac{(4 \times 2 - 1) \times 2.7004}{9} 191.1864 \times 1.3017}$$

$$= 3.1290$$

Then the limits can be computed as

$$L_1 = \frac{bL_B}{1 + (b-1)L_B} = \frac{4 \times 0.0143}{1 + (4-1)0.0143} = 0.0550$$

$$L_2 = \frac{bU_B}{1 + (b-1)U_B} = \frac{4 \times 3.1290}{1 + (4-1)3.1290} = 1.2050 = 1$$

Because the upper limit exceeds 1, it is set equal to 1.

The computations on the confidence interval for $\rho_w = \sigma^2/(\sigma^2 + \sigma_{BCA}^2 + \sigma_A^2)$ are more complicated. The following quantities must be computed first.

$$G_{12}^* = \left(1 - \frac{\nu_{among} + \nu_{subgr}}{X_7}\right)^2 \frac{(\nu_{among} + \nu_{subgr})^2}{\nu_{among}\nu_{subgr}} - \frac{\nu_{among}G_1^2}{\nu_{subgr}} - \frac{\nu_{subgr}G_2^2}{\nu_{among}}$$

$$= \left(1 - \frac{2+9}{21.920}\right)^2 \frac{(2+9)^2}{2 \times 9} - \frac{2 \times 0.7289^2}{9} - \frac{9 \times 0.5269^2}{2} = 0.3010$$

Box 10.1 (continued)

$$G_{13} = \frac{(F_5 - 1)^2 - G_1^2 F_5^2 - H_3^2}{F_5} = \frac{(5.0959 - 1)^2 - 0.7289^2 \times 5.0959^2 - 1.7249^2}{5.0959} = 7.0428 \times 10^{-4}$$

$$H_{13} = \frac{(1 - F_6)^2 - H_1^2 F_6^2 - G_3^2}{F_6} = \frac{(1 - 0.0254)^2 - 38.4979^2 \times 0.0254^2 - 0.4858^2}{0.0254} = -9.4639$$

$$G_{23} = \frac{(F_3 - 1)^2 - G_2^2 F_3^2 - H_3^2}{F_3} = \frac{(3.4358 - 1)^2 - 0.5269^2 \times 3.4358^2 - 1.7249^2}{3.4358} = -0.0929$$

$$H_{23} = \frac{(1 - F_4)^2 - H_2^2 F_4^2 - G_3^2}{F_4} = \frac{(1 - 0.2585)^2 - 2.3329^2 \times 0.2585^2 - 0.4858^2}{0.2585} = -0.1930$$

$$k_1 = MS_{\text{among}} + (b - 1)MS_{\text{subgr}} = 332.8379 + (3 - 1)191.1864 = 906.3971$$

$$k_2 = MS_{\text{within}} = 1.3017$$

$$k_3 = G_1^2 MS_{\text{among}}^2 + (b - 1)^2 G_2^2 MS_{\text{subgr}}^2 + (b - 1)G_{12}^* MS_{\text{among}} MS_{\text{subgr}}$$
$$= 0.7289^2 \times 332.8379 + (4 - 1)^2 0.5269 \times 191.1864 + (4 - 1)0.3010 \times 332.8379 \times 191.1864 = 2.0765 \times 10^5$$

$$k_4 = G_{13} MS_{\text{among}} MS_{\text{within}} + (b - 1)G_{23} MS_{\text{subgr}} MS_{\text{within}}$$
$$= 7.0428 \times 10^{-4} \times 332.8379 \times 1.3017 + (4 - 1)(-0.0929)191.1864 \times 1.3017 = -69.0479$$

$$k_5 = H_3^2 MS_{\text{within}}^2 = 1.7249^2 \times 1.3017^2 = 5.0413$$

$$k_6 = H_1^2 MS_{\text{among}}^2 (b - 1)^2 H_2^2 MS_{\text{subgr}}^2 = 38.4979^2 \times 332.8379^2 (4 - 1)^2 \times 2.3329^2 \times 191.1864^2 = 1.6598 \times 10^8$$

$$k_7 = H_{13} MS_{\text{among}} MS_{\text{within}} + (b - 1)H_{23} MS_{\text{subgr}} MS_{\text{within}}$$
$$= (-9.4639)332.8379 \times 1.3017 + (4 - 1)(-0.1930)191.1864 \times 1.3017 = -4.2443 \times 10^3$$

$$k_8 = G_3^2 MS_{\text{within}}^2 = 0.5269^2 \times 1.3017^2 = 0.3998$$

$$\hat{\rho} = \frac{MS_{\text{among}} + (b - 1)MS_{\text{subgr}}}{MS_{\text{within}}} = \frac{332.8379 + (4 - 1)191.1864}{1.3017} = 696.3358$$

Box 10.1 (continued)

$$V_L = (2 + k_4/k_1k_2)^2 - 4(1 - k_5/k_2^2)(1 - k_3/k_1^2)$$

$$= (2 + (-69.0479)/(906.3971 - 1.3017))^2 - 4(1 - 5.0413/1.3017^2)(1 - 2.0765 \times 10^5/906.3971^2) = 9.6737$$

$$V_U = (2 + k_7/k_1k_2)^2 - 4(1 - k_8/k_2^2)(1 - k_6/k_1^2)$$

$$= (2 + (-4244.3)/(906.3971 - 1.3017))^2 - 4(1 - 0.3998/1.3017^2)(1 - 1.6598 \times 10^8/906.3971^2) = 616.9027$$

$$L_w = \hat{\rho}\,\frac{2 + k_4/k_1k_2 - \sqrt{V_L}}{2(1 - k_5/k_2^2)} = 696.3358\,\frac{2 + (-69.0479)/(906.3971 \times 1.3017) - \sqrt{9.6737}}{2(1 - 5.0413/1.3017^2)} = 206.0027$$

$$U_w = \hat{\rho}\,\frac{2 + k_7/k_1k_2 - \sqrt{V_U}}{2(1 - k_8/k_2^2)} = 696.3358\,\frac{2 + (-4.2443 \times 10^3)/(906.3971 \times 1.3017) - \sqrt{616.9027}}{2(1 - 0.3998/1.3017^2)} = 10{,}591.0$$

Then the confidence limits for ρ_w can be computed as

$$L_1 = \frac{bn}{b(n-1) + U_w} = \frac{4 \times 2}{4(2-1) + 1059.1} = 0.00075509$$

$$L_2 = \frac{bn}{b(n-1) + L_w} = \frac{4 \times 2}{4(2-1) + 206.0027} = 0.0381$$

entire dataset of abn items. Its df would necessarily be $abn - 1$, which can easily be shown to be the sum of the three separate degrees of freedom $a - 1$, $a(b - 1)$, and $ab(n - 1)$. As is our practice, general formulas are shown first, followed by the specific example. Notice especially the expected mean squares in a nested anova. The variance for each level above the error contains within it the variances of all levels below. Thus, the expected variance of subgroups within groups is $\sigma^2 + n\sigma^2_{BCA}$, which resembles the familiar expression from a single-classification anova. The only new aspect is that σ^2_B has changed to σ^2_{BCA} to symbolize that it is the variance of level B *within* level A. The expected mean square among groups contains the terms below it plus $nb\sigma^2_A$. From these expected mean squares, the tests of significance are obvious.

We first test MS_{subgr}/MS_{within} for the presence of σ^2_{BCA} and then test MS_{among}/MS_{subgr} for the existence of σ^2_A. In the example of Box 10.1, we find an added variance component among females within cages, with an F_s value of 146.88, corresponding to a P-value or type I error equal to 1.35×10^{-7}. When we test the cages over the MS_{subgr}, however, we find the F_s-ratio to be only 1.741, $P = 0.230$, which could easily have been obtained by chance. Thus, we have no reason to reject the null hypothesis H_0: $\sigma^2_A = 0$. Although the females differed from each other for genetic reasons, or possibly because of different environmental experiences before being brought into the laboratory or during development, the fact that they were reared in different cages did not add variation to their wing lengths.

It is instructive to rearrange these nested data in two ways, each time creating a different single-classification anova. In part I of Table 10.1, we show the first of these arrangements in which we ignore the fact that the data are divided into three cages but simply treat each of the 12 females as one group in a single-classification anova. Convince yourself that the 24 observations at the head of Table 10.1 are the same as those at the head of Box 10.1. Now we simply have $a = 12$ females, each of which was measured $n = 2$ times. The fact that these 12 females come from 3 cages is hidden by this arrangement of the data. The analysis of variance of this table is straightforward, and the quantities shown beneath it yield the ensuing anova table.

The error mean square in the resulting anova table represents variance within females, which estimates the variance of repeated measurements on each female. The upper mean square represents variance among females, but this arrangement of the data confounds two sources of variation: the differences among females within a cage and possible differences from cage to cage. Note that this mean square yields $F_s = 166.66$, which would occur with an infinitesimal probability if the null hypothesis were correct. Obviously, this mean square should have been broken down into its components, representing variation of females within a cage and that among cages. It is reassuring to find this mean square large because this tells us that there is evidence of some differentiation at either or both of the upper levels of variation in the study.

The second rearrangement of the data of Box 10.1 is shown in part II of Table 10.1. We can remove the division into subgroups and simply treat the data as

TABLE 10.1 Analyses of Rearranged Data for a Two-Level Nested Anova

Data from Box 10.1.

I. Rearrangement of Data: First Example

						a females (*a* = 12)						
	1	**2**	**3**	**4**	**5**	**6**	**7**	**8**	**9**	**10**	**11**	**12**
n (measurements) $\left\{\begin{array}{c} \\ \end{array}\right.$ ($n = 2$)	58.5	77.8	84.0	70.1	69.8	56.0	50.7	63.8	56.6	77.8	69.9	62.1
	59.5	80.9	83.6	68.3	69.8	54.5	49.3	65.8	57.5	79.2	69.2	64.5
$\sum\limits^{n} Y$	118.0	158.7	167.6	138.4	139.6	110.5	100.0	129.6	114.1	157.0	139.1	126.6
$\bar{Y}$	59.0	79.35	83.8	69.2	69.8	55.25	50.0	64.8	57.05	78.5	69.55	63.3

Anova table

Source of variation		*df*	*SS*	*MS*	F_s	*P*
Among groups (females)	$\bar{Y} - \bar{\bar{Y}}$	11	2386.3533	216.9412	166.66	2.3289×10^{-11}
Within groups (between measurements on each female)	$Y - \bar{Y}$	12	15.6200	1.3017		

II. *Rearrangement of Data: Second Example*

	a cages (a = 3)		
	1	**2**	**3**
	58.5	69.8	56.6
	59.5	69.8	57.5
	77.8	56.0	77.8
n measurements	80.9	54.5	79.2
(*n* = 8)	84.0	50.7	69.9
	83.6	49.3	69.2
	70.1	63.8	62.1
	68.3	65.8	64.5
$\sum^{n} Y$	582.7	479.7	536.8
$\overline{Y}$	72.8375	59.9625	67.1000

Anova table

Source of variation	df	SS	MS	F_s	P
$\overline{Y} - \overline{\overline{Y}}$ Among groups (cages)	2	665.6758	332.838	4.026	0.0331
$Y - \overline{Y}$ Within groups (among measurements on all females in a cage)	21	1736.2975	82.681		

a single-classification anova of items within the higher level of classification. This procedure yields eight measurements in each of three cages. Sample sizes a and n now assume new values and should not be confused with the same symbols used in part I of Table 10.1. The mean square among cages is moderately larger than the error mean square, with a P-value of 0.033. The error term in this anova represents a mixture of variances among measurements and among females in a cage. One way of interpreting a nested classification is as a further partitioning of the SS within groups of a single-classification anova. Thus, in this case we separate variation between measurements of the same female from measurements among different females though still in the same cage.

You may wonder why the MS among cages yields a very small probability in the second anova of Table 10.1 but does not do so in Box 10.1. The reason is that we tested it over a different, smaller error term. The error term in the second anova of Table 10.1 is an average of the mean squares from the two lower levels of the analysis of variance in Box 10.1. If you added the two sums of squares and their degrees of freedom, you would obtain 1736.2975, the SS_{within} of the second anova of Table 10.1, and 21 error degrees of freedom. However, this error term confounds two separate sources of variation: the variation among measurements with 12 degrees of freedom and a sum of squares of 15.6200, and differences among females within cages with 9 degrees of freedom and a sum of squares of 1720.6775. Clearly, if several females represent each cage, the mean square among females must be included in the error term.

This Model II nested anova is completed by estimating the three variance components as shown in Box 10.1. The operations are simple extensions of the method of estimating variance components presented in Chapter 9. When we divide the overall variation into percentages attributable to each level, the error variance (measurements within one female) represents only 1.14% of the total variation. The replicate measurements evidently did not differ much from each other, and in future work it would probably suffice to measure each wing only once. Most of the variation (83.32%) represents variation among females (within cages), presumably largely genetic but possibly also environmental, as already discussed.

You may question the 15.54% of variance among cages. Having decided earlier that there is no added variance component among cages, why did we estimate it at all? One reason is to provide a general outline of all the computations in the box so that you can use the example as a model for others you will carry out in your research. A more important reason is that, even though we were not able to reject the null hypothesis, we cannot be certain that there is not a small contribution to the variance from cage differences. With larger sample sizes (more cages or more females or both), we could perhaps have shown an added component of variance among cages. Thus, we may legitimately estimate the variance component and calculate its percentage contribution to the total variation because the computation gives the best estimate of this variance component if it does indeed exist. Formulas for finding confidence limits of variance components are laid out in the remainder of Box 10.1. They are somewhat complex and practical only using computer programs.

In the example in Box 10.1, the test for subgroups within groups was significant, so we used the mean square for subgroups as a divisor to test for differences among groups. Let us imagine another example for which we accept the null hypothesis of no differences among subgroups. From the expected mean squares in Box 10.1 we see that, in that case, the MS_{subgr} estimates σ^2, as does MS_{within}. In the past, such a situation suggested to researchers that they could pool these two estimates of σ^2 to obtain a better estimate of σ^2 based on more degrees of freedom. Previous editions of this book featured conservative rules for pooling mean squares in anova when an intermediate level MS is nonsignificant. Hines (1996) and Janky (2000) summarize recent work on this topic and present new findings. The emerging consensus does not recommend pooling in anova. We join current practice and have omitted from this text a discussion of pooling mean squares. Despite these cautions, if the reader wishes to pool mean squares, the pooling procedure can be looked up in earlier editions of this volume.

Our second illustration of nested anova is of a design that is perhaps the most common in the literature: a two-level mixed model with equal sample sizes, shown in Box 10.2. The groups represent different strains of houseflies, some known to be DDT resistant, others not. The investigators wanted to test whether there were differences in various morphological characters among these strains and to see whether these differences could be related to DDT resistance. Because the morphology of flies is generally quite labile, depending substantially on environmental influences on the immature stages, three jars of the same medium were prepared for each strain in order to separate the differences among strains from the variances among separate jars. Clearly, if each strain is reared in only one jar, we cannot distinguish differences among jars from differences among strains. The variable reported here is the number of setae on the third abdominal sternum.

The data in Box 10.2 are in abbreviated form. Setae on eight females were counted per jar, but only the means of these eight counts are given in the table; the individual observations are omitted to conserve space. Therefore, the sum of squares within jars (quantity **4**, provided) cannot be computed from the information furnished. This is a nested anova because the jars were allocated at random, three to a strain. The quantities necessary for the computation are shown in the box, as is the completed anova. The box shows that there is no evidence for an added variance component among the jars because the mean square among jars is less than that within jars. It appears therefore that differences in the microclimate of the jars do not affect setae number in these flies. This is not true, however, of other characters. Sokal and Hunter (1955) found that many measurements of lengths of structures, such as wing length, were strongly affected by variation among the jars within strains. We find the differences among strains much larger than expected if the null hypothesis were true ($P = 0.0022$). Because the highest level of classification is Model I, we would ordinarily complete the analysis of these data with multiple comparisons (Sections 9.4 or 9.5), using MS_{subgr} as the error variance for computing standard errors.

BOX 10.2 Two-Level Nested Anova: Equal Sample Sizes, Mixed Model

Eight housefly strains differing in DDT resistance were tested for morphological differences. Three jars of each strain were prepared, and all jars were incubated together. Eight females were taken from each jar after all the adults had emerged. The character analyzed here is setae number on the third abdominal sternum. To conserve space, original observations are not shown. We give only the means for each jar; n (number of items per subgroup) = 8; b (number of subgroups per group) = 3; a (number of groups) = 8; abn (total sample size) = 192.

Strain	Jar	Jar means	Strain means		Strain	Jar	Jar means	Strain means
LDD	1	27.000			LC	1	28.500	
	2	27.750				2	26.875	
	3	26.625	27.1250			3	27.000	27.4583
OL	1	33.375			RH	1	29.500	
	2	38.125				2	30.375	
	3	31.250	34.2500			3	28.250	29.3750
NH	1	27.500			NKS	1	30.125	
	2	26.625				2	29.625	
	3	28.500	27.5417			3	31.750	30.5000
RKS	1	31.750			BS	1	27.875	
	2	31.750				2	25.625	
	3	35.250	32.9167			3	27.500	27.0000

Box 10.2 (continued)

Preliminary Computations

1. $\overline{\overline{Y}} = 29.5208$
2. $SS_{\text{among}} = 1323.42$
3. $SS_{\text{subgr}} = 357.25$
4. $SS_{\text{within}} = 4663.25$

Anova table

	Source of variation	df	SS	MS	F_s	P	Expected MS
$\overline{Y}_A - \overline{\overline{Y}}$	Among strains (groups)	7	1323.42	189.06	8.47	0.00219	$\sigma^2 + n\sigma^2_{B\subset A} + nb\dfrac{\Sigma\alpha^2}{a-1}$
$\overline{Y}_B - \overline{Y}_A$	Among jars within strains (subgroups within groups)	16	357.25	22.33	0.804	0.6798	$\sigma^2 + n\sigma^2_{B\subset A}$
$Y - \overline{Y}_B$	Within jars (error)	168	4663.25	27.76			σ^2
$Y - \overline{\overline{Y}}$	Total	191	6343.92				

SOURCE: Data from Sokal and Hunter (1955).

We now take up a three-level nested anova. The example that we have chosen is a mixed model; the highest level of classification is Model I, and the next two levels of classification are randomly chosen subgroups and subsubgroups (Model II). Box 10.3 features data on the glycogen content of rat livers. The measurement is in arbitrary units. Duplicate readings were made on each of three preparations of rat livers from each of two rats for three different treatments, as shown in the box. The setup of the data should by now be familiar. Note that we have a new symbol, c, the number of subsubgroups per subgroup, which in this case equals 3. If you were to have yet lower level classifications, you could introduce symbols d, e, and so on.

The preliminary computations in Box 10.3 follow the same pattern as those in Box 10.1. Note that we find the grand mean in step **1**, followed by the SS of groups, subgroups within groups, subsubgroups within subgroups, and within subsubgroups in steps **2** to **5**. The formulas are given in the box. It is easy to construct additional formulas if yet another level of variation is added.

The number of degrees of freedom could be provided by formula as in Box 10.1. We feel it is more instructive, however, to reason it out for the individual case. There are $a = 3$ treatments and hence $a - 1 = 2$ degrees of freedom among groups (treatments). There are $b = 2$ rats in each treatment, corresponding to $b - 1 = 1$ degree of freedom for rats in each treatment. Because there are $a = 3$ treatments, we have $a(b - 1) = 3$ df for subgroups within groups (rats within treatments). Each rat resulted in $c = 3$ preparations, yielding $c - 1 = 2$ degrees of freedom for preparations within one rat. There were $ab = 6$ rats in the study, so we have $ab(c - 1) = 12$ df among subsubgroups within subgroups (preparations within rats). Finally, $n = 2$ readings per preparation yield $n - 1 = 1$ df per preparation, and because there are $abc = 18$ preparations in the study, there are $abc(n - 1) = 18$ df within subsubgroups, corresponding to the error term (readings within preparations). These degrees of freedom, shown in the anova table in Box 10.3, total 35, as they should because there are $abcn = 36$ readings in all.

The anova table is shown in Box 10.3. We proceed to test each mean square over the one immediately beneath it. The mean square of subsubgroups within subgroups (preparations within rats) is almost large enough to allow us to reject the null hypothesis ($P = 0.0502$), whereas the MS of subgroups within groups (rats within treatments) allows a clear rejection of the null hypothesis ($P = 0.0141$). Thus, appreciable variation is added by these two levels of influences in the experiment. The F-ratio for the treatment mean square is not large enough for us to reject the null hypothesis ($P = 0.1970$). We conclude that although there is evidence for an added variance among preparations within rats and among rats within treatments, we cannot say that there are differences among treatments. Note that this test is based on only 3 degrees of freedom in the denominator. We might suspect that there really are differences among the treatments but that we cannot pick them up with so few degrees of freedom within treatments. We would be tempted to repeat this experiment using more rats per treatment and then might well be able to reject the null hypothesis. We will return to this example in Chapter 12 to examine the efficiency of its design.

BOX 10.3 Three-Level Nested Anova: Equal Sample Sizes, Mixed Model

Glycogen content of liver in arbitrary units. Duplicate readings on each of three preparations of rat livers from each of two rats for each of three treatments.

Treatments (a = 3)	Control						Compound 217						Compound 217 plus sugar					
Rats (b = 2)	1			2			1			2			1			2		
Preparations (c = 3)	1	2	3	1	2	3	1	2	3	1	2	3	1	2	3	1	2	3
Readings (n = 2)	131	131	136	150	140	160	157	154	147	151	147	162	134	138	135	138	139	134
	130	125	142	148	143	150	145	142	153	155	147	152	125	138	136	140	138	127
Preparation sums	261	256	278	298	283	310	302	296	300	306	294	314	259	276	271	278	277	261
Preparation means	130.5	128.0	139.0	149.0	141.5	155.0	151.0	148.0	150.0	153.0	147.0	157.0	129.5	138.0	135.5	139.0	138.5	130.5
Rat means	132.5000			148.5000			149.6667			152.3333			134.3333			136.0000		
Treatment means	140.5000						151.0000						135.1667					
Grand sum							5120											

Preliminary Computations

1. $\bar{\bar{Y}} = \dfrac{1}{abcn}\displaystyle\sum^{a}\sum^{b}\sum^{c}\sum^{n} Y = \dfrac{1}{36}5120 = 142.2222$

2. SS_{among} (among groups) $= bcn\displaystyle\sum^{a}(\bar{Y}_A - \bar{\bar{Y}})^2 = 1557.55$

3. SS_{subgr} (subgroups within groups) $= cn\displaystyle\sum^{a}\sum^{b}(\bar{Y}_B - \bar{Y}_A)^2 = 797.67$

4. SS_{subsubgr} (subsubgroups within subgroups) $= n\displaystyle\sum^{a}\sum^{b}\sum^{c}(\bar{Y}_C - \bar{Y}_B)^2 = 594.00$

5. SS_{within} (within subsubgroups; error) $= \displaystyle\sum^{a}\sum^{b}\sum^{c}\sum^{n}(Y - \bar{Y}_C)^2 = 381.00$

Box 10.3 (continued)

Rules for working out the degrees of freedom are given in the text.

Anova table

Source of variation		df	SS	MS	F_s	P	Expected MS
$\bar{Y}_A - \bar{\bar{Y}}$	Among groups (treatments)	2	1557.55	778.78	2.93	0.1970	$\sigma^2 + n\sigma^2_{C\subset B} + nc\sigma^2_{B\subset A} + ncb\dfrac{\Sigma\alpha^2}{a-1}$
$\bar{Y}_B - \bar{Y}_A$	Among subgroups within groups (rats within treatments)	3	797.67	265.89	5.37	0.0141	$\sigma^2 + n\sigma^2_{C\subset B} + nc\sigma^2_{B\subset A}$
$\bar{Y}_C - \bar{Y}_B$	Among subsubgroups within subgroups (preparations within rats)	12	594.00	49.50	2.34	0.0502	$\sigma^2 + n\sigma^2_{C\subset B}$
$Y - \bar{Y}_C$	Within subsubgroups (error; readings within preparations)	18	381.00	21.17			σ^2
$Y - \bar{\bar{Y}}$	Total	35	3330.22				

Estimation of Variance Components

$$s^2_{B\subset A} = (265.89 - 49.50)/6 = 36.06 \text{ (rats within treatments)}$$

$$s^2_{C\subset B} = (49.50 - 21.17)/2 = 14.16 \text{ (preparations within rats)}$$

$$s^2 = 21.17 \text{ (readings within preparations)}$$

These can also be expressed as percentages:

$$\text{Since } s^2_{B\subset A} + s^2_{C\subset B} + s^2 = 71.40$$

$$s^2_{B\subset A} \text{ (rats within treatments)} = 100 \times 36.06/71.40 = 50.5\%$$

$$s^2_{C\subset B} \text{ (preparations within rats)} = 100 \times 14.16/71.40 = 19.8\%$$

$$s^2 \text{ (readings within preparations)} = 100 \times 21.17/71.40 = 29.6\%$$

10.3 Nested Anovas with Unequal Sample Sizes

Regrettably, we do not always have equal sample sizes in an experiment or a study. The basic principles of a nested anova are the same, even when sample sizes are unequal. However, there are four additional problems: (1) More complex and cumbersome notation is needed, (2) computations become much more tedious, (3) there are no exact tests of significance (and the approximate tests can be very inexact), and (4) for a fixed total sample size power will be reduced when sample sizes are unequal. These complications were sufficiently annoying in the past for us to urge researchers to design studies with equal sample sizes whenever possible. Nowadays, almost all such work is carried out by prepackaged computer programs, obviating the first two of the above-listed problems.

If we were to provide the fully correct mathematical symbolism, including subscripts for the observations and subscripts and superscripts for the summation signs, the formulas would look forbiddingly complex for what is essentially a simple computational setup. Let us look at an example of a two-level nested anova with unequal sample sizes to see what the notational and computational problems are. Box 10.4 presents such a study, carried out on the blood pH of mice. In this experiment, 15 female mice (dams) were successively mated over a period of time to either 2 or 3 males (sires). These sires were different for the 15 dams; a total of 37 sires were employed. The litters resulting from each mating were kept separate, and the blood pH of female members of these litters was determined. The number of female offspring in these litters ranged from 3 to 5.

The data are arranged in the data table in Box 10.4, which shows individual readings, as well as litter means (the means of the several readings for each litter), litter sizes, dam means (the means of the readings for the several litters of any one dam), the sample sizes on which these dam means are based (the sum of litter sizes), and finally the grand sum of all readings as well as the total sample size of 160 readings. There are $a = 15$ groups (dams) in this study, each of them containing b_i subgroups (sires), where the i refers to the dam number. Thus, b_7 is the number of sires of dam number 7, which equals 3. The number of replicates (individual mice in a litter) must be symbolized with two subscripts, n_{ij}, where i refers to the dam number and j to the sire number of dam i. Thus, $n_{7,2}$ is the sample size of the number of female offspring from the mating of dam 7 and sire 2, which equals 5. The number of replicates for the ith dam is given as n_i which is computed as $\Sigma^{b_i} n_{ij}$. Thus, the first dam has $4 + 4 = 8$ female offspring. To express symbolically how many sires there are in the study, we have to write $\Sigma^a b_i = 37$; to find out how many mice were measured in the entire study, we have to sum the number of mice in each litter for all sires and over all dams as $\Sigma^a \Sigma^{b_i} n_{ij} = 160$.

The computations are straightforward and are given in Box 10.4. The degrees of freedom can be worked out directly from the table. There are 15 dams, and hence 14 degrees of freedom among groups; there are 37 sires within dams, which lose 1 degree of freedom per dam, so their $df = 37 - 15 = 22$. The within-subgroups degrees of freedom are based on 160 mice belonging to 37 litters, and hence losing

BOX 10.4 Two-Level Nested Anova with Unequal Sample Sizes

Blood pH for 3 to 5 female mice within litters resulting from mating 2 or 3 different sires to each of 15 dams. The data have been coded by subtracting 7.0 and then multiplying by 100. $a = 15$ dams; $b_i = 2$ or 3 sires per dam; $n_{ij} = 3$, 4, or 5 mice per litter (from one sire) where $i = 1, \ldots, a$, and $j = 1, \ldots, b_i$. This is a Model II anova.

Dam number	Sire number	Blood pH readings of individual mice 1	2	3	4	5	Litter size n_{ij}	Litter mean $\bar{Y}_B = \frac{1}{n_{ij}}\sum^{b_{ij}} Y_{ijk}$	Sample size of dam mean $n_i = \sum^{b_i} n_{ij}$	Dam mean $\bar{Y}_A = \frac{1}{n_{ij}}\sum_{ij=1}^{b_i}\sum_{k=1}^{n_{ij}} Y_{ijk}$
1	1	48	48	52	54		4	50.50000		
	2	48	53	43	39		4	45.75000	8	48.12500
2	1	45	43	49	40	40	5	43.40000		
	2	50	45	43	36		4	43.50000	9	43.44444
3	1	40	45	42	48		4	43.75000		
	2	45	33	40	46		4	41.00000		
	3	40	47	40	47	47	5	44.20000	13	43.07692
4	1	38	48	46			3	44.00000		
	2	37	31	45	41		4	38.50000	7	40.85714
5	1	44	51	49	51	52	5	49.40000		
	2	49	49	49	50		4	49.25000		
	3	48	59	59			3	55.33333	12	50.83333
6	1	54	36	36	40		4	41.50000		
	2	44	47	48	48		4	46.75000		
	3	43	52	50	46	39	5	46.00000	13	44.84615

Box 10.4 (continued)

Group	Subgroup						n_{ij}	Subgroup mean	Group mean
7	1	41	42	36	47		4	41.50000	
	2	47	36	43	38	41	5	41.00000	
	3	53	40	44	40	45	5	44.40000	42.35714
8	1	52	53	48			3	51.00000	
	2	40	48	50	40	51	5	45.80000	47.75000
9	1	40	34	37	45		4	39.00000	
	2	42	37	46	40		4	41.25000	40.12500
10	1	39	31	30	41	48	5	37.80000	
	2	50	44	40	45		4	44.75000	40.88889
11	1	52	54	52	56	53	5	53.40000	
	2	56	39	52	49	48	5	48.80000	51.10000
12	1	50	45	43	44	49	5	46.20000	
	2	52	43	38	33		4	41.50000	44.11111
13	1	39	37	33	43	42	5	38.80000	
	2	43	38	44			3	41.66667	
	3	46	44	37	54		4	45.25000	41.66667
14	1	50	53	51	43		4	49.25000	
	2	44	45	39	52		4	45.00000	
	3	42	48	45	51	48	5	46.80000	47.00000
15	1	47	49	45	43	42	5	45.20000	
	2	45	42	52	51	32	5	44.40000	
	3	51	51	53	45	51	5	50.20000	46.60000

Grand sum $\displaystyle\sum_{i=1}^{a}\sum_{j=1}^{b_i}\sum_{k=1}^{n_{ij}} Y_{ijk} = 7197$

Total sample size $\displaystyle\sum_{i=1}^{a}\sum_{j=1}^{b_i} n_{ij} = 160$

Box 10.4 (continued)

The structure of the analysis is within litters, among litters due to different sires (sires within dams), and among dams.

Preliminary Computations

1. Grand mean $\bar{\bar{Y}} = \dfrac{1}{\sum^a \sum^{b_i} n_{ij}} \sum^a \sum^{b_i} \sum^{n_{ij}} Y = \dfrac{1}{160}\, 7197 = 44.98125$

2. SS_{among} (among groups) $= \sum^a n_i (\bar{Y}_A - \bar{\bar{Y}})^2 = 1780.174$

3. SS_{subgr} (subgroups within groups) $= \sum^a \sum^{b_i} n_{ij}(\bar{Y}_B - \bar{Y}_A)^2 = 800.237$

4. SS_{within} (within subgroups; error SS) $= \sum^a \sum^{b_i} \sum^{n_{ij}} (Y - \bar{Y}_B)^2 = 3042.533$

Anova Table

	Source of Variation	df	SS	MS (or MS')	F_s (or F'_s)	P	Expected MS
$\bar{Y}_A - \bar{\bar{Y}}$	Among groups (dams)	14	1780.174	127.155	3.496 (3.473)[a]	0.0043 0.0050	$\sigma^2 + n'_0\sigma^2_{B\subset A} + (nb)_0\sigma^2_A$
$\bar{Y}_B - \bar{Y}_A$	Among subgroups (sires within dams)	22	800.237	36.374	1.470 (36.616)	0.0968	$\sigma^2 + n_0\sigma^2_{B\subset A}$
$Y - \bar{Y}_B$	Within subgroups (error; among mice of one litter— that is, from one sire)	123	3042.533	24.736			σ^2
$Y - \bar{\bar{Y}}$	Total	159	5622.944				

[a] Because the coefficient n_0 is not equal to n'_0, there is no MS over which we can test MS_{among} exactly. See the following discussion on how this value of F'_s has been obtained.

Box 10.4 (continued)

Tests of Significance

As can be seen from the Expected *MS* column of the preceding anova table, MS_{subgr} is tested directly over MS_{within}. Its *P*-value equals 0.0968, and we accept the null hypothesis that mice from the same litter—that is, the offspring of the same sire—do not present an added variance component. The BIOMstat computer program furnishes the coefficients of the variance components as follows:

$$n_0' = 4.376;\ n_0 = 4.287;\ (nb)_0 = 10.625$$

Solving for estimates of the 3 unknown variance components:

$$MS_{among} = 127.155 = s^2 + 4.376s^2_{BCA} + 10.625s^2_A$$

$$MS_{subgr} = 36.374 = s^2 + 4.287s^2_{BCA}$$

$$MS_{within} = 24.736 = s^2$$

Therefore

$$s^2 = MS_{within} = 24.736$$

$$s^2_{BCA} = \frac{MS_{subgr} - MS_{within}}{4.287} = \frac{36.374 - 24.736}{4.287} = 2.715$$

$$s^2_A = \frac{MS_{among} - MS_{within} - 4.376s^2_{BCA}}{10.625} = \frac{127.155 - 24.736 - (4.376)(2.715)}{10.625} = 8.521$$

Before we can perform a test of the hypothesis that $\sigma^2_A = 0$ using the MS_{among}, we must check whether the given data satisfy the Gaylor and Hopper (1969) conditions under which the Satterthwaite approximation can be used. First, verify that $df_{subgr} < 100$ and that $df_{subgr} < 2df_{within}$. In this case, because $df_{subgr} = 22$ and $2 df_{within} = 246$, the conditions are satisfied. Next, test that the ratio R is greater than the criterion C.

Box 10.4 (continued)

$$R = \frac{n_0'}{n_0' - n_0} \times \frac{MS_{subgr}}{MS_{within}}$$

$$= \frac{4.376}{(4.376 - 4.287)} \times \frac{36.374}{24.736} = 72.302$$

$$C = F_{.025\,[df_{within},\,df_{subgr}]} \times F_{.5\,[df_{subgr},\,df_{within}]}$$

$$= 2.074 \times 0.975 = 2.022$$

Because $R > C$ by Gaylor and Hopper's criterion, it is safe to use Satterthwaite's approximation (the fact that the weight w_1, computed as follows, is negative indicated that caution should still be used). (If these conditions were not met, only the simple approximate test, $F_s = MS_{among}/MS_{subgr}$ could be made.) For Satterthwaite's approximation, one evaluates the following weights

$$w_2 = \frac{n_0'}{n_0} = 1.0208$$

$$w_1 = 1 - w_2 = -0.0208$$

which are used to compute a new, synthetic denominator:

$$MS'_{subgr} = w_1 MS_{within} + w_2 MS_{subgr}$$

$$= -0.0208(24.736) + 1.0208(36.374)$$

$$= 36.616$$

This mean square is equal to

$$s^2 + n_0' s_{BCA}^2 = 24.736 + 4.376(2.715) = 36.617$$

Box 10.4 (continued)

(within rounding error). The Satterthwaite formula for the degrees of freedom of the reconstituted mean square MS'_{subgr} is

$$df'_{subgr} = \frac{(MS'_{subgr})^2}{\dfrac{(w_1 MS_{within})^2}{df_{within}} + \dfrac{(w_2 MS_{subgr})^2}{df_{subgr}}}$$

$$= \frac{(36.616)^2}{\dfrac{(-0.0208 \times 24.736)^2}{123} + \dfrac{(1.0208 \times 36.374)^2}{22}}$$

$$= 21.39 \approx 21$$

Finally,

$$F'_s = \frac{MS_{among}}{MS'_{subgr}} = \frac{127.154}{36.617} = 3.473$$

There is evidence for a variance component among dams, but, as we have seen, not for an added variance among sires.

37 degrees of freedom, which yields 123 df. By formula, the degrees of freedom could be obtained as $a - 1$, $\Sigma^a(b_i - 1)$, and $\Sigma^a\Sigma^{b_i}(n_{ij} - 1)$, respectively.

The first test is of the MS_{subgr} to test for the presence of the subgroups variance component. Note that $F_s = 1.470$ yields $P = 0.0968$. We have inadequate evidence of an added variance component among sires within dams. The next test, MS_{among} over MS_{subgr}, is our initial encounter with the third of the problems listed above that beset nested analyses of variance with unequal sample sizes. Not only must average sample sizes be computed (using a method similar to the special formula learned in Box 9.2) but also a glance at the expected mean squares of the anova table in Box 10.4 shows that the average coefficients of the same variance components do not correspond at different levels. Notice that $\sigma^2_{B \subset A}$ has n_0 as a coefficient in MS_{subgr} but has another quantity, n'_0, as a coefficient in MS_{among}. The numerical values for these coefficients as worked out by the computer program BIOMstat are also shown in Box 10.4.

In this example, the coefficients have similar values—4.287 for n_0 and 4.376 for n'_0—but in other examples, they can differ considerably. Next, we estimate the three unknown variance components—within subgroups, among subgroups within groups, and among groups—by simple substitutions as shown in Box 10.4. Thus, a general problem arises from unequal sample sizes: There are no exact tests of significance for testing the upper $k - 1$ levels in a k-level nested anova (although a resampling method could be used). Box 10.4 shows that the ratio $F_s = MS_{among}/MS_{subgr}$ does not simply estimate $(nb)_0 \sigma^2_A$ because $n_0 \neq n'_0$. We can perform the test by synthesizing a new denominator mean square (employing what is commonly referred to as the *Satterthwaite approximation*) against which to test the mean square of groups. This denominator mean square, MS'_{subgr}, has $s^2_{B \subset A}$ multiplied by $n'_0 = 4.376$ instead of $n_0 = 4.287$ as before.

We now obtain a variance ratio $F'_s = 3.473$ whose probability must be evaluated. The degrees of freedom of the denominator mean square are then calculated. First, we rewrite the new mean square as a linear combination of the previous mean squares, as shown in Box 10.4. Then we calculate df'_{subgr} from the coefficients of the old mean squares, which in this case is 21.39. To be conservative, we use 21 degrees of freedom for the denominator mean square and compare the $F_s = 3.473$ value to an $F_{[14,21]}$ distribution. We find it to be 0.005,04 and thus reject the null hypothesis that the added variance component is zero.

The test based on the Satterthwaite approximation can be unreliable when n_0 is different from n'_0 and $\sigma^2_{B \subset A}$ is small in comparison to σ^2 (Gaylor and Hopper, 1969; Boardman, 1974). Khuri (1995) warns that the approximation does not hold well when some of the w_i used to compute a synthetic MS are negative. Gaylor and Hopper (1969) suggest two rules for deciding when the Satterthwaite approximation can be used. These are given in Box 10.4. The conditions implied by these rules are satisfied for the example in Box 10.4. If these conditions are not met, then probably the only test one can do is the ordinary F-test (MS_{among}/MS_{subgr}). Using the Satterthwaite approximation when the required conditions are not satisfied tends to yield excessively conservative tests (in some Monte Carlo simulations, we have observed a type I error of 0.4% at an intended 5% level). This subject requires more theoretical work so that at least good approximate tests can be made in all cases.

The biological interpretation of this example is interesting. Because sires within dams did not show an added variance component, differentiation does not seem to depend on the sires. Either the sires came from an inbred strain and were identical genetically, or the variance among dams, but not among sires, can be explained through maternal effects. In this way, the analysis of variance can be of great help, suggesting further research in problems of genetics.

The next example is a three-level nested anova with unequal sample sizes. This analysis is carried out not for a planned experiment but to analyze a series of data that had accumulated in the laboratory. The data are diameters of pollen grains of a species of tree that were deposited in deep layers at the margin of a marsh. Samples of this pollen were analyzed; the variable measured was mean diameter of 100 pollen grains found on one slide. That the basic "observation" in this study is a mean need not concern us here. The pollen samples were grouped into three depths—the highest-level category in this study—which clearly are fixed treatment effects. For the first two depths, two core samples each had been obtained, but three were available for the third depth. Varying numbers of preparations had been made from these core samples and varying numbers of slides examined from these preparations. The structure of the analysis is shown in the table in Box 10.5; to conserve space, the data themselves are not given. The core samples within depths and the preparations within these samples can be considered to have been obtained at random; for this reason, the analysis becomes a mixed-model nested anova.

Although not shown in Box 10.5, the computations (given the data) could easily be derived from a combination of Boxes 10.3 and 10.4. The result of these computations is the anova table in Box 10.5. The purpose of this box is to show the analysis of this table because of the complexities of significance tests and estimating variance components in an example with unequal sample sizes. The first test, MS of subsubgroups over the error MS, is exact and yields $F_s = 1.712, P = 0.1324$, which does not permit rejection of the null hypothesis that the added variance component of preparations within core samples is zero.

Both $MS'_{subsubgr}$ and MS'_{subgr} satisfy the Gaylor and Hopper criteria, so the Satterthwaite approximation can be used with MS_{subgr} and MS_{among} to test for corresponding variance components or treatment effects (only the former allowed us to reject the null hypothesis). Again, we find that F and F' are similar. A Monte Carlo study of this design shows that if $\sigma^2_{B \subset A}$ were equal to zero, then the F'-test would have been excessively conservative. Out of 1000 replications of this design, the null hypothesis was rejected only 0.4% of the time at the 5% level, and none were rejected at the 1% level. The F-test yielded the correct probabilities. Thus, the Satterthwaite approximation cannot be used uncritically. The estimation of the variance components is also shown in detail in Box 10.5. No new procedures are introduced; they simply become more involved because of the extra level of variation. Because this is a mixed model, the highest level is Model I, and we do not estimate a variance component for it but simply perform a test of significance for added treatment effects. The box shows computation of a variance component for this level to complete the outline of all necessary computations for other examples that might be pure Model II.

BOX 10.5 Three-Level Nested Anova with Unequal Sample Sizes

Diameter of pollen grains of a species of tree from a paleobotanical study. All core samples were collected from the same locality at the margin of a marsh. Several core samples were obtained at each of three depths; varying numbers of preparations were made from each core sample and varying numbers of slides examined from each preparation. "Individual observations" are means of diameters of 100 pollen grains on a slide. Actual data are not shown here. This is a mixed-model anova.

The structure of the analysis and the sample sizes

Depths (a)		1			2				3		
Core samples (b_i)	1		2		1		2		1	2	3
Preparations (c_{ij})	1	2	1	2	1	2	3	1	2	1	2	1

Number of slides per preparation (n_{ijk})	4	5	4	4	5	4	4	5	5	3	4	2	2	5
Number of slides in core sample (n_{ij})		9		8			13		10		7	4	5
Number of slides per depth (n_i)			17				23				16		
Total sample size (n)							56						

From the original data, the various sums of squares were computed following the procedures in Boxes 10.3 and 10.4. The results of these computations are given in the following table.

Anova Table

	Source of variation	df	SS	MS (or MS')	F_s (or F_s')	P
$\overline{Y}_A - \overline{\overline{Y}}$	Among groups (depths)	2	396.64	198.32	1.239	0.3813
					(1.131)	0.4080
$\overline{Y}_B - \overline{Y}_A$	Among subgroups within groups (core samples within depths)	4	640.44	160.11	4.194	0.0480
					(175.40)	(4.466)	0.0416
$\overline{Y}_C - \overline{Y}_B$	Among subsubgroups within subgroups (preparations within core samples)	7	249.41	35.63	1.712	0.1324
					(35.85)		
$\overline{Y} - \overline{Y}_C$	Within subsubgroups (error; within preparations)	42	874.02	20.81		
$Y - \overline{\overline{Y}}$	Total	55	2160.51			

Note: The F_s' values are not computed as simple variance ratios of the mean squares in the table because of unequal coefficients in the expected mean squares. Their computation is explained in the following discussion.

Box 10.5 (continued)

As in Box 10.4, only the next-to-the-lowest MS (MS_{subsubgr}) may be tested exactly. In the current case, we are unable to reject the null hypothesis of no added variance of subsubgroups within subgroups

For the approximate tests of significance involving the other mean squares, we must calculate the coefficients of the variance components, estimate the components, and then synthesize new mean squares to use as the denominators of the F_S ratios if the Satterthwaite approximation is used.

The expected mean squares

Source of variation	Expected MS for a Model II
Among groups	$\sigma^2 + n_0''\sigma^2_{CCB} + (nc)_0'\sigma^2_{BCA} + (ncb)_0\sigma^2_A$
Subgroups within groups	$\sigma^2 + n_0'\sigma^2_{CCB} + (nc)_0\sigma^2_{BCA}$
Subsubgroups within subgroups	$\sigma^2 + n_0\sigma^2_{CCB}$
Within subsubgroups (error)	σ^2

In our case, we have a mixed model. The highest level is differentiated by fixed treatment effects. The top line of the expected mean squares (MS_{among}) should therefore read

$$\sigma^2 + n_0''\sigma^2_{CCB} + (nc)_0' \sigma^2_{BCA} + (ncb)_0 \frac{\sum \alpha^2}{a - 1}$$

Again, we obtain the coefficients of the variance components by a computer program.

$$n_0'' = 4.160; \; n_0' = 3.985; \; n_0 = 3.927$$
$$(nc)_0' = 8.425; \; (nc)_0 = 7.538; \; (ncb)_0 = 18.411$$

To solve for estimates of the four unknown variance components, we must solve the four simultaneous equations in which the four variances are the unknowns.

$$MS_{\text{among}} = 198.32 = s^2 + 4.160s^2_{CCB} + 8.425s^2_{BCA} + 18.411s^2_A$$
$$MS_{\text{subgr}} = 160.11 = s^2 + 3.985s^2_{CCB} + 7.538s^2_{BCA}$$
$$MS_{\text{subsubgr}} = 35.63 \quad = s^2 + 3.927s^2_{CCB}$$
$$MS_{\text{within}} = 20.81 \quad = s^2$$

Therefore,

$$s^2 = MS_{\text{within}} = 20.81$$

$$s^2_{CCB} = \frac{MS_{\text{subsubgr}} - MS_{\text{within}}}{3.927} = \frac{35.63 - 20.81}{3.927} = 3.774$$

$$s^2_{BCA} = \frac{MS_{\text{subgr}} - MS_{\text{within}} - 3.985s^2_{CCB}}{7.538}$$

$$= \frac{160.11 - 20.81 - [(3.985)(3.774)]}{7.538} = 16.485$$

Variance components may also be expressed as percentages:

Box 10.5 (continued)

$$\text{Since } s_{BCA}^2 + s_{CCB}^2 + s^2 = 41.069$$

$$s_{BCA}^2 \text{ (core samples within depths)} = 100 \times 16.485/41.069 = 40.1\%$$

$$s_{CCB}^2 \text{ (preparations within core samples)} = 100 \times 3.774/41.069 = 9.2\%$$

$$s^2 \text{ (slides within preparations)} = 100 \times 20.81/41.069 = 50.7\%$$

In this example, there is no s_A^2 because the highest-level classification is due to fixed treatment effects. However, we show the computation of s_A^2 here to provide a complete pattern for all computations in a Model II three-level anova.

$$s_A^2 = \frac{MS_{among} - MS_{within} - 4.160s_{CCB}^2 - 8.425s_{BCA}^2}{18.411}$$

$$= \frac{198.32 - 20.81 - [(4.160)(3.774)] - [(8.425)(16.485)]}{18.411} = 1.245$$

If the design is entirely Model II, then it would be appropriate to include s_A^2 in the sum of the variance components in the denominator of the fraction that evaluates the percentage variation. This would yield $100 \times 1.245/42.314 = 2.9\%$ for s_A^2 and 39.0%, 8.9%, and 49.2% for s_{BCA}^2, s_{CCB}^2, and s^2, respectively.

Tests of Significance

$MS_{subsubgr}$ may be tested directly over MS_{within}:

$$F_s = \frac{35.63}{20.81} = 1.712 \qquad P = 0.1324$$

Before we test MS_{subgr} we check with the Gaylor and Hopper (1969) criteria whether the Satterthwaite approximation can be used, as was done in Box 10.4.

First we verify that $df_{subsubgr} = 7 < 2\, df_{within} = 2(42) = 84$. Then we compute the ratio.

$$R = \left(\frac{n_0'}{n_0' - n_0}\right)\frac{MS_{subsubgr}}{MS_{within}} = \left(\frac{3.985}{3.985 - 3.927}\right)\frac{35.63}{20.81} = 117.657$$

This value is to be compared to the criterion

$$C = F_{.025[df_{within},\, df_{subsubgr}]} \times F_{.5[df_{subsubgr},\, df_{within}]}$$

$$= F_{.025[42,7]} \times F_{.5[7,42]}$$

$$= 4.301(0.921) = 3.961$$

Because $R > C$, it is safe by the Gaylor and Hopper (1969) criteria to use the Satterthwaite approximation given below. However, the negative value for w_1 implies that caution should be used.

If these conditions were not met, then one would have to carry out the approximate test $F_s = MS_{subgr}/MS_{subsubgr}$ (which, in this case, happens to give very similar results).

Box 10.5 (continued)

For the Satterthwaite approximation, compute the weights:

$$w_2 = \frac{n_0'}{n_0} = \frac{3.985}{3.927} = 1.0148$$

$$w_1 = 1 - w_2 = -0.0148$$

The new synthesized mean square is

$$MS_{subsubgr}' = w_1 MS_{within} + w_2 MS_{subsubgr}$$

$$= -0.0148(20.81) + 1.0148(35.63)$$

$$= 35.8493$$

Its degrees of freedom are computed as

$$df_{subsubgr}' = \frac{(MS_{subsubgr}')^2}{\dfrac{(w_1 MS_{within})^2}{df_{within}} + \dfrac{(w_2 MS_{subsubgr})^2}{df_{subsubgr}}}$$

$$= \frac{(35.8493)^2}{\dfrac{[-0.0148(20.81)]^2}{42} + \dfrac{[1.0148(35.63)]^2}{7}} = 6.88 \approx 7$$

$$F_s' = \frac{MS_{subgr}}{MS_{subsubgr}'} = \frac{160.11}{35.8493} = 4.466 \quad P = 0.0416$$

We reject the null hypothesis that the variance component among subgroups (among core samples within a depth) is zero.

Similar procedures are used to test MS_{among}. First, check that $df_{among} = 2 < 2\, df_{subgr} = 2(4) = 8$. Then compute the now much more complex ratio

$$R = \frac{n_0(nc)_0' MS_{subgr}}{[n_0''(nc)_0 + (nc)_0' n_0 - n_0(nc)_0 - (nc)_0' n_0'] MS_{within} + [(nc)_0' n_0' - n_0''(nc)_0] MS_{subgr}}$$

$$= \frac{3.927(8.425)160.11}{[4.160(7.538) + 8.425(3.927) - 3.927(7.538) - 8.425(3.985)]20.81 + [8.425(3.985) - 4.160(7.538)]35.63}$$

$$= 50.296$$

The criterion C is now

$$C = F_{.025[df_{(within+subsubgr)},\, df_{subgr}]} \times F_{.5[df_{subgr},\, df_{(within+subsubgr)}]}$$

$$= F_{.025[49,4]} \times F_{.5[4,49]}$$

$$= 8.323(0.851) = 7.134$$

Box 10.5 (continued)

Because $R > C$, we may use the Satterthwaite approximation again (however, the negative w_3 and w_4 values computed as follows imply that caution should be used). Compute the weights:

$$w_5 = \frac{(nc)_0'}{(nc)_0} = \frac{8.425}{7.538} = 1.1177$$

$$w_4 = \frac{n_0''}{n_0} - \frac{w_5 n_0'}{n_0} = \frac{4.160}{3.927} - \frac{(1.1177)3.985}{3.927} = -0.0749$$

$$w_3 = 1 - w_5 - w_4 = 1 - 1.1177 - (-0.0749) = -0.0428$$

$$MS'_{subgr} = w_3 MS_{within} + w_4 MS_{subsubgr} + w_5 MS_{subgr}$$

$$= -0.0428(20.81) - 0.0749)35.63 + 1.1177(160.11)$$

$$\approx 175.3956$$

The F'_s value is therefore

$$F'_s = \frac{MS_{among}}{MS'_{subgr}} = \frac{198.32}{175.3956} = 1.131 \quad P = 0.4080,$$

It is unlikely that an adjustment of the degrees of freedom will enable us to reject the null hypothesis for such a small F-value, but we compute the degrees of freedom as an illustration.

$$df'_{subgr} = \frac{(MS'_{subgr})^2}{\dfrac{(w_3 MS_{within})^2}{df_{within}} + \dfrac{(w_4 MS_{subsubgr})^2}{df_{subsubgr}} + \dfrac{(w_5 MS_{subgr})^2}{df_{subgr}}}$$

$$= \frac{(175.3956)^2}{\dfrac{[-0.0428(20.81)]^2}{42} + \dfrac{[-0.0749(35.63)]^2}{7} + \dfrac{[1.1177(160.11)]^2}{4}}$$

$$= 3.842 \simeq 4$$

Therefore, the degrees of freedom are unchanged, as is the P-value.

The conclusions from the analysis are that there is an added variance component among subgroups (core samples), while we are unable to demonstrate differences among preparations or depths. We therefore conclude that the diameter of the pollen grains shows added variance among core samples but not among preparations and does not differ for the depths recorded in this study. Thus, there must be some local heterogeneity in this variable, and different core samples appear to yield differing pollen samples. This heterogeneity is reflected in the percentage of variation attributable to the several levels: 40.1% among core samples, 9.2% among preparations within core samples, and 50.7% among slides within preparations. The preparations seem to represent core samples with high replicability, but the techniques for making slides and measuring the pollen grains on them merit further examination and refinement in view of the high percentage of variation of that level. Similarly, these results show substantial variation among core samples. Each depth should be adequately sampled by cores.

> The general method of analysis of nested anovas is the same for any number of levels. To enable you to generate coefficients for variance components for cases with unequal samples sizes other than those discussed in this section, general rules are given in Raktoe et al. (1981). Most computer programs furnish these coefficients automatically. BIOMstat computes nested anovas up to 10 levels, and in the case of unequal sample sizes, with and without the Satterthwaite approximation.

EXERCISES 10

10.1 J. A. Weir (unpublished results) took blood pH readings on mice of two strains that had been selected for high and low blood pH and obtained the following data on male litter mates. Only litters with at least four males were considered, and four males were selected at random whenever more than four males were present in a litter. Data are presented on seven litters for each strain. Analyze the data, test whether the pHH (high) strain has a higher average pH value than the pHL (low) strain (a one-sided test), and compute variance components and percent variation among males within litters and litters within strains.

ANSWER: Percent variation of litters within strains = 23.40%, $F_s = 2.222$ with $df = 12$, 42.

Strain	Litter Code Number	pH Readings			
pHH	387	7.43	7.38	7.49	7.49
	388	7.39	7.46	7.50	7.55
	389	7.53	7.50	7.63	7.47
	401	7.39	7.39	7.44	7.55
	402	7.48	7.43	7.47	7.44
	404	7.43	7.55	7.44	7.50
	405	7.49	7.49	7.51	7.54
pHL	392	7.40	7.46	7.43	7.42
	408	7.35	7.40	7.46	7.38
	413	7.51	7.39	7.42	7.43
	414	7.46	7.53	7.49	7.45
	415	7.48	7.53	7.52	7.43
	434	7.43	7.40	7.48	7.47
	446	7.53	7.47	7.50	7.53

10.2 Hasel (1938) made a study of sampling practices in timber surveys. The observations of board feet of timber in a ponderosa pine forest. He studied 9 sections, each subdivided into 4 quarter sections [each of which can be subdivided into 4 forty-acre tracts, or

16 ten-acre tracts, or 32 five-acre tracts, or 64 basic plots (of two-and-a-half acres)]. If sampling and distribution of timber were entirely at random, no added variance should exist beyond the basic plots.

Interpret the analysis of variance given in the following table and isolate variance components. To keep the notation uniform, identify variance components as follows: s^2 for basic plots, s_5^2 for 5s, s_{10}^2 for 10s, s_{40}^2 for 40s, s_Q^2 for quarter sections, and s_S^2 for sections.

Source of variation	df	SS
Between sections	8	109,693.35
Quarter sections within sections	27	57,937.40
40s within quarter sections	108	66,569.53
10s within 40s	432	161,109.77
5s within 10s	576	121,940.82
Basic plots within 5s	1152	133,337.74
Total	2303	650,588.61

Retain the results obtained. You will need them for the exercises of Chapter 12 to come.

10.3 The following are unpublished results from a study by R. R. Sokal on geographic variation in the aphid *Pemphigus populitransversus*. Mean length for antennal segment IV of stem mothers was computed for each of 75 localities in eastern North America. The means were grouped and coded 1 through 8. The data were subdivided according to drainage system along which the localities occur. A hierarchic classification was set up as follows. The drainages were divided into 3 coasts: East Coast, Gulf Coast, and Great Lakes. Each coast was subdivided into a number of areas, one of which, the Mississippi area, was further subdivided into 10 river systems. Analyze the data to determine whether differences in the length of antennal segment IV occur among the coasts. At what hierarchic level is there the greatest amount of variation? Interpret your findings. This problem might appear to be a Model I anova throughout. For our purposes, however, the areas and river systems are considered random samples from the available equivalent sampling units, and the emphasis is on the relative amounts of variation at these hierarchic levels. The data are presented in summary form.

Coast	Area	River System	n_{ijk}	$\sum\limits^{n_{ijk}} Y_{ijkl}$
East	New England		3	13
	Hudson		2	12
	Chesapeake Bay		2	11
	North Carolina		1	4
	Florida		1	1
Gulf	East		4	14
	Central		4	12
	Mississippi	Red	2	6
		Arkansas	7	29
		Lower Mississippi	10	33

(*Table continued on page 317*)

Coast	Area	River System	n_{ijk}	$\sum^{n_{ijk}} Y_{ijkl}$
		Illinois	6	32
		Missouri	3	15
		Upper Mississippi	6	34
		Lower Ohio	5	13
		Tennessee-Cumberland	3	9
		Wabash	2	9
		Middle Ohio	7	29
Great Lakes			7	39

$$SS_{total} = \sum^a \sum^{b_i} \sum^{c_{ij}} \sum^{n_{ijk}} (Y_{ijkl} - \overline{\overline{Y}})^2 = 226.74667$$

These data will present some computational problems for most packaged computer programs, including BIOMstat, because sums are given rather than the raw data. We suggest tricking the program by first calculating the mean for each area or river system and entering it as many times as needed to satisfy the stated sample size. Thus, for the first line of the table (New England), you should furnish 3 values of $13/3 = 4.333, 4.333, 4.333$. The program will then compute all the mean squares required, except that the error MS will be zero. However, its numerical value can be obtained from the information given above.

ANSWER: $MS_{Areas \subset Coasts} = 4.0133$; EMS for this MS is $\sigma^2 + 2.5333\sigma^2_{RCA} + 3.544\sigma^2_{ACC}$; s^2_{ACC} is negative.

10.4 Hanna (1953) studied hair pigment concentration (measured with a spectrophotometer and expressed as optical densities) in 39 pairs of monozygous and 40 pairs of dizygous twins. Two hair samples were taken from each person and hydrolyzed separately. Three readings were taken for each sample. Throughout the study, samples were tested in a random sequence to avoid bias in reading optical densities.

Source of variation	Monozygous Twins		Dizygous Twins	
	df	MS	df	MS
Among pairs	38	2676.1734	39	2509.9174
Between twins within pairs	39	44.2005	40	307.0163
Between samples within twins	78	3.2133	80	2.5451
Among readings with samples (error)	312	0.0662	320	0.0334

Complete the analysis of variance and estimate the added variance components. Compare and interpret the differences in the estimated variance components for monozygous and dizygous twins.

ANSWER: s^2_{TCP} equals 6.8312 for the monozygous twins and 50.7452 for the dizygous twins.

10.5 In an experiment to test the reliability of fat content determinations in dried eggs, an investigator sent out samples for analysis to six commercial laboratories. The contents of a single can of dried egg powder were stirred thoroughly, and each lab was sent two samples from the can. The labs were told that the two samples represented two "types," but in fact they were merely random samples from the same can. Each lab appointed two technicians to make independent assays of both samples, and each technician made two readings of each sample. These readings (within samples and technicians) furnish the basic error variance of the study. The three null hypotheses being tested are that there are no added variance components at the levels of technicians, samples, and laboratories. The data (from Bliss, 1967) are given in the following table. Test the three null hypotheses and estimate the variance components.

Fat Content	Lab	Technician	Sample	Fat Content	Lab	Technician	Sample
62	I	1	1	18	IV	1	1
55	I	1	1	47	IV	1	1
34	I	1	2	53	IV	1	2
24	I	1	2	32	IV	1	2
80	I	2	1	40	IV	2	1
68	I	2	1	37	IV	2	1
76	I	2	2	31	IV	2	2
65	I	2	2	43	IV	2	2
30	II	1	1	35	V	1	1
40	II	1	1	39	V	1	1
33	II	1	2	37	V	1	2
43	II	1	2	33	V	1	2
39	II	2	1	42	V	2	1
40	II	2	1	36	V	2	1
29	II	2	2	20	V	2	2
18	II	2	2	41	V	2	2
46	III	1	1	37	VI	1	1
38	III	1	1	43	VI	1	1
27	III	1	2	28	VI	1	2
37	III	1	2	36	VI	1	2
37	III	2	1	18	VI	2	1
42	III	2	1	20	VI	2	1
45	III	2	2	26	VI	2	2
54	III	2	2	6	VI	2	2

11

Two-Way and Multiway Analysis of Variance

This chapter covers designs in which the effects of two or more factors are considered simultaneously. Single-factor analyses were discussed in Chapter 9. Although Chapter 10 tested subclasses representing a separate source of variation, these were hierarchic, or nested, within the major classification. In this chapter, the sources of variation are of equal rank. Thus, we can group the observations by two separate kinds of treatments—for example, different doses of a drug and different temperatures at which these drugs are tested. Section 11.1 gives further details of this design. The computation of such an anova for replicated subclasses (more than one observation per subclass or factor combination) is shown in Section 11.2, which also contains a discussion of the meaning of interaction as used in statistics.

Significance testing in a two-way anova is the subject of Section 11.3. Section 11.4 discusses two-way anovas without replication (only a single observation per subclass). The well-known method of paired comparisons is treated in Section 11.5 as a special case of a two-way anova without replication. Sections 11.6 to 11.9 extend the two-way analysis of variance of the previous sections to the simultaneous consideration of three or more factors. In Section 11.6, we briefly discuss the nature of the design and the new problems encountered during the analysis. A three-factor (three-way) anova is illustrated in Section 11.7. Section 11.8 treats multiway anovas with more than three main effects. Section 11.9 enumerates some of the more common experimental designs, other than those already learned, that might be employed by readers of this book. Appropriate instances for application are given and references to detailed expositions of each design are cited. Finally, in Section 11.10, we discuss computer methods for the analysis of variance.

11.1 Two-Way Anova: Design

From the single-classification anova of Chapter 9, we progress to the two-way anova of this chapter by a single logical step. Individual items may be grouped into classes representing the different possible combinations of two treatments or factors. For example, the experiment reported in Table 9.1 testing five sugar treatments on pea sections could be carried out at two different pH levels. In this case, we would want to know not only whether sucrose induced a different growth than the control medium did but also whether pH treatment 1 affected the pea sections differently from treatment 2. Obviously, each combination of factors should be represented by a randomly chosen sample of pea sections. Thus, for five sugars and two pH levels, we need at least $5 \times 2 = 10$ samples. Such a design is a **two-way analysis of variance** of the effects of sugars and of pH treatments.

This method of anova assumes that a given pH and a given sugar each contribute a certain amount to the growth of a pea section and that these two contributions add their effects without influencing each other. In the next section, we will see how departures from this assumption are measured. Consideration of the expression for decomposing observations in a two-way anova is also deferred to Section 11.2.

The two factors in the present design may represent either Model I or Model II effects or one of each, in which case we talk of a **mixed model**.

Beginners are often confused between two-level nested anovas and two-way anovas. For example, if we subject samples of five pea sections not only to different sugar treatments but also to two pH levels, in what way does this differ from a nested design in which each sugar solution is prepared twice, so there are two batches of sugar made up for each of the treatments? We represent the design of the two-way anova as follows. The asterisks represent the individual observations.

	Sugar Treatments				
	1	2	3	4	5
pH level 1	*	*	*	*	*
	*	*	*	*	*
	*	*	*	*	*
	*	*	*	*	*
	*	*	*	*	*
pH level 2	*	*	*	*	*
	*	*	*	*	*
	*	*	*	*	*
	*	*	*	*	*
	*	*	*	*	*

The nested anova, on the other hand, is represented by the following arrangement. Again, the asterisks represent the individual observations.

	Sugar Treatments									
	1		2		3		4		5	
Sugar Batches	1	2	1	2	1	2	1	2	1	2
	*	*	*	*	*	*	*	*	*	*
	*	*	*	*	*	*	*	*	*	*
	*	*	*	*	*	*	*	*	*	*
	*	*	*	*	*	*	*	*	*	*
	*	*	*	*	*	*	*	*	*	*

Why can't we rearrange these data into a two-way table with batches at right angles to sugar treatments as in the first table? And why not nest the pH levels inside sugar treatments as in the second table? The reason is that the first (two-way) arrangement implies that the two pH classes are *common* to the entire study—that is, that pH level 1 is logically the same for all the sugar treatments and so is pH level 2, although their effects may not be the same for all sugars for the variable tested. Arranging the pH levels nested within each sugar treatment would imply that the two pH levels per treatment were random samples from all possible such levels and that pH level 1 in treatment 1 is not the same as pH level 1 in treatment 2—which is, of course, nonsense for these data. Conversely, if we tried to arrange the data from the nested analysis as a two-way anova, we would imply that batches 1 and 2 had the same meaning for all the strains in the experiment. This is not so. Batch 1 for treatment 1 has no closer relation to batch 1 in treatment 2 than it does to batch 2 in that treatment. Batches 1 and 2 are simply arbitrary designations for the two randomly prepared sugar solutions that represent each treatment. By contrast, if all batches labeled 1 were prepared on one day and all batches labeled 2 were made on the following day, then the "1" and "2" would represent common information for the study that should properly be arranged as a two-way anova.

The critical question to be asked is always this: *Does the arrangement of the data into a two-way table correctly imply a correspondence across the classes?* If there is a correspondence across classes, the two-way design is appropriate. If there is no correspondence across classes and we recognize that the factor represents only random subdivisions of the classes of another factor (as the batches within treatments), then we have a nested anova. The nested factor must thus be Model II.

This example also illustrates the difficulty in carrying out the laboratory procedure in a manner compatible with the statistical design. To have two levels of pH for each sugar treatment, one must make up two batches. Thus, the random variation due to batches within a sugar treatment is confounded with the effect of pH. To be a proper two-way design, each pH–sugar combination would have to be prepared independently (and in a random order to guard against subtle biases).

We will now illustrate the computation of a two-way anova. You will obtain closer insight into the structure of this design as we explain the computations.

11.2 Two-Way Anova with Equal Replication: Computation

We will explain the computation of a two-way anova using the simplest possible example—two factors, each divided into two classes. In a study of the inactivation of the effects of vitamin A by rancid fat, it became important to study differences in food consumption when rancid lard was substituted for fresh lard in the diet of rats. Box 11.1 shows the results of an experiment conducted on 12 rats, 6 males and 6 females, in which 3 randomly selected individuals of each sex were fed fresh lard and the other 3 rancid lard. The following analysis of these data is designed to be as instructive as possible. It is not, however, the most concise method of carrying out the computational steps; such a method is illustrated in Box 11.2, which will be considered later in this section. Notice some minor changes in symbolism in Box 11.1. We have called the number of rows in the table r and the number of columns c. This

BOX 11.1 Two-Way Anova with Replication (2 × 2 Case)

Differences in food consumption when rancid lard was substituted for fresh lard in the diet of rats. Food eaten (in grams) during 73 days by 12 rats aged 30 to 34 days at the start of the experiment. The data are classified in two ways, by fat (fresh vs. rancid lard) and by sex (male vs. female). Number of rats per fat and sex combination, $n = 3$.

Sex $(r = 2)$	Fat $(c = 2)$ Fresh	Fat $(c = 2)$ Rancid	Row means $\bar{R}$
♂	709	592	
	679	538	
	699	476	
Subgroup means $\bar{Y}$	695.6667	535.3333	615.5000
♀	657	508	
	594	505	
	677	539	
Subgroup means $\bar{Y}$	642.6667	517.3333	580.0000
Column means $\bar{C}$	669.1667	526.3333	
Grand sum			7173

Preliminary computation

1. $\bar{\bar{Y}} = \dfrac{1}{rcn} \sum^{r} \sum^{c} \sum^{n} Y = \dfrac{1}{12} 7173 = 597.7500$

2. $SS_{subgr} = n \sum^{rc} (\bar{Y} - \bar{\bar{Y}})^2 = 65{,}903.5833$

3. $SS_{within} = \sum^{rc} \sum^{n} (Y - \bar{Y})^2 = 11{,}666.6667$

4. $SS_{total} = $ quantity **2** + quantity **3** $ = 77{,}570.2500$

5. SS_{rows} (SS due to sex) $= \sum^{r} cn(\bar{R} - \bar{\bar{Y}})^2 = 3780.7500$

6. $SS_{columns}$ (SS due to fat) $= \sum^{c} rn(\bar{C} - \bar{\bar{Y}})^2 = 61{,}204.0833$

7. $SS_{interaction}$ (SS due to sex × fat) = quantity **2** − quantity **5** − quantity **6** $= 918.7500$

Completed anova

	Source of variation	df	SS	MS	P
$\bar{R} - \bar{\bar{Y}}$	Between rows (sex)	$1\ (r - 1)$	3780.75	3780.75	0.1460
$\bar{C} - \bar{\bar{Y}}$	Between columns (fat)	$1\ (c - 1)$	61,204.08	61,204.08	0.0002
$\bar{Y} - \bar{R} - \bar{C} + \bar{\bar{Y}}$	Interaction (sex × fat)	$1\ (r - 1)(c - 1)$	918.75	918.75	0.4503
$Y - \bar{Y}$	Error	$8\ rc(n - 1)$	11,666.67	1458.33	
$Y - \bar{\bar{Y}}$	Total	$11\ rcn - 1$	77,570.25		

SOURCE: Data from Powick (1925). Note: explanations of the tests corresponding to the P-values are given in the next section.

notation will be easy to remember for now, although in Box 11.2 we will adopt a more general symbolism consistent with that of previous and later chapters.

We commence by computing a single-classification anova of the four subgroups or subclasses, each representing readings on three rats. If we had no further classification of these four subgroups by sex or freshness of lard, such an anova would test whether there is variation among the four subgroups over and above the variance within the subgroups. Because we have subdivisions by sex and type of fat, however, our only purpose here is to compute some quantities necessary for further analysis. The quantities **1** through **4** are computed in the familiar manner of Box 9.1, although the symbolism is slightly different because in place of *a* groups we now have *rc* subgroups.

We continue the computation by finding the sums of squares for rows and columns of the table (quantities **5** and **6**). First, we sum the squared differences between row means and the grand mean; then we perform the same operation for column means and the grand mean. We weight the squared differences by the sample sizes of the row and column means, respectively.

In steps **2** through **4**, we divided the total sum of squares into two parts: the sum of squares among the four subgroups and that within the subgroups, the error sum of squares. The new sums of squares pertaining to row and column effects clearly are not part of the error but must contribute to the differences that comprise the sum of squares among the four subgroups. We therefore subtract row and column *SS* from the subgroup *SS*, which is 65,903.5833. The row *SS* is 3780.75, and the column *SS* is 61,204.0833. Together they add up to 64,984.8333—almost but not quite the value of the subgroup sum of squares. The difference represents a third sum of squares, called the **interaction sum of squares**, whose value in this case is 918.7500. We will discuss the meaning of this new sum of squares below. At the moment, let us say only that it is almost always present (but not necessarily large enough to be important). The interaction sum need not be independently computed but may be obtained as illustrated above—by the subtraction of the row *SS* and the column *SS* from the subgroup *SS*. This procedure is shown graphically in Figure 11.1, which illustrates

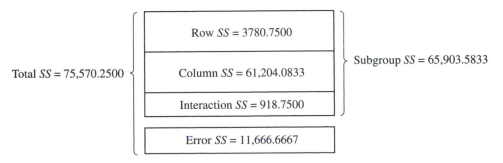

FIGURE 11.1   Diagram of the partitioning of the total sums of squares (*SS*) in a two-way orthogonal anova (Box 11.1). The areas of the subdivisions are not shown proportional to the magnitudes of the sums of squares.

TABLE 11.1   An Artificial Example to Illustrate the Meaning of Interaction

The readings for rancid and fresh lard of the male rats in Box 11.1 have been interchanged. Only subgroup and marginal means are given below.

		Fat	
Sex	**Fresh**	**Rancid**	$\overline{R}$
♂	535.3333	695.6667	615.5000
♀	642.6667	517.3333	580.0000
$\overline{C}$	589.0000	606.5000	$597.7500 = \overline{\overline{Y}}$

Completed Anova

Source of variation	df	SS	MS	P
Sex	1	3780.7500	3780.75	0.1460
Fat	1	918.7500	918.75	0.4503
S × F	1	61,204.0833	61,204.08	0.0002
Error	8	11,666.6667	1458.33	
Total	11	77,570.2500		

the decomposition of the total sum of squares into the subgroup SS and error SS, the former being subdivided into the row SS, column SS, and interaction SS. The relative magnitudes of these sums of squares differ from experiment to experiment. In Figure 11.1, they are not shown proportional to their actual values in the rat experiment; if they were, the area representing the row SS would have to be about $\frac{1}{16}$th of that allotted to the column SS.

Before we can test for the presence of effects due to sex or fat in this anova, we should understand the meaning of interaction. It measures the dependence of the effects of one factor on the levels of the other factor. We can best explain interaction by means of an artificial illustration based on the rat data. If we interchange the readings for rancid and fresh lard *for the male rats only*, we obtain Table 11.1. Only the means of the subgroups, rows, and columns are shown. We complete the analysis of variance in the manner presented already; note the results at the bottom of Table 11.1. The total, subgroup, and error SS are the same as before (see Box 11.1). This should not surprise you because we are using the same data. All we did was interchange the contents of the upper two cells of the table.

When we partition the subgroup SS, we do find some differences. Note that the SS between sexes (between rows) is unchanged. Because the change we made was within one row, the total for that row was not altered, and consequently the row SS did not change. The means of the columns, however, have been altered appreciably as a result of the interchange of the readings for fresh and rancid lard in the males.

The mean for fresh lard is now very close to that for rancid lard, and the difference between the fats, previously quite marked, is now no longer so. By contrast, the interaction *SS*, obtained by subtracting the sums of squares of rows and columns from the subgroup *SS*, is now a large quantity. Remember that the subgroup *SS* is the same in the two examples. In the first example, we subtracted sums of squares due to the effects of both fat and sex, leaving only a tiny residual representing the interaction. In the second example, these two *main effects* (fat and sex) account only for a small amount of the subgroup sum of squares, leaving the interaction sum of squares as a substantial residual. What is the essential difference between these two examples?

If we inspect the subgroup and marginal means for the original data from Box 11.1 and for the "doctored" data in Table 11.1, the original results are quite clear: Males eat slightly more than females, and this is true for both fresh and rancid fats. Note also that fresh lard is eaten in considerably greater quantity than rancid lard, and again this is true of males and females. Thus, our statements about differences due to sex or the freshness of the lard can be made independently of each other. When we interpret the artificial data (see Table 11.1), however, we note that although males still ate more fat than females (because row sums have not changed), this difference depends greatly on the nature of the fat. Males seem to eat considerably less fresh lard than the females, but of the rancid fat they appear to eat considerably more. Thus, we can no longer make an unequivocal statement about the amount of fat eaten by the two sexes. We have to qualify our statement by the type of lard that was consumed. For fresh fat, $\overline{Y}_\delta < \overline{Y}_\circ$, but for rancid fat, $\overline{Y}_\delta > \overline{Y}_\circ$. If we examine the effects of fresh versus rancid fats in the artificial example, we notice that slightly more rancid than fresh lard was consumed. However, we again have to qualify this statement by the sex of the consuming rat; the males ate more rancid fat, the females ate more fresh fat.

This dependence of the effect of one factor on the level of another factor is called **interaction**. It is a common and fundamental scientific idea. The sum of squares for interaction measures the departure of the subgroup means from the values expected on the basis of additive combinations of the row and column means. Any given combination of levels of factors such as fresh lard–male rats may result in a positive or negative deviation from the expected value based on the means for fresh lard and male sex. In common biological terminology, a large *positive* deviation of this sort is called **synergism**. When drugs act synergistically, the result of the interaction of the two drugs may be above and beyond the separate effects of each drug. When a combination of levels of two factors *inhibit* each other's effects, we call it **interference**. Synergism and interference both tend to magnify the interaction *SS*.

Testing for interaction is an important procedure in analysis of variance. If the artificial data of Table 11.1 were real, stating that rancid lard was consumed in slightly greater quantities than fresh lard would be of little value. This statement would cover up the important differences in the data, which are that males ate more rancid fat and females preferred fresh fat. Interactions are very common in biology.

We are now able to write an expression symbolizing the decomposition of a single observation in a two-way analysis of variance in the manner of Expressions (8.2) and (10.2) for single-classification and nested anovas. Expression (11.1) assumes that both factors represent fixed treatment effects, Model I. This assumption seems reasonable because sex and the freshness of the lard are fixed treatments. Observation Y_{ijk} is the kth item in the subgroup representing the ith group of treatment A and the jth group of treatment B. It is decomposed as follows:

$$Y_{ijk} = \mu + \alpha_i + \beta_j + (\alpha\beta)_{ij} + \epsilon_{ijk} \tag{11.1}$$

where μ equals the parametric mean of the population, α_i is the fixed treatment effect for the ith group of treatment A, β_j is the fixed treatment effect of the jth group of treatment B, $(\alpha\beta)_{ij}$ is the interaction effect in the subgroup representing the ith group of factor A and the jth group of factor B, and ϵ_{ijk} is the error term of the kth item in subgroup ij. We make the usual assumption that ϵ_{ijk} is distributed normally with a mean of 0 and a variance of σ^2. If one or both of the factors are Model II, we replace the α_i and β_j in the formula by A_i and B_j, respectively.

In previous chapters, we learned that each sum of squares represents a sum of squared deviations. What deviations does an interaction SS represent? Let us refer back to the anova of Box 11.1. The variation among subgroups is represented by $\overline{Y} - \overline{\overline{Y}}$, where $\overline{Y}$ stands for the subgroup means and $\overline{\overline{Y}}$ for the grand mean. When we subtract the deviations due to rows $\overline{R} - \overline{\overline{Y}}$ and columns $\overline{C} - \overline{\overline{Y}}$ from those of subgroups, we obtain

$$(\overline{Y} - \overline{\overline{Y}}) - (\overline{R} - \overline{\overline{Y}}) - (\overline{C} - \overline{\overline{Y}}) = \overline{Y} - \overline{\overline{Y}} - \overline{R} + \overline{\overline{Y}} - \overline{C} + \overline{\overline{Y}}$$

$$= \overline{Y} - \overline{R} - \overline{C} + \overline{\overline{Y}}$$

This complicated expression is the deviation due to interaction. When we evaluate one such expression for each subgroup, square it, sum these squares, and multiply the sum by n, we obtain the interaction SS. This partition of the deviations also holds for their squares because the sums of the cross products of the separate terms cancel out.

The deviations leading to the sums of squares for the artificial data of Table 11.1 are shown in Table 11.2. Calculate some of these deviations to confirm for yourself where they come from. Notice the large deviations due to interaction and their alternating signs—minus in the left upper and right lower quadrants and plus in the right upper and left lower quadrants. A simple method for revealing the nature of the interaction in the data is to inspect the means of the original data table. The original data (see Box 11.1), showing no interaction, would yield the following pattern of relative magnitudes (where F and R stand for fresh and rancid food, respectively):

$$♂F > ♂R$$

$$♀F > ♀R$$

The relative magnitudes of the means in the artificial example (see Table 11.1) yielding interaction can be summarized as follows:

$$♂F < ♂R$$

$$♀F > ♀R$$

TABLE 11.2 Deviations from Means

Artificial data of Table 11.1. Differences from values obtained in Box 11.1 and Table 11.1 are due to rounding errors.

Sex	Fresh		Rancid		$\overline{R} - \overline{\overline{Y}}$
♂	-62.4167		$+97.9167$		
		-71.4167		$+71.4167$	$+17.7500$
♀	$+44.9167$		-80.4167		
		$+71.4167$		-71.4167	-17.7500
$\overline{C} - \overline{\overline{Y}}$	-8.7500		$+8.7500$		

Left upper number in each quadrant: $\overline{Y} - \overline{\overline{Y}}$ (= subgroups). Right lower number in each quadrant: $\overline{Y} - \overline{R} - \overline{C} + \overline{\overline{Y}}$ (= interaction). $\overline{R} - \overline{\overline{Y}}$ (= rows); $\overline{C} - \overline{\overline{Y}}$ (= columns).

$$n\sum^{rc}(\overline{Y} - \overline{\overline{Y}})^2 = 65{,}903.5833 \qquad n\sum^{rc}(\overline{Y} - \overline{R} - \overline{C} + \overline{\overline{Y}})^2 = 61{,}204.0833$$

When the pattern of signs expressing relative magnitudes is not uniform, as in the second display here, interaction is indicated. As long as the pattern of means is uniform, as in the first table, interaction may not be present. However, interaction is often present without change in the *direction* of the differences; only the relative magnitudes may be affected. In any case, the statistical test needs to be performed to test whether the deviations are larger than can be expected from chance alone.

In summary, when the effect of two treatments applied together cannot be predicted from the average responses of the separate factors, statisticians call this phenomenon *interaction* and test for the presence of its effect by means of an interaction mean square. This is a very common scientific relationship. If we say that the effect of density on the fecundity or weight of a beetle depends on its genotype, we imply that a genotype × density interaction is present. If the geographic variation of a parasite depends on the nature of the host species it attacks, we speak of a host × locality interaction. If the effect of temperature on a metabolic process is independent of the effect of oxygen concentration, we say that temperature × oxygen interaction is absent.

The computations for a replicated two-way anova with more than two classes for one or both factors are illustrated in Box 11.2. In this example, the oxygen consumption of two species of limpets was measured at three concentrations of seawater. The two species represent one factor, and the three concentrations of seawater are the other. Because both factors are fixed treatment effects, this is a Model I anova. The outline of the computations, shown step by step in Box 11.2, follows the familiar format of previous analyses. In practice, the computation will be carried

BOX 11.2 Two-Way Anova with Replication (General Case: a Columns, b Rows)

Oxygen consumption rates of 2 species of limpets, *Acmaea scabra* and *A. digitalis*, at 3 concentrations of seawater. The variable measured is $\mu\ell$ O_2/mg dry body weight/min at 22°C. There are 8 replicates per combination of species and salinity ($n = 8$). This is a Model I anova.

Factor B: Seawater concentration ($b = 3$)	Factor A: Species ($a = 2$)				$\bar{Y}_B$
	Acmaea scabra		Acmaea digitalis		
100%	7.16	8.26	6.14	6.14	
	6.78	14.00	3.86	10.00	
	13.60	16.10	10.40	11.60	
	8.93	9.66	5.49	5.80	
	$\bar{Y} = 10.5612$		$\bar{Y} = 7.4288$		8.9950
75%	5.20	13.20	4.47	4.95	
	5.20	8.39	9.90	6.49	
	7.18	10.40	5.75	5.44	
	6.37	7.18	11.80	9.90	
	$\bar{Y} = 7.8900$		$\bar{Y} = 7.3375$		7.6138
50%	11.11	10.50	9.63	14.50	
	9.74	14.60	6.38	10.20	
	18.80	11.10	13.40	17.70	
	9.74	11.80	14.50	12.30	
	$\bar{Y} = 12.1738$		$\bar{Y} = 12.3262$		12.2500
$\bar{Y}_A$	10.2083		9.0308		461.74 = Grand sum

SOURCE: Data from F. J. Rohlf (unpublished results).

Box 11.2 (continued)

Preliminary computation

1. Grand mean $\bar{\bar{Y}} = \dfrac{1}{abn}\sum^a\sum^b\sum^n Y = \dfrac{1}{48}461.74 = 9.6196$

2. SS_A (SS of columns) $= nb\sum^a(\bar{Y}_A - \bar{\bar{Y}})^2 = 16.6381$

3. SS_B (SS of rows) $= na\sum^b(\bar{Y}_B - \bar{\bar{Y}})^2 = 181.3210$

4. $SS_{A\times B}$ (Interaction SS) $= n\sum^a\sum^b(\bar{Y} - \bar{Y}_A - \bar{Y}_B + \bar{\bar{Y}})^2 = 23.9262$

5. SS_{within} (Within subgroups; error SS) $= \sum^a\sum^b\sum^n(Y - \bar{Y})^2 = 401.5213$

Now fill in the anova table.

	Source of variation	df	SS	MS	Expected MS (Model I)
$\bar{Y}_A - \bar{\bar{Y}}$	A (columns)	$a-1$	2	$\dfrac{2}{(a-1)}$	$\sigma^2 + \dfrac{nb}{a-1}\sum^a\alpha^2$
$\bar{Y}_B - \bar{\bar{Y}}$	B (rows)	$b-1$	3	$\dfrac{3}{(b-1)}$	$\sigma^2 + \dfrac{na}{b-1}\sum^b\beta^2$
$\bar{Y} - \bar{Y}_A - \bar{Y}_B + \bar{\bar{Y}}$	A × B (interaction)	$(a-1)(b-1)$	4	$\dfrac{4}{(a-1)(b-1)}$	$\sigma^2 + \dfrac{n}{(a-1)(b-1)}\sum^{ab}(\alpha\beta)^2$
$Y - \bar{Y}$	Within subgroups	$ab(n-1)$	5	$\dfrac{5}{ab(n-1)}$	σ^2
$Y - \bar{\bar{Y}}$	Total	$abn-1$	$2+3+4+5$		

Because this example is a Model I anova for both factors, the expected MS above are correct. The following are the corresponding expressions for other models.

Box 11.2 (continued)

Source of variation	Model II	Mixed model (A fixed, B random)
A	$\sigma^2 + n\sigma_{AB}^2 + nb\sigma_A^2$	$\sigma^2 + n\sigma_{AB}^2 + \dfrac{nb}{a-1}\sum^{a}\alpha^2$
B	$\sigma^2 + n\sigma_{AB}^2 + na\sigma_B^2$	$\sigma^2 + na\sigma_B^2$
$A \times B$	$\sigma^2 + n\sigma_{AB}^2$	$\sigma^2 + n\sigma_{AB}^2$
Within subgroups	σ^2	σ^2

Anova table

Source of variation	df	SS	MS	F_s	P
A (columns; species)	1	16.6381	16.6381	1.740	0.1943
B (rows; salinities)	2	181.3210	90.6605	9.483	0.0004
$A \times B$ (interaction)	2	23.9262	11.9631	1.251	0.2967
Within subgroups (error)	42	401.5213	9.5600		
Total	47	623.4066			

Because this is a Model I anova, all mean squares are tested over the error *MS*. For hypothesis testing, see Section 11.3.

Conclusions

We are unable to reject the hypothesis that the two species of limpets have the same oxygen consumption, but we do conclude that O_2 consumption differs with salinity. At 50% seawater, the O_2 consumption is increased. Salinity appears to affect the two species equally because we are unable to reject the hypothesis of no species $\times$ salinity interaction.

out by means of a program, and no preliminary computations will show up in the output. We fill in the analysis of variance table and proceed to the tests of significance described in the next section.

11.3 Two-Way Anova: Hypothesis Testing

Before we can test hypotheses about the sources of variation isolated in Box 11.2, we must become familiar with the expected mean squares for this design. In the anova table of Box 11.2, we first show the expected mean squares for Model I; both species differences and seawater concentrations are fixed treatment effects. Incidentally, this model would also be appropriate for the earlier example of fat consumption in rats (sex and freshness of lard are fixed treatments). Note that the within-subgroups or error MS again estimates the parametric variance of the observations. The most important fact to remember about a Model I anova is that the mean square at each level of variation carries only the added effect due to that level of treatment; except for the parametric variance of the items, it does not contain any term from a lower line. Thus, the expected MS of factor A contains only the parametric variance of the items plus the added term due to factor A, but it does not also include interaction effects. In Model I, the significance test is therefore simple and straightforward. Any source of variation is tested by the variance ratio of the appropriate mean square over the error MS. Thus, for the appropriate tests, we employ variance ratios *A/Error*, *B/Error*, and *(A × B)/Error*, where each boldface term signifies a mean square. Thus, $A = MS_A$, $Error = MS_{within}$.

When we apply these tests to the example of Box 11.2, we are led to reject the null hypothesis that factor B, salinity, has no effect. However, we cannot similarly reject null hypotheses concerning factor A or the interaction. Using the common verbal shorthand of research publications, we find the "salinity effect significant, while those for species and salinity × species interaction are not." We conclude that the differences in oxygen consumption are induced by varying salinities, and there does not appear to be sufficient evidence for species differences in oxygen consumption. The following tabulation of the relative magnitudes of the means (in the manner of the previous section) shows that the pattern of signs in the two lines is identical.

	Seawater Concentrations (%)		
	100	75	50
A. scabra		>	<
A. digitalis		>	<

This representation is misleading, however, because the mean of *A. scabra* is far higher at 100% seawater than at 75%, but that of *A. digitalis* is only very slightly higher. Although the oxygen consumption curves of the two species appear far from parallel (Figure 11.2), this suggestion of a species × salinity interaction is not large enough in

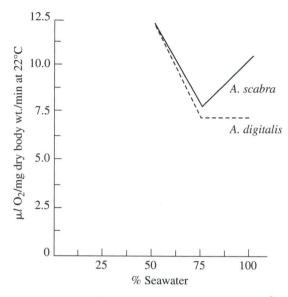

FIGURE 11.2 Oxygen consumption by two species of limpets at three salinities. Data from Box 11.2.

comparison to the within-subgroups variance to allow us to reject the null hypothesis of no interaction. Finding a significant difference among salinities does not conclude the analysis. The data suggest that at 75% seawater, oxygen consumption decreases. This can be tested by the methods of Section 9.5; however, a preferred method of analysis for response curves of this type will be presented in Section 16.9.

When we test the anova of fat consumption in rats (see Box 11.1), we notice again that we can only reject the null hypothesis for one mean square (between columns, representing fats). We conclude, therefore, that the apparent difference in fat consumption between males and females in rats is no larger than could be expected by chance, but that the higher consumption of fresh as compared with rancid lard is much larger than could be expected by chance. Absence of interaction means that this relation is the same in both sexes. When we analyze the results of the artificial example in Table 11.1, we find that only the interaction effect is larger than expected by chance. Thus, we would conclude that the response to freshness of lard differs in the two sexes. This conclusion would be supported by inspection of the data, which would show that males preferred rancid fat and females preferred fresh fat.

In the artificial example, we are unable to reject the null hypotheses for the effects of the two factors (main effects). However, many statisticians would not even test them after detecting the presence of an interaction effect because an overall statement of the effect for each factor would have little meaning in such a case. If the lines cross, then even a simple statement of preference for either rancid or fresh lard would be unclear. The presence of interaction makes us qualify our statements: Rancid fat is preferred by males, and fresh fat is preferred by females. Similarly, interaction in the example of Box 11.2 would have meant that the pattern of response to changes in salinity differed in the two species. We would consequently have to describe separate, nonparallel response curves for the two species.

In such a case, multiple-comparisons tests could be carried out (see Box 11.3) to test differences among subgroup means separately for each species. When interaction is present, the relative magnitudes of the sums of squares for the main effects and the interaction depend on the particular levels of the main effects chosen. In the limpet experiment illustrated in Figure 11.2, if we had chosen seawater concentrations ranging only between 50% and 75%, there would have been an increased effect of salinity, little if any species effect, and only a small interaction effect. Had we selected the range from 75% to 100%, the salinity factor would have been diminished, while species and interaction effects would have been large.

It may occasionally become important to test the main effects in a Model I anova in spite of the presence of interaction. Take, for example, the situation illustrated by Figure 11.3, which shows the effects of increasing proportions of two types of "conditioned" flour in fresh flour on the survival of wild-type *Tribolium castaneum*

BOX 11.3 Multiple Comparisons Among Cell Means in a Two-Way Anova with Interaction Present

Mean yields (in tons/acre) in an experiment growing three varieties of sugar cane (V_1, V_2, V_3) at three levels of nitrogen (in lb N per acre). There were $n = 4$ replications of the 3×3 design. This is a pure Model I anova.

		Varieties		
		V_1	V_2	V_3
Levels of nitrogen	150	66.525	61.450	68.600
	210	68.975	62.550	64.525
	270	75.950	70.430	57.900

Data from Cochran and Cox (1957).

Anova Table

Source	df	SS	MS	F	P
Varieties	2	319.3739	159.6869	3.6367	.0417
Nitrogen	2	56.5406	28.2703	0.6438	.5341
V×N	4	559.7878	139.9469	3.1871	.0311
Blocks	3	?	?		
Error	24	1053.8400	43.9100		

BOX 11.3 (continued)

There are 24 rather than 27 *df* for error (as one might expect) because the replications corresponded to blocks within which there was a 3×3 design with no replication. The *SS* for blocks was not published but is not needed for the analysis given here.

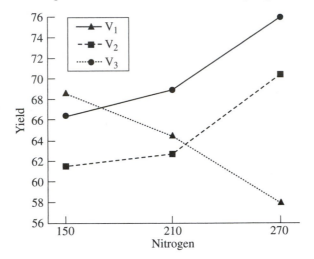

Testing the interaction term first, we conclude that the three varieties do not respond in the same way to the three levels of nitrogen. This is clear from a plot of the means. There is also evidence of differences in the overall mean yields for the three varieties, but that is of less interest because of the presence of an interaction effect. Although the plot is highly suggestive, unless there are prior hypotheses about differences among the cell means, multiple comparison tests for such differences should be performed. The maximum studentized range statistic of Copenhaver and Holland (1988) and Ferreira et al. (2007) can be used to test for differences among all pairs of variety means within each level of nitrogen.

At each level of nitrogen we can reject the null hypothesis of no difference for any pair of means for which the difference exceeds the following MSD value.

$$MSD = \sqrt{\frac{MS_{\text{error}}}{n}} q_{\alpha[r,c,\nu]}$$

where $q_{\alpha[r,c,\nu]}$ is the critical value from the maximum studentized range distribution for r groups, c treatments, and ν degrees of freedom (see Statistical Table **QQ** or the BIOMstat program). The $a = 3$ varieties are being compared at each of $b = 3$ levels of nitrogen, and $\nu = 24$ degrees of freedom. The critical value of $q_{.05[3,3,24]}$ was obtain here by computation.

$$MSD = \sqrt{\frac{MS_{\text{error}}}{n}} q_{.05[3,3,24]} = \sqrt{\frac{43.910}{4}} 4.20589 = 13.9351$$

Only the difference between the means for V_1 and V_3 for the highest level of nitrogen exceeds this MSD value. ($|\bar{Y}_{31} - \bar{Y}_{33}| = |75.950 - 57.900| = 18.050$) At the $P = 0.10$ level, varieties 2 and 3 would also differ in yield at the highest nitrogen concentration.

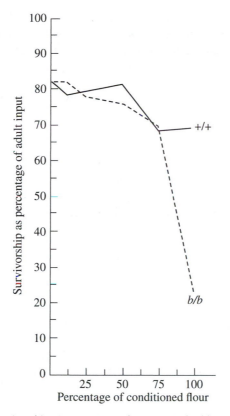

FIGURE 11.3   Adult survivorship as percentage of egg input of wild-type *Tribolium castaneum* (a beetle) reared on medium conditioned by either wild-type (+/+) or black (*b/b*) individuals in varying concentrations. (Extracted from Karten, 1965.)

beetles. Conditioned flour is medium that has been heavily contaminated by beetle waste products. The two types of flour were conditioned by two different strains of beetles. The shapes of the response curves are quite different. Undoubtedly, there is interaction between the strains conditioning the medium and the concentrations of the conditioned flour because these affect beetle survival. Yet the overall effects of increasing the proportion of conditioned flour are quite clear. As the concentration of conditioned flour increases, survival decreases in conditioned flour from both sources. To support this contention, we might wish to test the mean square among concentrations (over the error *MS*), regardless of whether an interaction effect is found.

Box 11.2 also lists expected mean squares for a Model II and a mixed model two-way anova. In the Model II anova, note that the two main effects contain the variance component of the interaction as well as their own variance component. Such an arrangement is familiar from nested analysis of variance. When performing tests in a Model II anova, we first test $(A \times B)/Error$. If the hypothesis of no interaction is rejected, we continue testing $A/(A \times B)$ and $B/(A \times B)$. When the effect of an $A \times B$ interaction is not detected, current practice is to nevertheless test a main effect MS over the interaction mean square. Only one type of mixed model is shown in which factor A is assumed to be fixed and factor B to be random. If the situation is reversed, the expected mean squares change accordingly. Here it is the mean square representing the fixed treatment that carries with it the variance component of the interaction, whereas the mean square representing the random factor contains only the error variance and its own variance component and does not include the interaction component. We therefore test the MS of the random main effect over the error but test the fixed treatment MS over the interaction.

You may wonder how to write the formulas for expected mean squares for cases other than those given in the boxes. We will learn general formulas for these in Section 11.7.

The expected mean squares for mixed-model anovas presented earlier are not the only ones suggested for this case. Alternative models do exist that are beyond the level of this book. The formulation chosen here is the one most frequently encountered in the literature. However, readers are warned that they need to know which formulation is employed by the computer program they are using: The various formulations will yield different results that could lead to different conclusions.

An example of a Model II two-way anova is featured in Table 11.3. The data represent measurements of blood pH of mice. Five inbred strains were measured repeatedly. In experiment I, the mice were tested every other day for seven test periods, and in experiment II they were tested at six weekly intervals. The purpose of these experiments was to estimate the proportion of the variation that is genetic. This is an example of a Model II analysis because the experimenter was not concerned with the differences between specific strains but was considering the five strains to be a "random" sample of the inbred mouse strains available. The sampling periods are considered to be random as well because the investigator did not wish to establish that day 1 produced blood pH values different from those of day 2 but merely tried to estimate the magnitude of fluctuations in blood pH over the period studied. There is no reason to believe in a directed trend of the blood pH over the period of the study.

Table 11.3 shows the expected mean squares of the two experiments, with the actual coefficients of the variance components. The symbolism differs somewhat from that given for Model II in Box 11.2. It is important that you become familiar with some variations in usage. Instead of using A and B as symbols for the two factors, we use M and D to represent mouse strains and days, respectively. Mnemonic notation like this is frequently used. In such cases, investigators often also use a different symbolism for the coefficients; m might be used for a and d for b (to correspond to M and D, respectively). Notice also that the source-of-variation column (copied from the original publication) does not have prepositions such as "among" or "between" and also that the error term is described as "mice treated alike," which is what the error term represents in the present experiment: variation among mice of

TABLE 11.3   A Model II Two-Way Anova with Replication

Strain differences and daily differences in blood pH for five inbred strains of mice. In experiment I, samples of two mice from each strain were tested every other day until seven such groups had been tested; in experiment II, samples of five mice from each strain were tested six times at one-week intervals. Only the anova tables are shown.

Experiment I ($a = 5$, $b = 7$, $n = 2$)

Source of variation	df	MS	F_s	P	Expected MS
Mouse strains	4	0.0427	11.24	2.79×10^{-5}	$\sigma^2 + 2\sigma^2_{MD} + 14\sigma^2_M$
Days of test	6	0.0148	3.89	0.0075	$\sigma^2 + 2\sigma^2_{MD} + 10\sigma^2_D$
Interaction	24	0.0038	1.03	0.4595	$\sigma^2 + 2\sigma^2_{MD}$
Mice treated alike (error)	35	0.0037			σ^2

Experiment II ($a = 5$, $b = 6$, $n = 5$)

Source of variation	df	MS	F_s	P	Expected MS
Mouse strains	4	0.0920	17.69	2.37×10^{-6}	$\sigma^2 + 5\sigma^2_{MD} + 30\sigma^2_M$
Days of test	5	0.0101	1.94	0.1324	$\sigma^2 + 5\sigma^2_{MD} + 25\sigma^2_D$
Interaction	20	0.0052	1.53	0.0833	$\sigma^2 + 5\sigma^2_{MD}$
Mice treated alike (error)	120	0.0034			σ^2

Variance components

	Experiment I		Experiment II	
Strains	0.00278	36.4%	0.00289	42.2%
Days	0.00110	14.4%	0.00020	2.9%
Interaction	0.00005	0.7%	0.00036	5.3%
Error	0.0037	48.5%	0.0034	49.6%

SOURCE: Data from Weir (1949).

a given strain within a given day. You will have to become used to such differences in naming the sources of variation.

Statistical testing in this example proceeds as follows. We first test interaction over the error; the null hypothesis is accepted in both experiments. We continue to test the main effects over the interaction. In experiment I, mouse strains and days of test yield very small probabilities for the tests of the variance components. In experiment II, however, only the test for mouse strains has such a small probability. The differences between the two experiments are reflected in the magnitudes of the estimated variance components given at the bottom of Table 11.3 as absolute values and as percentages. Confidence intervals can be computed for the variance components using the methods described in Section 9.2.

The error variance is remarkably alike in the two experiments, showing that the techniques were well controlled and gave comparable results in both experiments. The major difference between the experiments is that a substantial percentage (14.4%) of variation is among days in experiment I, but this variance component comprises only 2.9% of the overall variation in experiment II. Clearly, there are more fluctuations when readings are taken every two days. A far-fetched explanation might be that there is a rhythmic fluctuation in blood pH values with regular cycles every seven days. In such a case, samples taken every other day would fluctuate widely, while those taken once a week would be very constant. There is, however, no evidence for such a phenomenon in these mice. Experiments I and II were run during different seasons, however, and it is quite possible that during the 14 days necessary for experiment I there were considerable day-to-day fluctuations in climate. An inquiry to the investigator revealed that the mice were not maintained in temperature-controlled rooms; hence daily fluctuations during experiment I might have resulted in this large value of the variance. During the 42 days of experiment II, the weather might have been considerably more constant. Although these explanations may be incorrect, they illustrate how an anova furnishes insight into the structure of an experiment as well as clues to further questions to be asked and experiments to be undertaken.

A simple rearrangement of the data of Table 11.3 will provide some insight into the relations between two-way and nested anovas. In the design of experiment II in this table, five strains were observed on six separate days. In each strain, five mice had their blood pH measured on one day. This design is shown at the top of Table 11.4. We arrange it as a two-way anova because the days are common factors throughout the analysis. Day 1 is the same day for strain 1 as it is for strain 3. The anova table for this experiment is reproduced once more in Table 11.4, showing also the numerical values for the sums of squares.

Suppose, however, that we had misinterpreted this experiment. We might have thought that the days were nested within the strains, as shown in the next design in Table 11.4. Such a design implies that we randomly chose six days on which to measure the blood pH of each strain and that the six days chosen for the tests on strain 1 were different from the test days for strain 2, and so on. If we had randomly chosen the days without any thought of making day 1 the same for all the strains, a nested analysis would have been appropriate. Taking the same data and analyzing them (incorrectly, in view of the actual nature of design) as a two-level nested anova, we obtain the anova table shown at the bottom of Table 11.4. Note that the subordinate level (subgroups within groups, which is days within strains in this example) is the summation of two levels from the two-way anova. Degrees of freedom and sums of squares of "days of test" and "interaction" add to the corresponding values for "days within strains." Thus, in a hierarchic analysis of variance the subordinate level represents not only differences among the subgroups but also the interaction term. If differences among days exist but the nature of these differences varies from strain to strain, we would have a substantial contribution to interaction. This would not

TABLE 11.4 Experiment II of Table 11.3 Reanalyzed to Show Relationships Between a Two-Way Anova and a Nested Anova

Correct design (two-way anova with replication = 5 mice [*] in each cell)

	Strains				
Days	1	2	3	4	5
1	*	*	*	*	*
2	*	*	*	*	*
3	*	*	*	*	*
4	*	*	*	*	*
5	*	*	*	*	*
6	*	*	*	*	*

Anova table

Source of variation	df	SS	MS	F_s	P
Mouse strains	4	0.3680	0.0920	17.69	2.37×10^{-6}
Days of test	5	0.0505	0.0101	1.94	0.1324
Interaction	20	0.1040	0.0052	1.53	0.0833
Error	120	0.4080	0.0034		
Total	149	0.9305			

Incorrect design (nested anova—days within strains: 5 mice [*] in each cell)

	Strains				
	1	2	3	4	5
Days	1 2 3 4 5 6	1 2 3 4 5 6	1 2 3 4 5 6	1 2 3 4 5 6	1 2 3 4 5 6
Mice	* * * * * *	* * * * * *	* * * * * *	* * * * * *	* * * * * *

Anova table

Source of variation	df	SS	MS	F_s	P
Mouse strains	4	0.3680	0.0920	14.84	2.45×10^{-6}
Days within strains	25	0.1545	0.0062	1.82	0.0175
Error	120	0.4080	0.0034		
Total	149	0.9305			

be separately detectable in the nested anova but would contribute to the magnitude of the subgroups-within-groups *SS*. When we incorrectly consider this example as a nested anova, an effect of days within strains is found at the 2% level; but in the two-way anova, the null hypotheses for neither days of test nor interaction could be rejected, although the appropriate F_s-ratios are suspiciously large.

When there is no interaction and the null hypothesis for the treatment effects in a model I or a mixed-model two-way anova is rejected, one can carry out multiple comparisons in the manner of Section 9.5. This applies as well to the multiway analyses of variance discussed later in this chapter. Hochberg and Tamhane (1983) discuss such cases in detail. The error MS to be used is that used in the denominator of the *F*-test for the overall test of the main effects. When interaction is present, multiple comparisons can be made among the treatment means within each level of the other factor. An example was given in Box 11.3. The use of the maximum studentized range distribution allows one to make all pairwise comparisons among means within one of the factors of a two-way table, a total of $ba(a - 1)/2$ comparisons. For these data, it is clear that variety 3 did not respond in the same way as the other two varieties to increasing levels of nitrogen.

11.4 Two-Way Anova without Replication

In many experiments, there will be no replication for each combination of factors represented by a cell in the data table. In such cases, we cannot easily talk of "subgroups" because each cell contains a single reading only. At times it may be too difficult or too expensive to obtain more than one reading per cell, or the measurements may be known to be so repeatable that there is little point in estimating their error. As we will see in this section, a two-way anova without replication can be properly applied only with certain assumptions. For some models and tests in anova, we must assume that no interaction is present.

Our illustration for this design is taken from limnology. In Box 11.4 we show temperatures of a lake taken at about the same time on four successive summer afternoons. These temperature measurements were made at 10 different depths and were known to be highly repeatable. Therefore, only single readings were taken at each depth on any one day. What is the appropriate model for this anova? Clearly, the depths are Model I. The four days, however, are open to interpretation. If the days had been selected from among a wider span of time, say, April to October, we surely would have been interested in specific differences among the days because these would reflect seasonal changes in temperature profiles. In such cases, days would represent fixed treatment effects as well. These four readings were taken during a period of stable midsummer weather, however, so differences among the days are not likely to be of specific interest. It is improbable that an investigator would ask whether the water was colder on July 30 than on July 31. A more meaningful way to look at this problem would be to consider these four summer days as random samples that enable us to estimate the day-to-day variability of the temperature stratification in the lake during this climatically stable period.

BOX 11.4   Two-Way Anova without Replication

Temperatures (°C) of Rot Lake on 4 early afternoons of the summer of 1952 at 10 depths. This is a mixed model anova.

Factor B: Depth in meters (b = 10)	Factor A: Days (a = 4)					$\overline{Y}_B$
	29 July	30 July	31 July	1 August		
0	23.8	24.0	24.6	24.8		24.300
1	22.6	22.4	22.9	23.2		22.775
2	22.2	22.1	22.1	22.2		22.150
3	21.2	21.8	21.0	21.2		21.300
4	18.4	19.3	19.0	18.8		18.875
5	13.5	14.4	14.2	13.8		13.975
6	9.8	9.9	10.4	9.6		9.925
9	6.0	6.0	6.3	6.3		6.150
12	5.8	5.9	6.0	5.8		5.876
15.5	5.6	5.6	5.5	5.6		5.575
$\overline{Y}_A$	14.890	15.140	15.200	15.130		603.6 = Grand sum

SOURCE: Data from Vollenweider and Frei (1953).

The four sets of readings are treated as replications (blocks) in this analysis. Depth is a fixed treatment effect, and days are considered random effects; hence this is a mixed model anova.

Box 11.4 (continued)

Preliminary computation

1. Grand mean $\bar{\bar{Y}} = \dfrac{1}{ab}\sum\limits^{a}\sum\limits^{b} Y = \dfrac{1}{40}\,603.6 = 15.090$

2. SS_A (SS of columns) $= b\sum\limits^{a} (\bar{Y}_A - \bar{\bar{Y}})^2 = 0.5620$

3. SS_B (SS of rows) $= a\sum\limits^{b} (\bar{Y}_B - \bar{\bar{Y}})^2 = 2119.6510$

4. SS_{error} (remainder; discrepance) $= \sum\limits^{a}\sum\limits^{b} (Y - \bar{Y}_A - \bar{Y}_B + \bar{\bar{Y}})^2 = 2.2430$

Anova table

	Source of variation	df	SS	MS	F_s	P	Expected MS
$\bar{Y}_A - \bar{\bar{Y}}$	A (columns; days)	3	0.5620	0.1873	2.255	0.1048	$\sigma^2 + b\sigma_A^2$
$\bar{Y}_B - \bar{\bar{Y}}$	B (rows; depths)	9	2119.6510	235.5168	2835.0	$\ll 10^{-16}$	$\sigma^2 + \sigma_{AB}^2 + \dfrac{a}{b-1}\sum\beta^2$
$Y - \bar{Y}_A - \bar{Y}_B + \bar{\bar{Y}}$	Error (remainder; discrepance)	27	2.2430	0.0831			$\sigma^2 + \sigma_{AB}^2$
$Y - \bar{\bar{Y}}$	Total	39	2122.4560				

Conclusions

Large decreases in temperature occur as depth increases. The null hypothesis that depth has no effect on temperature can be rejected because the probability of a type I error is infinitesimal. To test days, we must assume interaction between days and depths to be zero. No added variance component among days can be demonstrated.

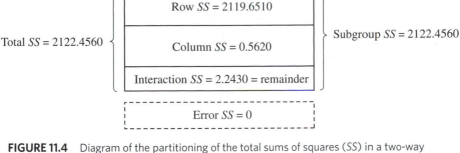

FIGURE 11.4 Diagram of the partitioning of the total sums of squares (*SS*) in a two-way orthogonal anova without replication (Box 11.4). The areas of the subdivisions are not shown proportional to the magnitudes of the sums of squares.

The computations, shown in Box 11.4, are the same as those in Box 11.2 except that the expressions to be evaluated are simpler. The subgroup sum of squares in this example is the same as the total sum of squares. If this is not immediately apparent, then consult Figure 11.4, which, when compared with Figure 11.1, illustrates that the error sum of squares based on variation within subgroups is missing in this example. Thus, after we subtract the sum of squares for columns (factor A) and for rows (factor B) from the total *SS*, we are left with only a single sum of squares; this is the equivalent of the previous interaction *SS* but is now the only source for an error term in the anova. This *SS* is known as the **remainder *SS*** or the **discrepance**.

If you refer to the expected mean squares for the two-way anova in Box 11.2, you will discover why we made the statement earlier that for some models and tests in a two-way anova without replication, we must assume that there is no interaction. If interaction is present, only a Model II anova can be entirely tested, whereas in a mixed model only the fixed level can be tested over the remainder mean square. In a pure Model I anova, or for the random factor in a mixed model, however, it would be improper to test the main effects over the remainder unless we could reliably assume that no added effect because interaction is present. General inspection of the data in Box 11.4 convinces us that the temperature profile for any one day is faithfully reproduced on the other days. Thus, interaction is unlikely to be present. If, for example, a severe storm had churned up the lake on one day, changing the temperature relationships at various depths, then interaction would have been apparent, and the test of the mean square among days carried out in Box 11.4 would not have been appropriate.

Because we assume no interaction, the row and column mean squares are tested over the error *MS*. The results are not surprising; casual inspection of the data would have predicted our findings. Added variance among days is no more than what could be expected by chance, but the differences in temperature due to depths are especially large, yielding a value of $F_s = 2835.0$. It is extremely unlikely that chance alone could produce such differences.

A common application of two-way anova without replication is the **repeated testing of the same individuals,** also known as **repeated measures design**. By this we mean that the same group of individuals is tested repeatedly over a period of time. The individuals are one factor (usually considered as random and serving as replication), and the time dimension is the second factor, a fixed treatment effect. For example, we might measure growth of a structure in 10 individuals at regular intervals. When testing for the presence of an added variance component (because of the random factor), such a model again must assume that there is no interaction between time and the individuals; that is, the responses of the several individuals are parallel through time.

This design is also used in physiological and psychological experiments in which we test the same group of individuals for the appearance of a response after treatment. Examples include increasing immunity after antigen inoculations, altered responses after conditioning, and measures of learning after a number of trials. For instance, we may study the speed with which 10 rats, repeatedly tested on the same maze, reach the end point. The fixed treatment effect would be the successive trials to which the rats have been subjected. The second factor, the 10 rats, is random, presumably representing a random sample of rats from the laboratory population. When the response variables are two-state attributes rather than being measurable, repeated testing of the same individuals is carried out by the method of Section 17.7.

The method of **randomized complete blocks** is a type of two-way anova (usually without replication) that is especially common not only in agricultural research but also in other biological field and laboratory experiments. When we apply a fertilizer to a small plot of land, we usually get a single reading expressed as yield of crop. Similarly, an application of an insecticide to a plot of land is measured as a single mortality among the insects in the plot or a single value for yield of crop. To test differences among five types of fertilizers (or insecticide treatments), we must have at least five plots available. The five resulting readings, however, would not yield an error variance useful for testing differences among treatments. Thus, we need replication of the treatments to provide an error term.

Let us assume for the sake of this discussion that we wish to replicate each treatment four times. We therefore need $5 \times 4 = 20$ plots to test five treatments. An area encompassing 20 plots of land is quite large relative to basic plot size and therefore is likely to be heterogeneous in its soil and in other microclimatic conditions. As a general rule, adjacent plots are more like each other. We therefore do not wish to put all the replicates of treatment A into the same general area; otherwise, what may appear to be a significant effect of treatment A may actually turn out to be due to the general area in which the plots of A are situated.

The method of randomized complete blocks is designed to overcome this difficulty. In our specific example, we partition the land available for our experiment into four blocks, hoping that the environment within each of the blocks is as homogeneous as possible. We then subdivide each block into five plots and allocate the treatments at random to these plots. Such an arrangement is shown diagrammatically in Figure 11.5. The five treatments are called A, B, C, D, and E. The random arrangement of the treatments

B	A	D	E	C	Block 1

B	C	E	D	A	Block 2

D	B	E	A	C	Block 3

C	B	A	D	E	Block 4

FIGURE 11.5 A randomized-complete-blocks design. The field has been arranged into four blocks of five treatments each.

within each block can be obtained by permuting the five letters. There are $5! = 120$ permutations of the letters A through E. We can number these sequentially; by consulting a table of random numbers, we can choose four permutations. The random arrangement of the treatments among the plots is necessary so that any given treatment does not always maintain the identical relative position in the field.

For analysis, the data are entered in a treatments $\times$ blocks table, all the treatments of one kind being entered in the appropriate block. If the $b = 4$ blocks are considered as rows (factor B) and the $a = 5$ treatments as columns (factor A), an anova table of this example would be constructed as follows:

Source of Variation	df	Expected MS
Blocks	$b - 1$	$\sigma^2 \qquad\qquad + a\sigma_B^2$
Treatment	$a - 1$	$\sigma^2 + [\sigma_{AB}^2] + \dfrac{b}{a-1}\sum^{a}\alpha^2$
Error	$(a-1)(b-1)$	$\sigma^2 + [\sigma_{AB}^2]$

The blocks are considered random effects. An individual observation in a randomized complete blocks design can therefore be decomposed as

$$Y_{ij} = \mu + \alpha_i + B_j + [(\alpha B)_{ij}] + \epsilon_{ij} \qquad (11.2)$$

where α_i is the treatment effect and B_j is the random blocks effect. Note that in this expression and in the expected mean squares above, the interaction term has been placed within brackets. The brackets indicate that we assume the interaction to be absent. Testing of the treatment effect is not dependent on the validity of this assumption because in the randomized-complete-blocks design we simply test the treatment MS over the error MS. The block effect is rarely tested, and we would have to assume that interaction is not present before we could test it. If the effect of blocks cannot be detected, there was less heterogeneity among the blocks than anticipated or the blocks were laid out in such a way that most of the heterogeneity was within blocks instead of among blocks. Finding added variance among the blocks does not disturb us because we expected to find such differences. In fact, if we found that the differences among blocks were small, then in future experiments we might want to use not a randomized-blocks design, but rather a single-classification anova.

The randomized-complete-blocks design may also be applied to experiments other than the agricultural ones for which it was first developed. If an experiment is too big to be undertaken at one time, it may have to be carried out over a period of days (each day representing a block), or it may have to be carried out in different laboratories (each serving as a block). When treatments are applied to different age classes of animals, these classes can become blocks. Whenever the environment in which the experiment is to be carried out or the research materials may be heterogeneous, these can be subdivided into more-homogeneous blocks and the experiment carried out and analyzed as randomized complete blocks.

One reason that the effects of blocks are rarely tested has already been explained: Unless we can be reasonably certain that there is no interaction variance component in our error term, we have no denominator mean square over which to test the block *MS*. But there is a second reason for our hesitancy. In our discussions so far, we have not emphasized an important consideration of two-way anova. In strict application, a two-way anova should be completely randomized. For example, to carry out the two-way anova on pea sections discussed earlier (see Section 11.1), we should assemble 100 dishes of pea sections and randomly allocate each to one of five sugar treatments and two pH levels so as to achieve 10 replicates for each treatment combination. Contrast this with the agricultural experiment. Before we apply our five fertilizer treatments to each of the four blocks of land, each block will have to be prepared by plowing, sowing seed, dressing soil, and so forth. This operation is unique for each block; given the vagaries of human performance, it cannot be exactly repeated for the same block in succeeding years. Thus, there is a random error deviation attached to each of the blocks because of the single and distinct instance of preparation each represents in the study.

On assigning the various plots to the study, we are free to randomly assign a plot to any one treatment, but we are not free to randomly assign a plot to any one block because the five plots within any one block are limited to that block. This restriction on the variation causes a random error deviation peculiar to each block, which in turn generates an error variance component that attaches itself to the block means square. This added error variance has been called **restriction error** by Anderson and McLean (1974). The block means square is then represented by $\sigma^2 + a\sigma_R^2 + a\sigma_B^2$, and there is no suitable denominator mean square over which to test it even if the interaction variance component is assumed to be absent. Thus, caution should be exercised whenever a randomized-complete-blocks experiment is analyzed. Do not test the block *MS* when restriction error is present. The restriction error inherent to agricultural experiments does not arise in the typical greenhouse experiment where each pot (equivalent to a plot) is separately prepared for the study.

Restriction error is not limited to randomized complete blocks but also applies to two-way anovas with replication. In such cases, the variance component due to restriction error will attach itself to the mean square of the main effect within which the restriction error operates. Such a main effect can be either a fixed treatment or a random effect. Let us review the cases that we have encountered so far. The hypothetical example of the pea sections should not have any restriction error if carried out properly as discussed earlier. It is unlikely in most experimental setups, however, that the test conditions will be replicated 100 times. Hence, there will be a restriction error attached to one of the mean squares in the study. The example of food consumption by male and female rats (see Box 11.1) has the possibility of restriction error. One cannot randomly assign a rat to one type of fat and one sex. Rats come prepackaged by nature into males and females; thus, any randomization of the rats for the experiments must be carried out within each sex, not across the entire study. However, we would expect that restriction error in this example would be negligible. There is little error in choosing a male or a female, and there should be little error in choosing males and females repeatedly, unlike error in the repeated tilling of the same block of agricultural land. The same sort of reasoning applies to the example from Box 11.2, in which two species of limpets were subjected to three salinities. Randomization can take place only within any one species. Nevertheless, again, there should be little if any restriction error due to sampling within one species. We would not hesitate, therefore, to test species as well as salinities, as we have done.

The analysis in Table 11.4 studying blood pH in mice is also a similar design. Assuming that all mice were living at the beginning of the experiment, any one mouse could be randomly assigned to the various time periods (days) when it is to be tested, but randomization could be only within any one strain. Assuming that the genetic factors determining blood pH are stable in these strains, we can ignore restriction error. But what if this were not the case? In a model such as this example, the presence of restriction error variance component would simply be confounded with the strain variance component and would inflate our estimate of the latter. There might be more hesitation in testing a fixed treatment main effect confounded with a restriction error variance component because in an extreme case an apparent treatment effect might be entirely due to restriction error, with no differences in means resulting from treatment differences.

In the example of water temperature for different depths on different days in Rot Lake (see Box 11.4), the error is restricted by depths having to be tested on any one day. Had a different test been run on any one date, different results would probably have been obtained. Because days are a random component, however, we need not be overly concerned about restriction error here other than to realize that our estimate of the variance component among days may be inflated by the presence of restriction error. In this particular example, an added variance component among days could not be demonstrated, so the restriction error, if any, cannot be very large.

It is important to note that each treatment must be present in every block. If this cannot be done, or when one has too many treatments, different combinations of treatments are applied to a series of blocks. In such cases, a more complex analysis must be carried out. Examples of such designs, called **incomplete blocks**, are described in Cox and Reid (2000). A more extensive account is furnished in Hinkelmann and Kempthorne (2005).

BOX 11.5   Example of a Randomized-Complete-Blocks Experiment

Mean dry weights (in milligrams) of 3 genotypes of beetles, *Tribolium castaneum*, reared at a density of 20 beetles per gram of flour. Four series of experiments represent blocks.

b series (*b* blocks; *b* = 4)	*a* genotypes (*a* = 3)			
	+ +	+*b*	*bb*	$\overline{Y}_B$
1	0.958	0.986	0.925	0.9563
2	0.971	1.051	0.952	0.9913
3	0.927	0.891	0.829	0.8823
4	0.971	1.010	0.955	0.9787
$\overline{Y}_A$	0.9568	0.9845	0.9153	11.426 = Grand sum

Anova table

Source of variation		df	SS	MS	F_s	P
MS_B	Series	3	0.021391	0.007130	10.23	0.0090
MS_A	Genotypes	2	0.009717	0.004859	6.97	0.0117
$MS_{E(RB)}$	Error	6	0.004184	0.000697		

SOURCE: Data from Sokal and Karten (1964).

An example of a randomized-complete-blocks experiment is shown in Box 11.5. This example is based on data extracted from a more extensive experiment by Sokal and Karten (1964) in which dry weights of three genotypes of *Tribolium castaneum* beetles were recorded. The experimental work involved measuring other variables and was very laborious. Therefore, the requisite amount of replication could not be carried out at one time, and four series of experiments, several months apart, were undertaken. Within one series of experiments, the order of weighing different genotypes was random, thus yielding a randomized-complete-blocks design with four blocks (series) and three treatments (genotypes).

The anova table in Box 11.5 shows that there is a difference in weight among the genotypes: The heterozygotes are the heaviest, followed by the wild type, with the homozygous mutant weighing the least. In Box 11.5, we have not actually tested the differences between pairs of means by multiple-comparison tests. From the analysis shown, we know only that the three means jointly are heterogeneous. Sokal and Karten, however, tested these differences and were able to reject the null hypothesis of no differences between all pairs of genotypes. In analogy to our earlier reasoning for the days mean square in the blood pH of mouse strains (see Table 11.3), we would consider series to be a random factor. The restriction error, if any, would be within genotypes, because samples of beetles cannot be randomly assigned to any one genotype. However, as in our earlier reasoning on sexes of rats, species of limpets, and strains of mice, we assume that the restriction error due to genotypes would be negligible. Beetles would be assigned to any one series on the assumption that the supply of beetles of any one genotype remains homogeneous throughout the study.

The significant effect of series implies that environmental factors among the series, minor changes in technique, or other factors were such that differences in weight resulted. By separating out this heterogeneity, we were able to perform a much more sensitive test. In Section 12.8, we shall revisit this example and learn a method for quantifying the increase in sensitivity for a randomized complete blocks design over a completely randomized design as in Chapter 9.

The computer program BIOMstat can be used to perform the computations described in Sections 11.2 and 11.4.

11.5 Paired Comparisons

The randomized-complete-blocks and repeated measures designs of the previous section may differ in the number of rows (blocks) and treatment classes. One special case, however, is common enough to merit discussion in a separate section. This special case is that of randomized complete blocks in which there are only two treatments or groups ($a = 2$) and is known as **paired comparisons** because each observation for one treatment is paired with one for the other treatment. The pair is composed of the same individuals tested twice or of two individuals with common experiences so that we can legitimately arrange the data as a two-way anova.

Let us elaborate on this point. Suppose we test the muscle tone of a group of individuals, subject them to severe physical exercise, and measure their muscle tone once more. Because the same group of individuals will have been tested twice, we can arrange our muscle tone readings in pairs, each pair representing readings on one individual (before and after exercise). Such data are appropriately treated by a two-way anova without replication, which in this case would be a paired-comparison test because there are only two treatment classes. This "before-and-after treatment" comparison is a frequent design leading to paired comparisons. Another design simply measures two stages in the development of a group of organisms, with time serving as the treatment intervening between the two stages. The example in Box 11.6 is of this nature. This study measures lower face width in a sample of girls five years old and in the same sample of girls when they are six years old. The paired comparison is for each individual girl, between her face width when she is five years old and her face width at six years.

Paired comparisons often result from dividing an organism or other individual unit so that one-half receives treatment 1 and the other half treatment 2, which may be the control. For example, if we wish to test the strength of two antigens or allergens in a random sample of volunteers, we might inject one into each arm of a single individual and measure the diameter of the red area produced. It would not be wise from the point of view of experimental design to test antigen 1 on half the individuals and antigen 2 on the other half. These individuals may be differentially susceptible

BOX 11.6   Paired Comparisons (Randomized Blocks with $a = 2$)

Lower face width (skeletal bigonial diameter in centimeters) for 15 North American white girls measured at 5 and again at 6 years old.

Individuals	(1) 5-year-olds	(2) 6-year-olds	(3) $\bar{Y}_B$	(4) $D_i = Y_{i2} - Y_{i1}$ (difference)
1	7.33	7.53	7.430	0.20
2	7.49	7.70	7.595	0.21
3	7.27	7.46	7.365	0.19
4	7.93	8.21	8.070	0.28
5	7.56	7.81	7.685	0.25
6	7.81	8.01	7.910	0.20
7	7.46	7.72	7.590	0.26
8	6.94	7.13	7.035	0.19
9	7.49	7.68	7.585	0.19
10	7.44	7.66	7.550	0.22
11	7.95	8.11	8.030	0.16
12	7.47	7.66	7.565	0.19
13	7.04	7.20	7.120	0.16
14	7.10	7.25	7.175	0.15
15	7.64	7.79	7.715	0.15
$\bar{Y}_A$	7.4613	7.6613	$\sum_a^b \sum Y = 226.84$	$\bar{D} = 0.20$

SOURCE: Data from a larger study by Newman and Meredith (1956).

Box 11.6 (continued)

I. Analysis as a randomized-complete-blocks design

This example is a mixed model with the $a = 2$ ages as fixed treatment effects and the $b = 15$ individuals constituting the randomly chosen blocks.

Preliminary computation

1. Grand mean $\bar{\bar{Y}} = \dfrac{1}{ab}\sum\limits^{a}\sum\limits^{b} Y = \dfrac{1}{30}\,226.84 = 7.5613$

2. SS_A (SS of columns; treatments) $= b \sum\limits^{a} (\bar{Y}_A - \bar{\bar{Y}})^2 = 0.3000$

3. SS_B (SS of rows; blocks) $= a \sum\limits^{b} (\bar{Y}_B - \bar{\bar{Y}})^2 = 2.63675$

4. SS_{error} (remainder; discrepance; residual SS) $= \sum\limits^{a}\sum\limits^{b} (Y - \bar{Y}_A - \bar{Y}_B + \bar{\bar{Y}})^2 = 0.01080$

Anova table

Source of variation	df	SS	MS	F_s	P	Expected MS
Ages (columns; factor A)	1	0.30000	0.300,000	388.89	$\ll 10^{-15}$	$\sigma^2 + \sigma_{AB}^2 + \dfrac{b}{a-1}\sum \alpha^2$
Individuals (rows; factor B)	14	2.63675	0.188,339	244.14	$\ll 10^{-15}$	$\sigma^2 + a\sigma_B^2$
Remainder	14	0.01080	0.000,771,4			$\sigma^2 + \sigma_{AB}^2$
Total	29	2.94755				

Box 11.6 (continued)

Conclusions

There is only a minute probability of obtaining such a result if the two age classes have the same lower face width. We conclude that faces of 6-year-old girls are wider than those of 5-year-olds. If we are willing to assume that the interaction is zero, then we may test for an added variance component among individual girls and would find a minuscule probability that it is zero.

II. The same analysis carried out using the t-test for paired comparisons

$$t_s = \frac{\bar{D} - (\mu_1 - \mu_2)}{s_{\bar{D}}}$$

where $\bar{D}$ is the mean difference between the paired observations

$$\bar{D} = \frac{\sum D}{b} = \frac{3.00}{15} = 0.20$$

and $s_{\bar{D}} = s_D/\sqrt{b}$ is the standard error of $\bar{D}$ calculated from the observed differences in column (4).

$$s_D = \sqrt{\frac{\sum\limits^{b} (D - \bar{D})^2}{b - 1}} = \sqrt{\frac{0.0216}{14}} = 0.039,279,2$$

$$s_{\bar{D}} = \frac{s_D}{\sqrt{b}} = \frac{0.039,279,2}{\sqrt{15}} = 0.010,141,9$$

We assume that the true difference between the means of the two groups, $\mu_1 - \mu_2$, equals zero.

$$t_s = \frac{\bar{D} - 0}{s_{\bar{D}}} = \frac{0.20 - 0}{0.010,141,9} = 19.7203$$

With $b - 1 = 14$ df, yields $P = 1.30 \times 10^{-11}$. $t_s^2 = 388.89$, which equals the previous F_s.

to these antigens, and we may learn little about the relative potency of the antigens because this would be confounded by the differential responses of the subjects. This (undesirable) design would consequently result in a large variance, which in turn would make it difficult to reject the null hypothesis of no difference, unless that difference was very large.

A much better design is to inject antigen 1 into the left arm and antigen 2 into the right arm of a group of n individuals and to analyze the data as a randomized-complete-blocks design with n rows (blocks) and 2 columns (treatments). It is probably immaterial whether an antigen is injected into the right or left arm, but if we were designing such an experiment and knew little about the reaction of humans to antigens, we might, as a precaution, randomly allocate antigen 1 to the left or right arm for different subjects and antigen 2 to the opposite arm. Note, however, that such a design should not be used if one needs to test whether there is some interaction effect due to both antigens being injected in the same person. There could, for example, be a difference depending on which antigen is injected first. A similar example is the testing of some plant viruses by rubbing a certain concentration of the virus over the surface of a leaf and counting the resulting lesions. Because different leaves are differentially susceptible, a conventional way of measuring the strength of the virus is to wipe it over half of the leaf on one side of the midrib, rubbing the other half of the leaf with a control or standard solution.

Another design leading to paired comparisons occurs when the treatment is given to two individuals sharing a common experience, whether genetic or environmental. For example, a drug or a psychological test might be given to groups of twins or sibs, one of each pair receiving the treatment, the other one not.

Finally, the paired-comparison technique may be used when the two individuals to be compared share a single experimental unit and are thus subjected to common environmental experiences. If we have a set of rat cages, each of which holds two rats, and we are trying to compare the effect of a hormone injection with that of a control, we might inject one of each pair of rats with the hormone and use its cage mate as a control. This design would yield a $2 \times n$ anova in which the n cages would represent the blocks.

The paired-comparison case in Box 11.6 analyzes face widths of 5- and 6-year-old girls, as already mentioned. The question posed is whether the faces of 6-year-old girls are wider than those of 5-year-old girls. The data for 15 individual girls are shown in columns (1) and (2) of Box 11.6. The marginal means are furnished in column (3) and below columns (1) and (2). The computations for the two-way anova without replication in Box 11.6 are the same as those already shown for Box 11.4. Remember that because there is no replication, the total sum of squares is identical to the subgroup sum of squares. The anova table shows that we can easily reject the null hypothesis of no differences in face width between the two age groups. If interaction is assumed to be zero, there is a large added variance component among the individual girls, undoubtedly representing genetic as well as environmental differences.

One reason for featuring the paired comparisons test in a separate section is that it alone among the two-way anovas without replication has an equivalent and traditional alternative method of analysis, the well-known **t-test for paired comparisons**. It is simple to apply and is illustrated in part II of Box 11.6. This method tests whether the mean of sample differences between pairs of readings in the two columns is different from a hypothetical mean, zero under the null hypothesis. The standard error over which this mean is tested is the standard error of the mean difference. The difference column [column (4) of the data table] must be calculated. The computations are straightforward, and the conclusions are the same as for the two-way anova above. This is another instance in which we obtain the value of F_s when we square the value of t_s (see Appendix A.7 for a proof).

Although the paired-comparisons t-test is the traditional method of solving this type of problem, we prefer the two-way anova. Its computation is no more time consuming and has the advantage of providing a measure of the variance component among the blocks. This is useful knowledge because if there is no added variance component among blocks, one might simplify the analysis and design of future, similar studies by employing a completely randomized anova. This example will also be revisited in Section 12.8 to show how we can estimate the increase in sensitivity for a paired comparisons design over a completely randomized design as in Chapter 9.

11.6 The Factorial Design

There is no reason to restrict the design of an anova to a consideration of only two factors. Three or more factors may be analyzed simultaneously using a method often called **factorial analysis of variance**. In the example in the next section, we will analyze the effects of temperature, oxygen concentration, and cyanide ion concentration on survival time in minnows. The number of main effects is theoretically unlimited. An analysis involving as many as five main effects, however, is a rarity because even without replication within a subgroup, the number of experimental units necessary becomes very large, and it is frequently impossible or prohibitive in cost to carry out such an experiment. Imagine an experiment with only three factors. If each of the three factors were divided into four classes or levels, we would have to carry out the experiment on $4 \times 4 \times 4 = 64$ experimental units to represent each combination of factors. This is a substantial number. You will realize that this would not permit any measurement of the basic experiment error, and we would have to employ an interaction term as an estimate of experimental error (on the assumption that no added interaction effect is present). In the minimal case of two levels per factor, a five-factor factorial anova would require $2^5 = 32$ experimental units, even for a single replicate.

There are also logistical difficulties with such large experiments. It may not be possible to run all the tests in one day or to hold all of the material in a single controlled environmental chamber. Thus, treatments may be confounded with undesired effects if different treatments are applied under not quite the same experimental conditions.

Another problem that accompanies a factorial anova with several main effects is the large number of possible interactions. We saw that a two-way anova (two-factor factorial) had only one interaction, $A \times B$. A three-factor factorial has three

first-order interactions, $A \times B$, $A \times C$, and $B \times C$; it also has a **second-order interaction**, $A \times B \times C$. A four-factor factorial has six first-order interactions, $A \times B$, $A \times C$, $A \times D$, $B \times C$, $B \times D$, and $C \times D$; four second-order interactions, $A \times B \times C$, $A \times B \times D$, $A \times C \times D$, and $B \times C \times D$; and one **third-order interaction**, $A \times B \times C \times D$. These numbers go up rapidly. In a k-factor factorial, $k!/[(m + 1)!(k - m - 1)!]$ mth-order interactions exist, representing the number of combinations of $m + 1$ items out of the k factors. Not only is the computation of these interactions tedious but also testing them and, more importantly, interpreting them become exceedingly complex.

The assumptions underlying multiway anova are the same as those for two-way anova. The expected value for a single observation in a (replicated) three-factor case in which all three main effects, A, B, and C, are fixed treatment effects (Model I) is

$$Y_{ijkl} = \mu + \alpha_i + \beta_j + \gamma_k + (\alpha\beta)_{ij} + (\alpha\gamma)_{ik} + (\beta\gamma)_{jk} + (\alpha\beta\gamma)_{ijk} + \epsilon_{ijkl}$$

where μ equals the parametric mean of the population; α_i, β_j, and γ_k are the fixed treatment effects for the ith, jth, and kth groups of treatments A, B, and C, respectively; $(\alpha\beta)_{ij}$, $(\alpha\gamma)_{ik}$, and $(\beta\gamma)_{jk}$ are first-order interaction effects in the subgroups represented by the indicated combinations of the ith group of factor A, the jth group of factor B, and the kth group of factor C; $(\alpha\beta\gamma)_{ijk}$ is the second-order interaction effect in the subgroup representing the ith, jth, and kth groups of factors A, B, and C, respectively; and ϵ_{ijkl} is the error term of the lth item in subgroup ijk. This expression is analogous to Expressions (8.2), (10.2), and (11.1). As before, all the main effects may be random (Model II), or only some of them may be, in which case the factorial anova is a mixed model.

In the next section, we will analyze a three-way factorial to illustrate the procedures and the problems encountered.

11.7 A Three-Way Factorial Design

The example we describe here comes from a study of pollution by factory effluents. The investigators measured the effects of different concentrations of cyanide ion on the survival time of minnows. Five concentrations of cyanide were combined with three concentrations of oxygen and three temperatures. The details of the experiment are given in Box 11.7. For each combination of factors, 10 replicate fishes were examined. The data shown here, however, are only the means of the 10 readings and thus do not provide an estimate of within-subgroup variance.

The basic data are tabulated in the second table of Box 11.7. Note that because this is a three-factor example, the third factor (oxygen concentration) is arranged within each class or level of the first factor (temperature) because we are limited by the two-dimensionality of the printed page. This arrangement, however, does not imply a nested anova; the factors define a three-dimensional table.

Next, we arrange the data as three two-way tables of means. We will symbolize the means of these tables by $\overline{Y}_{AB}$, $\overline{Y}_{AC}$, and $\overline{Y}_{BC}$. These tables are produced by summing all levels of the third factor for each two-factor combination and dividing

BOX 11.7 Three-Way Anova without Replication (a 5×3×3 Factorial Anova)

Time to intoxication by cyanide in *Phoxinus laevis*, a European minnow, using 5 concentrations of CN^- ion, 3 oxygen concentrations, and 3 temperatures. The variable is a transformation into logarithms of readings in minutes of survival time, coded and summed for 10 replicate fishes. Instead of presenting means, we furnish the sums, representing means coded × 10, as given in the original publication. In this example, all three main effects are fixed treatments; it is therefore a Model I anova.

Factor A (a = 3) Temperatures (in °C)	Factor C (c = 3) O_2-concentrations (in mg O_2/l)	Factor B (b = 5) CN^--concentrations (in mg CN^-/l)				
		Cy_1 0.16	Cy_2 0.8	Cy_3 4.0	Cy_4 20.0	Cy_5 100.0
T_1 5°C	O_1	201	150	131	130	97
	O_2	246	164	138	136	102
	O_3	271	170	149	127	99
		718	484	418	393	298
T_2 15°C	O_1	124	104	86	89	60
	O_2	158	111	99	91	74
	O_3	207	117	81	87	72
		489	332	266	267	206
T_3 25°C	O_1	79	63	50	51	32
	O_2	129	54	51	52	46
	O_3	142	93	62	51	52
		350	210	163	154	130

$O_1 = 1.5$, $O_2 = 3.0$, $O_3 = 9.0$

SOURCE: Data by Wuhrmann and Woker (1953).

Box 11.7 (continued)

For ease in computation, rearrange the data in three two-way tables of means as shown here. The value in each cell of a two-way table represents the mean of the items for the particular two-factor combination over all levels of the third factor. Thus, the value 239.3333 in the A×B table represents the combination $T_1 \times Cy_1$ averaged for the three oxygen concentrations O_1, O_2, and O_3 as follows: $(201 + 246 + 271)/3 = 239.3333$. The marginal means are computed from the bodies of the two-way tables and are used in computation.

Two-way tables of means

$A \times B$ ($T \times Cy$ in this example)

	Cy_1	Cy_2	Cy_3	Cy_4	Cy_5	$\overline{Y}_A$
T_1	239.3333	161.3333	139.3333	131.0000	99.3333	154.0667
T_2	163.0000	110.6667	88.6667	89.0000	68.6667	104.0000
T_3	116.6667	70.0000	54.3333	51.3333	43.3333	67.1333
$\overline{Y}_B$	173.0000	114.0000	94.1111	90.4444	70.4444	108.4000

$B \times C$ ($Cy \times O$ in this example)

	O_1	O_2	O_3	$\overline{Y}_B$
Cy_1	134.6667	177.6667	206.6667	173.0000
Cy_2	105.6667	109.6667	126.6667	114.0000
Cy_3	89.0000	96.0000	97.3333	94.1111
Cy_4	90.0000	93.0000	88.3333	90.4444
Cy_5	63.0000	74.0000	74.3333	70.4444
$\overline{Y}_C$	96.4667	110.0667	118.6667	108.4000

$A \times C$ ($T \times O$ in this example)

	O_1	O_2	O_3	$\overline{Y}_A$
T_1	141.8000	157.2000	163.2000	154.0067
T_2	92.6000	106.6000	112.8000	104.0000
T_3	55.0000	66.4000	80.0000	67.1333
$\overline{Y}_C$	96.4667	110.0667	118.6667	108.4000

Box 11.7 (continued)

The expressions for first-, second-, and third-order interaction SS in a four-factor factorial anova are

$$SS_{AB} = ncd \sum^{a} \sum^{b} (\bar{Y}_{AB} - \bar{Y}_A - \bar{Y}_B + \bar{\bar{Y}})^2$$

$$SS_{ABC} = nd \sum^{a} \sum^{b} \sum^{c} (\bar{Y}_{ABC} - \bar{Y}_{AB} - \bar{Y}_{AC} - \bar{Y}_{BC} + \bar{Y}_A + \bar{Y}_B + \bar{Y}_C - \bar{\bar{Y}})^2$$

$$SS_{ABCD} = n \sum^{a} \sum^{b} \sum^{c} \sum^{d} (\bar{Y}_{ABCD} - \bar{Y}_{ABC} - \bar{Y}_{ABD} - \bar{Y}_{ACD} - \bar{Y}_{BCD} + \bar{Y}_{AB} + \bar{Y}_{AC} + \bar{Y}_{AD} + \bar{Y}_{BC} + \bar{Y}_{BD} + \bar{Y}_{CD}$$
$$- \bar{Y}_A - \bar{Y}_B - \bar{Y}_C - \bar{Y}_D + \bar{\bar{Y}})^2$$

Note the regular alternation of signs before the terms. Note also, that because the example in this box is a three-factor anova, only the first two expressions are used. In a three-factor anova, omit the coefficient d in SS_{AB} and SS_{ABC}.

In the formulas below we include n, sample size per subgroup, to make the presentation general. Because in the present example $n = 1$, it could have been omitted.

Preliminary computation

1. Grand mean $\bar{\bar{Y}} = \dfrac{1}{abcn} \sum^{a} \sum^{b} \sum^{c} = 4878 \div 45 = 108.4000$

2. SS_A (SS of temperatures) $= nbc \sum^{a} (\bar{Y}_A - \bar{\bar{Y}})^2 = 57{,}116.1333$

3. SS_B (SS of cyanide concentrations) $= nac \sum^{b} (\bar{Y}_B - \bar{\bar{Y}})^2 = 55{,}545.4667$

Box 11.7 (continued)

4. SS_C (SS of oxygen concentrations) $= nab \sum^{c} (\bar{Y}_C - \bar{\bar{Y}})^2 = 3758.8000$

5. SS_{AB} (temperature $\times$ cyanide interaction SS) $= nc \sum^{a} \sum^{b} (\bar{Y}_{AB} - \bar{Y}_A - \bar{Y}_B + \bar{\bar{Y}})^2 = 3685.8667$

6. SS_{AC} (temperature $\times$ oxygen interaction SS) $= nb \sum^{a} \sum^{c} (\bar{Y}_{AC} - \bar{Y}_A - \bar{Y}_C + \bar{\bar{Y}})^2 = 97.0667$

7. SS_{BC} (cyanide $\times$ oxygen interaction SS) $= na \sum^{b} \sum^{c} (\bar{Y}_{BC} - \bar{Y}_B - \bar{Y}_C + \bar{\bar{Y}})^2 = 5264.5333$

8. SS_{ABC} (temperature $\times$ cyanide $\times$ oxygen interaction SS)

$$= n \sum^{a} \sum^{b} \sum^{c} (\bar{Y}_{ABC} - \bar{Y}_{AB} - \bar{Y}_{AC} - \bar{Y}_{BC} + \bar{Y}_A + \bar{Y}_B + \bar{Y}_C - \bar{\bar{Y}})^2 = 1034.9333$$

9. SS_{within} (within subgroups; error SS). Since this example is an anova without replication, there is no SS_{within}. However, if there had been replication, we could have calculated such an SS as follows:

$$SS_{\text{within}} = \sum^{a} \sum^{b} \sum^{c} \sum^{n} (\bar{Y} - \bar{Y}_{ABC})^2$$

The computation of these quantities is self-evident.

Box 11.7 (continued)

Now fill in the anova table.

Source of variation	df	SS	MS	Expected MS (Model I)
Main effects				
A $\bar{Y}_A - \bar{\bar{Y}}$	$a-1$	2	$\dfrac{2}{(a-1)}$	$\sigma^2 + \dfrac{bc}{a-1}\sum^{a}\alpha^2$
B $\bar{Y}_B - \bar{\bar{Y}}$	$b-1$	3	$\dfrac{3}{(b-1)}$	$\sigma^2 + \dfrac{ac}{b-1}\sum^{b}\beta^2$
C $\bar{Y}_C - \bar{\bar{Y}}$	$c-1$	4	$\dfrac{4}{(c-1)}$	$\sigma^2 + \dfrac{ab}{c-1}\sum^{c}\gamma^2$
First-order interactions				
$A \times B$ $\left(\bar{Y}_{AB} - \bar{Y}_A - \bar{Y}_B + \bar{\bar{Y}}\right)$	$(a-1)(b-1)$	5	$\dfrac{5}{(a-1)(b-1)}$	$\sigma^2 + \dfrac{c}{(a-1)(b-1)}\sum^{ab}(\alpha\beta)^2$
$A \times C$ $\left(\bar{Y}_{AC} - \bar{Y}_A - \bar{Y}_C + \bar{\bar{Y}}\right)$	$(a-1)(c-1)$	6	$\dfrac{6}{(a-1)(c-1)}$	$\sigma^2 + \dfrac{b}{(a-1)(c-1)}\sum^{ac}(\alpha\gamma)^2$
$B \times C$ $\left(\bar{Y}_{BC} - \bar{Y}_B - \bar{Y}_C + \bar{\bar{Y}}\right)$	$(b-1)(c-1)$	7	$\dfrac{7}{(b-1)(c-1)}$	$\sigma^2 + \dfrac{a}{(b-1)(c-1)}\sum^{bc}(\beta\gamma)^2$
Second-order interaction				
$A \times B \times C$ $\left(\bar{Y}_{ABC} - \bar{Y}_{AB} - \bar{Y}_{AC} - \bar{Y}_{BC} + \bar{Y}_A + \bar{Y}_B + \bar{Y}_C - \bar{\bar{Y}}\right)$	$(a-1)(b-1)(c-1)$	8	$\dfrac{8}{(a-1)(b-1)(c-1)}$	$\sigma^2 + \dfrac{1}{(a-1)(b-1)(c-1)}\sum^{abc}(\alpha\beta\gamma)^2$

Box 11.7 (continued)

If the anova had been replicated, all expected mean squares would have their added components multiplied by n. We would also have had an error term as follows:

$Y - \bar{Y}$	Within subgroups	$abc(n-1)$	$\dfrac{9}{abc(n-1)}$ σ^2

$\bar{\bar{Y}} - \bar{Y}$	Total	$abcn - 1$	$\displaystyle\sum (2 \dots 9)$

The expected mean squares shown above are for a Model I anova. We now show expected mean squares for Model II and mixed models. Because it is the most general case, the replicated design is shown. If you have a case without replication, simply eliminate the within-subgroups (error) term and set $n = 1$.

Expected mean squares for a three-way factorial anova (other models)

Source of variation	Model II	Mixed Model (A and B fixed, C random)
A	$\sigma^2 + n\sigma^2_{ABC} + nc\sigma^2_{AB} + nb\sigma^2_{AC} + nbc\sigma^2_A$	$\sigma^2 + nb\sigma^2_{AC} + \dfrac{nbc}{a-1}\displaystyle\sum^a \alpha^2$
B	$\sigma^2 + n\sigma^2_{ABC} + nc\sigma^2_{AB} + na\sigma^2_{BC} + nac\sigma^2_B$	$\sigma^2 + na\sigma^2_{BC} + \dfrac{nac}{b-1}\displaystyle\sum^b \beta^2$
C	$\sigma^2 + n\sigma^2_{ABC} + nb\sigma^2_{AC} + na\sigma^2_{BC} + nab\sigma^2_C$	$\sigma^2 + nab\sigma^2_C$
$A \times B$	$\sigma^2 + n\sigma^2_{ABC} + nc\sigma^2_{AB}$	$\sigma^2 + n\sigma^2_{ABC} + \dfrac{nc}{(a-1)(b-1)}\displaystyle\sum^{ab} (\alpha\beta)^2$
$A \times C$	$\sigma^2 + n\sigma^2_{ABC} + nb\sigma^2_{AC}$	$\sigma^2 + nb\sigma^2_{AC}$
$B \times C$	$\sigma^2 + n\sigma^2_{ABC} + na\sigma^2_{BC}$	$\sigma^2 + na\sigma^2_{BC}$
$A \times B \times C$	$\sigma^2 + n\sigma^2_{ABC}$	$\sigma^2 + n\sigma^2_{ABC}$
Within subgroups	σ^2	σ^2

Box 11.7 (continued)

Anova table

Source of variation		df	SS	MS	F_s	P
A	Temperature	2	57,116.1333	28,558.0667		
B	Cyanide	4	55,545.4667	13,886.3667		
C	Oxygen	2	3,758.8000	1,879.4000		
$A \times B$	$T \times Cy$	8	3,685.8667	460.7333	7.123	0.00046
$A \times C$	$T \times O$	4	97.0667	24.2667	0.3751	0.8230
$B \times C$	$Cy \times O$	8	5,264.5333	658.0667	10.174	5.46×10^{-5}
$A \times B \times C$	$T \times Cy \times O$	16	1,034.9333	64.6833		
Total		44	126,502.8000			

Because this is a Model I anova, the presence of each effect is tested by the ratio of its mean square over MS_{ABC}, which represents the error variance σ^2, on the assumption that the temperature $\times$ cyanide $\times$ oxygen interaction is zero.

Conclusions

The null hypotheses of no temperature $\times$ cyanide interaction and no cyanide $\times$ oxygen interaction can be rejected; that is, the effect of cyanide ion on survival time depends on the oxygen concentration and the temperature of the water. When, as in this instance, interactions involving all main effects are present, it is usually of little interest to test the main effect mean squares. A statement about the differences due to cyanide will not mean much unless we qualify it by specifying temperature and oxygen concentration. An examination of plots such as in Figures 11.6 and 11.7 can be very helpful for such interpretations. If we wish to assess the gross effects of these three variables, however, we can test them over the error MS; in this example, we would reject their null hypotheses also.

by the number of levels. Thus, the value for temperature level 2 against cyanide level 3 is the sum of all the oxygen concentrations for that combination of temperature and cyanide, divided by 3, the number of oxygen concentrations. In this case, $\bar{Y}_{A_2B_3} = \bar{Y}_{T_2Cy_3} = (86 + 99 + 91) \div 3 = 88.6667$. The means along the margins of these tables symbolized by $\bar{Y}_A$, $\bar{Y}_B$, and $\bar{Y}_C$ are the means for each level of the indicated factor, averaged over all levels of the other two factors. They can be most easily obtained by averaging the means of type over the appropriate row or column of the two-way tables. Necessarily, the marginal means for the same factor are identical in different tables. Finally, we need a symbol for the observations in the body of the three-way table. Because they are means, we label them $\bar{Y}_{ABC}$.

The computations in Box 11.7 are longer than those of previous anovas but are not more complicated. The deviations for the sums of squares are obtained from values in both the original data tables and the two-way tables. Note that in these data as presented here, there is no replication for each subgroup; hence there will be no error mean square and, as in the example of Box 11.4, the subgroup SS is the same as the total SS. Thus, step **9** is not carried out in the current example but would be undertaken in a three-way anova *with* replication.

The layout of the anova table is shown next in Box 11.7. Note that the source of variation is subdivided into three main effects, A, B, and C; three first-order interactions, $A \times B$, $A \times C$, and $B \times C$; and one second-order interaction, $A \times B \times C$. Had there been replication in these data, we would also have had a within-subgroups (error) line below the second-order interaction. The experiment with the minnows is clearly Model I. Therefore, the expected mean squares in the table consist simply of the error variance and the added effect due to treatment for the particular source of variation. Hypothesis testing is quite simple. The mean square for any source of variation is tested over the error mean square. Because we do not have a pure measure of error in this example, we must use the second-order interaction mean square on the assumption that the added effect due to the $A \times B \times C$ interaction, $\Sigma^{abc}(\alpha\beta\gamma)^2/(a - 1)(b - 1)(c - 1)$, is zero. The meaning of a second-order interaction will be explained later in this section. As mentioned earlier in this chapter, main effects in a Model I anova are rarely tested if one concludes that interaction effects are present, but if desired, such a test can be carried out as simply the ratio of the main effect mean square over the error mean square.

Box 11.7 also shows expected mean squares of a three-factor factorial, assuming Model II and a mixed model in which factors A and B are fixed and factor C is random. In these expected mean squares, we have taken the general case in which there is replication within subgroups. For anovas without replication, one simply eliminates the last line of the table (the within-subgroups mean square) and sets the coefficient $n = 1$ throughout the table.

In Model II and mixed model anovas, statistical tests are more complicated than in Model I. Let us look at a Model II anova. Inspection of the expected mean squares in Box 11.7 shows that $(A \times B \times C)/Error$ is the appropriate test for the second-order interaction. If second-order interaction is present, we test first-order interactions over the second-order interactions. However, there is no exact test for the main effects. There is no mean square in the table that, when subtracted from the mean square of a main effect, will leave only the added variance component due to the main effect. Thus, to test main effects we have to create a synthetic denominator mean square, which can be done in several ways. For example, to test the mean square of A we could construct a denominator MS as the sum of the mean square of $A \times B$ and $A \times C$ minus the mean square of $A \times B \times C$ (in order to subtract the extra values of $\sigma^2 + n\sigma^2_{ABC}$ from the denominator). Significance tests involving such synthetic mean squares are carried out by the approximate method learned in Chapter 10 (see Boxes 10.4 and 10.5) and are subject to the same limitations. We will illustrate such a test in the next section. In a mixed model, some mean squares can be tested directly over an appropriate MS, but others may need construction of a synthetic denominator mean square.

The analysis of the minnow data is shown in an anova table at the end of Box 11.7. Results of the significance tests lead us to accept the hypothesis that there is a temperature cyanide interaction, as well as a cyanide–oxygen interaction, but little evidence of a temperature oxygen interaction. These relationships are shown in Figure 11.6, in which the data from the three two-factor tables of Box 11.7 have been graphed. Note that lines depicting the relation between cyanide and temperature and cyanide and oxygen are not parallel, but those depicting the temperature–oxygen relationship are more or less parallel. At the low concentrations of cyanide, the effects of oxygen and temperature are much more marked than they are at the higher concentrations. This illustrates the meaning of interaction in these data. We could test the three main effects, the null hypotheses for which would all be strongly rejected, but in view of

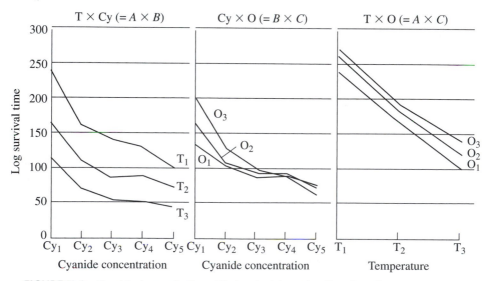

FIGURE 11.6 Graphic demonstration of first-order interaction. Data from Box 11.7.

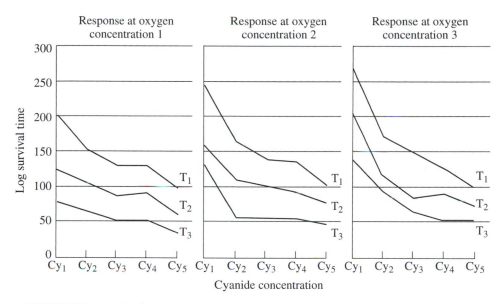

FIGURE 11.7 Graphic demonstration of second-order interaction. Data from Box 11.6.

the interactions this result would not be particularly meaningful. From Figure 11.6 it is clear that there is a substantial overall effect, at least of cyanide and temperature. As cyanide concentration increases, survival time of the fish decreases; also, as temperature increases, survival time decreases.

The meaning of a second-order interaction requires some explanation. Interaction $A \times B \times C$ can be interpreted as indicating that interaction $A \times B$ differs, depending on the level of factor C (or $A \times C$, depending on level of factor B, and so on). We can see this in Figure 11.7. The three panels of this figure depict the relationship between cyanide concentration and temperature separately for each oxygen concentration. Contrast this figure with the first panel of Figure 11.6, in which these relations were summarized for the whole experiment. One way of interpreting second-order interactions is to compare the three panels of Figure 11.7. Strong interaction would be reflected in marked differences in the patterns of the three temperature lines among the panels. Although some slight differences are evident, we assume that they were not of sufficient magnitude to have led to the conclusion that second-order interaction effects were present. Of course, we have no way of testing this assumption without an estimate of the error variance.

11.8 Higher-Order Factorial Anovas

It follows from the previous section that raising the number of factors above three will increase the complexity of the analysis considerably. Not only must there be more experimental units to provide at least one replicate per experimental condition but also there will be many interaction terms to be evaluated and interpreted. Writing down the expected mean square and testing each source of variation in the anova is more complex than in any example that we have tackled so far. If you carefully follow the rules we will describe, however, you should not have too much difficulty.

Assume that you have a four-factor factorial without replication. We label the factors A, B, C, and D. Assume further that factors A, B, and C are fixed treatment effects (Model I) and factor D is random. The number of levels for each corresponding main effect will be a, b, c, and d.

One problem is to write down the expected mean squares for any model anova involving any number of factors. You were shown these mean squares for three cases of a three-way anova in Box 11.7. We have found the following scheme helpful for writing down the expected mean squares, as well as for subsequent significance testing. The arrangement is illustrated in Table 11.5, and the steps are as follows. First, write the sources of variation along the left margin of the intended table, leaving three lines free at the top. Next, copy the same sources *in reverse order* across the top line labeled "Identification," starting with the error at the left and going to the topmost main effect at the right. Each symbolizes a component of variation for either an error, an interaction, or a main effect. Note that we left out the multiplication sign between letters in interaction terms for simplicity of notation. Block out by means of heavy lines the various levels of the sources of variation (main effects, first-order interactions, second-order interactions, third-order interaction, and error) in columns as well as in rows of the table.

The second line across the top of the table indicates whether the component in question is an added component due to treatment (symbolized by Σ^2) or a variance component (symbolized by σ^2). In a pure Model I anova, all components, except the one representing error, are Σ^2. In a pure Model II, all components are σ^2. In a mixed model, any component whose identification does not contain a letter representing a random effect is an added component due to treatment; all others are variance components. Thus, in the example of Table 11.5, because the interaction component AC does not contain the random effect D, it is an added component due to treatment and is identified by Σ^2. By contrast, interaction component ACD, which contains D, is a variance component identified by σ^2.

Next, we work out the coefficients for each component other than the error variance. These coefficients can be written down by a simple rule. The maximum possible coefficient is $nabcd$. Omit from this formula any letters whose corresponding capitals occur in the first line. Thus, for the component ABD, the coefficient is nc. Similarly, for the component of B, the coefficient is $nacd$. In anovas without replication, we can set $n = 1$ and ignore it when writing down the coefficients. We are now able to write down any component in its proper form with the correct coefficients. Thus, the component of the interaction ACD from Table 11.5 is $nb\sigma^2_{ACD}$, and the component for the main effect C is $nabd\Sigma\gamma^2/(c-1)$. Remember that added components due to treatment are labeled by Greek letters and are always divided by the corresponding degrees of freedom.

TABLE 11.5 Expected Mean Squares in a Four-Factorial Anova

Mixed model: A, B, C are fixed treatments; D is random.

Identification[a]	Error	ABCD	BCD	ACD	ABD	ABC	CD	BD	BC	AD	AC	AB	D	C	B	A
Component	σ^2	σ^2	σ^2	σ^2	σ^2	Σ^2	σ^2	σ^2	Σ^2	σ^2	Σ^2	Σ^2	σ^2	Σ^2	Σ^2	Σ^2
Coefficient	1	n	na	nb	nc	nd	nab	nac	nad	nbc	nbd	ncd	nabc	nabd	nacd	nbcd
Source of variation																
A	*									*						*
B	*							*							*	
C	*						*							*		
D	*												*			
AB	*				*							*				
AC	*			*							*					
AD	*									*						
BC	*		*						*							
BD	*							*								
CD	*						*									
ABC	*	*				*										
ABD	*				*											
ACD	*			*												
BCD	*		*													
ABCD	*	*														
Error	*															

[a]Identifying letters are used as subscripts in variance components and as equivalent lower case Greek letters squared in treatment effects.

We now determine the complete expected mean square for any source of variation by considering the following five rules. (1) *All sources of variation contain the error variance* as shown by an asterisk in that column. (2) *They also contain the component corresponding to their name.* For example, the main effect *B* contains the component for *B*, and interaction $A \times B \times C$ contains the component for *ABC*. The appropriate asterisks form a diagonal across the table starting at the lower left with the error mean square, and no mean square will have any component to the right of this diagonal.

Next, examine all components located between the error component and the final component in each line. (3) *Consider only those in blocks to the left of the block of the final component.* Thus, if you are working out a first-order interaction, consider only second- and higher-order interactions, and so on. (4) *In the blocks of interest, consider only the components that contain within them all the letters identifying the final component.* Thus, for main effect *B*, consider only *AB*, *BC*, and *BD*, as well as all higher-order interactions containing *B*; do not consider *AC*, *AD*, or *CD*. For interaction *BC*, consider only *ABC*, *BCD*, and *ABCD*. The final rule: (5) *If any of the letters in the interaction terms being considered (other than those identifying the final component in each line) represent a fixed treatment effect, this component is not included in the expected MS; if all letters represent random effects, include the component.*

Now let us apply these rules. In a pure Model I anova, all letters represent fixed effects. It is obvious, therefore, by rule (5) that all components except the final component in the line [rule (2)] and the error variance [rule (1)] will be omitted from each expected mean square. The result is the characteristic expected mean squares of a Model I anova containing only the error variance and the component representing a given source of variation (see Box 11.7). By contrast, a pure Model II anova contains all possible interaction terms. In a four-factor Model II anova with replication, the expected mean square for factor *B* would therefore read

$$\sigma^2 + n\sigma^2_{ABCD} + na\sigma^2_{BCD} + nc\sigma^2_{ABD} + nd\sigma^2_{ABC} + nac\sigma^2_{BD}$$
$$+ nad\sigma^2_{BC} + ncd\sigma^2_{AB} + nacd\sigma^2_B$$

Expected mean squares of mixed model anovas depend on the particular combination of fixed and random factors. We will compute several mean squares in the mixed model example of Table 11.5 to make certain that the rules are understood. The mean square of *A*, a fixed treatment effect, will lead to consideration of three first-order interactions containing an *A* [rules (3) and (4)]. *AB* and *AC* include letters representing fixed factors *B* and *C* and hence are omitted [rule (5)], but because *D* is a random factor, the *AD* term is retained. Any second- or third-order interaction including *A* also contains at least one letter representing a fixed effect. Thus, *ABD* contains *BD* in addition to *A*; although *D* is random, *B* is fixed, so the component is excluded [rule (5)]. Therefore, the expected mean square of A (EMS_A) consists of three components, as indicated by the asterisks in Table 11.5.

$$EMS_A = \sigma^2 + nbc\sigma^2_{AD} + \frac{nbcd}{a-1}\sum^a \alpha^2$$

The expected mean square for main effect D contains only the error variance and the final component because all other letters represent fixed effects and their components would be omitted [rule (5)]. Finally, let us work out the expected mean square for the first-order interaction $A \times B$. We find AB contained in the second order terms ABC and ABD [rule (4)]. In ABC, the letter C represents a fixed effect; hence the term is omitted [rule (5)]. In ABD, the D is a random factor, so the term is included.

All included terms are represented by asterisks in Table 11.5. When filled in completely, the table is useful for pointing out how significance tests for given sources of variation should be carried out. From Table 11.5, it is clear that we would test the main effect of C over the $C \times D$ mean square or the interaction $B \times C$ over the $B \times C \times D$ interaction MS. The mixed model in Table 11.5 is a relatively simple case. If two of the main effects had been random, the situation would have been more complicated. One of the problems frequently encountered in Model II or mixed factorials is that there is no appropriate error term against which to test a given mean square. For instance, in the expected mean squares of the Model II anova shown near the end of Box 11.7, there is no exact test for the expected mean square of main effect B. The appropriate significance test must be carried out by using the Satterthwaite approximation to estimate an appropriate mean square to use in the denominator. An estimate of $\sigma^2 + n\sigma^2_{ABC} + nc\sigma^2_{AB} + na\sigma^2_{BC}$ can be obtained by a combination of the other mean squares yielding the following F-ratio:

$$F_s = \frac{MS_B}{MS_{AB} + MS_{BC} - MS_{ABC}}$$

where the appropriate degrees of freedom for the denominator mean square have to be calculated, as in Box 10.4.

Another complication in calculating multifactor factorials is the computation of the higher-order interactions. We saw in Box 11.7 how this is done for a three-way anova. For higher-order interactions, similar arrangements into subtables of the original data table may be useful for interpreting direct computational formulas for the interaction sums of squares. Box 11.7 gives formulas for calculating first- through third-order interactions (in a four-factor factorial), from which the reader should have no difficulty generating yet higher-order interactions in factorial anovas of any number of factors. Note the regular alternation of plus and minus signs.

The following shorthand notation for indicating the nature of the factorial design is frequently used. A three-factor factorial with each factor expressed at two levels is indicated as a $2 \times 2 \times 2$ factorial or 2^3 factorial. A two-way anova with each factor at three levels is a 3^2 factorial. A $2 \times 3 \times 4$ factorial has three factors—one at two levels, another at three levels, and the third at four levels. For many standard factorials, especially efficient computational methods have been developed, although their importance has lessened with the increasing use of computers. Factorial anovas involving treatments at only two levels can be computed with special elegance. Complete computations for a 2×2 factorial two-way anova, each factor at two levels with replication, are illustrated in Steel and Torrie (1980, Section 15.3). Raktoe et al. (1981) discuss the general 2^k case, as do Dunn and Clark (1974, Chapter 8).

11.9 Other Designs

The analyses of variance that we have studied up to now are only an introduction to the field of **experimental design**, which covers a great variety of designs developed in response to the needs of experimental scientists. A detailed consideration of these is beyond the scope of this book. Therefore, if your experiments do not seem to fit the simple designs already discussed, we strongly urge you to consult a professional statistician. However, we will briefly outline some of the more common designs, the reasons for using them, and sources for in-depth discussion and computational outlines.

In Section 11.4, we first became acquainted with the randomized-complete-blocks design, in which the rows of a two-way anova represent blocks of an experimental layout. In many instances, it becomes necessary to lay out two-dimensional blocks. The most obvious cases are agricultural experiments, where fields may be heterogeneous not only in a north–south direction but also in an east–west direction. In such a case, to test a single treatment (factor) at k levels, the **Latin square design** is especially appropriate. We divide the field into k row blocks and k column blocks as shown in Figure 11.8. Each

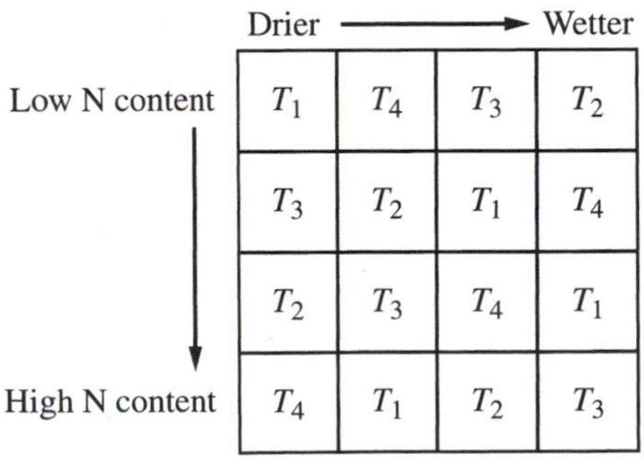

FIGURE 11.8   An example of four chemical treatments (T_1 through T_4) laid out in a field according to a Latin-square design to control known nitrogen content and moisture gradients.

level of the treatment T_i is replicated only once in a given row or column. For analysis of a Latin square experiment, see Steel and Torrie (1980, Sections 9.10 to 9.14) and Dunn and Clark (1974, Chapter 6).

This design need not be restricted to agricultural research. Suppose we wish to make quantitative chemical analyses of four related substances in a search for chemical differences among them. The analysis of a single sample is a tedious procedure that takes all day and ties up one analytical apparatus. Assuming that we have four such apparatuses in our laboratory, we can now run our analyses over four days, arranging the four treatments in the manner of a Latin square; that is, any one substance would be tested only once on any one day and only once in any one apparatus. The resulting anova table would yield effects representing day-to-day differences as well as differences among apparatuses, but the mean square of greatest interest would be that relating to differences among the four substances.

Another common design is the **split-plot design**, which is for two (or more) factors. Each level of one factor (factor A) is assigned to a whole plot within a block, and each plot is subdivided into as many subplots as there are levels in B, which are then randomly allocated to these subplots. Two different error mean squares are computed: one, expected to be larger, for the plots, and a smaller one for the subplots. This design yields more precise information on the factor allocated to the split plots or subplots at the expense of losing information on the factor assigned to the whole plots. For instance, we may be studying the differences between five genetic strains at three temperatures. The three temperatures may be produced in different temperature chambers in one laboratory. The five genetic strains may be in separate culture bottles in the chambers. The reader may ask why such an analysis could not be done as a simple factorial of strains against temperatures. It could indeed, but it would require 15 different temperature chambers, one for each combination of temperature and strains. The discussion of restriction errors from the previous chapter applies here (see Section 11.4). Split-plot designs are quite often incorrectly analyzed as factorial anovas. They are discussed in many texts. We find the discussion in Snedecor and Cochran (1989, section 16.15) quite clearly presented, Steel and Torrie (1980) devote a chapter (16) to this subject, and the book by Federer and King (2007) covers this and related designs.

Many other designs are found in textbooks of experimental design. For example, when in a design for which a randomized block is appropriate we find it impossible to include all treatments in every block because of the amount of work or space, we must use an **incomplete block design.** If only a fraction of combinations of factors (not all of them) are tested in a factorial design, then it is called a **fractional factorial.** The proper design of such experiments is complex, requiring certain assumptions for the analyses to be meaningful. Restriction errors (see Section 11.4) are especially important in more complicated designs. A simple book on experimental design is Pearce (1965). Several recent texts written for biologists are Glass (2006), Quinn and Keough (2002), Ruxton and Colegrave (2006), and Underwood (1997). Ryan (2007) is a more general text. Montgomery (2004) gives an extensive treatment of fractional factorial analyses. Some additional texts are designed to be used with particular statistical software.

11.10 Anova by Computer

By now you should be impressed by the number of possible designs in analysis of variance. For most of the common designs, the computations are relatively simple and straightforward when there are equal sample sizes. All of the examples of two-way and multiway anovas that we have considered thus far have had equal sample sizes for each subclass. Even with the best of intentions, however, an investigator often simply cannot produce such a balanced design. A given single replicate may be lost or ruined, or the data may vary naturally beyond the control of the investigator. For instance, in an experiment on the effects of treatment of parents on weight of their offspring, the number of offspring per parent is likely to differ among parents and will not be under the control of the investigator.

The computations needed for these analyses become more difficult when the sample sizes in the different cells are unequal. As will be described in Chapter 16, methods need to be used that are directly based on linear models such as shown in Equation 11.1. Because we have not yet covered the necessary background material, we must defer discussion of the details to Section 16.7. This approach necessitates the use of computer software even for relatively small experiments. Some software is written just for the analysis of particular types of designs. Such software is usually very easy to use but greatly limited in the types of analyses for which it can be used. Other software packages are able to handle a wide variety of types of designs by requiring the user to specify the appropriate linear model for the analysis to be performed. For example, for a two-way anova with interaction, one might enter a statement such as $Y = A + B + A*B$ (other statements will also have to be entered to indicate whether each factor is a fixed or random effect). Comprehensive programs include BMDP (www.statsol.ie/html/bmdp/bmdp_home.html), SAS (www.sas.com/), SYSTAT (www.systat.com/), and STATA (www.stata.com/). Although convenient, these programs are expensive, and for that reason researchers may wish to use the methods described in Section 16.7 that can be performed using more widely available software for multiple regression analysis—even though it requires more effort to set up the proper matrix of independent variables.

EXERCISES 11

11.1 Swanson et al. (1921) determined soil pH for various soil samples from Kansas. An extract of their data (acid soils) is shown here.

County	Soil Type	Surface pH	Subsoil pH
Finney	Richfield silt loam	6.57	8.34
Montgomery	Summit silty clay loam	6.77	6.13
Doniphan	Brown silt loam	6.53	6.32
Jewell	Jewell silt loam	6.71	8.30
Jewell	Colby silt loam	6.72	8.44
Shawnee	Crawford silty clay loam	6.01	6.80
Cherokee	Oswego silty clay loam	4.99	4.42

County	Soil Type	Surface pH	Subsoil pH
Greenwood	Summit silty clay loam	5.49	7.90
Montgomery	Cherokee silt loam	5.56	5.20
Montgomery	Oswego silt loam	5.32	5.32
Cherokee	Bates silt loam	5.92	5.21
Cherokee	Cherokee silt loam	6.55	5.66
Cherokee	Neosho silt loam	6.53	5.66

Do subsoils differ in pH from surface soils (assume that there is no interaction between localities and depth for pH reading)?

ANSWER: $F_s = 0.894$, with 1 and 12 degrees of freedom.

11.2 The following data were extracted from a Canadian record book of purebred dairy cattle. Random samples of 10 mature (five-year-old and older) and 10 two-year-old cows were taken from each of 5 breeds (honor roll, 305-day class). The average butterfat percentages of these cows were recorded, yielding a total of 100 butterfat percentages, broken down into 5 breeds and into 2 age classes. The 100 butterfat percentages are given below. Analyze and discuss your results.

					Breed				
Ayrshire		Canadian		Guernsey		Holstein-Friesian		Jersey	
Mature	2-yr	Mature	2-yr	Mature	2-yr	Mature	2-yr	Mature	2-yr
3.74	4.44	3.92	4.29	4.54	5.30	3.40	3.79	4.80	5.75
4.01	4.37	4.95	5.24	5.18	4.50	3.55	3.66	6.45	5.14
3.77	4.25	4.47	4.43	5.75	4.59	3.83	3.58	5.18	5.25
3.78	3.71	4.28	4.00	5.04	5.04	3.95	3.38	4.49	4.76
4.10	4.08	4.07	4.62	4.64	4.83	4.43	3.71	5.24	5.18
4.06	3.90	4.10	4.29	4.79	4.55	3.70	3.94	5.70	4.22
4.27	4.41	4.38	4.85	4.72	4.97	3.30	3.59	5.41	5.98
3.94	4.11	3.98	4.66	3.88	5.38	3.93	3.55	4.77	4.85
4.11	4.37	4.46	4.40	5.28	5.39	3.58	3.55	5.18	6.55
4.25	3.53	5.05	4.33	4.66	5.97	3.54	3.43	5.23	5.72

11.3 King et al. (1964) gave the following results for a study of the amount of cotton (in grams) used for nesting material in both sexes of two subspecies of the deer mouse *Peromyscus maniculatus*. Because methods for unequal n are not given until Chapter 16, perform a conservative analysis using $n = 24$ for all samples.

		P.m. gracilis		*P.m. bairdii*	
Male	$\overline{Y}$	2.9	$\overline{Y}$	1.7	
	s	1.4	s	0.9	
	n	24	n	24	
Female	$\overline{Y}$	2.6	$\overline{Y}$	2.1	
	s	1	s	1	
	n	26	n	26	

Analyze and interpret. Note that you will have to compute the error mean square as a weighted average of the individual variances.

ANSWER: $MS_{sex} = 0.0624$, $MS_{within} = 1.1845$.

11.4 Blakeslee (1921) studied length–width ratios of second seedling leaves of two types of Jimsonweed called *globe* (*G*) and *nominal* (*N*). Three seeds of each type were planted in 16 pots. Is there sufficient evidence to conclude that globe and nominal differ in length–width ratio?

Pot Identification Number							Types
		G			*N*		
16533	1.67	1.53	1.61	2.18	2.23	2.32	
16534	1.68	1.70	1.49	2.00	2.12	2.18	
16550	1.38	1.76	1.52	2.41	2.11	2.60	
16668	1.66	1.48	1.69	1.93	2.00	2.00	
16767	1.38	1.61	1.64	2.32	2.23	1.90	
16768	1.70	1.71	1.71	2.48	2.11	2.00	
16770	1.58	1.59	1.38	2.00	2.18	2.16	
16771	1.49	1.52	1.68	1.94	2.13	2.29	
16773	1.48	1.44	1.58	1.93	1.95	2.10	
16775	1.28	1.45	1.50	1.77	2.03	2.08	
16776	1.55	1.45	1.44	2.06	1.85	1.92	
16777	1.29	1.57	1.44	2.00	1.94	1.80	
16780	1.36	1.22	1.41	1.87	1.87	2.26	
16781	1.47	1.43	1.61	2.24	2.00	2.23	
16787	1.52	1.56	1.56	1.79	2.08	1.89	
16789	1.37	1.38	1.40	1.85	2.10	2.00	

11.5 The mean length of developmental period (in days) for three strains of houseflies at seven densities is given below (data from Sullivan and Sokal, 1963).

Density per Container	Strains		
	OL	BELL	bwb
60	9.6	9.3	9.3
80	10.6	9.1	9.2
160	9.8	9.3	9.5
320	10.7	9.1	10.0
640	11.1	11.1	10.4
1280	10.9	11.8	10.8
2560	12.8	10.6	10.7

Do these flies differ in developmental period with density and among strains? You may assume lack of strain × density interaction.

ANSWER: $MS_{error} = 0.3426$, $MS_{strains} = 1.3943$.

11.6   The following data were randomly selected from a larger dataset presented by French (1976) in a study on energy utilization of the pocket mouse (*Perognathus longimembris*) during hibernation at different temperatures. The selection was carried out to equalize subsample sizes for this homework exercise. We shall return to the full dataset when we discuss how to handle 2- and 3-way datasets with unequal subgroup sizes in Chapter 16. Because of our modification of the data, whatever conclusions you reach as the result of carrying out this exercise do not necessarily correspond to the actual findings by French (1976) based on the full dataset.

	Restricted food				Ad libitum food		
	8°C		18°C		8°C		18°C
Animal No.	Energy Used (kcal/g)	Animal No.	Energy Used (kcal/g)	Animal No.	Energy Used (kcal/g)	Animal No.	Energy Used (kcal/g)
1	62.69	5	72.60	13	95.73	17	101.19
2	54.07	6	70.97	14	63.95	18	76.88
3	65.73	10	59.10	15	144.30	19	74.08
4	62.98	12	61.89	16	144.30	22	84.83
$\overline{Y}$	61.37	$\overline{Y}$	66.14	$\overline{Y}$	112.07	$\overline{Y}$	84.13
(SE)	(2.53)	(SE)	(3.32)	(SE)	(19.71)	(SE)	(6.09)

Is there evidence that the amount of food available affects the amount of energy consumed at different temperatures during hibernation?

11.7   In a study of the infestation of potato tubers by the aphid *Rhopalosiphoninus latysiphon*, Haine (1955) investigated the following 14 varieties of potatoes, which differ in their time of maturation.

No.	Variety	Time of Maturation
1	Biene	Medium late
2	Heida	Medium late
3	Böhms Mittelfrühe	Medium early
4	Monika	Medium late–late
5	Johanna	Medium early–medium late
6	Condor	Medium early–medium late
7	Ostbote	Medium late
8	Falke	Medium late–late
9	Bevelander	Medium early–medium late
10	Erdgold	Medium late
11	Ackersegen	Late
12	Primula	Very early
13	Sieglinde	Early
14	Vera	Very early

Mean counts of aphids [transformed as $\log(Y - 1)$; see Section 13.6] at eight times during one year are given in the table below.

Varieties	Oct.	Nov.	Dec.	Jan.	Feb.	Mar.	Apr.	May
1	0.23	1.06	1.63	1.76	1.60	2.04	2.04	1.93
2	0.00	0.31	0.63	0.66	0.93	1.51	2.67	2.31
3	0.34	0.55	0.40	0.55	0.36	0.64	2.09	2.14
4	1.13	1.43	1.40	1.45	1.51	2.13	1.92	2.02
5	0.21	1.22	1.28	1.10	0.86	1.36	2.26	2.44
6	0.00	0.00	0.08	0.24	0.24	1.16	1.79	2.01
7	0.12	0.65	0.37	0.60	0.94	1.41	2.30	1.98
8	0.11	0.41	0.87	0.88	1.28	1.68	1.87	2.17
9	0.06	0.38	0.04	0.27	0.26	0.70	2.20	2.34
10	0.06	0.60	0.28	0.63	0.61	0.97	2.17	2.28
11	0.22	0.66	1.02	1.08	0.91	1.31	1.93	1.77
12	0.75	1.48	1.42	1.39	1.18	1.79	1.93	2.09
13	0.23	0.79	0.37	0.43	0.65	1.03	2.34	2.25
14	1.19	1.62	1.90	1.65	1.90	2.02	2.27	2.22

Carry out an analysis of variance. Design an orthogonal set of comparisons to test whether potatoes differing in time of maturation differ in their aphid infestations.

11.8 Price (1954) investigated the survival of *Bacterium tularense* in lice under three relative humidities (RH 100%, 50%, and 0%), four temperatures (37, 29, 20, and 4°C), and on three substrates (killed lice, starved lice, and louse feces). Two replicate experiments were run. Write out the expected mean squares and make the appropriate tests of significance. Interpret.

ANSWER: EMS for MST is $\sigma^2 + hs\sigma_{ET}^2 + ehs\sum\tau^2/(t-1)$.

Analysis of variance		
Source of variation	df	MS
Main effects		
Experiments (E)	1	0.1108
Humidities (H)	2	0.7613
Substrates (S)	2	0.0376
Temperatures (T)	3	3.1334
First-order interactions		
$E \times H$	2	0.1185
$E \times S$	2	0.1266
$E \times T$	3	0.0060
$H \times S$	4	0.1985
$H \times T$	6	0.0666
$S \times T$	6	0.0654
Second-order interactions		
$E \times H \times S$	4	0.0130
$E \times H \times T$	6	0.0601
$E \times S \times T$	6	0.0206
$H \times S \times T$	12	0.0475
Third-order interaction		
$E \times H \times S \times T$	12	0.0517

11.9 Analyze the data of Box 11.1 as if the experiment had been in the form of a three-factor anova without replication; that is, let three rats in each subgroup represent a dummy factor. In the resulting anova table, show which SS and df must be pooled in order to obtain the correct results shown at the end of Box 11.1.

11.10 Analyze the data given in Box 10.1 as if the experiment followed the form of a three-factor anova without replication. To do this, you must introduce the dummy factors "females" (with four levels) and "measurements" (with two levels). In the resulting anova table, indicate which SS and df must be pooled to obtain the correct anova.

ANSWER: SS among females within cages $= SS$ females $+ SS_{F \times C}$.

11.11 Write out the expected mean squares for a five-factor experiment in which factors A, B, and C are random and factors D and E are fixed treatment effects. Let $n = 1$.

11.12 Skinner and Allison (1923) studied the effect on cotton of date of planting and of the addition of borax to fertilizer. For each treatment combination, a single measurement was made of the weight of the green plants in pounds. Analyze and interpret. For the analysis, assume that the second-order interaction has negligible effects. Did the addition of borax affect the weight of harvested plants?

ANSWER: $MS_{error} = 16.8741$, $MS_{borax} = 167.6296$.

		Method of application		
		I	II	III
Date of planting	Amount of borax (pounds)	In drill; seed planted 1 week later	In drill; seed planted immediately	Broadcast; seed planted immediately
June 2	0	61	60	73
	5	56	59	80
	10	58	61	55
June 9	0	67	72	72
	5	69	62	79
	10	67	58	68
June 18	0	62	69	68
	5	62	70	71
	10	57	72	64
July 7	0	40	45	54
	5	37	49	51
	10	29	33	41
July 15	0	26	28	35
	5	26	24	36
	10	21	23	33
Aug. 3	0	10	13	17
	5	10	11	14
	10	8	12	14

11.13 The oven-dry weights (in grams) of new growth in hybrid poplars grown in concrete soil frames and treated with lime (L), nitrogen (N), phosphorus (P), and potassium (K) are given below. The frames were laid out in three blocks. These data can be interpreted as a five-way orthogonal anova without replication. Carry out the analysis and test significance. Data from Lunt (1947). O indicates absence of treatment (controls).

				Treatments				
Blocks	O	P	PK	K	NK	N	NP	NPK
1	13.9	14.2	14.7	13.6	31.7	57.9	49.5	49.7
2	14.3	22.8	12.8	12.7	25.6	21.7	35.5	38.1
3	15.8	22.1	13.3	15.5	25.7	31.0	30.7	36.3

				Treatments				
Blocks	L	LP	LPK	LK	LNK	LN	LNP	LNPK
1	15.3	11.8	17.8	16.6	41.2	43.0	63.8	53.4
2	19.4	23.2	21.4	20.1	59.3	62.5	59.7	53.5
3	15.9	22.7	20.6	15.1	32.0	37.1	41.3	58.5

12 Statistical Power and Sample Size in the Analysis of Variance

This chapter covers planning for sample sizes sufficient to ensure that important differences are likely to be detected in an *anova*. The emphasis will be on tests of significance, but the same considerations also apply to the construction of confidence intervals. Section 12.1 describes general methods for defining effect size (the magnitude of the differences among two or more means in an anova). Section 12.2 presents methods for computing confidence limits of effect sizes using noncentral *t*- and *F*-distributions. In Section 12.3, we extend the Section 7.6 discussion of power to include anova. Section 12.4 shows how one can invert the power computations to estimate the sample size to be used in a future analysis. Section 12.5 performs a related computation—given a specified sample size, what is the smallest difference one is likely to be able to detect? In Section 12.6, we discuss a common misapplication of power analysis. Section 12.7 presents ways in which a nested anova design can be modified so as to reduce the expected size of the error mean square and thereby increase power or to decrease the sample size while keeping the same power. Finally, Section 12.8 has a similar discussion but for randomized blocks and multiway designs.

12.1 Effect Size

To estimate the power of a planned test, one has to specify the magnitude of the expected difference between the parameters being tested and the parameter specified by the null hypothesis. In the examples of Section 7.5, they were differences such as $\mu_1 - \mu_0$. In a Model II single-classification anova, it would be the difference between σ_A^2 and its value of zero under the null hypothesis. Such differences are called **effect sizes**. Because true effect sizes are generally unknown, one customarily specifies the smallest difference that one would like to be able to detect. In some areas of research, it is possible to determine the smallest difference that would yield an economic benefit and would thus be worth detecting. In other areas, one can specify what effect size would be clinically important. In most applications, that is not possible. An investigator may desire to detect if there are *any* differences at all, but that is an impractical goal because it may take extremely large sample sizes in order to detect very small differences. Even if sufficient resources were available to collect enough data to detect extremely small differences, it may not be worth it. In some fields, standards have been developed for what constitutes small or large effects, but investigators still have to decide how small a difference they wish to be able to detect.

To develop general rules and compare different studies, it is convenient to express effect sizes as dimensionless numbers. For example, when testing differences between two means, one can use the *standardized effect size* Δ (a difference divided by a standard deviation). Values of Δ of about 0.2 or less are often considered "small effects," values around 0.8 or greater are considered "large effects," and values in between are considered "medium effects" (see Cohen, 1988). These definitions are, of course, arbitrary, but they may be helpful as guidelines. Note that the difference is divided by the standard deviation rather than the standard error of the mean. Note also that these terms refer to the size of effect in standard deviation units, but not to the biological, medical, or economic importance of such differences. They also do not measure the same thing as the *P*-value, which depends strongly on sample sizes. Investigators should consider the consequences of detecting or failing to detect differences of various magnitudes when deciding on an appropriate effect size for a particular study. It may be useful to examine the size of differences detected in related studies that tested similar hypotheses. Although tests of larger differences are more likely to yield smaller probabilities, obtaining a very small probability in a test of significance does not imply that the difference found is large or important. With very large sample sizes (common in large surveys, certain types of genetic experiments, or where measurements can be made automatically), even trivial differences may result in very small probabilities.

Several measures of effect size have been proposed. One of the simplest is $\Delta = (\mu_1 - \mu_2)/\sigma$, mentioned above. Perhaps the most commonly used measure is *Cohen's f*. Procedures based on it are well developed, and it is applicable in many situations, not just for anova. In terms of either a Model I or a Model II single-classification anova, it is

$$f = \sqrt{\frac{\sum \alpha_i^2}{(a-1)\sigma^2}} \quad \text{or} \quad f = \sqrt{\frac{\sigma_A^2}{\sigma^2}} \tag{12.1}$$

respectively. For Model I, it is the square root of the sum of the squared fixed treatment effects divided by its degrees of freedom and by the within-group variance, whereas for Model II it is the square root of the ratio of the among-group variance component to the within-group variance. The quantity f^2 has been called a *signal to noise ratio*. The modifications for a Model I anova are shown in Table 12.1. When comparing just two means, the investigator likely wishes to express the effect size in terms of standardized differences between the means, $\Delta = (\mu_1 - \mu_2)/\sigma$. As shown in that table, the measure of effect size in this case is $f = \Delta/2$. When there are more than two means, the measure of effect size depends on what assumptions are made about the relative magnitudes of the other means (equal to the overall mean, equally spaced, or all at the extreme limits). Values of f around 0.1 or less are considered small effects, those around 0.4 or greater are considered large effects, and values in between are usually considered medium sized effects.

TABLE 12.1 Different Versions of Cohen's Measures of Effect Size, f

Design	Estimate of Effect Size	Requirements
Anova	$f = \sqrt{\dfrac{\sum \alpha_i^2 / (a-1)}{\sigma^2}}$	Model I anova
Anova	$f = \sqrt{\dfrac{\sigma_A^2}{\sigma^2}}$	Model II anova
Pair of means	$f = \dfrac{\Delta_{\min}}{2}$	$a = 2,\ \Delta = (\mu_1 - \mu_2)/\sigma$
Based on largest standardized difference expected among the a means	$f = \dfrac{\Delta_{\min}}{\sqrt{2a}}$	$\Delta_{\min} = (\mu_{\max} - \mu_{\min})/\sigma$, other means equal to μ
	$f = \dfrac{\Delta_{\min}}{2} \sqrt{\dfrac{a+1}{3(a-1)}}$	Other means equally spaced between the two extreme means
	$f = \dfrac{\Delta_{\min}}{2}$	Other means equally distributed at the two extremes, a even
	$f = \Delta_{\min} \dfrac{\sqrt{a^2 - 1}}{2a}$	Other means equally distributed at the two extremes, a odd

$\Delta_{\min}$ is the smallest maximal difference between two means that one wishes to detect.

Another index that is used for a Model I anova can be written as

$$\omega^2 = \frac{\sigma_A^2}{\sigma^2 + \sigma_A^2} \tag{12.2}$$

This index is identical to r_I, the intraclass correlation coefficient as described in Section 9.2. Cohen (1988) called it η^2 and also PV, the *proportion of the variance*. It is also called the *correlation ratio*. The use of ω^2 is equivalent to f because one index can be converted into the other using the following relationships: $f = \sqrt{\dfrac{\omega^2}{1 - \omega^2}}$ and $\omega^2 = \dfrac{f^2}{1 + f^2}$. A convenience of the ω^2 index is that it ranges from 0 to 1 (whereas f ranges from 0 to infinity).

In the sampling experiment described in Section 8.4, the true effects, α_i, and σ^2 are known so that the effect size can be computed directly using Expression (12.1), $f = \sqrt{\sum \alpha_i^2/(a-1)\sigma^2} = \sqrt{56/([7-1]15.21)} = 0.7833$, which is considered a large effect size. For the same model, $\omega^2 = 0.3803$. Of course, σ^2 and σ_A^2 or $\sum \alpha_i^2/(a-1)$ are usually not known, so sample estimates must be used.

$$\hat{f} = \sqrt{\frac{(MS_A - MS_{\text{within}})/n}{MS_{\text{within}}}} \tag{12.3}$$

$$\hat{\omega}^2 = \frac{SS_A - v_A MS_{\text{within}}}{SS_{\text{total}} + MS_{\text{within}}} \tag{12.4}$$

Using Expression (12.2), we find a range of effect sizes in the examples in Chapter 9. The f-values are 0.69 for Box 9.2, 0.63 for Box 9.3, 1.19 for Box 9.4, 0.55 for Box 9.5, and a large 2.30 for the data of Table 9.1. One could also compute ω^2-values if desired.

Generalizations of the f and ω^2 indices can easily be applied to more complex anovas. A partial ω^2, ω_p^2, is used in a model I multi-way anova. The sample estimate contains just the SS for the effect being tested rather than the total SS. For example, for factor A in a two-way anova $\hat{\omega}_p^2 = (SS_A - v_A MS_{\text{within}})/(SS_A + [\Sigma\Sigma n_{ij} - v_A]MS_{\text{within}})$. Note that in more complex anovas, the magnitude of the variance in the denominator can be changed by alterations in the design of the experiment. Examples of this are discussed in Sections 12.7 and 12.8. Using the f-index, the effect sizes for the nested anova of Box 10.1 are a modest 0.30 for cages but a huge 8.54 for females within cages. For the two-way anova of Box 11.1, the f-values are 0.52 for sexes and 2.61 for fats. Because the estimated component for interaction is negative in this example, the estimated effect size is zero.

Another index for effect size has been proposed by Murphy et al. (2009). Unfortunately, it is also called PV, the proportion of variance. For a single-classification anova, it is

$$PV = \frac{v_1 F}{v_1 F + v_2}$$

The meaning of this index is becomes clearer if it is rewritten as $PV = SS_{\text{Among}}/(SS_{\text{Among}} + SS_{\text{Within}})$. Thus, it is actually a proportion of sums of squares rather than variances. Although this index is convenient in that it can be computed using just the degrees of freedom and F-ratios, it seems less useful as a measure of effect size because it is also a function of a and n.

12.2 Noncentral T- and F-Distributions and Confidence Limits for Effect Sizes

When sample statistics are used to estimate effect size, one obtains just that—an estimate of the effect size. It is important to set confidence limits to such estimates because they can be subject to considerable error when sample sizes are not large. The required computations are relatively straightforward, but they require access to software, such as BIOMstat, that computes probabilities for the *noncentral t-* and *F-distributions*.

As you may recall, the ordinary or *central t*-distribution, $t_{[v]}$, is the distribution of statistics such as $t = (\bar{Y} - \mu_0)/s_{\bar{Y}}$ when the samples are, in fact, taken from a population whose true mean is equal to μ_0. However, when that is not true, the statistic follows the noncentral t-distribution, $t_{[v,\lambda]}$. In addition to its degrees of freedom, this distribution has another parameter, λ, called its **noncentrality parameter**. It is equal to

$$\lambda = \frac{\mu - \mu_0}{\sigma/\sqrt{n}} \tag{12.5}$$

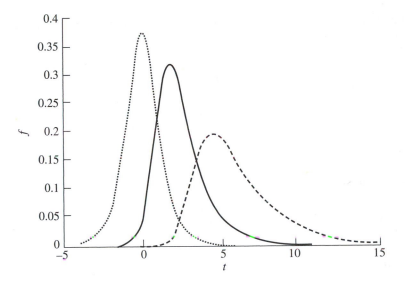

FIGURE 12.1 Examples of central and noncentral *t*-distributions. All distributions are for $\nu = 4$ degrees of freedom. From left to right, the distributions correspond to the central *t*-distribution and the noncentral *t*-distributions with the noncentrality parameter, λ, equal to 2 and 5, respectively.

where μ is the actual mean of the population from which the samples were drawn. In Figure 12.1, we contrast a central with two noncentral *t*-distributions for the same degrees of freedom. As discussed in Chapter 7, in place of $\overline{Y}$ and μ_0, one could substitute other statistics such as a difference between two means, or other normally distributed statistics. In Figure 7.10 for the normal distribution, the effect of including true differences between the means was just to shift the mean of the distribution—the shape of the distribution remained the same. However, in the case of the *t*-distribution (and the *F*-distribution discussed next), the *shape* as well as the *location* of the distribution changes. This is because the noncentral *t*-distribution takes into account differences between means as a ratio to their sample standard error. The fact that noncentral distributions change their shape as a function of their noncentrality parameter implies that to be useful, printed tables would have to be very extensive. The noncentral *t*-distribution is used in Box 12.1.

In a single-classification anova, the usual (central) *F*-distribution is the distribution of the ratio $(\Sigma n_i (\overline{Y}_i - \overline{\overline{Y}})^2 / a - 1) / MS_{\text{within}}$, assuming that the samples are actually drawn from populations with the same mean. If that is not the case, then in a Model I anova the ratio estimates

$$\frac{\sigma^2 + n\Sigma\alpha_i^2/(a-1)}{\sigma^2} = 1 + n\frac{\Sigma\alpha_i^2/(a-1)}{\sigma^2}$$

and follows the noncentral *F*-distribution, $F_{[\nu_1, \nu_2, \lambda]}$, with a noncentrality parameter:

$$\lambda = n\frac{\Sigma\alpha_i^2}{\sigma^2} = n(a-1)f^2 \tag{12.6}$$

BOX 12.1 **Estimating Effect Sizes and Their Confidence Limits**

I. t-Test

For the two samples of the Egyptian head-breadth data in Section 7.1, the following statistics were computed:

	Sample 1	Sample 2
n	8	10
$\overline{Y}$	131.75	135.90
s^2	10.7857	40.1000

Using Expression (9.3), we obtain $t_s = -1.6752$ with 16 degrees of freedom. For a 2-tailed test, this corresponds to $P = 0.0567$, which does not quite reach the 0.05 level.

The effect size for the difference in means is $\Delta = \dfrac{\mu_1 - \mu_2}{\sigma} = \dfrac{131.75 - 135.90}{5.22255} =$

-0.7946. This is considered a "large effect" even though it is not significantly different from zero for these data.

Confidence limits for Δ are computed as follows:

First we compute the noncentrality parameter, λ, for a noncentral t-distribution. For the difference between two means it is just the sample t-value, thus $\lambda = t_s = -1.6752$.

Next, we must use software for the noncentral t-distribution to find the noncentrality parameter, λ_1, of a noncentral t-distribution for which the observed t_s cuts off $\alpha/2$ of its right tail. Then we must find the noncentrality parameter, λ_2, of the noncentral t-distribution for which t_s cuts off $\alpha/2$ of its left tail. The figure below shows the observed value of the noncentrality parameter as a vertical dotted line and the two distributions for which t_s cuts off $\alpha/2$ of their left or right tails. The distribution on the left has a non-centrality value of -3.6941 and the one on the right a value of 0.3915. These are the 95% confidence limits for λ.

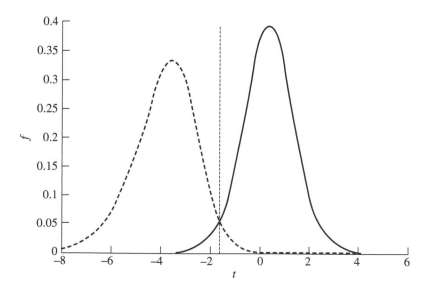

Box 12.1 (continued)

These limits are then back-transformed into effect sizes using the relationship

$$\Delta = \lambda \sqrt{\frac{n_1 + n_2}{n_1 n_2}}.$$

This yields 95% confidence limits $\Delta_1 = -3.6941\sqrt{\dfrac{8 + 10}{8 \times 10}} = -1.7523$ and $\Delta_2 = 0.3915\sqrt{\dfrac{8 + 10}{8 \times 10}} = 0.1857$. The interval includes zero because the *t*-test was not significant at the 0.05 level. Note that even though we were not able to reject the null hypothesis of no difference between the means, the confidence limits show that we cannot exclude the possibility that there is a relatively large (-1.7523) effect size, which suggests the possibility that $\mu_1 < \mu_2$.

II. Model I Single-Classification Anova

We shall use the transformed pea-section data of Table 9.1 as an example. There are $a = 5$ groups, each with $n = 10$ observations. $MS_{\text{within}} = 0.00297289$ and $MS_{\text{among}} = 0.160193$, $F_s = 53.885$. Following Expression (12.3), the effect size is

$$\hat{f} = \sqrt{\frac{(MS_{\text{among}} - MS_{\text{within}})/n}{MS_{\text{within}}}} = \sqrt{\frac{(0.160193 - 0.00297289)/10}{0.00297289}} = 2.2997$$

Converting it to ω^2 as formulated in Expression (12.2) and the text immediately following its display, $\omega^2 = \dfrac{f^2}{1 + f^2} = \dfrac{5.28846}{1 + 5.28846} = 0.84098$. Expressed in either form, these values represent very large effect sizes.

The noncentrality parameter $\lambda = n\dfrac{\Sigma \alpha_i^2}{\sigma^2}$ [Expression (12.6)] can be estimated by

$$\hat{\lambda} = (a - 1)\frac{MS_{\text{among}} - MS_{\text{within}}}{MS_{\text{within}}} = (5 - 1)\frac{0.160193 - 0.00297289}{0.00297289} = 211.53850,$$ as

shown in the text beneath Expression (12.6).

We must use software for the noncentral *F*-distribution to find the noncentrality parameter, λ_1, of a noncentral *F*-distribution for which the observed F_s cuts off $\alpha/2$ at its right tail. Then we must find the noncentrality parameter, λ_2, of the noncentral *F*-distribution for which the F_s cuts off $\alpha/2$ at its left tail. The figure below shows the value of $F_s = 53.885$ as a vertical dotted line and the two distributions for which F_s cuts off $\alpha/2$ at their left or right tails. The distribution on the left has a noncentrality value of 119.27839, and the one on the right a value of 330.26415. These are the 95% confidence limits for the noncentrality measure λ.

Box 12.1 (continued)

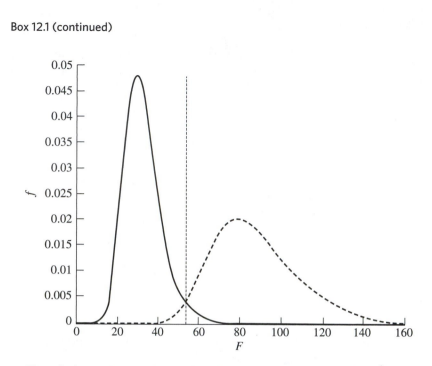

These limits are then transformed into effect sizes using the relationship $f^2 = \lambda/an$. The 95% confidence limits for f^2 are $f_1^2 = 119.27839/(5 \times 10) = 2.38557$ and $f_2^2 = 330.26415/(5 \times 10) = 6.60528$. The interval does not include zero because the F-test was able to reject the null hypothesis at the 0.05 level. These limits can also be transformed into confidence limits for the effect size index ω^2 by formulas shown in Section 12.1:

$$\omega_1^2 = \frac{f_1^2}{1 + f_1^2} = \frac{2.38557}{1 + 2.38557} = 0.70463$$

and

$$\omega_2^2 = \frac{f_2^2}{1 + f_2^2} = \frac{6.60528}{1 + 6.60528} = 0.86851$$

These values are large, and the 95% confidence intervals of the effect sizes are relatively narrow.

III. Model II Single-Classification Anova

a. For equal sample sizes. In general, this model is much more complicated, but an exact solution is possible when the sample sizes are equal (an approximate solution is given under **b** below for unequal sample sizes). In the data from Box 9.3, fertility in drosophila females, we have $a = 7$ groups with $n = 3$ observations in each group. $MS_{\text{among}} = 9.111$, $MS_{\text{within}} = 4.142$, $s_A^2 = 1.6563$, and $F_s = 2.20$. The estimated squared

Box 12.1 (continued)

effect size is $f^2 = s_A^2/MS_{\text{within}} = 1.6563/4.142 = 0.39988$. The confidence limits for f^2 can be computed without the use of special software because the required critical values, such as $F_{\alpha/2[v_1,v_2]}$, are for the central F-distribution. The 95% confidence limits for these squared effect size indices are computed as follows.

1. Lower confidence limit:

$$f_1^2 = \frac{(F_s/F_{\alpha/2[v_1,v_2]}) - 1}{n} = \frac{(2.20/3.50136) - 1}{3} = -0.12389 = 0 \text{ (the negative esti-}$$

mated limit is nonsensical and is replaced by zero)

2. Upper confidence limit:

$$f_2^2 = \frac{(F_s/F_{1-\alpha/2[v_1,v_2]}) - 1}{n} = \frac{(2.20/0.18879) - 1}{3} = 3.55100$$

3. These confidence limits can be transformed into an estimate of and confidence limits for ω^2 as follows:

$$\omega^2 = \frac{f^2}{1 + f^2} = \frac{0.39988}{1 + 0.39988} = 0.28565$$

$$\omega_1^2 = \frac{f_1^2}{1 + f_1^2} = \frac{0}{1 + 0} = 0 \text{ and } \omega_2^2 = \frac{f_2^2}{1 + f_2^2} = \frac{3.55100}{1 + 3.55100} = 0.78027$$

Because the F-test did not reject the null hypothesis, one expects the confidence limits for the effect sizes to include zero. What might be surprising is that the limits are so broad that they do not exclude the possibility that the effect size could actually be quite large. The confidence limits for Model II effects are usually much broader than if the same data were treated as a Model I.

b. *For unequal sample sizes.* Approximate confidence limits for the effect size indices f^2 and ω^2 can be computed using a method presented by Burdick et al. (2006). It is based on the modified large sample (MLS) method for approximate limits to σ_A^2 described in Box 9.2. These limits tend to be conservative (i.e., very broad), especially when the sample sizes are very unequal or ω^2 is small. Burdick et al. (2006) also describe methods that are more exact, but their computation is more complicated.

We apply the simpler methods to the data from Box 9.1 (scutum widths of tick larvae), which consist of $a = 4$ groups with sample sizes of 8, 10, 13, and 6. For these, we note their minimum ($n_{\min} = 6$) and maximum ($n_{\max} = 13$). Their harmonic mean ($H_n = 8.5363$), unweighted MS_{among} ($\widetilde{MS}_{\text{among}} = 576.2263$), and the $MS_{\text{within}} = 114.4849$ can be found in Box 9.2, which explains their computation.

1. The lower $(1 - \alpha)100\%$ confidence limit for f^2 is

$$f_1^2 = \frac{\widetilde{MS}_{\text{among}}}{H_n MS_{\text{within}} F_{\alpha/2[v_1,v_2]}} - \frac{1}{n_{\min}} = \frac{576.2263}{8.5363(114.4849)3.54287} - \frac{1}{6} = -0.00024009$$

Box 12.1 (continued)

Negative estimates should be set equal to zero. Because the null hypothesis for these data was rejected in Box 9.1, a lower limit of zero was not expected. This discrepancy reflects the fact that this method provides a conservative approximation to the limits.

2. The upper confidence limit is

$$f_2^2 = \frac{\widetilde{MS}_{\text{among}}}{H_n MS_{\text{within}} F_{1-\alpha/2[\nu_1,\nu_2]}} - \frac{1}{n_{\text{max}}} = \frac{576.2263}{8.5363(114.4849)0.071101} - \frac{1}{13} = 8.21593$$

3. The confidence limits for ω^2 can then be computed using the relationship $\omega^2 = f^2/(1 + f^2)$:

$$\omega_1^2 = \frac{f_1^2}{1 + f_1^2} = \frac{0}{1 + 0} = 0$$

$$\omega_2^2 = \frac{f_2^2}{1 + f_2^2} = \frac{8.21593}{1 + 8.21593} = 0.89149$$

The lower limit is zero because the lower limit for f^2 was zero. As in the case for equal sample sizes, we find that the confidence limits are very broad because of the small sample sizes.

IV. More Complex Designs

The procedures described above for a Model I single-classification anova can be adapted for more complex designs such as two-way or multiway Model I anovas. One replaces the MS_{among} in the formulas by the MS for the factor or interaction being tested, and the MS_{within} is replaced by the MS in the denominator of the F-test for that factor. In addition, partial effect sizes may be computed.

The two-way anova in Box 11.2 (oxygen consumption in two limpet species) can be used as an example. By generalizing Expression (12.3), the estimates of effect size indices f^2 for the effects A, B, and A×B can be computed as follows:

$$\hat{f}_A^2 = \frac{(MS_A - MS_{\text{within}})/nb}{MS_{\text{within}}} = \frac{(16.63810 - 9.56003)/(8 \times 3)}{9.56003} = 0.03085$$

$$\hat{f}_B^2 = \frac{(MS_B - MS_{\text{within}})/na}{MS_{\text{within}}} = \frac{(90.66050 - 9.56003)/(8 \times 2)}{9.56003} = 0.53021$$

$$\hat{f}_{AB}^2 = \frac{(MS_{AB} - MS_{\text{within}})/n}{MS_{\text{within}}} = \frac{(11.96310 - 9.56003)/8}{9.56003} = 0.03142$$

Estimates of their noncentrality parameters are

$$\hat{\lambda}_A = n(a - 1)f_A^2 = 8(2 - 1)0.03085 = 0.24679$$

$$\hat{\lambda}_B = n(b - 1)f_B^2 = 8(3 - 1)0.53021 = 8.48329$$

$$\hat{\lambda}_{AB} = n(a - 1)(b - 1)f_A^2 = 8(2 - 1)(3 - 1)0.059589 = 0.95342$$

Box 12.1 (continued)

As was done in part **II** of this box, the confidence limits for the noncentrality parameters are computed using the noncentral *F*-distribution to find values for λ such that $\alpha/2$ is cut off from their left or right tails. The following 95% confidence limits were obtained:

$$\lambda_{A1} = -0.24679 \text{ (set to zero because it is negative)}$$

$$\lambda_{A2} = 10.83434$$

$$\lambda_{B1} = 3.94097$$

$$\lambda_{B2} = 41.10623$$

$$\lambda_{AB1} = -0.95342 \text{ (set to zero because it is negative)}$$

$$\text{and } \lambda_{AB2} = 11.17952$$

We transformed these confidence limits for λ into those for f^2 and ω^2 using the relationships $f^2 = \lambda/an$ and $\omega^2 = f^2/(1 + f^2)$, respectively. The results are in the following table. The confidence limits for the partial effect sizes are more difficult to compute.

	f_1^2	f_2^2	ω_1^2	ω_2^2
A	0	0.246240	0	0.19758
B	0.087580	0.91347	0.08052	0.47739
A×B	0	0.24843	0	0.19900

The effect size for factor B (salinity) is clearly much larger than that for A (limpet species) or their interaction A×B.

Figure 12.2 shows an example of a central and a noncentral *F*-distribution with identical degrees of freedom. The noncentral distribution is displaced to the right because there actually are differences among the sample means in the simulation on which this plot is based (the sampling experiments described in Sections 8.1 and 8.4). In that experiment, there were $a = 7$ samples of $n = 5$ observations each. When the null hypothesis is true, the expected distribution of F_s-values is given by the *F*-distribution with 6 and 28 degrees of freedom. It corresponds to the curve at the left of Figure 12.2 (also shown in Figure 8.2, but for differently scaled axes). The upper 5% critical value for the central *F*-distribution is 2.45 and is indicated by the dashed line. The curve to the right is an example of the noncentral distribution expected when the null hypothesis is false due to the addition of treatment effects, α_i, to the means as described in Section 8.4. For this simulation, $\Sigma\alpha_i^2 = 56$.

For a single-classification anova, the noncentrality parameter is estimated as $\hat{\lambda} = (a - 1)(MS_A - MS_{\text{within}})/MS_{\text{within}}$ (note that the n appears to have been canceled out, but it is present implicitly because $MS_A - MS_{\text{within}}$ estimates $n\sigma_A^2$).

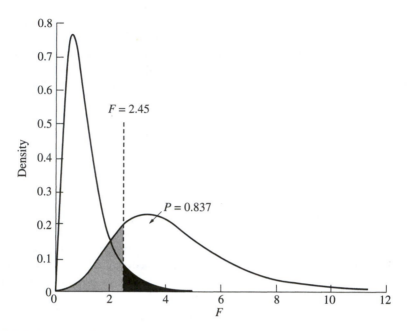

FIGURE 12.2   Probability density functions for the sampling experiment of Sections 8.1 and 8.4. The curve at the left is for the central F-distribution with 6 and 28 degrees of freedom. It gives the expected distribution of F_s when the null hypothesis is true. The vertical dashed line at the critical value of $F_{0.05[6,28]} = 2.45$ cuts off the upper 5% of that distribution (the black shaded area to its right). The curve at the right is for the noncentral F-distribution with noncentrality parameter $\lambda = 18.4089$ and with 6 and 28 degrees of freedom. The gray shaded area to its left corresponds to β, the remaining area to its right is 0.837.

The first part of Box 12.1 shows how confidence limits are computed for the estimated effect size for the Egyptian head-breadth data described in Section 7.1. The second and third parts of Box 12.1 show how to set confidence limits to effect size indices f and ω^2 by first estimating the confidence limits for the noncentrality parameter. The procedure for a Model I anova is analogous to that for testing the difference between just two means but uses the noncentral F-distribution rather than the noncentral t-distribution. The problem of setting confidence limits to effect size is more complex for Model II anovas—especially when sample sizes are not equal. Only approximate confidence limits are given for that case. Burdick and Graybill (1992) and Burdick et al. (2006) published surveys of the available methods for Model II anovas. Smithson (2003) furnishes a general account of the interpretation and computation of such confidence intervals.

12.3 Power in an Anova

As defined in Section 7.6, statistical power is the probability of correctly rejecting a false null hypothesis—that is, not making a type II error when the alternative hypothesis is true. It is equal to $1 - \beta$, where β is the probability of making a type II error. Several examples for small and large differences between two means are shown in Figure 7.10. Whenever there are more than just two means, probabilities must be computed using the F-distribution. For the sampling experiment of Section 8.4, $\lambda = n\Sigma\alpha_i^2/\sigma^2 = 5(56)/15.21 = 18.4089$. The curve on the right in Figure 12.2 illustrates the expected distribution for this simulation.

Power is computed as the area under the noncentral F-distribution to the right of the critical value found using the central F-distribution, $F_{0.05[6,28]} = 2.45$ in the current example. The computation of critical values or probabilities for the noncentral F-distribution is more complex than that for the central F-distribution. Software for their computation is available in the BIOMstat software and other statistical packages. Using the BIOMstat software, power is found to be 0.8373 for this example. The available tables for the noncentral F-distribution tend to be quite voluminous because for each value of λ, critical values or probabilities must be given for a wide range of combinations of upper and lower degrees of freedom. An alternative is to use power and sample size graphs such as Statistical Table **GG**. To use these, first select the graph corresponding to the numerator degrees of freedom and the desired type I error rate. Then select the curve for the sample size used and read off the power from the scale along the ordinate corresponding to the effect size, f, along the abscissa. Note that graphs available from other sources are used in a similar fashion, but the curves correspond to the denominator degrees of freedom, ν_2, and the scale along the abscissa is for $\phi = \sqrt{n\delta^2/2a\sigma^2} = f\sqrt{n/a}$ rather than for f.

An examination of the equation for the mean of the noncentral F-distribution is instructive.

$$\overline{F} = \frac{\nu_2}{\nu_2 - 2}\left(1 + \frac{(\nu_1 + 1)\lambda}{\nu_1 a}\right) \tag{12.7}$$

As one might expect, the mean F-ratio increases as the noncentrality parameter, λ, increases. Because of the term for the number of groups, a, in the denominator, $\overline{F}$ decreases as the number of groups increases; thus, using more groups than necessary implies a decrease in power. However, if expressed in terms of effect size, using Expression (12.6), $\lambda = naf^2$, we find that the number of groups cancels and for a fixed effect size power actually increases when a larger number of groups are used.

How large of a value for power should an investigator try to achieve? A value of at least 0.80 is often specified as the goal. The choice is, of course, arbitrary, just as using type I error rates of 5% or 1% is arbitrary. Murphy et al. (2009) points out that using a power of 0.80 implies that if the alternative hypothesis is true, then accepting the alternative hypothesis is four times as likely as rejecting it. Higher or lower values could be used, depending on the importance put on the decision to accept or reject the alternative hypothesis.

12.4 Sample Size in an Anova

Researchers frequently ask how large a sample should be for an anova. Regrettably, no answer is possible without the questioner first specifying three parameters.

1.   An important parameter is, of course, the minimum effect size that the researcher is interested in detecting. This requires indicating the smallest difference or σ_A^2 of interest, as well as an estimate of the error variance. As discussed in Section 12.1, one needs either reliable information about variation in prior studies or else an arbitrary decision must be made about the size of the effect that one would like to detect—for example, whether it is "small" or "large."

2.   The minimally acceptable statistical power must also be specified.

3.   In addition, one needs to state the type I error rate to be used and to ensure that the various assumptions of anova hold (see Chapter 13).

Box 12.2 provides several ways to find the required sample size for a single-classification anova (including the special case for just $a = 2$ means). If appropriate software that makes use of the noncentral F-distribution is not available, then special graphs, such as those in Statistical Tables **GG**, make approximate solutions quite easy. One simply selects the appropriate graphs based on the number of groups and the type I error rate and then sees which sample size curve comes closest to the intersection of the desired f along the abscissa and power, P, along the ordinate.

For the special case of just two means, a relatively simple formula (simple in comparison to that for the noncentral F-distribution) is available:

$$n \geq \frac{1}{2f^2}(t_{\alpha[\nu]} + t_{2(1 - P)[\nu]})^2 \tag{12.8}$$

For just two means, $f = \delta/2\sigma$. This equation must, however, be solved in an iterative fashion because the degrees of freedom for the two t-values depend on the sample size. Thus, one must first make an initial guess of the sample size and then look up the corresponding t-values and compute n. One then uses this value of n to look up new t-values and compute a new estimate of n. If the new estimate is not sufficiently close to the prior one, then the process is repeated. Fortunately, few iterations are usually needed unless the initial guess is very bad. Good initial approximations can be obtained using the special charts unless the point f, P is beyond the range covered by the charts.

A little experimentation with Expression (12.8) will show that it is not too sensitive to changes in α and P but is very sensitive to changes in the effect size. This indicates that large sample sizes may be required in order to detect small differences, as one might expect. These methods estimate the sample size needed in order to have a good chance for rejecting the null hypothesis of no differences among the means. Such sample sizes may not be sufficient to obtain tight confidence intervals around each mean. Thus, much larger sample sizes may be needed if the purpose of the study is to accurately estimate the means rather than simply demonstrating that they are likely to be different.

We can also see that any refinement of experimental technique that reduces σ will decrease the sample sizes required or will increase the possibility of detecting smaller differences. In Sections 12.7 and 12.8, we will show how changes in the experimental design can greatly affect the size of β and thus change f and the expected power.

BOX 12.2 Required Sample Size and Power for an Anova

I. Finding the Required Sample Size for an Anova

Several methods can be used to estimate the sample size needed for an anova. For all of them we must first compute the minimum effect size that we wish to be able to detect using a specified type I error rate and power.

First, let us imagine that the investigator who carried out the pea-section experiments of Table 9.1 wishes to do a new series using $a = 6$ treatments and wants to know how many replicates are needed to achieve a power of 0.80 using a type I error rate of $\alpha = 0.05$. The investigator assumes that in the new experiment the minimum effect size that needs to be detected is just one-fifth of what was observed in the original experiment.

Using Expression (12.3) for the estimated effect size in a single-classification anova, we have

$$\hat{f} = \sqrt{\frac{(MS_{among} - MS_{within})/n}{MS_{within}}} = \sqrt{\frac{0.16019 - 0.00297}{10(0.00297)}} = 2.30078$$

which is very large. One-fifth of that would be 0.46, which is still considered relatively large. Because the effect size is based on a sample estimate, one may wish to compute its confidence limits (1.544 to 2.570) and use the procedures given below to bracket a range of possible sample sizes that should be used.

Graphic Method

We can employ the special graphs given in Statistical Table **GG**. Consulting the graph for $\nu_1 = a - 1 = 5$ degrees of freedom and the panel for $\alpha = 0.05$, we find that the curve for $n = 12$ approximately intersects the point corresponding to a value of $P = 0.8$ along the ordinate and $f = 0.46$ along the abscissa. Thus, the new experiment should be performed with at least 12 replicates in each treatment group.

Note: previous editions of *Biometry* and many other texts featured the use of a set of graphs from Pearson and Hartley (1951). They are used in a similar manner, but the scale along the abscissa was for the quantity $\phi = \sqrt{\dfrac{n\Sigma\alpha_i^2}{a\sigma^2}}$ (which can also be written as $\phi = \sqrt{n}f$). The curves on those graphs were labeled in terms of the error degrees of freedom. Unlike Statistical Table **GG,** those graphs have to be used iteratively. One first guesses a value for n in order to compute both ϕ and ν_2. One then uses the graphs to find the value for power. If it is higher or lower than intended, then one must try another value for n until a sample size that gives the desired power is found.

Using the Noncentral F-Distribution

If software is available for the noncentral F-distribution (as in the BIOMstat package), then the required sample size can easily be estimated by trying different sample sizes until the desired power is achieved.

If it is not available, then a fairly good approximation from Patnaik (1949) can be used (see Winer et al., 1991, for a more accessible description). Power is estimated by looking up or computing the probability of the quantity $F = \dfrac{\nu_1 F'}{\nu_1 + \lambda}$ using the ordinary

Box 12.2 (continued)

central F-distribution with $v_1' = \dfrac{(v_1 + \lambda)^2}{v_1 + 2\lambda}$ and v_2 degrees of freedom. The quantity F' is the critical value from the central F-distribution with v_1 and v_2 degrees of freedom, and $\lambda = \dfrac{na\sigma_A^2}{\sigma_\epsilon^2} = naf^2$ is the noncentrality parameter. This approximation is accurate to at least one decimal place, which is probably sufficient for many practical applications.

For the current example, we can check the results given above using $a = 6$, $n = 12$, $v_1 = 5$, $v_2 = 66$, and $f = 0.46$. We first compute the noncentrality parameter $\lambda = naf^2 = 12(6)0.46^2 = 15.2352$, followed by the modified degrees of freedom and

F-values $\quad v_1' = \dfrac{(v_1 + \lambda)^2}{v_1 + 2\lambda} = \dfrac{(5 + 15.2352)^2}{5 + 2(15.2352)} = 11.5489, \quad F' = F_{0.05[5,66]} = 2.3538,$

and $F^* = \dfrac{v_1 F'}{v_1 + \lambda} = \dfrac{5(2.3538)}{5 + 15.2352} = 0.5813.$

This F^*-value is to be looked up in a table of the central F-distribution with $11.5489 \approx 12$ over 66 degrees of freedom. Because this value is less than 1, it will be easier to look up its reciprocal in Statistical Table **F**. It is 1.720 with 66 over 12 degrees of freedom. By computation, using fractional degrees of freedom, this is 0.1558. Because we used its reciprocal, this corresponds to a power of 0.8442, which is quite close to the exact value of 0.8394. If v_1' is rounded to 12, a value of 0.8495 would be obtained that is sufficiently close for practical applications.

II. Special Case of Just Two Means

Although the methods described in the preceding can be used when planning a test for the difference between just two means, one can take advantage of a computational simplification that is possible because the relevant distribution becomes the noncentral t-distribution rather than the noncentral F-distribution.

As in the general case, the procedure for computing the required sample size is iterative, but for two means Expression (12.8) directly produces an improved estimate of n given an initial estimate

$$n \geq \frac{1}{2f^2}(t_{\alpha[v]} + t_{2(1 - P)[v]})^2$$

where n is the new estimate for the number of replications, f is the minimum effect size to be detected (usually expressed in terms of the smallest difference between two means that one would like to detect, $f = \dfrac{\delta}{2\sigma}$), and $t_{\alpha[v]}$ and $t_{2(1 - P)[v]}$ are values from a two-tailed t-table (such as Statistical Table **B**) with $v = 2(n - 1)$ degrees of freedom and probabilities α and $2(1 - P)$, respectively. The expression for P, power, is more complex than one might expect in order to compensate for the fact that a one-tailed probability is required but a two-tailed table is being used. *Note:* If $P = 1/2$, then $t_{1[v]} = 0$.

For example, from previous studies you know that the coefficient of variation of wing length of a species of bird is about 6%. You plan to study two populations. How many measurements need be made from each population to have an 80% chance of detecting a 5% difference between two means while allowing for a 1% chance of a type I error?

Box 12.2 (continued)

Iterative Solution

We start by trying $n = 20$ as a reasonable initial guess. Then $v = 2(n - 1) = 2(20 - 1) = 38$. Because $V = 6\%$, $s = 6\bar{Y}/100$. We wish δ to be 5% of the mean; that is, $\delta = 5\bar{Y}/100$. Using s as an estimate of σ, we obtain $f = \dfrac{\delta}{2\sigma} = \dfrac{5\bar{Y}/100}{2(6\bar{Y}/100)} = \dfrac{5}{2(6)} = 0.41667$. Thus,

$$n \geq \frac{1}{2(0.41667)^2}(t_{.01[38]} + t_{2(1 - .80)[38]})^2$$
$$= 2.88000(2.7116 + 0.8512)^2 = 36.6$$

Next, we try $n = 37$, making $v = 2(37 - 1) = 72$:

$$n \geq 2.88000(t_{.01[72]} + t_{2(1 - .80)[72]})^2$$
$$= 2.88000(2.6479 + 0.8468)^2 = 35.2$$

We round $n = 35.2$ up to 36 to be conservative. Using 36, we obtain $n = 35.2$ again, indicating that we have iterated to stability. It appears that 36 replications per group are necessary.

Graphic Solution

The preceding obtained result is consistent with the graph in Statistical Table **GG** for $v_1 = 1$ and $\alpha = 0.01$, where we can see that for $f = 0.41667$ a sample size somewhat larger than 30 would be necessary to achieve a power of 0.80.

12.5 Minimum Detectable Difference

Another useful computation, when already given a sample size, is to determine the smallest effect size one is likely to be able to detect.

For two means, one can rearrange Expression (12.8) to give

$$f \geq \sqrt{\frac{1}{2n}}(t_{\alpha[v]} + t_{2(1 - P)[v]}) \tag{12.9}$$

Using the relationship $f = \Delta/2 = \delta/(2\sigma)$ from Table 12.1, this can also be expressed as a minimum detectable difference between two means:

$$\delta \geq \sqrt{\frac{2}{n}}\sigma(t_{\alpha[v]} + t_{2(1 - P)[v]}) \tag{12.10}$$

These equations can be solved directly (i.e., without iteration) because the value of n is specified.

For more than two means, one can use Statistical Table **GG** to find the value of f at which the curve for the specified sample size intersects the point along the ordinate corresponding to the desired power. This minimum detectable effect size can be converted to a minimum detectable σ_A by multiplying it by an estimate for σ.

12.6 Post Hoc Power Analysis

Many authors have suggested that it is useful to compute the **observed power** after an experiment has been completed. In such computations, the true difference and error variance are assumed to be equal to the observed difference and variance—that is, the true effect size is assumed to be equal to the observed effect size. Using those values, the computed probability of rejecting the null hypothesis is called the *observed power.* If one did this and found that the observed power was large when the test was nonsignificant, then that was taken as evidence in favor of the null hypothesis—rather than considering the nonsignificance to be merely a consequence of having too small a sample size n or too large of an error variance. Hoenig and Heisey (2001) have shown that such computations provide no additional information because the observed power is a relatively simple function of the observed P-value. Figure 12.3 illustrates this for a two-tailed test for the difference between two means. When the

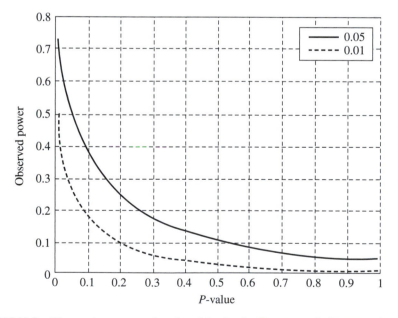

FIGURE 12.3   Observed power as a function of the P-value for a two-tailed t-test at the 0.01 and 0.05 type I error levels and assuming very large sample sizes. Based on a similar figure in Hoenig and Heisey (2001).

observed P-value is equal to the desired type I error rate, the observed power is approximately equal to 0.5. Smaller P-values yield larger observed power values and larger P-values yield smaller values. Observed power curves for other tests would be different but would still be functions of P.

Another type of computation that is sometimes done is to calculate the hypothetical true difference that would have to exist in order to have a high level of power (e.g., 0.9) for detection in the present test. This is done to determine the detectable effect size—what is assumed to be an upper bound on the true effect size. The closer this value is to that of the null hypothesis, the stronger the evidence is considered to be for the null hypothesis to be true. A related computation is the biologically significant effect size. For this, one computes the power using some minimal effect size that is still considered to be important. When the null hypothesis is accepted, higher values from this power calculation are interpreted as evidence that the true difference is close to that of the null hypothesis. Hoenig and Heisey (2001) point out that there are a number of problems with these procedures, and they show that once one has constructed a confidence interval for a statistic, no additional insights are gained by these methods and thus they should not be used. Additional arguments against the use of post hoc power methods are given by Levine and Ensom (2001), and a contrary view is given by Onwuegbuzie and Leech (2004), who apparently were unaware of the earlier Hoenig and Heisey (2001) paper.

An alternative is to treat the current data as a pilot experiment in order to obtain estimates of the variance and likely effect sizes for future studies. However, Kraemer et al. (2005) discuss potential problems with performing power analyses on data from pilot experiments. Their main concern is that pilot experiments are often based on relatively small samples. Therefore, the estimates of effect sizes and power are subject to considerable sampling error (as illustrated in several examples in Box 12.1). In such cases, they must be interpreted with caution.

The preceding sections have focused on the effect of different sample sizes on power, but changes in the error variance are equally important. In single-classification designs, the use of controlled environments or better measuring devices can reduce the error variance and thus enable one to achieve greater power or achieve the same power with smaller sample sizes. In more complex experiments, the size of the error variance can be reduced by the use of more efficient experimental designs.

12.7 Optimal Allocation of Resources in a Nested Design

In the earlier sections, there were just two aspects of the experimental design that we could control—the number of groups to be tested and the sample sizes within each group. In nested designs, this problem is more complex. Before we can determine how many replications to use, we must decide how these replications should be structured. Take the experiment of Box 10.3 as an example. We wish to test for differences among treatments, and each treatment is represented by one or more rats. But how should the data for each rat be structured? Recall that $c = 3$ preparations were made from each rat liver and $n = 2$ readings were taken on each preparation. Would it have been wiser to take four readings per preparation but only one preparation per rat, or would it have been better to take five preparations from one rat liver but only one reading per preparation? Only after making such decisions can we solve for required sample size (number of rats per treatment) in the manner of Box 12.2.

We approach this problem of optimal allocation of resources by determining the expected variance of means for various possible designs (ncb times this quantity will be used as the estimate of the error variance for estimating power and sample size when testing differences among treatment means). In a Model I single-classification analysis of variance (as in Box 9.1), the anova is structured as follows:

$$\text{Treatments (groups)} \qquad \sigma^2 + n\frac{\sum \alpha^2}{a-1}$$

$$\text{Replicates (within groups)} \qquad \sigma^2$$

The estimated error variance is s^2, and the expected variance of a group mean is $s_{\bar{Y}}^2 = s^2/n$, which is simply the familiar formula for the variance of the mean.

In a mixed-model two-level nested anova (as in Box 10.2), the expected mean squares are structured as follows:

$$\text{Treatments (groups)} \qquad \sigma^2 + n\sigma_{BCA}^2 + nb\frac{\sum \alpha^2}{a-1}$$

$$\text{Subgroups within groups} \qquad \sigma^2 + n\sigma_{BCA}^2$$

$$\text{Replicates within subgroups} \qquad \sigma^2$$

and the variance of group means can be estimated as MS_{subgr}/nb, which yields

$$s_{\bar{Y}}^2 = \frac{s^2}{nb} + \frac{s_{BCA}^2}{b} \tag{12.11}$$

where s^2 and s_{BCA}^2 are estimates of σ^2 and σ_{BCA}^2, respectively. Note that the error variance is now divided by nb, the total number of observations on which a subgroup mean is based, while the variance component of subgroups within groups is divided only by b, the number of subgroups constituting a mean.

In a mixed-model three-level nested anova (as in Box 10.3), expected mean squares have the following structure:

$$\text{Treatments (groups)} \qquad \sigma^2 + n\sigma_{CCB}^2 + nc\sigma_{BCA}^2 + ncb\frac{\sum \alpha^2}{a-1}$$

$$\text{Subgroups within groups} \qquad \sigma^2 + n\sigma_{CCB}^2 + nc\sigma_{BCA}^2$$

$$\text{Subsubgroups within subgroups} \qquad \sigma^2 + n\sigma_{CCB}^2$$

$$\text{Replicates within subsubgroups} \qquad \sigma^2$$

with s^2, $s^2_{C \subset B}$, and $s^2_{B \subset A}$ estimating variance components for increasing hierarchic levels. The expected variance of a group mean based on b subgroups, c subsubgroups, and n replicates within subsubgroups is MS_{subgr}/ncb, which yields

$$s^2_{\bar{Y}} = \frac{s^2}{ncb} + \frac{s^2_{C \subset B}}{cb} + \frac{s^2_{B \subset A}}{b} \tag{12.12}$$

The variances of group means are expectations of the variances among means sampled from identical populations. They do not therefore include added treatment or random effects among groups. Thus, we always divide the MS just below MS_{among} by the number of replicates on which a group mean is based, yielding $s^2_{\bar{Y}}$, as shown in Expressions (12.11) and (12.12).

The following example will make these formulas more meaningful to you. In a hypothetical pharmacological experiment testing the effects of various drugs, each drug was tested on several rats. Several replicated readings of the chemical composition of each rat's blood were made. A mixed-model two-level nested anova (among treatments, among rats within treatments, readings within rats) resulted in the following variance component estimates:

$$\text{Among rats within treatments} \qquad s^2_{B \subset A} = 25$$

$$\text{Readings within rats} \qquad s^2 = 100$$

We will now contrast two of many possible designs for further research with these drugs. In design (1), we propose to employ $b = 10$ rats per treatment and to make $n = 2$ readings on each rat. The expected variance of treatment means (if no added treatment effects are present) from Expression (12.11) is:

$$s^2_{\bar{Y}}(1) = \frac{100}{2 \times 10} + \frac{25}{10} = 7.50$$

Design (2) might consist of two rats per treatment but with nine readings on each rat. Its expected variance would be:

$$s^2_{\bar{Y}}(2) = \frac{100}{9 \times 2} + \frac{25}{2} = 18.06$$

Thus, design (2) would have a considerably greater variance than design (1) and, all other considerations being equal, would be less desirable because it would be far less sensitive than design (1) to differences among treatment means.

To compare two designs, we compute the **relative efficiency** (*RE*) of one design with respect to the other. *RE* is a ratio, usually expressed as a percentage, of the variances resulting from the two designs. The design whose variance is in the denominator is the one whose relative efficiency is being evaluated. Thus, we measure the relative efficiency of design (2) with respect to design (1) by

$$RE = \frac{s^2_{\bar{Y}}(1)}{s^2_{\bar{Y}}(2)} \times 100 \tag{12.13}$$

In the previous example, the relative efficiency of design (1) with respect to design (2) is

$$RE = \frac{100s_{\bar{Y}}^2(2)}{s_{\bar{Y}}^2(1)} = \frac{100 \times 18.06}{7.50} = 240.8\%$$

The increase in efficiency on changing from design (2) to design (1) is

$$240.8\% - 100\% = 140.8\%$$

The meaning of relative efficiency can be explained as follows. If we take the design whose relative efficiency is being measured and divide its number of replications by RE (expressed as a ratio), the resulting design will be as sensitive as the original design with which it is being compared. Thus, in the pharmacological experiment, if we take $10/2.408 = 4.153$ rats per treatment (with 2 readings per rat), we will get a denominator MS for testing the treatment MS approximately equal to that of design (2), 2 rats with 9 readings per rat. The equality is only approximate because the denominator mean squares are based on different degrees of freedom that affect the sensitivity of the design. Furthermore, we obviously cannot apportion 4.153 rats per treatment but would need to settle for 4 rats (although one might wish to be conservative and round up to 5).

The relative efficiency of one design with respect to another is not very meaningful, however, unless the relative costs of obtaining the two designs are taken into consideration. Clearly, if one design is twice as efficient as another (that is, has half the other's variance) but at the same time is 10 times as expensive to achieve, we might not choose it. To introduce the idea of cost, we construct a cost function, which tells us the cost per treatment class. For a two-level nested design,

$$C = bc_{BCA} + nbc_{RCB} \tag{12.14}$$

where C is the cost per treatment class, c_{BCA} is the cost of one subgroup unit, c_{RCB} is the cost of one replicate reading within a subgroup, and b and n have their familiar meanings.

Let us assume for the pharmacological experiment that the cost of one subgroup unit (the cost of buying and feeding one rat) is $c_{BCA} = \$150$. Similarly, suppose the cost of one replicate within the subgroup (the price of one blood test on a rat) is $c_{RCB} = \$30$; then the total cost of design (1) by Expression (12.14) will be

$$C(1) = (10 \times \$150) + (10 \times 2 \times \$30) = \$2100$$

and of design (2) will be

$$C(2) = (2 \times \$150) + (2 \times 9 \times \$30) = \$840$$

Although more efficient than design (2), design (1) costs considerably more: It is $2100/840 = 2.50$ times more expensive. Thus, the cost efficiency of design (1) with respect to design (2) is $2.408/2.50 = 0.9632$—that is, the first design is slightly less efficient than the second.

We now ask two questions: (a) For a given amount of money per treatment class, what is the optimal (most efficient) design? (b) What is the least expensive design for obtaining a given variance of a treatment mean? To answer these questions, we need one more formula:

$$n = \sqrt{\frac{c_{BCA}s^2}{c_{RCB}s^2_{BCA}}} \qquad (12.15)$$

which yields the number of replicates n per subgroup unit that corresponds to the optimal trade-off between cost and variance—that is, it is a number that will result in the minimal cost and minimal variance.

Expression (12.15) shows that the optimal number of replicates per subgroup depends on the ratio of costs of the two levels, as well as on the ratio of their variance components. If the costs (= effort) of obtaining a replication at either hierarchic level are the same, Expression (12.15) becomes the ratio of the standard deviations corresponding to the variance components. Expression (12.15) may yield $n < 1$. In such a case, we set $n = 1$ because obviously we must have at least one reading per subgroup. A rat will be of no use to us in the pharmacological experiment unless we take at least one blood sample from it! Cases in which the optimal number of replicates per subgroup is <1 are those in which variance among subgroups is great and the error variance is relatively small. In our problem,

$$n = \sqrt{\frac{150 \times 100}{30 \times 25}} = 4.472$$

Therefore, the optimal number of readings per rat is four, ignoring the decimal fraction. Thus, we answer questions (a) and (b) as follows:

(a) Assume that we wish to spend the same amount of money as in design (1), that is, \$2100. What is the optimal design at this cost? We substitute the known values in Expression (12.14) and solve for b:

$$C = bc_{BCA} + nbc_{RCB}$$

$$2100 = (b \times 150) + (4 \times b \times 30) = 270b$$

$$b = 2100/270 = 7.778 \approx 8$$

Thus, 8 rats with 4 readings per rat should yield the most efficient way of spending approximately \$2100. Note that we said "approximately"; obviously, we cannot use a fractional number of rats or blood samples, which would be necessary to yield an exact cost of \$2100. To check our results, we calculate the new cost of the design, which is

$$C = (8 \times \$150) + (4 \times 8 \times \$30) = \$2160$$

This is close to the figure of \$2100 we aimed at. To show that the new design is more efficient than the old one, we calculate the new variance of a treatment mean based on 8 rats and 4 blood tests per rat by substituting in Expression (12.11):

$$s_{\bar{Y}}^2 = \frac{100}{8 \times 4} + \frac{25}{8} = 6.25$$

The new variance is smaller than that of design (1) at nearly the same cost. The relative efficiency,

$$RE = \frac{7.50 \times 100}{6.25} = 120\%$$

has increased by 20%, but the increase in cost is only 2.9%. Clearly, the new design is preferable, even though design (1) was relatively not a bad design to start with.

The process may also be shown graphically. The solid line in Figure 12.4 is a graph of the cost equation $2100 = $150b + $30nb$ and represents all combinations of n and b that would yield an experiment costing $2100. The dotted lines represent the equation $s_{\bar{Y}}^2 = 100nb + 25/b$ for various values of $s_{\bar{Y}}^2$. All intersections between the line for cost $= 2100 and equations for the various $s_{\bar{Y}}^2$'s give the $s_{\bar{Y}}^2$, n, and b for all possible experiments costing $2100. What we want in this case is the one with the smallest $s_{\bar{Y}}^2$. One can see that no line representing an experiment with $s_{\bar{Y}}^2 < 6.41$ can intersect the cost equation, so $s_{\bar{Y}}^2 = 6.41$ is the best we can expect. These two lines intersect at $n = 4$ and $b = 7.8$ (≈ 8), which is the best combination of n and b for our requirements.

(b) If we wish to achieve a given variance, as, for example, the value of 18.06 resulting from design (2), what is the least expensive way to do so? We substitute the known and given terms in Expression (12.11) and solve for b:

$$s_{\bar{Y}}^2 = \frac{s^2}{nb} + \frac{s_{BCA}^2}{b}$$

$$18.06 = \frac{100}{4b} + \frac{25}{b}$$

$$b = \frac{100 + 100}{4 \times 18.06} = 2.8 \approx 3$$

Thus, 3 rats with 4 readings per rat will give the most economic design for a treatment mean square approximating 18.06. When we check this result, we find

$$s_{\bar{Y}}^2 = \frac{100}{(3 \times 4)} + \frac{25}{3} = 16.67$$

This is a value less than the aimed-at variance (because we used whole numbers of replicates), but the new cost would be

$$C = (3 \times $150) + (4 \times 3 \times $30) = $810$$

Thus, a 6% decrease in cost would be accompanied by an 8% increase in relative efficiency. This result may also be determined graphically by finding the smallest cost equation that intersects the line of the equation for the desired $s_{\bar{Y}}^2$.

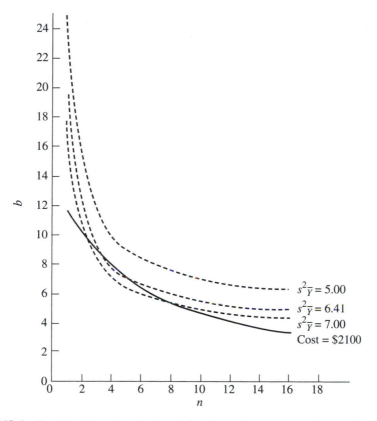

FIGURE 12.4 Equations for cost and estimated variance of a group mean in a two-level anova. Solid line: $b = 2100/(150 + 30n)$ because $\$2100 = 150b + \$30nb$. Dashed lines: $b = (100/s_{\bar{Y}}^2)/n - 25/s_{\bar{Y}}^2$ because $s_{\bar{Y}}^2 = 100/nb - 25/b$.

Box 12.3 summarizes the computations of optimal allocation of resources for two and three levels of sampling. The formulas in the latter case are similar to those in the former and should be self-explanatory. Once one determines the optimum values for n and b in a two-level anova, one can use nb $s_{\bar{Y}}^2$ as the error variance in Expression (12.1) to compute the effect size. In Box 12.1, the sample size estimated is that of the product nb. Because the value for n has already been optimized using Expression (12.15), one only adjusts b to obtained the desired level of power. The computations for a three-level nested anova are similar. One uses ncb $s_{\bar{Y}}^2$ as the estimated error variance to compute the effect size. In Box 12.2, the sample size corresponds to the product ncb. Because the values of n and c have been optimized as shown in Box 12.3, one treats them as fixed and only varies b to achieve the desired level of power.

We should point out that whenever the cost of an experiment is of little consideration, but the design of the most efficient experiment (lowest variance) for a given

---

BOX 12.3 Optimal Allocation of Resources in Two- and Three-Level Sampling: Outline of Procedure

Two Levels

In a study of variation in the aphid *Pemphigus populitransversus*, Sokal (1962) measured the forewing length of each of 2 alates for each of 15 galls from each of 23 localities in eastern North America: $n = 2$; $b = 15$. The estimates of the variance components were:

$$\text{Galls within localities} \qquad s^2_{BCC} = 10.507$$

$$\text{Within galls} \qquad s^2 = 6.813$$

In this instance the costs of obtaining another gall or another alate in the same gall are roughly equal. Therefore, $c_{RCB} = c_{BCA}$.

Optimum sample size n per subgroup is computed from Expression (12.15):

$$n = \sqrt{\frac{c_{BCA}s^2}{c_{RCB}s^2_{BCA}}} = \sqrt{\frac{6.813}{10.507}} = 0.805$$

Although $n < 1$, we must use at least one alate per gall in further studies. The number of galls to be measured depends on the desired variance of locality means. It must satisfy the following equation:

$$s^2_{\bar{Y}} = \frac{s^2}{nb} + \frac{s^2_{BCA}}{b} = \frac{6.813}{1 \times b} + \frac{10.507}{b}$$

For example, if an $s^2_{\bar{Y}}$ less than or equal to 1.7 were desired, one could solve for b and obtain the following result:

$$b = \frac{6.813 + 10.507}{1.7} = 10.188 \approx 10$$

Three Levels

In Box 10.3, we studied the effect of 3 treatments on glycogen content in rat livers. The experiment was structured into b rats within treatments ($s^2_{BCA} = 36.1$), c preparations within rats ($s^2_{CCB} = 14.2$), and n readings within preparations ($s^2 = 21.1$). Let us assume that a rat costs $c_{BCA} = \$50$, a preparation $c_{CCB} = \$200$, and a reading $c_{RCC} = \$20$. Extending Expression (12.14) for two-level samples, the cost C of a treatment class for a given design is

$$C = bc_{BCA} + cbc_{CCB} + ncbc_{RCC}$$

where b, c, and n are the number of replicates at the B, C, and lowest levels, and c_{BCA}, c_{CCB}, and c_{RCC} are the costs per replicate at these levels, respectively.

Expression (12.15) yields

$$c = \sqrt{\frac{c_{BCA}s^2_{CCB}}{c_{CCB}s^2_{BCA}}} \qquad n = \sqrt{\frac{c_{CCB}s^2}{c_{RCC}s^2_{CCB}}}$$

Box 12.3 (continued)

as the optimal number of replicates c and n in the subgroups and subsubgroups, respectively.

Solving for our case, we get

$$c = \sqrt{\frac{5 \times 14.2}{20 \times 36.1}} = 0.3136$$

$$n = \sqrt{\frac{20 \times 21.2}{2 \times 14.2}} = 3.864$$

Thus, employing four readings per preparation and one preparation per rat will give us the most efficient design.

If we can spend \$2000 per treatment, we substitute this amount in the cost formula above and solve for b:

$$2000 = (b \times 50) + (1 \times b \times 200) + (4 \times 1 \times b \times 20) = 330b$$

$$b = \frac{2000}{330} = 6.06$$

We should use six rats per treatment, make one preparation from each rat, and carry out four readings on each preparation.

If we wish to obtain the most efficient design for a given variance—say, 22—we proceed as follows. Substituting in Expression (12.12) and solving for b, we get

$$s_{\bar{Y}}^2 = \frac{s^2}{ncb} + \frac{s_{CCB}^2}{cb} + \frac{s_{BCA}^2}{b}$$

$$22 = \frac{21.2}{4 \times 1 \times b} + \frac{14.2}{1 \times b} + \frac{36.1}{b}$$

$$b = \frac{222.4}{4 \times 22} = 2.53$$

We decide to take three rats per treatment.

Thus, our expected variance of a treatment mean based on three rats, one preparation, and four readings is

$$s_{\bar{Y}}^2 = \frac{21.2}{4 \times 1 \times 3} + \frac{14.2}{1 \times 3} + \frac{36.1}{3} = 18.533$$

This value is not exactly 22 because $b = 3$ is fairly far from the exact solution to the equation ($b = 2.53$). Cost for this design would be

$$C = (3 \times 50) + (1 \times 3 \times 200) + (4 \times 1 \times 3 \times 20) = \$990$$

Let us compare this with the original design in Box 10.3, in which we employed two rats per treatment, three preparations per rat, and two readings per preparation. By Expression (12.12), this yields the following expected MS of means:

$$\frac{21.2}{2 \times 3 \times 2} + \frac{14.2}{3 \times 2} + \frac{36.1}{2} = 22.183$$

Box 12.3 (continued)

When we multiply by $ncb = 12$, we obtain $MS_{subgr} = 266.20$, which differs from the actually calculated value of 265.89 because of rounding errors. The relative efficiency of the optimal design computed above is $RE = 22.183/18.533 = 119.7\%$. The cost of the design in Box 10.3 is thus

$$C = (2 \times 50) + (3 \times 2 \times 200) + (2 \times 3 \times 2 \times 20) = \$1540$$

This is 55.6% higher than the optimal design. Clearly, one preparation per rat and four readings per preparation are a more efficient allocation of resources for this study.

total number of observations is important, the most efficient procedure will always be to take a single observation for each of the subgroups; that is, set $n = 1$ in a two-level anova and both n and $c = 1$ in a three-level anova. This design is also often used when the cost of making measurements (such as concentrations of various metals in environmental monitoring programs) is very expensive. But, of course, such studies would yield no information on the magnitude of variation at lower levels nor information on the statistical distribution of the variable.

12.8 Randomized Blocks and Other Two-Way and Multiway Designs

In the previous section, we saw how a restructuring of the pattern of subsamples could reduce the error mean square in testing the effect of some treatment. In a two- or more-way design, this reduction can be quite dramatic. The data in Box 11.6 serve as a good example. The investigator wanted to see whether growth in lower face width could be detected in young girls. This study could have been carried out using a design in which a single visit to a school resulted in two samples of 15 subjects being collected—one of five-year-olds and another of six-year-olds. Instead, two visits were made. During the first, measurements were made on 15 five-year-olds and then a year later the same individuals would be measured again to provide a sample of six-year-olds. The single-visit design would be simpler and therefore less expensive because only a single visit would be necessary. However, the repeated-visits design proved to yield a statistically more powerful result because the variation among girls was separated out from the possible variation due to growth.

The result was that the error variance was only a small fraction of what it would have been if the single-visit design had been used. It is possible to quantify the comparisons of these two designs. The single-visit design is called a **completely randomized design** because the null hypothesis assumes that there is a single pool of possible subjects who are randomly selected and assigned to the two treatment groups. The repeated-visits design is called a **randomized-blocks design** because the subjects come in blocks (in this case, individual girls) selected at random from some population. Within each block, treatments (ages in this case) are assumed to be assigned at random to the observations. This design could also be called a **randomized-complete blocks-design** because each block contains all treatments. It is also an example of a paired-comparisons design because each block contains only two treatments.

Had the investigator used the single-visit design, which ignored differences among girls and simply analyzed these data using a single-classification anova, the expected error variance would have been

$$MS_{E(CR)} = \frac{b(a-1)MS_{E(RB)} + (b-1)MS_B}{ab-1} \tag{12.16}$$

In this expression, $MS_{E(CR)}$ is the expected error mean square for a completely randomized design, $MS_{E(RB)}$ is the observed error mean square in the randomized-blocks design, and MS_B is the observed mean square among blocks. This formula is derived in Appendix A.6. For the values in Box 11.6, this is $MS_{E(CR)} = 0.091318$, which is considerably larger than the error mean square error, $MS_{E(RB)} = 0.0007714$, used in Box 11.6.

We can now calculate the **relative efficiency** of the randomized-blocks design relative to the completely randomized design by computing the ratio of their error mean squares and taking into account the differences in their degrees of freedom.

$$RE_{RB/CR} = \frac{MS_{E(CR)}(df_{RB} + 1)(df_{CR} + 3)}{MS_{E(RB)}(df_{RB} + 3)(df_{CR} + 1)} \times 100$$

$$= \frac{0.09131\ (14+1)(28+3)}{0.0007714(14+3)(28+1)} \times 100 = 11,164.64\% \tag{12.17}$$

This means that the sample size would have to be 111.6464 times larger in a completely randomized design in order to have the same ability to test for differences between ages as in the randomized-blocks design. Unless the costs of a second visit to the school were extremely expensive, it is clearly much more efficient to use the randomized-blocks design.

The second factor in expression (12.17) is the correction for the differences in the degrees of freedom in the two designs proposed by Cochran and Cox (1957). The completely randomized design will always have larger degrees of freedom than the randomized-blocks design. The correction will have relatively little effect on the results unless the degrees of freedom are small—for example, less than 20.

Box 11.5 shows another example of an experiment that used a randomized-complete-blocks design. In these data, the blocks represent replications of an experiment performed several months apart because the experimental protocol was so laborious. Within each block, all three treatments were represented. The alternative would have been to break up the experiment by testing the three treatments separately but including all replications of the same treatment at one time. This would be a completely randomized design because the variation due to replications within each treatment would be independent. Using Expression (12.16), the estimate of $MS_{E(CR)}$ is 0.002452, which is larger than the observed $MS_{E(RB)}$ of 0.000697. If we use Expression (12.17), we find that the randomized-blocks design has a relative efficiency of 328.2%.

Section 11.9 describes several other types of experimental designs. The purpose of most of them is to enable a researcher to perform experiments that are more efficient or less costly or both.

EXERCISES 12

12.1 If a study such as that given in Exercise 9.5 were to be repeated with only two samples (one a control and the other representing pigeons fed a special diet), what sample size should be used to be 80% certain of observing a true difference between two means as small as a 10th of a millimeter using a type I error rate of 5%? Assume that the error variance in this new experiment would be the same as in the previous one.

12.2 Estimate f^2 and ω^2 for the data of Exercise 9.6. Also compute their 95% confidence limits.

12.3 What sample size should be used in a new density experiment based on Exercise 9.10? The investigator plans to use 5 densities, wishes to establish whether there is a difference of at least 1.0 between the most distant means, and desires a power of at least 80% using a type I error rate of 5%.

ANSWER: A sample size of $n = 5$ yields power of about 93%.

12.4 Estimate f^2, ω^2, and their 95% confidence limits for the data presented in Exercise 9.2.

12.5 Hasel (1938) made a study of sampling practices in timber surveys. The variable studied is board feet of timber in a ponderosa pine forest. He studied 9 sections, each subdivided into 4 quarter sections [each of which can be subdivided into 4 forty-acre tracts, or 16 ten-acre tracts, or 32 five-acre tracts, or 64 basic plots (of two-and-a-half acres)]. If sampling and distribution of timber were entirely at random, no added variance should exist beyond the basic plots.

Interpret the analysis of variance given below and isolate variance components. To keep the notation uniform, identify variance components as follows: s^2 for basic plots, s_5^2 for fives, s_{10}^2 for tens, s_{40}^2 for forties, s_Q^2 for quarter sections, and s_S^2 for sections.

Source of variation	df	SS
Between sections	8	109,693.35
Quarter sections within sections	27	57,937.40
Forties within quarter sections	108	66,569.53
Tens within forties	432	161,109.77
Fives within tens	576	121,940.82
Basic plots within fives	1152	133,337.74
Total	2303	650,588.61

12.6 Using the data on geographic variation in the aphid *Pemphigus populitransversus* given in Exercise 10.3, design a new study with equal sample sizes that would be expected to achieve the same standard error for a coast mean. Comment on the total numbers of samples required by the new study in comparison to the original.

13 Assumptions of Analysis of Variance

In Chapters 8 to 11, we blithely proceeded to study design after design in analysis of variance without bothering to discuss the assumptions required for carrying out anova and its attendant tests of significance. In Chapter 12, we dealt with problems of effect size, power, and the determination of sufficient sample sizes for these tests. This chapter will examine the underlying assumptions of the analysis of variance, methods for testing whether these assumptions are valid, the consequences if the assumptions are violated, and steps to be taken if the assumptions cannot be met. By taking up the fundamental assumptions of anova last in our consideration of the subject, we may seem to be building the superstructure before the foundation. Rigorous mathematical treatment clearly demands statement of the model prior to a treatment of the subject. Yet we have found that the generally nonmathematical audiences we have been addressing learn the concepts more easily if they first understand the structure and purpose of analysis of variance and are able to carry out the computations. We should stress, however, that before carrying out an anova on an actual research problem you should reassure yourself that the assumptions listed in this chapter seem reasonable. If they are not, you should carry out one of several possible alternative steps to remedy the situation.

Section 13.1 refreshes your memory of a fundamental assumption: random sampling. In each of the next three sections, we explain a special assumption of analysis of variance, describe a procedure for testing it, briefly discuss the consequences if the assumption does not hold, and give instructions on how to proceed if it does not. Section 13.2 treats the assumption of independence, Section 13.3 the homogeneity of variance, and Section 13.4 normality.

In many cases, departure from the assumptions of analysis of variance can be rectified by transformation of the original data into a new scale. Section 13.5 gives a brief general introduction to transformations. The next four sections discuss some useful transformations. Section 13.6 describes the logarithmic transformation, which for some types of variables is able to make a sample closer to a normal distribution and to equalize its variances. In addition, for two-way and higher-order anova designs, it is sometimes able to simplify an analysis by making it additive by the elimination of interaction terms. Section 13.7 describes the square root transformation, Section 13.8 the Box–Cox transformation, and Section 13.9 the arcsine or angular transformation.

When transformations are unable to make the data conform to the assumptions of analysis of variance, other techniques of analysis, analogous to the intended anova, must be employed. These are the nonparametric or distribution-free techniques, which are sometimes used by preference even when the parametric method (anova, in this case) can be legitimately employed. Ease of computation and a preference for the simpler assumptions of the nonparametric analyses cause many research workers to turn to them. When the assumptions of the anova are met, however, these methods are less powerful than analysis of variance. Section 13.10 examines several nonparametric methods in lieu of single-classification anova, and Section 13.11 features nonparametric methods in place of two-way anova.

13.1 A Fundamental Assumption

All anovas require that sampling of individuals be random. Thus, in a study of the effects of three doses of a drug (plus a control) on five rats each, the five rats allocated to each treatment must be selected at random. If the five rats employed as controls are either the youngest or the smallest or the heaviest rats, whereas those allocated to some other treatment are selected in some other way, the results are not apt to yield an unbiased estimate of the true treatment effects. Nonrandomness of sample selection may well be reflected in lack of independence of the items (see Section 13.2), in heterogeneity of variances (Section 13.3), or in nonnormal distribution (Section 13.4). Adequate safeguards are essential to ensure random sampling during the design of an experiment or when sampling from natural populations.

13.2 Independence

An assumption stated in each explicit expression for the expected value of an observation [for example, Expression (8.2) was $Y_{ij} = \mu + \alpha_i + \epsilon_{ij}$] is that the error term ϵ_{ij} is a normally distributed random variable with a mean of zero. In addition, for completeness we should add that it is assumed that the ϵ's are independent of one another and have the same variance (see Section 13.3).

Thus, if the observations within any one group are arranged in the order in which the measurements were obtained, we expect the ϵ_{ij}'s to succeed each other in a random sequence. Consequently, we assume a long sequence of large positive values followed by an equally long sequence of negative values to be quite unlikely. We would also not expect positive and negative values to alternate with regularity.

How could departures from independence arise? An obvious example is an experiment in which the experimental units are plots of ground laid out in a field. In such a case, adjacent plots of ground often give similar yields due to spatial nonindependence. It would thus be important not to group all the plots containing the same treatment into an adjacent series of plots but to randomize the allocation of treatments among the experimental plots. The physical process of randomly allocating the treatments to the experimental plots ensures that the ϵ's will be independent.

Lack of independence of the ϵ's can also result from correlation in time rather than in space. In an experiment, we might measure the effect of a treatment by recording weights of 10 individuals. The balance we use may suffer from a maladjustment that yields successively greater underestimates, resulting in a declining trend of the weights. Occasional recalibrations of the balance will introduce large apparent increases of the weights. Differences in weights among several treatments may therefore not be entirely due to added treatment effects but could be due to the biases of the balance and the order in which the individuals were weighed. Conversely, compensation by the operator of the balance may result in regularly alternating over- and underestimates of the true weight. Here again randomization may overcome the problem of nonindependence of errors. For example, we may determine the sequence in which individuals of the various groups are weighed according to some random procedure.

Both of these examples—the spatial and the temporal—are instances of positive **autocorrelation**, the self-similarity of observations adjacent in space or time. Regular alternation of positive and negative errors is a manifestation of negative autocorrelation.

Independence of errors in a sequence of observations of a continuous variable may be tested as first proposed by the well-known mathematician John von Neumann (von Neumann et al., 1941), with critical values tabulated by Young (1941). The test is based on successive differences between normally distributed observations, $d_i = Y_{i+1} - Y_i$. In the case of independent errors, the ratio $\eta = \Sigma d^2 / \Sigma y^2$ should approximate 2 (note: this is not the same η sometimes used as a measure of effect size). In Section 15.4, you will learn why the expected sum of such squared differences is twice the sum of squares of variable Y if the observations are independent. If there are sequences of similar readings, their differences will be less than what they would have been if the observations were randomly ordered, and the ratio η will be less than 2. Conversely, if there is a nonrandom alternation of the magnitudes of the observations, the variance of the differences will be greater than expected and η will be greater than 2. In Statistical Table **HH**, we expanded a shorter table of critical values of $|1 - \eta/2|$ by Young (1941) up to a sample size of $n = 50$. When $n > 50$, we can calculate

$$t_s = \frac{|1 - \eta/2|}{\sqrt{(n - 2)/(n^2 - 1)}}$$

and compare this to $t_{\alpha[\infty]}$, the normal distribution, which this expression approximates.

The computations are summarized in Box 13.1, where we examine the sequence of 25 numbers representing the aphid stem mother femur lengths from Box 2.1. We compute first differences to match all but the last observation; then we square and sum these d's. The result is an estimate of $\Sigma d^2 = 9.3700$. When we divide this value by the sum of squares of the femur lengths ($\Sigma y^2 = 3.2096$), we obtain $\eta = 2.9194$. Because $\eta > 2$, a nonrandom alternation of observations is indicated. Computing $|1 - \eta/2|$, we obtain 0.459683, which in Statistical Table **HH** yields a two-tailed $0.01 < P < 0.02$ for $n = 25$. Had this example been based on more than 50 observations, we could have tested it by using the normal approximation. If we do so in any case, we obtain $t_s = 2.394$, which yields $P = 0.0167$, consistent with the result from Statistical Table **HH**. We conclude that the observations occur in a sequence

---

BOX 13.1   Test for Serial Independence of a Continuous Variable

Twenty-five aphid stem mother femur lengths. Data from Box 2.1.

Y_i	$d_i^2 = (Y_{i+1} - Y_i)^2$
3.8	0.04
3.6	0.49
4.3	0.64
3.5	0.64
4.3	1.00
3.3	1.00
4.3	0.16
3.9	0.16
4.3	0.25
3.8	0.01
3.9	0.25
4.4	0.36
3.8	0.81
4.7	1.21
3.6	0.25
4.1	0.09
4.4	0.01
4.5	0.81
3.6	0.04
3.8	0.36
4.4	0.09
4.1	0.25
3.6	0.36
4.2	0.09
3.9	
$\Sigma y^2 = 3.2096$	$\Sigma d^2 = 9.3700$

Computation

1. Make a column of the observations. Construct a second column of first differences between the observations and square them as shown.

2. Compute the sum of squares of the observations and the sum of the squared differences (shown at the bottom of the columns).

3. Compute $\eta = \Sigma d^2/\Sigma y^2 = 9.3700/3.2096 = 2.9194$.

4. Evaluate $|1 - \eta/2| = 0.459683$. If $n \le 50$, consult Statistical Table **HH** for the critical values. In our case, the two-tailed probability is $0.01 < P < 0.02$. For illustrative purposes, we also evaluate

$$\frac{|1 - \eta/2|}{\sqrt{(n - 2)/(n^2 - 1)}}$$

Box 13.1 (continued)

and compare it with $t_{\alpha[\infty]}$. This is the approximation we would use with $n > 50$. Because n for this example is 25, the approximation should be close. We find that $t_s = 0.459683/\sqrt{23(25^2 - 1)} = 2.394348$, which for a normal distribution yields $P = 0.0167$. The observations are not serially independent. The fact that η is greater than 2 suggests a nonrandom alternation of the observations. Values of $\eta < 2$ indicate serial correlation (= positive autocorrelation) between adjacent variates.

We used a two-tailed significance test here because we had no a priori notion of the nature of the departure from serial independence. In some instances, our alternative hypothesis would be one-tailed, in which case we should halve the probabilities of Statistical Table **HH**, or, when using the normal approximation, halve the P-values for $t_{\alpha[\infty]}$ in Statistical Table **B** or those obtained by computer programs such as BIOMstat.

that appears to be nonrandom, so we question the assumption of independence in these data. Later work indicated that the femur lengths may have come from a dimorphic sample. Possibly the technician mounting the aphids on slides for measurement alternated between the two types of aphids in a conscious (but misguided) attempt to strike a balance. A ratio of η less than 2 would have indicated a positive serial correlation (= positive autocorrelation); succeeding observations would be more similar to each other because of technician or instrument bias.

When serial independence is tested in an anova with two or more groups, the test is applied separately to the observations within each group. For a nonparametric serial correlation test in a continuous variable, or when the observations are nominal, employ a runs test (see Section 18.2).

There is no simple adjustment or transformation to overcome the lack of independence or errors. The basic design of the experiment or the way in which it was performed must be changed. We have seen how a randomized-blocks design often overcomes lack of independence of error by randomizing the effects of differences in soils or cages. Similarly, in the experiment with the biased balance, we could obtain independence of errors by redesigning the experiment, using different times of weighing as blocks. Of course, if a source of error is suspected or known, attempts can be made to remove it; if we know, for example, that the balance is biased, we may have it fixed. If the ϵ's are not independent, the validity of the usual F-test of significance can be seriously impaired.

13.3 Homogeneity of Variances

In Section 9.4 and Box 9.4, in which we described the t-test for the difference between two means, we said that the statistical test was valid only if we could assume that the variances of the two samples were equal. Although we have not stressed it so far, this assumption that the ϵ_{ij}'s have identical variances also underlies the equivalent anova test for two samples—and, in fact, for any type of anova. *Equality of variances* in a set of samples is an important precondition for several statistical tests. Synonyms for this condition are *homogeneity of variances* or **homoscedasticity**, a jawbreaker that makes students in any biometry class sit up and take notice. The term is coined from Greek roots meaning "equal scatter." The converse condition (inequality of variances among samples) is called **heteroscedasticity**. Because we assume that each sample variance is an estimate of the same parametric error variance, the assumption of homogeneity of variances makes intuitive sense.

We have already seen how to test whether two samples are homoscedastic prior to a *t*-test of the differences between two means or a two-sample analysis of variance: We use an *F*-test for the hypotheses $H_0: \sigma_1^2 = \sigma_2^1$ and $H_1: \sigma_1^2 \neq \sigma_2^1$ (see Section 8.3 and Box 8.1). Markowski and Markowski (1990) point out that the reliability of these tests depends on the normality of the underlying distributions. A test for differences in dispersion that does not depend on normality is the interquantile range test (Shoemaker, 1995). We feature this nonparametric test in a part of Box 13.8 later in this chapter. When the two sample sizes are equal, Markowski and Markowski also state that the *t*-test is insensitive to the heterogeneity of the variances, making preliminary tests of equality of the two variances unnecessary.

When there are more than two groups in the anova, there is a "quick and dirty" test for homoscedasticity that, although not quite as efficient as the methods presented below, is preferred by many because of its simplicity. This method is Hartley's (1950) F_{max}-test. It uses a statistic that is the ratio of the largest to the smallest of several sample variances. Critical values are given in Statistical Table G. The F_{max}-test is shown in Box 13.2. Despite their great apparent differences, we cannot conclude that the variances of the four samples of ticks are heterogeneous.

The appropriate test for homogeneity of variances when the number of samples $a > 2$ has long been a contentious subject among statisticians. For a useful review of this topic, see Boos and Brownie (2004). Current opinion appears to favor **Levene's test for homogeneity of variances**, proposed by Levene (1960). The principle behind the test is quite simple. For each observation, one calculates a deviation $D_{ij} = |Y_{ij} - St_i|$, where the subscripts index the rows and columns of an anova and St_i is a statistic of location for group i. The absolute deviation is employed to avoid positive and negative deviations canceling each other. Deviations from samples with greater variances will have higher means than those with lesser variances, and Levene's test is simply a single-classification anova of the deviations. This anova tests for differences among means, hence among variances, of samples (groups). Levene (1960) originally proposed the arithmetic mean for the statistic St_i, but the recent consensus has been to use the median M_i for each group i. The test can be carried out directly by performing a single-classification anova of the aforementioned deviations. A preferable test mode, now that software for resampling approaches is becoming readily available, is to carry out a bootstrapping experiment with the data and compare the variance ratio obtained for the actual data to the distribution of such ratios resulting from numerous bootstrapped samples. For a refresher on bootstrapping, see Section 7.13.

BOX 13.2 Tests of Homogeneity of Variances Among $a > 2$ Groups

---

We apply these tests to the scutum widths of tick larvae from Box 9.1.

The F_{max}-Test

We inspect the variances of the $a = 4$ groups. They are not shown in Box 9.1 but can be easily obtained by dividing the furnished sum of squares of each group, $\sum^{n_i} y_i^2$, by its sample size less one, $n_i - 1$, shown in column (1) of the following table. The sample variances, s_i^2, are shown in column (2).

Groups ($a = 4$)	(1) $df = n_i - 1$	(2) s_i^2	(3) $\ln s_i^2$
1	7	54.2143	3.9929
2	9	142.0444	4.9561
3	12	79.5641	4.3766
4	5	233.0667	5.4513
	33		

Find the greatest variance, max $s_i^2 = 233.0667$, and the smallest, min $s_i^2 = 54.2143$. Then compute the maximum variance ratio: $F_{max} = $ max s_i^2/min $s_i^2 = 233.0667/54.2143 = 4.2990$. Statistical Table **G** (cumulative probability distribution of $F_{max\ \alpha[a, n-1]}$ assumes that the dfs of all the variances are equal, but we can make a conservative, approximate test using the lesser of the degrees of freedom of the two variances making up the variance ratio ($n_4 = 5$). In Statistical Table **G**, we find $F_{max\ .05[4,5]} = 13.7$. We cannot reject the null hypothesis that the variances of the four samples are homogeneous, despite their great apparent differences.

Levene's Test for Homogeneity of Variances

This test is carried out on absolute deviations of observations from measures of location. The most generally accepted variation of this test (see Section 13.3) is to apply it to absolute deviations of observations Y_{ij} from their group medians M_i. These deviations are defined as $D_{ij} = |Y_{ij} - M_i|$, where the subscripts i index the number of the group, $i = 1, 2, \ldots, a$, and the subscripts j identify the observations within the group, $j = 1, 2, \ldots, n_i$. These deviations, D_{ij}, are arrayed as in an ordinary single-classification anova. For the scutum widths, the first two observations, 380 and 376, would be transformed into $D_{11} = |380 - 373| = 7$ and $D_{12} = |376 - 373| = 3$. The medians of the four groups have not been calculated before. They are 373, 353, 354, and 366 for groups 1, 2, 3, and 4, respectively. The last deviation is $D_{46} = |360 - 366| = 6$.

Next, this table of differences is subjected to a single-classification anova, which will test whether there are differences in magnitude among the means of the absolute measures of variability and, by extension, whether the variances within the groups differ among themselves. The results of the anova of the D_{ij}-values are as follows.

Box 13.2 (continued)

	df	MS	F_s	P
Among samples	3	12.3843	0.8230	0.4906
Within samples	33	15.0478		

There is no evidence that the dispersions of the four groups differ, which is in agreement with the results of the earlier F_{max}-test.

Given the result just obtained, one would not attempt any refinement of Levene's test, but when the outcome of the test is critical and the P-value is borderline, it is worthwhile to attempt a resampling approach. Bootstrapping the Levene test, as suggested by various authors (see Boos and Brownie, 2004), has achieved improved results by several criteria. At this stage, we ask readers to review the meaning and mode of bootstrapping in Section 7.13. This cannot be carried out without a computer program, so we shall not give an explicit algorithmic account of applying the method to the scutum widths but instead will describe verbally what is done by the program and furnish the results of a bootstrap experiment we conducted for these data.

Random samples (with replacement) of the 37 observations are taken and arranged in 4 groups with the same sample sizes as in the original data. Each sample is analyzed by Levene's test of the deviations taken from the group median, as shown earlier for the single example. We carried out about 9999 such samples and computations, with the actual sample added to yield the 10,000th replication. We find that the observed F_s-value of 0.8230 is the 874th item in the array of 10,000 numbers, meaning that 8.74% of the items are equal to or greater than the observed value. Again, we find little support to reject the null hypothesis of homogeneity of dispersions among groups.

Bartlett's Test for Homogeneity of Variances

1. Convert the $a = 4$ variances to their natural logarithms. This has already been done for you in column (3) of the table early on in this box.

2. The sum of the degrees of freedom of the $a = 4$ groups is 33, shown at the bottom of column (2) of the table.

3. Compute a weighted average variance

$$s^2 = \frac{\sum_{}^{a}(n_i - 1)s_i^2}{\sum_{}^{a}(n_i - 1)} - \frac{7(54.2143) + 9(142.044) + \cdots + 5(5.4513)}{33} = 114.4849$$

and find its natural logarithm, ln 114.4849 = 4.7404.

4. $\sum_{}^{a}(n_i - 1)\ln s_i^2 = 7(3.9929) + 9(4.9561) + \cdots + 5(5.4513) = 152.3309$

5. $X^2 = \left[\sum_{}^{a}(n_i - 1)\right]\ln s^2 - \sum_{}^{a}(n_i - 1)s_i^2$

   = (quantity **2** $\times$ quantity **3**) − quantity **4**

   = 33(4.7404) − 152.3309 = 4.1037

Box 13.2 (continued)

6. Correction factor $C = 1 + \dfrac{1}{3(a-1)}\left[\displaystyle\sum^{a}\dfrac{1}{n_i - 1} - \dfrac{1}{\displaystyle\sum^{a}(n_i - 1)}\right]$

$$= 1 + \dfrac{1}{3(4-1)}\left[\dfrac{1}{7} + \dfrac{1}{9} + \cdots + \dfrac{1}{5} - \dfrac{1}{33}\right] = 1.05633317$$

Adjusted $X^2 = \dfrac{X^2}{C} = \dfrac{4.1037}{1.0563} = 3.8850$, $df = a - 1 = 3, P = 0.2742$.

We conclude that the variances of the four samples are homogeneous. The three tests featured in this box are in agreement for this dataset.

In Box 13.2, we examined the homogeneity of variances of the scutum widths of ticks, first encountered in Box 9.1. There are $a = 4$ samples, with varying numbers n_i of individuals in each. Obtain the deviations from the group means and then carry out an anova of these deviations. The results are clear-cut. There is no evidence for heterogeneity among the variances of the four groups, confirming the results of the F_{max}-test. In such an outcome, there would be no reason to pursue the matter further by bootstrapping the data, but we carried out a bootstrapping experiment anyway to illustrate the procedure. Again, we find no evidence to reject the null hypothesis of homogeneity of variances.

An alternative test for $a > 2$ groups is **Bartlett's test for homogeneity of variances**. Although it has fallen out of favor in recent years, we include it here for the sake of completeness. The computation is explained near the end of Box 13.2, where the method is applied to the scutum widths. The final test statistic is X^2, which is distributed as chi squared for $a - 1$ degrees of freedom. The samples again give no evidence for rejection of the null hypothesis, and the variances appear to be homogeneous. Regrettably, Bartlett's test is unduly sensitive to departures from normality in the data (see Section 13.4), and a large X^2 may therefore indicate nonnormality rather than heteroscedasticity. For this reason, many statisticians no longer recommend Bartlett's test when the normality of the distribution is in doubt, although this is less of a problem when sample sizes are equal. An advantage of Bartlett's test (shared by the F_{max}-test) is that, unlike the Levene test, it is carried out using only the variances of the a groups, not the individual observations. It is possible, therefore, to apply this test to data tables found in the literature that feature only sample sizes and variances of some variable for a groups, a frequent occurrence. The BIOMstat package of computer programs carries out the tests of homogeneity of variances discussed here.

What could be the reasons for heteroscedasticity had it been found? Some of the samples may inherently vary more than others, with some populations relatively uniform for one character and others quite variable for the same character. In an anova representing the results of an experiment, one sample may well have been obtained under less standardized conditions than the others and hence has a greater variance. There are also many cases in which the heterogeneity of variances is a function of an improper choice of measurement scale. With some measurement scales, variances vary as functions of means. Thus, differences among means bring about heterogeneous variances. For example, in variables following the Poisson distribution, the variance is equal to the mean, so populations with greater means consequently have greater variances. Such departures from the assumption of homoscedasticity are often easily corrected by a suitable transformation, as discussed later in this chapter.

A very rapid first inspection for heteroscedasticity checks for correlation between the means and variances. An indication of such correlation is that the ratios $s^2/\overline{Y}$ or $s/\overline{Y} = V$ will be approximately constant for the samples. If means and variances are independent, these ratios will vary widely.

The consequences of moderate heterogeneity of variances are not too serious for the overall test of significance, but single degree-of-freedom comparisons may be far from accurate.

What can we do if our data are inherently heteroscedastic? We may carry out tests for the equality of means that are less sensitive to heterogeneity of variances. Using equal sample sizes reduces the problem also. Ramsey et al. (2010) found that most of the standard multiple-comparison methods had inflated type I error rates when the variances were heterogeneous and the distributions deviated from normality. They found the T-method and the Games and Howell (1976) method to be less sensitive to these problems—at least for the equal sample size case that they investigated. However, because the type I error rates tended to be too large, they recommended that an overall anova F-test be applied first. If the null hypothesis is accepted, then no further tests are made. If it is rejected, then the Games and Howell method shown in Box 13.3 is applied. This method performs multiple comparisons between pairs of means using a studentized range with specially weighted average degrees of freedom and a standard error based on the averages of the variances of the means. We find that means for samples 1 and 8 differ from each other and from the other means (samples 2 to 7), which do not differ among themselves. The BIOMstat package of computer programs carries out this test.

When only two means are to be tested (see Section 9.4), we frequently test the hypothesis of equality of means by a t-test. Such a test assumes the equality of the two sample variances. When this assumption is not valid, we can perform **Welch's approximate t-test** (Welch, 1951), which is illustrated in Box 13.4. This test uses an approximate t-value, t'_s, for which the critical value is calculated as a weighted average of the critical values of t based on the corresponding degrees of freedom of the two samples. The computation is quite simple. Gans (1991) recommends using Welch's test in cases of two-sample tests, whether heteroscedastic or not, and Markowski and Markowski (in a comment published below Gans's letter) concur that this test might be advisable for normally distributed data but that other tests are better when the parent distributions differ from the normal.

BOX 13.3 Approximate Test of Equality of Means when the Variances Are Heterogeneous. Unplanned Comparisons Among Pairs of Means Using the Games–Howell Method

Length of the third molar of eight species of the condylarth *Hyopsodus*. Samples are listed by magnitude of $\bar{Y}$.

(1)	(2)	(3)	(4)	(5)	(6)
Samples $(a = 8)$	n_i	$\bar{Y}_i$	s_i^2	$s_{\bar{Y}_i}^2 = \dfrac{s_i^2}{n_i}$	$\dfrac{(s_{\bar{Y}_i}^2)^2}{(n_i - 1)}$
1	18	3.88	.0707	.00393	.000,000,907
2	13	4.61	.1447	.01113	.000,010,325
4	16	4.73	.0836	.00522	.000,001,820
3	17	4.79	.0237	.00139	.000,000,121
5	8	4.92	.2189	.02736	.000,106,958
6	11	4.96	.1770	.01609	.000,025,892
7	10	5.20	.0791	.00791	.000,006,952
8	10	6.58	.2331	.02331	.000,060,373

SOURCE: Data from Olson and Miller (1958).

Computation

Before the Games and Howell (1976) procedure is carried out, we first perform a single-classification anova to test for the equality of all of the means. Using the preceding data, we obtain $MS_{among} = 7.08166$, $MS_{within} = 0.11245$, $F_s = 62.9739$, with 7 over 95 degrees of freedom. This corresponds to a probability of less than 10^{-32}, so we proceed to carry out the Games and Howell procedure as described below.

1. Make a table giving n, $\bar{Y}$, and s^2 for each sample as shown above in columns (2) through (4). It is most convenient to arrange the samples according to the magnitude of their means.

2. Compute the variance of each mean, $s_{\bar{Y}}^2$, and the quantity $(s_{\bar{Y}}^2)^2/(n - 1)$ for each sample as shown in columns (5) and (6).

3. The minimum significant difference, MSD_{ij}, between any pair (i, j) of means is given by the following expression (from Games and Howell, 1976).

$$MSD_{ij} = Q_{\alpha[k, \nu^*]} \left(s_{\bar{Y}_i}^2 + s_{\bar{Y}_j}^2 \right)^{1/2}$$

where

$$\nu^* = \frac{\left(s_{\bar{Y}_i}^2 + s_{\bar{Y}_j}^2 \right)^2}{\dfrac{(s_{\bar{Y}_i}^2)^2}{(n_i - 1)} + \dfrac{(s_{\bar{Y}_j}^2)^2}{(n_j - 1)}}$$

Box 13.3 (continued)

is the special weighted average degrees of freedom based on the quantities in columns (5) and (6) above. The critical value $Q_{\alpha[k,\nu^*]}$ is obtained from the table of the studentized range, Statistical Table **J**.

For the comparison $\overline{Y}_8 - \overline{Y}_1 = 6.58 - 3.88 = 2.70$ (the most deviant pair of means) the $MSD_{1,8}$ is computed as follows:

$$\nu^* = \frac{0.00393 + 0.02331)^2}{(0.000,000,907 + 0.000,060,373)}$$

$$= \frac{(0.027424)^2}{0.000,061,280}$$

$$= 12.11$$

In Table **J** we interpolate to obtain $Q_{.05[8,12.11]} = 5.111$. The MSD is then

$$MSD_{1,8} = 5.119(0.00393 + 0.02331)^{1/2}$$

$$= 0.8435$$

The means of samples 1 and 8 are clearly different with an experimentwise type I error rate of < 0.05.

4. One may systematically test the differences between all pairs of means, most conveniently by making a table of the difference between each pair of means and then comparing it with a table of MSD_{ij} values, as was done in Box 9.11.

					j			
	1	2	4	3	5	6	7	8
1	0	.5851	.4403	.3417	.9609	.7060	.5548	.8435
2	.73*	0	.6098	.5528	1.0043	.7867	.6750	.9088
4	.85*	.12	0	.3879	.9806	.7150	.5735	.8529
i 3	.91*	.18	.06	0	.9861	.6879	.5281	.8338
5	1.04*	.31	.19	.13	0	1.0402	.9907	1.1120
6	1.08*	.35	.23	.17	.04	0	.7628	.9576
7	1.32*	.59	.47	.41	.28	.24	0	.8867
8	2.70*	1.97*	1.85*	1.79*	1.66*	1.62*	1.38*	0

The differences $\overline{Y}_i - \overline{Y}_j$ are given below the diagonal, and the corresponding MSD_{ij} values for an experimentwise type I error rate of 0.05 are given above the diagonal. Differences greater than their corresponding MSD are indicated with an asterisk.

From this table we conclude that the means for samples 1 and 8 differ from each other and from all other means and that all pairs of means for samples 2 through 7 do not differ at the $\alpha = 0.05$ experimentwise type I error rate.

BOX 13.4 Welch's Approximate t-Test of Equality of the Means of Two Samples Whose Variances Are Assumed to Be Unequal

---

Comparison of Chemical Composition of the Urine of Apes

Milligrams of glutamic acid per milligram of creatinine.

	n	$\bar{Y}$	$s_{\bar{Y}}$
Chimpanzees	37	0.115	0.017
Gorillas	6	0.511	0.144

SOURCE: Data from Gartler et al. (1956).

First we reconstruct the variances of the two samples: $s^2 = s_{\bar{Y}}^2 \times n$

$$\text{Chimpanzees} \quad s^2 = (0.017)^2 \times 37 = 0.010693$$

$$\text{Gorillas} \quad s^2 = (0.144)^2 \times 6 = 0.124416$$

By the method of Box 8.1, $F_s = 0.124416/0.010693 = 11.64$. Because this is a 2-tailed test, the usual 1-tailed probability, $P = 9.79 \times 10^{-7}$, must be doubled to yield 1.9579×10^{-6}. It is most improbable that the two samples came from populations with the same variances. We therefore cannot employ the t-tests of Sections 9.4 and 9.6 and use the following approximate test that allows for unequal variances:

$$t_s' = \frac{(\bar{Y}_1 - \bar{Y}_2) - (\mu_1 - \mu_2)}{\sqrt{\dfrac{s_1^2}{n_1} + \dfrac{s_2^2}{n_2}}},$$

where t_s' is approximately distributed as t with v^* degrees of freedom (defined below). Note that Ruxton (2006) warns against deciding to use Welch's test based on the results of a test for equality of variances, as we have done here. He, therefore, suggests that Welch's test always be used for a t-test of the difference between two means instead of the method we show in Box 9.4, unless one has prior information that the variances are very likely to be equal.

Usually, the formula for v^* will yield a fractional number of degrees of freedom. This is not a problem for many computer programs of Student's distribution, such as the one in BIOMstat. If no such program is available, a conservative step is to round v^* down to the nearest integer value.

We compute

$$t_s' = \frac{0.115 - 0.511}{\sqrt{0.017^2 + 0.144^2}} = -2.73$$

Box 13.4 (continued)

Next, we compute its degrees of freedom, v^*. Its formula, together with the numerical values for the present example is shown below. The ratio of variances is $u = 0.124416/0.010693 = 11.635275$ and

$$v^* = \frac{\left(\dfrac{1}{n_1} + \dfrac{u}{n_2}\right)^2}{\dfrac{1}{n_1^2(n_1 - 1)} + \dfrac{u^2}{n_2^2(n_2 - 1)}} = \frac{\left(\dfrac{1}{37} + \dfrac{11.635275}{6}\right)^2}{\dfrac{1}{37^2(37 - 1)} + \dfrac{11.635275^2}{6^2(6 - 1)}} = 5.1402.$$

Using the t-distribution with v^* degrees of freedom, t_s' corresponds to a probability of $P = 0.04005$. We decide to reject the null hypothesis that the two means are identical and conclude that the amount of glutamic acid in the urine is not the same in gorillas and chimpanzees (it seems to be larger in gorillas).

A related test based on fiducial probability is the **Behrens–Fisher test**, which can be found in the explanatory comments for Table VI in Fisher and Yates (1963). Unplanned multiple-comparison tests based on unequal variances are presented by Hochberg (1976), Games and Howell (1976), and Fligner and Policello (1981).

You may wish to test a set of coefficients of variation for homogeneity. Such a test has been proposed by Feltz and Miller (1996) and is illustrated in Example 10.14 in Zar (2010).

13.4 Normality

We have assumed that the error terms, ϵ_{ij}, of the observations in each sample will be independent, that the variances of the error terms of the several samples will be equal, and, finally, that the error terms are distributed normally. One method to test this last assumption is to compute normal expected frequencies from the observed data and test the departure of the observed from the expected frequencies. In Section 17.2, you will be introduced to two tests, the G-test and the Kolmogorov–Smirnov test for goodness of fit, either of which could routinely be applied to each sample in an anova. The BIOMstat package of computer programs provides such tests. Alternatively, a graphic test (as illustrated in Section 6.7) might be applied to each sample separately. Testing for $\gamma_1 = 0$ and $\gamma_2 = 0$ (see Box 7.2) is a partial test for normality. When the size of each sample is quite small, one can test normality for the entire analysis by computing the standardized deviates $(Y_{ij} - \overline{Y}_i)/s_i$ separately for each sample i, pooling the resulting deviates across samples, and testing them for normality by any of the techniques mentioned here.

A more powerful test for normality, which has become widely adopted, was developed by Shapiro and Wilk (1965). The method depends on some theoretical and computational topics that have not yet been covered. Some of them (e.g., regression coefficients, covariance matrices) are taken up in later chapters; others (e.g., order statistics) are beyond the purview of this introductory text. For each sample in a proposed anova, the method computes a coefficient W, defined as

$$W = \frac{\left(\sum\limits_{i}^{n} a_i Y_{(i)}\right)^2}{\sum\limits_{i}^{n} (Y_i - \bar{Y})^2}$$

where Y_i is the ith observation in the sample and $Y_{(i)}$ is the ith order statistic of the sample—that is, the expected value of the ith observation when the sample is ordered by ascending magnitude. The a_i's are special coefficients that either have to be looked up in a table such as Statistical Table **LL** or computed by a somewhat complicated algorithm.

This test can be carried out using BIOMstat or other statistical software. The coefficient can be interpreted as approximately the squared correlation (see Section 15.2) between the ordered observations and the expected quantiles for a normal distribution. It is a measure of the straightness of the normal quantile plot (see Section 6.7). The sum of squares in the denominator is the observed sum of squares for the sample. This yields observed values of W less than 1. The critical values for W are estimated by simulations. Statistical Table **MM** provides critical values up to $n = 100$ that are more accurate than those in the original publication. The null hypothesis of normality is rejected if the value of W is appreciably below 1. For larger sample sizes, a computer program such as BIOMstat is essential, because tables for the a_i-coefficients are available only up to $n = 50$ and our table of critical values are limited to $n \leq 100$. We furnish a computational outline for the Shapiro–Wilk test in Box 13.5.

When examining a sample for normality, an investigator may notice one or more extreme observations, or **outliers**. These may be the result of errors of measurement, recording or transcription errors, admixture of one or more individuals from a population different from the one under study, and so forth. If the investigator can assume that the population is normally distributed, then tests are available for detecting outliers. For sample sizes up to $n = 25$, we can use a test devised by Dixon (1950). We show how to apply this test using the data from Exercise 9.5. We discover that 3.8 is the smallest of the $n = 20$ observations in sample 1, considerably less than most other observations in the sample. Might this be an outlier?

To carry out Dixon's test, order the n observations from low to high or from high to low so that the first observation, Y_1, is the suspected outlier. Then consult Statistical Table **CC**. Here we find that for samples of size 20, the critical ratio should be computed as

$$r_{22} = \frac{Y_3 - Y_1}{Y_{n-2} - Y_1}$$

where Y_1 is the suspected outlier and the other subscripted values of Y are the respective observations in an ordered array, with the first item being the suspected outlier. For sample 1, we obtain

$$r_{22} = \frac{4.5 - 3.8}{5.8 - 3.8} = 0.350$$

BOX 13.5 The Shapiro-Wilk Test for Normality

Sample A of Nymphs of the Chigger Trombicula lipovskyi *from Box 13.8.*

The variable measured is length of cheliceral base stated as micrometer units.

(1) Y_i	(2) i	(3) $Y_{(i)}$	(4) a_i
118	1	104	−0.5056
128	2	109	−0.3295
128	3	112	−0.2514
128	4	114	−0.1939
118	5	116	−0.1448
112	6	118	−0.1007
109	7	118	−0.0594
123	8	119	−0.0196
119	9	121	0.0196
126	10	123	0.0594
121	11	125	0.1007
116	12	126	0.1448
125	13	126	0.1939
104	14	128	0.2514
126	15	128	0.3295
114	16	128	0.5056

Sample A has a sample size (n) of 16. The observations are listed in column (1). The rank orders from 1 to 16 are given in column (2). The flow of computations is as follows:

1. Reorder the observations in column (1) from smallest to largest and enter them in column (3).

2. Calculate the sum of squares of the observations $Y_{(i)}$ for the denominator of the W coefficient. The SS is the same whether or not the observations are ordered.

$$SS = 799.4375$$

3. Construct a column of n a_i-coefficients as shown in column (4) by copying the k coefficients from the tabled values in Statistical Table **LL** but preceding each value with a negative sign. Samples whose sample size n is an odd number will have a zero as the $(k + 1)$th coefficient. The a_i-coefficients $k + 2, \ldots, 2k = n$ are the same coefficients again, but they are *positive* and in *reverse order*—that is, they increase in magnitude. The beginning and end of the column of coefficients should thus have the identical largest coefficients but of *unlike sign*. Then compute the sum of the n products $a_i Y_{(i)}$ for the numerator of the W coefficient. We obtained 27.2320.

4. The Shapiro–Wilk statistic W is computed as

$$W = \frac{\left(\sum_{}^{n} a_i Y_{(i)}\right)^2}{SS} = \frac{27.2320^2}{799.4375} = 0.9276.$$

Box 13.5 (continued)

We enter Statistical Table **MM** at $n = 16$ and find $P > 0.05$. We cannot reject the null hypothesis that the measurements of sample A are normally distributed.

If we carry out a Shapiro–Wilk test on sample B from Box 13.8, we obtain $W = 0.9322$. In Statistical Table **MM** for $n = 10$, we find $P > 0.05$ and arrive at the identical conclusion as for sample A. Thus, we would conclude that we cannot reject the hypothesis of normality for either of these samples.

For $\alpha = 0.10$, Statistical Table **CC** shows that the critical value of this ratio is 0.401. Hence, we conclude that the suspected outlier is not sufficiently deviant to justify its deletion from the sample. Statistical Table **CC** is one-tailed. For a two-tailed test, double the values of α.

For sample sizes beyond the scope of Statistical Table **CC** ($n > 25$), we can use the test $(Y_1 - \bar{Y})/s$, where Y_1 is the suspected outlier, $\bar{Y}$ is the sample mean, and s is the sample standard deviation. Critical values for this ratio have been computed by Grubbs (1969) and can be looked up in Statistical Table **DD**. As an example, we may examine the 37 urine samples of chimpanzees from Exercise 4.2. Is either the smallest or the largest extreme observation an outlier? We test $(0.008 - 0.115)/0.10404 = -1.028$ and $(0.440 - 0.115)/0.10404 = 3.124$. The second of these deviations is larger than expected using a two-tailed probability of 0.05.

When outliers have been detected, they can be removed if external consideration of the sampling procedure and of the experimental design warrant it; an alternative procedure would be to **Winsorize** the data. In Winsorization, the outliers in an ordered array are replaced by their neighboring values. Thus, in the chimpanzee example, the abnormally high reading of 0.440 from Exercise 4.2 would be replaced by the next highest adjacent value, 0.370. The mean is then computed from the Winsorized sample. It has been shown that for samples from normal distributions, the relative efficiency of estimating means by this technique is more than 90% whenever sample sizes are greater than 5. However, as Dorfman (1996) points out, excluding observations from the sampling scheme without a valid reason for suspecting that they are true outliers may yield false conclusions. Such procedures may lead to pseudorandom sampling designs and loss of power and size (see Chapter 12) of the hypothesis tests based on such data.

The consequences of nonnormality of error are not too serious because means will follow the normal distribution more closely than the distribution of the observations themselves (a consequence of the central limit theorem; see Section 7.2). Only highly skewed distributions would have a marked effect on the type I error rate of the F-test or on the efficiency of the design. The best way to correct for lack of normality is to carry out a transformation that will make the data normally distributed, as explained in later sections of this chapter. If no simple transformation is satisfactory, a nonparametric test, as carried out in Sections 13.10 and 13.11, should be substituted for the analysis of variance.

13.5 Transformations

If the evidence indicates that the assumptions for an analysis of variance or a t-test cannot be maintained, two courses of action are open to us. We may carry out a different test not requiring the rejected assumptions, such as the distribution-free tests in lieu of anova discussed at the end of this chapter; use resampling methods; or we may transform the variable to be analyzed in such a manner that the resulting transformed observations meet the assumptions of the analysis. Let us look at a simple example of what transformation will do. A single observation of the simplest type of anova (completely randomized, single classification, Model I) decomposes as follows: $Y_{ij} = \mu + \alpha_i + \epsilon_{ij}$. In this model, the components are additive with the error term ϵ_{ij} distributed normally. However, we might encounter a situation in which the components were multiplicative in effect, where $Y_{ij} = \mu\alpha_i\epsilon_{ij}$, the product of these three terms. If we fitted the standard anova model as given here to these data, the observed deviations from the group means would lack normality and homoscedasticity.

The general parametric mean μ is constant in any one anova, but the treatment effect α_i differs from group to group. Clearly, the scatter among the observations Y_{ij} would double in a group in which α_i was twice as great as in another. Assume that $\mu = 1$, the smallest $\epsilon_{ij} = 1$, and the greatest 3; then, if $\alpha_i = 1$, the range of the Y's will be $3 - 1 = 2$. When $\alpha_i = 4$, however, the corresponding range will be four times as wide, from $4 \times 1 = 4$ to $4 \times 3 = 12$, a range of 8. Such data will be heteroscedastic. To correct this situation, we simply change our model into an additive one by transforming the data using logarithms. We would therefore obtain $\log Y_{ij} = \log \mu + \log \alpha_i + \log \epsilon_{ij}$, which is additive and homoscedastic. The entire analysis of variance would then be carried out on the transformed observations.

At this point, many of you might feel more or less uncomfortable about what we have done. Transformation seems too much like "data grinding." When you learn that often the results of a statistical test may be declared "significant" after transformation of a set of data, though it would not have been without such a transformation, you may feel even more suspicious. What is the justification for transforming the data? It takes some getting used to the idea, but there is really no scientific necessity to employ the common linear or arithmetic scale to which we are accustomed. If a relation is multiplicative on a linear scale, it may make much more sense to think of it as an additive system on a logarithmic scale.

The square root of a variable is another frequent transformation. In some cases, the square root of the surface area of an organism may be a more appropriate measure of the fundamental biological variable subjected to physiological and evolutionary forces than is the area. This concept is reflected in the normal distribution of the square root of the variable as compared to the skewed distribution of areas. In many cases, experience has taught us to express experimental variables not in linear scale but as logarithms, square roots, reciprocals, or angles. Thus, pH values are logarithms, and dilution series in microbiological titrations are expressed as reciprocals. As soon as you are ready to accept the idea that the scale of measurement is arbitrary, you simply have to look at the distributions of transformed observations to decide which transformation most closely satisfies the assumptions of the analysis of variance before carrying out an anova.

A fortunate fact about transformations is that frequently several departures from the assumptions of anova are cured simultaneously by the same transformation to a new scale. Simply by making the data homoscedastic, we often also make them approach normality and ensure additivity of the treatment effects.

When a transformation is applied, tests of significance are performed on the transformed data, but estimates of means are usually rendered in the familiar untransformed scale. Because the transformations discussed in this chapter are nonlinear, confidence limits computed in the transformed scale and changed back to the original scale would be asymmetrical. Stating the standard error in the original scale would therefore be misleading. In reporting results of research with variables that require transformation, you must furnish means in the untransformed scale followed by their (asymmetrical) confidence limits rather than by their standard errors.

An easy way to find out whether a given transformation will yield a distribution satisfying the assumptions of the anova is to plot the cumulative distributions of the several samples against a probability scale. By changing the scale of the axis used for the variable from linear to logarithmic, square root, or any other scale, we can see whether a previously curved line, indicating skewness, straightens out to indicate normality (you may wish to refresh your memory on these graphic techniques, which we studied in Section 6.7).

We can look up class limits on transformed scales. In earlier days, statisticians used a variety of available probability graph papers with variable axes in logarithmic, angular, or other scales. Nowadays, many computer programs, BIOMstat included, will plot the data as probability plots for a wide variety of transformations. Thus, we not only test whether the data become closer to a normal distribution through transformation but also can get an estimate of the standard deviation under transformation as measured by the slope of the estimated line. The assumption of homoscedasticity implies that the slopes for the several samples should be the same. If the slopes are very heterogeneous, homoscedasticity has not been achieved. Alternatively, we can examine goodness-of-fit tests for normality (see Chapter 17) for the samples under various transformations. The transformation that yields the best fit over all samples will be chosen for the anova. It is important that the transformation not be selected because it gives the best anova results: Such a procedure would distort the type I error rate.

Four transformations will be discussed here: the logarithmic transformation (Section 13.6), the square root transformation (Section 13.7), the Box–Cox transformation (Section 13.8), and the angular or arcsine transformation (Section 13.9). Two other transformations will be taken up in connection with regression analysis in Section 14.10.

13.6 The Logarithmic Transformation

The most common transformation is the conversion of all observations into logarithms, usually common logarithms. Whenever the mean is positively correlated with the variance (greater means are accompanied by greater variances), the logarithmic transformation is likely to remedy the situation and make the variance independent of the mean. Frequency distributions skewed to the right are often made more symmetrical by transformation to logarithmic scale. We shall learn later that logarithmic transformation is also called for when effects in anova are multiplicative. It is also commonly needed for the analysis of morphological measurements.

Table 13.1 illustrates the effect of a logarithmic transformation. These data are lengths of juvenile silver salmon, checked in a downstream trap during four different two-week periods. Note that the samples with larger means also have larger variances. This relation could have been foreseen from the increase in range of each sample with increase in mean if samples were compared based on approximately the same number of fish, such as the first and third samples. As soon as the observations are transformed to logarithms, the relationship no longer holds. In most cases, this transformation also removes heteroscedasticity; in this particular example, the heterogeneity among the variances is reduced but not eliminated—differences among the periods could be tested by the nonparametric techniques discussed in Section 13.10. To report these data, you should transform the means back into linear scale using the inverse log function (also called the *antilogarithm*; for natural logs this is the exponential function). To indicate the reliability of your sample mean, confidence limits are computed in logarithmic scale and then transformed back to linear scale. Thus, for the first sample ($n = 421$), we would report the 95% confidence limits as follows:

$$L_1 = \text{antilog}(\overline{\log Y} - t_{.05[420]}\sqrt{s^2_{\log Y}/n}) = \text{antilog}\,[1.9068 - 1.966(0.001768)]$$

$$= 80.038$$

$$L_2 = \text{antilog}\,[1.9068 + 1.966(0.001768)] = 81.329$$

The back-transformed mean is also the geometric mean of the sample. $GM_Y = \text{antilog}(\overline{\log Y}) = \text{antilog}\,(1.9068) = 80.681$. It will always be less than the ordinary mean, here 80.962, unless the observations are all identical. This difference has been called the *logarithmic transformation bias* (Smith, 1993). There will be a similar discrepancy between the mean and the back-transformed mean for all of the nonlinear transformations discussed in this chapter. If the results are to be expressed on the original linear scale, then the ordinary mean and the back-transformed confidence limits should be reported.

Note that the preceding confidence limits are asymmetrical around the geometric mean: $L_2 - GM_Y = 81.329 - 80.681 = 0.648$, $GM_Y - L_1 = 80.681 - 80.038 = 0.643$. These limits differ from confidence limits computed from the original measurements, which would be the following symmetrical limits:

$$\overline{Y} = 80.962$$

$$L_1 = \overline{Y} - t_{.05[420]}\sqrt{s^2/n} = 80.962 - 1.966(0.3303) = 80.313$$

$$L_2 = 80.962 + 1.966(0.3303) = 81.611$$

TABLE 13.1 An Application of the Logarithmic Transformation

Length of the juvenile silver salmon *Oncorhynchus kisutch* checked in a downstream trap during four different two-week periods (pooled over period 1933 to 1942.)

(1) Length in mm Y	(2) log of length log Y	(3) Nov. 12–25 f	(4) Jan. 7–20 f	(5) Apr. 1–14 f	(6) Apr. 29– May 12 f
60	1.778	1	2	—	—
65	1.813	4	—	—	—
70	1.845	33	1	—	—
75	1.875	94	4	—	1
80	1.903	125	2	1	—
85	1.929	98	12	5	5
90	1.954	44	5	6	20
95	1.978	18	5	5	105
100	2.000	3	—	16	328
105	2.021	1	—	25	757
110	2.041	—	—	45	1371
115	2.061	—	—	57	1562
120	2.079	—	—	87	1372
125	2.097	—	—	65	781
130	2.114	—	—	20	388
135	2.130	—	—	5	159
140	2.146	—	—	5	51
145	2.161	—	—	5	18
150	2.176	—	—	1	6
155	2.190	—	—	—	4
160	2.204	—	—	1	1
165	2.217	—	—	—	1
n		421	31	349	6930

Untransformed variable				
$\overline{Y}$	80.962	83.710	116.977	115.551
s^2	45.176	83.2796	124.6719	82.5407
log transformation				
$\overline{\log Y}$	1.9068	1.9200	2.0661	2.0614
$s^2_{\log Y}$	0.001316	0.002603	0.001825	0.001166
Back-tranformed means				
antilog $(\overline{\log Y})$	80.681	83.181	116.427	115.194

SOURCE: Data from Shapovalov and Taft (1954).

When the observations to be transformed include zeros, a problem arises, because the logarithm of zero is negative infinity. In such cases, the use of the transformation $\log (Y + 1)$ avoids the problem. When values include numbers between 0 and 1, it may be desirable to code the observations by multiplying by 10,000, or some higher power of 10 to avoid negative characteristics in the logarithms. Although logarithms to any base are acceptable, common or natural logarithms may be simplest.

Logarithmic transformations are frequently needed in the analysis of variables related to the size or growth of organisms, such as the example in Table 13.1. The use of the logarithmic transformation may also simplify the results of a two-way or higher-order anova by eliminating some or all of the interactions. This is especially useful in an anova without replication in which it is necessary to assume that interaction is not present if one wishes to make tests of the main effects in a Model I anova. This assumption of no interaction in a two-way anova is sometimes also referred to as the assumption of **additivity** of the main effects. By this we mean that any single observed observation can be decomposed into additive components representing the treatment effects of a particular row and column, as well as a random term special to it. If interaction is present, the F-test will be very inefficient and possibly misleading if the effect of the interaction is very large. A check of this assumption requires either more than a single observation per cell (so that an error mean square can be computed) or an independent estimate of the error mean square from previous *comparable* experiments.

Interactions can result from a variety of causes. Most frequently, interaction results from a given treatment combination (such as level 2 of factor A when combined with level 3 of factor B) that makes an observation deviate from the expected sum of the treatment effects. Such a deviation is regarded as an inherent property of the natural system under study, as in examples of synergism or interference (see Section 11.2). Similar effects occur when a given replicate is quite aberrant, as may happen if an exceptional plot is included in an agricultural experiment, if a diseased individual is included in a physiological experiment, or if by mistake an individual from a different species is included in a biometric study. Finally, an interaction term will result if the effect of the two factors A and B on the response variable Y are multiplicative rather than additive. An example will make this clear.

In Table 13.2, we show the additive and multiplicative treatment effects in a hypothetical two-way anova. Let us assume that the expected population mean μ is zero. Then the mean of the sample subjected to treatment 1 of factor A and treatment 1 of factor B should be 2 by the conventional additive model. This is so because each factor at level 1 contributes unity to the mean. Similarly, the expected subgroup mean subjected to level 3 for factor A and level 2 for factor B is 8 because the respective contributions to the mean are 3 and 5. If the process is multiplicative rather than additive, however, as occurs in a variety of physicochemical and biological phenomena, the expected values are quite different. For treatment A_1B_1, the expected value

TABLE 13.2 Additive and Multiplicative Effects

Factor B	Factor A $\alpha_1 = 1$	$\alpha_2 = 2$	$\alpha_3 = 3$	
$\beta_1 = 1$	2	3	4	Additive effects
	1	2	3	Multiplicative effects
	0	0.30	0.48	Log of multiplicative effects
$\beta_2 = 5$	6	7	8	Additive effects
	5	10	15	Multiplicative effects
	0.70	1.00	1.18	Log of multiplicative effects

equals 1, which is the product of 1 and 1. For treatment A_3B_2, the expected value is 15, the product of 3 and 5.

If we were to analyze multiplicative data of this sort by a conventional anova, we would find that the interaction sum of squares would be greatly augmented because of the nonadditivity of the treatment effects. In this case, there is a simple remedy. By transforming the variable into logarithms (see Table 13.2), we are able to restore the additivity of the data. The third item in each cell gives the logarithm of the expected value, assuming multiplicative relations. Notice that the increments are strictly additive again ($SS_{A\times B} = 0$). As a matter of fact, on a logarithmic scale we could simply write $\alpha_1 = 0$, $\alpha_2 = 0.30$, $\alpha_3 = 0.48$, $\beta_1 = 0$, $\beta_2 = 0.70$. This is a good illustration of how transformation of scale helps us meet the assumptions of analysis of variance.

If we wish to ascertain whether the interaction found in a given set of data can be explained in terms of multiplicative main effects, we can perform a test devised by Tukey (1949), which is illustrated in Box 13.6. This test is also useful when testing for nonadditivity in a two-way Model I anova without replication in experiments where it is reasonable to assume that interaction, if present at all, could only result from multiplicative main effects. Tukey's test partitions the interaction sum of squares into one degree of freedom due to multiplicative effects of the main effects and a residual sum of squares to represent the other possible interactions or to serve as error in case the anova has no replication. The computations are straightforward. We find that in this example (oxygen consumption of two species of limpets), for which we did not reject the null hypothesis of no interaction, the single-degree-of-freedom SS for nonadditivity because of multiplicative effects is very small (the F-ratio is even less than 1). If the SS for nonadditivity had been large and the residual SS nonsignificant, then an analysis of log oxygen consumption would be preferred because it would yield a simpler analysis. The BIOMstat program contains Tukey's test for nonadditivity.

BOX 13.6 Tukey's Test for Nonadditivity

Oxygen Consumption of Two Species of Limpets (Acmaea) *at Three Salinities*

Data are from Box 11.2. Because only the subgroup cell means are needed for computation, the individual observations are not given here. Each subgroup cell mean is based on $n = 8$ observations.

Factor B: Seawater concentration ($b = 3$)	(1) Factor A: Species ($a = 2$) A. scabra	(2) A. digitalis	(3) $\overline{Y}_B$	(4) $n\overline{Y}_B - n\overline{\overline{Y}}$
100%	10.5613	7.4288	8.9950	−4.9967
75%	7.8900	7.3375	7.6138	−16.0467
50%	12.1738	12.3263	12.2500	21.0433
$\overline{Y}_A$	10.2083	9.0308	$9.6196 = \overline{\overline{Y}}$	
$n\overline{Y}_A - n\overline{\overline{Y}}$	4.7100	−4.7100		

Computation

1. Form a two-way table of subgroup (cell) means. Compute row means divided by a, column means divided by b, and the grand mean divided by ab. These calculations yield values of $\overline{Y}_B$, $\overline{Y}_A$, and $\overline{\overline{Y}}$, respectively.

2. Evaluate the following expressions:

$$Q = n^3 \sum_{}^{a} \sum_{}^{b} \overline{Y}_{ij}(\overline{Y}_{Ai} - \overline{\overline{Y}})(\overline{Y}_{Bj} - \overline{\overline{Y}}) = -1044.7515$$

$$K = n^4 \sum_{}^{a} (\overline{Y}_{Ai} - \overline{\overline{Y}})^2 \sum_{}^{b} (\overline{Y}_{Bj} - \overline{\overline{Y}})^2 = 32{,}179.5485$$

In these expressions $\overline{Y}_{ij}$ stands for the subgroup mean for the ith column and the jth row, and $\overline{Y}_{Ai}$ and $\overline{Y}_{Bj}$ represent the $\overline{Y}_A$ mean for column i and the $\overline{Y}_B$ mean for column j, respectively.

3. *SS* for nonadditivity $SS_{\text{nonadd}} = \dfrac{Q^2}{Kn} = \dfrac{(-1044.7515)^2}{(32{,}179.5485 \times 8)} = 4.2399$

This *SS* with one degree of freedom is a part of the interaction *SS* (calculated in Box 11.2)

Source of variation	df	SS	MS
$A \times B$	2	23.9262	
Nonadditivity	1	4.2399	4.2399
Residual	1	19.6863	19.6863

Box 13.6 (continued)

$F_s = MS_{\text{nonadd}}/MS_{\text{resid}} = 4.2399/19.6863 = 0.2154$, $P = 0.7234$. It is clearly not very reasonable to reject the null hypothesis. We have little evidence of nonadditivity (multiplicative effects of the factors A and B) in these data.

These quantities may appear to have been pulled out of a hat. Actually, the SS for nonadditivity is the explained sum of squares for the regression of the subgroup means on the deviations of their marginal means. The meaning of this sentence may not be clear until you have mastered the regression chapter through Section 14.3.

13.7 The Square Root Transformation

When the data are counts—insects on a leaf or blood cells in a hemacytometer, for example—we frequently find the square root transformation of value. Remember that such distributions are likely to be Poisson rather than normally distributed and that the variance is the same as the mean in a Poisson distribution. Therefore, the mean and variance cannot be independent but will vary identically. Transforming the observations to square roots generally makes the variances independent of the means. When the counts include many zero values, it is often desirable to code all observations by adding 0.5. The transformation then is $\sqrt{Y + \frac{1}{2}}$. Improved transformations have been suggested by Anscombe (1948; $Y' = \sqrt{Y + \frac{3}{8}}$) and by Freeman and Tukey (1950; $Y' = \sqrt{Y} + \sqrt{Y + 1}$), but they are less widely used.

Table 13.3 shows an application of the square root transformation. Again, the sample with the greater mean has a greater variance prior to transformation. After transformation, the variances no longer appear different. To report means, the transformed means are squared again and confidence limits are reported in lieu of standard errors.

If there are only two counts that need to be compared, one cannot test means of transformed variables. In such a case, we take advantage of the fact that the difference between two independent observations sampled from the same distribution has a variance twice that of the observations (see Section 15.4). We set

$$t_{\alpha[\infty]} = \frac{Y_1 - Y_2}{\sqrt{Y_1 + Y_2}} \tag{13.1}$$

In this equation, Y_1 and Y_2 are the two counts; they are also the best estimates of the parametric means μ_1 and μ_2 of counts from which the individual counts were sampled. Recall that for Poisson distributions, $\mu = \sigma^2$; thus, Expression (13.1) becomes

$$t = \frac{Y_1 - Y_2}{\sqrt{2s^2}}$$

TABLE 13.3 An Application of the Square Root Transformation

The data represent the number of adult *Drosophila* emerging from 15 single-pair cultures for two different medium formulations (medium A contained DDT).

(1) Number of flies emerging Y	(2) Square root of number of flies $\sqrt{Y}$	(3) Medium A f	(4) Medium B f
0	0.000	1	—
1	1.000	5	—
2	1.414	6	—
3	1.732	—	—
4	2.000	3	—
5	2.236	—	—
6	2.449	—	—
7	2.646	—	2
8	2.828	—	1
9	3.000	—	2
10	3.162	—	3
11	3.317	—	1
12	3.464	—	1
13	3.606	—	1
14	3.742	—	1
15	3.873	—	1
16	4.000	—	2
		15	15

Untransformed variable

$\overline{Y}$		1.933	11.133
s^2		1.495	9.410

Square root transformation

$\overline{\sqrt{Y}}$		1.299	3.307
$s^2_{\sqrt{Y}}$		0.2634	0.2099

Tests of equality of variances

$$F_s = \frac{s^2_2}{s^2_1} = \frac{9.410}{1.495} = 6.294, P = 0.000729$$

$$F_s = \frac{s^2_{\sqrt{Y_1}}}{s^2_{\sqrt{Y_2}}} = \frac{0.2634}{0.2099} = 1.255, P = 0.3383$$

Table 13.3 (continued)

Back-transformed (squared) means		
$(\overline{\sqrt{Y}})^2$	1.687	10.937
95% confidence limits		
$L_1 = \overline{\sqrt{Y}} - t_{.05}s_{\overline{\sqrt{Y}}}$	$1.299 - 2.145\sqrt{\dfrac{0.2634}{15}}$	$3.307 - 2.145\sqrt{\dfrac{0.2099}{15}}$
	$= 1.015$	$= 3.053$
$L_2 = \overline{\sqrt{Y}} + t_{.05}s_{\overline{\sqrt{Y}}}$	1.583	3.561
Back-transformed (squared) confidence limits		
L_1^2	1.030	9.324
L_2^2	2.507	12.681

SOURCE: Data from R. R. Sokal (unpublished results).

Let us illustrate this with an example. During a given week, two counties with approximately equal populations reported 17 and 32 new influenza cases. Do these figures indicate different morbidities in these two counties? Applying Expression (13.1), we obtain

$$t = \frac{17 - 32}{\sqrt{17 + 32}} = -15/7 = -2.143$$

The probability of obtaining such an extreme value is 0.0321 (two-tailed test).

13.8 The Box–Cox Transformation

Often one has no a priori reason for selecting a specific transformation. Rather than simply trying various transformations to find out which one works best, Box and Cox (1964) developed a procedure for estimating the best transformation to normality within the family of power transformations:

$$Y' = (Y^\lambda - 1)/\lambda \text{ (for } \lambda \neq 0) \tag{13.2}$$

$$Y' = \ln Y \text{ (for } \lambda = 0) \tag{13.2a}$$

The generality of the Box–Cox transformation is evident from the fact that setting $\lambda = 1$ gives a simple linear transformation (which does not change the shape of the distribution) and that $\lambda = \frac{1}{2}$ is equivalent to the square root transformation, $\lambda = 0$ corresponds to the logarithmic transformation, and $\lambda = -1$ is equivalent to the reciprocal transformation. The optimum value, $\hat{\lambda}$, is that which maximizes

$$L = -\frac{v}{2}\ln s_T^2 + (\lambda - 1)\frac{v}{n}\sum \ln Y \tag{13.3}$$

a **log-likelihood function**. This yields the λ-value that gives the best transformation to normality within this family of transformations. In this expression, s_T^2 is the variance of the *transformed* λ values (based on v degrees of freedom). For a single sample, s_T^2 would be the usual sample variance, but for a series of samples it could be the error mean square. The second term in the equation involves the sum of the natural logarithms of the original *untransformed* observations. The other terms are v and n, which are degrees of freedom and sample sizes, as usual. Because $\hat{\lambda}$ must be found by an iterative procedure, the transformation is feasible only if performed by computer (it is included as an option in the BIOMstat package of computer programs).

For the milk-yield data of Table 6.1, the Box–Cox transformation after several iterations yields an estimate of $\hat{\lambda} = -2.1089$. Figure 13.1 shows L as a function of λ; L is maximized at $\hat{\lambda} = -2.1089$. The Box–Cox algorithm also estimates confidence limits to $\hat{\lambda}$. Because $-2L_{\max}$ approximately follows the $\chi_{[1]}^2$ distribution, the lower and upper confidence limits (λ_1 and λ_2) correspond to the pair of values of λ that have L values equal to $L_{\max} - \frac{1}{2}\chi_{\alpha[1]}^2$. Solutions for λ_1 and λ_2 can be obtained by solving for the two roots of $L - (L_{\max} - \frac{1}{2}\chi_{\alpha[1]}^2)$ iteratively. For the milk-yield data, we obtain the limits $\lambda_1 = -3.420$ and $\lambda_2 = -0.834$ (see Figure 13.1). Because these limits exclude $\lambda = 1.0$ (linear transformation), we conclude that a transformation should be applied.

Note that the optimal transformation $\hat{\lambda} = -2.1089$ is approximately an inverse square transformation. There is no theoretical reason to suggest an inverse square law for milk yields—it is simply an empirical result. Its effectiveness is demonstrated by the finding that it reduces the $g_1 = 0.941$ for the untransformed milk yields to $g_1 = -0.108$ after the Box–Cox transformation. A test of the null hypothesis $g_1 = 0$ no longer results in a probability < 0.05. Figure 13.2 shows the data before and

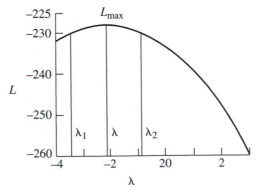

FIGURE 13.1 Plot of the log-likelihood function L versus the Box–Cox parameter λ for the milk-yield data of Table 6.1. The 95% confidence limits (λ_1, λ_2) are also shown.

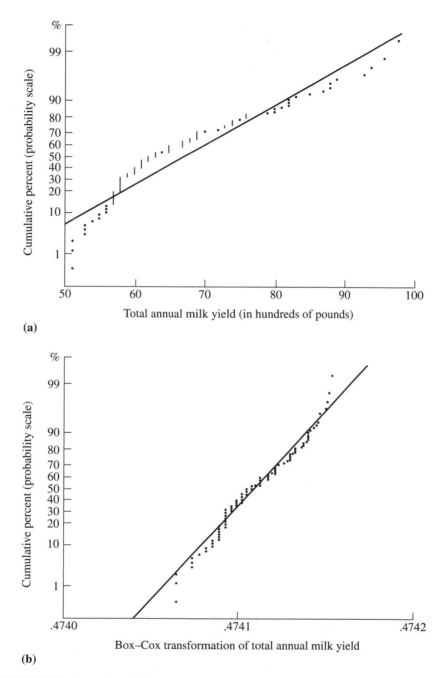

FIGURE 13.2 Normal probability plots for the milk-yield data of Table 6.1. (a) Untransformed data. (b) Data transformed using the Box-Cox transformation. Straight lines fitted by eye.

after transformation plotted on arithmetic probability paper in the manner of Section 6.7. It is quite clear that the transformation has made the sample approximate the normal distribution better than it did before. The Kolmogorov–Smirnov test for goodness of fit to a normal distribution (see Section 17.2)—which previously yielded a value of $D_{max} = 0.150$ ($P \leq 9.371 \times 10^{-6}$), implying that the milk yields are not normally distributed—after transformation yielded $D_{max} = 0.0939$ ($P = 0.0299$), a clear improvement.

One may wish to transform data not only to achieve normality but also to induce homogeneity of variances. The Box–Cox transformation can be extended simply by defining a new log-likelihood function:

$$L' = L - \tfrac{1}{2}X^2 \tag{13.4}$$

where L is as defined earlier and X^2 is the test statistic defined in Box 13.2 for Bartlett's test of homogeneity of variances (the preferred Shapiro–Wilk statistic is not used here because the Box–Cox transformation requires a likelihood statistic). The value of λ that maximizes L' corresponds to the transformation that attempts to normalize the distribution as well as to equalize the variances. An iterative procedure for finding the value of λ that maximizes L' is included in the BIOMstat package of computer programs. When we applied this procedure to the *Hyopsodus* data from Box 13.2, we found that L' is maximized at $\lambda = 0.1098$. The 95% confidence limits for λ and $\lambda_1 = -0.543$ and $\lambda_2 = 0.728$. Because λ is so close to 0 (and the confidence limits include the value of 0), we are tempted to use the logarithmic transformation, as is often done on such morphometric data. The transformed data yield an adjusted X^2 value of 16.731, $P = 0.0192$. The untransformed data yield an adjusted X^2 value of 20.920, $P = 0.0039$ (see Box 13.2). The F_{max} value has also been reduced and now has a probability > 0.05. The amount of heterogeneity has been reduced, although perhaps not sufficiently to warrant carrying out an anova.

If a program for the Box–Cox transformation is not available, the following rule of thumb may be helpful. Try the series of power transformations $1/\sqrt{Y}$, $\sqrt{Y}$, $\ln Y$, $1/Y$ for samples skewed to the right (g_1 positive) and the transformation Y^2, Y^3, ... for samples skewed to the left (g_1 negative).

Note: Box and Cox (1964) also give an alternative form of the Box–Cox transformation using the geometric mean. It is not given here because it leads to the same value for λ.

13.9 The Arcsine Transformation

This transformation, also known as the *angular transformation*, is especially appropriate to percentages and proportions. You may remember from Section 5.2 that the standard deviation of a binomial distribution is $\sigma = \sqrt{pq/k}$. Because $\mu = p$, $q = 1 - p$, and k is constant for any one problem, in a binomial distribution the variance is a function of the mean. The arcsine transformation reduces this dependence.

The arcsine transformation finds $\theta = \arcsin \sqrt{p}$, where p is a proportion. Thus, it is actually the arcsine of the square root of a proportion even though the name of the transformation does not mention the square root. The term *arcsin* is synonymous with inverse sine or $\sin^{-1}$, which stands for the angle whose sine is the given quantity. Thus, for the proportion 0.431, we find 41.03°, the angle whose sine is $\sqrt{0.431}$. The arcsine transformation stretches out both tails of a distribution of percentages or proportions and compresses the middle. An example of its application is shown in Table 13.4, where percent fertilities in a sample of 100 vials of 10 eggs each are shown for a strain of *Drosophila*. Figure 13.3 shows that the distribution is not normal (it is platykurtic; see Section 6.6). When the percentages are transformed to angles in column (2) of Table 13.4, the distribution closely approximates the normal. The expected variance of such a distribution of arcsines is $\sigma_\theta^2 = 180^2/4\pi^2 n = 820.7/n$, where θ represents the angles in which the arcsines are expressed. When θ is measured in radians rather than degrees, $\sigma_\theta^2 = 1/4n$. To report means, we need to convert them back to proportions or percentages. This conversion is accomplished by converting the angle θ to the proportion p by $p = [\sin (\theta)]^2$. When the percentages in the original data fall between 30 and 70%, it is generally not necessary to apply the arcsine transformation.

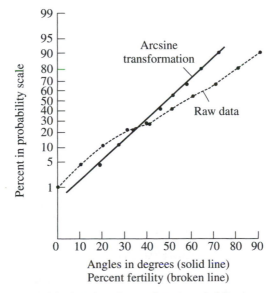

FIGURE 13.3 Data from Table 13.4 plotted on arithmetic probability paper to show the effect of the arcsine transformation. Angles in degrees (solid line).

TABLE 13.4 An Application of the Arcsine Transformation

Percent fertility of eggs of the CP strain of Drosophila melanogaster *raised in 100 vials of 10 eggs each.*

(1)	(2)	(3)	(4)[a]
			Cumulative
% fertility			**frequencies**
100p	**arcsin $\sqrt{p}$**	**f**	**F**
0	0	1	1
10	18.43	3	4
20	26.57	8	12
30	33.21	10	22
40	39.23	6	28
50	45.00	15	43
60	50.77	14	57
70	56.79	12	69
80	63.43	13	82
90	71.57	9	91
100	90.00	9	100
		100	

SOURCE: Data from Sokal (1966); see Figure 13.3.
[a]Needed for preparation of Figure 13.3.

Improved angular transformations for proportions have been suggested by Johnson and Kotz (1969) and Freeman and Tukey (1950). These transformations are, respectively,

$$\theta = \arcsin\sqrt{\frac{Y + 3/8}{n + 3/4}}$$

and

$$\theta' = 2\sqrt{n}\left[\arcsin\sqrt{\frac{Y + 3/8}{Y + 3/4}} - \arcsin\sqrt{p}\right]$$

where Y is the observed number possessing the property and n is the sample size.

13.10 Nonparametric Methods in Lieu of Single-Classification Anova

If we are unable to transform the data to a semblance of normality by any of the preceding methods, the best approach to testing for group differences is to use a randomization test by either complete enumeration or Monte Carlo sampling. We have already studied these methods in Section 7.1, and any reader who needs a refresher should refer back to that section. The example we wish to analyze is that of the pea-section data of Table 9.1. Note that these are the original, untransformed observations, whereas the values reported in Table 9.1 have already been (reciprocally) transformed. The assumption we make here is that the observations in the several groups of the anova are interchangeable. This also implies that the observations in each group are homoscedastic. Recall that these data are 50 observations arranged as 5 groups of 10, representing the 5 sugar treatments (including the controls). We cannot attempt complete enumeration because the total number of possible permutations is $>10^{31}$. We undertook 100,000 random partitions of the total dataset into 5 groups of 10 readings each. For the criterion, whose distribution over this large number of partitions we wish to examine, we specify the familiar sum of squares among groups,

$$SS_{among} = \sum_{i=1}^{a} n_i(\overline{Y}_i - \overline{\overline{Y}})^2,$$ described in Section 9.1 for the analysis of variance. Note

that we do not have to compute its ratio to an error MS, because the SS_{total} is a constant over all permutations so that the SS_{within} is then just a function of the SS_{among} for each random partition. None of the random permutations yielded an SS_{among} that was equal to or greater than the value we would obtain if we analyzed the untransformed observations of the actual experiment ($SS_{among} = 1077.32$). The null hypothesis that all five groups were sampled from the same population is, therefore, very improbable with a P-value of 10^{-5}.

Research workers may not wish to run a randomization test each time they have a dataset for which none of the transformations already described manage to make the data meet the assumptions of analysis of variance. In such a case, they may resort to an analogous **nonparametric method**. These techniques are also called *distribution-free methods* because they are not dependent on a given distribution (such as the normal in the case of anova) but usually will work for a wide range of different distributions. They are called *nonparametric methods* because their null hypothesis is not concerned with specific parameters (such as the mean in analysis of variance) but only with the distribution of the observations. Nonparametric methods are popular, because being based on ranks, they are simpler to compute than anovas and permit freedom from worry about many of the distributional assumptions of parametric methods. Yet in cases where the assumptions of analysis of variance hold entirely or even approximately, anova is generally the more powerful statistical procedure for detecting departures from the null hypothesis. Another disadvantage of these methods is that they cannot provide estimates of the magnitude of any differences because that information is lost when the observed data values are replaced by their ranks.

The easiest way to get a feel for these nonparametric tests is to become acquainted with one. The first tests to be taken up are all based on the idea of *ranking* the observations in an example after pooling all groups and considering them as a single sample for purposes of ranking. A convenient method for arraying the observations in rank order is by the stem-and-leaf displays introduced in Section 2.5. The current section considers analogues of single-classification analysis of variance. For the general case with a samples and n_i observations per sample, the **Kruskal–Wallis test** (Kruskal and Wallis, 1952) is widely recommended.

Box 13.7 illustrates the Kruskal–Wallis test applied to the rate of growth of pea sections, introduced in Table 9.1 and already analyzed by a permutational test earlier in this section. This is an example of a single classification with equal sample sizes, but the computational formulas shown in Box 13.7 are general and apply to varying sample sizes as well. First, we rank all observations from the smallest to the largest, ignoring the division into groups. The largest rank will be N, the total sample size of the entire study, $\overset{a}{\Sigma}n_i$. During this procedure, ties are a frequent problem. You will note several ties for almost every value on the measurement scale of the variable. For

---

BOX 13.7 **Kruskal-Wallis Test (a Test for Differences of Location in Ranked Data Grouped by Single Classification)**

Effect of Different Sugars on Growth of Pea Sections

These are the *original, untransformed values* for which we show the transformed observations in Table 9.1: $a = 5$ groups; n_i = number of items in group i (in this example, sample sizes are equal, $n_i = n = 10$); N = total sample size of the entire study, $\overset{a}{\Sigma}n_i = 50$. In this equal-sample-size case, the formula for N can be simplified to $a \times n$.

			Treatments ($a = 5$)		
Observations	Control	2% Glucose Added	1% Glucose + 2% Fructose Added	1% Fructose Added	2% Sucrose Added
1	75	57	58	58	62
2	67	58	61	59	66
3	70	60	56	58	65
4	75	59	58	61	63
5	65	62	57	57	64
6	71	60	56	56	62
7	67	60	61	58	65
8	67	57	60	57	65
9	76	59	57	57	62
10	68	61	58	59	67

Computation

1. Rank all observations from smallest to largest when pooled together into a single sample. In case of ties, compute the average ranks. For example, the 4 observation $Y = 59$ represent ranks 18, 19, 20, and 21. Their average rank therefore equals $(18 + 19 + 20 + 21)/4 = 19.5$.

Box 13.7 (continued)

Average Rank	Rank	Y	Average Rank	Rank	Y	Average Rank	Rank	Y
2	1	56	19.5	18	59	37.5	35	64
	2	56		19	59		36	65
	3	56		20	59		37	65
7	4	57		21	59		38	65
	5	57	23.5	22	60		39	65
	6	57		23	60		40	66
	7	57		24	60	42.5	41	67
	8	57		25	60		42	67
	9	57	27.5	26	61		43	67
	10	57		27	61		44	67
14	11	58		28	61		45	68
	12	58		29	61		46	70
	13	58	31.5	30	62		47	71
	14	58		31	62	48.5	48	75
	15	58		32	62		49	75
	16	58		33	62		50	76
	17	58		34	63			

2. Replace each observation in the original data table by its rank or average rank.

				Treatments					
Control		2% glucose added		2% fructose added		1% glucose + 1% fructose added		2% sucrose added	
Y	Rank	Y	Rank	Y	Rank	Y	Rank	Y	Rank
75	48.5	57	7	58	14	58	14	62	31.5
67	42.5	58	14	61	27.5	59	19.5	66	40
70	46	60	23.5	56	2	58	14	65	37.5
75	48.5	59	19.5	58	14	61	27.5	63	34
65	37.5	62	31.5	57	7	57	7	64	35
71	47	60	23.5	56	2	56	2	62	31.5
67	42.5	60	23.5	61	27.5	58	14	65	37.5
67	42.5	57	7	60	23.5	57	7	65	37.5
76	50	59	19.5	57	7	57	7	62	31.5
68	45	61	27.5	58	14	59	19.5	67	42.5
$\left(\sum^{n_j} R\right)_i$	450.0		196.5		138.5		131.5		358.5
$\overline{R}_i$	45.00		19.65		13.85		13.15		35.85

Box 13.7 (continued)

3. Sum the ranks separately for each group. Enter in row $(\Sigma^{n_i} R)_i$. For example,

$$\left(\sum^{n_i} R\right)_1 = 48.5 + 42.5 + \cdots + 50 + 45 = 450.0.$$

Then compute the means of ranks of each treatment group as $\left(\sum^{n_i} R\right)_i \div n_i$. For example, $450.0 \div 10 = 45.00$, $196.5 \div 10 = 19.65$, and so on. Finally, compute the grand mean of the ranks $\overline{\overline{R}} = \sum^a \left(\sum^{n_i} R\right)_i \div N$ or, more simply, by formula as $(N + 1)/2 = 25.5$.

4. Calculate the sum of squares among groups for the ranks, $SS_{among} = \sum_{i=1}^{a} n_i(\overline{R}_i - \overline{\overline{R}})^2$

$$= (50 - 1)[10(45.00 - 25.50)^2 + 10(19.65 - 25.50)^2 + \ldots + 10(35.85 - 25.50)^2]$$
$$= 8098.40$$

5. Compute Expression (13.5). The number 12 is a constant in this expression.

$$H = \frac{SS_{among}}{\sigma^2} = \frac{\sum_{i=1}^{a} n_i(\overline{R}_i - \overline{\overline{R}})^2}{N(N + 1)/12} = \frac{\text{quantity 4}}{50(51)/12} = \frac{8098.40}{212.50} = 38.110$$

6. Because there were ties, this H value must be corrected by dividing it by

$$D = 1 - \frac{\sum^m T_j}{(N - 1)N(N + 1)}$$

where T_j is a function of the t_j, the number of observations tied in the jth group of ties. (This t has no relation to Student's t.) The function is $T_j = t_j^3 - t_j$, computed most easily as $(t_j - 1)t_j(t_j + 1)$. Because in most cases the tied group will range from $t = 2$ to 10 ties, we furnish a small table of T over this range; the summation of T_j is over the m different ties.

t_j	2	3	4	5	6	7	8	9	10
T_j	6	24	60	120	210	336	504	720	990

For example, for the first tied group in the table of ranks, $t_j = 3$, because there are 3 observations of equal magnitude.

The t_j's for the present example are shown below, together with the corresponding T_j's:

t_j	3	7	7	4	4	4	4	4	4	2
T_j	24	336	336	60	60	60	60	60	60	6

Box 13.7 (continued)

$$\sum_{j}^{m} T_j = 24 + 336 + \cdots + 6 = 1062$$

$$D = 1 - \frac{1062}{(50 - 1)50(50 + 1)} = 0.99150$$

Adjusted $H = \dfrac{H}{D} = \dfrac{38.110}{0.99150} = 38.437, \quad P[\chi^2_{[4]} \geq H_{adj}] = 9.1046 \times 10^{-8}$

If the null hypothesis (that the a groups do not differ in "location") is true, H (or H_{adj}) will be distributed approximately as $\chi^2_{[a-1]}$. The minute P-value permits us to reject the null hypothesis with confidence and conclude that different sugars affect the rate of growth of pea sections differentially. Box 13.9 gives a method for nonparametric unplanned testing of differences among treatments.

these ties, we calculate the average of the ranks occupied by the tied values, as shown in Box 13.7. Next, we reconstitute the original data table but replace each original observation by its rank or average rank, as appropriate. At the bottom of each column representing a treatment, compute the mean of the n_i ranks for that treatment.

Research workers wishing to compute the Kruskal–Wallis statistic on their own may find themselves confused by several different looking formulas. Most but not all of these yield the same numerical value. Expression (13.5) is built on knowledge you have already acquired. We first calculate the sum of squares among groups for the ranks, which is $\sum_{i=1}^{a} n_i(\bar{R}_i - \bar{\bar{R}})^2$. Note that this is the same quantity we employed several paragraphs ago for the randomization approach to these data, except that ranks are now used. As shown in Box 13.7, we obtain 8098.40 for the SS_{among} of the ranks. How can we test whether the observed SS_{among} is larger than what one would expect if the differences among the mean ranks resulted just from chance? In Section 7.8, we showed that a sum of squares divided by a parametric variance is distributed as χ^2 with the same degrees of freedom as the sum of squares. Assuming the null hypothesis is true, the MS_{total} can be used as the variance. It is a parametric variance because the sample variance of the ranks from 1 to N is simply a function of N, $\sigma^2 = N(N + 1)/12$. The numeral 12 in these formulas is a constant. The Kruskal–Wallis statistic is

$$H = \frac{SS_{among}}{\sigma^2} = \frac{\sum_{i=1}^{a} n_i(\bar{R}_i - \bar{\bar{R}})^2}{N(N + 1)/12} \tag{13.5}$$

For the current example, $N(N + 1)/12 = 212.5$ and $SS_{among} = 8098.40$, yielding $H = 38.1101$ with $a - 1 = 4$ df. There is a probability of 1.06×10^{-7}—about 1 in

10 million—that we could obtain such a result by chance if the null hypothesis were true. The null hypothesis is decisively rejected.

An alternative formula, frequently published but less easily understood, yields exactly the same numerical value, 38.1101.

$$H = \left[\frac{12}{\sum^{a} n_i \left(\sum^{a} n_i + 1 \right)} \sum^{a} \frac{\left(\sum^{n_i} R \right)^2_i}{n_i} \right] - 3 \left(\sum^{a} n_i + 1 \right) \qquad (13.5a)$$

The statistic H, as shown by Expressions (13.5) or (13.5a), is appropriate for data without ties but must be divided by a correction factor D when ties are present. This correction factor is computed using a formula shown in Box 13.7. When this formula is applied to our example, the resultant adjusted H-value increases slightly to 38.437, making the null hypothesis even less likely at $P\ [\chi^2_{[4]} \geq H_{adj}] = 9.1046 \times 10^{-8}$.

For tests at $\alpha = 0.10$ or $\alpha = 0.05$, the χ^2 approximation is very good even with n as small as 5. For $\alpha = 0.01$, however, the test is conservative for small values of n; it rejects less than 1% of the tests if the hypothesis is true (personal communication from K. R. Gabriel, based on an unpublished Monte Carlo study). When $a = 3$ and the sample sizes n_i are each less than 5, the statistic H is not well approximated by the χ^2 distribution and Table O in Siegel and Castellan (1988), which gives exact critical values for such cases, should be consulted.

If the populations are not different from each other, we expect their rank sums to be approximately the same (allowing for differences in sample size where such exist). Note in the example of Box 13.7 that we decisively reject the null hypothesis, which is that the true "location" of the several populations is the same. By location in this context, we refer to the rank-ordered positions of the individual observations along the measurement- or Y-axis. The null hypothesis is that each of the a groups contains a random allocation of all available ranks. The alternative hypothesis is that the ranks differ among the groups. However, the Kruskal–Wallis test is *not* a test of the equality of medians of the a groups, as is sometimes stated in various publications. The conclusion reached by the Kruskal–Wallis test for these data is the same as that reached by the regular anova in Table 9.1. The BIOMstat program performs the Kruskal–Wallis test.

When the test is between only two samples (in parametric tests, such a design would give rise to a t-test or anova with two classes), we employ either of two nonparametric tests: the **Mann–Whitney U-test** or the **Wilcoxon two-sample test**. These tests start out from different premises but yield the same statistic and give the same results. The null hypothesis is that the two samples come from populations having the same "location." The meaning of this term is the same as for the Kruskal–Wallis test and was discussed in the preceding paragraph. Each of these tests is applied in Box 13.8 to data that are morphological measurements on two samples of chigger nymphs. The Mann–Whitney U-test, as illustrated in Box 13.8,

---

BOX 13.8 Mann–Whitney U Test and Wilcoxon Two-Sample Test for Two Samples (Ranked Observations, Not Paired)

Two Samples of Nymphs of the Chigger Trombicula lipovskyi

The variable measured is the length of the cheliceral base stated as micrometer units.

(1)	(2)	(3)	(4)
Sample A		Sample B	
Y	Rank (R)	Y	Rank (R)
104	2	100	1
109	7	105	3
112	9	107	4.5
114	10	107	4.5
116	11.5	108	6
118	13.5	111	8
118	13.5	116	11.5
119	15	120	16
121	17.5	121	17.5
123	19.5	123	19.5
125	21		$91.5 = \overset{n_2}{\sum} R$
126	22.5		
126	22.5		
128	25		
128	25		
128	25		
	$259.5 = \overset{n_1}{\sum} R$		

SOURCE: Data from D. A. Crossley (unpublished results).

Designate sample size of the larger sample as n_1 and that of the smaller sample as n_2. In this case, $n_1 = 16$, $n_2 = 10$. If the two samples are of equal size, it does not matter which is designated as sample 1.

There are two equivalent procedures for carrying out a test of equality of "location" of two samples.

The Mann–Whitney U-test

1. List the observations from the smallest to the largest in such a way that the two samples may be compared easily. A convenient method is to make a graph as shown here.

Box 13.8 (continued)

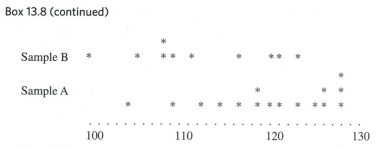

2. For each observation in one sample (it is convenient to use the smaller sample, but it does not matter which is selected), count the number of observations in the other sample that are lower in value (to the left). Count one-half for each tied observation. For example, there are zero observations in sample A less than the first observation in sample B; one observation less than the second, third, fourth, and fifth observations in sample B; two observations in A less than the sixth in B; 4 observations in A less than the seventh in B, but one is equal (tied) with it, so we count $4\frac{1}{2}$. Continuing in a similar manner, we obtain counts of 9, $8\frac{1}{2}$, and $9\frac{1}{2}$. The sum of these counts $C = 36\frac{1}{2}$. The Mann–Whitney statistic U_s is the greater of the two quantities C and $n_1 n_2 - C$, in this case $36\frac{1}{2}$ and $(16 \times 10) - 36\frac{1}{2} = 123\frac{1}{2}$.

The Wilcoxon two-sample test

1. Rank all of the observations together from low to high, as shown in columns (2) and (4) above. Give average ranks in the case of ties.

2. Sum the ranks of the smaller sample (sample size n_2).

3. Compute the Wilcoxon statistic as

$$C = n_1 n_2 + \frac{n_2(n_2 + 1)}{2} - \sum^{n_2} R = 16(10) + \frac{10(10 + 1)}{2} - 91.5 = 123.5$$

where n_1 is the size of the larger sample and n_2 is the size of the smaller. If the two sample sizes are equal, it does not matter which sample is labeled 1 and which 2. This statistic must be compared with $n_1 n_2 - C$ and the greater of the two quantities chosen as a test statistic U_S. In this case, $160 - 123.5 = 36.5$, so we use 123.5. This is identical to the value obtained in the Mann–Whitney test.

Obtaining the Probability of a U_S-Value

No tied observations in samples (or observations tied within *one or both groups only):* When $n \leq 20$, compare U_s, with the critical value for $U_{\alpha[n_1, n_2]}$ in Statistical Table **U**. The null hypothesis is rejected if the observed value is too large.

In cases where $n_1 > 20$, calculate the following quantity:

$$t_s = \frac{\left(U_s - \dfrac{n_1 n_2}{2}\right)}{\sqrt{\dfrac{n_1 n_2(n_1 + n_2 + 1)}{12}}}$$

Box 13.8 (continued)

which is approximately normally distributed. The denominator 12 is a constant. One can look up the critical values of $t_{\alpha[\infty]}$ in Statistical Table **B** to compare against the observed t_s value for a one-tailed test or two-tailed test as required by the hypothesis.

Tied observations (ties occur across both samples): There is no exact test. When $n_1 \leq 20$, the critical values in Statistical Table **U** are conservative; that is, the actual probability value will be less than the tabled one. Our example is a case in point. Ranks for observations 116, 121, and 123 are tied across samples (ranks for observations 107, 118, 126, and 128 are tied within samples). We consult Statistical Table **U** and find $U_{.025[16,10]} = 118$ and $U_{.01[16,10]} = 124$. Hence, the two samples are different at least at $0.05 > P > 0.02$ (we double the probabilities because this is a two-tailed test).

For sample sizes $n_1 > 20$, and when an approximation to the correct probability value is desired, evaluate

$$t_s = \frac{\left(U_s - \dfrac{n_1 n_2}{2} \right)}{\sqrt{\left(\dfrac{n_1 n_2}{(n_1 + n_2)(n_1 + n_2 - 1)} \right)\left(\dfrac{(n_1 + n_2)^3 - (n_1 + n_2) - \sum\limits^{m} T_j}{12} \right)}}$$

where $\sum^{m} T_j$ has the same meaning as in Box 13.7. Compare t_s with $t_{\alpha[\infty]}$. For our example, we compute

$$t_s = \frac{\left(123.5 - \dfrac{16 \times 10}{2} \right)}{\sqrt{\left(\dfrac{16 \times 10}{(16+10)(16+10-1)} \right)\left(\dfrac{(16 + 10)^3 - (16+10) - (6+6+6+6+6+24)}{12} \right)}}$$
$$= 2.297, P = 0.0216.$$

We reject the null hypothesis that the two samples do not differ in location. The corresponding value of t_s without the correction for ties is 2.293, $P = 0.0218$, so in this case the conclusions are the same.

A Test for Different Dispersions in Two-Sample Data

For the tests described above to be valid, one must assume equal dispersion of the observations in the two samples. A test proposed by Shoemaker (1995) addresses this assumption. It is a nonparametric test that examines the membership in the outlier groups when the two samples are combined. Before we can combine the two samples, they have to be scaled so the spreads of the two sets of readings can be fairly compared. To accomplish this, we evaluate the deviations of each observation from the median of each group. We find the median of sample A to be 120 and that of sample B to be 109.5. Next, we amalgamate the two samples of deviates and order them from smallest (most negative) to largest (most positive). We take care to tag the deviates that originally came from the smaller, second sample. We show the amalgamated and ordered sample of

Box 13.8 (continued)

deviates in the following three columns. The 10 deviates from sample B are each marked with a dagger. All these operations can be easily carried out on a spreadsheet.

-16	$-2.5^\dagger$	$1.5^\dagger$	8
-11	$-2.5^\dagger$	3	8
$-9.5^\dagger$	-2	5	$10.5^\dagger$
-8	-2	6	$11.5^\dagger$
-6	$-1.5^\dagger$	6	$13.5^\dagger$
$-4.5^\dagger$	-1	$6.5^\dagger$	
-4	1	8	

Next, we have to define the outlier groups. Shoemaker's analysis has shown that an interquantile range leaving 1/16th of the distribution in each tail yields a satisfactory screen. This results in either two or four deviates in each tail being considered outliers for combined samples ranging from 20 to 60 deviates. Obviously, the number of outliers considered has to be rounded to the nearest integer.

The final step in the analysis is to calculate the probability of the distribution of the observed counts of deviates from samples A or B beyond the interquantile range in the combined sample of deviates. The size of sample A, $n_1 = 16$, and that of sample B, $n_2 = 10$. Thus, their combined sample size $N = 26$ and the number of outliers in each tail of the combined sample is $26/16 = 1.62$ rounded up to 2. Note that the divisor of N and sample size n_1 are both 16 by happenstance in this example. In other problems n_1 is likely to differ from the constant divisor. We can easily see that the total number of outliers delimited in both tails of the combined sample is $k = 4$ and that the number of outliers from sample A is $r = 2$. We have to evaluate the probability that by chance we have arrived at the observed outcome. The appropriate model here is the hypergeometric distribution defined earlier by Expression (5.13). If we set $n_1 = pN$ and $n_2 = qN$, we have all the quantities necessary to evaluate the expression for our data. It gives

$$\frac{\binom{n_1}{r}\binom{n_2}{k-r}}{\binom{N}{k}} = \frac{\binom{16}{2}\binom{10}{4-2}}{\binom{26}{4}} = 0.3612$$

We have little reason to reject the null hypothesis that the two samples have equal dispersion. The Mann–Whitney U-test or the Wilcoxon two-sample test, therefore, satisfy their assumptions and their employment is justified.

is a semigraphical test and is quite simple to apply. It is especially convenient when the data are already graphed and there are not too many items in each sample. The display of data for anovas as a series of dots has other advantages as well. It permits the investigator to determine at a glance whether the assumptions of normality and homoscedasticity are reasonable and also whether the null hypothesis of identical group means appears to hold. Systematic testing of data by such an approach (especially for randomized-blocks designs) has been suggested by Fawcett (1990), which should be consulted for further details.

Note that the methods of Boxes 13.7 and 13.8 do not require that each individual observation represent a precise measurement. As long as you can order the observations, you are able to perform these tests. Thus, for example, suppose you placed some meat out in the open and studied the arrival times of individuals of two species of blowflies. You could record exactly the time of arrival of each fly, starting from a point zero in time when the meat was set out. On the other hand, you might simply rank arrival times of the two species, noting that one individual of species B came first, two individuals from species A next, then three individuals of B, followed by the simultaneous arrival of one of each of the two species (a tie), and so forth. Although such ranked or ordered data cannot be analyzed by the parametric methods studied earlier, the techniques of Boxes 13.7 and 13.8 are entirely applicable.

The method of calculating the sample statistic U_s for the Mann–Whitney and the Wilcoxon tests is straightforward, as shown in Box 13.8. When there are no tied observations across the two groups, the critical values for $U_{\alpha[n_1,n_2]}$ are given by Statistical Table U, which is adequate for cases in which the larger sample size $n_1 \leq 20$. The probabilities in Statistical Table U assume a one-tailed test. For a two-tailed test, you should double the value of the probability shown in that table. When $n_1 > 20$, compute the first expression shown in the section on *Obtaining the Probability of a U_S-Value* in Box 13.8. Because this expression is distributed as a normal deviate, consult the table of t (Statistical Table **B** or computer software for the t-or normal distributions) for $t_{\alpha[\infty]}$, using one- or two-tailed probabilities, depending on your hypothesis. A further complication arises from tied observations across the two groups. There is no exact test. For sample sizes $n_1 \leq 20$ use Statistical Table **U**, which will then be conservative. Larger sample sizes require the more elaborate formula shown at the bottom of Box 13.8. It takes a substantial number of ties to affect the outcome of the test appreciably. Corrections for ties increase the t_s-value slightly; hence, the uncorrected formula is more conservative.

It is desirable to obtain an intuitive understanding of the rationale behind these tests. What we are doing in the Mann–Whitney test is counting the number of individuals in the opposite group, which the members of the first group would have to overtake in order to separate the two samples completely along the axis of the variable under study. We can conceive of two extreme situations: In one case, the two samples overlap and coincide entirely; in the other, they are quite separate. In the latter case, if we take the sample with the lower-valued observations, there will be no points of the contrasting sample to the left of it; that is, we can go through every observation in the lower-valued sample without having any items of the higher-valued one to the left of it. Thus, $C = 0$. Conversely, all the points of the lower-valued sample would be to the left of every point of the higher-valued one if we had started out with the latter. Our total count would therefore be the total count of one sample multiplied by every observation in the second sample, which yields $n_1 n_2$. Thus, because we are told to take the greater of the two values, the sum of the counts C or $n_1 n_2 - C$, our result in this case would be $n_1 n_2$. But if two equal-sized samples coincide completely, then for each point in one sample we would have those points below it plus a half point for the tied value representing the observation in the second sample that is at exactly the same level as the observation under consideration. A little experimentation will show C to be $[n(n-1)/2] + (n/2) = n^2/2$. Clearly, the range of possible U-values must be between this and n_1n_2, and the critical value must be somewhere within this range.

In computing the Wilcoxon statistic, the magnitude of the sum of the ranks of sample 2 determines the outcome of the test. If this sum is unusually large or unusually small, the sample under consideration is near the upper or lower extreme of the overall distribution of ranks of the two samples considered together. This is the place to remove a misconception prevalent among some users of the Mann–Whitney–Wilcoxon two-sample test (and even disseminated by some textbooks), namely, that these tests are the reduction of the Kruskal–Wallis test to $a = 2$ samples, analogous to the relation between single-classification anova of $a \geq 3$ groups and the two-sample t-test or its anova version. The supposed analogy holds only for the design, not for the structure of the tests. The H-statistic is based on an among-groups sum of squares of the ranks, whereas the U-statistic is a measure of the overlap of the ranks of the two samples.

Our conclusion as a result of the tests in Box 13.8 is that the two samples differ in the distribution of cheliceral base length. The chiggers of sample A have longer cheliceral bases than those of sample B.

Even the Mann–Whitney U-test and the Wilcoxon two-sample test are not entirely free of assumptions. The test assumes equal dispersions in both samples. A test for this assumption has been developed by Shoemaker (1995). We illustrate this test and apply it to the chigger nymph measurements in the last section of Box 13.8. The principle behind the test is to look at the outliers in a combined sample of n_1 and n_2 observations and to note whether the bracketing outliers of the combined sample all belong to one or the other of the two samples. The probability of this occurring is evaluated based on the number of outliers studied and the sample sizes involved. We are not able to show that samples A and B differ in their dispersions.

Alternatively, the nonparametric test may be targeted at a comparison of the medians of the two samples only. Fligner and Policello (1981) describe such a test as a modification of the Mann–Whitney–Wilcoxon test. Tests on differences among medians can also be carried out by means of two-way tests of independence (see Section 17.4).

Now that you have learned techniques for testing two groups, we can return to the case of several groups and show how we could carry out multiple comparisons between pairs of treatments based on a nonparametric test of significance. This is the nonparametric analogue to the multiple-comparison tests we studied in Sections 9.4 and 9.5. One approach illustrated here is the simultaneous test procedure of Dwass (1960) as further developed by K. R. Gabriel (unpublished). This technique, described in Box 13.9, finds the Mann–Whitney–Wilcoxon U-statistic for pairwise comparisons for each of the treatments and compares these with a critical value found by the formula shown in Box 13.9. The computation of the statistic, U_s, is quite simple. We again take the greater of two quantities C or $(n^2 - C)$. In Box 13.8, in place of n^2 we used n_1n_2, but the *STP* (simultaneous test procedure) method of Box 13.9 applies only to equal sample sizes; hence n_1n_2 becomes n^2. The method

---

BOX 13.9 Nonparametric Multiple Comparisons by *STP* (an Unplanned Test for Equal Sample Sizes)

This method is based on U, the Mann–Whitney statistic, and requires equal sample sizes, with $n > 8$.

The technique is applied to the data on pea sections in Box 13.6. As in the ordinary Mann–Whitney test, it is simplest to prepare a graph of the data as shown in the following (see also Box 13.8).

For each pair of samples, compute U_s as follows. For each observation in one sample, count the number of observations in the other samples that are lower in value (to the left). Count a $\frac{1}{2}$ for each tied observation. For example, control (C) versus glucose (G): 10 observations in G are below the first observation in C, 10 observations in G are less than the second observation in C, and so on. The sum of the counts is $C(C, G) = 10 + 10 + \cdots + 10 = 100$. Obviously, the same sum will be found for $C(C, F)$ and $C(C, G + F)$.

$$C(C, S) = 6\frac{1}{2} + 9\frac{1}{2} + 9\frac{1}{2} + 9\frac{1}{2} + 10 + 10 + 10 + 10 + 10 + 10 = 95$$
$$C(G, F) = 3 + 3 + 5\frac{1}{2} + 7 + 7 + 7\frac{1}{2} + 7\frac{1}{2} + 7\frac{1}{2} + 9 + 10 = 67$$
$$C(G, G + F) = 2\frac{1}{2} + 2\frac{1}{2} + 5\frac{1}{2} + 8 + 8 + 9 + 9 + 9 + 9\frac{1}{2} + 10 = 73$$
$$C(G, S) = 0 + 0 + 0 + 0 + 0 + 0 + 0 + 0 + 0 + 1\frac{1}{2} = 1\frac{1}{2}$$
$$C(F, G + F) = \frac{1}{2} + \frac{1}{2} + 2\frac{1}{2} + 2\frac{1}{2} + 5\frac{1}{2} + 5\frac{1}{2} + 5\frac{1}{2} + 9 + 9\frac{1}{2} + 9\frac{1}{2} = 50\frac{1}{2}$$

$C(F, S)$: All points of F are below those of S, so the count must be the minimum possible; namely, zero. This is also true of $C(G + F, S)$.

The greater of the two quantities C or $(n^2 - C)$ is now entered as U_s into the table below (n is the sample size for each group).

Box 13.9 (continued)

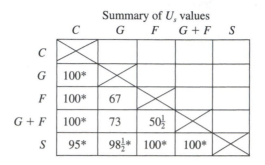

Summary of U_s values

	C	G	F	G + F	S
C					
G	100*				
F	100*	67			
G + F	100*	73	$50\frac{1}{2}$		
S	95*	$98\frac{1}{2}$*	100*	100*	

*Null hypothesis rejected at an experimentwise error rate of 0.05 by test described below

The critical value for U_s is:

$$U_{\alpha[a,n]} = \frac{n^2}{2} + Q_{\alpha[a,\infty]}n \sqrt{\frac{2(10) + 1}{24}}$$

where a is the number of samples, n is the number of observations within each sample, and $Q_{\alpha[a,\infty]}$ is the α significance level for the studentized range for a groups and ∞ df (see Statistical Table **J**).

In our case

$$U_{.05[5,10]} = \frac{10^2}{2} + (3.858)10 \sqrt{\frac{2(10) + 1}{24}} = 86.09$$

The null hypothesis is rejected for all U_s-values equal to or greater than 86.09. Such values have been marked with an asterisk in the preceding table. All sets of groups containing a U_s greater than this critical value between at least one pair are declared heterogeneous. Thus, we find that the set (G, F, G + F) is not heterogeneous. Because C and S were heterogeneous with any other treatment, any set containing either C or S (or both) would be heterogeneous. The data therefore can be broken up into the following partition:

$$(G, F, G + F) \quad (S) \quad (C)$$

This is the same conclusion that was reached in Box 9.7 and Section 9.7, although such complete agreement cannot always be expected.

When there are tied observations across groups as in the example here, the formula for the critical value is not quite correct. The square rooted terms should be changed to

$$\sqrt{\frac{(2n)^3 - 2n - \sum_{j}^{m} T_j}{2n(2n - 1)24}}$$

requiring a recomputation of each value of $U_{\alpha[a,n]}$ for each pair of samples. However, because the correction due to ties is generally quite small, and the critical values computed by the simpler formula are conservative, we did not adjust for ties.

is recommended only for n equal to or greater than 8. Another procedure is the test developed by Nemenyi (1963) for comparisons of all pairs of mean ranks. The test is carried out using the same procedures as the T-method described in Box 9.8, but on mean ranks. It uses the standard error $SE = \sqrt{a(N + 1)/24}$, where a is the number of groups and $N = an$ is the total sample size. The same table of critical values is used to compute the MSD, but the $Q_{\alpha[k,\nu]}$ value is looked up using infinite degrees of freedom. Cook and Wheater (2000) show an example of the application of this test.

In our discussion of comparisons among means in Section 9.4, we pointed out that although the experimentwise type I error is α, tests of subsets have actual error rates of $\alpha' < \alpha$. Thus, in this test, the 5% type I error probabilities for tests involving any set of four samples is 0.0323; that for any set of three samples is 0.0175, and that for a pair of samples is 0.0064. Our tests of pairs are therefore quite conservative. The computations are simple and lead to the conclusion that glucose, fructose, and the mixed sugar are not different from each other but are different from sucrose, which in turn is different from the control. This is the same conclusion that was reached in Box 9.7, but of course this nonparametric method need not always give the same conclusion as a corresponding anova.

Two additional points should be noted. We only tested all pairwise comparisons in this example because nonparametric *STP* test have the following properties. If any pair of samples within a larger set is found to be larger than the criterion, then the larger set must also be declared "significant" at the same probability level. Conversely, all pairs of samples within a set declared significant are not necessarily significantly different; that is, there must be at least one significant pair leading to the rejection of the hypothesis of homogeneity of the larger set. Therefore, by knowing which pairs are nonsignificant, we can construct the maximally nonsignificant sets, as we have done in this example. The second point relates to an issue we took up a few paragraphs ago. Although the Kruskal–Wallis test of Box 13.7 is a generalization of the Wilcoxon two-sample test employed in the *STP*, it is not based on the same test criterion. Therefore, although this is unlikely, a nonsignificant Kruskal–Wallis test for differences among several groups might yield at least some significant pairs when investigated by the nonparametric *STP* of Box 13.9.

The Kruskal–Wallis and Mann–Whitney tests are based on ranks and measure differences in location. A nonparametric test that tests differences between two distributions is the **Kolmogorov–Smirnov two-sample test**. Its null hypothesis is that the two samples are distributed identically; thus, the test is sensitive to differences in location, dispersion, skewness, and so forth. This test is quite simple to carry out. It is based on the unsigned differences between the relative cumulative frequency distributions of the two samples. Expected critical values can be looked up in a table or approximated. Comparison between observed and expected values leads to decisions about whether the maximum difference between the two cumulative frequency distributions is larger than would be expected by chance. Box 13.10 gives some examples.

First, we show the application of the method to small samples (both n_1 and $n_2 \leq 25$. We apply the test to the example from Box 13.8, where two samples of chigger nymphs were measured for length of cheliceral base. We use the symbol

---

BOX 13.10 Kolmogorov–Smirnov Two-Sample Test, Testing Differences in Distributions of Two Samples of Continuous Observations

I. *Test for Two Small Samples (Both n_1 and $n_2 \leq 25$)*

Data on length of cheliceral base in chigger nymphs from Box 13.8. The sample sizes are $n_1 = 16$, $n_2 = 10$.

(1) Y	(2) Sample A F_1	(3) Sample B F_2	(4) $\dfrac{F_1}{n_1}$	(5) $\dfrac{F_2}{n_2}$	(6) $d = \left| \dfrac{F_1}{n_1} - \dfrac{F_2}{n_2} \right|$
100	0	1	0	0.100	0.100
101	0	1	0	0.100	0.100
102	0	1	0	0.100	0.100
103	0	1	0	0.100	0.100
104	1	1	0.062	0.100	0.038
105	1	2	0.062	0.200	0.138
106	1	2	0.062	0.200	0.138
107	1	4	0.062	0.400	0.338
108	1	5	0.062	0.500	0.438
109	2	5	0.125	0.500	0.375
110	2	5	0.125	0.500	0.375
111	2	6	0.125	0.600	0.475 ← D
112	3	6	0.188	0.600	0.412
113	3	6	0.188	0.600	0.412
114	4	6	0.250	0.600	0.350
115	4	6	0.250	0.600	0.350
116	5	7	0.312	0.700	0.388
117	5	7	0.312	0.700	0.388
118	7	7	0.438	0.700	0.262
119	8	7	0.500	0.700	0.200
120	8	8	0.500	0.800	0.300
121	9	9	0.562	0.900	0.338
122	9	9	0.562	0.900	0.338
123	10	10	0.625	1.000	0.375
124	10	10	0.625	1.000	0.375
125	11	10	0.688	1.000	0.312
126	13	10	0.812	1.000	0.188
127	13	10	0.812	1.000	0.188
128	16	10	1.000	1.000	0

Box 13.10 (continued)

Procedure

1. Form cumulative frequencies, F, of the items in samples 1 and 2. Thus in column (2) note that there are 3 measurements in sample A at or below 112.5 micrometer units. By contrast there are 6 such measurements in sample **B** [column (3)].

2. Compute relative cumulative frequencies by dividing frequencies in columns (2) and (3) by m_1 and n_2, respectively, and enter in columns (4) and (5).

3. Compute d, the absolute value of the difference between the relative cumulative frequencies in columns (4) and (5), and enter in column (6).

4. Locate the largest unsigned difference D. It is 0.475.

5. Multiply D by $n_1 n_2$. We obtain $(16)(10)(0.475) = 76$.

6. Compare $n_1 n_2 D$ with its critical value in Table **W**, where we obtain a value of 84 for $P = 0.05$. We accept the null hypothesis that the two samples have been taken from populations with the same distribution. The results of the Kolmogorov–Smirnov two-sample test contrast with the moderately significant difference by the Mann–Whitney U-test in Box 13.7. The Kolmogorov–Smirnov test is less powerful than the Mann–Whitney U-test with respect to the alternative hypothesis of the latter, i.e., differences in location. However, the Kolmogorov–Smirnov test examines differences in both shape and location of the distributions and is thus a more comprehensive test.

    An alternative computational scheme, which is more efficient but perhaps less instructive for the novice, was proposed by Gideon and Mueller (1978). Pool both samples into a single array. For the chigger data just considered this would yield

Y	Running sum		Y	Running sum
100	-16		118	-42
104	-6		119	-32
105	-22		120	-48
107	~~-38~~		121	~~-38~~
107	-54		121	-54
108	-70		123	~~-44~~
109	-60		123	-60
111	-76		125	-50
112	-66		126	~~-40~~
114	-56		126	-30
116	~~-46~~		128	~~-20~~
116	-62		128	~~-10~~
118	~~-52~~		128	0

Note that all members of sample A have been underlined for convenience

    Next, compute a running sum for which you add n_2 whenever an item from sample A is encountered and subtract n_1 whenever the item is from sample B. The running sum is shown alongside the arrayed observations. Cross out running sums for all but the last of each set of tied values. The largest running sum (ignoring sign) is $n_1 n_2 D = 76$, as before.

Box 13.10 (continued)

II. *Approximate Test for Two Larger Samples (Both Sample Sizes Between 26 and 40, or n_1 or $n_2 > 40$).*

Compute D as shown in part **I** and test as follows. Using Expression (13.6) compute

$$D_\alpha = K_\alpha \sqrt{\frac{n_1 + n_2}{n_1 n_2}}$$

where

$$K_\alpha = \sqrt{\frac{1}{2}\left[-ln\left(\frac{\alpha}{2}\right)\right]}$$

For equal sample sizes the formula for D_α simplifies to $K_\alpha \sqrt{2/n}$. These critical values are for two-tailed tests.

Whenever $n_1 = n_2 = n$ and $26 < n \le 40$, the exact distribution of D can be looked up in Table 15.4 of Owen (1962). The quantity tabled there is actually the cumulative probability distribution corresponding to values of nD (called nr in Owen's table). This table is intended for a two-tailed test.

III. *Approximate Test for Two Frequency Distributions (Large Sample Sizes)*

Length of silver salmon at two time periods. Data from Table 13.1.

$n_1 = 421$, $n_2 = 349$

(1) Length in mm	(2) Upper class limit	(3) Nov. 12–25 F_1	(4) Apr. 1–14 F_2	(5) $\dfrac{F_1}{n_1}$	(6) $\dfrac{F_2}{n_2}$	(7) $d = \left\|\dfrac{F_1}{n_1} - \dfrac{F_2}{n_2}\right\|$
60	62.5	1	0	0.002	0	0.002
65	67.5	5	0	0.012	0	0.012
70	72.5	38	0	0.090	0	0.090
75	77.5	132	0	0.314	0	0.314
80	82.5	257	1	0.610	0.003	0.607
85	87.5	355	6	0.843	0.017	0.826
90	92.5	399	12	0.948	0.034	0.914
95	97.5	417	17	0.990	0.049	0.941 ← D
100	102.5	420	33	0.998	0.095	0.903
105	107.5	421	58	1.000	0.166	0.834
110	112.5	421	103	1.000	0.295	0.705
115	117.5	421	160	1.000	0.458	0.542
120	122.5	421	247	1.000	0.708	0.292
125	127.5	421	312	1.000	0.894	0.106
130	132.5	421	332	1.000	0.951	0.049
135	137.5	421	337	1.000	0.966	0.034
140	142.5	421	342	1.000	0.980	0.020
145	147.5	421	347	1.000	0.997	0.003
150	152.5	421	348	1.000	0.997	0.003
155	157.5	421	348	1.000	0.997	0.003
160	162.5	421	349	1.000	1.000	0

Box 13.10 (continued)

Procedure

1. Form cumulative frequency distributions F from the frequencies in the two samples (shown in Table 13.2). Column (2) gives the upper class limits. In column (3) we note 5 fish at or below 67.5 mm. These were obtained by adding 1 (for 60 mm) and 4 (for 65 mm).

2. Compute relative cumulative frequencies by dividing frequencies in columns (3) and (4) by n_1 and n_2, respectively, and enter in columns (5) and (6).

3. Compute d, the absolute difference between the relative cumulative frequencies in columns (5) and (6) and enter in column (7).

4. Locate the largest unsigned difference D. In this case, $D = 0.941$.

5. Critical values D_α for D are found from Expression (13.6) as given in part **II**.

$$D_\alpha = K_\alpha \sqrt{\frac{n_1 + n_2}{n_1 n_2}}$$

where

$$K_\alpha = \sqrt{\frac{1}{2}\left[-\ln\left(\frac{\alpha}{2}\right)\right]}$$

Thus for $\alpha = 0.05$, $K_{.05} = 1.35810$, and for $\alpha = 0.01$, $K_{.01} = 1.62762$. In this case,

$$D_{.05} = 1.35810\sqrt{\frac{421 + 349}{421(349)}} = 0.98316$$

$$D_{.01} = 1.62762(0.072392) = 0.11783$$

The unsigned difference $D = 0.941 \gg D_{.001} = 0.118$, so we conclude that the two samples come from populations with (very) different distributions. The critical value D_a is intended for a two-tailed test. When $n_1 = n_2 = n$ the formula for D_α simplifies to

$$D_\alpha = K_\alpha \sqrt{\frac{2}{n}}$$

F for cumulative frequencies, which are summed with respect to the class marks shown in column (1) and give the cumulative frequencies of the two samples in columns (2) and (3). Relative expected frequencies are obtained by division by the respective sample sizes in columns (4) and (5); column (6) features the absolute difference between relative cumulative frequencies. The maximum unsigned difference is 0.475, which is multiplied by $n_1 n_2$, yielding 76. The critical value for this statistic can be found in Statistical Table **W**, which furnishes critical values for the two-tailed, two-sample Kolmogorov–Smirnov test. We obtain $n_1 n_2 D_{.10} = 76$ and $n_1 n_2 D_{.05} = 84$.

There is a 10% probability of obtaining the observed difference by chance alone, and we conclude that the two samples do not differ in their distributions by more than one would usually expect by chance. This contradicts the findings of a difference by the Mann–Whitney U-test in Box 13.8. However, the two tests differ in their sensitivities to different alternative hypotheses—the Mann–Whitney U-test is sensitive to the number of interchanges in rank necessary to separate the two samples (shift in location), whereas the Kolmogorov–Smirnov test measures differences in the entire distributions of the two samples and is thus less sensitive to differences in location only.

A more efficient algorithm for the same test is featured next (see Box 13.10). The instructions are given for the exact test and for an approximate test that can be used when sample sizes exceed 25. An approximate two-tailed critical value for the test statistic D can be computed as

$$D_\alpha = K_\alpha \sqrt{\frac{n_1 + n_2}{n_1 n_2}} \tag{13.6}$$

where

$$K_\alpha = \sqrt{\frac{1}{2}\left[-\ln\left(\frac{\alpha}{2}\right)\right]}$$

The final example in Box 13.10 is for large samples arrayed in frequency distributions. The data are the lengths of silver salmon from Table 13.1. We have chosen the samples for two time periods, November and April, and are testing whether these differ in distribution of body length. The columns are computed as in the earlier example, and the critical value D_α is computed by Expression (13.6). We see from Box 13.10 that the maximum unsigned difference $D = 0.941 \gg D_{.01} = 0.118$ and thus we reject the null hypothesis that the two samples come from populations with the same distribution. An underlying assumption of all Kolmogorov–Smirnov tests (including goodness-of-fit tests by means of this statistic that are treated in Section 17.2) is that the variables studied are continuous.

13.11 Nonparametric Methods in Lieu of Two-Way Anova

Nonparametric tests exist for randomized blocks, the most common design for two-way analysis of a variance. The tests are for designs in which, if they were to be done by anova, one factor would be Model I, and the other factor would represent the blocks in the experiment. A Model II design, estimating the variance components, would clearly not be appropriate in a test based on ranks because we are concerned only with location of the various samples in the overall ranking.

The test illustrated in Box 13.11 is **Friedman's method for randomized blocks**. This example involves the lake-temperature data from Box 11.3, where they were used to illustrate the computation of a randomized-blocks anova without replication. Recall that the four dates are considered blocks and the depths are the treatment effects. In Friedman's method, the observations are ranked *within each block*, as

BOX 13.11 Friedman's Method for Randomized Blocks

*Temperatures (°C) of Rot Lake on 4 Early Afternoons of the Summer of 1952 at 10
Depths*

Data are from Box 11.4.

Depth in meters ($a = 10$)	Days ($b = 4$)			
	29 July	30 July	31 July	1 August
0	23.8	24.0	24.6	24.8
1	22.6	22.4	22.9	23.2
2	22.2	22.1	22.1	22.2
3	21.2	21.8	21.0	21.2
4	18.4	19.3	19.0	18.8
5	13.5	14.4	14.2	13.8
6	9.8	9.9	10.4	9.6
9	6.0	6.0	6.3	6.3
12	5.8	5.9	6.0	5.8
15.5	5.6	5.6	5.5	5.6

Procedure

The four sets of readings are treated as blocks in this analysis.

1. Assign ranks to items within each of the $b = 4$ blocks (columns in this case) separately, resulting in the table that follows. If there are ties, treat them in the conventional manner (by average ranks). There are $a = 10$ treatments (depths; rows in this case).

Depth in meters	Days				$\sum^{b} R_{ij}$
	29 July	30 July	31 July	1 August	
0	10	10	10	10	40
1	9	9	9	9	36
2	8	8	8	8	32
3	7	7	7	7	28
4	6	6	6	6	24
5	5	5	5	5	20
6	4	4	4	4	16
9	3	3	3	3	12
12	2	2	2	2	8
15.5	1	1	1	1	4

2. Sum the ranks for each of the $a = 10$ groups (rows in the present case) $= \sum^{b} R_{ij}$. This yields values such as 40, 36, 32, ...

Box 13.11 (continued)

3. Compute

$$X^2 = \left[\frac{12}{ab(a+1)} \sum^a \left(\sum^b R_{ij} \right)^2 \right] - 3b(a+1)$$

$$= \frac{12}{10(4)(10+1)} (40^2 + 36^2 + \cdots + 8^2 + 4^2) - 3(4)(10+1) = 36.000$$

This value is to be compared with $\chi^2_{[a-1]}$. We find $P\left[\chi^2_9 \geq X^2\right] = 3.9647 \times 10^{-5}$ and reject the null hypothesis that water depth does not affect its temperature. This outcome was to be expected in view of the consistency of the data.

Multiple comparisons

We can again carry out multiple comparisons among the treatment means. The procedure is that of Box 13.8. The only difference is that in the formula for the critical value of U, we replace the quantity n by bn, where b is the number of blocks and n is the number of replicates per subgroups (block per treatment). In the frequent cases where each treatment–block combination is not replicated (as in the example of this box), $n = 1$ and b replaces the n of the formula in Box 13.8. In this example the formula for the critical value of U_s is

$$U_{\alpha[a,b]} = \frac{b^2}{2} + Q_{\alpha[a,\infty]} b \sqrt{\frac{2b+1}{24}}$$

$$= \frac{4^2}{2} + 4.474(4) \sqrt{\frac{2(4)+1}{24}} = 18.9590$$

Inspection of the values in the table reveals that the observations are virtually nonoverlapping between adjacent depths. Therefore, the value of C is $4 + 4 + 4 + 4 = 16$ between each pair [except for $C(9,12)$, which is $3\frac{1}{2} + 3\frac{1}{2} + 4 + 4 = 15$]. It therefore appears that the null hypothesis cannot be rejected for any of the pairwise comparisons despite the firm rejection for the overall analysis.

---

shown in Box 13.11. Note that because of the regularity of the thermal stratification of the lake, the ranking for the four days is entirely uniform. We sum the ranks for each of the treatment groups and compute a statistic X^2, identical to H in Box 13.7. Again, X^2 is distributed as $\chi^2_{[a-1]}$, where a is the number of treatments (depths in this example). This value will be maximal in our example because the ranks are ordered uniformly in each of the blocks. In an example in which the observations were not as uniformly ranked in each block, the marginal sums (row sums in this case) would not be as high or as low as in this instance but would tend more toward intermediate values. Because these sums are being squared, the large values exert a disproportionate influence upon the sample statistic X^2.

We conclude that the effect of depth on water temperature is larger than we would expect by chance, which agrees with our earlier finding by analysis of variance. At the end of Box 13.11, we also show how to carry out multiple comparisons among the groups, and we find that, despite the very small overall probability, none of the individual comparisons has a probability less than 0.05. This result illustrates the sharp decrease in power because of both the recoding of the observations as ranks and the decrease in pairwise α in the multiple-comparisons test. Had we carried out the multiple-comparisons test parametrically using the T-method, all but 4 of the 45 pairwise comparisons would have yielded probabilities <0.01.

A special case of randomized blocks is the paired-comparisons design, discussed in Section 11.5 and illustrated in Box 11.5. This design can be carried out by Friedman's test, but a more widely used method is **Wilcoxon's signed-ranks test**, which is applied in Box 13.12 to a new example. The data are mean litter sizes in two strains of guinea pigs kept in large colonies during the years 1916 through 1924. Each value is the average of many litters. Note the parallelism in the changes in the variable in the two strains. During 1917 and 1918 (war years for the United States), a shortage of caretakers and food resulted in a decrease in the number of offspring per litter. As soon as better conditions returned, the mean litter size increased again. Notice that a subsequent drop in 1922 is again mirrored in both lines, suggesting that these fluctuations are environmentally caused. It is therefore appropriate that the data be treated as randomized blocks, with years as blocks and the strain differences as the fixed treatments.

Column (3) in Box 13.12 lists the differences on which a conventional paired-comparisons t-test could be performed. For Wilcoxon's test, these differences are ranked *without regard to sign,* so the smallest absolute difference is ranked 1, and the largest absolute difference (of the nine differences) is ranked 9. Tied ranks are computed as averages as usual. After the ranks have been computed, the original sign of each difference is assigned to the corresponding rank. The sum of the positive or of the negative ranks, whichever one is smaller in absolute value, is then computed (it is labeled T_s) and is compared with the critical value T in Statistical Table **V** for the corresponding sample size. In view of the low probability of the rank sum, it is clear that strain B has a litter size different from that of strain 13. This test is simple to carry out, but it is, of course, not as efficient as the corresponding parametric t-test, which is preferred if the necessary assumptions hold. Note that one needs at least six differences in order to carry out Wilcoxon's signed-ranks test. In six paired comparisons, all differences must be of like sign for the test to be able to reject the null hypothesis at the 5% level.

For a large sample, an approximation to the normal curve is available, which is given in Box 13.12. Note that the absolute magnitudes of these differences play a role only as they affect the ranks of the differences.

A still simpler test is the **sign test**, in which we count the number of positive and negative signs among the differences (omitting all differences of zero). We then test the hypothesis that the n plus and minus signs are sampled from a population in which the two kinds of signs are present in equal proportions, as might be expected if

BOX 13.12 | Wilcoxon's Signed-Ranks Test for Two Groups Arranged as Paired Observations

Mean Litter Size of Two Strains of Guinea Pigs, Compared Over n = 9 Years

Year	(1) Strain B	(2) Strain 13	(3) D	(4) Rank (R)
1916	2.68	2.36	+0.32	+9
1917	2.60	2.41	+0.19	+8
1918	2.43	2.39	+0.04	+2
1919	2.90	2.85	+0.05	+3
1920	2.94	2.82	+0.12	+7
1921	2.70	2.73	−0.03	−1
1922	2.68	2.58	+0.10	+6
1923	2.98	2.89	+0.09	+5
1924	2.85	2.78	+0.07	+4
	Absolute sum of negative ranks			1
	Sum of positive ranks			44

SOURCE: Data from S. Wright (unpublished results).

Procedure

1. Compute the differences between the n pairs of observations. These are entered in column (3), labeled D.

2. Rank these differences from the smallest to the largest *without regard to sign*.

3. Assign the original signs of the differences to the ranks.

4. Sum the positive and negative ranks separately. The sum that is smaller in absolute value, T_s, is compared with the values in Statistical Table **V** for $n = 9$.

$T_s = 1$, which is equal to or less than the entry for one-tailed $\alpha = 0.005$ in the table. The null hypothesis of no difference can be rejected at the 1% level, because for our example a two-tailed test seems appropriate. Litter size in strain B is different from that of strain 13.

For large samples ($n > 50$), compute

$$t_s = \frac{T_s - \dfrac{n(n+1)}{4}}{\sqrt{\dfrac{n(n+\frac{1}{2})(n+1)}{12}}}$$

where T_s is as defined in step **4** above. The values 4 and 12 are constants. Compare the computed value with $t_{\alpha[\infty]}$ in Statistical Table **B**.

there were no true difference between the two paired samples. Such sampling should follow the binomial distribution, and the test of the hypothesis that the parametric frequency of the plus signs is $\hat{p} = 0.5$ can be made in a number of ways. Let us learn these methods by applying the sign test to the guinea pig data of Box 13.12. There are nine differences, of which eight are positive and one is negative. We could follow the methods of Section 5.2 (illustrated in Table 5.3) in which we calculate the expected probability of sampling one minus sign in a sample of nine on the assumption of $\hat{p} = \hat{q} = 0.5$. The probability of such an occurrence and all "worse" outcomes equals 0.0195. Because we have no a priori notions that one strain should have a greater litter size than the other, this is a two-tailed test and we double the probability to 0.0390. Clearly, this is an improbable outcome, and we reject the null hypothesis that $\hat{p} = \hat{q} = 0.5$.

Because the computation of the exact probabilities is tedious, a second approach is to make use of Statistical Table **Q**, which furnishes critical values for selected parametric proportions. The first part of Statistical Table **Q** gives values for $p = 0.5$, appropriate for the sign test. We reduce sample size n by 1 for each omitted zero-difference sign. There are none in our example. We find for $n = 9$ that one minus and eight plus signs are critical values at 5%. To yield a 2% type I error level, we would have required no minus and nine plus signs. We conclude that the guinea pig strains differ in mean litter size at the 5% level. The percentages at the head of Statistical Table **Q** are for a two-tailed test. Had our results been one plus sign and eight minus signs, we could have employed the table in an identical manner. If our alternative hypothesis had been that strain B has larger litters than strain 13, we could have halved the percentages at the head of the table, used the critical value for the appropriate tail of the distribution, and recorded a one-tailed probability of 2.5%. Obviously, such a one-tailed test would be carried out only if the results were in the direction of the alternative hypothesis. Thus, if the alternative hypothesis were that strain 13 in Box 13.12 had greater litter size than strain B, we would not have bothered testing this example at all because the observed proportion of years showing this relation was less than half. For samples greater than 100, one may need to interpolate in Statistical Table **Q**.

For larger samples, we can also use the normal approximation to the binomial distribution as follows:

$$t_s = \frac{Y - \mu}{\sigma_Y} = \frac{Y - kp}{\sqrt{kpq}}$$

substituting the mean and standard deviation of the binomial distribution (see Section 5.2). In our case, we let n stand for k and assume that $\hat{p} = \hat{q} = 0.5$. Therefore,

$$t_s = \frac{Y - \frac{1}{2}n}{\sqrt{\frac{1}{4}n}} = \frac{Y - \frac{1}{2}n}{\frac{1}{2}\sqrt{n}}$$

The value of t_s is then compared with $t_{\alpha[\infty]}$ in Statistical Table **B**, using one tail or two tails of the distribution as warranted. When the sample size $n \geq 12$, this approximation is satisfactory. For a more accurate approximation of the binomial confidence limits, consult Burstein (1973).

A third approach would be to test the departure from expectation $\hat{p} = \hat{q} = 0.5$ by one of the methods of Section 17.1.

We can employ the sign test also for slightly more complicated hypotheses. If the two columns of data in a paired-comparisons case are labeled Y_1 and Y_2, respectively, we can test whether the response Y_1 is greater than Y_2 by K units. Simply evaluate the difference $D = Y_1 - (Y_2 + K)$ and carry out a sign test on the signs of the D-values. A second hypothesis might be that the response Y_1 is greater than Y_2 by $100p\%$. In such a case, the sign test should be performed on the differences $D = Y_1 - (1 + p)Y_2$.

When the two-way anova is replicated and both factors are fixed treatment effects, we can employ an extension of the Kruskal–Wallis test devised by Scheirer et al. (1976). The procedure is simple. Treat all the observations in the basic data table as a single array and rank them. Then replace the ranks in the data table and carry out a two-way anova on the ranks as though they were the observations. The expected variance of any array of n ranks is $n(n + 1)/12$. Convince yourself that this statement is true by calculating the sum of squares of a small array and dividing it by $n - 1$ to obtain the mean square. Because the MS computed by this formula is constant for any given number n of ranks, we can consider it a parameter. Recall from Section 7.8 that the chi-square distribution is the ratio of a sum of squares divided by a parametric variance. Therefore, we can test the sums of squares that emerged from the anova of the ranks by dividing them by the total MS of the ranks and considering the ratio H as a χ^2-distributed variable with the degrees of freedom pertaining to the

---

BOX 13.13   A Two-Way Anova Design for Ranked Data. The Scheirer–Ray–Hare Extension of the Kruskal–Wallis Test

Food consumption related to fat quality and sex in rats. Data from Box 11.1. $a = 2$ (fats), $b = 2$ (sexes), $n = 3$ (number of rats per subgroup).

Sex $(b = 2)$	Fresh		Rancid	
	Y	Rank	Y	Rank
♂	709	12	592	6
	679	10	538	4
	699	11	476	1
♀	657	8	508	3
	594	7	505	2
	677	9	539	5

Fats $(a = 2)$

Box 13.13 (continued)

Computation

1. Rank all observations from smallest to largest when pooled together into a single sample. Replace each variate in the original data table by its rank.

2. Carry out an analysis of variance of the ranks following the outline in Box 11.1. The following table results.

Source of variation	df	SS	MS
A (columns; fats)	1	108.0000	108.0000
B (rows; sexes)	1	8.3333	8.3333
$A \times B$ (interaction)	1	5.3333	5.3333
Error	8	21.3334	2.6667

3. Compute the total MS as $abn(abn + 1)/12$, 13 in the present case. If there are ties in the ranks it is easiest to compute the total MS from the total SS, by adding all sums of squares in the table and dividing by the total degrees of freedom. Alternatively, we could compute the correction factor D from Box 13.7 and use it to adjust H (computed in step **4** below).

4. Compute H as SS/MS_{total}

For factor **A** $H = 108.0000/13.0000 = 8.3077, P[\chi^2_{[1]} \geq H] = 0.003948$

For factor **B** $H = 8.3333/13.0000 = 0.6410, P[\chi^2_{[1]} \geq H] = 0.4233$

For factor **AB** $H = 5.3333/13.0000 = 0.4103, P[\chi^2_{[1]} \geq H] = 0.5218$

The probability for H is computed as a χ^2-variable, with the degrees of freedom pertaining to the SS being tested. In this example, $df = 1$ for all three tests.

We are able to reject the null hypothesis of no effect only for factor A, the quality of the fat consumed.

---

sum of squares in the numerator. This is entirely analogous to what the Kruskal–Wallis test does for the single-classification design, and we could have computed the probability for a Kruskal–Wallis test by using the procedures outlined here.

In Box 13.13, we employ the "rats and fats" example of Box 11.1 to illustrate the method just described. When analyzed nonparametrically, the results do not differ from those of the earlier anova. We can reject the null hypothesis for the effect of fats, but not for either sex or interaction. This approach can be extended to multiway anovas in a straightforward manner. A general but computationally more complex formula for dealing with such cases is given by Groggel and Skillings (1986).

EXERCISES 13

13.1 In a study of flower color in butterfly weed (*Asclepias tuberosa*), Woodson (1964) obtained the following results:

Geographic region	$\overline{Y}$	n	s
C1	29.3	226	4.59
SW2	15.8	94	10.15
SW3	6.3	23	1.22

The variable recorded was a color score (ranging from 1 for pure yellow to 40 for deep orange red) obtained by matching flower petals to sample colors in Maerz and Paul's *Dictionary of Color*. Analyze and interpret (note that the standard deviations appear unequal).

ANSWER: $v^*_{1,2} = 108.8$; all have probabilities less than 0.05 using the method of Games and Howell (see Box 13.3).

13.2 The number of bacteria in 1 cm^3 of milk from three cows was counted at three intervals (data from Park et al., 1924), yielding the following results:

	At time of milking	After 24 hours	After 48 hours
Cow No. 1	12,000	14,000	57,000
2	13,000	20,000	65,000
3	21,500	31,000	106,000

(a) Calculate means and variances for the three periods and examine the relation between these two statistics. Transform the observations to logarithms and compare means and variances based on the transformed data. Discuss.

(b) Carry out an anova on transformed and untransformed data. Discuss your results.

(c) Analyze the data by Friedman's method for randomized blocks.

13.3 Allee and Bowen (1932) studied survival time of goldfish (in minutes) when placed in colloidal silver suspensions. Experiment 9 involved 5 replications, and experiment 10 involved 10 replicates. Do the results of the two experiments differ? Addition of urea, NaCl, and Na_2S to a third series of suspensions apparently prolonged the life of the fish.

Colloidal silver		Urea and salts added
Experiment No. 9	Experiment No. 10	
210	150	330
180	180	300
240	210	300
210	240	420
210	240	360
	120	270

| | Colloidal silver | | Urea and |
Experiment No. 9	Experiment No. 10		salts added
	180		360
	240		360
	120		300
	150		120

Analyze and interpret. Check assumptions of normality and equality of variances. Compare anova results with those obtained using the Kruskal–Wallis test.

ANSWER: $H = 12.0047$

13.4 In a study of a sea cucumber (*Stichopus* sp.), measurements were made of the lengths of "buttons" (in micrometer units) in samples of tissue from three body regions for two specimens (data from David Olsen, unpublished results).

	Body regions										
Specimen		Dorsal				Lateral		Ventral			
No. 2	115	91	102	89	92	90	96	90	72	93
	88	84	80	73	90	91	98	114	89	81
	100	98	77	100	90	91	80	90	91	124
	96	87	91	111	93	82	93	87	111	92
	88	91	85	94	85	91	96	128	91	108
No. 9	81	108	92	88	113	112	91	104	96	107
	75	90	90	110	92	90	117	105	85	98
	90	96	100	91	96	105	104	91	90	80
	95	119	100	97	85	91	91	85	82	83
	99	88	91	95	88	92	81	80	89	90

Are the variances homogeneous? Are the distributions normal? Make histograms for the data in each cell of the two-way table and perform graphic analyses. Analyze by a two-way anova. Why can you show the effect of body regions? Is there anything disturbing about these data?

13.5 Test for a difference in surface and subsoil pH in the data of Exercise 11.1, using Wilcoxon's signed-ranks test.

ANSWER: $T_s = 38; P > 0.10$

13.6 Reanalyze the birth-weight data of Exercise 9.3 by using the Kolmogorov–Smirnov two-sample test.

13.7 Reanalyze the fecundity data of Box 9.5 by nonparametric methods (Mann–Whitney *U*-test and Kolmogorov–Smirnov two-sample test). Do your conclusions differ from those arrived at in Box 9.5?

ANSWER: No; $U_s = 31$ and $n_1 n_2 D = 21$

13.8 Check whether the butterfat data of Exercise 4.3 should be transformed by using the Box–Cox procedure. Perform a graphic analysis on both the original and the transformed data.

13.9 Allee et al. (1934) studied the rate of growth of *Ameiurus melas* in conditioned and unconditioned well water and obtained the following results for the gain in average length of sample fish. Although the original observations are not available, we may still test for differences between the two treatment classes.

	Average gain in length (in millimeters)	
Replicate	Conditioned water	Unconditioned water
1	2.20	1.06
2	1.05	0.06
3	3.25	3.55
4	2.60	1.00
5	1.90	1.10
6	1.50	0.60
7	2.25	1.30
8	1.00	0.90
9	-0.09	-0.59
10	0.83	0.58

ANSWER: $T_s = 3$, $P = 0.0049$ using Wilcoxon's signed-ranks test

13.10 Olson and Miller (1958) have published data on the lengths of the third molar of eight species of the condylarth *Hyopsodus*. Before an anova of their means (not shown) can be undertaken, we have to be certain that the variances within samples are homogenous. Carry out the possible tests, relying only on the information furnished in the following table.

(1) Groups ($a = 8$)	(2) n_i	(3) s_i^2
1	18	0.0707
2	13	0.1447
3	16	0.0836
4	17	0.0237
5	8	0.2189
6	11	0.1770
7	10	0.0791
8	10	0.2331

ANSWER: For Bartlett's test, adjusted $X^2 = 20.914$, $P = 0.0039$.

14 Linear Regression

We now turn to the simultaneous analysis of two variables. Even though we may have considered more than one variable at a time in the studies we have described thus far (e.g., seawater concentration and oxygen consumption in Box 11.2, or depth and water temperature in Box 11.3), our actual analyses were of only one variable. Frequently, however, we measure two or more variables on each individual; we would therefore like to be able to express more precisely the nature of the relationships between these variables. Thus, we come to the subjects of **regression** and **correlation.** In regression, we estimate the relationship of one continuous variable with another by expressing one in terms of a linear (or a more complex) function of the other. We also use regression to predict values of one variable as a function of the other. In correlation analysis, which is sometimes confused with regression, we estimate the degree to which two variables vary *together*, that is, covary. Chapter 15 deals with correlation, and we will postpone our effort to clarify the relation and distinction between regression and correlation until then. The variables involved in regression and correlation are continuous or, if meristic, are treated as though they were continuous (see Chapter 2). If the variables are qualitative (that is, if they are attributes), then the methods of regression and correlation cannot be used.

In Section 14.1, we review mathematical functions and introduce the new terminology required for regression analysis. Section 14.2 follows with a discussion of the appropriate statistical models for regression analysis. The basic computations in simple linear regression are shown in Section 14.3, which outlines the computation for the case of one dependent observation for each independent observation. Hypothesis tests and computation of confidence intervals for regression problems are discussed in Section 14.4, where we also discuss how to fit a regression line through the origin. Section 14.5 presents the case with several dependent observations for each independent observation.

Section 14.6 summarizes regression and discusses the various uses of regression analysis in biology. In this section, we also show how to estimate the relative efficiency of experimental designs in which more than one variable is measured. Section 14.7 examines a procedure known as *inverse prediction*, in which the most probable value of the independent observation is estimated from an observed dependent variable. In Section 14.8, we test differences between two regression coefficients.

In Section 14.9, we relate regression to the analysis of variance, showing how the sums of squares and degrees of freedom among groups can be partitioned into single-degree-of-freedom contrasts, and we demonstrate that these contrasts are really regression computations. Linear regression can be treated as such a contrast,

471

and the residual degrees of freedom can be used to test for departures from linearity in the regression relationship. Section 14.10 describes how to examine residuals from regression to detect outliers or systematic departures from the regression line. We also show how transformation of scale can straighten out curvilinear relationships for ease of analysis. When transformation cannot linearize the relation between variables, an alternative approach is a nonparametric test for regression. Such tests are illustrated in Section 14.11.

Most methods we have described here assume that the independent variable is controlled by the investigator or is measured without error. When both variables are subject to natural variation and the purpose of the analysis is to estimate the functional relationship between the two variables, then different techniques, usually referred to as Model II regression methods, are needed. We will cover these methods in Section 14.12.

Finally, in Section 14.13, we discuss methods for measuring effect sizes, power, and sample size in regression analysis.

14.1 Introduction to Regression

Much scientific thought concerns the relations between pairs of variables hypothesized to be in a cause-and-effect relationship. We will not discuss here the philosophical requirements for establishing whether the relationship between two variables is really one of cause and effect. We will be content with establishing the form and significance of *functional relationships* between two variables, leaving the demonstration of cause-and-effect relationships to the established procedures of the scientific method. A **function** is a mathematical relationship that enables us to predict what values of variable Y correspond to given values of a variable X. Such a relationship, generally written as $Y = f(X)$, is familiar to all of us from our general scientific and cultural experience. Nevertheless, as an introduction to the subject of regression, we will briefly review functions.

The simplest type of regression follows the equation $Y = X$, as illustrated in Figure 14.1. This graph shows the number of growth rings on a tree as a function of age in years (by drawing this as a continuous function, we are assuming we can estimate fractions of growth rings corresponding to fractions of years). Although the graph is self-explanatory, let us nevertheless examine the properties of the function $Y = X$. Whatever the value of X (age in years), the value of Y (number of growth rings) will be of corresponding magnitude. Thus, if the tree is 10 years old, it will have 10 growth rings. When $X = 0$, Y also equals zero, and the line describing the function goes through the origin of the coordinate system illustrated in Figure 14.1, indicating that a tree zero years old has no growth rings, a reasonable fact in terms of the biology of trees. Clearly, with the relationship $Y = X$, we can predict accurately the number of growth rings, given the age of a tree. We call the variable Y the **dependent variable** and X the **independent variable.** The magnitude of Y depends on the magnitude of X and can therefore be predicted from the independent variable, which presumably is free to vary. Remember that although a cause is always considered an

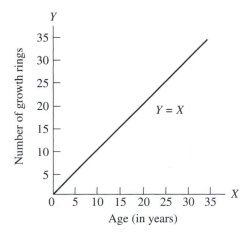

FIGURE 14.1 Number of growth rings on a tree as a function of age. An example of the simplest type of regression.

independent variable and an effect a dependent variable, a functional relationship observed in nature might not be a cause-and-effect relationship. It is conventional in statistics to label the dependent variable Y and the independent variable X.

Figure 14.2 shows another functional relationship, indicated by the equation $Y = bX$. Here the independent variable X is multiplied by a coefficient b, a slope factor. In this example, which relates the height of a plant in centimeters to its age in days, the slope factor $b = \frac{1}{7} = 0.143$, which means that for an increase of 7 units of X, there will be an increase of 1 unit of Y. Thus, a 14-day-old plant will be 2 cm tall. For any fixed scale, the slope of the function line depends on the magnitude of b, the slope factor. As b increases, the slope of the line becomes steeper. Note again that when $X = 0$, the dependent variable also equals zero. This makes sense; a plant that is zero days old has no height.

In biology, such a relationship is appropriate over only a limited range of values of X. Negative values of X are meaningless, and it is unlikely that the plant will continue to grow at a uniform rate. Quite probably the slope of the functional relationship will flatten out as the plant gets older and approaches its mature size. For a

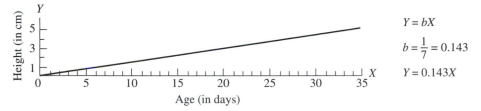

$Y = bX$

$b = \frac{1}{7} = 0.143$

$Y = 0.143X$

FIGURE 14.2 Height of a plant as a function of age.

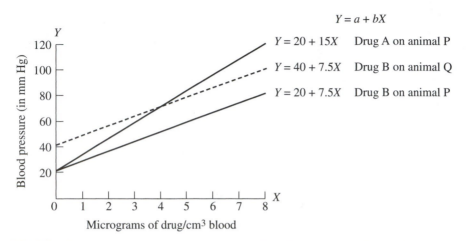

FIGURE 14.3 Blood pressure of an animal as a function of drug concentration in the blood.

limited portion of the range of variable X (age in days), however, the linear relationship $Y = bX$ may be an adequate description of the functional dependence of Y on X.

A third type of function is depicted in Figure 14.3, which illustrates the effect of two drugs on blood pressure in two species of animals. The relationships shown in this graph can be expressed by the formula $Y = a + bX$. The highest line is of the relationship $Y = 20 + 15X$, which represents the effect of drug A on animal P. The quantity of drug is measured in micrograms, the blood pressure in millimeters of mercury. From the equation, it is easy to calculate the blood pressure to be expected in the animal after a given amount of drug has been administered. Thus, after 4 μg of the drug have been given, the blood pressure will be $Y = 20 + (15)(4) = 80$ mm Hg. Note that by this formula, when the independent variable equals zero, the dependent variable does not also equal zero, but equals a. When $X = 0$, the blood pressure is 20 mm, which is the normal blood pressure of animal P in the absence of the drug. It would not be biologically reasonable to assume that the animal had no blood pressure in the absence of the drug. The magnitude 20 mm can be obtained by solving the equation after substituting zero for X or by examining the graph for the intersection of the function line with the Y-axis. This point yields the magnitude of a, which is therefore called the **Y-intercept.**

The two other functions in Figure 14.3 show the effects of varying both a, the Y-intercept, and b, the slope. In the lowest line, $Y = 20 + 7.5X$, the Y-intercept remains the same, but the slope has been halved. This line represents the effect of a different drug, B, on the same organism, P. Obviously, when no drug is administered, the blood pressure should be at the same Y-intercept because the identical organism is being studied. However, different drugs are likely to exert different hypertensive effects, as reflected by the different slopes. The third relationship (the broken line) also describes the effect of drug B, which is assumed to remain the same, but the experiment is carried out on a different species, Q, whose normal blood pressure is assumed to be 40 mm Hg. Thus, the equation for the effect of drug B on species Q is written as $Y = 40 + 7.5X$. Note that this line is exactly parallel to the one studied previously because it has the same slope.

You have probably noticed that the equation represented in Figure 14.3, $Y = a + bX$, is the most general linear equation and contains within it the two simpler versions from Figures 14.1 and 14.2. In $Y = bX$, we simply assume that $a = 0$; in $Y = X$, $a = 0$ and $b = 1$.

Those of you familiar with analytical geometry will have recognized the slope factor b as the *slope* of the function $Y = a + bX$, generally symbolized by m. If you know calculus, you will, of course, recognize b as the *derivative* of that same function ($dY/dX = b$). In biometry, b is called the **regression coefficient** and the function is called a **regression equation.** The origin of the word "regression" lies in biometric studies made by Francis Galton, in which the literal meaning of regression as "returning to" or "going back to" was appropriate; however, the accepted meaning in statistics now relates to the dependence of the means of variable Y on the independent variable X by some mathematical equation. When we wish to stress that the regression coefficient is of variable Y on variable X, we write $b_{Y \cdot X}$. If ever we wish to regress X on Y (when this is legitimate), the proper symbol for the regression coefficient is $b_{X \cdot Y}$.

14.2 Models in Regression

In any real example, observations do not lie perfectly along a regression line but scatter along both sides of the line. This scatter usually results from inherent, natural variation of the items (genetically and environmentally caused) and to measurement error. Thus, in regression a functional relationship does not mean that given X the value of Y must be $a + bX$, but rather that the mean (or expected value) of Y is $a + bX$.

The appropriate computations and significance tests in regression involve the following two models. The more common of these, **Model I regression,** is especially suitable in experimental situations. It is sometimes referred to as *ordinary least-squares* (OLS) *regression.* It is based on four assumptions:

1. The independent variable X is measured without error. We therefore say that the X's are "fixed," which means that whereas the dependent variable Y is a random variable, X does not vary at random, but rather is under the control of the investigator. In the example of Figure 14.3, for instance, the dose of drug which we varied at will was X, whereas the variable whose response we studied, blood pressure, was the random variable, Y. In this respect, we can manipulate X in a manner similar to that for the treatment effect in a Model I anova. However, as Warton et al. (2006) have pointed out, there is no inherent correspondence between Models I and II in anova and identically numbered models in regression, despite some superficial similarities.

2. The expected value for the variable Y for any given value X is described by the linear function $\mu_Y = \alpha + \beta X$. This relation is the same function we encountered in the previous section, but here we use Greek letters for a and b because we are describing a parametric relationship. Another way of stating this assumption is that the parametric means μ_Y of the values of Y are a function of X and lie on a straight line described by this equation.

3. For any given value X_i of X, the Y's are independently and normally distributed. This relationship can be represented by the equation $Y_{ij} = \alpha + \beta X_i + \epsilon_{ij}$, where the ϵ_{ij}'s are assumed to be normally distributed error terms with a mean of zero. Figure 14.4 illustrates this concept with a regression line similar to the ones in Figure 14.3. A given experiment can be repeated several times. Thus, for instance, we could administer 2, 4, 6, 8, and 10 μg of the drug to each of 20 individuals of an animal species and obtain a frequency distribution of blood pressure responses Y to the independent treatment values $X = 2, 4, 6, 8,$ and 10 μg. The inherent variability of biological material means that the responses to each dosage would not be the same in every individual; thus, we would obtain a distribution of values of Y (blood pressure) around the expected value. Assumption 3 states that these sample values will be independently and normally distributed, as is indicated by the normal curves superimposed about several points in the regression line in Figure 14.4. A few curves are shown to give you an idea of the scatter about the regression line. In actuality, there is, of course, a continuous scatter, as though these separate normal distributions were stacked right next to each other because there is an infinite number of possible intermediate values of X between any two dosages. In the rare cases in which the independent variable is discontinuous, the distributions of Y are physically separate from each other and occur only along the points of the abscissa that correspond to independent treatment values. An example of such a case is the weight of offspring (Y) in litters of various sizes in mice. There may be three or four offspring per litter, but there would be no intermediate value of X representing 3.25 mice per litter.

Not every experiment has more than one value of Y for each value of X. In fact, the basic computations we will learn in the next section apply to only one value of Y per X because this is the more common case. Even in such instances, however, the basic assumption of Model I regression is that the single observation of Y corresponding to the given value of X is a sample of size 1 from a population of independently and normally distributed observations.

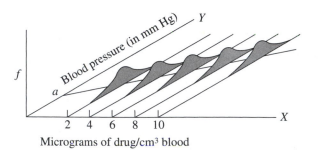

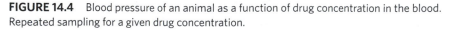

FIGURE 14.4 Blood pressure of an animal as a function of drug concentration in the blood. Repeated sampling for a given drug concentration.

4. The final assumption is familiar: The samples along the regression line are homoscedastic; that is, they have a common variance, σ^2, which is the variance of the ϵ's in the expression for assumption 3. Thus, we assume that the variance around the regression line is constant and hence is independent of the magnitude of X or Y.

Many regression analyses in biology do not meet the assumptions of Model I regression. Frequently, both X and Y are subject to natural variation and/or measurement error. Also the variable X may not be fixed, that is, under control of the investigator. Suppose we sample a population of female flies and measure wing length and total weight of each individual. We might be interested in studying wing length as a function of weight or we might wish to predict wing length for a given weight. In this case, the weight, which we treat as an independent variable, is not fixed and is certainly not the "cause" of differences in wing length. The weights of the flies vary for genetic and environmental reasons and are also subject to measurement error. In these cases, one is usually interested in good estimates of the slope itself rather than prediction or just showing that the slope is different from zero. **Model II regression** methods are used on datasets in which both variables show random variation. Some of the computations for estimating functional relationships and prediction in this model differ from the ones of Model I. The why's and how's of these differences will be taken up in Section 14.12.

As we shall see, in Model II regression the appropriate regression line may vary depending on whether functional relationship or prediction is the aim of the investigator. It is therefore important to keep this aim clearly in mind. In Model I regression (the familiar "regression" of statistics texts and research articles), the same equation serves both purposes. In the following sections, which are devoted to Model I regression, we therefore do not stress the distinction between these aims. We now turn to the rationale behind the computational steps in Model I regression.

14.3 The Linear Regression Equation

To learn the basic computations of a Model I linear regression, we choose an example with only one Y-value per independent variate X because this example is computationally simpler than dealing with samples and is by far the more common case encountered in research work. The extension to a sample of values of Y for each X is explained in Section 14.5.

The data for our discussion of regression come from a study of water loss in *Tribolium confusum,* the confused flour beetle. Nine batches of 25 beetles were weighed (individual beetles could not be weighed with available equipment), kept at different relative humidities, and weighed again after six days of starvation. Weight loss in milligrams was computed for each batch. This example is a Model I regression in which the weight loss is the dependent variable Y and the relative humidity is the independent variable X, a fixed treatment effect under the control of the experimenter. The purpose of the analysis is to establish whether the relationship between relative humidity and weight loss can be described adequately by a linear regression of the general form $Y = a + bX$. The original data are shown in columns (1) and (2) of

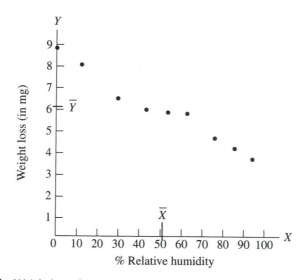

FIGURE 14.5 Weight loss of nine batches of 25 *Tribolium* beetles after six days of starvation at nine different relative humidities. Data from Table 14.1.

Table 14.1. They are plotted in Figure 14.5, from which it appears that a negative relationship exists between weight loss and humidity: as humidity increases, weight loss decreases. The means of weight loss and relative humidity, $\overline{Y}$ and $\overline{X}$, respectively, are marked along the coordinate axes. The average humidity is 50.39%, and the average weight loss is 6.022 mg. How can we fit a regression line to these data, permitting us to estimate a value of Y for a given value of X?

Unless the observations lie exactly on a straight line, we need a criterion for determining the best possible placing of the regression line. Statisticians generally follow the principle of least squares, which we first encountered in Chapter 4. If we were to draw a horizontal line through $\overline{X}, \overline{Y}$ (that is, a line parallel to the X-axis at the level of $\overline{Y}$), then deviations to that line drawn parallel to the Y-axis would represent the deviations from the mean for these observations with respect to variable Y (as in Figure 14.6). We learned in Chapter 4 that the sum of these observations, $\Sigma(Y - \overline{Y}) = \Sigma y = 0$. The sum of squares of these deviations, $\Sigma(Y - \overline{Y})^2 = \Sigma y^2$, is less than those from any other horizontal line. Any horizontal line drawn through the data at a point other than $\overline{Y}$ would yield a sum of deviations other than zero and a sum of deviations squared greater than Σy^2. Therefore, a mathematically correct but impractical method for finding the mean of Y would be to draw a series of horizontal lines across a graph, calculate the sum of squares of deviations from it, and choose the line yielding the smallest sum of squares.

In linear regression, we again draw a straight line through our observations, but the line is not necessarily horizontal. A sloped regression line indicates for each value of the independent variable X_i an estimated value of the dependent variable.

TABLE 14.1 Basic Computations in Regression

Weight loss (in milligrams) of nine batches of 25 Tribolium beetles after six days of starvation at nine different humidities.

(1) Percent relative humidity X	(2) Weight loss in mg Y	(3) $x = (X - \bar{X})$	(4) $y = (Y - \bar{Y})$	(5) x^2	(6) xy	(7) y^2	(8) $\hat{Y}$	(9) $d_{Y \cdot X} = (Y - \hat{Y})$	(10) $d^2_{Y \cdot X}$	(11) $\hat{y} = (\hat{Y} - \bar{Y})$	(12) $\hat{y}^2$
0	8.98	−50.39	2.958	2539.1521	−149.0536	8.7498	8.7038	0.2762	0.0763	2.6818	7.1921
12	8.14	−38.39	2.118	1473.7921	−81.3100	4.4859	8.0652	0.0748	0.0056	2.0432	4.1747
29.5	6.67	−20.89	0.648	436.3921	−1.5367	0.4199	7.1338	−0.4638	0.2151	1.1118	1.2361
43	6.08	−7.39	0.058	54.6121	−0.4286	0.0034	6.4153	−0.3353	0.1124	0.3933	0.1547
53	5.90	2.61	−0.122	6.8121	−0.3184	0.0149	5.8831	0.0169	0.0003	−0.1389	0.0193
62.5	5.83	12.11	−0.192	146.6521	−2.3251	0.0369	5.3776	0.4524	0.2047	−0.6444	0.4153
75.5	4.68	25.11	−1.342	630.5121	−33.6976	1.8010	4.6857	−0.0057	0.0000	−1.3363	1.7857
85	4.20	34.61	−1.822	1197.8521	−63.0594	3.3197	4.1801	0.0199	0.0004	−1.8419	3.3926
93	3.72	42.61	−2.302	1815.6121	−98.0882	5.2992	3.7543	−0.0343	0.0012	−2.2677	5.1425
Sum 453.5	54.20	−0.01	0.002	8301.3889	−441.8176	24.1307	54.1989	0.0011	0.6160	0.0009	23.5130
Mean 50.39	6.022						6.022				
Sum/$(n-1)$				1037.6736	−55.2272	3.0163			0.0880[a]		

SOURCE: Nelson (1964).

[a] Sum divided by $n - 2$.

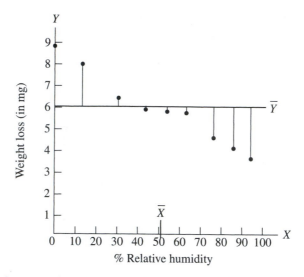

FIGURE 14.6   Deviations from the mean of Y for the data of Figure 14.5.

To distinguish between the estimated values of Y_i and the observed values, we will hereafter designate the former as $\hat{Y}$ (read: Y-hat or Y-caret); the latter will remain Y_i. The regression equation therefore should read

$$\hat{Y} = a + bX \tag{14.1}$$

which indicates that for given values of X, this equation provides estimated values $\hat{Y}$ (as distinct from the observed values Y in any actual case). The deviation of an observation Y_i from the regression line is $(Y_i - \hat{Y}_i)$ and is generally symbolized as $d_{Y \cdot X}$. These deviations are drawn parallel to the Y-axis, and they meet the sloped regression line at an angle (Figure 14.7). The sum of these deviations is again zero ($\Sigma d_{Y \cdot X} = 0$), and the sum of their squares yields a quantity $\Sigma(Y - \hat{Y}_2) = \Sigma d_{Y \cdot X}^2$ analogous to the sum of squares Σy^2. For reasons that will become clear later, $\Sigma d_{Y \cdot X}^2$ is called the unexplained or residual sum of squares. The **least squares linear regression line** through a set of points is defined as the straight line that results in $\Sigma d_{Y \cdot X}^2$ being at a minimum. Geometrically, the basic idea is that we prefer using a line that is in some sense close to as many points as possible. For ordinary Model I regression analysis, it is most useful to define closeness in terms of the vertical distances from the points to a line, using the line that makes the sum of these squared deviations as small as possible.

A convenient consequence of this criterion is that the line must pass through the point $\overline{X}, \overline{Y}$. It would be possible but impractical to calculate the correct regression slope by pivoting a ruler around the point $\overline{X}, \overline{Y}$ and calculating the unexplained sum of squares, $\Sigma d_{Y \cdot X}^2$, for each of the innumerable possible positions. Whichever position gave the minimal value of $\Sigma d_{Y \cdot X}^2$ would be the least squares regression line. The usual derivation of the formula for the slope of a line yielding a minimal value of

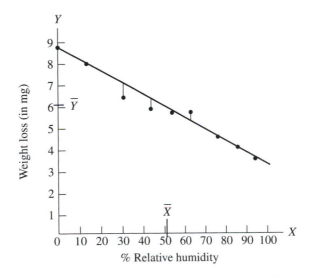

FIGURE 14.7   Deviations from the regression line for the data of Figure 14.5.

$\Sigma d_{Y\cdot X}^2$ uses calculus (see Appendix A.8). We can obtain equations for b and a by plain algebra as well (see Appendix A.9). Either results in the formula for the regression coefficient of Y on X, $b_{Y\cdot X} = \Sigma xy/\Sigma x^2$. We will now calculate this quantity for our weight loss data.

We first compute the deviations from the respective means of X and Y as shown in columns (3) and (4) of Table 14.1. The sums of these deviations, Σx and Σy, differ slightly from their expected value of zero because of rounding errors. The squares of these deviations yield sums of squares and variances in columns (5) and (7). In column (6) are the products xy, which in this example are all negative because the x- and y-deviations are of unlike sign. The sum of these products, $\Sigma''xy$, is a new quantity, called the **sum of products.** This poor but well-established term refers to Σxy, the sum of the products of the deviations, rather than ΣXY, the sum of the products of the observations. Recall that Σy^2 is called the *sum of squares,* whereas ΣY^2 is the sum of the squared observations. The sum of products is analogous to the sum of squares. When divided by the degrees of freedom, the sum of products yields the **covariance,** just as the variance results from a similar division of the sum of squares. (We first encountered covariances in Section 8.4.) Note that the sum of products can be negative as well as positive. A negative sum of products indicates a negative slope of the regression line: as X increases, Y decreases. In this respect, the sum of products differs from a sum of squares, which can be only positive. From Table 14.1 we find that $\Sigma xy = -441.8176$, $\Sigma x^2 = 8301.3889$, and $b = \Sigma xy/\Sigma x^2 = -0.05322$. Thus, for a one-unit increase in X, there is a decrease of 0.05322 units in Y. In terms of the present example, a 1% increase in relative humidity indicates a weight loss of 0.05322 mg.

You may wish to convince yourself that the formula for the regression coefficient is intuitively reasonable. This formula is the ratio of the sum of *products* of deviations for X and Y over the sum of *squares* of deviations for X. The product for X_i, a single value of X, is $x_i y_i$. Similarly, the squared deviation for X_i is x_i^2 or $x_i x_i$. The ratio $x_i y_i / x_i x_i$ thus reduces to y_i / x_i. Although $\Sigma xy / \Sigma x^2$ only approximates the average of y_i / x_i for the n values of X_i, the latter ratio indicates the direction and magnitude of the change in Y for a unit change in X. Thus, if y_i on the average equals x_i, b will equal 1. When $y_i = -x_i$, $b = -1$. Also, when $|y_i| > |x_i|$, $b > |1|$, and conversely when $|y_i| < |x_i|$, $b < |1|$.

How can we complete the equation $\hat{Y} = a + bX$? We have stated that the regression line passes through the point $\overline{X}, \overline{Y}$. At $\overline{X} = 50.39$, $\hat{Y} = 6.022$; that is, we use $\overline{Y}$, the observed mean of Y, as an estimate $\hat{Y}$ of the mean. We can substitute these means into Expression (14.1):

$$\hat{Y} = a + bX$$

$$\overline{Y} = a + b\overline{X}$$

$$a = \overline{Y} - b\overline{X}$$

$$a = 6.022 - (-0.05322)(50.39)$$

$$= 8.7038$$

Therefore, $\hat{Y} = 8.7038 - 0.05322X$.

This equation predicts weight loss from relative humidity. Note that when X is zero (0% relative humidity), the estimated weight loss is greatest, equal to $a = 8.7038$ mg. As X increases to a maximum of 100, however, the weight loss decreases to 3.3818 mg.

We can use the regression formula to draw the regression line. Simply estimate $\hat{Y}$ at two convenient points of X such as $X = 0$ and $X = 100$ and draw a straight line between them. This line, added to the observed data, is shown in Figure 14.8. Note that it goes through the point $\overline{X}, \overline{Y}$. In fact, to draw the regression line, we frequently use the intersection of the two means and one other point.

Now bear with us through some very elementary algebra. Because

$$a = \overline{Y} - b\overline{X}$$

we can write Expression (14.1), $\hat{Y} = a + bX$, as

$$\hat{Y} = (\overline{Y} - b\overline{X}) + bX$$
$$= \overline{Y} + b(X - \overline{X})$$

Therefore,

$$\hat{Y} = \overline{Y} + bx \tag{14.2}$$

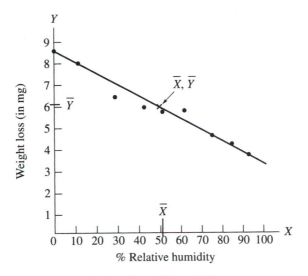

FIGURE 14.8 Linear regression fitted to data of Figure 14.5.

Also,

$$\hat{Y} - \bar{Y} = bx$$

$$\hat{y} = bx \qquad (14.3)$$

where $\hat{y}$ is defined as the deviation $\hat{Y} - \bar{Y}$. Next, using Expressions (14.1) or (14.2), we estimate $\hat{Y}$ for every one of our given values of X. The estimated values $\hat{Y}$ are shown in column (8) of Table 14.1. Compare them with the observed values of Y in column (2). Overall agreement between the two columns of values is good. Note that except for rounding errors, $\Sigma\hat{Y} = \Sigma Y$ and hence $\bar{\hat{Y}} = \bar{Y}$. Observed Y-values, however, usually are different from the estimated values $\hat{Y}$ because of individual variation around the regression line.

The deviations of each observed Y-value from its estimated value $(Y - \hat{Y}) = d_{Y \cdot X}$ are listed in column (9) of Table 14.1. Note that these deviations exhibit one of the properties of deviations from a mean: They sum to zero except for rounding errors. Thus, $\Sigma d_{Y \cdot X} = 0$, just as $\Sigma y = 0$. Next, we compute in column (10) the squares of these deviations and sum them to give a new sum of squares, $\Sigma d_{Y \cdot X}^2 = 0.6160$. A comparison of $\Sigma(Y - \bar{Y})^2 = \Sigma y^2 = 24.1307$ with $\Sigma(Y - \hat{Y})^2 = \Sigma d_{Y \cdot X}^2 = 0.6160$ shows that the new sum of squares is much less than the previous one. What has caused this reduction? Allowing for its relationship with X has eliminated most of the variance of Y from the sample. Remaining is the *residual* or **unexplained sum of squares** $\Sigma d_{Y \cdot X}^2$, which expresses that portion of the total SS of Y which is not accounted for by differences in X. This value is unexplained with respect to X. The difference between the total SS, Σy^2, and the unexplained SS, $\Sigma d_{Y \cdot X}^2$, is (not surprisingly) called the **explained sum of squares** or the *sum of squares due to regression*, $\Sigma\hat{y}^2$, and is based on the deviations $\hat{y} = \hat{Y} - \bar{Y}$. The computation of these deviations and their squares is shown in columns (11) and (12) of Table 14.1. Note that $\Sigma\hat{y}$ approximates zero and that $\Sigma\hat{y}^2 = 25.5130$. Adding the unexplained SS (0.6160) to this value results in $\Sigma y^2 = \Sigma\hat{y}^2 + \Sigma d_{Y\cdot X}^2 = 24.1290$, which is equal (except for rounding errors) to the independently calculated value of 24.1307 in column (7).

We will return to the meaning of the unexplained and explained sums of squares later. We retain the terms "explained" and "unexplained" for regression sums of squares in this book because they are well established, but we must strongly caution the reader not to read too much into them. The explanation refers to how well the function $\hat{Y} = a + bX$ fits the observed data, not necessarily to any causal explanation of variable Y by variable X. One sometimes sees the phrase "sum of squares accounted for by X" used instead.

For the regression equation in cases where there is a single value of Y for each value of X, the regression coefficient $\Sigma xy/\Sigma x^2$ can be computed as follows:

$$b_{Y\cdot X} = \frac{\sum_{}^{n}(X - \bar{X})(Y - \bar{Y})}{\sum_{}^{n}(X - \bar{X})^2} \tag{14.4}$$

We have already learned that the numerator of this expression is called the sum of products of X and Y, and we recognize the denominator as the sum of squares of X.

To compute regression statistics, we need eight quantities initially: n, ΣX, ΣY, $\bar{X}$, $\bar{Y}$, Σx^2, Σy^2, and Σxy. From these quantities the regression equation is calculated as shown in Box 14.1, which also illustrates computation of the explained sum of squares, $\Sigma\hat{y}^2 = \Sigma(\hat{Y} - \bar{Y})^2$, and the unexplained sum of squares, $\Sigma d_{Y\cdot X}^2 = \Sigma(Y - \hat{Y})^2$. There are also convenient formulas for these two quantities, but they have to be used with caution because they may be sensitive to rounding errors. That

$$\sum d_{Y\cdot X}^2 = \sum y^2 - \frac{\left(\sum xy\right)^2}{\sum x^2} \tag{14.5}$$

is demonstrated at the end of Appendix A.9. The term subtracted from Σy^2 is the explained sum of squares, as we show here:

$$\sum \hat{y}^2 = \sum b^2 x^2 = b^2 \sum x^2 = \frac{\left(\sum xy\right)^2}{\left(\sum x^2\right)^2}\sum x^2$$

$$\sum \hat{y}^2 = \frac{\left(\sum xy\right)^2}{\sum x^2} \tag{14.6}$$

The BIOM-pc package of computer programs calculates linear regression with one or more values of Y per value of X.

BOX 14.1 Computation of Regression Statistics: Single Value of Y for Each Value of X

Data from Table 14.1.

Weight loss in mg (Y)	8.98	8.14	6.67	6.08	5.90	5.83	4.68	4.20	3.72
Percent relative humidity (X)	0	12.0	29.5	43.0	53.0	62.5	75.5	85.0	93.0

Computation

1. Compute sample size, sums, means, sums of squares, and the sum of products.

$$n = 9$$

$$\sum X = 453.5 \qquad \sum Y = 54.20$$
$$\overline{X} = 50.389 \qquad \overline{Y} = 6.022$$
$$\sum x^2 = 8301.3889 \qquad \sum y^2 = 24.1306$$
$$\sum xy = \sum (X - \overline{X})(Y - \overline{Y}) = -441.8178$$

2. The regression coefficient is

$$b_{Y \cdot X} = \frac{\sum xy}{\sum x^2} = \frac{-441.8178}{8301.3889} = -0.05322$$

3. The Y-intercept is

$$a = \overline{Y} - b_{Y \cdot X}\overline{X} = 6.022 - (-0.05322)(50.389) = 8.7040$$

4. The explained sum of squares is

$$\sum \hat{y}^2 = \sum (\hat{Y} - \overline{Y})^2 \quad \text{or}$$
$$\frac{(\sum xy)^2}{\sum x^2} = \frac{(-441.8178)^2}{8301.3889} = 23.5145$$

5. The unexplained sum of squares is

$$\sum d_{Y \cdot X}^2 = \sum (Y - \hat{Y})^2 = 0.6161$$

14.4 Hypothesis Testing in Regression

Up to now we have interpreted regression as a method for providing an estimate, $\hat{Y}_i$, given a value of X_i. Regression can also be interpreted as a method for accounting for some of the variation of the dependent variable Y in terms of variation of the independent variable X. The SS of a sample of Y-values, $\sum y^2$, is computed by summing and squaring deviations, $y = Y - \overline{Y}$. Figure 14.9 shows that the deviation y can

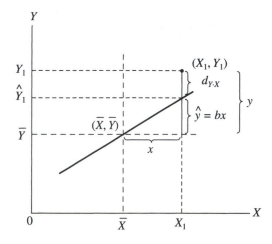

FIGURE 14.9 Schematic diagram showing the relations involved in partitioning the sum of squares of the dependent variable.

be decomposed into two parts, $\hat{y}$ and $d_{Y \cdot X}$. Figure 14.9 also shows that the deviation $\hat{y} = \hat{Y} - \overline{Y}$ represents the deviation of the estimated value $\hat{Y}$ from the mean of Y. The height of $\hat{y}$ is clearly a function of x. We have already seen that $\hat{y} = bx$ [see Expression (14.3)]. Those of you familiar with analytical geometry will recognize this expression as the point-slope form of the equation. If b, the slope of the regression line, were steeper, $\hat{y}$ would be relatively larger for a given value of x. The remaining portion of the deviation y, $d_{Y \cdot X}$, represents the residual variation of the variable Y after the explained variation has been subtracted. We can see that $y = \hat{y} + d_{Y \cdot X}$ by writing out these deviations explicitly: $Y - \overline{Y} = (\hat{Y} - \overline{Y}) + (Y - \hat{Y})$.

For each of these deviations, we can compute a corresponding sum of squares. From Expression (14.5) we obtain

$$\sum y^2 = \frac{\left(\sum xy\right)^2}{\sum x^2} + \sum d_{Y \cdot X}^2$$

Of course, $\sum y^2$ corresponds to y, $\sum d_{Y \cdot X}^2$ to $d_{Y \cdot X}$, and

$$\sum \hat{y}^2 = \frac{\left(\sum xy\right)^2}{\sum x^2}$$

corresponds to $\hat{y}$ (as shown in the previous section). Thus, we are able to partition the sum of squares of the dependent variable in regression in a way analogous to the partition of the total SS in analysis of variance (see Section 8.5). You may wonder how the additive relation of the deviations can be matched by an additive relation of their squares without the presence of any cross products. Simple algebra (see Appendix A.10) shows that the cross products cancel out. The magnitude of the unexplained deviation $(d_{Y \cdot X})$ is independent of the magnitude of the explained deviation $(\hat{y})$ just as in anovas the magnitude of the deviation of an item from the sample mean is independent of the magnitude of the deviation of the sample mean from the grand mean. This relationship between regression and analysis of variance can be carried further. We can undertake an analysis of variance of the partitioned sums of squares as follows:

	Source of Variation	df	SS	MS	Expected MS
$\hat{Y} - \bar{Y}$	Explained (estimated Y from mean of Y)	1	$\Sigma \hat{y}^2 = \Sigma(\hat{Y} - \bar{Y})^2$	$s_{\hat{Y}}^2$	$\sigma_{Y \cdot X}^2 + \beta^2 \Sigma x^2$
$Y - \hat{Y}$	Unexplained, error (observed Y from mean of Y)	$n - 2$	$\Sigma d_{Y \cdot X}^2 = \Sigma(Y - \hat{Y})^2$ $= \Sigma y^2 - \Sigma \hat{y}^2$	$s_{Y \cdot X}^2$	$\sigma_{Y \cdot X}^2$
$Y - \bar{Y}$	Total (observed Y from mean of Y)	$n - 1$	$\Sigma y^2 = \Sigma(Y - \bar{Y})^2$	s_Y^2	

The **explained mean square,** or *mean square due to linear regression,* measures the amount of variation in Y accounted for by linear variation in X. The explained mean square is tested over the **unexplained mean square,** which measures the residual variation and is used as an error *MS*. The mean square due to linear regression, $s_{\hat{Y}}^2$, is based on one degree of freedom, and consequently $n - 2$ df remain for the error *MS* because the total sum of squares possesses $n - 1$ degrees of freedom. The test is of the null hypothesis $H_0: \beta = 0$. Carrying out such an anova on the weight loss data of Box 14.1 produces the following results:

Source of Variation	df	SS	MS	F_s	P
Explained: due to linear regression	1	23.5145	23.5145	267.18	7.816×10^{-7}
Unexplained: error around regression line	7	0.6161	0.08801		
Total	8	24.1306			

The hypothesis test is $F_s = s_{\hat{Y}}^2 / s_{Y \cdot X}^2$. The observed value of F_s shows that a large portion of the variation of Y has been accounted for by regression on X. We can estimate this portion by the fraction $\Sigma \hat{y}^2 / \Sigma y^2 = 23.5145/24.1306 = 0.9745$. Thus, 97.45% of the variation of Y can be accounted for by the variation of X. This ratio is known as the *coefficient of determination* (see Sections 15.2 and 15.9) and is symbolized by r^2 or R^2. The probability that this coefficient has been sampled from a population with a true value of 0 approaches one in a million.

We now proceed to the computation of the standard errors for various regression statistics, their employment in tests of hypotheses, and the computation of confidence limits. Box 14.2 lists these standard errors in two columns. Column (3) is for the case of more than one Y-value for each value of X, which will be discussed in Section 14.5. Column (4) is for the case with a single Y-value for each value of X. The first row of the table gives the *standard error of the regression coefficient,* which is simply the square root of the quotient of the unexplained variance divided by the sum

BOX 14.2 Standard Errors of Regression Statistics and Their Degrees of Freedom

For explanation of this box, see Sections 14.4 and 14.5; v = degrees of freedom; a = number of values of X when there are n_i Y-values for each X; n = sample size when there is a single Y-value for each value of X.

(1) Statistic	(2)	(3) More than one Y-value for each value of X	(4) Single Y-value for each value of X
Regression coefficient b_{YX} (derivation in appendix Section A.11)	s_b	$\sqrt{\dfrac{s_{Y \cdot X}^2}{\sum x^2}}$ $\qquad v = a - 2$	$\sqrt{\dfrac{s_{Y \cdot X}^2}{\sum x^2}}$ $\qquad v = n - 2$
Sample mean $\bar{Y}_i$ or $\bar{Y}$	$s_{\bar{Y}}$	$\bar{Y}_i$ for group i $\sqrt{\dfrac{MS_{\text{within}}}{n_i}}$ $\qquad v = \sum n_i - a$	$\bar{Y}$ at $\bar{X}$ $\sqrt{\dfrac{s_{Y \cdot X}^2}{n}}$ $\qquad v = n - 2$
Estimated Y for a given value $X_i\,\hat{Y}_i$ (derivation in appendix Section A.12)	$s_{\hat{Y}}$	$\sqrt{s_{Y \cdot X}^2 \left[\dfrac{1}{\sum n_i} + \dfrac{(X_i - \bar{X})^2}{\sum x^2}\right]}$ $\left(v = \sum n_i - 2 \text{ if pooled } s_{Y \cdot X}^2 \text{ is employed}\right)$	$\sqrt{s_{Y \cdot X}^2 \left[\dfrac{1}{n} + \dfrac{(X_i - \bar{X})^2}{\sum x^2}\right]}$ $\qquad v = n - 2$

Box 14.2 (continued)

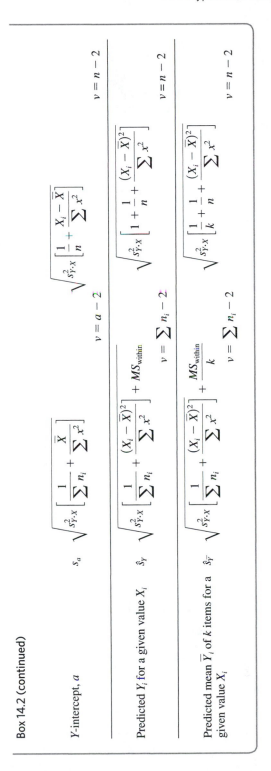

Y-intercept, a

$$s_a \quad \sqrt{s^2_{Y\cdot X}\left[\frac{1}{\sum n_i}+\frac{\bar X}{\sum x^2}\right]} \qquad \sqrt{s^2_{Y\cdot X}\left[\frac{1}{n}+\frac{X_i-\bar X}{\sum x^2}\right]} \qquad \nu = n-2$$

$$\nu = a - 2$$

Predicted Y_i for a given value X_i

$$\hat s_Y \quad \sqrt{s^2_{Y\cdot X}\left[\frac{1}{\sum n_i}+\frac{(X_i-\bar X)^2}{\sum x^2}\right]+MS_{within}} \qquad \sqrt{s^2_{Y\cdot X}\left[1+\frac{1}{n}+\frac{(X_i-\bar X)^2}{\sum x^2}\right]} \qquad \nu = n-2$$

$$\nu = \sum n_i - 2$$

Predicted mean $\bar Y_i$ of k items for a given value X_i

$$\hat s_{\bar Y} \quad \sqrt{s^2_{Y\cdot X}\left[\frac{1}{\sum n_i}+\frac{(X_i-\bar X)^2}{\sum x^2}\right]+\frac{MS_{within}}{k}} \qquad \sqrt{s^2_{Y\cdot X}\left[\frac{1}{k}+\frac{1}{n}+\frac{(X_i-\bar X)^2}{\sum x^2}\right]} \qquad \nu = n-2$$

$$\nu = \sum n_i - 2$$

of squares of X. Note that the unexplained variance $s^2_{Y \cdot X}$ is a fundamental quantity that is a part of all standard errors in regression. The standard error of the regression coefficient permits us to test various hypotheses and to set confidence limits to our sample estimate of β because it is normally distributed if the $d_{Y \cdot X}$ are normally distributed. The computation of s_b is illustrated in step **1** of Box 14.3, using the weight loss example of Box 14.1.

BOX 14.3 **Hypothesis Tests and Computation of Confidence Limits of Regression Statistics: Single Value of Y for Each Value of X**

Based on standard errors and degrees of freedom of Box 14.2; using example of Box 14.1.

$$n = 9 \qquad \bar{X} = 50.389 \qquad \bar{Y} = 6.022$$

$$b_{Y \cdot X} = -0.05322 \qquad \sum x^2 = 8301.3889$$

$$s^2_{Y \cdot X} = \frac{\sum d^2_{Y \cdot X}}{(n-2)} = \frac{0.6161}{7} = 0.08801$$

$$a = \bar{Y} - b\bar{X} = 6.022 - (-0.05322)50.389 = 8.7037$$

Computation

1. Standard error of the regression coefficient:

$$s_b = \sqrt{\frac{s^2_{Y \cdot X}}{\sum x^2}} = \sqrt{\frac{0.08801}{8301.3889}} = 0.003{,}256{,}1$$

2. Testing whether regression coefficient $\beta = 0$:

$$t_s = \frac{(b-0)}{s_b} = \frac{-0.053{,}22}{0.003{,}2561} = -16.345$$

$$t_{.001[7]} = 5.408 \qquad P = 7.819 \times 10^{-7}$$

3. 95% confidence limits for regression coefficient:

$$t_{.05[7]}s_b = 2.365(0.003{,}256{,}1) = 0.00770$$

$$L_1 = b - t_{.05[7]}s_b = -0.05322 - 0.00770 = -0.06092$$

$$L_2 = b + t_{.05[7]}s_b = -0.5322 + 0.00770 = -0.04552$$

4. Standard error of the sampled mean $\bar{Y}$ (at $\bar{X}$):

$$s_{\bar{Y}} = \sqrt{\frac{s^2_{Y \cdot X}}{n}} = \sqrt{\frac{0.08801}{9}} = 0.098{,}888{,}3$$

5. 95% confidence limits for the mean μ_Y corresponding to $\bar{X}$ ($\bar{Y} = 6.022$):

$$t_{.05[7]}s_{\bar{Y}} = 2.365(0.098{,}888{,}3) = 0.233871$$

$$L_1 = \bar{Y} - t_{.05[7]}s_{\bar{Y}} = 6.022 - 0.2339 = 5.7881$$

$$L_2 = \bar{Y} + t_{.05[7]}s_{\bar{Y}} = 6.022 + 0.2339 = 6.2559$$

Box 14.3 (continued)

6. Standard error of $\hat{Y}$, an estimated Y for a given value of X_i:

$$s_{\hat{Y}} = \sqrt{s_{Y \cdot X}^2 \left[\frac{1}{n} + \frac{(X_i - \bar{X})^2}{\sum x^2} \right]}$$

For example, for $X_i = 100\%$ relative humidity,

$$s_{\hat{Y}} = \sqrt{0.08801 \left[\frac{1}{9} + \frac{(100 - 50.389)^2}{8301.3889} \right]} = 0.18940$$

7. 95% confidence limits for μ_{Y_i} corresponding to the estimate $\hat{Y}_i = 3.3817$ at $X_i = 100\%$ relative humidity:

$$t_{.05[7]}s_{\hat{Y}} = 2.365(0.18940) = 0.44793$$

$$L_1 = \hat{Y}_i - t_{.05[7]}s_{\hat{Y}} = 3.3817 - 0.4479 = 2.9338$$

$$L_2 = \hat{Y}_i + t_{.05[7]}s_{\hat{Y}} = 3.3817 + 0.4479 = 3.8296$$

8. Standard error of a predicted mean $\bar{Y}_i$ to be obtained in a new experiment run at $X_i = 100\%$ relative humidity. Our best prediction for this mean would be $\bar{Y}_i = \hat{Y}_i = 3.3817$. If the new experiment were based on a sample size of $k = 5$, the standard error of the predicted mean would be

$$\hat{s}_{\bar{Y}} = \sqrt{s_{Y \cdot X}^2 \left[\frac{1}{k} + \frac{1}{n} + \frac{(X_i - \bar{X})^2}{\sum x^2} \right]}$$

$$= \sqrt{0.08801 \left[\frac{1}{5} + \frac{1}{9} + \frac{(100 - 50.389)^2}{8301.3889} \right]} = 0.23124$$

9. 95% prediction limits for a sample mean of 5 weight losses at 100% relative humidity (using the standard error computed above):

$$t_{.05[7]}\hat{s}_{\bar{Y}} = 2.365(0.23124) = 0.5469$$

$$L_1 = \hat{Y}_i - t_{.05[7]}\hat{s}_{\bar{Y}} = 3.3817 - 0.5469 = 2.8348$$

$$L_2 = \hat{Y}_i + t_{.05[7]}\hat{s}_{\bar{Y}} = 3.3817 + 0.5469 = 3.9286$$

10. Standard error of the Y-intercept, a:

This is a special case of the standard error for an estimated Y at $X = 0$.

$$s_a = \sqrt{s_{Y \cdot X}^2 \left[\frac{1}{n} + \frac{\bar{X}^2}{\sum x^2} \right]}$$

$$= \sqrt{0.08801 \left[\frac{1}{9} + \frac{(50.389)^2}{8301.3889} \right]} = 0.19156$$

11. Testing whether the Y-intercept $a = 0$:

$$t_s = \frac{a - 0}{s_a} = \frac{8.7037}{0.19156} = 45.43589 \qquad P < 10^{-15}$$

The hypothesis test illustrated in step **2** tests the *"significance" of the regression coefficient*; that is, it tests the null hypothesis that the sample value of b comes from a population with a parametric value $\beta = 0$ for the regression coefficient. This is a t-test, the appropriate degrees of freedom being $n - 2 = 7$. If we cannot reject the null hypothesis, there is no evidence that the regression is significantly deviant from zero in either the positive or negative direction. Our conclusions for the weight loss data are that a highly significant negative regression is present. In Section 9.4, we saw that $t^2 = F$. The square of $t_s = -16.345$ from Box 14.3 is 267.16, which (within rounding error) equals the value of F_s found in the anova earlier in this section. The associated probabilities of t and F are also identical within rounding error. The significance test in step **2** of Box 14.3 could, of course, also have been used to test whether b is significantly different from a parametric value β other than zero.

Setting confidence limits to the regression coefficient presents no new features. The computation is shown in step **3** of Box 14.3. Because s_b is small, the confidence interval is narrow. The confidence limits are shown in Figure 14.10 as dashed lines representing the 95% bounds of the slope. Note that the regression line, as well as its confidence limits, pass through the means for X and Y. Variation in b therefore rotates the regression line about the point $\overline{X}, \overline{Y}$.

Next, we calculate a *standard error for the observed sample mean* $\overline{Y}$. Recall from Section 7.1 that $s_{\overline{Y}}^2 = s_Y^2/n$. Now that we have regressed Y on X, however, we are able to account for (that is, hold constant) some of the variation of Y in terms of the variation of X. The variance of Y around the point $\overline{X}, \overline{Y}$ on the regression line is less than s_Y^2; it is $s_{Y\cdot X}^2$. At $\overline{X}$ we may therefore compute confidence limits of $\overline{Y}$, using as a standard error of the mean $s_{\overline{Y}} = \sqrt{s_{Y\cdot X}^2/n}$ with $n - 2$ degrees of freedom. This

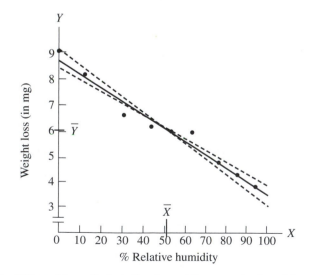

FIGURE 14.10   95% confidence limits to the slope of the regression line of Figure 14.8.

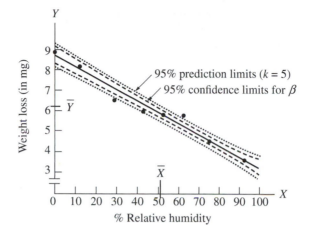

FIGURE 14.11   95% confidence limits to regression estimates for data of Figure 14.8.

standard error is computed in step **4** of Box 14.3, and 95% confidence limits for the sampled mean $\bar{Y}$ at $\bar{X}$ are calculated in step **5**. These limits (5.7881–6.2559) are considerably narrower than the confidence limits for the mean based on the conventional standard error $s_{\bar{Y}}$, which would be from 4.687 to 7.357. Thus, knowing the relative humidity greatly reduces the uncertainty in weight loss.

The standard error for $\bar{Y}$ is only a special case of the *standard error for any estimated value $\hat{Y}$ along the regression line*. A new factor now enters the error variance. The magnitude of this factor is in part a function of the distance of a given value X_i from its mean $\bar{X}$. The farther away X_i is from its mean, the greater will be the error of estimate. This factor is shown in the third row of Box 14.2 as the deviation $X_i - \bar{X}$, squared and divided by the sum of squares of X. The standard error for an estimate $\hat{Y}_i$ for a relative humidity $X_i = 100\%$ is given in step **6** of Box 14.3. The 95% confidence limits for $\mu_{\hat{Y}_i}$, the parametric value corresponding to the estimate $\hat{Y}_i$, are shown in step **7**. Note that the width of the confidence interval is $3.8296 - 2.9338 = 0.8958$, considerably wider than the confidence interval at $\bar{X}$ calculated in step **5**, which was $6.2559 - 5.7881 = 0.4678$. If we calculate a series of confidence limits for different values of X_i, we obtain a biconcave confidence belt (see Figure 14.11). The farther away from the mean we get, the less reliable are our estimates of Y because of the uncertainty about the true slope, β, of the regression line.

Furthermore, the linear regressions that we fit are often only rough approximations to more complicated functional relationships between biological variables. Very often there is an approximately linear relation along a certain range of the independent variable, beyond which range the slope changes rapidly. For example, the pulse of a poikilothermic animal is directly proportional to temperature over a range of tolerable temperatures, but beneath and above this range the pulse eventually decreases as the animal freezes or suffers heat prostration. Hence common sense indicates that one should be cautious about extrapolating from a regression equation, especially if one has any doubts about the linearity of the relationship.

The confidence limits for α, the parametric value of a, are a special case ($X_i = 0$) of the confidence limits for $\mu_{\hat{Y}_i}$, and the standard error of a is therefore

$$s_a = \sqrt{s_{Y\cdot X}^2 \left[\frac{1}{n} + \frac{\bar{X}^2}{\sum x^2} \right]}$$

In certain situations, we may require that a regression line pass through the origin. This would be proper for cases in which the parametric Y-intercept α is really zero, as in growth curves where the size of a structure is effectively zero at age zero or in biochemical work where none of a substance is produced by time zero. Fitting a regression through the origin would be supported if the confidence limits for α include zero. In such a case, the computations are modified so that the regression line, instead of passing through the bivariate mean $(\bar{X}, \bar{Y})$, is forced to pass through the origin of the coordinate system $(0, 0)$. The deviations in the regression formulas, therefore, are taken from the origin; thus, a deviation such as $Y - \bar{Y}$ becomes $Y - 0 = Y$. The formula for the regression coefficient consequently modifies Expression (14.4) as follows:

$$b_{Y\cdot X} = \frac{\sum XY}{\sum X^2} \tag{14.7}$$

and the regression equation becomes $\hat{Y} = b_{Y\cdot X}X$. For the unexplained sum of squares in regression through the origin, we modify Expression (14.5):

$$\sum d_{Y\cdot X}^2 = \sum Y^2 - \frac{(\sum XY)^2}{\sum X^2} \tag{14.8}$$

and for the explained SS, we do the same with Expression (14.6):

$$\sum \hat{y}^2 = \frac{(\sum XY)^2}{\sum X^2} \tag{14.9}$$

The total SS for Y necessarily becomes $\sum Y^2$. Because the sums of squares do not center on the mean, we do not lose a degree of freedom for that parameter. Thus, for regression through the origin, the three mean squares s_Y^2, $s_{\hat{Y}}^2$, and $s_{Y\cdot X}^2$ possess n, 1, and $n - 1$ degrees of freedom, respectively.

To evaluate standard errors for regression statistics when the regression line is forced to run through the origin, modify the formulas in Box 14.2 by replacing $\sum x^2$ and $\bar{X}$ with $\sum X^2$ and 0, respectively, and employ them with $n - 1$ degrees of freedom. Keep in mind the following caution about using the explained proportion of the total sum of squares in cases where the intercept is zero: The statistic has to be computed in a different way, with the formula $\sum \hat{Y}^2 / \sum Y^2$, and cannot be compared with its counterpart for the model with a nonzero intercept. For more on this topic, see Kvånalseth (1985).

The last two standard errors in Box 14.2 have somewhat different applications. They relate to *predications made on the basis of the regression equation*. Once you have established a functional relationship between two variables, as between weight loss and relative humidity in this example, you can use this equation to predict the outcome of future experiments. If, for example, you wish to run another batch of *Tribolium* beetles at 100% relative humidity, your best estimate for the weight loss would be $\hat{Y} = 3.3817$ mg, based on the regression equation of Box 14.1. You should expect a standard error greater than that of the estimate $\hat{Y}$ because the error variance of weight losses within the new sample at $X_i = 100\%$ relative humidity has been added. The formula in the fourth row of Box 14.2 is what is given most frequently in statistics texts, but it is of limited usefulness because it describes the expected variance of the sample items—that is, your prediction of the variance of future individual weight loss readings.

More useful is the formula in the last row of the box, which describes the *standard error of the predicted sample mean based on k items*. Thus, if you were to repeat the experiment on weight loss and run five batches of beetles at 100% relative humidity, you could expect the standard error of your sample mean of the five batches to be 0.23124 as calculated in step **8** of Box 14.3. The prediction limits for the sample mean are shown in step **9**. Note that they are wider than the confidence limits of the corresponding estimated value $\hat{Y}$ for the relative humidity $X_i = 100\%$. The biconcave prediction belts, graphed in Figure 14.11, are wider than the confidence belts of the estimated Y-values. As the size of the intended sample increases, the standard error diminishes and the prediction limits approach the confidence limits shown in step **7**. When the sample size decreases to unity, one computes the prediction limits from the standard error in the fourth row in Box 14.2. These prediction limits yield bounds for the sample items.

In the next section, we show how to compute regression when there is more than one Y-value per value of X. We also show the computation of standard errors for such cases based on the formulas in the third column of Box 14.2.

14.5 More Than One Value of Y for Each Value of X

We now take up the case of Model I regression as originally defined in Section 14.2 and illustrated by Figure 14.4. For each value of the treatment X, we sample Y repeatedly, obtaining a sample distribution of Y-values at each of the chosen points of X. The example we use here is an experiment from Sokal's laboratory in which *Tribolium* beetles were reared from eggs to adulthood at four different densities. The percentage survival to adulthood was calculated for varying numbers of replicates at these densities. Following Section 13.9, these percentages were given arcsine transformations, which are listed in Box 14.4. These values are more likely to be normal and homoscedastic than are percentages. The arrangement of these data is very much

> **BOX 14.4** Computation of Regression with More Than One Value of Y per Value of X: General Case with Unequal Sample Sizes

The observations Y are arcsine transformations of the percentage survival of the beetle *Tribolium castaneum* at 4 densities (X = number of eggs per gram of flour medium).

| | Density = X | | | |
| | ($a = 4$) | | | |
	5/g	20/g	50/g	100/g
	61.68	68.21	58.69	53.13
	58.37	66.72	58.37	49.89
Survival (in degrees)	69.30	63.44	58.37	49.82
	61.68	60.84		
	69.30			
n_i	5	4	3	3
$\bar{Y}_i$	64.0660	64.8025	58.4767	50.9467
$\displaystyle\sum_{a} n_i = 15$				

SOURCE: Data from Sokal (1967).

Anova computation

The steps are the same as those described in Box 9.1.

1. Grand mean $\displaystyle \bar{\bar{Y}} = \frac{1}{\sum_{a} n_i} \sum^{a} \sum^{n_i} Y = 60.5207$

2. $\displaystyle SS_{among} = \sum^{a} n_i(\bar{Y}_i - \bar{\bar{Y}})^2 = 423.7016$

3. $\displaystyle SS_{within} = \sum^{a} \sum^{n_i} (Y_{ij} - \bar{Y}_i)^2 = 138.6867$

4. SS_{total} = quantity **2** + quantity **3** = 562.3883

Anova table

Source of variation	df	SS	MS	F_s	P
$\bar{Y} - \bar{\bar{Y}}$ Among groups	3	423.7016	141.2339	11.20	0.0011
$Y - \bar{Y}$ Within groups	11	138.6867	12.6079		
$Y - \bar{\bar{Y}}$ Total	14	532.3883			

The null hypothesis H_0: $\mu_5 = \mu_{20} = \mu_{50} = \mu_{100}$ is rejected.

Box 14.4 (continued)

We proceed to test whether the differences among the survival values can be accounted for by linear regression on density. If

$$F_s < \frac{1}{a-1} F_{\alpha[1,\sum_{}^{a} n_i - a]}$$

we cannot reject the null hypothesis that the slope is zero.

Computation for regression analysis

5. Sum of squares of $X = \sum x^2 = \sum\limits^{a} n_i(X_i - \bar{X})^2 = 18{,}690$

6. Sum of products $= \sum xy = \sum\limits^{a} n_i(X_i - \bar{X})(\bar{Y}_i - \bar{\bar{Y}}) = -2747.62$

7. Explained sum of squares $= \sum \hat{y}^2 = \sum\limits^{a} n_i(\hat{Y}_i - \bar{\bar{Y}})^2 = \dfrac{(\sum xy)^2}{\sum x^2}$

$$= \frac{(\text{quantity } \mathbf{6})^2}{\text{quantity } \mathbf{5}} = \frac{(-2747.62)^2}{18{,}690} = 403.9281$$

8. Unexplained sum of squares $= \sum d^2_{Y \cdot X}$

$$= \sum\limits^{a} \sum\limits^{n_i} (Y_{ij} - \hat{Y}_i)^2 = SS_{\text{among}} - \sum \hat{y}^2$$

$$= \text{quantity } \mathbf{2} - \text{quantity } \mathbf{7}$$

$$= 423.7016 - 403.9281 = 19.7735$$

Completed anova table with regression

Source of variation	df	SS	MS	F_s	P	Expected MS
$\bar{Y} - \bar{\bar{Y}}$ Among densities (groups)	3	423.7016	141.2339	11.20	0.0011	$\sigma^2 + \dfrac{n_0}{a-1}\sum \alpha^2$
$\hat{Y} - \bar{\bar{Y}}$ Linear regression	1	403.9281	403.9281	40.86	0.0236	$\sigma^2 + n_0\sigma_D^2 + \beta^2 \sum x^2$
$\bar{Y} - \hat{Y}$ Deviations from regression	2	19.7735	9.8868	0.7842	0.4804	$\sigma^2 + n_0\sigma_D^2$
$Y - \bar{Y}$ Within groups	11	138.6867	12.6079			σ^2
$Y - \bar{\bar{Y}}$ Total	14	562.3883				

Box 14.4 (continued)

In addition to the familiar mean squares, MS_{among} and MS_{within}, we now have the mean square due to linear regression, $MS_{\hat{Y}}$, and the mean square for deviations from regression, $MS_{Y \cdot X}$ ($= s^2_{Y \cdot X}$). To test whether the deviations from linear regression are significant, compare the ratio $F_s = MS_{Y \cdot X}/MS_{within}$ with $F_{\alpha[a-2, \sum n_i - a]}$. Because $P = 0.4804$, we accept the null hypothesis that the deviations from linear regression are zero.

To test for linear regression, we test $MS_{\hat{Y}}$ over the mean square of deviations from regression $s^2_{Y \cdot X}$, and since $F_s = 403.9281/9.8868 = 40.86$, which yields $P = 0.0236$, we reject our hypothesis that there is no regression, or that $\beta = 0$.

9. Regression coefficient (slope of regression line) $= b_{Y \cdot X} = \dfrac{\sum xy}{\sum x^2}$

$$= \frac{\text{quantity } \mathbf{6}}{\text{quantity } \mathbf{5}} = \frac{-2747.62}{18,690} = -0.14701$$

10. Y-intercept $= a = \bar{\bar{Y}} - b_{Y \cdot X} \bar{X}$

$$= \text{quantity } \mathbf{1} - \left(\text{quantity } \mathbf{9} \times \frac{\sum\limits^{a} n_i X}{\sum\limits^{a} n_i} \right)$$

$$= 60.5207 - (-0.14701)(37.0000) = 65.96004$$

Thus, the regression equation is $\hat{Y} = 65.96004 - 0.14701X$.

like that of a single-classification Model I anova. There are four different densities and several replicated survival values at each density. We would like to determine whether there are differences in survival among the four groups and whether we can establish a regression of survival on density.

Our first approach is to carry out an analysis of variance, using the methods of Section 9.2 and Table 9.1. Various possible outcomes are illustrated in Figure 14.12. If the outcome is as in Figure 14.12a, it is likely that we would accept the null hypothesis that the four means do not differ from each other. It would thus seem unlikely that a regression line fitted to these data would have a slope significantly different from zero. However, although both analysis of variance and linear regression test the same null hypothesis—equality of means—the regression test is more powerful (less type II error; see Section 7.1) against the alternative hypothesis that there is a linear relationship between the group means and the independent variable X. Thus, when the means increase or decrease slightly as X increases, they may not be different enough for the mean square among groups to be significant by an anova, yet a significant regression could be found. When we find a marked regression of

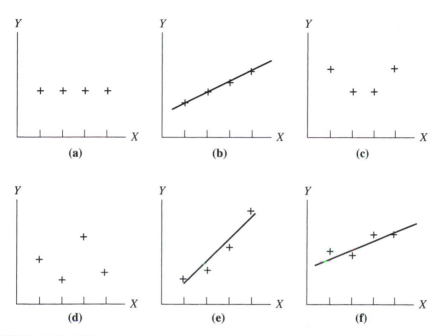

FIGURE 14.12 Differences among means and linear regression. General trends only are indicated by these figures. Significance of any of these would depend on the outcomes of appropriate tests. (For further explanation, see text.)

the means on X (Figure 14.12b), we usually find a significant difference among the means by an anova.

We cannot turn this argument around, however, and say that a significant difference among means as shown by an anova necessarily indicates that a significant linear regression can be fitted to these data. In Figure 14.12c, the means follow a U-shaped function (a parabola). Although the means would likely be significantly different from each other, a straight line fitted to these data would be a horizontal line halfway between the upper and the lower points. For such data, *linear* regression can explain only very little of the variation of the dependent variable. A curvilinear regression (see Section 16.9), however, would fit these data and remove most of the variance of Y.

A similar case is shown in Figure 14.12d, in which the means describe a periodically changing phenomenon, rising and falling alternatively. Again, the regression line for these data has slope zero. A curvilinear (cyclical) regression could also be fitted to such data, but our main purpose in showing this example is to indicate that there could be heterogeneity among the means of Y unrelated linearly to the magnitude of X. Remember that in real examples, you rarely get a regression as clear-cut as the linear case in Figure 14.12b or the curvilinear one in Figure 14.12c; nor will you necessarily get heterogeneity of the type shown in Figure 14.12d, in which any straight line fitted to the data would be horizontal. You are more likely to get data in which linear regression can be demonstrated, but which will not fit a straight line well. The residual deviations of the means around linear regression might be removed by changing from linear to curvilinear regression (as is suggested by the pattern of points in Figure 14.12e) or they remain as inexplicable residual heterogeneity around the regression line, as suggested in Figure 14.12f.

We carry out the computations by first following the familiar outline for analysis of variance and obtain the anova table in Box 14.4. The three degrees of freedom among the four groups yield a mean square that would be highly significant if tested over the within-groups mean square. Next, we carry out the additional steps for the regression analysis. We compute the sum of squares of X, the sum of products of X and Y, the explained sum of squares of Y, and the unexplained sum of squares of Y. The formulas may look unfamiliar because of the complication of the several Y's per value of X. The computations for the sum of squares of X involve the multiplication of X by the number of items in the study. Thus, although there may appear to be only four densities, there are as many densities (although of only four magnitudes) as there are values of Y in the study. Having completed these computations, we present the results in another anova table (see Box 14.4). Note that the major quantities in this table are the same as in a single-classification anova, but in addition, we now have a sum of squares representing linear regression, which is always based on one degree of freedom. This sum of squares is subtracted from the SS among groups, leaving a residual sum of squares (of two degrees of freedom in this case) representing the deviations from linear regression.

We should understand what these sources of variation represent. The linear model for regression with replicated Y per X is derived directly from Expression (8.2), which is

$$Y_{ij} = \mu + \alpha_i + \epsilon_{ij}$$

The treatment effect $\alpha_i = \beta x_i + D_i$, where βx is the component due to linear regression and D_i is the deviation of the mean $\overline{Y}_i$ from regression, which is assumed to have a mean of zero and a variance of σ_D^2. Thus, we can write

$$Y_{ij} = \mu + \beta x_i + D_i + \epsilon_{ij} \tag{14.10}$$

The SS due to linear regression represents the portion of the SS among groups that can be explained by linear regression on X. The SS due to deviations from regression represents the residual variation or scatter around the regression line (see Figure 14.12). The SS within groups is a measure of the variation of the items around each group mean. The expected mean squares for each source of variation are given in the completed anova table in Box 14.4.

We first test whether the mean square for deviations from regression ($MS_{Y \cdot X} = s_{Y \cdot X}^2$) is larger than the within-groups MS by computing the variance ratio of $MS_{Y \cdot X}$ over the within-groups MS. In our case, the deviations from regression are clearly not significant because the mean square for deviations is less than that within groups. We next test the mean square for regression, $MS_{\hat{Y}}$, over the mean square for deviations from regression and find $MS_{\hat{Y}}$ to be significantly greater. Thus, linear regression on density has clearly removed some of the variation of survival values. We should caution you again about confusing significance and importance when reporting a test result. If the result is not (statistically) significant, then you have not substantiated that any difference exists. If it is significant, it is important only if its effect is large. In this case, the effect of density is both significant and important because most of the variance of percentage survival is accounted for by the variation in density.

Finding a small probability for the mean square for deviations from regression could mean either that Y is a curvilinear function of X or that there is a large amount of random heterogeneity around the regression line (see Figure 14.12). Actually, a mixture of both conditions may prevail. If the $MS_{Y \cdot X}$ is large and curvilinearity is suspected, a curvilinear regression (see Section 16.9) may be fitted. If heterogeneity is responsible for a large $MS_{Y \cdot X}$, we use it as the denominator mean square against which to test for the presence of linear regression.

You may wonder how the terms in this example are related to those in the earlier example with a single Y for each value of X (see Box 14.1). Obviously, the single degrees of freedom for sum of squares due to linear regression correspond in the two examples. However, there is no replication in the earlier example. As a result, we have no independent estimate of error of the Y's for a given value of X. The unexplained sum of squares with $n - 2$ degrees of freedom in the nonreplicated example includes the deviation sum of squares with $a - 2$ degrees of freedom as well as the error sum of squares with $\Sigma^a n_i - a$ degrees of freedom in the replicated case. The total sum of squares with $n - 1$ degrees of freedom in the nonreplicated case is equivalent to the sum of squares among groups with $a - 1$ degrees of freedom in the replicated example. Thus, in the nonreplicated case, we are unable to separate the two sources of variation—deviations from regression and error among Y's for a given value of X—just as in an analogous case in a two-way anova, we were unable to separate interaction and error variance in a completely randomized blocks design, whereas we were able to achieve this separation in a two-way anova with replication (see Chapter 11). These relations are shown in Figure 14.13.

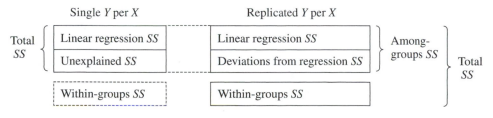

FIGURE 14.13 Diagrammatic representation of the partitioning of the total sums of squares (SS) in two regression designs—single Y per X and more than one Y per X—corresponding to Boxes 14.1 and 14.4. The areas of the subdivisions are not shown proportional to the magnitudes of the sums of squares.

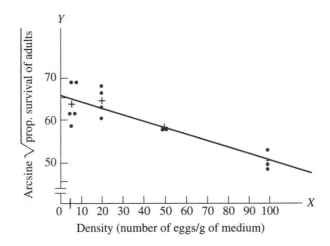

FIGURE 14.14 Linear regression fitted to data of Box 14.4. Sample means are identified by + signs.

We complete computation of the regression coefficient and regression equation as shown at the end of Box 14.4. Our conclusions are that as density increases, survival decreases and that this relationship can be expressed by a linear regression of the form $\hat{Y} = 65.9601 - 0.14701X$, where X is density per gram and $\hat{Y}$ is the arcsine transformation of percentage survival (Figure 14.14).

We will now compute standard errors and confidence limits for the example we have just discussed. The pertinent standard errors are found in column (3) of Box 14.2. We compare the mean square for deviations from regression with its 2 degrees of freedom to the within-groups mean square, MS_{within}, with its 11 df. The appropriate degrees of freedom for the standard errors are given in Box 14.2. The computation of the *standard error of the regression coefficient* and the 95% *confidence limits of β* is shown in steps **1** and **2** of Box 14.5. We calculate the *standard error of the sample mean* $\overline{Y}_i$ for any one group i based on the MS_{within}. Thus, in step **3** we compute the standard error for the mean survival of 64.0660 at density 5/g as 1.58795 (expressed in degrees; remember this is an arcsine transformation), and the 95% confidence limits computed in step **4** are 60.5709 to 67.5611.

Next, we estimate the mean survival $\hat{Y}_i$ for another value of the density, say $X_i = 59$/g. Note that this value of X_i was not one of the four employed in the original experiment. As long as X (density) is a continuous variable, we can estimate Y for any biologically reasonable value of X that does not involve undue extrapolation beyond the range of observed X-values. The estimate of $\hat{Y}_i$ for $X_i = 59$/g can be obtained from the regression equation and yields a survival value of 57.2865°. In step **5**, we calculate the *standard error of the estimated mean*, followed by the *confidence limits for the estimated mean* (53.1702 to 61.4028) in step **6**. Next, we predict the mean that we would obtain in another experiment. At density 59/g, our best estimate for such a survival value is, of course, again the regression estimate of 57.2865°. However,

BOX 14.5 Computation of Standard Errors and Confidence Limits of Regression Statistics: More Than One Value of Y for Each Value of X

Data from Box 14.4.

$$b_{Y \cdot X} = -0.14701 \qquad s_{Y \cdot X}^2 = 9.8868$$
$$\Sigma x^2 = 18,690 \qquad MS_{\text{within}} = 12.6079$$

Computation

In all of these computations the *MS* for deviations from regression ($s_{Y \cdot X}^2 = 9.8868$ with 2 *df*) is used. The formulas for the standard errors are from column (3) of Box 14.2.

1. Standard error of the regression coefficient:

$$s_b = \sqrt{\frac{s_{Y \cdot X}^2}{\Sigma x^2}} = \sqrt{\frac{9.8868}{18,690}} = 0.02300$$

2. 95% confidence limits for regression coefficient β:

$$t_{.05[2]} s_b = 4.303(0.02300) = 0.09897$$
$$L_1 = b - t_{.05[2]} s_b = -0.14701 - 0.09897 = -0.24598$$
$$L_2 = b + t_{.05[2]} s_b = -0.14701 + 0.09897 = -0.04804$$

3. Standard error of a sample mean $\overline{Y}_i$ for any group i:

$$s_{\overline{Y}} = \sqrt{\frac{MS_{\text{within}}}{n_i}}$$

For the first sample at density $X_1 = 5/g$, we find $n_1 = 5$. The standard error would be

$$s_{\overline{Y}} = \sqrt{\frac{12.6079}{5}} = 1.58795$$

4. The 95% confidence interval for the mean of the first sample ($X_1 = 5/g$; $\overline{Y}_1 = 64.0660$):

$$t_{.05[11]} s_{\overline{Y}} = 2.201(1.58795) = 3.4951$$
$$L_1 = \overline{Y}_1 - t_{.05[11]} s_{\overline{Y}} = 64.0660 - 3.4951 = 60.5709$$
$$L_2 = \overline{Y}_1 + t_{.05[11]} s_{\overline{Y}} = 64.0660 + 3.4951 = 67.5611$$

5. Standard error of $\hat{Y}_i$, an estimated mean for a given value of X_i. As an example we will compute $\hat{Y}_i$ for $X_i = 59/g$. Since $\hat{Y} = a + bx$, for $X_i = 59, \hat{Y}_i = 65.9601 - 0.14701(59) = 57.2865$. We use the following formula for the standard error:

$$s_{\hat{Y}} = \sqrt{s_{Y \cdot X}^2 \left[\frac{1}{\Sigma n_i} + \frac{(X_i - \overline{X})^2}{\Sigma x^2} \right]}$$

Box 14.5 (continued)

$\overline{X}$ in this formula refers to

$$\frac{\sum\limits_{a}^{a} n_i X}{\sum\limits_{a} n_i} = \frac{555}{15} = 37.0 \qquad \text{(using values from Box 14.4)}$$

For a density of $X_i = 59/g$

$$s_{\hat{Y}} = \sqrt{9.8868 \left[\frac{1}{15} + \frac{(59 - 37)^2}{18,690} \right]} = 0.95662$$

6. 95% confidence interval for μ_{Y_i} corresponding to the estimate $\hat{Y}_i = 57.2865$ at $X_i = 59/g$:

$$t_{.05[2]}s_{\overline{Y}} = 4.303(0.95662) = 4.1163$$

$$L_1 = \hat{Y}_i - t_{.05[2]}s_{\overline{Y}} = 57.2865 - 4.1163 = 53.1702$$

$$L_2 = \hat{Y}_i + t_{.05[2]}s_{\overline{Y}} = 57.2865 + 4.1163 = 61.4028$$

95% of such intervals should contain the true mean μ_{Y_i} for $X_i = 59$. These limits are plotted in Figure 14.15 for values of X_i from 0 to 100.

7. Standard error of a predicted mean $\overline{Y}_i$ to be obtained in a new experiment run at a density of $X_i = 59/g$. Our best prediction for this mean would be $\overline{Y}_i = \hat{Y} = 57.2865$. If the new experiment were based on a sample of size $k = 10$, the standard error of the predicted mean would be

$$\hat{s}_{\overline{Y}} = \sqrt{s_{Y \cdot X}^2 \left[\frac{1}{\sum n_i} + \frac{(X_i - \overline{X})^2}{\sum x^2} \right] + \frac{MS_{\text{within}}}{k}}$$

$$= \sqrt{9.8868 \left[\frac{1}{15} + \frac{(59 - 37)^2}{18,690} \right] + \frac{12,6079}{10}} = 1.47510$$

8. 95% prediction limits for the mean of a sample of 10 survival readings at density 59/g (using the standard error computed above):

$$t_{.05[13]}\hat{s}_{\overline{Y}} = 2.160(1.4751) = 3.1862$$

$$L_1 = \hat{Y}_i - t_{.05[13]}\hat{s}_{\overline{Y}} = 57.2865 - 3.1862 = 54.1003$$

$$L_2 = \hat{Y}_i + t_{.05[13]}\hat{s}_{\overline{Y}} = 57.2865 + 3.1862 = 60.4727$$

Although the standard error of these prediction limits is greater than that of the estimated $\hat{Y}_i$ in step **5**, these limits (also shown in Figure 14.15) are narrower than those found in step **6** because of the greater number of degrees of freedom and consequent decrease in magnitude of t.

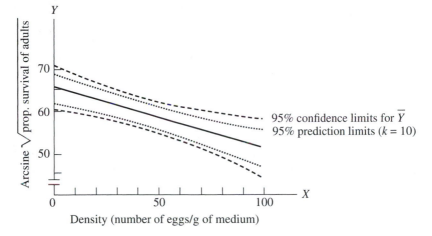

FIGURE 14.15 Confidence limits to regression estimates for data of Figure 14.14. Note the unusual circumstance in which the confidence limits are outside the prediction limits, a phenomenon that occurs here because in this problem the latter are based on substantially more degrees of freedom.

the standard error is as shown in the last row of the table in Box 14.2. Let us assume that the mean to be obtained in our next experiment will be based on 10 replicates. In steps **7** and **8**, we obtain the *standard error for the predicted mean* ($\hat{s}_{\bar{Y}} = 1.4751$) and *prediction limits* of 54.1003 and 60.4727, which in this unusual case are narrower than the confidence limits for the estimated mean obtained previously because of the greater number of degrees of freedom of the standard error and consequent decrease of t. Confidence and prediction bands around the regression line are shown in Figure 14.15.

If the sample sizes for the separate values of X are equal, we can simplify the formulas for the various sums of squares. For example, the formula for the sum of products (step **6** in Box 14.4) becomes $n\Sigma^a(X_i - \bar{X})(\bar{Y}_i - \bar{\bar{Y}})$. Almost all such computations are carried out by computers nowadays, making these simplifications largely irrelevant, but significance tests for the various regression statistics could be carried out by using the formulas in Box 14.2, simplifying whenever possible to take advantage of the equality of sample sizes.

There are many important extensions to the usual methods of regression analysis. **Weighted regression** allows one to take into account the fact that observations may differ in their reliability. When we regressed $\bar{Y}$ on X, the means were weighted by their sample sizes. In an analogous fashion, one can also weight inversely by the variance within different samples. For more information, see, for example, Ryan (1997) or Wolberg (2006).

14.6 The Uses of Regression

We have been so busy learning the mechanics of regression analysis that we have not given much thought to the applications of regression. In this section, we will discuss six more or less distinct types of applications. All are discussed in terms of Model I regression. When applying these techniques, be careful that the four assumptions listed in Section 14.2 are correct. From a logical and philosophical point of view, some of these applications also fit a Model II case. The computational procedures would then differ as discussed in Section 14.12.

The first application of regression is the *study of causation.* If we wish to know whether variation in a variable Y is caused by changes in another variable X, we manipulate X in an experiment to see whether we can find statistical evidence for a linear relation of Y on X. Causation is a complex, philosophical idea that we will not go into here. You have undoubtedly been cautioned from your earliest scientific experience not to confuse correlation with causation. When variables vary together, this covariation may be by chance or both variables may be functions of another variable. These cases are usually examples of Model II regression. When we manipulate one variable and find that such manipulations affect a second variable, we generally are satisfied that the variation of the independent variable X is the cause of the *variation* of the dependent variable Y (not the cause of the variable!). Even here, however, it is best to be cautious. When we find that heart rate in a cold-blooded animal is a function of ambient temperature, we may conclude that temperature is one of the causes of differences in heart rate. Other factors may also be affecting heart rate.

Sometimes we make the mistake of inverting the cause-and-effect relationship. Such a mistake in this case is unlikely; suggesting that heart rate affects the temperature of the general environment seems immediately unreasonable. If we were testing two chemical substances in the blood, however, we might more easily be mistaken about the cause-and-effect relationship. If these two chemicals are related by a feedback relationship, then administering chemical A, the precursor, might increase production of chemical B, the product. Injection of chemical B, however, might inhibit production of the precursor. If we did not know which was the precursor and which the product, we might be confused about cause-and-effect relationships here. Despite these cautions, regression analysis is a common device for screening out causal relationships. Although a significant regression of Y on X does not prove that changes in X are the cause of variations in Y, the converse statement is true. When we find no significant linear or curvilinear regression of Y on X, we may infer that variation in X does not affect Y. Note, however, the possibility of type II error. There may indeed be an effect of X on Y that cannot be substantiated because the sample size of the study is inadequate.

The *description of scientific laws* is a second general application of regression analysis. The aim of science is mathematical description of relations between variables in nature, and regression analysis permits us to estimate functional relationships between variables. These functional relationships do not always have clearly interpretable biological meaning. Thus, in many cases it may be difficult to assign a biological interpretation to the statistics *a* and *b,* or their corresponding parameters α and β. When we are able to do so, we have a **structural mathematical model,** one whose components have clear scientific meaning. Mathematical curves that are not structural models are also of value in science. Most regression lines are **empirically fitted curves,** in which the functions simply represent the best mathematical fit (by a criterion such as least squares) to an observed set of data. The constants necessary to fit these curves may not possess any clear inherent meaning. Biologists generally do not like empirically fitted curves. When they adequately describe the relations between natural phenomena, however, empirically fitted curves are of value as temporary devices until enough insight into the phenomena is obtained to postulate a hypothesis from which a structural model can be constructed. Structural models are preferred for many reasons, although they may actually not be markedly better predictors than are empirically fitted curves over the observed range in which the line is being fitted.

Successful *prediction* of an expected value of *Y* for a given value of *X* is a third aspect of regression analysis and has been suggested as one justification for employing empirically fitted curves. Prediction is frequently an objective of applied research. Associated with prediction is estimating the difference between observed and expected values in the dependent observation; these differences can lead to the discovery of atypical values, that is, outliers, which result from inhomogeneity of samples or measurement errors. Although correct structural mathematical models should be better at prediction than empirically fitted curves are—especially for extrapolation beyond the range of *X*'s observed in the sample—empirical curves are frequently of great utility. Examples from the history of science are not hard to find. Good empirical formulas for the motions of the heavenly bodies were available in the time of Copernicus and Kepler; the Newtonian ideas on gravitation altered the nature of the formulas but improved the prediction only relatively mildly. Subsequent improvements in the structural formulas have continued to lessen the error of prediction. A biological analogue relates to the use for many years of heart-muscle extract as a simulated antigen in the Wassermann test for syphilis. Although there seems to be no logical or biological reason why such an extract should provide an antigen for a spirochete-induced disease, the method proved to be reasonably reliable. It is with the same sort of philosophy that we often use regression equations to describe scientific laws, knowing full well that the structure of the equations may not represent the workings of nature.

Comparison of dependent observations is another application of regression. As soon as it is established that a given variable is a function of another one, as in Box 14.4, where we found the survival of beetles to be a function of density, we can ask how strong this relation is. In the beetle example, for instance, we ask to what degree the observed difference in survival between two samples of beetles is a function of the density at which they have been raised. Comparing beetles raised at very high density (and expected to have low survival) with those raised under optimal conditions of low density would be unfair. This same point of view makes us hesitate to compare the mathematical knowledge of a fifth-grader with that of a college student. Because we could undoubtedly obtain a regression of mathematical knowledge on years of schooling in mathematics, we should be comparing how far a given individual deviates from his expected value based on such a regression. Thus, relative to other classmates and to age group, the fifth-grader may be far better than is the college student relative to his/her peer group. This consideration suggests that we calculate *adjusted Y-values* that allow for the magnitude of the independent variable X. A conventional way of calculating such adjusted Y-values is to estimate the Y-value one would expect if the independent variable were equal to its mean $\overline{X}$ and the observation retained its observed deviation $(d_{Y \cdot X})$ from the regression line. Because $\hat{Y} = \overline{Y}$ when $X = \overline{X}$, the adjusted Y-value can be computed as

$$Y_{adj} = \overline{Y} + d_{Y \cdot X} = Y - bx \qquad (14.11)$$

Let us apply this formula to the data of Table 14.1. Comparing the weight loss in the *Tribolium* sample raised at 29.5% relative humidity ($Y = 6.67$ mg) and that at 62.5% relative humidity ($Y = 5.83$ mg), we are led to conclude that weight loss in the more humid environment was less by 0.84 mg. According to our regression line, we expect 7.1338 mg weight loss at 29.5% relative humidity (RH), whereas at 62.5% RH we expect a loss of 5.3776 mg. Thus, the expected difference should be greater (and it is, 1.76 mg). To calculate adjusted Y-values for these two relative humidities, we need the following quantities:

X_i	29.5	62.5	%RH
Y_i	6.67	5.83	mg
$\hat{Y}_i$	7.1338	5.3776	mg
$d_{Y \cdot X}$	−0.4638	0.4524	mg
$Y_{adj} = \overline{Y} + d_{Y \cdot X}$	5.558	6.474	mg

Remembering that $\overline{Y} = 6.022$, we obtain the adjusted Y-values using Expression (14.11). These adjusted values lead to interpretations different from those based on the observed Y-values. We now find a higher adjusted weight loss at the higher humidity. What does this mean? It indicates that the sample used in the experiment at the higher humidity happened to deviate strongly above the regression line. It lost more weight than was expected. On the other hand, the sample at 29.5% RH had a lower weight loss than expected. Thus, although overall weight loss was greater at the lower humidity, adjusted weight loss allowing for the regression relationship is quite the reverse for these two samples: The sample of beetles tested at 62.5% RH lost relatively more weight than the sample at the lower humidity.

In this manner, adjusted means are useful for making comparisons among dependent observations after allowing for differences among independent observations that affect them. For the standard error of an adjusted mean, we use the standard error of estimated Y for a given value of X_i, shown in the third row of Box 14.2. Sometimes we do not even bother to calculate adjusted Y. We can simply state $d_{Y \cdot X}$ and compare the unexplained deviations among the observations. A large and positive unexplained deviation means that this individual or sample lies considerably above expectation by regression; a negative value of $d_{Y \cdot X}$ indicates the converse. Because the deviations $d_{Y \cdot X}$ are in absolute measurement units, it may be desirable to divide them by $s_{Y \cdot X}$ to obtain standardized scores. These are quite frequently used in educational statistical research to provide an estimate in standard deviation units of an individual's relative standing with respect to his/her group, allowing for an independent variable such as age or amount of education.

Differences among adjusted Y-values are not necessarily scientifically meaningful. They may merely represent random error around the regression line—the result of unknown and probably unknowable factors differentiating the responses of the individuals concerned. Such differences may, however, reflect biologically meaningful distinctions among the individuals leading to recognition of new and important causal factors. Statistical analysis cannot by itself distinguish between these alternatives. Further experimentation and analysis are required to reveal these new insights.

Statistical control is an application of regression that is not widely known among biologists and represents a scientific philosophy that is not well established in biology outside agricultural circles. Biologists frequently categorize work as descriptive or experimental, implying that only the latter can be analytical. Statistical approaches applied to descriptive work can in some instances, however, take the place of experimental techniques quite adequately—occasionally they are even to be preferred. These approaches are attempts to substitute statistical manipulation of a concomitant variable for control of the variable by experimental means. An example will clarify this technique.

Assume that we are studying the effects of diet on blood pressure in rats. We find that the variability of blood pressure in our rat population is considerable, even before we introduce differences in diet. Further study reveals that the variability is due largely to differences in age among the rats of the experimental population. This relationship can be demonstrated by a significant linear regression of blood pressure on age. Thus, to reduce the variability of blood pressure in the population, we should keep the age of the rats constant. The reaction of most biologists at this point would be to repeat the experiment using rats of only one age group; this is a valid, common-sense approach, which is part of the experimental method. An alternative approach using the analysis of covariance (see Section 16.8) is superior in cases in which holding the variable constant is impractical or too costly. For example, we could continue to use rats of variable ages and simply record the age of each rat as well as its blood pressure. Then we could regress blood pressure on age and use an adjusted mean as the basic blood pressure reading for each individual. We could then evaluate the effect of differences in diet on these adjusted means, or we could analyze the effects of diet on unexplained deviations, $d_{Y \cdot X}$, after the experimental blood pressures were regressed on age (which amounts to the same thing).

What are the advantages of such an approach? Sometimes it is impossible to secure adequate numbers of individuals all of the same age. By using regression we can use all the individuals in the population. Employing statistical control assumes that recording the independent variable X is relatively easy and that this variable can be measured without error, which would be generally true of such a variable as age of a laboratory animal. Statistical control may also be preferable because we obtain information over a wider range of both Y and X and because we add to our knowledge about the relations between these two variables, which would not happen if we restricted ourselves to a single age group. Furthermore, when the assumed additivity of treatment effects does not hold, this fact can be learned only over a range of age groups. Suppose, for example, that the effect of diet on blood pressure is noticeable only in older rats and not in younger ones. Clearly, we would not discover this unless we used rats of different age groups.

Regression permits us to predict what the variance of an organism would be under statistical or experimental control. Instead of s_Y^2, the previous variance of Y, we can now use the unexplained residual variance $s_{Y \cdot X}^2$. By regressing Y on X, we increase the information per individual by $100(s_Y^2/s_{Y \cdot X}^2 - 1)\%$. This quantity is very much like the relative efficiency in design discussed in Section 10.4. In calculating relative efficiency in regression, we customarily take the ratio of total variance over unexplained variance.

Let us work out an example with the survival values of Box 14.4. Because these data are structured into variance among densities and within densities, we have to be careful which variance we use here. The variance among densities was reduced by regression on density, and we should compare it with the unexplained mean square around regression, referred to in Box 14.4 as "the mean square of deviations from regression." The relative efficiency is computed as $RE = (MS_{\text{among}} \times 100)/MS_{Y \cdot X} = (141.2339 \times 100)/9.8868 = 1428.5\%$. We could improve the efficiency of our design by $1428.5 - 100 = 1328.5\%$ by recording the density at which the beetles were raised and by regressing survival on density. If information on density were easily available, this procedure would be well worthwhile. A similar increase in efficiency would result from using only beetles reared at the previous mean density.

Finally, *substitution of variables* is a special application of regression that may occasionally be useful. Suppose we are interested in a response variable that is very difficult or expensive to measure. For example, assume that we wish to measure blood pressure in mice, which might be quite complex with the equipment at our disposal. Assume further that after much effort, we have accurately measured the blood pressures of 25 mice of known, but differing ages. After regressing blood pressure on age, we find that a substantial portion of the variance of blood pressure was a function of age. Instead of continuing to measure blood pressure, we could simply record the ages of the mice and predict blood pressure from them. Obviously, this approach would not be as efficient as measuring the blood pressure directly. If our initial experiment had been carried out properly, however, we might be able to predict to a satisfactory degree of accuracy what the blood pressures of mice should be, given their age distribution. We might, therefore, have to use a greater number of mice to predict blood pressure from age than if we measured the blood pressure, but it may be much more economical to record the ages of large samples of mice than to measure the blood pressures of smaller samples.

Whenever the dependent variable is difficult and costly to measure, consider substituting variables.

14.7 Estimating *X* From *Y*

Occasionally, we have a problem in which we know the value of *Y* for an individual and wish to estimate the corresponding value of *X*. This problem may appear simple, requiring only that we reverse the regression equation to write an equation of the type $\hat{X} = a' + b_{X \cdot Y}Y$, where $a \neq a'$ and $b_{X \cdot Y} = \Sigma xy / \Sigma y^2$. Such an approach would be improper, however, because our initial assumptions were that *X* is measured without error and *Y* is the dependent, random, and normally distributed variable. Unless the situation is reversed, it is not legitimate to regress *X* on *Y*. The appropriate procedure is simple, although the computations (outlined in Box 14.6) are tedious.

Because $\hat{Y}_i = a + b_{Y \cdot X}X_i$, we can estimate X_i by rearranging this equation to yield $\hat{X}_i = (Y_i - a)/b_{Y \cdot X}$. In Box 14.6, we assume that the weight loss in a sample of 25 *Tribolium* beetles is 7 mg. We wish to estimate the relative humidity $\hat{X}_i$ at which these beetles were kept, assuming that the experimental setup was the same as before. Following the formula just given, we estimate the relative humidity at 32.0124%. It is not appropriate to assign standard errors to such an estimate, but there is a method for providing it with confidence limits. Because computational formulas for these limits are unwieldy, we divide the computation into three steps. First, we compute quantity *D* as defined in Box 14.6; then we proceed to a second quantity, *H*. The limits are relatively simple functions of *D* and *H*, as shown in the box. However, note that these confidence limits are unusual in one respect—they are *not* symmetrical around the estimate $\hat{X}_i$ but are symmetrical around another value, $\bar{X} + [b_{Y \cdot X}(Y_i - \bar{Y})/D]$, which is close to $\hat{X}_i$ but not identical to it. By the way, all values of *t* in Box 14.6 are shown as $t_{.05}$ to provide 95% confidence limits. For other $100(1 - \alpha)\%$ confidence limits, you should use t_α. Note the wide confidence interval: 17.3 to 45.9% relative humidity after rounding (the estimate $\hat{X}_i$ should also be rounded to 32.0% RH).

Box 14.6 also provides a formula that is appropriate when the estimate of *X* is based on an initial study of replicated *Y*-values per single value of *X* and there is more than one value of *Y* from which to estimate $\hat{X}_i$. This method of estimating *X* from *Y*, also called **inverse prediction,** is applied frequently in the statistical analysis of dosage-mortality problems in bioassay. Such a study involves a regression of cumulative mortalities of organisms on dosage of a substance. Thus, dosage X_1 will cause a mortality of Y_1 percent; dosage $X_2 > X_1$, a mortality of $Y_2 > Y_1$ percent; $X_3 > X_2$, a mortality of $Y_3 > Y_2$ percent; and so forth. At a certain dosage, mortality is 100%; the entire sample of organisms is killed. Frequently, such data are transformed—the mortalities to a so-called probit scale (see Section 14.10) and the dosages to logarithms to make the regression of mortality on dosage linear. A common measure of the potency of the substance (or of the tolerance of the organisms) is the dosage required to kill 50% or 95% of the organisms. Such a point is called an LD_{50} or LD_{95} of the organisms, the 50% or 95% lethal dose. This problem is clearly one

---

BOX 14.6 Estimating X from Y

Data from Table 14.1 and Box 14.1.

Given a weight loss reading of $Y_i = 7$ mg for a sample of 25 *Tribolium* beetles, what can we infer about the relative humidity under which they were kept (assuming that the experimental setup was identical to that previously used)?

Since $\hat{Y}_i = a + b_{Y \cdot X} X_i$,

$$\hat{X}_i = \frac{(Y_i - a)}{b_{Y \cdot X}} = \frac{7.0 - 8.7037}{-0.05322} = 32.0124$$

The 95% confidence limits of this estimate are computed as follows:

We define a quantity D as

$$D = b_{Y \cdot X}^2 - t_{.05[n-2]}^2 s_b^2 = (-0.05322)^2 - (2.365)^2 (0.0032561)^2 = 0.002773$$

where $n = 9$, the number of samples (of beetles) in the previous study (see Table 14.1), and another quantity H as

$$H = \frac{t_{.05[n-2]}}{D} \sqrt{s_{Y \cdot X}^2 \left[D \left(1 + \frac{1}{n} \right) + \frac{(Y_i - \bar{Y})^2}{\sum x^2} \right]}$$

$$= \frac{2.365}{0.002773} \sqrt{0.08801 \left[(0.002773) \left(1 + \frac{1}{9} \right) + \frac{(7.0 - 6.022)^2}{8301.3889} \right]}$$

$$= 14.30428$$

The 95% confidence limits are

$$L_1 = \bar{X} + \frac{b_{Y \cdot X}(Y_i - \bar{Y})}{D} - H$$

$$= 50.389 + \frac{(-0.05322)(7 - 6.022)}{0.002773} - 14.30428 = 17.315$$

$$L_2 = L_1 + 2H = 17.315 + 2 \times 14.30428 = 45.924$$

Note that the limits are symmetrical about $\bar{X} + [b_{Y \cdot X}(Y_i - \bar{Y})/D]$, not about $\hat{X}_i = (Y_i - a)/b_{Y \cdot X}$.

If we wish to estimate X_i for a $\bar{Y}_i$ based on a sample of size n in a regression analysis with more than one Y-value per value of X (as in Box 14.4, for example), H becomes

$$H = \frac{t_{\alpha[v]}}{D} \sqrt{s_{Y \cdot X}^2 \left[D \left(\frac{1}{n} + \frac{1}{a} \right) + \frac{(\bar{Y}_i - \bar{\bar{Y}})^2}{\sum x^2} \right]}$$

where α is the significance level chosen, $s_{Y \cdot X}^2$ is the *MS* of deviations around regression, a is the number of groups in the anova, and v is the number of degrees of freedom for $s_{Y \cdot X}^2$.

of inverse prediction: Given a value of *Y,* namely 50% or 95% mortality, estimate a corresponding value of *X,* the dosage. Confidence limits to the estimate are set in the manner indicated above.

> If the inverse prediction is applied to a regression equation through the origin, the formula changes. Consult Seber (1977, Section 7.3) for such cases.
>
> Sometimes the interest is directly that of regressing the proportions of observations falling below various thresholds onto an independent variable. An example would be regressing the median (the 50th percentile) of several samples on an independent variable. This method is called **quantile regression** and requires more complex computational methods. For more information, see, for example, Cade and Noon (2003), He (1997), or Koenker (2005).

14.8 Comparing Two Regression Lines

Often an investigator obtains two regression lines from similar data and wishes to know whether the functional relationships described by the regression equations are the same. For example, the scientist may have established a regression of blood pressure on age in a sample of animals and may now wish to compare this regression equation with those in another sample, which has been subjected to a different diet or drug. The basic design of such a test is the single-classification analysis of variance. There will be $a = 2$ samples, representing the treatment group and the control. There is one major new aspect, however. In previous analyses, we encountered only one variable, *Y*; in this example, *Y* would be the blood pressure. In addition, however, for each reading of *Y* we also have a reading of *X,* the age of the animal. Thus, two separate analyses of variance are possible, one for each variable, as well as a joint analysis—the analysis of the covariance between *X* and *Y.* Such an analysis in its complete form is called the *analysis of covariance* and is discussed in Section 16.8.

In this section, we test two regression lines for the homogeneity of their slopes. Why would we be interested in testing differences between regression slopes? We might find that different toxicants yield different dosage-mortality curves or that different drugs yield different relationships between dosage and response (see, for example, Figure 14.3). In the example analyzed below, genetically differing cultures yielded potentially different responses to increasing density, an important fact in understanding the effect of natural selection in these cultures. The regression slope of one variable on another is as fundamental a statistic of a sample as is the mean or the standard deviation; when comparing samples, it may be as important to compare regression coefficients as it is to compare these other statistics. Another reason for testing differences among slopes is that the homogeneity of regression slopes is an important assumption in the analysis of covariance.

The example we have chosen is based on data we have already encountered in Figure 7.15. In this experiment, mean dry weights of adult houseflies reared from different densities of eggs per constant weight of medium were compared for four

BOX 14.7 Test for the Equality of Slopes of Two Regression Lines for Single Values of Y for Each Value of X

The effect of varying density (X = number of eggs per 36 g of medium) on mean adult dry weights (Y) of two strains of houseflies. Because the seven densities were arrayed in a series in which each term after the first is twice the density of the previous value, all densities are transformed to logarithms.

Preliminary computations

The original data values are not shown here to conserve space. The regression computations for the two samples were carried out by simple linear regression methods, as shown in Box 14.1. In the following table, we show only the summary statistics necessary for the test at hand. [Data excerpted from Sullivan and Sokal (1963).]

Sample	Strain	$b_{Y \cdot X}$	n_i	s_X^2	Σx^2	$\Sigma d_{Y \cdot X}^2$	$\Sigma \hat{y}^2$	$P(H_0: b_i = 0)$
1	OL	−2.2162	7	0.42289	2.53734	0.6843	12.4622	0.000214
2	bwb	−1.7915	7	0.42289	2.53734	0.7879	8.1432	0.000811

The two P-values clearly suggest that each of the two regression coefficients differs from zero and that there is a significant negative regression of dry weight on log density. The mean weights and their regression lines are shown in the figure below. We next proceed to a test of the principal question asked of these data.

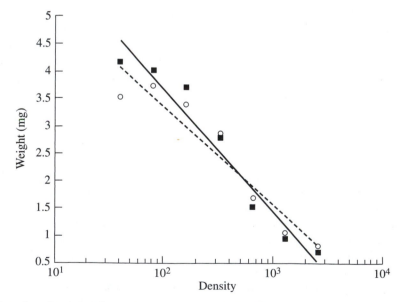

Plot of weight against density (on log scale). Strain OL: filled squares and solid regression line. Strain *bwb*: open circles and dashed regression line.

Box 14.7 (continued)

F-tests for difference between two regression coefficients

The formula for $k = 2$ regression coefficients is as follows:

$$F_s = \frac{(b_1 - b_2)^2}{\dfrac{\sum x_1^2 + \sum x_2^2}{\sum x_1^2 \sum x_2^2}\left(\dfrac{\sum d_{Y \cdot X(1)}^2 + \sum d_{Y \cdot X(2)}^2}{n_1 + n_2 - 4}\right)}$$

Substituting the appropriate values from the table above into this formula, we obtain

$$F_s = \frac{[(-2.2162) - (-1.7915)]^2}{\dfrac{2.53734 + 2.53734}{2.53734 \times 2.53734}\left(\dfrac{0.6843 + 0.7879}{7 + 7 - 4}\right)} = 1.5543$$

For degrees of freedom 1 and 10, this yields $P = 0.2409$. Thus, despite the apparent substantial difference between the two regression slopes as shown in the figure, we cannot reject the null hypothesis that they are sampled from the same slope.

genetically differing strains. For our analysis here, we compare the regression line of mean dry weights of strain OL on 7 logarithmically transformed densities to a similar line for strain *bwb*. The results of the preliminary regression computations are shown in Box 14.7. The rest of the box shows a simple test of the null hypothesis that there is *no difference between two regression coefficients*. This can be tested either by a *t*-test or by an *F*-test. The test shown in Box 14.7 is an *F*-test. The resulting *P*-value (0.240873) indicates that we have no reason for rejecting the null hypothesis of no difference between the regression coefficients of mean dry weights of adults emerging from different densities. Tests for comparing $k > 2$ regression slopes are illustrated in Section 16.8, where we shall learn to carry these out with more powerful and elegant procedures.

14.9 Linear Comparisons in Anovas

In previous sections, we noted the close relationship between regression and analysis of variance. The closeness of this relationship will become even more obvious here, as our discussion relates to tests that were introduced in Section 9.4. Recall that we were able to subdivide the sum of squares and the degrees of freedom among groups into separate sums of squares, each possessing a single degree of freedom and furnishing certain desired information about the overall differences in the analysis of variance. The sum of squares among groups (treatments) could be decomposed into a set of orthogonal single-degree-of-freedom comparisons—that is, their *SS* and *df* were independent of each other and additive (added up to the *SS* and *df* among groups). An important idea, put forth by R. A. Fisher, is that all degrees of freedom and sums of squares in an anova eventually can be partitioned into single degrees of freedom with corresponding sums of squares that can be identified with certain contrasts. Not all of these contrasts will be scientifically meaningful, and it is generally not practical to carry out so complete a decomposition of the sums of squares; yet the general idea has considerable attraction, and, as we will see in this section and in Section 16.7, in many cases we can probe deeply into the nature of the treatment and interaction sums of squares by means of individual-degree-of-freedom comparisons.

We will now provide a more general method, related to regression, for performing such individual-degree-of-freedom comparisons, at the same time laying the groundwork for the study of the general linear hypothesis and orthogonal polynomials (see Section 16.7). If the subject matter of Section 9.4 has by now become vague, we urge you to reread it before proceeding.

In Section 14.5 (replicated Y's for each value of X), we studied a single-classification anova in which survival of *Tribolium* beetles was regressed on density. The density classes were the groups of the anova, but at the same time these classes had numerical values of their own. Thus, one density was 5/g, another 20/g, and so forth. Quite naturally, therefore, we considered density an independent variable X, and we regressed the survival values Y on density X. However, we can also carry out a regression of response variable Y in an anova in which the groups are not expressly quantified. We do this by employing dummy variables to symbolize the groups and then regressing the response means $\bar{Y}_i$ on these dummy variables. This sounds like a futile exercise, but establishing a significant regression of Y on a carefully chosen dummy variable can yield important analytical insights. An example will clarify this point.

Refer to the pea section data that were introduced in Table 9.1. The means for these data together with coefficients of linear comparison for two orthogonal contrasts, control versus treatments (sugars) and mixed versus pure sugars, are shown in Table 9.3. We will employ these coefficients as dummy variables. Tests of the null hypotheses for these contrasts, which in Section 9.4 were carried out as conventional analyses of variance, are now carried out as regressions on these dummy variables. Instead of the usual formula for the explained sum of squares, $(\Sigma xy)^2/\Sigma x^2$, we can write an analogous formula for an explained SS for regression of $\bar{Y}$ on the dummy variable c, omitting the subscripts from c_{ij} for simplicity:

$$\frac{\left(\sum^a c\bar{y}\right)^2}{\sum^a c^2}$$

Note that the dummy variable c is already written in lower case; it is a deviate as well as a coefficient of linear comparison because its sum and mean are zero. Because $\Sigma^a c = 0$, $(\Sigma^a c\bar{y})^2/\Sigma^a c^2$ can be written as

$$\frac{\left[\sum^a c(\bar{Y} - \bar{\bar{Y}})\right]^2}{\sum^a c^2} = \frac{\left[\sum^a c\bar{Y} - \bar{\bar{Y}}\sum^a c\right]^2}{\sum^a c^2} = \frac{\left(\sum^a c\bar{Y}\right)^2}{\sum^a c^2}$$

The final step is to multiply the sum of squares by n to make it equivalent to a sum of squares of *groups,* not of *means,* thereby making it compatible with the rest of the anova to which it belongs. Therefore, the final formula for the explained *SS* based on a linear comparison is

$$\frac{n\left(\sum\limits^{a} c\bar{Y}\right)^2}{\sum\limits^{a} c^2} \qquad (14.12)$$

Now we apply Expression (14.12) to the two planned comparisons of Table 9.3. We compute

$$10[(4 \times 70.1) + (-1 \times 59.3) + (-1 \times 58.2) + (-1 \times 58.0) + (-1 \times 64.1)]^2$$
$$\div [4^2 + (-1)^2 + (-1)^2 + (-1)^2 + (-1)^2]$$
$$= 832.32$$

and

$$10[(0 \times 70.1) + (-1 \times 59.3) + (-1 \times 58.2) + (3 \times 58.0) + (-1 \times 64.1)]^2$$
$$\div [0^2 + (-1)^2 + (-1)^2 + 3^2 + (-1)^2]$$
$$= 48.13$$

If you check back, you will find that these sums of squares for the difference between treatments and the control and between the pure and mixed sugars are the same as were found in Section 9.4. These values are incorporated into the anova as was shown there. We are led to reject the null hypotheses for both linear comparisons.

Why do we go to all this trouble when we were able to carry out such an analysis by the more conventional method of obtaining sums of squares learned in Section 9.4? Although explaining how to evaluate a sum of squares from a linear comparison of treatments as shown here took some time, the setup of the data and the computation are generally simpler than the corresponding techniques of Box 9.5, once the idea has been mastered. A major advantage of the new technique is that it permits a comprehensive view of all comparisons through the table of coefficients, which often helps avoid inconsistencies. Furthermore, the coefficients enable us to test the orthogonality of the comparisons, as illustrated in Section 9.4. Finally, understanding the use of these coefficients will facilitate part of our study of advanced regression in Chapter 16.

Box 14.8 shows some examples of individual-degree-of-freedom comparisons that have been completely worked out. These are all independent tests whose probabilities are evaluated on a per-comparison basis. The first example, the per-diem fecundity in three selected lines of *Drosophila*, comes from Box 9.5. First, we test selected lines against nonselected lines. The coefficients are 1 for both RS and SS because there is one line in the contrasted set. The NS line, however, has a coefficient of 2, because there are two lines, RS and SS, in its contrasted set. We arbitrarily

BOX 14.8 Individual-Degree-of-Freedom per-Comparison Testing of Means in Analysis of Variance

Per diem fecundity in Drosophila melanogaster

Data from anova of Box 9.5 (equal n): $a = 3$; $n = 25$.

Comparisons		Resistant (RS) Line	Susceptible (SS) Line	Nonselected (NS) Line	(1) $\sum^a c\bar{Y}$	(2) $\sum c^2$	(3) $\sum \hat{y}^2 = n(1)^2/(2)$
	$\bar{Y}$	25.256	23.628	33.372			
Selected vs. nonselected	c_{i1}	+1	+1	-2	-17.860	6	1329.08
Resistant vs. susceptible	c_{i2}	+1	-1	0	1.628	2	33.13

Anova table

Source of variation	df	SS	MS	F_s	P
Strains	2	1362.21	681.10	8.665	0.000425
Selected vs. nonselected	1	1329.08	1329.08	16.909	0.000103
Resistant vs. susceptible	1	33.13	33.13	0.422	0.5180
Error	72	5659.02	78.60		

Computation

See text for method of assigning correct coefficients of linear comparison.

1. $\sum^a c\bar{Y} = (1 \times 25.256) + (1 \times 23.628) + (-2 \times 33.372) = 1.628$

2. $\sum c^2 = (1)^2 + (1)^2 + (-2)^2 = 6$ and $(1)^2 + (-1)^2 = 2$

3. $\sum \hat{y}^2 = n(\text{quantity } 1)^2/\text{quantity } 2 = 25(-17.860)^2/6 = 1329.08$ and $25(1.628)^2/2 = 33.13$

Box 14.8 (continued)

Partitioning degrees of freedom in a larger example (degrees of freedom for strains not completely subdivided)

Abdominal bristle number in housefly strains. Strains are subdivided on the basis of resistance to DDT: $a = 8$ strains; $n = 24$ flies per strain.

Meaningful comparisons		Group 1 Strongly resistant	Group 2 Slightly resistant		Group 3 Normal			Group 4 Susceptible		(1) $\overset{a}{\sum} c\bar{Y}$	(2) $\sum c^2$	(3) $n(1)^2/(2)$
		OL	LDD	RKS	RH	LC	BS	NKS	NH			$\sum \hat{y}^2 =$
$\bar{Y}$		34.2500	27.1250	32.9167	29.3750	27.4583	27.0000	30.50000	27.5417			
Resistant vs. nonresistant strains	c_{i1}	+5	+5	+5	−3	−3	−3	−3	−3	45.8335	120	420.14
Group 1 vs. group 2	c_{i2}	+2	−1	−1	0	0	0	0	0	8.4583	6	286.17
LDD vs. RKS (within group 2)	c_{i3}	0	+1	−1	0	0	0	0	0	−5.7917	2	402.53
Group 3 vs. group 4	c_{i4}	0	0	0	+2	+2	+2	−3	−3	−6.4585	30	33.37
Among strains of group 3 (SS for 2 df; compute in conventional manner as shown below)												
NKS vs. NH (within group 4)	c_{i5}	0	0	0	0	0	0	+1	−1	2.9583	2	105.02

Among strains of group 3
$$SS = \sum^{k} n_i(\bar{Y}_i - \bar{\bar{Y}})^2 = 24(29.3750 - 27.9444)^2 + 24(27.4583 - 27.9444)^2 + 24(27.000 - 27.9444)^2 = 76.20$$

Note that 27.9444 is the mean of group 3.

SOURCE: Data from Sokal and Hunter (1955).

Box 14.8 (continued)

Anova table

Source of variation	df	SS	MS	F_s	P
Strains	7	1323.42	189.06	8.467	0.000220
Resistant vs. nonresistant	1	420.14	420.14	18.815	0.000509
Group 1 vs. group 2	1	286.17	286.17	12.815	0.002504
LLD vs. RKS (within group 2)	1	402.52	402.52	18.026	0.000617
Group 3 vs. group 4	1	33.37	33.37	1.494	0.2393
Among strains of group 3	2	76.20	38.10	1.706	0.2130
NKS vs. NH (within group 4)	1	105.02	105.02	4.703	0.0455
Jars within strains	16	357.25	22.33	0.804	0.6798
Within jars	168	4663.25	27.76		

Strains and all linear contrasts are tested over the mean square for jars within strains.

Individual-degree-of-freedom comparisons in an anova with unequal n

Length of larval period (in hours) in lines of *Drosophila melanogaster* selected for short (SL) and long (LL) larval period compared with the control strain (CS).

Box 14.8 (continued)

Comparisons		SL	CS	LL	(1) $\sum\limits^{a} n_i c_i' \bar{Y}_i$	(2) $\sum\limits^{a} n_i c_i'^2$	(3) $\sum \hat{y}^2 = \dfrac{(1)^2}{(2)}$
$\dfrac{n_i}{\bar{Y}_i}$		80 100.9750	69 105.6667	33 110.3030			
Effect of selection (shortest vs. longest; linear)	Unweighted c_{i1} Weighted c_{i1}'	(−1) −33	0 0	(+1) +80	24,625.9200	298.320	2032.84
Symmetry of selection (ends vs. middle; quadratic)	Unweighted c_{i2} Weighted c_{i2}'	(−1) −69	+2 +113	(−1) −69	15,341.3289	1,419.054	165.85

SOURCE: Data from Hunter (1959).

See text for assignment of correct coefficients of linear comparison. Computation as before except that in step **3** it is not necessary to multiply (quantity **1**)^2 by n because quantity **1**, $\sum\limits^{a} n_i c_i' \bar{Y}_i$, is already weighted by n_i.

Anova table

Source of variation	df	SS	MS	F_s	P
Among lines	2	2198.70	1099.35	27.895	2.848×10^{-11}
Effect of selection	1	2032.84	2034.84	51.633	$<1 \times 10^{-15}$
Symmetry of selection	1	165.85	165.85	4.208	0.0417
Error	179	7055.25	39.41		

assign a plus sign to one of the contrasted sets and a minus sign to the other. In testing resistant lines against susceptible lines, we set RS = +1 and SS = −1 and assign a coefficient of zero to the nonselected line, which is not involved in this test. The computations are straightforward, as outlined in Box 14.8. The results are the same as in Box 9.5. We reject the null hypothesis of no difference in fecundity because of selection but accept the H_0 of no difference in fecundity between the two selected lines.

The next analysis in Box 14.8 is a more extensive study in which the number of abdominal bristles in eight strains of houseflies differing in DDT resistance is analyzed. The strains are arranged in four groups according to their degree of resistance to the toxicant. This study was undertaken in the hope of finding a morphological correlate for DDT resistance, such as bristle number, that could be used (by inverse prediction) to indicate the resistance status of a housefly population. This example is a nested analysis of variance in which differences among jars within strains, as well as differences among strains, were tested. The overall analysis of variance shows, however, that can we only reject the null hypothesis with confidence for differences among strains; the number of bristles was not affected by environmental variations in the culture medium among different jars. We test the linear comparisons over the mean square of jars within strains. We now wish to partition the SS among strains based on seven degrees of freedom into separate single-degree-of-freedom comparisons to learn more about the differences in this study. For example, the first comparison suggested by the design of the study is resistant versus nonresistant strains—that is, groups 1 and 2 versus groups 3 and 4. The coefficients for this comparison are 5 for members of groups 1 and 2 because there are five strains comprising the contrasted set, groups 3 and 4; and, conversely, 3 for members of groups 3 and 4 because there are three strains in groups 1 and 2. Similar considerations govern the choice of coefficients for the other comparisons. The computations at the right side of the table in Box 14.8, which are no different from the previous calculations, are not discussed in detail.

The P-values in the anova table indicate that resistant flies differ from nonresistant ones in bristle number, that the very resistant group 1 differs from the slightly resistant group 2, and that group 2 is heterogeneous because strain LDD differs from strain RKS. We cannot reject the null hypothesis of no difference between groups 3 and 4. The investigators had no logical basis for designing contrasts within group 3, consisting of three strains with two degrees of freedom among them. They therefore calculated the sum of squares pertaining to the two degrees of freedom among the three strains in the conventional manner, as shown at the bottom of the table in Box 14.8. The investigators could not reject the null hypothesis of no differences in abdominal bristle numbers among the three strains of group 3. The P-value of 0.0455 for the test between the two strains of group 4 is borderline. We cannot reject the H_0 of no difference between them with confidence.

We may conclude that selection for resistance has affected abdominal bristle number variously in different strains, apparently raising it the most in the highly resistant strain, whereas among the nonresistant strains there are no differences for this variable. We might wish to consider one further contrast, between groups 2 and 3, which cannot be orthogonal to the previous contrasts because they already used up the available 7 degrees of freedom. Because the data have already been divided into groups 1 and 2 versus groups 3 and 4, a contrast between groups 2 and 3 can no longer be orthogonal. We can easily convince ourselves of this fact by accumulating the products of the coefficients of the first row with those for the new contrast (let us call them c_{i6}): $\sum c_{i1}c_{i6} = (5 \times 0) + (5 \times 3) + (5 \times 3) + (-3 \times -2) + (-3 \times -2) + (-3 \times -2) + (-3 \times 0) + (-3 \times 0) = 48$. Because these products do not sum to zero, they are not orthogonal. A test of a set of such nonorthogonal contrasts can be carried out by the procedures described in Box 9.7, part III.

The last example in Box 14.8 shows the minor complications that arise when sample sizes are unequal. The data are length of larval period in three strains of *Drosophila melanogaster*, one selected for short larval period (SL), one for long larval period (LL), and a control strain (CS). The sample sizes and sums are given in the first table. The simplest way of carrying out the computations for such a case is first to write down the "unweighted" coefficients as if the sample sizes were the same to give us a general idea of the contrasts we would like to compute. Note that the first contrast, strain SL versus strain LL, tests the overall effect of selection. If the strains selected for short and long larval period were not different, selection clearly would have been ineffective. The second comparison tests the symmetry of the results of selection. If selection for short larval period was as effective as selection for long larval period, then the control (unselected) strain should be exactly halfway between the two. This symmetry can be tested by comparing the control strain with the average of the short and long strains.

Having established the contrasts, we now replace the unweighted coefficients by others weighted according to sample sizes. The weighted coefficients, c'_{i1} and c'_{i2}, for each member of one set is the *total sample size* of the contrasted set. Thus, for the first comparison, the weighted coefficient for SL is the sample size of the contrasted LL; conversely, for the LL it is the sample size of the contrasted SL. In the second comparison, the weighted coefficient for both SL and LL is 69, the sample size of the contrasted CS, and the weighted coefficient for CS is the sum of the sample sizes of the contrasted lines, SL and LL, $80 + 33 = 113$. The signs of these coefficients are the same as those of the unweighted coefficients. The rest of the computations are as before, except that in step **3** we need not multiply the numerator by n because the coefficients are already weighted by sample size. The resulting anova table shows that there is a clear difference in length of larval period between the strains selected in opposite directions for this variable, and that there is a suggestion ($P = 0.0417$) that their distance from the nonselected control strain is not symmetrical.

We can estimate multiple confidence limits for means and contrasts by means of coefficients of linear comparison. Such a set of confidence limits is computed with an experimentwise error rate α so that all k confidence intervals that the investigator intends to construct contain the parameter with an overall probability of $1 - \alpha$. Note that these are confidence limits. They are thus not intended for significance tests based on testing overlaps between intervals, as are the comparison limits of Section 9.5. To set these multiple confidence limits for contrasts, employ Expression (14.13) for equal sample sizes:

$$\sum_{i=1}^{a} c_i \bar{Y}_i \pm t_{\alpha'[v]} \sqrt{\left(\frac{1}{n}\sum_{i=1}^{a} c_i^2\right) MS_{\text{within}}} \tag{14.13}$$

and Expression (14.14) for unequal sample sizes:

$$\sum_{i=1}^{a} c'_i \bar{Y}_i \pm t_{\alpha'[v]} \sqrt{\left(\sum_{i=1}^{a} \frac{c'^2_i}{n_i}\right) MS_{\text{within}}} \tag{14.14}$$

Multiple confidence limits for means are set as

$$\bar{Y}_i \pm t_{\alpha'[v]} \sqrt{\frac{MS_{\text{within}}}{n_i}} \tag{14.14m}$$

For the special case of the contrast between two means based on equal sample sizes, Expression (14.13) simplifies to

$$(\bar{Y}_1 - \bar{Y}_2) \pm t_{\alpha'[a(n-1)]} \sqrt{\frac{2}{n} MS_{\text{within}}} \tag{14.15}$$

In all four cases, $\alpha' = 1 - (1 - \alpha)^{1/k}$ where k is the number of contrasts intended. We do not have to compute α' but can look up $t_{\alpha'}$ directly in Statistical Table C as $t_{\alpha[k,v]}$.

Note that the width of the confidence interval differs for different contrasts and sample sizes. As an example, let us set confidence limits to the difference between the selected and nonselected lines (the first contrast in the first example of Box 14.9). We assume that we had planned to set two confidence intervals at a confidence level of 95%. From Statistical Table C, we find that $t_{.05[2,72]} = 2.284$. Following Expression (14.13), we compute

$$[(+1)(25.256) + (+1)(23.628) + (-2)(33.372)]$$

$$\pm 2.284 \sqrt{\left[\frac{1}{25}(1^2 + 1^2 + (-2)^2\right]78.60} = -17.860 \pm 9.920$$

14.10 Examining Residuals and Transformations in Regression

In Section 14.5, we stressed that the nature of the departure of points from linear regression gives us a clue to interpreting the goodness of fit. The modern tendency has been to carry out some examination of residuals in addition to the hypothesis tests that we have already learned. Such an examination may detect outliers in a sample. Removal of such outliers may improve the regression fit considerably. One might also detect systematic departures from regression that can be adjusted by transformation of scale or by fitting a curvilinear regression line.

Two common procedures are used for such an examination. The first is to compute **leverage coefficients** (also known as the diagonal elements of the hat matrix;

Hoaglin and Welsch, 1978) to match each dependent observation Y_i. These coefficients are computed as

$$h_i = \frac{1}{n} + \frac{(X_i - \overline{X})^2}{\sum x^2}$$

for $i = 1 \ldots n$. The values of h_i, which range from near 0 to 1, give an indication of the *leverage* of a given value of X—that is, the influence of the corresponding response value Y_i on the estimated value $\hat{Y}_i$. Note that the quantities h_i are the same quantities as those in square brackets in the formula for $s_{\overline{Y}}$ in Box 14.2. They are functions of the magnitude of the departure for any given value X_i from its mean; thus, values of Y_i corresponding to deviant observations X_i (deviant from their mean $\overline{X}$, that is) are given greater weight in determining the fit of the line. For more than one value of Y per X, a single h_i is computed per X_i, with $1/\sum n_i$ replacing $1/n$ in the formula for h_i.

Examining the h_i values, however, is not sufficient. The second procedure is to compute and plot the **standardized residuals** from the regression line, which are the unexplained deviations $d_{Y \cdot X}$ divided by $s_{Y \cdot X}(1 - h_i)^{1/2}$, in order to put them into a standard deviation scale. Both sets of quantities have been computed for the weight loss data of Box 14.1 and are shown in Figure 14.16. Although several of the residuals seem suspiciously large, suggesting that a parabola might have fit the data better, by the criteria that we will describe shortly, none of the h_i or $d_{Y \cdot X}$ are high or outliers, respectively. In examining such a figure, we look first at high values of h_i, which indicate points with potential high leverage. If such points also show high residuals, they may be affecting the slope of the line unduly. High values of h_i with low residuals are not a problem; they simply indicate consistency of the observations with the regression model. Large residuals for a point with a low leverage value are

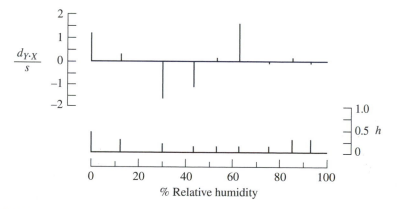

FIGURE 14.16   Examination of residuals in regression. Weight loss data of Box 14.1 and Figure 14.8. The ordinate of the upper graph represents deviation from regression in standard deviation scale, that of the lower graph h, the leverage coefficients (diagonal elements of the hat matrix).

relatively unimportant because they do not influence the regression line very much. When outliers are detected by this method, the regression may be recomputed after the outliers have been omitted.

Hoaglin and Welsch (1978) recommend that values of h_i greater than $4/n$ be considered high. The standardized residuals can be tested by comparing them to $t_{\alpha[n-2]}$, although the nominal probability levels in such an exhaustive data analysis are only suggestive. Hoaglin and Welsch also refine the computation of the standardized residual for the ith observation by recomputing $s_{Y \cdot X}$, leaving that observation out of the computation, so that the observation being tested does not contribute to the unexplained error estimate. In practice, such a strategy can be carried out only by computer.

In transforming either or both variables in regression, we aim at achieving a normal and homoscedastic distribution of points around the regression line. As a by-product of such a procedure, we simplify a curvilinear relationship to a linear one, thereby usually increasing the proportion of the variance of the dependent variable explained by the independent variable. Rather than fit a curvilinear regression to points plotted on an arithmetic scale (see Section 16.9), it may be more expedient to compute a simple linear regression for observations plotted on a transformed scale. A general test of whether transformation will improve linear regression is to graph the points to be fitted on an arithmetic scale as well as on the scale suspected to improve the relationship. These operations can be carried out either by computer graphics or with suitably scaled graph paper. If the function straightens out and the systematic deviation of points around a visually fitted line is reduced, the transformation is worthwhile. If a suitable transformation cannot be found, methods of nonlinear least squares analysis may have to be employed.

Such procedures are more complex and beyond the scope of this book because they usually have no explicit solution and there may be multiple solutions that minimize the least-squares criterion. See Wolberg (2006), Seber and Wild (2003), or Ritz and Streibig (2008) for more information. Nonlinear models can be fit using such software as Mathematica (Wolfram Research, 2008), Prism (GraphPad Software), MATLAB (The MathWorks, 2009), R (R Development Core Team, 2009), or even websites such as statpages.org/nonlin.html.

We will discuss briefly a few of the transformations common in regression analysis. Square root and arcsine transformations (see Sections 13.7 and 13.9, respectively) are not mentioned here, but they are also effective in regression cases involving data suited to such transformations, such as counts or percentages.

The **logarithmic transformation** is the most frequently used. Usually we transform the dependent variable Y. This transformation is indicated when percentage changes in the dependent variable vary directly with changes in the independent variable. Such a relationship is represented by the equation $\hat{Y} = ae^{bX}$, where a and b are constants and e is the base of the natural logarithm. Transformation results in the equation $\widehat{\log Y} = \log a + b(\log e)X$. In this expression, $\log e$ is a constant that, when multiplied by b, yields a new constant factor b', which is a regression coefficient in the new scale. Similarly, $\log a$ is a new Y-intercept, a'. We can then simply

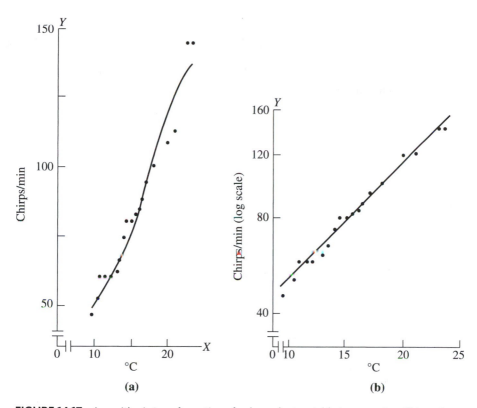

FIGURE 14.17 Logarithmic transformation of a dependent variable in regression. Chirp rate as a function of temperature in males of the tree cricket *Oecanthus fultani*. Each point represents the mean chirp rate per minute for all observations at a given temperature in °C. **(a)** Original data. **(b)** *Y*'s plotted on logarithmic scale. (Data from Block, 1966.)

regress log *Y* on *X* to obtain the function $\log \hat{Y} = a' + b'X$ and obtain all our prediction equations and confidence intervals in this form. Figure 14.17 shows an example of transforming the dependent observations to logarithmic form, which results in considerable straightening of the response curve.

A logarithmic transformation of the independent variable *X* in regression is effective when proportional changes in the independent variable produce linear responses in the dependent variable. An example is the decline in weight of an organism as density increases, where the successive increases in density need to be in a constant ratio to effect equal decreases in weight. This is an instance of a well-known class of biological phenomena, another example of which is the Weber–Fechner law in physiology and psychology, which states that a stimulus has to be increased by a constant proportion to produce a constant increment in response. Figure 14.18 illustrates how logarithmic transformation of the independent variable results in the straightening of the regression line. For computation, we transform *X* into logarithms.

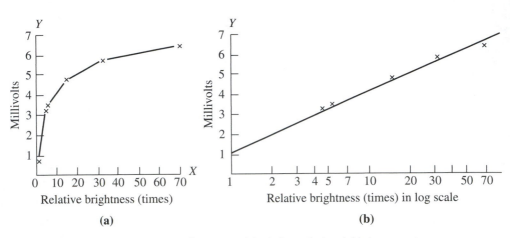

FIGURE 14.18   Logarithmic transformation of the independent variable in regression. Electrical response as a function of illumination in cephalopod eyes. **(a)** Original data. **(b)** *X*'s plotted on logarithmic scale. A proportional increase in *X* (relative brightness) produces a linear electrical response *Y*. (Data from Fröhlich, 1921.)

Logarithmic transformation for both variables applies to situations in which the true relationship can be described by the formula $\hat{Y} = aX^b$. The curve described by this relationship becomes straight when both variables are transformed to the logarithmic scale, as Figure 14.19 shows. The regression equation is $\widehat{\log Y} = \log a + b \log X$, and the computation is carried out in the conventional manner. This equation is the well-known **allometric growth curve,** applicable in many organisms where the ratio between increments in structures of different size remains roughly constant, yielding a relatively great increase of one variable with respect to the other on a linear scale. Examples are the greatly disproportionate growth of organs in some organisms, such as the antlers of deer or horns of stage beetles, with respect to their general body sizes. Note that the symbolism we use here is the converse of the conventional symbolism for allometric growth because we think it is important to retain the symbol *a* for *Y*-intercept and *b* for the regression coefficient. For comparison of structures subject to allometric growth, it is best to compare adjusted and transformed means after the allometric organ has been regressed on some measure of general body size.

The analysis of allometric growth relationships is a subject of considerable depth that we cannot pursue here. The seminal work in this field is by Huxley (1932). Simpson et al. (1960) present an elementary discussion of this topic; Teissier (1960) and Gould (1966, 1975) give early reviews of this topic. Mosimann (1970) developed important theories about measures of size and shape. A useful review is furnished by Klingenberg (1996). Warton et al. (2006) review the statistical methods used in studies of allometry.

With models that are more complicated, one may transform the variables, as well as rearrange the equation slightly to put it into a form amenable to ordinary linear regression analysis. For example, studies of population growth sometimes involve a

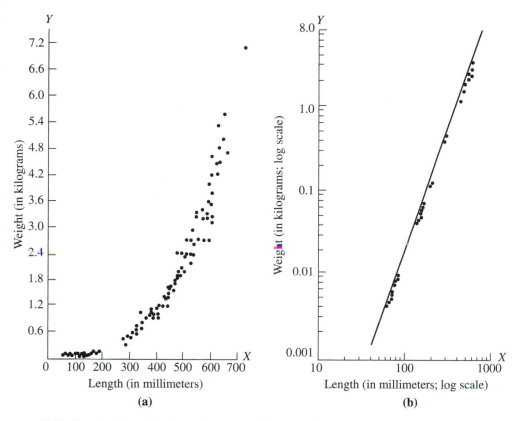

FIGURE 14.19 Logarithmic transformation of both variables in regression. Weight versus length in the cabezon *Scorpaenichthys marmoratus.* **(a)** Original data. **(b)** X's and Y's plotted on logarithmic scale. (Data from O'Connell, 1953.)

slightly more complicated functional relationship that requires a double logarithmic transformation. The Ricker function $\hat{Y} = aXe^{-bX}$ (Ricker, 1954) relates the expected number of recruits Y into a population to the size of the parental population X. Taking logarithms of both sides of this equation results in $\ln \hat{Y} = \ln a + \ln X - bX$. If we move $\ln X$ to the left side, we can rewrite the equation as $\ln \hat{Y} - \ln X = \ln a - bX$. In regression analysis, we assume that the independent variable is measured without error, so we may consider regressing a new artificial variable $Y' = \ln Y - \ln X$ on the independent variable X, using simple linear regression techniques to estimate the slope, $-b$, and the intercept, $\ln a$. A complication sometimes overlooked is that the proportion of the variance of Y' as explained by regression on X should not be considered a measure of fit of this model. Because Y' is simply the difference between $\ln Y$ and $\ln X$, it is expected to be rather highly correlated with X even if there is no relationship between Y and X. To investigate the degree of fit of this model, one must see how well the predicted values $\widehat{\ln Y}$ match the logarithms of the observed value, $\ln Y$.

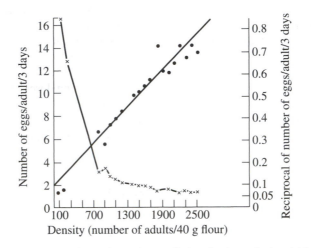

FIGURE 14.20 The reciprocal transformation applied to the dependent variable. Fecundity in *Tribolium* beetles expressed as number of eggs per adult per three days graphed as a function of density. Transforming the dependent variable into reciprocal scale changes the relationship from a hyperbolic to a linear one. X-marks are plots in original scale, dots in reciprocal scale. (Data from R. R. Sokal, unpublished results.)

Many rate phenomena (a given performance per unit of time or per unit of population), such as wing beats per second or number of eggs laid per female, yield hyperbolic curves when plotted in original measurement scale. Thus, they form curves described by the general mathematical equations $bXY = 1$ or $(a + bX)Y = 1$. From these we can derive $1/Y = bX$ or $1/Y = a + bX$. Transforming the dependent variable into its reciprocal frequently results in a straight-line regression. The **reciprocal transformation** is illustrated in Figure 14.20.

We may also apply the Box–Cox transformation to either the dependent or the independent variables or to both of them simultaneously. In these cases, the log-likelihood function to be maximized is as given earlier (see Section 13.8). In the present case, s_T^2 is the variance of the deviations from regression, $s_{Y \cdot X}^2$, from the regression analysis carried out on the transformed data. Estimating $\hat{\lambda}$ and setting confidence limits are as described earlier. If we have more than one Y for each X and we are concerned with heteroscedasticity, we can maximize L' using Expression (13.4).

Some cumulative curves can be straightened by the **probit transformation.** Refresh your memory on the cumulative normal curve shown in Figure 6.5 and discussed in Section 6.7. Remember that by changing the ordinate into normal equivalent deviates (NEDs) or normal quantiles, we were able to make the cumulative distribution function of a normally distributed variable linear. In the older literature, **probits** were quantiles coded by the addition of 5.0 to avoid negative values. Thus, a probit value of 5.0 would correspond to a quantile of zero and a cumulative frequency of 50%, probit value 6.0 corresponds to a quantile of 1 and a cumulative frequency of 84.13%, and probit value 3.0 corresponds to a quantile of -2 and a cumulative frequency of 2.27%. Probit tables, giving the probit equivalents of cumulative percentages, are available in Fisher and Yates (1963) and Pearson and Hartley (1958). Probit values can also be found by inverse lookup in Statistical Table **A**.

Figure 14.21 shows mortality percentages for increasing doses of an insecticide. These percentages represent differing points of a cumulative frequency distribution. With increasing dosages, an ever greater proportion of the sample dies, until at a high enough dose the entire sample is killed. If the doses of toxicants are transformed into logarithms, the tolerances of many organisms to these poisons are distributed approximately normally. These transformed doses are technically called **dosages.** Increasing dosages lead to a cumulative normal distribution of mortalities, called a **dosage–mortality curve.** These curves are the subject of an entire field of biometric analysis, **bioassay,** to which we can refer only in passing here. The most common technique in this field is **probit analysis.** Graphic approximations can be carried out on *probit paper,* which is probability graph paper in which the abscissa has been transformed into logarithmic scale. A regression line is fitted by eye to dosage–mortality data graphed on probit paper (see Figure 14.21). From the line fitted by eye, the 50% kill is estimated by inverse prediction (see Section 14.7).

Govindarajulu (2000) gives a general account of methods for bioassay. Robertson et al. (2007) includes examples of the use of the POLO series of programs. The by now classic reference is Finney (1971).

Another cumulative distribution that superficially resembles the cumulative normal curve is the **logistic equation.** This distribution ranges from a proportion of zero to 1 and is especially suitable for proportions near 0 and 1. By transforming proportions p to **logits**—that is, $\ln[p/(1 - p)]$, we can make the logistic function linear and

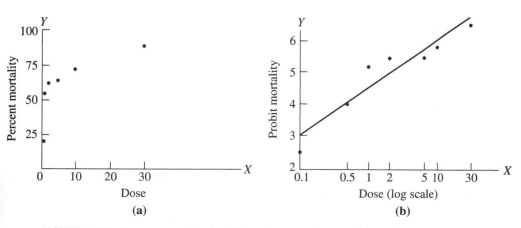

FIGURE 14.21 Dosage–mortality data illustrating an application of the probit transformation. Data are mean mortalities for two replicates. Twenty *Drosophila melanogaster* per replicate were subjected to seven doses of an "unknown" insecticide in a class experiment. **(a)** Original data. **(b)** Probit transformation. The points at dose 0.1 that yielded 0% mortality have been assigned a probit value of 2.5 in lieu of $-\infty$, which cannot be plotted.

compute regressions of data for proportions or occurrences. This method, which has become quite popular in recent years, may be the optimal technique for regressing proportions. Percentages, which are readily expressed as proportions, can, of course, also be transformed to logits. Because this method is intimately related to the analysis of frequencies, we feature it in Chapter 17.

Finally, there is an iterative algorithm that transforms both dependent and independent variables to yield regressions with minimal unexplained sums of squares. Known as the *ACE algorithm* (Breiman and Friedman, 1985), this program finds optimal nonlinear transformations. The user can restrict these to monotone transformations if so desired. The data may be continuous, ordered, or categorical, making this a very powerful transforming instrument. Because the ACE algorithm fits the dependent variable of a sample to its independent variable(s) empirically, it does not yield a general transformation formula, as do the other transformations we have considered. Adding to or subtracting from the data points in the sample would yield a different transformation. The ACE method is widely available (e.g., acepack in R). Hastie and Tibshirani (1990) includes examples and comparisons to other methods.

When transforming variables, you should avoid the pitfall of comparing the goodness of fit of regressions based on untransformed and transformed variables by means of the coefficient of determination R^2. In Section 14.4, we defined this statistic only as $\Sigma \hat{y}^2 / \Sigma y^2$, but Chapter 15 will give other formulations that are mathematically identical in the case of linear regressions with an intercept a. When the model is nonlinear or lacks a Y-intercept, however, these mathematical identities break down. The R^2 computed from transformed variables holds only for the linearized space resulting from the transformation and cannot be compared with an R^2 computed from the untransformed data. Kvålseth (1985) suggests computing R^2 as $1 - \Sigma(Y - \hat{Y}')^2 / \Sigma(Y - \overline{Y}')^2$ where $\hat{Y}'$ is the backtransformed value of the estimated Y in the transformed space and $\overline{Y}'$ is the backtransformed mean. Kvålseth (1985) and Scott and Wild (1991) discuss these problems in some detail.

14.11 Nonparametric Tests for Regression

When transformations are unable to make the relationship between the dependent and independent variables linear, the investigator may wish to carry out a simpler nonparametric test in lieu of regression analysis. We show **Kendall's robust line-fit method** in Box 14.9 (Kendall and Gibbons, 1990, Section 13.4). We order the X- and Y-pairs by the magnitude of X. Then we compute a slope S_{ji} for every pair of X-values, X_i and X_j, by the following formula:

$$S_{ji} = \frac{Y_j - Y_i}{X_j - X_i}$$

where i ranges from 1 to $n - 1$ and $j > i$. There will be $n(n - 1)/2$ such slope estimates S_{ji}. The nonparametric estimate b_K of the slope is the median of the S_{ji}-values. To estimate the Y-intercept, compute the n values of $Y_i - b_K X_{ii}$ and again choose

BOX 14.9 Kendall's Robust Line-Fit Method for Nonparametric Regression

Weight loss data of Box 14.1.

Weight loss in mg (Y)	8.98	8.14	6.67	6.08	5.90	5.83	4.68	4.20	3.72
Percent relative humidity (X)	0.0	12.0	29.5	43.0	53.0	62.5	75.5	85.0	93.0

The X-values are already arranged in ascending order of magnitude. If they had not been, then one should order them together with their paired Y-values.

Computation

1. Calculate $S_{ji} = (Y_j - Y_i)/(X_j - X_i)$ for all pairs X_i, X_j, where $j > i$.

$$S_{21} = (8.14 - 8.98)/(12.0 - 0.0) = -0.07000$$
$$S_{32} = (6.67 - 8.14)/(29.5 - 12.0) = -0.08400$$
$$\cdot$$
$$\cdot$$
$$\cdot$$
$$S_{31} = (6.67 - 8.98)/(29.5 - 0.0) = -0.07831$$
$$\cdot$$
$$\cdot$$
$$\cdot$$
$$S_{91} = (3.72 - 8.98)/(93.0 - 0.0) = -0.05656$$

   The median of the $n(n - 1)/2$ slopes is $b = -0.05436$.

2. Estimate the Y-intercept: $Y_i - bX_i$ for $i = 1$ is 8.98000, for $i = 2$ is 8.79226, and so on

   The median of the n intercepts is 8.78382.

3. The nonparametric regression equation is $Y = 8.78382 - 0.05436X$.

the median. Using this approach, we find a regression equation of $Y = 8.78382 - 0.05436X$ for the weight loss/relative humidity data of Box 14.1. This equation is close to the estimate given in that box.

An appropriate significance test for b_K is the **ordering test** (Quenouille, 1952), which is equivalent to Kendall's rank correlation coefficient (see Box 15.4) and can be carried out most easily as such. According to this procedure, we first rank observations X and Y. Then we arrange the independent variable X in increasing order of ranks and calculate τ, the Kendall rank correlation of Y with X. The computational steps are shown in Box 15.4. If we carry out this computation for the weight loss data of Box 14.1 (reversing the order of percent relative humidity, X, which is negatively related to weight loss, Y), we obtain a quantity $N = 72$, yielding a value of $\tau = 1.0000$ and leading us to reject the H_0 of no rank correlation between these two variables. The probability of type I error is $P < 0.01$ (see Statistical Table **S**). We conclude that weight loss is a function of relative humidity. The ranks of the weight losses are a perfect monotonic function of the ranks of the relative humidities.

Noether (1985) described a direct hypothesis test for b_K, but because it too is related to Kendall's rank correlation, we limit ourselves to presenting the latter method. If we do not need either a prediction equation or a functional relationship but simply want to know whether the dependent variable Y is a monotonically increasing (or decreasing) function of the independent variable X, we can simply carry out the rank correlation test. Such a test could also be applied to means in a single-classification analysis of variance that are assumed to represent a trend on X, such as the data in Box 14.4, where survival was considered a function of density. Note, however, that this particular example cannot be tested by this method because there are only four densities and the minimum number of points required for significance testing by the rank correlation method is five.

When the data are presented as ranks in a randomized-complete-blocks design, such as the data analyzed by Friedman's test in Box 13.11, a monotonic trend among the treatment effects can be tested by **Page's L-test for ordered alternatives** (Page, 1963). This test can be performed easily using the layout of Box 13.11 for Friedman's method. Order the treatments in such a way as to produce a monotonically increasing response according to your alternative hypothesis (the null hypothesis, of course, is no trend). Then multiply the treatment rank sums $\Sigma^b R_{ij}$ by the rank order of the particular treatment. In the Rot Lake temperature data in Box 13.11, for example, the alternative hypothesis is that increases in depth produce decreases in temperature. It is therefore advisable to rank the greatest depths (15.5 m) as rank 1 and to rank the other depths successively until rank 10, which is the shallowest depth (0 m). Page's L-statistic is computed as

$$L = \sum_{i=a}^{a} i\left(\sum R_{ij}\right) \tag{14.16}$$

where i refers to the rank number of the treatment level. Applying this formula to the data in Box 13.11, we obtain $L = 1(4) + 2(8) + 3(12) + \ldots + 10(40) = 1540$. The critical value of L is given in Statistical Table **Z**, where a and b refer to the number of treatments and of blocks as before. The table furnishes critical values of L for a from 3 to 10 and for b from 2 to 24. For values beyond these arguments, use the following expression:

$$L_\alpha = \frac{b(a^3 - a)}{12}\left[\frac{t_{(2\alpha)[\infty]}}{\sqrt{b(a-1)}} + \frac{3(a+1)}{a-1}\right] \tag{14.17}$$

Looking up L_α for these data, we obtain 1382 for $L_{.001}$. Because our observed L is much greater than this value, there is little doubt that there is a monotonic trend of temperature in relation to depth.

An alternative nonparametric test for trend in ordered observations is the runs-up-and-down test discussed in Section 18.2 and Box 18.8.

14.12 Model II Regression

Having learned a good deal about regression, all of it in a Model I context, the reader is ready to tackle Model II regression. This subject has aroused a good deal of confusion and controversy in the past, compounded by a terminological quagmire. The thorough review of the topic by Warton et al. (2006) is helpful. Do not be misled by its title, as it covers a broader topic than just allometry. The review has an ample bibliography for those interested in exploring the subject in depth. The appropriate analysis for Model II is determined jointly by the types of data at hand as well as by the questions to which the investigator wishes to have answers. Table 14.2 lists the cases of interest to us, as well as the recommended solutions. More detail on some of the issues involved is provided below.

As we have seen, there are two reasons to fit a line to bivariate data: (1) to predict Y for a given X_i; and (2) to estimate the functional relationship between Y, the dependent, and X, the independent variable and to perform tests of hypotheses about the slopes and location statistics. In Model I regression, both prediction and estimation of functional relationships are carried out best by means of linear regression as described above, but when both variables are subject to natural variation and measurement error, the appropriate method depends on the nature of the data and the intentions of the investigator. We therefore need to examine in some detail the possible types of situations subsumed under Model II regression.

The typical case of Model II regression involves two continuous variables distributed according to the bivariate normal distribution. This distribution will be discussed in Section 15.2. It will suffice for now to state that in a bivariate normal distribution, both variables are normally distributed and the correlation (association) between the two variables is determined by the magnitude of the parameter ρ (see Figures 15.1 and 15.2). Because neither X or Y are assumed to be fixed, we assign the symbols Y_1 and Y_2 for these, respectively. Examples of such data include the following measurements: the length of both left and right forewings in each individual of a sample of bees; the height of each person, as well as her score on an achievement test, in a sample of college women; and the strength of muscle contraction and the naturally varying concentration of a particular cation in the blood of a sample of frogs. In all these cases, the variables vary naturally (individual differences are genetic and/or are environmentally caused).

Note that in the statistical literature, this type of variation is sometimes called *equation error* (e.g., Warton et al. 2006, Carroll and Ruppert 1996). In addition, there will likely be error in making the measurements. The proportion of measurement error varies with the variable and depends on the circumstances of the study. Rarely do we encounter no measurement error. Frequently, the dependent variable Y_2 may not be normally distributed (and hence Y_1, Y_2 will not follow a bivariate normal distribution) for either or both of two reasons. First, if the underlying parent population is not a bivariate normal distribution, the sample will reflect this departure from normality.

TABLE 14.2 | Analyzing Bivariate Data

Types of Data

(A) X can be determined exactly or is an intended or nominal value fixed by the investigator rather than a measured value subject to natural variation or subject to measurement error (the Berkson case).

(B) X is subject to natural variation and to measurement error as well. (X and Y change to Y_1 and Y_2, respectively.) Both variables are in comparable units of measurement originally or after transformation.

(C) X is subject to natural variation and measurement error as well. (X and Y change to Y_1 and Y_2, respectively.) The two variables are *not* in comparable units of measurement.

(D) X is not subject to natural variation and is subject to measurement error only.

Purpose of the Analysis

1. Find a prediction equation for Y
2. Determine slope (functional relationship) of Y on X
3. Test H_0: $\beta = 0$, testing for a nonzero slope
4. Test H_0: $\alpha = 0$, testing for a nonzero Y-intercept
5. Test H_0: $\beta_1 = \beta_2$, testing for the difference between two or more slopes
6. Test H_0: $\alpha_1 = \alpha_2$, testing for the difference between two or more Y-intercepts

Recommended Solutions

For		
	A1, B1, C1, D1	Employ OLS (Box 14.1 or 14.4)
	A2	OLS (Box 14.1 or 14.4)
	A3, A4	OLS (Boxes 14.2 & 14.3)
	B2	MA (Box 14.10, part **II**)
	B3, B4	MA (Box 14.10 part **II**)
	C2	SMA (Box 14.10, part **III**)
	C3, C4	Combination of OLS & SMA (Box 14.10, part **III**)
	D2, D3, D4	See Carroll and Rulpert (1996).
	B5, B6, D5, D6	See Appendix D in Warton et al. (2006). Warton and Weber (2002) have software in R.

Acronyms: OLS is Ordinary Least Squares (the usual Model I regression method); MA is Major Axis regression; and SMA is Standard Major Axis regression.

In many cases, suitable transformations can correct this problem. Second, sampling may not have been random but rather biased to include more extreme values of the independent variable. For example, in a study relating chest width of men to their height, we may set up classes for the height distribution and sample approximately equal frequencies of men for each height class in an attempt to have equal replication over the range of the observed data. Again, the proportion of measurement error varies for different variables and studies.

Cases in which both variables lack natural variability and are subject to measurement error only are quite rare in the biological sciences. We therefore do not treat them in this book. Most such instances come from the physical sciences; an example is measurements of the same piece of metal for conductivity at different controlled temperatures in an experiment to establish a functional relation between these two variables.

There is one special case of apparent Model II regression that permits us to apply Model I methods for tests of significance. This is the so-called *Berkson case,* in which independent variables are subject to error but are controlled by the experimenter. This is case A in Table 14.2, a situation that occurs frequently in experimental work. You may have been uneasy earlier about applying Model I regression to cases in which different temperature settings of an environmental chamber, doses of a hormone, or different densities are applied to organisms. How can we be sure that such independent variables are applied without error? The controls on the environmental chamber may not be accurate. There can also be an error in administering the dose of hormone or reading the density. In addition, the effective dose or density (the dose or density directly interacting with the organism) will not necessarily be the ones that we intend. Thus, there is an error δ that attaches itself to each of our intended independent observations.

We can now write $X = \zeta + \delta$, where X is the intended or nominal value of the independent observation, ζ is the actual or effective value, and δ is the error term making the difference between the intended and actual observations. In this model, X and δ are not expected to be correlated because there is no reason to suppose, according to our model, that the magnitude of the intended observation and its error of application should be correlated. We may, therefore, use the ordinary Model I regression procedures to estimate slope without bias if we use the intended values of the independent variable; in such cases, prediction and functional relationship can be obtained using simple linear regression. If, on the other hand, we use a measured temperature within the chamber or actual readings of the amount of a hormone circulating in the blood of the organisms, then model II regression methods must be used.

If the regression line is being fitted mainly for purposes of prediction, the current consensus is to employ simple linear regression techniques (the Model I design already studied). Some further remarks on the distinction between prediction and functional relationship are, however, in order. Because in Model I regression the least squares simple regression equation $\hat{Y} = a + bX$ is appropriate for both purposes (prediction and determining functional relationship), investigators are tempted to use the line obtained in Model II cases interchangeably as well. If one has an equation relating Y_2 to Y_1, why not predict Y_2 for a given value of Y_1? Unbiased estimates of Y_2 given Y_1, and minimal-width confidence belts around the estimates of Y_2, however, are not generally given by functional equations that best describe the joint variation of two random variables. Thus, investigators who wish to determine the mutual slope of two random variables and the Y-intercept of that slope (i.e., their functional relation), should not attempt to use the resulting equation to predict values of Y_2 given Y_1.

The reason that special techniques are necessary for Model II regression is that even though Model I methods still provide the best prediction of the dependent variable, the estimates of the slopes are biased and do not represent the trend in the

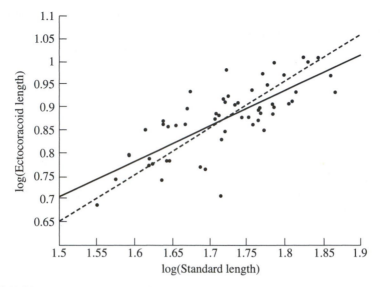

FIGURE 14.22   Log10 of ectocoracoid lengths plotted against log10 of standard length for the stickleback data of Box 14.10. Solid line is the usual least-squares regression. The major axis and standard major axis lines are indistinguishable and are shown by the dashed line.

bivariate scatter—the slope is too shallow (reflecting the property of regressing towards the mean mentioned earlier). Several approaches have been suggested to overcome this problem. When both variables are in the same units of measurement, determining the **slope of the major axis** (or *principal axis*) of the bivariate sample has been suggested. This method provides a single axis to represent the trend expressed by the scattergram and is preferred for other reasons as well, several of which we will mention.

Our first task, therefore, is to find the slope and equation of the major axis from a sample. A criterion of fitting a regression line through bivariate data is that it must pass through $(\overline{X}, \overline{Y})$ and that the SS of the deviations of the observed Y-values from regression is minimal (see Section 14.3). For simple linear regression, these deviations are parallel to the Y-axis (see Figure 14.7). A corresponding criterion for locating the principal axis is that the SS of deviations, which represent perpendiculars to the principal axes, be at a minimum. These perpendiculars take into consideration the deviation of a given point with respect to both $\overline{Y}_1$ and $\overline{Y}_2$ (Figure 14.22). By contrast, in simple linear regression the deviation was measured with respect only to Y, not to X.

The equation of the major axis defined in this way is $Y_2 = \overline{Y}_2 + b(Y_1 - \overline{Y}_1)$. The expression involves the means of the two variables and b_1, the slope of the major axis. This slope is computed from $b_1 = s_{12}/(\lambda_1 - s_2^2)$, where s_{12} is the covariance

$\Sigma y_1 y_2/(n-1)$, s_2^2 is the variance of Y_2, and λ_1 is a new quantity defined as follows in terms of variances and covariance of Y_1 and Y_2:

$$\lambda_1 = \frac{1}{2}[s_1^2 + s_2^2 + \sqrt{(s_1^2 + s_2^2)^2 - 4(s_1^2 s_2^2 - s_{12}^2)}]$$

The slope, b_2, of the minor axis, which is at right angles to the slope of the major axis, can be found easily as $b_2 = -1/b_1$. Details of the computation are given in Box 14.10.

To describe the shape of an ellipse, we must also know the ratio of the lengths of its major and minor axes. This ratio is $\sqrt{\lambda_1/\lambda_2}$, where $\lambda_2 = s_1^2 + s_2^2 - \lambda_1$. In mathematics and statistics, λ_1 and λ_2 are important quantities, known by many names. Among other terms, they are called **eigenvalues,** *latent roots,* or *characteristic roots* of the variance–covariance matrix of Y_1 and Y_2. In the type of problem we are discussing here, the eigenvalues are quantities analogous to variances, λ_1 and λ_2 measure variability along the major and minor axes, respectively. The major and minor axes are known as eigenvectors, latent vectors, or characteristic vectors. In data that are highly correlated and consequently represented by a very narrow, elongated ellipse, most of the variance can be accounted for by the major axis; in this case, the value of λ_1 would be very great with respect to the magnitude of λ_2. If the quantities λ_1 and λ_2 and are equal, the major and minor axes would be of equal length, the data would be represented by a circle, and hence there would be no correlation between the variables. Part **II** of Box 14.10 illustrates the computation of approximate confidence limits to the slope of the principal axis. This calculation is based on the approximate procedure for small samples proposed by Jolicoeur (1968), exact only for large *n*. With small *n* and low correlation *r*, *H* may be >1, and as a result, the limits will not exist.

When the two variables have different units of measurement, the slope of the major axis is generally meaningless and another technique for Model II regression should be employed. When the units of measurement are predetermined and nonarbitrary, however, this criticism does not apply and the slope of the major axis might be a suitable measure of functional relationship among the variables. In morphometric work, logarithmically transformed variables are often employed, and their functional relationship is estimated by the slope of the major axis.

Another way to overcome the scale dependence of the major axis is to standardize variables before the slope is computed—that is, to transform each of the two variables so that it has a mean of zero and a standard deviation of one. The principal axis of these standardized variables is known by several names: the *reduced major axis* (Kermack and Haldane, 1950), *relation d'allométrie* (Teissier, 1948), *geometric mean (GM) regression* (Ricker, 1973), or **standard major axis** (SMA, Jolicoeur, 1975). We now prefer the latter because the term reduced major axis was based on a mistranslation (see Jolicoeur, 1975). The slope of this line is simply

$$v_{2\cdot 1} = \pm\sqrt{\frac{\sum y_2^2}{\sum y_1^2}} = \pm\frac{s_2}{s_1} \tag{14.18}$$

BOX 14.10   Model II Regression

Standard lengths (Y_1—from the upper jaw to end of the last vertebra) of fossil three-spine sticklebacks, *Gasterosteus doryssus,* and lengths of their ectocoracoid bones (Y_2) in a random sample of size, $n = 55$.

Standard length in mm, Y_1	Ectocoracoid length in mm, Y_2	Standard length in mm, Y_1	Ectocoracoid length in mm, Y_2
57.21	7.27	59.79	8.86
59.07	9.38	63.02	9.29
72.42	9.30	43.93	6.08
65.23	8.55	66.64	10.17
53.01	8.39	63.75	8.05
52.54	7.04	69.75	10.16
58.11	7.82	35.53	4.89
46.65	7.87	37.56	5.54
51.56	7.62	54.14	8.03
43.07	5.54	54.53	8.11
52.39	8.27	64.50	8.17
48.54	5.89	43.33	7.40
39.11	6.26	59.20	7.10
73.45	8.53	51.09	7.70
49.30	5.81	41.62	6.14
40.99	7.10	58.25	7.86
57.08	8.64	67.70	9.94
58.59	7.69	41.91	5.99
44.00	7.23	52.52	8.16
60.68	8.06	60.98	9.91
41.58	5.96	58.08	7.44
52.72	9.54	55.26	7.54
44.14	6.09	46.35	7.29
51.81	5.09	47.07	8.57
52.01	6.74	56.49	7.52
43.29	7.31	61.08	7.92
45.05	7.26	60.70	7.69
50.94	7.48		

Unpublished data by M. A. Bell, F. J. Rohlf, and M. Travis.

Both variables are subject to natural variation as well as measurement error. It is desired to establish a functional relationship between ectocoracoid length and standard length to determine whether variation is isometric (slope $= 1$) or if there is a positive or negative allometric relationship. Linear measurements, such as those shown above, are frequently lognormally distributed. We have therefore transformed them to common logarithms. As a space-saving measure, we do not show the transformed numbers in the data table. To simplify the appearance of the formulas, we have omitted the log prefix in all expressions below.

Box 14.10 (continued)

I. *Linear regression*

The usual Model I regression line is computed first for comparison with the major axis and standard major axis methods illustrated in parts **II** and **III**. Some of the quantities computed here are also needed for these other methods.

$$n = 55$$

$$\sum y_1^2 = 0.308263 \qquad \sum y_2^2 = 0.309378 \qquad \sum y_1 y_2 = 0.233217$$
$$\bar{Y}_1 = 1.72151 \qquad \bar{Y}_2 = 0.87580 \qquad b_{2\cdot1} = 0.75655$$
$$a = -0.42661 \qquad s_{2\cdot1} = 0.05008 \qquad s_b = 0.090204$$

95% confidence limits for β: $L_1 = 0.5756$, $L_2 = 0.9375$

II. *Principal axes and confidence limits for the slope of the major axis*

1. From the basic quantities provided in part **I**, compute additionally s_1^2, s_2^2, and s_{12}. The quantity s_{12} is an abbreviated symbol for the covariance of Y_1 and Y_2, computed as shown. Using the sums of squares and cross products given in part **I**, we find:

$$s_1^2 = \frac{\sum y_1^2}{n-1} = \frac{0.308263}{54} = 0.005709$$

$$s_2^2 = \frac{\sum y_2^2}{n-1} = \frac{0.309378}{54} = 0.005729$$

$$s_{12} = \frac{\sum y_1 y_2}{n-1} = \frac{0.233217}{54} = 0.004319$$

2. The eigenvalues (see text) are found as follows (note that s_{12}^2 is the square of the covariance s_{12}):

Let

$$D = \sqrt{(s_1^2 + s_2^2)^2 - 4(s_1^2 s_2^2 - s_{12}^2)}$$

$$= \sqrt{(0.005709 + 0.005729)^2 - 4(0.005709 \times 0.005729 - 0.004319^2)} = 0.008638$$

Then

$$\lambda_1 = \frac{s_1^2 + s_2^2 + D}{2} = \frac{0.005709 + 0.005729 + 0.008638}{2} = 0.010038$$

and

$$\lambda_2 = \frac{s_1^2 + s_2^2 - D}{2} = \frac{0.005709 + 0.005729 - 0.008638}{2} = 0.001400$$

(Can also be computed as $\lambda_2 = s_1^2 + s_2^2 - \lambda_1$).

3. The slope of the principal axis is

$$b_1 = \frac{s_{12}}{\lambda_1 - s_1^2} = 0.99761$$

Box 14.10 (continued)

The equation of this axis is

$$\hat{Y}_1 = \bar{Y}_1 + b_1(Y_2 - \bar{Y}_2) = 0.84780 + 0.99761 Y_2$$

and is plotted in Figure 14.22.

4. The slope of the minor axis (at right angles to the principal axis) is

$$b_2 = -\frac{1}{b_1} = 1.00239$$

The equation of the minor axis is thus

$$\hat{Y}_1 = \bar{Y}_1 + b_2(Y_2 - \bar{Y}_2) = 2.59941 - 1.00239 Y_2$$

5. 95% confidence limits for β_1, the slope of the principal axis:

Let

$$H = \frac{F_{.05[1,n-2]}}{[(\lambda_1/\lambda_2) + (\lambda_2/\lambda_1 - 2)](n-2)}$$

$$= \frac{4.023017}{[0.010038/0.001400 + 0.001400/0.010038 - 2]53} = 0.014296$$

and

$$A = \sqrt{\frac{H}{1-H}} = 0.120430$$

The 95% confidence limits to the slope β_1 are

$$L_1 = \frac{b_1 - A}{1 + b_1 A} = \frac{0.99761 - 0.120430}{1 + 0.99761(0.120430)} = 0.78309$$

$$L_2 = \frac{b_1 + A}{1 - b_1 A} = \frac{0.99761 + 0.120430}{1 - 0.99761(0.120430)} = 1.27073$$

Shortcut using trigonometric functions:

$$L_1 = \tan[\arctan(b_1) - \tfrac{1}{2}\arcsin(2\sqrt{H})]$$

$$L_2 = \tan[\arctan(b_1) + \tfrac{1}{2}\arcsin(2\sqrt{H})]$$

III. *Standard major axis (reduced major axis) regression*

$$v_{2 \cdot 1} = \sqrt{\textstyle\sum y_2^2 / \sum y_1^2} = \sqrt{0.30826/0.30938} = 0.99820$$

$$a_v = 0.84729, \quad s_v = s_b = 0.089879$$

95% confidence limits for v: $L_1 = 0.81792$, $L_2 = 1.17847$.

The slopes from the major axis and standard major axis methods are very similar and consistent with an isometric model. As expected, linear regression yields a much lower slope.

This slope can be shown to be the geometric mean of the linear regression coefficient of Y on X and of the reciprocal of the regression coefficient of X on Y:

$$\pm \sqrt{b_{2 \cdot 1} \frac{1}{b_{2 \cdot 1}}} = \pm \sqrt{\frac{\sum y_1 y_2}{\sum y_1^2} \cdot \frac{\sum y_2^2}{\sum y_1 y_2}} = \pm \sqrt{\frac{\sum y_2^2}{\sum y_1^2}} = \pm \frac{s_2}{s_1}$$

which explains the name *geometric mean regression,* coined by Ricker (1973).

Note that $v_{2 \cdot 1}$, unlike $b_{Y \cdot X}$, is independent of any association between the two variables. It is merely the ratio of two standard deviations. This property of the standard major axis and other aspects of its behavior have led to serious criticisms of the method by Jolicoeur (1975) and Kuhry and Marcus (1977). The SMA regression of Y_1 on Y_2, $v_{2 \cdot 1}$, is obviously the same line, but because the axes are reversed, its slope is $1/v_{2 \cdot 1}$. The SMA regression line is computed as in linear regression, by forcing the line to pass through the bivariate mean $\overline{Y}_2, \overline{Y}_1$. Thus, the Y_2-intercept a_2 can be obtained from $\overline{Y}_2 - v\overline{Y}_1$ by analogy with the formula for a. Note that the sign of v cannot be obtained from its formula; it is the sign of the sum of products $\sum y_1 y_2$. When the sum of products equals zero, there is no association between X and Y, the sign of v is undefined, and there is little point in fitting a regression line. The slope of the SMA regression line will always be greater than that of the usual regression line, because $v_{2 \cdot 1} = b_{2 \cdot 1} / r_{12}$. (See Chapter 15 for the correlation coefficient r_{XY}.)

The standard error of the SMA regression slope, s_v, can be approximated by s_b, the standard error of the linear regression slope, and can be used to set confidence limits to v in the same way as for b. There is one difference, however. Whereas a test of the null hypothesis $H_0: \beta = 0$ was appropriate for linear regression and tested for the existence of any association between Y and X, a similar test is inappropriate for v because this is a ratio of the two standard deviations and will not be zero or undefined unless one of the standard deviations equals zero (an unsuitable case for a regression problem). Alternatively, one could use Kendall's robust line-fit method (see Section 14.11).

We illustrate the computation of Model II regression in Box 14.10. The data are from a sample of 55 fossil stickleback fish for which the length of the ectocoracoid bone and the "standard length" (from the upper jaw to the end of the last vertebra) had been recorded in millimeters. We want to estimate the functional relationship between ectocoracoid length (Y_1) and standard length (Y_2). These variables both vary naturally among individuals and are subject to measurement error. Box 14.10 shows three possible ways of estimating their functional relations computations, whose results we will compare shortly. The computation for the SMA regression is very simple. The regression line is fitted to the data plotted in Figure 14.22. Methods for setting confidence limits to the regression coefficient are also shown in Box 14.10. These are tedious to compute but contain no new complexities.

We can now compare the regression equations obtained by the four methods discussed in this chapter.

Linear regression	$\hat{Y}_2 = -0.42661 + 0.75655Y_1$
Major axis	$\hat{Y}_2 = 0.84780 + 0.99761Y_1$
Standard major axis	$\hat{Y}_2 = 0.84729 + 0.99820Y_1$
Kendall's robust line-fit	$\hat{Y}_2 = -0.4158 + 0.74349Y_1$

As is often the case, the slopes of the two Model II methods are very similar. As expected, the slope from linear regression is much lower. Because the two variables happen to have very similar variances, the slope is close to their correlation, 0.755. The Kendall's robust line-fit method gives an estimate of the slope similar to that found by linear regression. As mentioned earlier, the problem of an optimal Model II strategy is still an open question. Recommendations for showing an optimal functional relationship may yet be revised.

14.13 Effect Size, Power, and Sample Size in Regression

Because the magnitude of a regression coefficient depends in part on the units in which the dependent and independent variables are expressed, it is convenient to standardize it. This can be done by multiplying $b_{Y·X}$ by s_X/s_Y, the ratio of the standard deviation of the independent and dependent variables. The result is r_{XY}, the product-moment correlation coefficient between the two variables. For that reason, the methods for computing power and sample sizes for regression are the same as for correlations. This coefficient will be discussed in Chapter 15. Methods for computing confidence limits to estimated effect sizes and for estimating sample sizes needed to detect specified minimal effect sizes are described in Section 15.9.

EXERCISES 14

14.1 The following temperatures (Y) were recorded in a rabbit at various times (X) after being inoculated with rinderpest virus. (Data from Carter and Mitchell, 1958.)

Time after Injection (Hours)	Temperature (°F)
24	102.8
32	104.5
48	106.5
56	107.0
72	103.9
80	103.2
96	103.1

Graph the data. Clearly, the last three data points represent a different phenomenon from that of the first four pairs. *For the first four points:* (a) Calculate b. (b) Calculate the regression equation and draw the regression line on your graph. (c) Test the hypothesis that $\beta = 0$ and set 95% confidence limits. (d) Set 95% confidence limits to your estimate of the rabbit's temperature 50 hours after the injection.

ANSWER: $b = 0.1300$, $\hat{Y}_{50} = 106.5$

14.2 The following table is extracted from data by Sokoloff (1955). Adult weights of female *Drosophila persimilis* reared at 24°C are affected by their density as larvae. Carry out an anova among densities, then calculate the regression of weight on density and partition the sums of squares among groups into that explained and that unexplained by linear regression. Graph the data with the regression line fitted to the means. Interpret your results.

Larval Density	Mean Weight of Adults (in mg)	s of Weights (not $s_{\bar{Y}}$)	n
1	1.356	0.180	9
3	1.356	0.133	34
5	1.284	0.130	50
6	1.252	0.105	63
10	0.989	0.130	83
20	0.664	0.141	144
40	0.475	0.083	24

14.3 Davis (1955) reported the following results in a study of the amount of energy metabolized by the English sparrow, *Passer domesticus*, under various constant temperature conditions and a 10-hour photoperiod.

Temperature (°C)	Calories $\bar{Y}$	n	s
0	24.9	6	1.77
4	23.4	4	1.99
10	24.2	4	2.07
18	18.7	5	1.43
26	15.2	7	1.52
34	13.7	7	2.70

Analyze and interpret.

ANSWER: $b_{Y \cdot X} = -0.35558$, $F_s = 80.002$

14.4 Using the complete data given in Exercise 14.1, calculate the regression equation and compare it with the one obtained in that exercise. Discuss the effect of including the last three points in the analysis.

14.5 The following results were obtained in a study of oxygen consumption (microliters/milligrams dry weight/hour) in the corn earworm *Heliothis zea* under controlled temperatures and photoperiods. (Data from Phillips and Newsom, 1966.)

Temperature (°C)	Photoperiod (Hours)	
	10	14
18	0.51	1.61
21	0.53	1.64
24	0.89	1.73

Compute regression for each photoperiod separately and test for homogeneity of slopes.

ANSWER: $s^2_{Y \cdot X} = 0.019267$ and 0.00060 for 10- and 14-hour photoperiods, respectively.

14.6 Length of developmental period (in days) of the potato leafhopper, *Empoasca fabae,* from egg to adult at various constant temperatures. (Data from Kouskolekas and Decker, 1966.) The original data were weighted means, but for this analysis we will consider them ordinary means.

Temperature (°F)	Mean Length of Developmental Period (in Days) $\overline{Y}$	Number Reared N
59.8	58.1	10
67.6	27.3	20
70.0	26.8	10
70.4	26.3	9
74.0	19.1	43
75.3	19.0	4
78.0	16.5	20
80.4	15.9	21
81.4	14.8	16
83.2	14.2	14
88.4	14.4	27
91.4	14.6	7
92.5	15.3	4

Analyze and interpret. Compute deviations from the regression line $(\overline{Y}_i - \hat{Y}_i)$ and plot against temperature.

14.7 Webber (1955) studied the relation between pupal weight (in milligrams) and the number of ovarioles in females of the blowfly *Lucilia cuprina* and obtained the following data.

Pupal Weight	Number of Ovarioles
6.30	62
7.91	76
8.20	81
8.25	80
8.65	83
9.48	105
16.08	142
18.15	152
21.70	184
23.12	186
21.88	182
26.20	230
28.82	193
32.00	260

Both variables are subject to error, but the count of the number of ovarioles is probably fairly accurate and could be used as the independent variable in a Model I regression. The equation needed, however, is a prediction equation for the number of ovarioles, given the pupal weight, because it is technically far easier to weigh a pupa than to dissect an adult and count the number of ovarioles. Estimate the number of ovarioles expected in a female pupa weighing 25 mg and set 95% confidence limits to your estimate.

ANSWER: 95% confidence limits are 173.2 and 236.0.

14.8 Given the lower face widths (skeletal bigonial diameter in centimeters) for the 15 North American white females at ages 5 and 6 (see Box 11.5), evaluate their functional relationships by a Model II regression analysis. Also compute a prediction equation for face width at age 6 in terms of face width at age 5.

14.9 The experiment in Exercise 14.3 was repeated using a 15-hour photoperiod, yielding the following results.

Temperature (°C)	Calories $\overline{Y}$	n	s
0	24.3	6	1.93
10	25.1	7	1.98
18	22.2	8	3.67
26	13.8	10	4.01
34	16.4	6	2.92

Test for the equality of slopes of the regression lines for the 10- and 15-hour photoperiods. Comment on the results.

ANSWER: $F_s = 0.003$

14.10 Regress the men's winning long jump distance (in inches) against year for the data provided below. To what extent does the unusual value for 1968 (in Mexico City) influence the result? Values from Olympic website. Note that the last 4 values were converted from meters to inches to match the other records.

Year	Long Jump Distance (Inches)
1900	282.875
1904	289.000
1908	294.500
1912	299.250
1920	281.500
1924	293.125
1928	304.750
1932	300.750
1936	317.3125
1948	308.000
1952	298.000
1956	308.250
1960	319.750
1964	317.750
1968	350.500
1972	324.500
1976	328.500

Year	Long Jump Distance (Inches)
1980	336.250
1984	336.250
1988	343.337
1992	341.369
1996	334.675
2000	336.644
2004	338.219
2008	328.375

14.11 The following data were sampled at random from a larger study by Prange et al. (1979). They reinvestigated the question of whether birds have proportionately lighter skeletons than mammals because of their adaptation for flight. Compare the slopes for the estimated relationship between log skeletal mass and log body mass for the two samples. All weights are in kilograms. Compare the results one obtains using different methods for estimating structural relationships. Plot the data and superimpose the usual Model I regression line and the major axis and standard major axis lines.

Birds		Mammals	
Skeletal Mass Y	Body Mass X	Skeletal Mass Y	Body Mass X
1.995	40.667	0.193	3.35
0.072	1.225	0.227	3.915
0.0054	0.163	0.0003	0.0063
0.203	2.504	0.039	0.790
0.043	0.701	0.027	0.820
0.027	0.416	0.244	4.836
0.186	2.379	0.002	0.030
0.0058	0.124	0.015	0.275
0.028	0.427	0.020	0.365
0.00174	0.031	0.0025	0.030
0.00182	0.029	0.0076	0.115
0.00102	0.020	1.146	22.7
0.024	0.383	0.748	11.950
0.00618	0.144	0.250	3.395
0.00184	0.038	0.107	2.460
0.00297	0.069	0.224	4.260
0.00183	0.045	0.233	4.210
0.00076	0.013	0.0173	0.350
0.00128	0.023	0.270	4.450
0.00049	0.0087	0.448	6.725
0.00062	0.0126	0.135	1.560
0.00061	0.0092	0.342	2.931
0.00117	0.019	4.120	20.000
0.00051	0.013	12.161	67.310
0.0011	0.032	1782.	6600.

ANSWER: For the bird data, $b_{Y \cdot X} = 1.03059$ and $v_{Y \cdot X} = 1.07284$; for mammals, the corresponding values are 1.10295 and 1.11272. We cannot reject the null hypothesis of no difference in slope because $P > 0.05$.

14.12 Carry out a nonparametric test for regression on the data in Exercise 14.7.

14.13 J. Calaprice (unpublished results) studied the effect of crowding on weight (in grams) of *Salmo trutta*, the brown trout. The fish were subjected to 4 densities: 175, 350, 525, and 700 per container. From each container, 25 fish were sampled at random and weighed. The mean weights at the four densities are given below. A preliminary analysis of variance is shown, below. Can we reject the null hypothesis of no difference among men weights? Calculate the regression of weight on density for these trout and test whether trends are supported statistically.

Density	175	350	525	700
Mean weight	3.51224	3.71020	2.92680	2.53960

Anova Table

Source of variation	df	SS	MS	F_s
Among densities	3	21.63688	7.21229	6.944
Within densities	96	99.71534	1.03870	
Total	99	121.35222		

15 Correlation

In this chapter, we continue our discussion of bivariate statistics. Chapter 14 dealt with the functional relation of one variable to the other; here we discuss the measurement of the amount of association between two variables. This general topic is called *correlation analysis.*

It is not always obvious which type of analysis—regression or correlation—should be employed in a given problem. This topic has produced considerable confusion not only in the minds of investigators but also in the literature. We will try to make the distinction between these two approaches clear at the outset in Section 15.1. In Section 15.2, you will be introduced to the product–moment correlation coefficient, the common correlation coefficient of the literature. We will derive a formula for this coefficient and present some of its theoretical background. Additionally, we consider the close mathematical relationship between regression and correlation analysis. A fair amount of algebra is involved, but none of it is difficult, and its mastery will improve your understanding of several topics in this and previous chapters. The actual computation of a product–moment correlation coefficient is the subject of Section 15.3. In the following section, 15.4, we make a detour illustrating the algebraic relations between the *t*-test for paired comparisons and correlation.

Section 15.5 presents tests of significance involving correlation coefficients. After this introduction to correlation coefficients, we discuss their applications in Section 15.6. Section 15.7 features several nonparametric methods for association to be used when the necessary assumptions for tests involving correlation coefficients do not hold, or when quick but less than fully efficient tests are preferred for reasons of convenience. Section 15.8 shows how to construct confidence regions for the mean in normally distributed bivariate data. The construction of equal frequency ellipses is also shown. Section 15.9 discusses several measures of effect size for comparing correlations and methods for determining sample sizes that should be used in order to detect a specified minimum effect size with sufficient power.

When variables are nominal or categorical, the measures of association discussed in this chapter cannot be applied. Near the end of Section 17.4, we show how to compute a coefficient of association, the phi coefficient, between two categorical variables, each with two classes.

15.1 Correlation Versus Regression

Correlation and regression are often confused. Quite frequently, correlation problems are treated as regression in the scientific literature, and the converse is equally true. There are several reasons for this confusion. First, there are close mathematical relationships between the two methods of analysis. Algebraically, one can easily move from the formula for one to the formula for the other; hence, the temptation to do so is great. Second, earlier texts did not make the distinction between the two approaches sufficiently clear, and the resulting confusion persists. Finally, although the approach chosen by an investigator may be correct for his or her intentions, the data available for analysis may make one or the other of the techniques inappropriate.

Let us examine these points at length. The many and close mathematical relations between regression and correlation will be detailed in Section 15.2. It suffices for now to state that most of the computational steps are the same, whether one carries out a regression or a correlation analysis. Recall that the fundamental quantity required for regression analysis is the sum of products. This same quantity is the base for computing the correlation coefficient. Because there are some simple mathematical relations between regression coefficients and their corresponding correlation coefficients, the temptation exists to compute a correlation coefficient corresponding to a given regression coefficient. As we will see later, this approach would be wrong unless our intention at the outset was to study association and the data were appropriate for such a computation.

An analogy from analysis of variance comes to mind. Recall the fundamental distinctions in purpose of Model I and Model II anovas (see Chapter 8). The computations for the two models, however, were almost the same, in some cases up to and including the initial hypothesis tests. Only subsequent to these preliminary tests did the procedural paths separate. In Model I, we estimated means, subdivided the differences among treatments into separate comparisons, and established confidence limits. In Model II, we estimated variance components. Although it would be perfectly possible to carry out the computation for estimating a variance component in a Model I anova, the quantity obtained should be interpreted not as an added variance component but rather as a fixed sum of squares whose magnitude is determined by the choice of included treatments. Calling it a variance component would serve no useful purpose. Conversely, testing the differences between the means of random samples in a Model II anova by individual degree-of-freedom comparisons might yield some statistically supported pairs, but establishing such differences would not be useful in a Model II because of the random sampling of the groups.

In a similar manner, we can easily compute a correlation coefficient from data that were properly analyzed by Model I regression, but such a coefficient is meaningless as an estimate of any population correlation coefficient because of its strong dependence on the choice of values for the independent variable. Conversely, data for which correlations could have been properly computed easily yield a Model I regression equation of one variable on the other, but such a slope would be biased; the appropriate equation depends on the purpose of the investigator and the nature of the variables.

Let us then look at the intentions or purposes behind the two types of analyses. In regression, we intend to describe the dependence of a variable Y on an independent variable X so that we can predict Y given a value for X. As we have seen, we employ regression equations to support hypotheses regarding the possible causation of changes in Y by changes in X, to predict Y in terms of X, and to explain some of the variation of Y by X, by using the latter variable as a statistical control. Studies of the effects of temperature on heart rate, nitrogen content of soil on growth rate in a plant, age of an animal on its blood pressure, or dose of an insecticide on mortality of an insect population are typical examples of regression for these purposes.

In correlation, by contrast, we generally want to determine whether variation in the two variables is interdependent—that is, whether they vary together or *covary*. We do not express one as a function of the other. Thus, there are no independent or dependent variables. In the pair of variables whose correlation is studied, one may well be the cause of the other, but we neither know nor assume this. A more typical (but not essential) assumption is that variation in both variables is caused by variation in other variables. What we wish to estimate is the degree to which these variables vary together. For example, we might be interested in the correlation between arm length and leg length in a population of mammals, between body weight and egg production in female blowflies, or between color intensity and nitrogen content in the leaves of a plant. Reasons for demonstrating and measuring association between pairs of variables need not concern us yet. We will discuss this topic in Section 15.6. All we need to know now is that when we wish to establish the degree of association between pairs of variables in a population sample, correlation analysis is the proper approach.

Even if we attempt the correct method for our purposes, we may run afoul of the nature of the data. For example, in an attempt to establish cholesterol content of blood as a function of weight using regression, we may take a random sample of men of the same age group, obtain each individual's cholesterol content and weight, and regress the former on the latter. However, both these variables will have been measured with error. Individual observations of the supposedly independent variable X were not deliberately chosen or controlled by the experimenter. The underlying assumptions of Model I regression do not hold, and fitting a Model I regression to the data is not legitimate, although you will have no difficulty finding instances of such improper practices in the published research literature.

As discussed in Section 14.12, if what we want is an equation describing the dependence of Y on X, then we should carry out a Model II regression. If it is the degree of association between the variables (interdependence) that is of interest, however, we should carry out a correlation analysis, for which the data in this example are suitable. The converse difficulty is trying to obtain a correlation coefficient from data that are properly computed as a regression—that is, when X is fixed. An example is heart rate of a poikilotherm as a function of temperature, where several temperatures have been applied in an experiment. Such a correlation coefficient is easily obtained mathematically but would simply be a numerical value, not an estimate of a parametric measure of correlation. The square of the correlation coefficient can be interpreted in a way that is relevant to a regression problem, but it is not an estimate of a squared parametric correlation.

This discussion is summarized in Table 15.1, which shows the relations between correlation and regression. The two columns of the table indicate the two conditions

TABLE 15.1 The Relations Between Correlation and Regression

The correct computation for any combination of purposes and variables is shown.

Purpose of investigator	Nature of the two variables	
	Y random, *X* fixed	Y_1, Y_2 both random
Establish and estimate dependence of one variable upon another (describe functional relationship and/or predict one in terms of the other).	Model I regression.	Model II regression. (Model I generally inappropriate. For exceptions, the Berkson case and some instances of prediction, see Section 14.12)
Establish and estimate association (interdependence) between two variables.	Meaningless for this case. If desired, an estimate of the proportion of the variation of *Y* explained by *X* can be obtained as the square of the correlation coefficient between *X* and *Y*.	Correlation coefficient. (Hypothesis tests entirely appropriate only if Y_1, Y_2 are distributed as bivariate normal variables.)

of the pair of variables: in one case, one random and measured with error, the other fixed; in the other case, both variables random. We have departed from the usual convention of labeling the pair of variables Y and X or X_1, X_2 for both correlation and regression analyses. In regression, we continue the use of Y for the dependent variable and X for the independent variable, but in correlation both of the variables are random variables, which we have throughout the text designated as Y. We therefore refer to the two variables as Y_1 and Y_2. The rows of the table indicate the intention of the investigator in carrying out the analysis, and the four quadrants of the table indicate the appropriate procedures for a given combination of intention of investigator and nature of the pair of variables.

15.2 The Product–Moment Correlation Coefficient

There are many correlation coefficients in statistics. The most common is the **product–moment correlation coefficient,** the current formulation of which was worked out by Karl Pearson. We will derive its formula using an intuitive approach similar to the one we used to introduce the regression coefficient (see Section 14.3).

Because it is a measure of covariation, the sum of products is the basic quantity from which the formula for the correlation coefficient is derived. We will label the variables whose correlation is to be estimated as Y_1 and Y_2. Their sum of products, therefore, will be $\Sigma y_1 y_2$ and their covariance $\Sigma y_1 y_2 / (n - 1) = s_{12}$. The latter quantity is analogous to a variance—that is, a sum of squares divided by its degrees of freedom.

A standard deviation is expressed in original measurement units such as inches, grams, or cubic centimeters. Similarly, a regression coefficient is expressed as units of Y per unit of X, such as 5.2 grams per day. A measure of association, however, should be independent of the original scale of measurement so that we can compare the degree of association in one pair of variables with that in another. One way to make the covariance scale independent is to divide it by the standard deviations of variables Y_1 and Y_2:

$$r_{Y_1 Y_2} = \frac{s_{12}}{s_{Y_1} s_{Y_2}} \tag{15.1}$$

This expression is the formula for the product–moment correlation coefficient $r_{Y_1 Y_2}$ between variables Y_1 and Y_2. Simplifying the symbolism, we obtain $r_{Y_1 Y_2} = r_{12}$. This notation is adequate unless we have two-digit variable numbers, in which case commas are necessary. For example, $r_{12,14}$ is the correlation between variables Y_{12} and Y_{14}.

By multiplying the covariance in the numerator and the product of the two standard deviations in the denominator of Expression (15.1) by $n - 1$, we can rewrite it as

$$r_{12} = \frac{\sum y_1 y_2}{\sqrt{\sum y_1^2 \sum y_2^2}} \tag{15.2}$$

To state Expression (15.1) more generally for variables Y_j and Y_k, we can write it as

$$r_{jk} = \frac{s_{jk}}{s_j s_k} \tag{15.3}$$

The correlation coefficient r_{jk} can range from $+1$ for perfect positive association to -1 for perfect negative association. The meaning of these values is intuitively obvious when we consider the correlation of a variable Y_j with itself. Expression (15.3) would then yield

$$r_{jj} = \frac{\sum y_j y_j}{\sqrt{\sum y_j^2 \sum y_j^2}} = \frac{\sum y_j^2}{\sum y_j^2} = 1$$

a perfect correlation of $+1$. If deviations in one variable were paired with opposite but equal deviations representing another variable, the result would be a correlation of -1 because the sum of products in the numerator would be negative. Proof that the correlation coefficient is bounded by $+1$ and -1 will be given shortly.

We can compute r_{jk} for any set of paired values of Y_j and Y_k, no matter what their underlying distribution is and even if they properly should be analyzed by regression. However, when regression is the correct approach, the correlation coefficient is simply a mathematical index, not a sample statistic estimating an unknown parameter, and would be of little interest. But when the observations follow a specified distribution, the **bivariate normal distribution** (Figure 15.1), the correlation coefficient

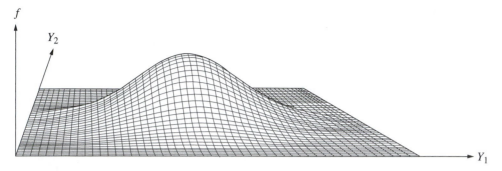

FIGURE 15.1   Bivariate normal frequency distribution. The parametric correlation ρ between variables Y_1 and Y_2 equals zero. The frequency distribution may be visualized as a bell-shaped mound.

r_{jk} will estimate a parameter of that distribution symbolized by ρ_{jk}. Actually, r_{jk} is slightly biased and underestimates ρ_{jk} in small samples. Kendall and Stuart (1961, Section 26.17) suggest the following corrected formula:

$$r_{jk}^* = r_{jk}\left[1 + \frac{1 - r_{jk}^2}{2(n - 4)}\right] \tag{15.4}$$

where r_{jk}^* is the unbiased estimate of ρ_{jk}, and r_{jk} and n have their previous meanings. This bias can be ignored when sample size $n > 42$, because in these cases r and r^* are identical to two decimal places.

The probability density function for the bivariate normal distribution has a formidable-looking formula that we will not reproduce here. We instead approach the distribution empirically. Suppose you have sampled 100 items and measured two variables on each item, obtaining 100 paired observations in this manner. If you plot these 100 items on a graph in which the variables Y_1 and Y_2 are the coordinates, you will obtain a scattergram of points (such as in Figure 15.3a below). Let us assume that both variables, Y_1 and Y_2, are normally distributed and independent of each other. The fact that one individual is greater than the mean for variable Y_1 thus has no effect on its value for Y_2. This same individual may therefore be greater or less than the mean for variable Y_2. If there is absolutely no relation between Y_1 and Y_2 and if the two variables are standardized to make their variances the same, the outline of the scattergram is roughly circular.

Of course, for a sample of 100 items, the circle would be only imperfectly outlined; the larger the sample, the more clearly one could discern a circle with the central area around the point $\overline{Y}_1, \overline{Y}_2$ heavily darkened because of the aggregation there of many points. Continued sampling requires superimposing new points on previous points. If you visualize these points in a physical sense, such as grains of sand, a mound peaked in the shape of a bell would gradually form. This three-dimensional realization of a normal probability density function is shown in perspective in Figure 15.1.

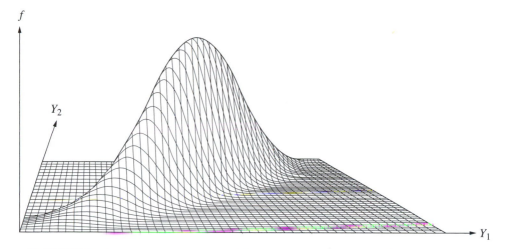

f

Y_2

Y_1

FIGURE 15.2 Bivariate normal frequency distribution. The parametric correlation ρ between variables Y_1 and Y_2 equals 0.9. The bell-shaped mound of Figure 15.1 has become elongated.

Regarded from either coordinate axis, the mound has a two-dimensional appearance, with the outline of a normal distribution curve, the two views giving the distributions of Y_1 and Y_2, respectively.

If we assume that the two variables Y_1 and Y_2 are not independent but are correlated positively to some degree, then an individual that has a large value of Y_1 is likely to have a large value of Y_2 as well. Similarly, a small value of Y_1 will likely be associated with a small value of Y_2. Were you to sample items from such a population, the resulting scattergram (such as that shown in Figure 15.3**d**) would be elongated in the form of an ellipse because the parts of the circle that formerly included individuals high for one variable and low for the other (and vice versa) would now be sparsely represented. Continued sampling (with the sand-grain model) yields a three-dimensional elliptical mound (Figure 15.2). The ridge of high density is along the major axis we learned about in Section 14.12. If correlation is perfect, then all the data would fall along a single line, which is the major axis (the identical line would describe the regression of Y_1 on Y_2 and of Y_2 on Y_1). In a physical model, they would be represented by a flat, essentially two-dimensional normal curve lying on this major axis line.

The shape of the outline of the scattergram and of the resulting mound is a function of the degree of correlation between the two variables, which is the parameter ρ_{jk} of the bivariate normal distribution. By analogy with Expression (15.1), the parameter ρ_{jk} can be defined as

$$\rho_{jk} = \frac{\sigma_{jk}}{\sigma_j \sigma_k} \tag{15.5}$$

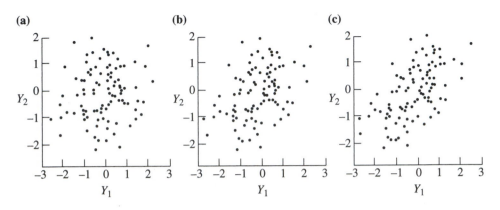

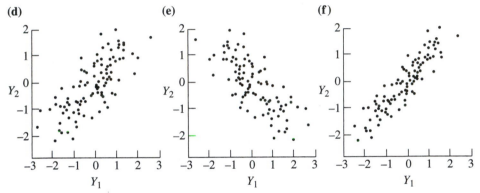

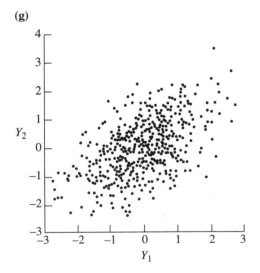

FIGURE 15.3 Random samples from bivariate normal distributions with varying values of the parametric correlations coefficient ρ. Sample size $n = 100$ in all graphs except part (**g**), where $n = 500$. (**a**) $\rho = 0$. (**b**) $\rho = 0.3$. (**c**) $\rho = 0.5$. (**d**) $\rho = 0.7$. (**e**) $\rho = -0.7$. (**f**) $\rho = 0.9$. (**g**) $\rho = 0.5$.

where σ_{jk} is the parametric covariance of variables Y_j and Y_k, and σ_j and σ_k are the parametric standard deviations of variables Y_j and Y_k, respectively, as before. When two variables are distributed according to the bivariate normal distribution, a sample correlation coefficient r_{jk} estimates their parametric correlation coefficient ρ_{jk}. We can make some statements about the sampling distribution of ρ_{jk} and set confidence limits to it.

Regrettably, the elliptical shape of scattergrams of correlated variables is not usually very clear unless either large samples have been taken or the parametric correlation ρ_{jk} is high. To illustrate this point, we show in Figure 15.3 several scattergrams resulting from samples of 100 items from bivariate normal populations with differing values of ρ_{jk}. Note that when $\rho_{jk} = 0$ (Figure 15.3a), the circular distribution is only very vaguely outlined. A much larger sample is required to demonstrate the circular shape of the distribution more clearly. No substantial difference is visible in Figure 15.3b, where $\rho_{jk} = 0.3$. Because we know that this graph depicts a positive correlation, we can visualize a positive slope in the scattergram, but without this knowledge the positive slope would be difficult to detect visually. The shape in Figure 15.3c, where $\rho_{jk} = 0.5$, is clearer but still does not exhibit an unequivocal trend. In general, correlation cannot be inferred from inspecting scattergrams based on samples from populations with ρ_{jk} between -0.5 and $+0.5$ unless the sample is very large, say, 500. The graph in Figure 15.3g, also sampled from a population with $\rho_{jk} = 0.5$ but based on a sample of 500, illustrates this point. Here the positive slope and elliptical outline of the scattergram are quite evident. Figure 15.3d, based on $\rho_{jk} = 0.7$ and $n = 100$, shows the trend. (Figure 15.3e, based on the same magnitude of ρ_{jk}, represents the equivalent negative correlation.) Finally, Figure 15.3f, representing a correlation of $\rho_{jk} = 0.9$, shows a tight association between the variables, reasonably approximating an ellipse of points.

Now let us return to the equation for the sample correlation coefficient shown in Expression (15.2). Squaring both sides results in

$$r_{12}^2 = \frac{\left(\sum y_1 y_2\right)^2}{\sum y_1^2 \sum y_2^2}$$

$$= \frac{\left(\sum y_1 y_2\right)^2}{\sum y_1^2} \cdot \frac{1}{\sum y_2^2}$$

Look at the left term of the last expression. It is the square of the sum of products of variables Y_1 and Y_2, divided by the sum of squares of Y_1. If this were a regression problem, this term would be the formula for the explained sum of squares of variable Y_2 on variable Y_1, $\sum \hat{y}_2^2$. Using the symbolism that we used for regression (see Chapter 14), we would get $\sum \hat{y}^2 = (\sum xy)^2/\sum x^2$. Thus, we can write

$$r_{12}^2 = \frac{\sum \hat{y}_2^2}{\sum y_2^2} \tag{15.6}$$

The square of the correlation coefficient, therefore, is the ratio of the explained sum of squares of variable Y_2 divided by the total sum of squares of variable Y_2 or, equivalently,

$$r_{12}^2 = \frac{\sum \hat{y}_1^2}{\sum y_1^2} \tag{15.6a}$$

which could be derived as easily as Expression (15.6) was. Remember that we are not really regressing one variable on the other, so explaining Y_1 by Y_2 is just as legitimate as the other way around. This ratio is a proportion between zero and one. This range becomes obvious after a little contemplation of the meaning of this formula. The explained sum of squares of any variable must be less than or equal to its total sum of squares. In other words, if all the variation of a variable has been explained, its explained sum of squares can be as great as the total sum of squares, but no greater. When none of the variance can be explained by the other variable with which the covariance has been computed, the explained sum of squares is at its lowest value, zero. Thus, we obtain an important measure of the proportion of the variation of one variable determined by the variation of the other. This quantity, the square of the correlation coefficient, r_{12}^2, is called the **coefficient of determination.**

We encountered the coefficient of determination in Section 14.4. It ranges from 0 to 1 and must be positive regardless of whether the correlation coefficient is negative or positive. Incidentally, here is proof that the absolute value of the correlation coefficient cannot exceed 1: Its square is the coefficient of determination and we have just shown that the bounds of the latter are 0 to 1.

The coefficient of determination is also useful when one is considering the relative importance of correlations of different magnitudes. As Figure 15.3 shows, the rate at which the distributions of the scatter diagrams vary from a circular outline to an elliptical outline seems to be more directly proportional to r^2 than to r. Thus, in Figure 15.3b, with $\rho^2 = 0.09$, the correlation is difficult to detect visually. By the time we reach Figure 15.3d, with $\rho^2 = 0.49$, however, the presence of correlation is very apparent.

The use of the coefficient of determination was also discussed in Section 14.4. Recall that in a regression, we used an anova to partition the total sum of squares into explained and unexplained sums of squares. After such an analysis of variance has been carried out, one can obtain the ratio of the explained sums of squares over the total SS as a measure of the proportion of the total variation that has been explained by the regression. As we explained in Section 15.1, however, the square root of such a coefficient of determination would not be meaningful as an estimate of the parametric correlation of these variables.

We can invert Expression (15.6) and use it to compute the explained and unexplained sum of squares in cases where we know the coefficients of correlation or determination. Thus, given a sample in which you know that the correlation between two variables is r_{12}, and given the total sum of squares of variable Y_1 as $\sum y_1^2$, you can compute the explained SS as $\sum \hat{y}_1^2 = r_{12}^2 \sum y_1^2$ and the unexplained SS as $\sum d_{1 \cdot 2}^2 = (1 - r_{12}^2) \sum y_1^2$ and subject these to an analysis of variance. The quantity

$1 - r_{12}^2$ is sometimes known as the **coefficient of nondetermination** because it expresses the proportion of variance of a variable that has *not* been explained by the other variable. The square root of this coefficient, $\sqrt{1 - r_{12}^2}$, is known as the **coefficient of alienation,** measuring the lack of association between variables Y_1 and Y_2.

We will now examine some mathematical relations between the coefficients of correlation and regression. At the risk of sounding redundant, we stress again that the ease of converting one coefficient into the other does not mean that the two types of coefficients can be used interchangeably on the same type of data. An important relationship between the correlation coefficient and the regression coefficient can be derived as follows from Expression (15.1):

$$r_{12} = \frac{s_{12}}{s_1 s_2} = \frac{s_{12}}{s_1} \cdot \frac{1}{s_2}$$

Multiplying the numerator and the denominator of this expression by s_1, we obtain

$$r_{12} = \frac{s_{12}}{s_1 s_2} \cdot \frac{s_1}{s_2} = \frac{s_{12}}{s_1^2} \cdot \frac{s_1}{s_2} = b_{21} \cdot \frac{s_1}{s_2} \tag{15.7}$$

Similarly, we could demonstrate that

$$r_{12} = b_{1 \cdot 2} \frac{s_2}{s_1} \tag{15.7a}$$

and hence

$$b_{2 \cdot 1} = r_{12} \frac{s_2}{s_1}$$

and

$$b_{1 \cdot 2} = r_{12} \frac{s_1}{s_2} \tag{15.7b}$$

In these expressions, $b_{2 \cdot 1}$ is the regression coefficient for variable Y_2 and Y_1. The correlation coefficient is thus the regression slope multiplied by the ratio of the standard deviations of the variables. The correlation coefficient may therefore be regarded as a standardized regression coefficient. If the two standard deviations are identical, the regression and correlation coefficients will also be numerically identical.

A second relation between the two coefficients is the following: Multiplying together the $b_{2 \cdot 1}$ and $b_{1 \cdot 2}$ terms of Expression (15.7b), we obtain

$$b_{2 \cdot 1} b_{1 \cdot 2} = r_{12} \frac{s_2}{s_1} r_{12} \frac{s_1}{s_2} = r_{12}^2$$

Therefore,

$$r_{12} = \pm \sqrt{b_{2 \cdot 1} b_{1 \cdot 2}} \tag{15.8}$$

Thus, if we know the two regression coefficients, we can easily compute the correlation coefficient as their geometric mean (see Section 4.2). The sign of the correlation coefficient will be that of the regression coefficients and is, of course, determined by the sign of the sum of products $\Sigma y_1 y_2$.

If the two regression lines are drawn on the same graph, correlation is a measure of the angle between them. If this angle is a right angle, then the correlation is zero; the result is a circular scattergram. If this angle is very small, then correlation is very high. In the case of perfect correlation, the two regression lines coincide, and the angle between them is zero.

In the next section, we discuss methods of computing correlation coefficients.

15.3 Computing the Product–Moment Correlation Coefficient

The computation of a product–moment correlation coefficient is quite simple. The basic quantities needed are the same as those required for computing the regression coefficient (see Section 14.3). Box 15.1 illustrates the computation. The example is based on a sample of 12 crabs in which gill weight, Y_1, and body weight, Y_2, have been recorded. We wish to know whether there is a correlation between the weight of the gill and that of the body, the latter representing a measure of overall size. The existence of a positive correlation might lead you to conclude that a bigger-bodied crab, with its resulting greater amount of metabolism, would require larger gills to provide the necessary oxygen. The high positive correlation coefficient computed in Box 15.2 ($r_{12} = 0.865$) corroborates the clear slope and narrow elliptical outline of the scattergram for these data (Figure 15.4).

This example has another interesting feature: It is an example of a **part–whole correlation.** Although only a small fraction of the total, gill weight is a part of body weight, and the correlation is therefore between two variables, one of which is a part of another. We can rewrite these variables as follows: Y_1 is gill weight as before, but Y_2 (the body weight) is made up of Y_0, body weight *minus* gill weight, and Y_1, gill weight. Because $Y_2 = Y_0 + Y_1$, Y_1 and Y_2 should be correlated positively to some degree because Y_1 is part of Y_2 and will clearly be correlated positively with itself. The correlation between Y_1 and Y_2 depends on the degree of correlation between Y_0 and Y_1 and their variances. The correlation can be predicted from the formula

$$r_{12} = \frac{s_1 + r_{01}s_0}{\sqrt{s_0^2 + 2r_{01}s_0s_1 + s_1^2}} \tag{15.9}$$

which is derived in Section 16.4. For now, we shall treat this equation as a given. Thus, the correlation of a part with a whole is a function of s_1 and s_0, the standard deviations of the part and of the whole minus that part, as well as of r_{01}, the correlation between these variables. When the correlation $r_{01} = 0$, the formula for a part–whole correlation simplifies to

$$r_{12} = \frac{1}{\sqrt{1 + (s_0^2/s_1^2)}} \tag{15.9a}$$

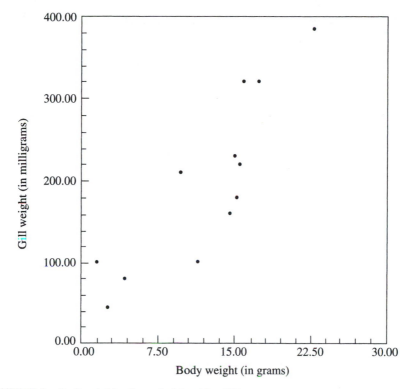

FIGURE 15.4 Scatterplot for the crab data of Box 15.1.

and the correlation is then simply a function of the ratio of the variances. If, in addition to no correlation between the part and its complement, the variances were equal, then the formula for the part–whole correlation would simplify to $r_{12} = 1/\sqrt{1 + 1}$ $= 1/\sqrt{2} = 0.707$, which shows that when two parts are equally variable, the correlation of one part with their sum will be quite high. In the past, such correlations were considered spurious. They are not spurious, however, but rather are logical consequences of the way in which the variables were defined. As long as we keep this in mind, there is nothing necessarily wrong with calculating such correlations, except that one should not be surprised to find a sizable correlation of a part with a whole. When the part has a much smaller variance than the whole, Expression (15.10) shows that the expected value of the part–whole correlation is not likely to be great unless r_{01} is substantial.

Many correlation programs exist. The BIOMstat computer program performs this operation. We should stress, however, that it is important to examine a scattergram of the points as part of the analysis. Inspection may reveal that the scatter of points, rather than circular or elliptical, shows a peculiar shape that might be improved by transformation. Most computer programs display a bivariate scattergram.

BOX 15.1 Computing the Product–Moment Correlation Coefficient

Relationship between gill weight and body weight in the crab *Pachygrapsus crassipes;* $n = 12$.

(1) Y_1 Gill weight in milligrams	(2) Y_2 Body weight in grams
159	14.40
179	15.20
100	11.30
45	2.50
384	22.70
230	14.90
100	1.41
320	15.81
80	4.19
220	15.39
320	17.25
210	9.52

SOURCE: Data from L. Miller (unpublished results).

Computation

1. $\bar{Y}_1 = 195.58333$, $\bar{Y}_2 = 12.04750$

2. Sum of squares of $Y_1 = \sum y_1^2 = \sum (Y_1 - \bar{Y}_1)^2 = 124{,}368.9111$

3. Sum of squares of $Y_2 = \sum y_2^2 = \sum (Y_2 - \bar{Y}_2)^2 = 462.47817$

4. Sum of products $= \sum y_1 y_2 = \sum (Y_1 - \bar{Y}_1)(Y_2 - \bar{Y}_2) = 6561.61737$

5. Product-moment correlation coefficient [by Expression (15.3)]:

$$r_{12} = \frac{\sum y_1 y_2}{\sqrt{\sum y_1^2 \sum y_2^2}} = \frac{\text{quantity } 4}{\sqrt{\text{quantity } 2 \times \text{quantity } 3}}$$

$$= \frac{6561.61737}{\sqrt{(124{,}368.9111)(462.47817)}} = 0.86519$$

In addition, the discussion of leverage coefficients in Section 14.10 in conjunction with regression applies also to correlation. Outliers seriously affect values of a sample correlation coefficient.

The advantage of the computer becomes especially noticeable when more than two variables have been recorded for each item. For p variables, the outcome of the study will be a $p \times p$ **correlation matrix**—that is, a table of correlation coefficients of variable 1 against variable 2, variable 1 against variable 3, and so on until variable 1 is correlated with variable p. This process is followed by the correlation of 2 against 3, 2 against 4, and so on up to 2 against p. The final value in the correlation matrix is the correlation of variable $p - 1$ against variable p. Computing such correlation matrices in the precomputer days was a very difficult and tedious procedure, often taking several months for sizable studies. The ease with which correlation matrices can now be obtained has given impetus to considerable research employing multivariate analysis, some of which is discussed in Chapter 16.

15.4 The Variance of Sums and Differences

Having engaged in Section 15.2 in more than the usual amount of algebra for this book, we hope the reader is in a frame of mind to tolerate a little more. Now that we know about the coefficient of correlation, some of the earlier work on paired comparisons (see Section 11.5) can be put into proper perspective. In Section A.13 of the appendix, we show that the variance of a sum of two variables is

$$s^2_{Y_1 + Y_2} = s^2_1 + s^2_2 + 2r_{12}s_1s_2 \tag{15.10}$$

where s_1 and s_2 are standard deviations of Y_1 and Y_2, respectively, and r_{12} is the correlation coefficient between these variables. Similarly, for a difference between two variables, we obtain the following equation for its variance:

$$s^2_{Y_1 - Y_2} = s^2_1 + s^2_2 - 2r_{12}s_1s_2 \tag{15.10a}$$

Expression (15.10) indicates that if we make a new composite variable that is the sum of two other variables, the variance of this new variable will be the sum of the variances of the variables of which it is composed, plus an added term, which is a function of the standard deviations of these two variables and of the correlation between them. Appendix A.13 shows that this added term is twice the covariance of Y_1 and Y_2. When the two variables being summed are uncorrelated, this added covariance term will also be zero, and the variance of the sum will simply be the sum of the variances of the two variables. This is why, in an anova or in a t-test of the difference between the two means, we had to assume the independence of the two variables to permit us to add their variances. Otherwise, we would have had to allow for a covariance term.

By contrast, in the paired-comparisons technique, we expect correlation between the variables, because each pair shares a common experience. The guinea pig data in Table 15.2, which we analyzed by a nonparametric technique in Box 13.12, make this point clearer. First we calculate the correlation coefficient between variables Y_1 and Y_2 and find it to be quite high, 0.882. If these data were (improperly)

TABLE 15.2	Relationships Between the Correlation Coefficient and the Paired-Comparisons Test

Litter size in guinea pigs. (Data from Box 13.12.)

Year	Strain B Y_1	Strain 13 Y_2	$D = Y_1 - Y_2$
1916	2.68	2.36	+0.32
1917	2.60	2.41	+0.19
1918	2.43	2.39	+0.04
1919	2.90	2.85	+0.05
1920	2.94	2.82	+0.12
1921	2.70	2.73	-0.03
1922	2.66	2.58	+0.08
1923	2.98	2.89	+0.09
1924	2.85	2.78	+0.07
$\bar{Y}$	2.749	2.646	+0.1033

Computation

1. Computing the correlation coefficient between Y_1 and Y_2:

$$n = 9$$

$$\sum y_1^2 = 0.2619 \qquad \sum y_2^2 = 0.3638$$

$$s_1^2 = 0.03274 \qquad s_2^2 = 0.04548$$

$$\sum y_1 y_2 = 0.2723 \qquad \sqrt{\sum y_1^2 \sum y_2^2} = 0.3087$$

$$r_{12} = 0.882$$

2. *t*-test for difference of means (in the manner of Box 9.6):

$$t_s = \frac{(\bar{Y}_1 - \bar{Y}_2) - (\mu_1 - \mu_2)}{\sqrt{\frac{1}{n}(s_1^2 + s_2^2)}} = \frac{0.1033}{0.093226} = 1.108, \quad df = 2(n-1) = 16, \quad P = 0.2842$$

3. *t*-test for paired comparisons (in the manner of Box 11.5):

$$s_D^2 = 0.01015 \qquad s_{\bar{D}} = 0.03358$$

$$t_s = \frac{\bar{D} - 0}{s_{\bar{D}}} = \frac{0.1033}{0.03358} = 3.077, \quad df = n - 1 = 8, \quad P = 0.0151$$

The difference in outcome of the two tests in steps **2** and **3** shows presence of correlation. We can calculate r_{12} in the usual way (step **1**) or as follows:

$$s_D^2 = s_{(Y_1 - Y_2)}^2 = s_1^2 + s_2^2 - 2r_{12}s_1s_2 \qquad \text{[from Expression (15.10a)]}$$

$$2r_{12}s_1s_2 = s_1^2 + s_2^2 - s_D^2$$

$$r_{12} = \frac{s_1^2 + s_2^2 - s_D^2}{2s_1s_2} = \frac{0.03274 + 0.04548 - 0.01015}{2(0.1809)(0.2132)} = 0.882$$

analyzed by a t-test for difference of means (see Section 9.4), the result, as shown in step 2 of Table 15.2, would not lead to a rejection of H_0 because the standard error for the t-test is too large, being based on the sum of the variances of the two variables without subtracting a term to account for the correlation. The paired-comparison test (see Section 11.5 and Box 11.6) automatically subtracts such a term, resulting in a smaller standard error and consequently in a larger value of t_s, because the numerator of the ratio remains the same. Thus, whenever correlation between two variables is positive, the variance of their differences will be smaller than the sum of their variances; this is why with positive correlation the paired-comparison test should be used in place of the t-test for difference of means. These considerations are equally true for the corresponding analyses of variance: single-classification and two-way anova.

At the bottom of Box 15.1, we show how to compute the correlation coefficient from the variances of the variables and of their differences. Although this method is not recommended, it does provide further insight into the relationships that we have just discussed.

Finally, you might ask how negative correlations would affect the considerations of the last four paragraphs. Such a situation might arise with a compensatory phenomenon. For instance, if two animals share a cage and the total amount of food available is constant, the more one animal eats, the less the other animal will get. If the two animals can be differentiated—for example, by sex—then over a series of cages there will be a negative correlation between the amount of food eaten by males and females. In such a case, the standard error of the paired-comparison test will be greater than that of the difference-of-means test because the covariance term is now added to the sum of the two variances (the negative sign before the covariance term and the negative sign of the covariance itself together yield a plus sign).

Therefore, with negative correlations, a paired-comparisons test will have less power and thus is less likely to lead to the rejection of the null hypothesis than is the test for difference of means. In this case, an overall test for difference of means of the variables (treatments) has little meaning. Also differences due to the classification factor grouping them into pairs will not mean much in view of the negative correlation. The important aspect of the analysis is the negative interaction between classification factor and food intake. To estimate such an interaction, however, one needs to have replication of each subgroup. Otherwise, the only way to demonstrate such a relationship is by computing a correlation coefficient between the variables.

15.5 Hypothesis Tests for Correlations

The most common significance test determines whether a sample correlation coefficient could have come from a population with a parametric correlation coefficient of zero. The null hypothesis is, therefore, $H_0: \rho = 0$. If the sample comes from a bivariate normal distribution and $\rho = 0$, the standard error of the correlation coefficient is

$$s_r = \sqrt{\frac{1 - r^2}{n - 2}}$$

This enables one to construct a *t*-test with $n - 2$ degrees of freedom:

$$t_s = \frac{r - 0}{\sqrt{\dfrac{1 - r^2}{n - 2}}} = r\sqrt{\frac{n - 2}{1 - r^2}}$$

Note that this standard error applies only when $\rho = 0$, so it cannot be applied to testing a hypothesis that ρ is a value other than zero. This *t*-test for r is mathematically equivalent to the *t*-test of the null hypothesis that $\beta = 0$. This equivalence is analogous to the situation in Model I and Model II single-classification anovas, where the same *F*-test is used regardless of the model.

Box 15.2 illustrates tests of the hypothesis H_0: $\rho = 0$, using Statistical Table **R**, as well as the *t*-test previously discussed.

For ρ close to ± 1.0, the distribution of sample values of r is markedly skewed, and, although a standard error has been found for r in such cases, it should not be applied unless the sample is very large ($n > 500$). To overcome this difficulty, we transform r to a function z, developed by R. A. Fisher. The formula for z is

$$z = \frac{1}{2} \ln \left(\frac{1 + r}{1 - r} \right) \tag{15.11}$$

You may recognize this expression as $z = \tanh^{-1} r$, the formula for the inverse hyperbolic tangent of r. This function is found on most pocket calculators and in spreadsheet programs. Inspection of Expression (15.11) will show that when $r = 0$, z also equals zero, because $\frac{1}{2} \ln 1$ equals zero. As r approaches 1, however, $(1 + r)/(1 - r)$ approaches infinity; consequently, z approaches infinity. Therefore, substantial differences between r and z occur at the higher values for r. Thus, when $r = 0.115$, $z = 0.11551$; for $r = -0.531$, we obtain $z = -0.59154$; and when $r = 0.972$, $z = 2.12730$. Note by how much z exceeds r in this last pair of values. The inverse function, $r = \tanh z$, yields values of r given a value of z. Thus, $z = 0.70$ corresponds to $r = 0.60437$, and $z = -2.76$ corresponds to $r = -0.99202$.

The advantage of the **z-transformation** is that although correlation coefficients are distributed in skewed fashion for values of $\rho \neq 0$, the values of z are approximately normally distributed for any value of its parameter, which we call ζ (zeta), following the usual convention. The expected variance of z is

$$\sigma_z^2 = \frac{1}{n - 3} \tag{15.12}$$

This approximation is adequate for sample sizes $n \geq 50$ and tolerable even when $n \geq 25$. An interesting aspect of the variance of z evident from Expression (15.12) is that it is independent of the magnitude of r and is simply a function of sample size n. Where the hypothesis tests or confidence limits are critical and a better approximation is desired, Hotelling (1953) recommends another transformation, z^*, which is computed as follows:

$$z^* = z - \frac{3z + r}{4(n - 1)} \tag{15.13}$$

BOX 15.2 Hypothesis Tests and Confidence Limits for Correlation Coefficients

Test of the null hypothesis H_0: $\rho = 0$ versus H_1: $\rho \neq 0$

The simplest procedure is to consult Statistical Table **R**, where the 5% and 1% critical values of r are tabulated for $df = n - 2$ from 1 to 1000. If the absolute value of the observed r is greater than the tabulated value in the column for 1 independent variable, then we reject the null hypothesis (the other columns are for testing multiple correlation coefficients as discussed in Chapter 16).

Examples: In Box 15.1, we found the correlation between body weight and gill weight to be 0.86519, based on a sample of $n = 12$. For 10 degrees of freedom, the critical values are 0.576 at the 5% level and 0.708 at the 1% level. Because the observed correlation is greater than both of these, we can reject the null hypothesis, H_0: $\rho = 0$, at $P < 0.01$.

In Box 14.10, we encountered paired measurements of 55 stickleback fishes as an illustration of model II regression. If we wish to estimate the association between these variables, we calculate their product–moment correlation coefficient following the procedures learned in Box 15.1. Remember that the two variables, standard length and ectocoracoid length, have been transformed into the logarithms of the original measurements. We obtain $r = 0.75519$, a value far greater than the critical values for $\nu = 50$ degrees of freedom (0.273 and 0.354) given in Statistical Table **R**. These are conservative values because the actual degrees of freedom for this example are $\nu = 55 - 2 = 53$. Again, we reject the null hypothesis that $\rho = 0$ at $P \ll 0.01$. Statistical Table **R** is based on the following test, which may be carried out when the table is not available or when an exact test is needed at probability levels or at degrees of freedom other than those furnished in the table. The null hypothesis is tested by means of the t-distribution (with $n - 2$ df) by using the standard error of r. When $\rho = 0$,

$$s_r = \sqrt{(1 - r^2)/(n - 2)}$$

Therefore,

$$t_s = \frac{(r - 0)}{\sqrt{(1 - r^2)/(n - 2)}} = r\sqrt{(n - 2)/(1 - r^2)}$$

For the body-weight and gill-weight data of Box 15.1, the value would be

$$t_s = 0.86519\sqrt{(12 - 2)/(1 - 0.86519^2)} = 5.45618, P = 0.000{,}278$$

For a one-tailed test, the 0.10 and 0.02 values of t should be used for 5% and 1% tests, respectively. Such tests would apply if the alternative hypothesis were either H_1: $\rho > 0$ or H_1: $\rho < 0$, rather than H_1: $\rho \neq 0$.

When n is greater than 50, we can also make use of the z-transformation described in the text. For the stickleback data of Box 14.10, because $\sigma_z = 1/\sqrt{n - 3}$, we test

$$t_s = \frac{z - 0}{1/\sqrt{n - 3}} = z\sqrt{n - 3}$$

Because z is normally distributed and we are using a parametric standard deviation, we compare t_s with $t_{\alpha(\infty)}$ or employ Statistical Table **A**, areas of the normal curve.

Box 15.2 (continued)

Previously in this box, we reported $r = 0.75519$. This yields $z = 0.98492$ and $t_{s[\infty]} = 7.10236$, $P = 1.22645 \times 10^{-12}$ or $P = 1.23479 \times 10^{-12}$, obtained from the t-distribution and the normal distribution functions, respectively, of the statistical calculator in BIOMstat. Clearly, there is only an infinitesimal probability that the null hypothesis $\rho = 0$ is true.

Test of the null hypothesis H_0: $\rho = \rho_1$, where $\rho_1 \neq 0$

To test this hypothesis, we cannot use Statistical Table **R** or the t-test based on the standard error of r but must make use of Fisher's z-transformation.

Let us imagine a morphogenetic law that, for our log-transformed stickleback data, would predict that the correlation between ectocoracoid length and standard length is 0.5. We would test such a prediction with a test of the null hypothesis H_0: $\rho = +0.5$ versus H_1: $\rho \neq +0.5$. We would use the following expression:

$$t_s = \frac{z - \zeta}{1/\sqrt{n - 3}} = (z - \zeta)\sqrt{n - 3}$$

where z and ζ are the z-transformations of r and ρ, respectively. Again we compare t_s with $t_{\alpha[\infty]}$ or look it up in Table **A**. Using Fisher's z $= \tanh^{-1}$ transformation, we find

$$\text{for } r = 0.75519 \qquad z = 0.98492$$
$$\text{for } \rho = 0.500 \qquad \zeta = 0.54931$$

Therefore,

$$t_s = (0.98492 - 0.54931)\sqrt{52} = 3.14123$$

We obtain $P = 0.001683$ from the t-distribution or the normal distribution functions of the statistical calculator in BIOMstat. It is unlikely that the null hypothesis $\rho = 0.5$ is true. Our findings do not support the prediction of the (imagined) morphogenetic law.

Confidence limits

If $n > 50$, we can set confidence limits to r using the z-transformation. First, we convert the sample r to z, set confidence limits to z, and then transform these limits back to the r-scale. The following calculations determine 95% confidence limits for the stickleback measurements already analyzed ($r = 0.75519$; $z = 0.98492$; $\alpha = 0.05$, $\sigma_z = 1/\sqrt{n - 3} = 1/\sqrt{52}$):

$$L_1 = z - t_{\alpha[\infty]}\sigma_z = 0.98492 - \frac{1.960}{\sqrt{52}} = 0.71312$$

$$L_2 = z + t_{\alpha[\infty]}\sigma_z = 0.98492 + \frac{1.960}{\sqrt{52}} = 1.25672$$

We retransform these z-values to the r-scale by means of the tanh function:

$$L_1 = 0.6126 \quad \text{and} \quad L_2 = 0.8502$$

are the 95% confidence limits around $r = 0.75519$.

Box 15.2 (continued)

The use of the z-transformation when sample sizes n < 50

For small samples, calculating exact probabilities is difficult. The following modified z-transformation has been suggested by Hotelling (1953) for use in small samples:

$$z^* = z - \frac{3z + r}{4(n - 1)}$$

$$\sigma_{z^*} = \frac{1}{\sqrt{n - 1}}$$

The distribution of z^* is closer to a normal distribution than is z. We do not know for how small a sample size this transformation is adequate, but most likely it should not be used for $n < 10$. As an example, z^* and σ_{z^*} are computed for a small sample, the gill weights versus body weights from Box 15.1.

$$r = 0.86519 \qquad n = 12 \qquad z = 1.31363$$

$$z^* = z - \frac{(3z + r)}{4(n - 1)} = 1.31363 - \frac{3(1.31363) + 0.86519}{4(11)} = 1.20440$$

$$\sigma_{z^*} = \frac{1}{\sqrt{n - 1}} = \frac{1}{\sqrt{11}} = 0.30151$$

The differences between z and z^* and between their standard errors are slight and are important only for critical tests near the borderline values of P. For example, to test the null hypothesis $H_0: \rho = 0$ versus $H_1: \rho \neq 0$,

$$t_s = (z^* - \zeta^*)\sqrt{n - 1}$$

where ζ^* is the z^*-transformation applied to the ρ of the null hypothesis:

$$\zeta^* = \zeta - \frac{(3\zeta + \rho)}{4n}$$

which equals zero because ζ and ρ are both zero. Again, we compare t_s with $t_{\alpha[\infty]}$ or look it up in Statistical Table **A**, or we compute it directly with a computer program. Applying this formula to the gill-weight–body-weight correlation of Box 15.1, we obtain

$$t_s = 1.20440\sqrt{11} = 3.99454, P = 6.48 \times 10^{-8}$$

If we had used the ordinary z-transformation, our result would have been

$$t_s = (z - \zeta)\sqrt{n - 3} = 1.31363\sqrt{9} = 3.94088$$

leading to the same conclusion as before.

We can use Expressions (15.13) and (15.14) to set confidence limits to ρ in small samples. A limitation of the z^*-transformation is that one cannot use it to back-transform confidence limits into correlation coefficients.

Its variance is given as

$$\sigma_{z*}^2 = \frac{1}{n-1} \tag{15.14}$$

This expression is actually another approximation for a more involved expression, but it should be satisfactory for sample sizes $n \geq 10$. Unfortunately, the $z*$ function cannot be inverted to provide a back transformation to r.

As Box 15.2 shows, for sample sizes greater than 50 we can also use the z-transformation to test the hypothesis H_0: $\rho = 0$ for a sample r. In the second section of Box 15.2, we show the test of a null hypothesis that $\rho \neq 0$. We may hypothesize that the true correlation between two variables is a given value ρ different from zero. Hypotheses about the expected correlation between two variables are frequent in genetic work, and also in morphometric studies. We may wish to test observed data against such a hypothesis. Although there is no a priori reason to assume that the true correlation between the two stickleback measurements is 0.5, we show the test of such a hypothesis to illustrate the method. Corresponding to $\rho = 0.5$ is ζ, the parametric value of z—that is, the z-transformation of ρ. Note that the probability that the sample r of 0.75519 could have resulted from a population with $\rho = 0.5$ is quite small.

Next in Box 15.2, we show how to assign confidence limits to a sample correlation coefficient r. We use the z-transformation, which will result in asymmetrical confidence limits when these are retransformed to r-scale, as when setting confidence limits with variables subjected to square root or logarithmic transformations.

Occasionally, we measure the same two variables in several samples and obtain correlation coefficients for each sample. It may be of interest to know whether these correlation coefficients can be considered samples from a population exhibiting a common correlation among the variables. One way of stating the null hypothesis is to say that the k sampled correlation coefficients are homogeneous and estimate a common parametric value of ρ. Such problems arise occasionally in systematics, where taxonomic differentiation results in differences in correlation between body structures. Thus, not only do characters vary among taxa, or geographically within a taxon, but also the correlation between pairs of characters changes in different populations.

The example in Box 15.3 illustrates the **test of homogeneity among correlation coefficients.** In this case, the variables are wing length and width of a band on the wing in females of 10 populations of a species of butterfly. These 10 populations were taken from different islands and localities in the Caribbean. Note that the sample correlation coefficients differ considerably, from a low of 0.29 to a high of 0.70. The computations are quite simple and in effect consist of calculating a weighted sum of squares of the z-values corresponding to the correlation coefficients. This sum of squares is called X^2 and is distributed as χ^2 with $k - 1$ degrees of freedom.

The reason that the weighted sum of squares of z should be distributed as χ^2 may not be immediately obvious. This sum of squares is the quantity

$$X^2 = \sum_{}^{k} \left(\frac{z - \bar{z}}{\sigma_z} \right)^2 = \sum_{}^{k} \left(\frac{z - \bar{z}}{1/\sqrt{n_i - 3}} \right)^2 = \sum_{}^{k} (n_i - 3)(z - \bar{z})^2 \tag{15.15}$$

BOX 15.3 Tests of Homogeneity Among Two or More Correlation Coefficients

Correlation coefficients between length of wing and width of band on wing in females of 10 populations of the butterfly *Heliconius charitonius* based on varying sample sizes.

(1) n_i	(2) $n_i - 3$	(3) r_i	(4) z_i	(5) Weighted $z_i = (n_i - 3)z_i$
100	97	0.29	0.2986	28.96093
46	43	0.70	0.8673	37.29392
28	25	0.58	0.6625	16.56157
74	71	0.56	0.6328	44.93116
33	30	0.55	0.6184	18.55144
27	24	0.67	0.8107	19.45786
52	49	0.65	0.7753	37.98964
26	23	0.61	0.7089	16.30519
20	17	0.64	0.7582	12.88895
17	14	0.56	0.6328	8.85966
423	393			241.8003

SOURCE: Data from Brown and Comstock (1952).

We wish to test whether the $k = 10$ sample r's could have been taken from the same population; if so, we will combine them into an estimate of ρ.

Computation

1. Compute average z:

$$\bar{z} = \frac{\sum^{k} \text{weighted } z_i}{\sum (n_i - 3)} = \frac{\text{sum of column } (5)}{\text{sum of column } (2)} = \frac{241.8003}{393} = 0.615268$$

2. To test that all r's are from the same population compute

$$X^2 = \sum^{k} (n_i - 3)(z_i - \bar{z})^2 = 15.26352$$

Compare X^2 with χ^2-values for $k - 1$ df in Statistical Table **D** or employ the statistical calculator in BIOMstat for the more precise value of $P = 0.08395$. We do not have sufficient evidence to reject the null hypothesis of no heterogeneity among the correlation coefficients.

3. Estimate common ρ: The r-value corresponding to $\bar{z} = 0.615268$ is $r = 0.547825$.

Box 15.3 (continued)

The special case of two correlation coefficients

When we have only two correlation coefficients we may test H_0: $\rho_1 = \rho_2$ versus H_1: $\rho_1 \neq \rho_2$ as follows:

$$t_s = \frac{z_1 - z_2}{\sqrt{\dfrac{1}{n_1 - 3} + \dfrac{1}{n_2 - 3}}}$$

Because $z_1 - z_2$ is normally distributed and we are using a parametric standard deviation, we compare t_s with $t_{\alpha[\infty]}$ or employ Statistical Table **A**, areas of the normal curve. For a more precise answer, use the statistical calculator in BIOMstat, or elsewhere.

For example, the correlation between body weight and wing length in *Drosophila pseudoobscura* was found by Sokoloff (1966) to be 0.552 in a sample of $n_1 = 39$ at the Grand Canyon and 0.665 in a sample of $n_2 = 20$ at Flagstaff, Arizona.

$$\text{Grand Canyon: } z_1 = 0.6213 \qquad \text{Flagstaff: } z_2 = 0.8017$$

$$t_s = \frac{0.6213 - 0.8017}{\sqrt{\frac{1}{36} + \frac{1}{17}}} = -0.6130$$

From the statistical calculator in BIOMstat, we obtain $P = 0.4601$, so we clearly have no evidence on which to reject the null hypothesis.

Confidence limits to the difference in z-values can be computed by inverting the t-test previously given. For the present data, the 95% confidence limits are

$$L_1 = (z_1 - z_2) - t_{\alpha[\infty]}\sqrt{\frac{1}{n_1 - 3} + \frac{1}{n_2 - 3}} = -0.1804 - 1.960(0.29428) = -0.75719$$

$$L_2 = (z_1 - z_2) + t_{\alpha[\infty]}\sqrt{\frac{1}{n_1 - 3} + \frac{1}{n_2 - 3}} = -0.1804 + 1.960(0.29428) = 0.39639$$

The interval includes zero because the t-test did not reject the null hypothesis.

Procedures to follow for small sample sizes are suggested in the text (Section 15.5).

Multiple comparisons among correlation coefficients

To carry out a Tukey–Kramer multiple comparisons test for correlation coefficients, we rank order the k correlation coefficients, r_i, by magnitude from a list such as the butterfly data shown earlier in this box. Next to the column with the correlations, we enter two columns of their associated z-transforms and sample sizes, n_i. The relevant minimum significant differences for unequal sample sizes, MSD_{ij}, are

$$MSD_{ij} = Q_{\alpha[k,\infty]}\sqrt{\frac{(n_i - 3)^{-1} + (n_j - 3)^{-1}}{2}}$$

Box 15.3 (continued)

Any absolute difference between two z_i-values for a pair of samples that equals or exceeds the associated *MSD* value will lead to a rejection of the null hypothesis of no difference for their two parametric correlations. When all sample sizes are the same, the *MSD* formula simplifies into

$$MSD = Q_{\alpha[k,\infty]}\sqrt{(n-3)^{-1}}$$

and needs to be computed just once for a given set of comparisons.

Let us apply these techniques to the correlation of wing measurements from 10 populations of butterflies previously analyzed. The computation is conveniently carried out within the framework of an all-pairs matrix layout, as presented in part **IV** of Box 9.7. By now this should be familiar. To conserve space, we will not furnish a fully filled in matrix here, but we present an outline of the computations sufficient to enable you to carry them out on your own.

We first test the pair of localities showing the greatest difference in correlations (hence in z-values). These happen to be the localities in the first two rows of the table. Their absolute difference is 0.5687. The *MSD* for these two localities is $5.157\sqrt{(97^{-1} + 43^{-1})/2} = 0.6681$. We cannot reject the null hypothesis that the two correlation coefficients were sampled from the same population, despite their apparent great differences. Rather than test all possible pairs, we experimented with these data and found that no combination of possible large differences and available large sample sizes would lead to the rejection of the null hypothesis. This set of correlation coefficients appears to be truly homogeneous. These results support the acceptance of overall homogeneity among the 10 populations with respect to the correlations under study reported in the earlier part of the box. In a multiple comparisons study with more structure, such as the tiger beetle study in Box 9.7 (for means, not r's), the interpretation of the results (i.e., that one or more difference-pairs lead to a rejection of the null hypothesis) will be more interesting than in the present case. In most computer programs (as in BIOMstat), it is more convenient to test all pairwise differences and let the user subsequently try to make sense out of the results.

If we had equal sample sizes, and thus equal standard deviations, we could rewrite this expression as $(1/\sigma_z^2)\Sigma^k(z - \bar{z})^2$. In this form, the expression is analogous to Expression (7.8), $(1/\sigma^2)\Sigma(Y_i - \bar{Y})^2$, which, as you may recall, followed the χ^2-distribution with $n - 1$ degrees of freedom. Thus, if the z's represent a random sample from a single normally distributed population, the value of X^2 obtained as shown here should follow the χ^2-distribution with $k - 1$ degrees of freedom.

In the example in Box 15.3, we find that X^2 is smaller than the 5% critical value of χ^2. We therefore do not reject the null hypothesis. The correlation coefficients in the 10 populations should thus be considered homogeneous (even though the $r = 0.29$ appears quite deviant from the rest). This result is rather typical; the r-values have to be very different or based on larger sample sizes before this test can detect differences. If the sample correlations are adjudged homogeneous, we can use them to estimate an overall average to obtain a common estimate of ρ. We estimate ρ by back-transforming the value of $\bar{z}$ to r. The estimated common correlation is 0.547825.

This average is weighted by the reciprocals of the variances, which are a function of sample size. If some samples are small and an especially precise estimate of ρ is desired, we should allow for a bias introduced in our sample estimate of ζ. An unbiased estimate, z_u, is given by

$$z_u = z - \frac{\rho}{2(n-1)}$$

Note that to obtain such an estimate we need to know ρ, which is usually impossible; if we did know ρ, then we would, of course, not be interested in z_u. We therefore obtain a more approximate estimate of ζ from the sample r by the following formula:

$$z_u = z - \frac{r}{(2n-5)}$$

These biases are of the same general type as the bias in estimating the population variance as $\Sigma y^2/n$, first encountered in Section 4.7. In practice, we would correct the estimate $\bar{z}$ by subtracting the following quantity from it:

$$\frac{\bar{r} \sum\limits^{k} \left(\dfrac{n_i - 3}{2n_i - 5} \right)}{\sum\limits^{k} (n_i - 3)}$$

For the example in Box 15.3, this correction amounts to 0.006848, which results in a new estimate of $\bar{z} = 0.615268 - 0.006848 = 0.60842$. The new $\bar{r}$ is, therefore, 0.543014, which is only slightly different from the original $\bar{r}$ of 0.547825. Although the literature cautions against applying the homogeneity test (and the two-sample test discussed next) to data with small samples, it gives no clear instructions on what constitutes a small sample or on how to proceed in such cases.

A test for the difference between two sample correlation coefficients is a special case of the test for homogeneity among r's and can be carried out in the manner just described. A simpler procedure is shown in the middle section of Box 15.3. A standard error for the difference between two z's is computed and is used to carry out a t-test of the difference. Because the appropriate degrees of freedom for z-tests are infinite, we compare t_s with a normal deviate; look up either Statistical Table **A** or **B** or use the statistical calculator of BIOMstat. In the example in Box 15.3, the correlation between body weight and wing length in two *Drosophila* populations was tested, and the null hypothesis of no difference in correlation coefficients between the two populations could not be rejected. The formula given is an acceptable approximation when the smaller of the two samples is greater than 25. It is also frequently used with even smaller sample size, as this example shows.

It might have occurred to you that it would be of interest to probe a little deeper into the relationship structure of the 10 butterfly correlations that tested homogeneous overall previously in this box. When we have correlation coefficients for $k > 2$ samples, we can adapt the Tukey–Kramer procedure from Section 9.5, testing for homogenous pairs of means, to pairs of correlation coefficients. The computations are outlined in the final section of Box 15.3. We conclude that none of the pairs of localities differs in their correlations of wing lengths with wing band widths. These results support our previously reported finding of overall homogeneity for these correlations among the 10 populations studied.

15.6 Applications of Correlation

The purpose of correlation analysis is to measure the intensity of association observed between two variables and to test whether this correlation is greater than could be expected by chance alone if $\rho = 0$. Once established, such an association is likely to lead to reasoning about causal relationships between the variables. Students of statistics are taught early not to confuse correlation with causation. We are also warned about so-called nonsense or spurious correlations, such as the well-known case of the positive correlation between the number of Baptist ministers and the total liquor consumption in cities with populations of more than 10,000 in the United States. Individual cases of correlation must be carefully analyzed before inferences are drawn from them. It is useful to distinguish correlations in which one variable is the entire, or more likely the partial, cause of another, from others in which the two correlated variables have a common cause and from more complicated situations involving both direct influence and common causes.

Almost all observed correlations can be described in terms of one of the structural designs in Figure 15.5 or in terms of a more complicated structure composed of these elements. There are probably few situations in nature in which two variables are totally correlated, either because one is the sole cause of the other (Figure 15.5a) or because both arise from a single common cause (Figure 15.5b). Exceptions include trivial cases such as age of tree and number of growth rings (an example of Figure 15.5a). The model described by Figure 15.5c is more common. The correlation between age (Y_2) and weight (Y_1) in an animal is likely to be of this type.

Factors other than age (Y_3, Y_4) also affect the animal, so the correlation between Y_1 and Y_2 is unlikely to be perfect. Weight (Y_1) and length (Y_2) of an animal usually have age (Y_4) as a common factor, leading to a model such as that in Figure 15.5d. Most correlations in nature are likely to be of the type shown in Figure 15.5e, in which two or more common causes (Y_4 and Y_5) lead to correlation of variables Y_1 and Y_2. Correlation between morphological parts in organisms is surely of this type, with many morphogenetic forces exerting an influence on the variation of the correlated variables. The next degree of complexity is shown in Figure 15.5f, where the correlation between Y_1 and Y_2 results from both the direct effect of one of the variables (Y_2) and a common cause (Y_4).

These relationships can be even more complicated. Adult weight and length of developmental period, for example, are correlated in many insects. There is a direct effect of length of developmental period on adult weight: The longer the immature insect feeds, the heavier the adult will be. The density at which the insects were

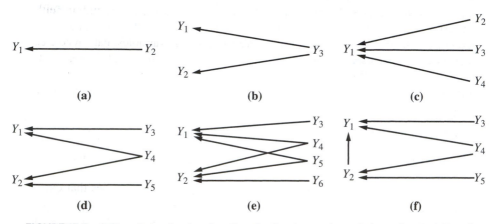

FIGURE 15.5 Different structural explanations for the observed correlation r_{12} between Y_1 and Y_2 (assuming only linear relations between variables). **(a)** Y_2 is the entire cause of the variation of Y_1. In such a case, r_{12} must equal 1, and r_{12}^2 must also equal 1. **(b)** The common cause of Y_3 totally determines variables Y_1 and Y_2. Again, r_{12} must be 1. **(c)** In this case, Y_2 is one of several causes of Y_1, and r_{12} will be less than 1. **(d)** The correlation between variables Y_1 and Y_2 is due to a common cause Y_4. Because other causes, Y_3 and Y_5, also determine Y_1 and Y_2, respectively, the correlation between these variables is not perfect. **(e)** The correlation between variables Y_1 and Y_2 is due to two common causes, Y_4 and Y_5. **(f)** The correlation between variables Y_1 and Y_2 is due to the direct effect of Y_2 on Y_1, as well as from a common cause, Y_4.

reared, however, affects both larval period and weight. A high density reduces the weight because the food supply decreases, but generally increases the developmental period. Thus, the same common cause increases one but lowers the other of the two correlated variables, hence decreasing the correlation caused by the direct effect of density on weight. Models of this type, including instances of mutual causation between two variables, can rapidly become quite complex.

The establishment of a significant correlation does not tell us which of these structural models is appropriate. Further analysis is needed to distinguish between models. Path analysis (see Section 16.3) is especially designed for the study of models such as those in Figure 15.5.

The traditional distinction of real versus nonsense or illusory correlation is of little use. In the supposedly legitimate correlations, the relations were of the types shown in Figure 15.5 and causal connections were known or at least believed to be clearly understood. In so-called illusory correlations, no reasonable connection between the variables can be found; or if one is demonstrated, it is of no real interest or may be shown to be an artifact of the sampling procedure. For example, the correlation between Baptist ministers and total liquor consumption is simply a consequence of city size. The larger the city, the more Baptist ministers it will contain on average and the greater the total amount of liquor consumption will be. Because city size determines both variables, this case is really an example of Figure 15.5d, correlation due to a common cause. The correlation is of little interest to anyone studying either the distribution of Baptist ministers or the consumption of alcohol. Correlations can also be expected as a consequence of the definition of the variables. Part–whole correlations were discussed earlier. Another example is the correlation between a ratio, Y_1/Y_2, and the variable, Y_2, in the denominator of the ratio. Although the expected value depends on the variances and covariances of the variables, one usually expects a rather large negative correlation. A well-known example is the ratio where Y_1 is the size of a species on an island and Y_2 is the size of its most closely related species on the mainland. Similarly, a rate of evolutionary change, the amount of change divided by the time period, would be expected to be negatively correlated with the time period over which the change is measured.

Some correlations have time as the common factor, and processes that change with time are frequently likely to be correlated, not because of any functional biological reasons but simply because the change with time in the two variables under consideration happens to be in the same direction. Thus, size of an insect population increasing through the summer may be correlated with the height of some weeds, but this association may simply be a function of the passage of time. There may be no ecological relation between the plant and the insects.

Perhaps the only correlations properly called *nonsense* or *illusory* are those assumed by popular belief or scientific intuition, for which we are unable to reject the null hypothesis that the correlations are zero, when they are tested by proper statistical methodology using adequate sample sizes (see Section 15.9). Thus, if we can show that there is no significant correlation between the amount of saturated fats eaten and the degree of atherosclerosis, we can consider this correlation to be illusory (assuming our study was carefully done and had sufficient sample sizes to be able to detect a biologically important correlation). Remember also that when performing hypothesis tests, you must allow for type I errors (see Section 7.1), which will result in a certain percentage of correlations to be judged statistically significant when in fact the parametric value of ρ equals zero.

Correlation coefficients have a history of extensive use and application dating back to the English biometric school at the turn of the century. Recently, this technique has been applied somewhat less frequently as biological research has become more experimental. In experiments in which one factor is varied and the response of another variable to the deliberate variation of the first is examined, the method of regression is more appropriate, as has already been discussed. Large areas of biology and of other sciences remain, however, where the experimental method is not suitable because variables cannot be brought under control of the investigator.

There are many areas of ecology, systematics, evolution, and other fields in which experimental methods are difficult to apply. As yet, the weather cannot be controlled, and historical evolutionary factors cannot be altered. Nevertheless, we need to understand the scientific mechanisms underlying these phenomena as much as we understand those in biochemistry or experimental embryology. In such cases, correlation analysis is a preliminary descriptive technique to estimate the degrees of association among the variables involved. Other techniques such as factor analysis, briefly discussed in Section 16.11, lead to additional analytical insights into the mechanisms of such phenomena, even without applying the experimental method.

15.7 Nonparametric Tests for Association

Remember that the product–moment correlation coefficient describes the strength of the linear component of relations between two variables. Depending on the non-linearity and the range of values used in computing the coefficient, even a strictly deterministic phenomenon may have a small linear correlation. For example, the correlation between X and Y for the relation $Y = (X - 5)^2$ in the range $1 \leq X \leq 10$ is $r = 0.37$. For the range $11 \leq X \leq 20$, however, $r = 0.99$.

When we know that our data do not conform to a bivariate normal distribution, but we wish to test for the presence of an association between two variables, one method of analysis is to rank the observations and calculate a coefficient of rank correlation. This approach belongs to the general family of nonparametric methods, encountered in Chapter 13, where we learned methods for analyses of ranked observations paralleling anovas. In other cases especially suited to ranking methods, we cannot measure the variable on an absolute scale, but only on an ordinal scale. This limitation is typical of data in which we estimate relative performance, as in assigning positions in a class. For example, we can say that A is the best student, B the second best student, C and D are both equal to each other and next best, and so on. Two instructors may independently rank a group of students, and we can then test whether these two sets of rankings are independent, which one would not expect if the judgments of the instructors are based on objective evidence.

Of greater biological interest are the following examples. We might wish to test whether the order of emergence in a sample of insects is associated with their ranking in size, or order of germination in a sample of plants with rank order of flowering. A geneticist might predict the rank order of performance of a series of n genotypes he synthesizes and wants to show the correlation of his prediction with the rank orders of the realized performance of these genotypes. A taxonomist might wish to array n organisms from those most like form X to those least like it. Will a similar array prepared by a second taxonomist be correlated with the first—that is, are the taxonomic judgments of the two observers related?

There are several rank correlation coefficients. In Box 15.4, we present **Kendall's coefficient of rank correlation,** generally symbolized by τ (the Greek letter tau), although it is a sample statistic, not a parameter. The formula for Kendall's coefficient of rank correlation is $\tau = N/n(n - 1)$, where n, as usual, is the sample size and N is a count of ranks that can be obtained in a variety of ways (we show two simple approaches in Box 15.4). If a second variable, Y_2, is perfectly correlated with the first variable, Y_1, then the Y_2-observations should be in the same order as the Y_1-observations. If the correlation is less than perfect, however, the order of the Y_2-observations will not entirely correspond to that of the Y_1. The quantity N measures how well the order of the second variable corresponds to the order of the first. It has a

BOX 15.4 Kendall's Coefficient of Rank Correlation, τ

Computation of a rank correlation coefficient between the total length (Y_1) of 15 aphid stem mothers and the mean thorax length (Y_2) of their parthenogenetic offspring (based on measurement of four alates, or winged forms); $n = 15$ pairs of observations. Measurements are in micrometer units.

(1) Stem mother	(2) Y_1	(3) R_1	(4) Y_2	(5) R_2
1	8.7	8	5.95	9
2	8.5	6	5.65	4
3	9.4	9	6.00	10
4	10.0	10	5.70	6.5
5	6.3	1	4.70	2
6	7.8	5	5.53	3
7	11.9	15	6.40	15
8	6.5	2	4.18	1
9	6.6	3	6.15	13
10	10.6	12	5.93	8
11	10.2	11	5.70	6.5
12	7.2	4	5.68	5
13	8.6	7	6.13	12
14	11.1	13	6.30	14
15	11.6	14	6.03	11

SOURCE: Data from a more extensive study by R. R. Sokal (unpublished results).

Computation

1. Rank variables Y_1 and Y_2 separately and then replace the original observations with the ranks (assign average ranks in the case of ties). These ranks are listed in columns (3) and (5) above. Because they were tied, the 4th and 11th observations of variable Y_2 were assigned an average rank of 6.5.

2. Write down the n ranks of one of the two variables in order, paired with the rank values assigned for the other variable (as shown in the next table). If only one variable has ties, then order the pairs by the variable without ties (as in the present example). If both variables have ties, it does not matter which of the variables is ordered. We feature two equivalent methods for computing the rank correlation coefficient.

3a. The conventional method is to obtain a sum of the counts C_i, as follows. Examine the first value in the column of ranks R_2 paired with the ordered column. In our case, this is rank 2. Count all ranks subsequent to it that are higher than the rank being considered. In this example, count all ranks greater than 2. There are 14 ranks

Box 15.4 (continued)

following the 2, and all of them except rank 1 are greater than 2. Therefore, we count a score of $C_1 = 13$. Now we look at the next rank (rank 1) and find that all 13 subsequent ranks are greater than it; therefore, C_2 is also equal to 13. C_3, however, is equal to 2 because only ranks 14 and 15 are higher than rank 13. Continue in this manner, taking each rank of the variable in turn and counting the number of higher ranks subsequent to it. This procedure can usually be done in one's head, but we show it here explicitly so that the method will be entirely clear. In the case of ties, count a $\frac{1}{2}$. Thus, for C_{10}, there are $4\frac{1}{2}$ ranks greater than the first rank of 6.5.

R_1	R_2	Subsequent ranks greater than pivotal rank R_2	Counts C_i
1	2	13, 5, 3, 4, 12, 9, 10, 6.5, 6.5, 8, 14, 11, 15	13
2	1	13, 5, 3, 4, 12, 9, 10, 6.5, 6.5, 8, 14, 11, 15	13
3	13	14, 15	2
4	5	12, 9, 10, 6.5, 6.5, 8, 14, 11, 15	9
5	3	4, 12, 9, 10, 6.5, 6.5, 8, 14, 11, 15	10
6	4	12, 9, 10, 6.5, 6.5, 8, 14, 11, 15	9
7	12	14, 15	2
8	9	10, 14, 11, 15	4
9	10	14, 11, 15	3
10	6.5	(6.5), 8, 14, 11, 15	$4\frac{1}{2}$
11	6.5	8, 14, 11, 15	4
12	8	14, 11, 15	3
13	14	15	1
14	11	15	1
15	15		0
			$78\frac{1}{2} = \sum_{}^{n} C_i$

We then compute the following quantity:

$$N = 4\sum_{}^{n} C_i - n(n-1) = 4(78\tfrac{1}{2}) - 15(14) = 104$$

3b. The following semigraphic procedure is often very convenient. Draw lines connecting identical ranks. In the case of ties, draw the lines from the tied ranks in such a way that they do not intersect. Count the number of intersections formed by the lines ($X = 26$ in the current case). The lines need not be straight but they should be drawn so as to avoid ambiguous multiple intersections.

Box 15.4 (continued)

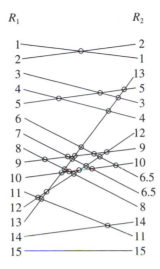

The quantity N previously found by counting ranks can now be computed as follows:

$$N = n(n - 1) - 4X - \sum_{}^{m} T_2$$

where $\Sigma^m T_2$ is a correction term needed to correct for the presence of ties in the R_2 ranks (it is zero if there are no ties). A T-value equal to $t(t - 1)$ is computed for each group of tied observations and summed over m such groups. In our case, there is $m = 1$ group of $t = 2$ tied observations; therefore,

$$\sum_{}^{m} T_2 = T = 2(1) = 2$$

Thus,

$$N = 15(14) - 4(26) - 2 = 104$$

If there is a negative correlation between the ranks, ranking one of the variables in reverse order will simplify the counting of intersections. The sign of N will then have to be reversed.

4. The Kendall coefficient of rank correlation, τ, can be found as follows:

$$\tau = \frac{N}{\sqrt{\left[n(n - 1) - \sum_{}^{m} T_1 \right]\left[n(n - 1) - \sum_{}^{m} T_2 \right]}}$$

where $\Sigma^m T_1$ and $\Sigma^m T_2$ are the sums of correction terms for ties in the ranks of variable Y_1 and Y_2, respectively, defined as in step **3b**. In our example $\Sigma^m T_1 = 0$, because there were no ties in the ranks R_1; $\Sigma^m T_2 = 2$ because of one group of $t = 2$ tied ranks

Box 15.4 (continued)

R_2. $T = t(t - 1) = 2(2 - 1) = 2$. Had there been more groups of ties, we would have summed the T's. Thus,

$$\tau = \frac{104}{\sqrt{[15(14)][15(14) - 2]}} = 0.4976$$

If there are no ties, the equation can be simplified to

$$\tau = \frac{N}{n(n - 1)}$$

J. S. Smart (personal communication) suggests the following treatment for ties. When it is believed that the ties are real, as for example scores of error-free integers, then the procedure described above is appropriate. When it is believed that the ties are only apparent, however, because differences in items are being obscured by limitations on precision of measurement or judgment, then it would be appropriate to average the correlation coefficient over all possible ways of assigning untied ranks to the tied observations. This result can be obtained directly by using formula $N/n(n - 1)$, where the numerator is the N corrected for ties as shown above and the denominator is the one appropriate for the case without ties. It could be argued that the present example falls into the second category. The aphids are probably not truly tied in length, but the measurement device was unable to discriminate at a more precise level. Thus $T = 104/15(14) = 0.4952$, a very slight decrease in the strength of the association.

5. To test τ for sample sizes > 40, we can use a normal approximation to test the null hypothesis that the true value of $\tau = 0$:

$$t_s = \frac{\tau}{\sqrt{2(2n + 5)/9n(n - 1)}} = \frac{0.4976}{\sqrt{2[2(15) + 5]/9(15)(14)}}$$

$$= 2.59 \text{ compared with } t_{\alpha[\infty]}$$

When we look up this value in Statistical Table **A** (areas of the normal curve) or using the statistical calculator of BIOMstat, we find the probability of such a t_s arising by chance to be 0.0096 (both tails).

When $n \le 40$, the approximation given here is not adequate, and Statistical Table **S** must be consulted. The table gives various (two-tailed) critical values for $n = 4$ to 40. These are exact only if there are no ties. If there are ties, a special table can be consulted (see Burr, 1960) but the computation is still complex.

maximal value of $n(n - 1)$ and a minimal value of $-n(n - 1)$. The following sample will make this clear.

Suppose we have a sample of five individuals that have been arrayed by rank of variable Y_1:

$$Y_1 \quad 1\,2\,3\,4\,5$$

$$Y_2 \quad 1\,3\,2\,5\,4$$

Note that the ranking by variable Y_2 is not totally concordant with that by Y_1. One of the techniques in Box 15.7 (step **3a**) is to count the number of higher ranks following any given rank, sum this quantity for all ranks, multiply the sum ($\Sigma^n C_i$) by 4, and subtract $n(n - 1)$ from it to obtain a statistic N. For variable Y_1, we find $\Sigma^n C_i = 4 + 3 + 2 + 1 + 0 = 10$; then we compute $N = 4\Sigma^n C_i - n(n - 1) = 40 - 5(4) = 20$ to obtain the maximum possible score $N = n(n - 1) = 20$. Obviously, Y_1 being ordered is always perfectly concordant with itself. For Y_2, however, we obtain $\Sigma^n C_i = 4 + 2 + 2 + 0 + 0 = 8$, and $N = 4(8) - 5(4) = 12$. Because the maximum score of N is $n(n - 1) = 20$ and the observed score 12, an obvious coefficient suggests itself as

$$\tau = \frac{N}{n(n - 1)} = \frac{4\sum_{}^{n} C_i - n(n - 1)}{n(n - 1)} = \frac{12}{20} = 0.6$$

Calculated in this manner, τ is actually a measure of the conformity to strict monotonicity of the Y_2 ranks when these are arrayed to match the Y_1 ranks. A graphic alternative computational approach for obtaining N is featured in step **3b** of Box 15.4. Kendall's τ is the converse of another coefficient called the **coefficient of disarray,** equal to $-N/n(n - 1)$, which measures the degree to which two sets of rankings are in disarray, or are not ordered in the same way. Frequently, ties present minor computational complications, which are dealt with in Box 15.4, where the correlation is between total body size of aphid stem mothers and mean thorax length of their offspring. In this case, there was no special need to turn to rank correlation except that there is some evidence that these data are bimodal and not normally distributed. The probability of τ for sample sizes greater than 40 can be obtained easily by using the standard error shown in Box 15.4. For sample sizes up to 40, look up critical values of N in Statistical Table **S**. This table is exact for samples without ties only. A table allowing for ties is given in Burr (1960). The normal approximation given in step **5** of Box 15.4 is not conservative and should be used with some caution.

Another rank correlation coefficient, **Spearman's coefficient of rank correlation,** is computed for data arranged in a manner similar to that described in Box 15.4. There is no simple mathematical relation between the two coefficients of rank correlation. Spearman's coefficient, r_s, can be computed directly from the differences between the ranks r_1 and r_2 of paired variables 1 and 2 as follows:

$$r_s = 1 - \frac{6\sum_{}^{n} (R_1 - R_2)^2}{n(n^2 - 1)}$$

Alternatively, r_s can be computed as the product–moment correlation coefficient of the two columns of ranks.

If the two variables Y_1 and Y_2 are independent, then the numerical value of Spearman's coefficient of rank correlation is highly correlated with that of Kendall's coefficient ($\rho_{r_s\tau} = 1$ for $n = 2$, decreases to 0.98 for $n = 5$, and then increases to 1.0 again as $n \to \infty$). If the true correlation is not zero, the two coefficients are sensitive to different types of departures from independence. Specifically, r_s gives greater weight to pairs of ranks that are farther apart, whereas τ weights each disagreement in rank equally. Therefore, r_s is more appropriate when there is less certainty about the reliability of close ranks. When $n \le 10$, special tables are needed for significance testing. Statistical Table **NN** provides critical values for r_s for one- and two-tailed tests for all sample sizes from 5 to 100. For $n > 25$, one can also test r_s as an ordinary product–moment correlation coefficient with relatively little error. However, the normal approximation for τ is considered more accurate than that for r_s.

When ranks of more than two variables are to be compared, one can obtain an overall measure of agreement among the rankings of the k variables. This value is

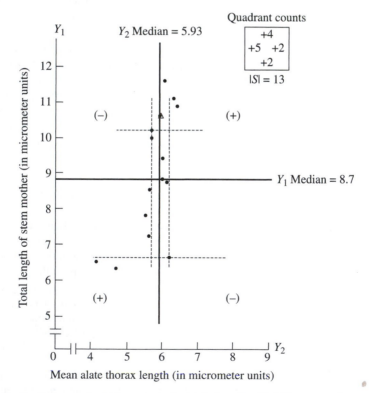

FIGURE 15.6 Scattergram of aphid morphological data from Box 15.4, illustrating the procedure for Olmstead and Tukey's corner test for association. (See Box 15.5 for explanation.)

W, **Kendall's coefficient of concordance.** The data would be set up as in the second table of Box 13.11 (Friedman's method for randomized blocks). We could ask the following question for that example: Do the temperature rankings shown in that box over the 10 depths agree for the four days of the study? The formula of the coefficient of concordance is

$$W = \frac{X^2}{k(n - 1)} \tag{15.16}$$

where X^2 is as defined in Box 13.11, k is the number of variables, and n is the number of items per variable. Note that in the symbolism of Box 13.11, $k = b$ and $n = a$. Thus, for the example of the temperatures of Rot Lake, we find $W = 36/4(9) = 1.000$. Values of W can range from 0 to 1. It should not surprise us that there is perfect concordance among the temperature readings because this fact is obvious on visual inspection of the data. The significance test for W is the same as that for Friedman's method: X^2 is compared with $\chi^2_{[n-1]}$.

It is sometimes desirable to test for an association between two variables not by computation, but by inspection of a scattergram. **Olmstead and Tukey's corner test for association** is designed for this purpose and permits a significance test for plotted points, regardless of whether the exact numerical values are known to the investigator. The method is described in Box 15.5 and illustrated in Figure 15.6.

BOX 15.5 Olmstead and Tukey's Corner Test for Association

A graphic "quick and dirty" method for ascertaining the presence but not the magnitude of correlation in a scattergram. Data from Box 15.4: $n = 15$.

1. Start with a scattergram of the data to be tested. The scattergram in Figure 15.6 shows total length of stem mother along the ordinate, Y_1, and mean thorax length of alates along the abscissa, Y_2.

2. Draw the medians of Y_1 and Y_2 into the scattergram. When n is an odd number, the median lines will run across the median item of each array. When n is an even number, the median lines will pass between the pair of central observations of each array. Label the upper right and lower left quadrants $+$ and the upper left and lower right quadrants $-$, respectively.

3. Apply a ruler to the left side of the scattergram. Note the leftmost point and the sign of the quadrant in which it is located. Now slowly move the rule to the right. Count 1 for each point, including the first, preceded by the sign of the quadrant in which it is located. Stop counting as soon as you reach a point that crosses the median perpendicular to your rule (the Y_1 median in this case). In other words, if your first point is $+1$, stop counting when you reach a negative point or vice versa. Enter the count ("quadrant count") of your points preceded by the appropriate sign at the left side of a square table.

Box 15.5 (continued)

Repeat this procedure starting at the bottom of the scattergram. Enter the quadrant count at the bottom of the table. Continue with the right side and top of the scattergram and enter quadrant counts at the right and top sides of the table, respectively.

Following this procedure on the scattergram in Figure 15.8, we obtain $+5$ for the left quadrant, $+2$ for the bottom, and $+2$ for the right. Obtaining the top quadrant count involves some complications. Because the sample size is odd in this example, two points will lie on the medians (unless the same point is the median observation for both variables). The fourth point from the top lies on the Y_2 median at $Y_1 = 10.6$. In such a case, we are told to ignore this point (as well as the point on the Y_1 median at $Y_2 = 5.95$) and replace them by a single point at $Y_1 = 10.6$ and $Y_2 = 5.95$. This new point is marked by a triangle in the figure whereas the points to be ignored are circles. The top quadrant count becomes $+4$. The broken lines in Figure 15.8 have been drawn to indicate the limits across which points change sign.

4. Sum the four quadrant counts, take the absolute value of this "quadrant sum," and read the probability from Statistical Table **T**. In the present example, the absolute value of the quadrant sum $|S| = 13$, which yields a P-value of 0.02. Thus, it is improbable that the association observed results from just sampling error.

NOTE: Another complication, not encountered in our example, occurs when on moving the rule inward, we are faced with a tie—two or more points on the same level and carrying opposite signs. In such a case, count the number of points in the tied group with the same sign as the quadrant counts and divide by (1 + the number of points with the opposite sign). In the case of two tied points, this will give $\frac{1}{2}$.

15.8 Major Axes and Confidence Regions

We have used the correlation coefficient as a measure of intensity of association between two variables in a bivariate scattergram of the type in Figure 15.3. A measure of the reliability of the bivariate mean is also desirable. Simple univariate confidence limits are not adequate because points on the scattergram can vary in two dimensions. Instead, we refer to **confidence regions** around the sample mean $(\overline{Y}_1, \overline{Y}_2)$. The major axis method described in Section 14.12 provides a single axis to represent the trend expressed by the scattergram. We noted in Section 14.12 that the slope of this axis has been suggested as one of the ways of estimating the slope of the functional relationship in Model II regression when both variables are in the same units of measurement.

We saw in Section 15.2 that the bivariate normal density function can be represented by concentric ellipses describing the topography of the surface of a bell-shaped mound. In the univariate work of Chapter 6, we assumed that observed samples had been taken from a normal distribution and, consequently, calculated expected frequencies (areas under the normal curve). Analogously, we will calculate ellipses in these data representing various volumes under the bell-shaped surface of the bivariate normal density function. In effect, the ellipses are the contour lines of the three-dimensional model of Figure 15.2.

In the univariate case, we assigned confidence limits to the mean and calculated expected frequencies for normal distributions whose mean and standard deviation were identical to the sample statistics. These calculations required two parameters: a fixed point, the mean μ for which our best estimate is $\overline{Y}$, and a measure of variation, the standard deviation σ, which we estimate as s. The corresponding values necessary in the bivariate case are the sample mean $(\overline{Y}_1, \overline{Y}_2)$, the standard deviations s_1 and s_2, and the covariance s_{12}. Recall from analytical geometry that an ellipse can be described as two **principal axes** that are perpendicular to each other—the major axis and the minor axis, with the major axis being the longest possible axis of the ellipse. Once the required parameters are estimated, we can draw ellipses around the mean.

Our first task, therefore, is to find the slope and equation of the major axis from a sample. Details of this computation are given in Box 14.10.

To describe the shape of an ellipse, we must also know the ratio of the lengths of its major and minor axes. This ratio is $\sqrt{\lambda_1/\lambda_2}$, where $\lambda_2 = s_1^2 + s_2^2 - \lambda_1$ and the λ values are computed using the formulas in Boxes 14.10 and 15.6. In mathematics and statistics, λ_1 and λ_2 are important quantities, known by many names. Among other terms, they are called **eigenvalues,** *latent roots*, or *characteristic roots* of the variance–covariance matrix of Y_1 and Y_2. In the type of problem we are discussing here, the eigenvalues are quantities analogous to variances; λ_1 and λ_2 measure variability along the major and minor axes, respectively. The major and minor axes are known as *eigenvectors, latent vectors,* or *characteristic vectors*. In data that are highly correlated and consequently represented by a very narrow, elongated ellipse, most of the variance can be accounted for by the major axis; in this case, the value of λ_1 would be very great with respect to the magnitude of λ_2. If the quantities λ_1 and λ_2 are equal, the major and minor axes would be of equal length, the data would be represented by a circle, and hence there would be no correlation between the variables.

Box 15.6 shows a method for finding the elliptical confidence region for the bivariate mean (μ_1, μ_2). Because two variables are involved, we cannot state univariate confidence intervals, but we do need to identify an elliptical confidence region around the point $(\overline{Y}_1, \overline{Y}_2)$. The shape of the ellipse is a function of the correlation between the variables, and the size (area) of the ellipse is a function of the confidence coefficient $1 - \alpha$. The equation describing the ellipse is shown in Box 15.6. We expect $100(1 - \alpha)\%$ of such ellipses to contain the true mean (μ_1, μ_2).

In addition, we have shown an equal-frequency ellipse that should, on average for bivariate normal samples, contain 95% of the observations. This is analogous to a range of $\pm 1.960s$ around a univariate sample mean. We simply use a different formula for C_α, which has the same relation to the previous C_α as the variance of items has to the variance of means (see Box 15.6 for the formula). Drawing an ellipse requires plotting a sufficient number of points to be able to draw a curve connecting them. By substituting our sample estimates of $\overline{Y}_1$ for μ_1 and $\overline{Y}_2$ for μ_2, and by evaluating C_α as shown in the box, we can solve for various sample values of Y_1, given possible values of Y_2. These equations are tedious to work out, so we try to plot only points for which most of the equation can be simplified considerably. We have labeled such points on the equal frequency ellipse with lowercase letters in

---

BOX 15.6 Bivariate Confidence Regions and Equal Frequency Ellipses

Data from Box 14.10.

Confidence regions for the bivariate mean (μ_1, μ_2)

The outer boundary of the confidence region is an ellipse around the point $(\bar{Y}_1, \bar{Y}_2)$. The equation for a $100(1 - \alpha)\%$ confidence region is

$$P[s_1^2(\bar{Y}_2 - \mu_2)^2 - 2s_{12}(\bar{Y}_1 - \mu_1)(\bar{Y}_2 - \mu_2) + s_2^2(\bar{Y}_1 - \mu_1)^2 \le C_\alpha] = 1 - \alpha$$

where μ_1 is the true mean for $\bar{Y}_1$, μ_2 is the true mean for $\bar{Y}_2$, and

$$C_\alpha = \frac{2\lambda_1\lambda_2(n - 1)}{n(n - 2)}F_{\alpha[2,n-2]}$$

As shown in Box 14.10, λ_1 and λ_2 are computed as follows: first, let

$$D = \sqrt{(s_1^2 + s_2^2)^2 - 4(s_1^2 s_2^2 - s_{12}^2)}$$

$$= \sqrt{(0.005709 + 0.005729)^2 - 4(0.005709 \times 0.005729 - 0.004319^2)} = 0.008638$$

then

$$\lambda_1 = (s_1^2 + s_2^2 + D)/2 = (0.005709 + 0.005729 + 0.008638)/2 = 0.010038$$
$$\text{and } \lambda_2 = s_1^2 + s_2^2 - \lambda_1 = 0.005709 + 0.005729 - 0.010038 = 0.001400$$

Using these formulas, the constant C_α for a 95% confidence region for the bivariate mean for our example would be computed as follows:

$$F_{.05[2,53]} = 3.17163$$

$$C_{.05} = \frac{2(0.010038)(0.001400)(55 - 1)}{55(55 - 2)} 3.17163 = 1.65138 \times 10^{-6}$$

As in the case of confidence intervals for a single mean, 95% of such bivariate confidence regions should contain the true bivariate mean.

The simplest way to make use of the confidence region is to graph its boundary. Several methods are possible. By computer, one can either convert the equation of the ellipse to polar coordinates or else express the equation for the ellipse as a function Y_1 and then solve a quadratic equation to obtain the corresponding values of Y_2.

In addition, it is also often useful to construct 95% equal frequency ellipses that should enclose 95% of the observations in a bivariate normally distributed sample. The procedure is the same except that the division by n in the formula for C_α is omitted. The new constant is

$$C_{.05} = \frac{2(0.010038)(0.001400)(55 - 1)}{(55 - 2)} 3.17163 = 9.08261 \times 10^{-5}$$

When sketching the ellipse by hand, it is easiest to compute the coordinates of particular points along the ellipse for which most of the preceding equation reduces to zero. These points, a through h, for a 95% equal frequency ellipse are shown in Figure 15.7 and are computed as follows.

a. $Y_2 = \bar{Y}_2 = 1.72151$

$$Y_1 = \bar{Y}_1 + \sqrt{\frac{C_\alpha}{s_2^2}} = 1.00171$$

b. $Y_2 = \bar{Y}_2 = 0.74989$

$$Y_1 = \bar{Y}_1 - \sqrt{\frac{C_\alpha}{s_2^2}} = 1.72151$$

Box 15.6 (continued)

c. $Y_2 = \bar{Y}_2 + \sqrt{\dfrac{C_\alpha}{s_1^2}} = 0.87580$

d. $Y_2 = \bar{Y}_2 - \sqrt{\dfrac{C_\alpha}{s_1^2}} = 0.87580$

$Y_1 = \bar{Y}_1 = 1.84765$

$Y_1 = \bar{Y}_1 = 1.59538$

e. $Y_2 = \bar{Y}_2 + \sqrt{\dfrac{C_\alpha}{\lambda_2(1 + b_1^2)}} = 1.05612$

f. $Y_2 = \bar{Y}_2 - \sqrt{\dfrac{C_\alpha}{\lambda_2(1 + b_1^2)}} = 0.69548$

$Y_1 = \bar{Y}_1 + b_1(Y_2 - \bar{Y}_2) = 1.90140$

$Y_1 = \bar{Y}_1 + b_1(Y_2 - \bar{Y}_2) = 1.54163$

g. $Y_2 = \bar{Y}_2 + \sqrt{\dfrac{C_\alpha}{\lambda_1(1 + b_2^2)}} = 0.80862$

h. $Y_2 = \bar{Y}_2 - \sqrt{\dfrac{C_\alpha}{\lambda_1(1 + b_2^2)}} = 0.94298$

$Y_1 = \bar{Y}_1 + b_2(Y_2 - \bar{Y}_2) = 1.78886$

$Y_1 = \bar{Y}_1 + b_2(Y_2 - \bar{Y}_2) = 1.65417$

Figure 15.7, using the data from Box 15.6. At points a and b, $Y_2 = \bar{Y}_2$; at points c and d, the ellipse is cut at $Y_1 = \bar{Y}_1$; at points e and f, the ellipse is cut by the major axis; and at points g and h, it is cut by the minor axis.

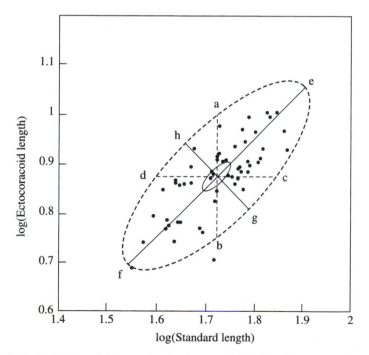

FIGURE 15.7 The 95% confidence region for the mean (solid ellipse), 95% equal frequency ellipse (dotted ellipse), with major and minor axes shown. The points a through h correspond to points computed in Box 15.6 to facilitate sketching the ellipses by hand.

All of these computations are performed by the correlation routine in the BIOMstat computer program. Altman (1978) presents another technique for plotting ellipses by hand by constructing 12 tangents around the ellipse.

The technique of principal axes is important in multivariate analysis, where instead of elliptical clouds of bivariate points, we encounter multidimensional clouds of observations describing hyperellipsoids in a multidimensional space. To simplify the description of these clouds of points, we calculate the principal axes through the hyperellipsoids. An important property that we have not yet emphasized is that the eigenvalues, which represent the variance along the principal axes, are such that as we determine successive principal axes representing the major axis, the second major axis, and so forth of the hyperellipsoid, we are successively finding the orthogonal axis that accounts for the greatest, second greatest, and successively smaller amounts of variation in the data. This technique is called *principal component analysis,* a branch of multivariate analysis (see Everitt and Dunn, 2001; Johnson and Wichern, 2007; Krzanowski, 1990; Manly, 2004).

15.9 Effect Size, Power, and Sample Size

When judging the strength of a correlation—that is, when comparing it to zero—the correlation coefficient is itself a convenient measure of effect size. It also is used as a standardized measure of effect size for the regression slope. As in the case of anova, Cohen (1988) provided subjective guidelines for interpreting effect sizes. He suggested that $r = 0.10$ would correspond to a small effect size, 0.3 to a medium effect, and 0.5 or larger to a large effect size. A re-examination of Figure 15.3 is useful to put these values into perspective. These guidelines seem very conservative because even a correlation of 0.5 is not very impressive and accounts for only 25% of the variance.

Some alternative measures of effect size have been proposed because the strength of a correlation is not a linear function of r. A popular index is r^2, the coefficient of determination discussed in Section 14.3. Others are $f^2 = \dfrac{r^2}{1 - r^2}$ (analogous to Cohen's measure of effect size introduced in Section 12.1. This implies that r^2 is equivalent to the ω^2 statistic also described in Section 12.1. Another measure is $E = 1 - \sqrt{1 - r^2}$, the complement of the coefficient of alienation mentioned in Section 15.2. These measures of effect size are equivalent in the sense that one can easily convert from one measure to another. If desired, the upper and lower limits for the confidence interval for a correlation coefficient can be transformed into limits for the other measures of effect size. Box 15.7 shows how to estimate the sample size needed in order to have sufficient power to detect a specified correlation. The procedure is analogous to that used in part 2 of Box 12.2 but simpler because a z-transformed correlation is approximately normally distributed.

BOX 15.7 Power and Sample Size

Sample size needed to detect an expected correlation

The required sample size to test the null hypothesis, H_0: $\rho = 0$, with a type I error rate of 1% and power of $P = 0.90$, given the smallest correlation, $|\rho_1|$, that one needs to detect can be found by the following formula:

$$n = \left(\frac{t_{\alpha[\infty]} + t_{2(1-P)[\infty]}}{\zeta_1}\right)^2 + 3$$

This is Expression (12.3.5) in Cohen (1988), where n is the desired sample size, $\alpha = $ type I error rate, $P = $ desired power if the absolute value of the true correlation is as small as ρ_1, and ζ_1 is the z-transformation of ρ_1. Because the critical t-values have infinite degrees of freedom, they can be looked up in either the bottom row of Statistical Table **B** for the t-distribution or in Table **A**, a table of areas for the normal distribution.

Again, we use the stickleback data. By consulting Statistical Table **B** for $t_{0.01[\infty]} = 2.576$ and $t_{0.2[\infty]} = 1.282$ and then computing $\zeta_1 = \tanh^{-1}\rho_1 = \tanh^{-1}(0.50) = 0.5493$, we find that

$$n = \left(\frac{2.576 + 1.282}{0.5493}\right)^2 + 3 = 52.33$$

We therefore need a sample of 53 paired observations to meet the specified and rather stringent conditions. The sample size available to us, 55 pairs of observations, slightly exceeded these requirements. We had sufficient power for our test of the null hypothesis $\rho_1 = 0.5$.

The use of the z^*-transformation yields a more accurate estimate of the required sample size, but one must employ an iterative solution because the z^*-transformation requires that one already has an estimate of the sample size:

$$n = \left(\frac{t_{\alpha[\infty]} + t_{2(1-P)[\infty]}}{\zeta_1^*}\right)^2 + 1$$

For the present example, it indicates that a sample size of about 51 is needed.

The graphs provided in Statistical Tables **KK** provide an easier, though more approximate, solution. To use them, select the graph corresponding to the desired type I error rate. The required sample size is that of the closest curve above the point intersecting the minimum correlation, ρ_1, along the abscissa and the desired power, P, along the ordinate. The graphs are based on the z^*-transformation but do not require an iterative solution.

A similar procedure is used to determine the sample sizes needed to detect a specified difference between two correlations:

$$n = \left(\frac{t_{\alpha[\infty]} + t_{2(1-P)[\infty]}}{|\zeta_1 - \zeta_2|}\right)^2 + 3$$

where n is the required sample size, $\alpha = $ type I error rate, $P = $ desired power if the absolute value of the true difference between the z-transformed correlations is as small as $|\zeta_1 - \zeta_2|$, and ζ_1 and ζ_2 are the z-transformations of ρ_1 and ρ_2, the correlations one

Box 15.7 (continued)

wishes to distinguish. Because the critical t-values have infinite degrees of freedom, they can be looked up in either the bottom row of Statistical Table **B** for the t-distribution or in Table **A**, a table of areas for the normal distribution.

As previously noted, using the z^*-transformation will yield a more accurate estimate of the required sample size, but an iterative solution must be employed because the z^*-transformation requires that one already has an estimate of the sample size:

$$n = \left(\frac{t_{\alpha[\infty]} + t_{2(1-P)[\infty]}}{|\zeta_1^* - \zeta_2^*|} \right)^2 + 1$$

The graphs provided in Statistical Tables **KK** provide an easier, though more approximate, solution. To use them, select the graph corresponding to the desired type I error rate. The required sample size is that of the closest curve above the point intersecting the minimum difference in z^*-values, $|\zeta_1^* - \zeta_2^*|$, along the abscissa and the desired power, P, along the ordinate. Note that the abscissa of these graphs are for differences in $\zeta_1^* - \zeta_2^*$ values, not for differences in $\rho_1 - \rho_2$ values because different sample sizes are required for differences in correlations, depending on their magnitudes. The graphs are based on the z^*-transformation but do not require an iterative solution.

When testing for the difference between two correlation coefficients, the observed difference, $r_1 - r_2$, is not useful as a measure of effect size because confidence intervals for correlations near zero are broader than those for larger correlations. Differences in terms of the r^2 (or ω^2), f^2, or E measures of effect size don't have this limitation, but their statistical distributions are complex. Cohen (1988) suggests using the difference between the z-transformed correlations, $q = z(r_1) - z(r_2)$ (the z^*-transformation could, of course, also be used). This coefficient has the advantage that it is approximately normally distributed and that its standard error is known; thus, its confidence limits can easily be computed (Box 15.3). A disadvantage is that differences in z-values are not in familiar units [although Cohen (1988) suggested the same subjective cutoffs for small, medium, and large effect sizes as for correlations]. Box 15.7 also shows how to estimate the sample size needed to detect that two z-transformed correlations differ by more than a specified amount. The approach is quite similar to that in Box 12.2, except that the solution is found directly, not iteratively, because the appropriate degrees of freedom for the z-transformation are infinite rather than dependent on the estimated sample size.

EXERCISES 15

15.1   Graph the following data in the form of a bivariate scatter diagram. Compute the correlation coefficient and set 95% confidence intervals to ρ for these data from a study of geographic variation in the aphid *Pemphigus populitransversus*. The values in the data

table represent locality means based on equal sample sizes for 23 localities in eastern
North America. The variables, extracted from Sokal and Thomas (1965), are expressed
in millimeters. Y_1 = tibia length; Y_2 = tarsus length. The correlation coefficient will
estimate correlation of these two variables over localities.

ANSWER: $r = 0.910$

Locality code number	Y_1	Y_2
1	0.631	0.140
2	0.644	0.139
3	0.612	0.140
4	0.632	0.141
5	0.675	0.155
6	0.653	0.148
7	0.655	0.146
8	0.615	0.136
9	0.712	0.159
10	0.626	0.140
11	0.597	0.133
12	0.625	0.144
13	0.657	0.147
14	0.586	0.134
15	0.574	0.134
16	0.551	0.127
17	0.556	0.130
18	0.665	0.147
19	0.585	0.138
20	0.629	0.150
21	0.671	0.148
22	0.703	0.151
23	0.662	0.142

IMPORTANT NOTE: The probability of this correlation, calculated in the usual way,
is likely to be understated because these data are spatially autocorrelated (Sokal and
Oden, 1978a, b); that is, individual localities are not truly independent observations,
but nearby localities tend to be similar. Clifford et al. (1989) present a method of
correcting the statistical significance of correlation coefficients based on spatially
autocorrelated data.

15.2 The data given below are the average weekly household spending, in British pounds, on
tobacco products and alcoholic beverages for each of 12 regions of Great Britain from
1999 to 2000. **(a)** Compute the correlation coefficient and test the null hypothesis $\rho = 0$ against the alternative hypothesis $\rho \neq 0$. **(b)** Are any outliers present in the data?
What is the correlation after such values are removed? **(c)** Compute the various mea-
sures of effect size.

Region	Alcohol	Tobacco
Northeast	14.4	6.2
Northwest	16.4	7.1
York & Humber	15.9	6.3
East Midlands	14.5	5.9
West Midlands	14.2	5.7
East	12.5	4.8
London	15.3	5.7
Southeast	13.8	5.1
Southwest	13.3	4.7
Wales	14.2	6.5
Scotland	14.7	8.3
Northern Ireland	12.4	8.7

From *Family Spending: A Report on the 1999–2000 Family Expenditure Survey*, Office for National Statistics.

15.3   The following data were extracted from a larger study of Brower (1959) on speciation in a group of swallowtail butterflies. Morphological measurements are in millimeters coded $\times$ 8. Compute the correlation coefficient separately for each species. Test for homogeneity of the four correlation coefficients.

ANSWER: For *Papilio rutulus*, $r = 0.1958$.

Species	Specimen number	Y_1 Length of 8th tergite	Y_2 Length of superuncus
Papilio	1	24.0	14.0
multicaudatus	2	21.0	15.0
	3	20.0	17.5
	4	21.5	16.5
	5	21.5	16.0
	6	25.5	16.0
	7	25.5	17.5
	8	28.5	16.5
	9	23.5	15.0
	10	22.0	15.5
	11	22.5	17.5
	12	20.5	19.0
	13	21.0	13.5
	14	19.5	19.0
	15	26.0	18.0
	16	23.0	17.0
	17	21.0	18.0
	18	21.0	17.0
	19	20.5	16.0
	20	22.5	15.5

Species	Specimen number	Y_1 Length of 8th tergite	Y_2 Length of superuncus
Papilio	21	20.0	11.5
rutulus	22	21.5	11.0
	23	18.5	10.0
	24	20.0	11.0
	25	19.0	11.0
	26	20.5	11.0
	27	19.5	11.0
	28	19.0	10.5
	29	21.5	11.0
	30	20.0	11.5
	31	21.5	10.0
	32	20.5	12.0
	33	20.0	10.5
	34	21.5	12.5
	35	17.5	12.0
	36	21.0	12.5
	37	21.0	11.5
	38	21.0	12.0
	39	19.5	10.5
	40	19.0	11.0
	41	18.0	11.5
	42	21.5	10.5
	43	23.0	11.0
	44	22.5	11.5
	45	19.0	13.0
	47	22.5	14.0
	48	21.0	12.5
Papilio	49	17.5	9.0
glaucus	50	19.0	9.0
	51	18.0	10.0
	52	19.0	8.0
	53	19.0	8.5
	54	21.0	8.5
	55	19.0	9.0
	56	21.0	9.5
	57	19.0	9.5
	58	17.5	9.0
	64	23.5	10.5
	65	16.0	9.5
	66	19.5	10.0
	67	17.0	7.0
	68	16.0	8.0
	69	18.0	9.0
	70	18.5	8.5

Species	Specimen number	Y_1 Length of 8th tergite	Y_2 Length of superuncus
Papilio	71	19 .0	8.5
glaucus	72	16.5	8.5
(continued)	73	18.5	9.5
	74	17.5	9.5
	75	19.5	8.0
	76	18.5	8.0
	77	19.0	9.0
	78	20.5	9.0
	79	17.0	9.5
	80	18.5	9.0
	81	16.5	9.0
	82	17.5	9.0
	83	19.0	9.0
	84	18.5	7.5
	85	18.5	8.5
	86	17.5	9.0
	87	18.5	9.5
	88	16.5	11.0
	89	23.0	9.0
	90	22.0	8.0
	91	18.0	10.0
	92	21.5	9.0
	93	18.5	9.0
	94	20.5	8.0
	95	21.0	10.0
	96	21.0	8.0
	97	22.5	9.5
	98	18.0	10.0
Papilio	99	17.5	11.5
eurymedon	100	17.5	11.5
	101	17.0	11.0
	102	18.5	11.5
	103	19.5	10.0
	104	16.0	11.5
	105	17.0	10.5
	106	17.5	10.5
	107	19.0	10.5
	108	19.0	12.0
	109	19.0	11.5
	110	21.0	11.0
	111	19.5	10.5
	112	19.5	10.5
	113	19.5	9.5

Species	Specimen number	Y_1 Length of 8th tergite	Y_2 Length of superuncus
Papilio	114	19.0	10.5
eurymedon	115	19.5	12.0
(continued)	116	21.5	11.5
	117	16.5	11.0
	118	21.5	13.5

15.4 Graph the data for *Papilio glaucus* in Exercise 15.3 in the form of a scatter diagram. (**a**) Compute the major and minor axes. (**b**) Draw the 95% equal frequency ellipse for the scatter diagram. (**c**) Draw the 95% confidence ellipse for the bivariate mean (μ_1, μ_2).

15.5 Test for the presence of association between tibia length and tarsus length in the data of Exercise 15.1 using (**a**) Olmstead and Tukey's corner test of association, (**b**) Kendall's coefficient of rank correlation, and (**c**) Spearman's rank correlation.

ANSWER: (**a**) $|S| = 27$.

15.6 The following four correlation coefficients are extracted from a paper by Clark (1941). They represent correlations between skull width and skull length in four samples of the deer mouse, *Peromyscus maniculatus sonoriensis,* from different localities. Can we consider the four sample r's to have come from a single population with parametric correlation ρ? If so, what is your estimate of ρ? Estimate and set confidence limits to the measure of effect size, q, for the difference between the first two correlations. (See the Important Note beneath Exercise 15.1.)

	n	r
Miller Canyon, Ariz.	71	0.68
Winslow, Ariz.	54	0.54
San Felipe, Calif.	83	0.56
Victorville, Calif.	57	0.43

15.7 The following table of data is from an unpublished morphometric study of the cottonwood, *Populus deltoides,* by T. J. Crovello. One hundred leaves from one tree were measured when fresh and after drying. The variables shown are fresh leaf width (Y_1) and dry leaf width (Y_2), both in millimeters. Calculate r. Discuss the implications of the findings.

ANSWER: $r = 0.974$.

Y_1	Y_2	Y_1	Y_2	Y_1	Y_2
90	88	94	87	86	80
88	87	83	81	98	90
55	52	90	88	100	98
100	95	90	84	70	65

Y_1	Y_2	Y_1	Y_2	Y_1	Y_2
86	83	91	90	100	96
90	88	88	86	105	98
82	77	98	94	95	88
78	75	98	94	95	90
115	109	89	85	93	89
100	95	90	86	95	97
110	105	105	102	95	90
84	78	95	90	104	99
76	71	100	93	81	78
100	97	104	99	85	83
110	105	84	78	108	103
95	90	85	81	112	106
99	98	89	92	96	90
104	100	118	110	98	93
92	92	104	97	83	78
80	82	105	100	103	98
110	106	113	108	107	104
105	97	106	100	72	67
101	98	87	83	105	100
95	91	100	96	80	75
80	76	108	103	99	95
103	97	101	97	108	102
108	103	92	90	86	84
113	110	84	85	94	92
90	85	89	85	106	104
97	93	120	111	98	104
107	106	96	86	100	96
112	115	97	94	95	90
101	98	86	83	108	102
95	91				

15.8 Buley (1936) studied the relative consumption of diatoms and dinoflagellates by the California sea mussel, *Mytilus californianus*. Test whether the four variables listed are concordant in their pattern of variation during the course of the study.

	Average numbers of diatoms and dinoflagellates			
	In plankton hauls		In mussel stomachs	
Date	Diatoms/ℓ seawater	Dinofl./ℓ seawater	Diatoms per mussel	Dinofl. per mussel
9/5	7,340	860	3,525	20,200
9/7	6,960	480	13,000	37,750
9/13	1,160	480	2,450	32,700

	Average numbers of diatoms and dinoflagellates			
	In plankton hauls		In mussel stomachs	
Date	Diatoms/ℓ seawater	Dinofl./ℓ seawater	Diatoms per mussel	Dinofl. per mussel
9/15	1,380	820	897	37,790
9/19	1,480	340	2,500	53,125
10/5	3,960	1,300	2,000	100,350
10/7	3,080	1,860	850	38,940
10/11	80	6,500	525	691,650
10/13	240	5,300	1,575	363,860
10/19	100	2,040	4,900	260,725
10/21	480	1,820	500	350,950
10/26	220	4,300	1,860	241,350
11/3	640	2,160	300	137,000
11/11	40	3,260	130	780,018
11/19	0	1,380	625	59,437
11/22	220	740	175	26,370
12/2	180	260	1,500	5,000
12/9	1,020	340	700	1,566
12/16	120	160	3,333	11,000
12/23	820	260	1,300	2,066
12/29	20	40	1,400	930
1/6	100	300	4,000	430
1/13	8,800	300	1,300	1,150
1/20	940	300	550	800
1/27	240	0	1,100	750
2/3	1,100	80	850	500
2/11	27,400	20	2,500	850
2/18	11,240	60	650	150
2/25	103,040	40	2,600	250
3/3	179,760	80	1,550	200
3/10	7,940	40	2,400	900
3/16	20	280	400	2,000
3/23	7,720	100	400	1,550
3/27	55,920	420	50	50
3/28	601,980	300	1,150	3,400
3/31	628,000	3,200	16,800	5,800
Total	1,663,740	40,220	80,345	3,271,557

Counts are averages for 25 liters of seawater and 5 mussel stomachs for each date.

16 Multiple and Curvilinear Regression

A corollary to the uses of regression for studying causation and prediction (as described in Section 14.6) is its application to reduce the unexplained error variance of a variable. Prior to regression on X, all of s_Y^2, the variance of Y, is "unexplained." After regression, the unexplained or residual variance, $s_{Y \cdot X}^2$, is smaller. We could visualize regressing the unexplained deviations $d_{Y \cdot X}$ on a third variable Z in order to remove more of the variance of Y. We would then obtain a new residual variance of Y, $s_{Y \cdot XZ}^2$, in which the portion of the fluctuations of Y determined by X and by Z has been removed. The new residual deviations $d_{Y \cdot XZ}$ would be those that neither X nor Z could explain. This process could go on indefinitely. In fact, this process of successively regressing a dependent variable Y on a series of independent variables, $X, Z, W, \ldots$, or more conventionally, $X_1, X_2, X_3, \ldots$, is analogous to the fundamental process of science. Scientists examine a phenomenon, study its variation, and successively reduce its unexplained variation as more and more of its causes are understood. We might state tongue-in-cheek that the aim of science is to reduce residual variances $\sigma_{Y \cdot X_{i \ldots}}^2$ to zero.

Regressing a variable Y on a series of independent variables could be done by successively regressing deviations in the manner we just described. Generally, however, if we suspect several variables of being functionally related to Y, we try to regress Y on all of them simultaneously. This technique is called **multiple regression.** It computes the least squares best-fitting linear function of two or more independent variables. Multiple regression is explained in Section 16.1, and detailed computations are illustrated for two independent variables. Tests of significance in multiple regression are discussed and illustrated in Section 16.2.

In Section 16.3, we discuss path analysis and structural equation modeling. These are important techniques that will help your understanding of various methods employed in this chapter. Path analysis is a method for studying the direct and indirect effects of one set of variables taken as causes on another set taken as effects. At the end of the section, there is a brief account of structural equation modeling. This approach, a generalization of path analysis, is illustrated with an example.

Just as we were able to extend the subject of regression to more than two independent variables, so are we able to study correlations between pairs of variables, other variables being held constant. In Section 16.4, we cover this topic, partial correlation, and a related topic, multiple correlation. In Section 16.5, we discuss how to select predictor variables in multiple-regression studies.

The next three sections deal with the general approach to computing multiple correlation and analysis of variance using matrix methods. Section 16.6 shows how to compute multiple-regression coefficients by matrix methods. In Section 16.7, we introduce the reader to the general linear hypothesis, which permits a general solution to all sorts of anova problems. We express analyses of variance in terms of multiple-regression problems, which can then be solved easily by matrix methods. Section 16.8 explains analysis of covariance and describes how to carry it out by matrix methods.

Section 16.9 introduces curvilinear regression for relationships between variables that cannot be expressed as straight lines. Effect size, power, and sample size in regression, bivariate as well as multiple, are covered in Section 16.10. Finally, in Section 16.11, we mention briefly some advanced topics in the general area of multivariate analysis, referring the reader to sources where these topics can be pursued in depth.

A full understanding of the material in this chapter and in some sections of Chapter 17 requires an elementary knowledge of matrix algebra. For those readers whose knowledge of matrices is rusty, and for those who have not been exposed to the subject in the past, we provide a minimalist primer in Appendix **B.** We urge you to review this material, as needed.

Although the distinction between Model I and II regression (see Section 14.2) also applies to the methods discussed in this chapter, techniques for handling Model II cases of multiple regression have not been fully developed. Therefore, exercise caution in interpreting the results obtained by the Model I techniques of this chapter when the assumptions of that model have not been met.

16.1 Multiple Regression: Computation

There are two main purposes of multiple-regression analysis. One is to establish a linear prediction equation that will enable a better prediction of a dependent variable Y than would be possible by any single independent variable X_j. Typically, one aims for that subset of all possible predictor variables that predicts an appreciable proportion of the variance of Y. We trade off adequacy of prediction against the cost of measuring more predictor variables. The second purpose of multiple regression is to estimate and fit a structural model to "explain" or account for variation in the observations of Y in terms of the independent variables X_j. Among a set of putative causal variables, which ones affect the dependent variable appreciably and what are the estimates of the relative magnitudes of the contributions of the independent variables? To examine these goals, consider the following example.

Table 16.1 shows data on air pollution in 41 American cities. The dependent variable is the annual arithmetic mean concentration of sulfur dioxide expressed in micrograms per cubic meter. When data were available, means for three years, 1969, 1970, and 1971, were averaged. The predictor variables are two human ecological variables—number of manufacturing enterprises employing 20 or more workers (X_2),

TABLE 16.1 Air Pollution (SO$_2$ Content of Air) in 41 U.S. Cities Associated with Six Environmental Variables

Cities	Y	X$_1$	X$_2$	X$_3$	X$_4$	X$_5$	X$_6$
Phoenix	10	70.3	213	582	6.0	7.05	36
Little Rock	13	61.0	91	132	8.2	48.52	100
San Francisco	12	56.7	453	716	8.7	20.66	67
Denver	17	51.9	454	515	9.0	12.95	86
Hartford	56	49.1	412	158	9.0	43.37	127
Wilmington	36	54.0	80	80	9.0	40.25	114
Washington	29	57.3	434	757	9.3	38.89	111
Jacksonville	14	68.4	136	529	8.8	54.47	116
Miami	10	75.5	207	335	9.0	59.80	128
Atlanta	24	61.5	368	497	9.1	48.34	115
Chicago	110	50.6	3344	3369	10.4	34.44	122
Indianapolis	28	52.3	361	746	9.7	38.74	121
Des Moines	17	49.0	104	201	11.2	30.85	103
Wichita	8	56.6	125	277	12.7	30.58	82
Louisville	30	55.6	291	593	8.3	43.11	123
New Orleans	9	68.3	204	361	8.4	56.77	113
Baltimore	47	55.0	625	905	9.6	41.31	111
Detroit	35	49.9	1064	1513	10.1	30.96	129
Minneapolis–St. Paul	29	43.5	699	744	10.6	25.94	137
Kansas City	14	54.5	381	507	10.0	37.00	99
St. Louis	56	55.9	775	622	9.5	35.89	105
Omaha	14	51.5	181	347	10.9	30.18	98
Albuquerque	11	56.8	46	244	8.9	7.77	58
Albany	46	47.6	44	116	8.8	33.36	135
Buffalo	11	47.1	391	463	12.4	36.11	166
Cincinnati	23	54.0	462	453	7.1	39.04	132
Cleveland	65	49.7	1007	751	10.9	34.99	155
Columbus	26	51.5	266	540	8.6	37.01	134
Philadelphia	69	54.6	1692	1950	9.6	39.93	115
Pittsburgh	61	50.4	347	520	9.4	36.22	147
Providence	94	50.0	343	179	10.6	42.75	125
Memphis	10	61.6	337	624	9.2	49.10	105
Nashville	18	59.4	275	448	7.9	46.00	119
Dallas	9	66.2	641	844	10.9	35.94	78
Houston	10	68.9	721	1233	10.8	48.19	103
Salt Lake City	28	51.0	137	176	8.7	15.17	89
Norfolk	31	59.3	96	308	10.6	44.68	116
Richmond	26	57.8	197	299	7.6	42.59	115
Seattle	29	51.1	379	531	9.4	38.79	164
Charleston	31	55.2	35	71	6.5	40.75	148
Milwaukee	16	45.7	569	717	11.8	29.07	123
Mean	30.049	55.763	463.10	608.61	9.4439	36.769	113.90
s	23.472	7.2277	563.47	579.11	1.4286	11.772	26.506

Notes:

Y − SO$_2$ content of air in micrograms per cubic meter.

X$_1$ − Average annual temperature in °F.

X$_2$ − Number of manufacturing enterprises employing 20 or more workers.

X$_3$ − Population size (1970 census); in thousands.

X$_4$ − Average annual wind speed in miles per hour.

X$_5$ − Average annual precipitation in inches.

X$_6$ − Average number of days with precipitation per year.

Data compiled from several government publications. Cities are ordered alphabetically by state.

and population size (X_3)—and four climatic averages for weather stations at these cities: average annual temperature (X_1), average annual wind speed (X_4), average annual precipitation (X_5), and average number of days with precipitation per year (X_6).

We can state the multiple-regression equation in two distinct but interrelated ways, the conventional and the standardized form. The conventional equation is

$$\hat{Y} = a + b_{Y1}.X_1 + b_{Y2}.X_2 + \cdots + b_{Yk}.X_k \tag{16.1}$$

where the estimate of the dependent variable $\hat{Y}$ is a function of k independent variables $X_1, X_2, \ldots, X_k$. A coefficient such as b_{Yj}. denotes the regression coefficient of Y on variable X_j that one would expect if all the other variables in the regression equation had been held constant experimentally. It is called a **partial regression coefficient.** The dot in the subscript of such coefficients separates the two labeled variables (dependent and independent variables) from other independent variables that are held constant. In those cases where the dot terminates the subscript, all other independent variables in the study are held constant. Note that to simplify the notation the symbol b_{Yj}. is used rather than $\hat{\beta}_{Yj}$. to indicate a sample estimate of β_{Yj}..

Figure 16.1 is a diagrammatic representation of a multiple regression of Y on two independent variables, X_1 and X_2. The shaded area represents the regression plane, which is a generalization of the regression line from Chapter 14. The plane is situated so that it minimizes the sum of squared vertical distances from the points to the plane. The partial regression coefficients determine the slope or tilt of the plane in the

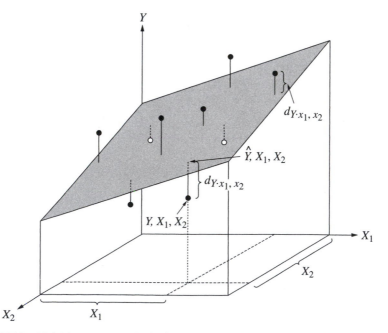

FIGURE 16.1 Multiple regression of Y on X_1 and X_2. Note that the regression line has become a plane.

X_1 and in the X_2 directions. Such coefficients express the rate of change of variable Y per unit of variable X_j with all other variables in the study held constant. Thus, if for the example in Table 16.1 we compute $b_{Y1 \cdot 2 \ldots 6}$, the partial regression coefficient yields the rate of change of SO_2 concentration as a function of average temperature (X_1), with variables X_2 through X_6 held constant. The coefficient would be expressed as $\mu g/m^3/°F$. Expression (16.1) is a direct extension of the by now familiar regression equation $\hat{Y} = a + bX$ from Chapter 14.

In the standardized form of the equation, we transform the variables to standard deviates by subtracting means and dividing by the standard deviation

$$y' = \frac{Y - \bar{Y}}{s_Y}$$

and

$$x'_j = \frac{X_j - \bar{X}_j}{s_{X_j}}$$

In standardized form, the multiple-regression equation is thus

$$\hat{y}' = b'_{Y1 \cdot} x' + b'_{Y2 \cdot} x'_2 + \cdots + b'_{Yk \cdot} x'_k \tag{16.2}$$

Here the coefficients b'_{Yj} are **standard partial regression coefficients,** also known as *b-primes*, *beta coefficients*, and *beta weights*. They are in a simple relation to the conventional partial regression coefficients b_{Yj}:

$$b'_{Yj \cdot} = b_{Yj \cdot} \frac{s_{X_j}}{s_Y} \tag{16.3}$$

Instead of expressing a rate of change in the original measurement units as in $b_{Yj \cdot}$, the standard partial regression gives the rate of change in standard deviation units of Y per one standard deviation unit of X_j (all other X variables held constant, of course). One advantage of standard partial regression coefficients is that their magnitudes can be compared directly to show the relative standardized strengths of the effects of several independent variables on the same dependent variable. This property eliminates the effect of differences in measurement scale for different independent variables. Standard partial regression implies that the range of variation in the independent variables is comparable to that expected in future studies.

In standard form, the familiar simple linear regression equation would be

$$\hat{y}' = b'_{Y \cdot X} x' \tag{16.4}$$

We saw in Section 15.2 that the standard regression coefficient $b'_{Y \cdot X}$ equals the correlation coefficient r_{XY} [see Expression (15.7)]. Thus, another way of looking at a correlation coefficient is as the slope in standard deviation units of Y on X or X on Y. Standard partial regression coefficients, however, are not equal to the partial *correlation* coefficients, which we will discuss in Section 16.4.

Many multiple-regression analyses require computation of both conventional and standard partial regression coefficients, and most computer programs routinely furnish both types of coefficients. Once the standard deviations of the variables are known, one type of coefficient is easily transformed into the other. Computing conventional and standard partial regression coefficients directly, however, requires different equations.

The computation of the b_{Yj}. is based on the same principles as those in ordinary regression. As in simple linear regression, we want the coefficients that minimize the sum of squared errors $Y - \hat{Y}$. The derivation of these coefficients involves the calculus of vector variables and results in a series of simultaneous linear equations. These equations are called **normal equations,** and they can be written as follows for the general case:

$$
\begin{array}{ccccc}
X_1 & X_2 & \cdots & X_k & Y \\
\end{array}
$$

$$
\begin{array}{l}
X_1 \quad s_1^2 b_{Y1}. + s_{12} b_{Y2}. + \cdots + s_{1k} b_{Yk}. = s_{1Y} \\
X_2 \quad s_{12} b_{Y1}. + s_2^2 b_{Y2}. + \cdots + s_{2k} b_{Yk}. = s_{2Y} \\
\quad \vdots \qquad \vdots \qquad \vdots \qquad \quad \vdots \qquad \vdots \\
X_k \quad s_{1k} b_{Y1}. + s_{2k} b_{Y2}. + \cdots + s_k^2 b_{Yk}. = s_{kY}
\end{array}
\tag{16.5}
$$

Note the regularities in these equations, presented so that the variables and coefficients line up as in a table or matrix. The rows of the table are labeled for the independent variables, as are the first k columns of the table. The last column is labeled for the dependent variable Y. The partial regression coefficients are lined up in columns, and there are k such values of b_{Yj}., which are the unknowns in a normal equation. The coefficients in the equations are variances, s_i^2, and covariances, s_{ij}, of the independent variables $X_1, \ldots , X_k$. The right-hand sides of the equations are covariances between variables X_i and Y. The variances s_i^2 along the diagonal correspond to the covariances of each variable with itself. Because there are k equations for k unknowns, we can solve for the regression coefficients using a variety of standard procedures.

When we want standardized partial regression coefficients, we can set up an analogous set of normal equations by replacing the b_{Yj}. with b'_{Yj}., setting all variances to unity, and replacing all covariances by correlation coefficients. Later we show how to calculate standard partial regression coefficients in problems with only two independent variables. This is simple enough to be done with a pocket calculator.

Although eventually we will employ all six predictor variables in the air pollution example of Table 16.1, for this demonstration we shall limit ourselves to accounting for the amount of pollution Y in terms of two variables—average temperature (X_1) and number of manufacturing enterprises (X_2). Problems involving three or more independent variables nowadays are almost never carried out without a computer, and the investigator usually need not be concerned whether the program solves the normal equations for variances and covariances or for correlations.

Computation of the multiple regression on two independent variables is explained in Box 16.1. The normal equations for the two variables in standardized form are:

$$b'_{Y1\cdot} + r_{12}b'_{Y2\cdot} = r_{1Y}$$

$$r_{12}b'_{Y1\cdot} + b'_{Y2\cdot} = r_{2Y} \qquad (16.6)$$

BOX 16.1 **Multiple Regression for Two Independent Variables**

Data from Table 16.1, where means and standard deviations are also furnished.
We propose to regress SO_2 content of air, Y, on average temperature, X_1, and number of
manufacturing enterprises, X_2.

1. Compute the correlations among the three variables concerned following the procedures of Box 15.2. We obtain

   $$r_{12} = -0.1900$$
   $$r_{1Y} = -0.4336$$
   $$r_{2Y} = 0.6448$$

2. Formulas for the standard partial regression coefficients are as shown below. These
   are solutions to the normal equations [Expression (16.6)]. As r_{12} approaches unity
   the denominator must be computed with more precision.

   $$b'_{Y1\cdot2} = \frac{(r_{1Y} - r_{2Y}r_{12})}{(1 - r_{12}^2)} = \frac{[-0.4336 - (0.6448)(-0.1900)]}{[1 - (-0.1900)^2]} = -0.3227$$

   $$b'_{Y2\cdot1} = \frac{(r_{2Y} - r_{1Y}r_{12})}{(1 - r_{12}^2)} = \frac{[0.6448 - (-0.4336)(-0.1900)]}{0.9639} = 0.5835$$

3. The multiple-regression equation in standard format is

   $$\hat{y}' = -0.3227x'_1 + 0.5835x'_2$$

4. Conventional partial regression coefficients, based on Expression (16.3):

   $$b_{Y1\cdot2} = b'_{Y1\cdot2}\frac{s_Y}{s_{X_1}} = -0.3227\left(\frac{23.472}{7.2277}\right) = -1.0480$$

   $$b_{Y2\cdot1} = b'_{Y2\cdot1}\frac{s_Y}{s_{X_2}} = 0.5835\left(\frac{23.472}{563.47}\right) = 0.02431$$

5. The Y-intercept is

   $$a = \bar{Y} - b_{Y1\cdot2}\bar{X}_1 - b_{Y2\cdot1}\bar{X}_2$$
   $$= 30.049 - (-1.0480)(55.763) - (0.02431)(463.10)$$
   $$= 77.231$$

   The prediction equation is

   $$\hat{Y} = 77.231 - 1.0480X_1 + 0.02431X_2$$

We can solve for the *b*-primes by the customary algebraic techniques for solving simultaneous equations. Box 16.1 gives explicit solutions for the standard partial regression coefficients in terms of the known correlation for two independent variables. These solutions are based on the determinantal method for solving simultaneous equations. Solution becomes more difficult as r_{12} approaches ± 1. In step **2**, we obtain $b'_{Y1\cdot2} = -0.3227$ and $b'_{Y2\cdot1} = 0.5835$. Because these are standard partial regression coefficients, they express the average change in standard deviation units of the dependent variable Y for one standard deviation unit of each independent variable, the other one being held constant. Thus, if we consider towns with the same number of manufacturing enterprises (variable X_2 held constant), $b'_{Y1\cdot2} = -0.3227$ means that for an increase of one standard deviation of annual temperature, there will be a decrease in the SO_2 pollution by roughly one-third of its standard deviation. By contrast, the number of factories increases pollution; when cities are compared at the same temperature (average annual temperature, X_1, held constant), an increase of one standard deviation in number of enterprises will result in a corresponding increase of 0.5835 standard deviations of SO_2. The increase in SO_2 from effects of manufacturing is greater in absolute magnitude than the decrease from the effects of temperature.

Because the standard partial regression coefficients are free of the original measurement scale, they are more convenient for estimating the relative importance of the influence of each of the set of independent variables being considered in the equation. In the next section, we will discuss testing of the difference between two *b*-primes. In Box 16.1 we write the multiple-regression equation in step **3** in standardized form.

$$\hat{y}' = -0.3227x'_1 + 0.5835x'_2$$

This is the equation that would have been obtained if all three variables had been transformed to standard deviates before the computation. Workers concerned with best estimates of the population standard partial regression coefficients, $\beta'_{Yj\cdot}$, may wish to apply a correction suggested by Mayer and Younger (1976).

Next, we obtain the conventional partial regression coefficients by inverting Expression (16.3), shown as step **4** of Box 16.1, which yields $b_{Y1\cdot2} = -1.0480$ and $b_{Y2\cdot1} = 0.02431$. Note that $b_{Y1\cdot}$ and $b_{Y2\cdot}$ are in original measurement units. Thus, $b_{Y1\cdot2} = -1.0480$ $\mu g/m^3/°F$. To complete the multiple-regression equation, we replace Y, X_1, and X_2 by $\overline{Y}$, $\overline{X}_1$, and $\overline{X}_2$, then solve for a, the Y-intercept. We obtain $a = 77.231$ in step **5** and can now write out the entire multiple-regression equation in terms of original measurement units as

$$\hat{Y} = 77.231 - 1.0480X_1 + 0.02431X_2$$

Note that unless the independent variables are measured in comparable units, you cannot infer the importance of an effect from conventional partial regression coefficients; numerically, X_1 now seems to have a greater effect than X_2 does, but we have seen that exactly the converse is true when *b*-primes are considered.

Next, we turn to a computation involving three independent variables. In addition to X_1 and X_2, we will include variable X_3 from Table 16.1. This variable is population size (to the nearest thousand) in the 1970 census. We aim to predict SO_2 content of the air in terms of this variable and the previous two. Although formulas for multiple-regression coefficients involving three or more independent variables could be given, they are much too tedious to evaluate without a computer program. We shall defer (to Section 16.6) consideration of how the computation is carried out and shall first focus on examining the results for three independent variables shown in the following.

The multiple-regression equation in standard format is

$$\hat{y}' = -0.1807x'_1 + 1.7111x'_2 - 1.1522x'_3$$

In conventional format (i.e., all measurements in original scale), the prediction equation is

$$\hat{Y} = 58.183 - 0.5868X_1 + 0.07128X_2 - 0.04670X_3$$

Inspection of the standard multiple-regression equation reveals three points of interest. First, standard partial regression coefficients can be greater than 1 (one standard deviation change in the independent variable may effect more than one standard deviation change in the dependent variable). Second, and more important, the value of b_{Yj} for Y and X_j depends on which other independent variables have been held constant. Introducing population size X_3 into the equation not only changed the partial regression coefficients on temperature and number of manufacturing establishments, but also altered their relative magnitudes. The reason for this effect is that the independent variables are correlated with each other. Later in this chapter, we will see exactly how this correlation affects the standard partial regression coefficients. For now, it is important to remember that as we add or drop independent variables, we have to recompute the entire problem with a new set of normal equations—a tedious task impractical without a computer when there are more than three independent variables.

The third point is that the sign of a partial regression coefficient need not be the same as that of the correlation between the two variables concerned. Note that $r_{3Y} = 0.4938$, while $b'_{Y3\cdot1,2} = -1.1522$. Such a disparity is not possible in simple linear regression, where the sign of the covariance in the numerator determines the sign of both the correlation and the regression coefficients. Again, this phenomenon depends on the pattern of correlations among the independent variables and will be discussed later in this chapter.

A corollary to the last two points is that partial regression coefficients may differ in sign and magnitude, depending on which independent variables are held constant. Readers with an appreciation for graphic rather than arithmetic relations may find the illustration in Mullett (1972) instructive. One can plot the three variables

Y	X_1	X_2
-5	-4	3
-7	-2	3
-1	-2	1
-3	0	1
3	0	-1
1	2	-1
7	2	-3
5	4	-3

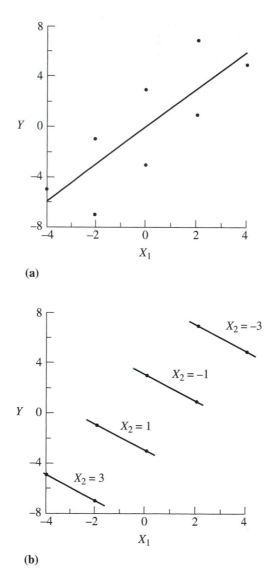

(a)

(b)

FIGURE 16.2 Comparison of simple linear and partial regression in an artificial data set.
(a) Simple linear regression of Y on X_1. **(b)** The partial regression of Y on X_1, with variable X_2 held constant for the four distinct values of X_2. The observations on which this figure is based are given in the text.

as simple linear regressions of Y on X_1 to obtain the graph of Figure 16.2a, which shows the least squares equation $\hat{Y} = 1.5X_1$ fitted to these data. If we compute the partial regression coefficient of Y on X_1 with X_2 fixed, we obtain a slope of $b_{Y1\cdot2} = -1.0$. We can easily compute the four regression lines of Y on X_1, corresponding to the four values of X_2. Line segments representing these regression lines are shown in Figure 16.2**b**. Note that the slopes of the partial regression coefficients are now negative, while the overall slope of Y on X_1 is positive. The least squares fit of Y on X_1 given X_2 constant is thus quite different from the least squares fit of Y on X_1 without any second independent variable held constant.

Returning to the air pollution example, the new equation based on three variables shows that number of factories still affects SO_2 content of air most strongly but that temperature is no longer important when factories *and* population size are held constant. However, the new variable, population size, now affects SO_2 importantly *in a negative direction*. This seems counterintuitive at first. Should we not expect more pollution in more populous cities? Remember, however, that this is a *partial* regression. In cities with the same average temperature and number of factories (i.e., with these variables held constant), an increase in population size probably implies a larger residential area, hence *less* SO_2 emissions per unit volume of air. We will return to the interpretation of this case later. The conventional multiple-regression equation permits prediction of the SO_2 content of air for a city with given values for the three variables. Thus, for Phoenix, where the average annual temperature is 70.3°F and there are 213 manufacturing establishments as well as 582,000 inhabitants (see Table 16.1), we would expect

$$\hat{Y} = 58.183 - (0.5868)(70.3) + (0.07128)(213) - (0.04670)(582)$$

$$= 4.9 \ \mu g \ SO_2/m^3$$

The observed value is $10 \ \mu g \ SO_2/m^3$. As in simple linear regression, some observations are closer to the predicted values, others farther away from them. The goodness of fit will be estimated by means of the unexplained sum of squares, as shown in Section 16.2.

The BIOMstat package of statistical computer programs computes standard and conventional partial regression coefficients, the Y-intercept, and other statistics necessary for interpreting the multiple-regression equation. Most computer programs solve the normal equations in one of two ways: either in the form already familiar (i.e., solving for b-primes when correlation coefficients are given); or in another form in which conventional partial regression coefficients b_{Yj}. replace the standard ones, sums of products $\Sigma x_i x_j$ and $\Sigma x_i y$ replace the correlations r_{ij} and r_{iY}, respectively, and sums of squares Σx_i^2 replace r_{ii} (equals unity) in the diagonal entries. Thus,

$$
\begin{array}{ccccccc}
& X_1 & X_2 & \cdots & X_k & Y \\
X_1 & b_{Y1}{\cdot}\sum x_1^2 & + \ b_{Y2}{\cdot}\sum x_1 x_2 & + \cdots + & b_{Yk}{\cdot}\sum x_1 x_k & = \sum x_1 y \\
X_2 & b_{Y1}{\cdot}\sum x_1 x_2 & + \ b_{Y2}{\cdot}\sum x_2^2 & + \cdots + & b_{Yk}{\cdot}\sum x_2 x_k & = \sum x_2 y \\
\vdots & \vdots & \vdots & \vdots & \vdots & \vdots \\
X_k & b_{Y1}{\cdot}\sum x_1 x_k & + \ b_{Y2}{\cdot}\sum x_2 x_k & + \cdots + & b_{Yk}{\cdot}\sum x_k^2 & = \sum x_k y
\end{array}
\tag{16.7}
$$

Expression (16.7) is simply Expression (16.5) with every term in the latter multiplied by $n - 1$ to change variances and covariances into sums of squares and products.

Expression (16.7) can be expressed using matrix algebra as

$$\mathbf{Ab} = \mathbf{c} \tag{16.8}$$

where $\mathbf{A}$ is the sum of squares–sum of products matrix (a square, symmetrical matrix of dimension k), $\mathbf{b}$ is the column vector of unknown partial regression coefficients, and $\mathbf{c}$ is the column vector of sums of products of the k independent variables with the dependent variable Y. Those of our readers who studied matrix algebra long ago or who never studied it at all should not panic at this stage. We furnish (in Appendix **B**) an introduction to matrix algebra that briefly covers all the material required for an understanding of the several sections of this chapter that employ matrix methods.

To solve for $\mathbf{b}$, we rewrite Expression (16.8) as

$$\mathbf{b} = \mathbf{A}^{-1}\mathbf{c} \tag{16.9}$$

Thus, we can find the regression coefficients by premultiplying $\mathbf{c}$ by the inverse of the sum of squares—sum of products matrix. The solution by Expression (16.9) is useful because, as you will see in the next section, the elements of the inverse matrix $\mathbf{A}^{-1}$ are needed to compute standard errors for most statistics in multiple regression. These elements of $\mathbf{A}^{-1}$ are usually called **Gaussian multipliers** and are symbolized by c_{ii} and c_{ij} for diagonal and off-diagonal elements, respectively. Because $\mathbf{A}^{-1}$ is also a square, symmetrical matrix of dimension k, there will be k diagonal elements, $c_{11}, \ldots, c_{kk}$, and $k(k - 1)/2$ different off-diagonal elements, $c_{ij} = c_{ji}$. Because calculating the inverse of a sizable matrix is practical only by computer, we will not concern ourselves here with the details of the computation.

16.2 Multiple Regression: Hypothesis Tests

In Section 14.4, we learned how to compute a sum of squares for the deviations $\hat{y} = \hat{Y} - \bar{Y}$ of the predicted values from the mean in a linear regression. This explained sum of squares, $\Sigma \hat{y}^2$, was used in hypothesis testing of the regression by analysis of variance; in Section 15.2, we saw that the ratio $\Sigma \hat{y}^2/\Sigma y^2$ expresses the proportion of the variation of Y due to variation in X. This proportion was called the coefficient of determination, which in simple linear regression was shown to be r_{XY}^2, the square of the correlation coefficient between the two variables.

We could predict a value of $\hat{Y}$ for each pair of values of X_1, X_2 of the example in Section 16.1, then evaluate the deviation $\hat{y} = \hat{Y} - \bar{Y}$ and compute $\Sigma \hat{y}_{1,2}^2$, the sum of squares of Y explained by variables X_1, and X_2. Such an explained sum of squares can obviously be computed for any k-tuple of independent variables $X_1, X_2, \ldots, X_k$. The ratio $\Sigma \hat{y}_{1\ldots k}^2/\Sigma y^2$ is known as the **coefficient of multiple determination.** It is generally symbolized by $R_{Y \cdot 1 \ldots k}^2$ and is an estimate of the proportion of the variation of Y jointly explained by variables X_1 through X_k. A more practical way of computing this coefficient is by the following formula, which we state without proof:

$$R_{Y \cdot 1 \ldots k}^2 = r_{1Y}b'_{Y1} + r_{2Y}b'_{Y2} + \cdots + r_{kY}b'_{Yk}. \tag{16.10}$$

Thus, for the two independent variables in Box 16.1, we get

$$R^2_{Y \cdot 1,2} = r_{1Y} b'_{Y1 \cdot 2} + r_{2Y} b'_{Y2 \cdot 1}$$

$$= (-0.4336)(-0.3227) + (0.6448)(0.5835) = 0.5162$$

This value of R^2 shows that almost 52% of the total variation of SO_2 in the air is explained by average annual temperature and number of manufacturing establishments (in the population of cities from which our 41 cities are a sample). Remember that the hypothesis test for linear regression was carried out by an anova and that in Section 15.2 we learned one of several ways to compute the explained and unexplained sums of squares. We showed there that $\Sigma \hat{y}^2 = r^2_{XY} \Sigma y^2$ and that $\Sigma d^2_{Y \cdot X} = (1 - r^2_{XY}) \Sigma y^2$. These formulas can easily be derived from Expression (15.6), $r^2_{12} = \Sigma \hat{y}^2_2 / \Sigma y^2_2 = \Sigma \hat{y}^2 / \Sigma y^2$. In a similar manner, we can compute the sum of squares of Y due to regression on $X_1, \ldots, X_k$:

$$\sum \hat{y}^2_{\cdot 1 \ldots k} = R^2_{Y \cdot 1 \ldots k} \sum y^2 \tag{16.11}$$

as well as the sum of squares unexplained by these variables:

$$\sum d^2_{Y \cdot 1 \ldots k} = (1 - R^2_{Y \cdot 1 \ldots k}) \sum y^2 \tag{16.12}$$

The square root of the coefficient of multiple determination is the coefficient of multiple correlation (see Section 16.4).

Now let us use these coefficients in an **overall hypothesis test of the multiple regression.** We proceed as in the analysis of variance for linear regression in Section 14.4. What will be the degrees of freedom for $\Sigma \hat{y}^2_{\cdot 1 \ldots k}$, the sum of squares of Y explained by variables X_1 through X_k? In the case of simple linear regression, we counted one degree of freedom for $\Sigma \hat{y}^2$ due to regression on X. Now that we have k separate X-variables, we allocate k degrees of freedom to $\Sigma \hat{y}^2_{\cdot 1 \ldots k}$. What remains is $\Sigma d^2_{Y \cdot 1 \ldots k}$ with $n - k - 1$ df, which agrees with the earlier instance in which the unexplained SS had $n - 2$ df, because in that case $k = 1$ (see Section 14.4). We will use these facts to test the null hypothesis of no multiple regression of SO_2 on temperature and manufacturing enterprises in Box 16.1.

Earlier in this section, we computed $R^2_{Y \cdot 1,2} = 0.5162$. Hence $1 - R^2_{Y \cdot 1,2} = 0.4838$. The total sum of squares for SO_2 is 22,038 (computation not shown). We can therefore obtain the explained and unexplained sums of squares by simple multiplication using Expressions (16.11) and (16.12): $\Sigma \hat{y}^2_{\cdot 1,2} = 0.5162 \times 22{,}038 = 11{,}376$ and $\Sigma d^2_{Y \cdot 1,2} = 0.4838 \times 22{,}038 = 10{,}662$. We display the computation in conventional anova format:

Source of variation	df	SS	MS	F_s	P
Explained—due to regression on X_1 and X_2	2	11,376	5688.00	20.27	1.02×10^{-6}
Unexplained—error around regression plane	38	10,662	280.58		

In view of the minute P-value, we can safely reject the null hypothesis of no dependence on temperature and number of manufacturing enterprises.

In multiple-regression analysis, we will want to repeatedly carry out tests such as the one just completed. Even for one and the same example, we will want to test the effects of a single independent variable, of two variables, and of k variables. A look at the F_s value just obtained will show us how we can simplify the computation further. We computed

$$F_s = \frac{\sum \hat{y}^2_{\cdot 1 \ldots k}/k}{\sum d^2_{Y \cdot 1 \ldots k}/(n - k - 1)}$$

From Expressions (16.11) and (16.12), however,

$$F_s = \frac{R^2_{Y \cdot 1 \ldots k} \sum y^2/k}{(1 - R^2_{Y \cdot 1 \ldots k}) \sum y^2/(n - k - 1)}$$

$$= \frac{R^2_{Y \cdot 1 \ldots k}/k}{(1 - R^2_{Y \cdot 1 \ldots k})/(n - k - 1)} \tag{16.13}$$

Thus, for hypothesis tests we need only compute the coefficients of multiple determination and their complements, the **coefficients of multiple nondetermination,** and use their ratio (properly weighted by degrees of freedom) to compute their probability using the $F_{\alpha[k, n - k - 1]}$-distribution.

These calculations are shown in step **1** of Box 16.2. Note that the values of the coefficients of determination in Box 16.2 were obtained by computer and that, because of rounding errors, the value for variables X_1 and X_2 differs slightly from that calculated earlier in this section immediately beneath Expression (16.10). Variable X_2, number of manufacturing enterprises, has the highest correlation of any independent variable with Y, the SO_2 content of air ($r_{2Y} = 0.6448$). We might try to explain Y entirely in terms of X_2. The coefficient of determination $r^2_{2Y} = 0.41573$ and the test of the hypothesis that it is zero yields a minute P-value by the F-test shown in Box 16.2. When we regress Y on X_1 as well as X_2, we obtain $R^2_{Y \cdot 1,2} = 0.51611$, an appreciable increase over the 0.41573 due to X_2 alone. Again, an F-test for this coefficient yields a very small P-value. In Box 16.2, we also show that including X_3 among the independent variables raises R^2 to 0.61255; including all available independent variables ($X_1 - X_6$) further raises the coefficient of determination to 0.66951. All regressions tested are highly significant.

What hypotheses are we testing with these tests? There are two mathematically equivalent ways of looking at this matter. We test $H_0: \rho^2_{Y \cdot 1 \ldots k} = 0$, which is to say that the parametric R^2 equals zero and none of the variation of Y can be explained by $X_1, \ldots, X_k$. Alternatively, we test $H_0: \beta_{Y1} = \beta_{Y2} = \cdots = \beta_{Yk} = 0$, that is, that all of the partial regression coefficients are equal to zero.

The tests carried out in step **1** of Box 16.2 raise new questions that are almost self-evident and lead to *hypothesis tests for additional independent variables.*

BOX 16.2 Hypothesis Testing in Multiple Regression

Based on data in Table 16.1, analyses in Box 16.1, and standard errors in Box 16.3.

$$n = 41 \qquad \sum y^2 = 22038$$

$$r^2_{2Y} = 0.41573$$

$$R^2_{Y \cdot 1,2} = 0.51611$$

$$R^2_{Y \cdot 1,2,3} = 0.61255$$

$$R^2_{Y \cdot 1 \ldots 6} = 0.66951$$

The coefficient of multiple determination for two independent variables ($R^2_{Y \cdot 12}$) could be obtained from Box 16.1 by means of Expression (16.10). Its value here, as well as the value of the coefficient for all six independent variables ($R^2_{Y \cdot 1 \ldots 6}$), were calculated by computer, which explains any discrepancies in terminal digits between these values and those calculated earlier in the text. The tests in this box are for a priori selected sets of variables.

1. Testing the null hypothesis for the regression on k independent variables. Compare

$$F_s = \frac{R^2_{Y \cdot 1 \ldots k}/k}{(1 - R^2_{Y \cdot 1 \ldots k})/(n - k - 1)}$$

with $F_{\alpha[k, n-k-1]}$.

$k = 1$; regression of Y on X_2:

$$F_s = \frac{r^2_{2Y}/1}{(1 - r^2_{2Y})/(41 - 1 - 1)} = \frac{0.41573}{0.58427/39} = 27.750, P = 5.36 \times 10^{-6}$$

$k = 2$; regression of Y on X_1 and X_2:

$$F_s = \frac{R^2_{Y \cdot 1,2}/2}{(1 - R^2_{Y \cdot 1,2})/(41 - 2 - 1)} = \frac{0.51611/2}{0.48389/38} = 20.265, P = 1.02 \times 10^{-6}$$

$k = 3$; regression of Y on X_1, X_2, and X_3:

$$F_s = \frac{R^2_{Y \cdot 1,2,3}/3}{(1 - R^2_{Y \cdot 1,2,3})/(41 - 3 - 1)} = \frac{0.61255/3}{0.38745/37} = 19.498, P = 9.49 \times 10^{-8}$$

$k = 6$; regression of Y on $X_1 \ldots X_6$:

$$F_s = \frac{R^2_{Y \cdot 1 \ldots 6}/6}{(1 - R^2_{Y \cdot 1 \ldots 6})/(41 - 6 - 1)} = \frac{0.66951/6}{0.33049/34} = 11.480, P = 5.42 \times 10^{-7}$$

We can reject the null hypotheses of there being no linear relationship between the dependent variable and the predictor variable X_2, and by X_2 in combination with other variables with a high degree of confidence. We conclude that number of manufacturing concerns, X_2, can be used to predict SO_2 content in air singly and even better when in combinations with other variables such as average temperature, X_1, and population size, X_3. Remember that adding more variables will, of course, improve prediction but we test below whether the increases are better than one would expect by chance.

Box 16.2 (continued)

2. Testing the increase in proportion of variance determined as the number of independent variables is augmented from k_1 to k_2. Compare

$$F_s = \frac{(R^2_{Y\cdot 1\ldots k_2} - R^2_{Y\cdot 1\ldots k_1})/(k_2 - k_1)}{(1 - R^2_{Y\cdot 1\ldots k_2})/(n - k_2 - 1)}$$

with $F_{\alpha[k_2-k_1, n-k_2-1]}$.

Testing the increment in determination due to addition of X_1 over determination by X_2 alone; $k_2 = 2$, $k_1 = 1$:

$$F_s = \frac{(R^2_{Y\cdot 1,2} - r^2_{2Y})/(2 - 1)}{(1 - R^2_{Y\cdot 1,2})/(41 - 2 - 1)} = \frac{(0.51611 - 0.41573)/1}{(1 - 0.51611)/38} = 7.883, P = 0.00783$$

Testing the increment in determination due to addition of X_3 over joint determination by X_1 and X_2; $k_2 = 3$, $k_1 = 2$:

$$F_s = \frac{(R^2_{Y\cdot 1,2,3} - R^2_{Y\cdot 1,2})/(3 - 2)}{(1 - R^2_{Y\cdot 1,2,3})/(41 - 3 - 1)} = \frac{(0.61255 - 0.51611)/1}{(1 - 0.61255)/37} = 9.210, P = 0.00439$$

Testing the increment in determination due to addition of X_4, X_5, and X_6 over joint determination by X_1, X_2, and X_3; $k_2 = 6$, $k_1 = 3$:

$$F_s = \frac{(R^2_{Y\cdot 1\ldots 6} - R^2_{Y\cdot 1,2,3})/(6 - 3)}{(1 - R^2_{Y\cdot 1\ldots 6})/(41 - 6 - 1)} = \frac{(0.66951 - 0.61255)/1}{(1 - 0.66951)/38} = 1.953, P = 0.140$$

We conclude that the null hypotheses of no increased determination by first adding X_1, then X_3, should both be rejected because chance increases of the observed magnitude are very unlikely. However, the small amount of added explained variation produced by the subsequent addition of variables X_4, X_5, and X_6 would occur 14% of the time by chance alone if the null hypothesis were true, leading to acceptance of the null hypothesis of no increased explanation by these three independent variables.

For the following tests, additional quantities are needed. All the examples are from the regression of Y on X_1 and X_2, as shown in Box 16.1. The unexplained standard deviation

$$s_{Y\cdot 1,2} = \left[\frac{(1 - R^2_{Y\cdot 1,2})\sum y^2}{(n - k - 1)}\right]^{1/2} = \left[\frac{(1 - 0.51611)22038}{38}\right]^{1/2} = 16.7520$$

Also needed are the Gaussian multipliers. Their computation for this simple case with two independent variables is shown in the text. For larger problems, one needs to invert the sum of squares–sum of products matrix of the independent variables. In this case, $c_{11} = 4.964,929,584 \times 10^{-4}$, $c_{22} = 8.168,974,959 \times 10^{-8}$, and $c_{12} = 1.210,292,356 \times 10^{-6}$.

3. Standard error of the partial regression coefficient:

$$s_{b_{Y1\cdot}} = s_Y.\sqrt{c_{11}} = 16.7520\sqrt{4.964,929,584 \times 10^{-4}} = 0.37327$$

$$s_{b_{Y2\cdot}} = s_Y.\sqrt{c_{22}} = 16.7520\sqrt{8.168,974,959 \times 10^{-8}} = 0.004,788,0$$

Box 16.2 (continued)

4. Testing the partial regression coefficients:

$$t_s = \frac{(b_{Y1\cdot} - 0)}{s_{b_{Y1\cdot}}} = \frac{-1.0480}{0.37327} = -2.808, P = 0.00783$$

$$t_s = \frac{(b_{Y2\cdot} - 0)}{s_{b_{Y2\cdot}}} = \frac{0.02431}{0.004,788,0} = 5.077, P = 1.04 \times 10^{-5}$$

5. 95% confidence limits for partial regression coefficients:

$$t_{.05[38]}s_{b_{Y1\cdot}} = 2.025(0.37327) = 0.75587$$

$$L_1 = b_{Y1\cdot} - t_{.05[38]}s_{b_{Y1\cdot}} = -1.0480 - 0.75587 = -1.8039$$

$$L_2 = b_{Y1\cdot} + t_{.05[38]}s_{b_{Y1\cdot}} = -1.0480 + 0.75587 = -0.2981$$

$$t_{.05[38]}s_{b_{Y2\cdot}} = 2.025(0.004,788,0) = 0.009,695,7$$

$$L_1 = b_{Y2\cdot} - t_{.05[38]}s_{b_{Y2\cdot}} = 0.02431 - 0.009,695,7 = 0.01461$$

$$L_2 = b_{Y2\cdot} + t_{.05[38]}s_{b_{Y2\cdot}} = 0.02431 + 0.009,695,7 = 0.03401$$

6. Standard error of standard partial regression coefficients:

We no longer furnish formulas for these standard errors for the following reasons: (1) The values for the t-tests for a partial regression coefficient and a standard partial regression coefficient are identical, and we can test the latter by testing the former. (2) To set confidence limits for the standard partial regression coefficients, special methods are required, which we shall discuss in Section 16.10.

7. Testing the difference between two partial regression coefficients.

$$s_{b_{Y1\cdot} - b_{Y2\cdot}} = s_Y \cdot \sqrt{c_{11} + c_{22} - 2c_{12}}$$

$$= 16.7520[4.964,929,584 \times 10^{-4} + 8.168,974,959 \times 10^{-8}$$

$$- 2(1.210,292,356 \times 10^{-6})]^{1/2}$$

$$= 0.37285$$

$$t_s = \frac{b_{Y1\cdot} - b_{Y2\cdot}}{s_Y \cdot \sqrt{c_{11} + c_{22} - 2c_{12}}} = \frac{-1.048 - 0.02431}{0.37285} = 2.876, P = 0.00657$$

This test is shown here for illustrative purposes. It make sense only when the regression coefficients being compared are in identical measurement units, which is not the case in this example.

8. Standard error of the sampled mean $\bar{Y}$ (at $\bar{X}_1, \bar{X}_2$):

$$s_{\bar{Y}} = \sqrt{\frac{s_{Y\cdot}^2}{n}} = \frac{16.7520}{\sqrt{41}} = 2.6162$$

9. 95% confidence limits for the mean μ_Y corresponding to $\bar{X}_1, \bar{X}_2$. ($\bar{Y} = 30.049$):

$$t_{.05[38]}s_{\bar{Y}} = 2.025(2.6162) = 5.29785$$

$$L_1 = \bar{Y} - t_{.05[38]}s_{\bar{Y}} = 30.049 - 5.298 = 24.751$$

$$L_2 = \bar{Y} + t_{.05[38]}s_{\bar{Y}} = 30.049 + 5.298 = 35.347$$

Box 16.2 (continued)

10. Standard error of $\hat{Y}$, an estimated Y for specified values of X_1 and X_2. We propose to estimate SO_2 content of air (Y) for a town with an annual average temperature of $50°F$ (X_1) and 1000 manufacturing enterprises (X_2).

$$s_{\hat{Y}} = s_{Y \cdot} \sqrt{\frac{1}{n} + \sum_{i=1}^{k} c_{ii} x_i^2 + 2 \sum_{ij} c_{ij} x_i x_j}$$

$$= s_{Y \cdot} \left[\frac{1}{n} + c_{11}(X_1 - \bar{X}_1)^2 + c_{22}(X_2 - \bar{X}_2)^2 + 2c_{12}(X_1 - \bar{X}_1)(X_2 - \bar{X}_2) \right]^{1/2}$$

$$= 16.7520 \left[\frac{1}{41} + 4.964,929,584 \times 10^{-4}(50 - 7.2277)^2 \right.$$

$$+ 8.168,974,959 \times 10^{-8}(1000 - 563.47)^2$$

$$\left. + 2(1.210,292,356 \times 10^{-6})(50 - 7.2277)(1000 - 563.47) \right]^{1/2}$$

$$= 16.6972$$

11. 95% confidence limits for μ_Y corresponding to the estimate $\hat{Y}$ for $X_1 = 50$ and $X_2 = 1000$. Employing the prediction equation from step **5** of Box 16.1,

$$\hat{Y} = 77.231 - 1.0480(50) + 0.0243(1000) = 49.141$$

$$t_{.05[38]} s_{\hat{Y}} = 2.025(16.6972) = 33.81183$$

$$L_1 = \hat{Y} - t_{.05[38]} s_{\hat{Y}} = 49.141 - 33.812 = 15.329$$

$$L_2 = \hat{Y} + t_{.05[38]} s_{\hat{Y}} = 49.141 + 33.812 = 82.953$$

12. Standard error for a predicted mean $\bar{Y}$ to be obtained in a new study of towns with $X_1 = 50°F$ and $X_2 = 1000$ manufacturing enterprises. Our best prediction for this mean would be $\bar{Y} = \hat{Y} = 49.141 \ \mu g/m^3$. If the new study were based on a sample size of $m = 5$, the standard error of the predicted mean would be

$$\hat{s}_{\bar{Y}} = s_{Y \cdot} \sqrt{\frac{1}{m} + \frac{1}{n} + \sum_{i=1}^{k} c_{ii} x_i^2 + 2 \sum_{ij} c_{ij} x_i x_j}$$

$$= 16.7520 \left(\frac{1}{5} + 0.99347 \right)^{1/2} \quad \text{(using numerical values obtained in step 10)}$$

$$= 16.7520(1.09246) = 18.3009$$

13. 95% prediction limits for a sample mean of 5 towns at $X_1 = 50°F$ and $X_2 = 1000$ manufacturing enterprises:

$$t_{.05[38]} \hat{s}_{\bar{Y}} = 2.025(18.3009) = 37.05934$$

$$L_1 = \hat{Y} - t_{.05[38]} \hat{s}_{\bar{Y}} = 49.141 - 37.059 = 12.082$$

$$L_2 = \hat{Y} + t_{.05[38]} \hat{s}_{\bar{Y}} = 49.141 + 37.905 = 86.200$$

Box 16.2 (continued)

14. Testing $R_{Y \cdot 1 \ldots k}$. For $k \leq 4$, use Statistical Table **R** and look up critical values for $v = n - k - 1$ degrees of freedom and k independent variables. *Example:* Test $R_{Y \cdot 1,2}$. Because $R^2_{Y \cdot 1,2} = 0.51611$, $R_{Y \cdot 1,2} = 0.7184$ this is greater than the 0.01 critical value for 35 df and $k = 2$, which is 0.481. The actual degrees of freedom, $v = n - k - 1 = 38$, are not listed in Statistical Table **R,** but there is no need to interpolate in view of the magnitude of the coefficient. We reject the null hypothesis that $\rho_{Y \cdot 1,2} = 0$. For any value of k, we can use the following relationship to find a desired critical value:

$$R_{Y \cdot 1 \ldots k(\alpha)} = \left[\frac{kF_{\alpha[k, n-k-1]}}{n - k - 1 + kF_{\alpha[k, n-k-1]}} \right]^{1/2}$$

This formula can also be inverted to yield a probability corresponding to a given value of R^2.

Because we rejected H_0 for the proportion of the variance of Y determined by X_2, we are not surprised that when using X_1 and X_2 together we also reject H_0. The explained sum of squares increases as additional independent variables are included in the regression equation. But is the observed increase in the explained sum of squares itself larger than the amount one would usually expect by chance?

If we compute an explained sum of squares based on a larger number, k_2, of independent variables and subtract from it an explained sum of squares based on a smaller number of variables, k_1, the resulting difference, $\Sigma \hat{y}^2_{\cdot 1 \ldots k_2} - \Sigma \hat{y}^2_{\cdot 1 \ldots k_1}$, is a sum of squares pertaining to the added explanation due to variables $X_{k_1 + 1}, \ldots, X_{k_2}$. This sum of squares possesses $k_2 - k_1$ degrees of freedom and can be tested against the unexplained sum of squares $\Sigma d^2_{Y \cdot 1 \ldots k_2}$ with $n - k_2 - 1$ df. We are thus partitioning the explained SS with the greater number of variables into two sums of squares—one due to the effect of the first k_1 variables and one due to the effect of the additional $k_2 - k_1$ variables, with the k_1 variables held constant. Rather than set up the tests as formal analyses of variance, we can again use the shortcut employed in Expression (16.13) and write

$$F_s = \frac{(R^2_{Y \cdot 1 \ldots k_2} - R^2_{Y \cdot 1 \ldots k_1})/(k_2 - k_1)}{(1 - R^2_{Y \cdot 1 \ldots k_2})/(n - k_2 - 1)} \qquad (16.14)$$

You should be able to derive this formula easily from the facts already given. These tests are carried out on the air pollution data from Table 16.1 in step **2** of Box 16.2. We reject the H_0 of no increase in determination for the additions of X_1 and then of X_3 to X_2. Adding variables X_4, X_5, and X_6, however, increases the coefficient of multiple determination only a little; with X_1, X_2, and X_3 only, $R^2 = 0.61255$; with all six X-variables, $R^2 = 0.66951$. A test of a H_0 for this increase yields $P = 0.140$. Thus, we cannot reject the hypothesis that $\beta_{Y4} = \beta_{Y5} = \beta_{Y6} = 0$, and we would be tempted to omit X_4, X_5, and X_6 from the prediction equation (although the best fit for the observed data is obtained by using all available variables).

All the other hypothesis tests familiar from simple linear regression apply to multiple regression as well. Remember that the key variance necessary in linear regression is $s_{Y \cdot X}^2$, the unexplained mean square with $n - 2$ degrees of freedom. The analogous quantity in multiple regression is the unexplained mean square $s_{Y \cdot 1 \ldots k}^2$, computed as follows:

$$s_{Y \cdot 1 \ldots k}^2 = \frac{(1 - R_{Y \cdot 1 \ldots k}^2) \sum y^2}{n - k - 1}$$

In addition to $s_{Y \cdot 1 \ldots k}^2$, we need the Gaussian multipliers, the elements of the inverse of the *SS–SP* (sum of squares–sum of products) matrix (see Section 16.1). Larger matrices, for cases with more than two independent variables, should be inverted by computer. Many computer programs for multiple-regression analysis automatically calculate various standard errors of interest, and the user may not see or need to employ the Gaussian multipliers explicitly. However, you should at least know of their existence and how they might be employed if the standard error you wish to use is not included in the program available to you.

As an example, we will employ the regression of Y on X_1 and X_2 from Box 16.1. At the beginning of Box 16.2, we computed $R_{Y \cdot 1,2}^2 = 0.51611$. From this value we can obtain

$$s_{Y \cdot 1,2} = \left[\frac{(1 - R_{Y \cdot 1,2}^2) \sum y^2}{n - k - 1} \right]^{1/2} = \left[\frac{(1 - 0.51611)22,038}{38} \right]^{1/2} = 16.7520$$

In this simple example with only two independent variables, one can easily obtain the Gaussian multipliers without a computer. These quantities are the elements of the inverse of the *SS–SP* matrix $\mathbf{A}$. In the current example,

$$\mathbf{A} = \begin{bmatrix} \sum x_1^2 & \sum x_1 x_2 \\ \sum x_1 x_2 & \sum x_2^2 \end{bmatrix}$$

$$= \begin{bmatrix} 2089.595122 & -30,958.85366 \\ -30,958.85366 & 12,700,115.61 \end{bmatrix}$$

These values were obtained by computation from the data in Table 16.1. In this 2×2 matrix, computing an inverse is a trivial task. First, we compute the determinant, $|\mathbf{A}| = \sum x_1^2 \sum x_2^2 - (\sum x_1 x_2)^2 = 2.557,964,901 \times 10^{10}$, and its reciprocal, $1/|\mathbf{A}| = 3.909,357,785 \times 10^{-11}$. Then we compute the elements of the inverse of $\mathbf{A}$ as follows:

$$\begin{bmatrix} c_{11} & c_{12} \\ c_{12} & c_{22} \end{bmatrix} = \frac{1}{|\mathbf{A}|} \begin{bmatrix} \sum x_1^2 & \sum x_1 x_2 \\ \sum x_1 x_2 & \sum x_2^2 \end{bmatrix}$$

$$= 3.909,357,785 \times 10^{-11} \begin{bmatrix} 12,700,115.61 & 30,958.85366 \\ 30,958.85366 & 2089.595122 \end{bmatrix}$$

$$= \begin{bmatrix} 4.964,929,584 \times 10^{-4} & 1.210,292,356 \times 10^{-6} \\ 1.210,292,356 \times 10^{-6} & 8.168,974,959 \times 10^{-8} \end{bmatrix}$$

Using the variance $s^2_{Y \cdot 1 \cdots k}$ and the Gaussian multipliers, we can set standard errors and confidence limits to various statistics related to multiple regression. The necessary standard errors are listed in Box 16.3.

The first row of Box 16.3 shows the *standard error of the partial regression coefficient* for the regression of Y on X_j, all other independent variables being held constant. Its computation is illustrated in step **3** of Box 16.2, while in step **4** of the same box we show the by now familiar *t*-test for *the partial regression coefficients*. The null hypothesis can easily be rejected for both coefficients: Temperature and manufacturing enterprises both explain SO_2 content of air, each appreciably affecting the dependent variable when the other is held constant. Note that in contrast to simple linear regression, the significance test of the regression coefficients does not follow from that of the overall regression (the anova test in step **1** of Box 16.2). The anova tests whether both X_1 and X_2 explain a significant proportion of the variance of Y. If this were the case, we still would not know whether X_1 or X_2 or both are related to Y.

For example, we saw in step **1** of Box 16.2 that $X_1, \ldots, X_6$ together reject the H_0 of not explaining Y. If we tested the partial regression coefficients, however, we would find that only those regressing on X_1, X_2, and X_3 reject the H_0. The standard errors of the partial regression coefficients also permit *setting of confidence limits for partial regression coefficients* in a manner by now familiar. The procedure is illustrated in step **5** of Box 16.2.

The *standard error of standard partial regression coefficients* is computed by the formula in the second row of Box 16.3. Sample computations are shown in step **6** of Box 16.2. Hypothesis tests (not shown) give identical results to those on partial regression coefficients. Thus, the standard errors of *b*-primes simply allow for the standardizations. The fact that the standard errors of $b'_{Y2 \cdot 1}$ in this case have identical numerical values (within rounding error) is a peculiarity of the case with two independent variables. When there are more than two such variables, the standard errors differ. We can set confidence limits to *b*-primes in the customary manner.

The factors $c_{jj}\Sigma x_j^2$ in the equations for $s_{b'_j}$ (see Box 16.3) are called the **variance inflation factors** (VIF) (Marquardt, 1970), because they represent the factor by which the standardized unexplained variance $(1 - R^2)/(n - k - 1)$ is inflated because of the pattern of intercorrelation among the independent variables. They can also be called measures of multicollinearity. If the independent variables are uncorrelated, the VIF will equal unity. In the current case (with $r_{12} = -0.19$), they are equal to 1.037,469,26. As the independent variables become more highly correlated (whether positively or negatively), the factors can become much larger. When the VIFs are greater than 100, the ordinary precision available on most computers will not permit a sufficiently accurate solution of the normal equations (Snee, 1973).

Although the test makes sense only when the independent variables concerned are in the same units of measurement, we may wish to *test differences between pairs of partial regression coefficients*. A standard error for such differences is shown in the third row of Box 16.3, and its use is illustrated in step **7** of Box 16.2—although its employment in this particular case is not of interest. Change in SO_2 per degree

BOX 16.3 Standard Errors of Multiple Regression Statistics

For explanation of this box, see Section 16.2; the degrees of freedom appropriate to these tests are $v = n - k - 1$, where n = sample size and k = number of independent variables in the study.

Statistic	s	Formula for standard error
$b_{Yj\cdot}$ (partial regression coefficient of Y on X_j)	$s_{b_{Yj\cdot}}$	$s_Y \cdot \sqrt{c_{jj}}$
$b'_{Yj\cdot}$ (standard partial regression coefficient of Y on X_j)	$s_{b'_{Yj\cdot}}$	$\sqrt{\dfrac{(1 - R^2_{Y\cdot 1 \ldots k})}{n - k - 1} c_{jj} \sum x_j^2}$
$b_{Yi\cdot} - b_{Yj\cdot}$ (difference between two partial regression coefficients)	$s_{b_{Yi\cdot} - b_{Yj\cdot}}$	$s_Y \cdot \sqrt{c_{ii} + c_{jj} - 2c_{ij}}$
$\overline{Y}$ (sample mean)	$s_{\overline{Y}}$	$\sqrt{\dfrac{s_{Y\cdot}^2}{n}}$
$\hat{Y}$ (estimated Y for given values of $X_1, X_2, \ldots, X_k$)	$s_{\hat{Y}}$	$s_Y \cdot \sqrt{\dfrac{1}{n} + \sum_{i=1}^{k} c_{ii} x_i^2 + 2 \sum_{ij} c_{ij} x_i x_j}$
Predicted Y for given values of $X_1, X_2, \ldots, X_k$	$\hat{s}_Y$	$s_Y \cdot \sqrt{1 + \dfrac{1}{n} + \sum_{i=1}^{k} c_{ii} x_i^2 + 2 \sum_{ij} c_{ij} x_i x_j}$
Predicted mean $\overline{Y}$ of m items for given values of $X_1, X_2, \ldots, X_k$	$\hat{s}_{\overline{Y}}$	$s_Y \cdot \sqrt{\dfrac{1}{m} + \dfrac{1}{n} + \sum_{i=1}^{k} c_{ii} x_i^2 + 2 \sum_{ij} c_{ij} x_i x_j}$

NOTES: All standard errors except that for $b'_{Yj\cdot}$ contain the unexplained variance $s^2_{Y\cdot 1 \ldots k}$ (shown in simplified symbolism as $s_{Y\cdot}^2$). This can be obtained from the anova testing overall regression. The terms c_{ii}, c_{jj}, and c_{ij} are the Gaussian multipliers, the elements of the inverse of the sum of squares-sum of products matrix.

In the last three rows the estimated $\hat{Y}$, Y, and $\overline{Y}$ are for a specified set of values of X_1, $X_2, \ldots, X_k$. The deviations under the radicals on the right are for these specified values. Thus $x_1 = X_1 - \overline{X}_1$, where X_1 is the specified observation of variable X_1. The summation symbolized by $\sum_{ij}$ is over all pairs of variables X_i, X_j, where $i < j$, i ranges from 1 to $k - 1$, and j ranges from $i + 1$ to k.

Fahrenheit cannot readily be compared with change per number of manufacturing enterprises.

The other standard errors in Box 16.3 correspond to those for simple linear regression in Box 14.2, which are illustrated in Box 14.3. A new, smaller *standard error of $\overline{Y}$* can be computed, leading to narrower confidence limits (see steps **8** and **9** of Box 16.2). When we wish to construct confidence "belts" around the plane, we need to take the deviation of the specified values of X_1 and X_2 from their respective means into consideration, as was done in simple linear regression. The formulas are more complicated, however, because the partial regression coefficients are not independent of each other and because in addition to the Gaussian multipliers that furnish their variance (remember that $s_Y^2 . c_{jj}$ is the variance for b_{Yj}.), we need those that provide their covariance ($s_Y^2 . c_{ij}$). The confidence surfaces are closest to the plane at $\overline{X}_1, \overline{X}_2$ and curve away from it as X_1 and X_2 depart from their respective means. The computations for the *standard error of an estimated Y for specified values of X_j* and those for the *standard error of predicted sample mean based on m items,* are shown in steps **10** and **12,** respectively, and computations of confidence or prediction limits based on the standard errors are given in steps **11** and **13,** respectively, of Box 16.2.

16.3 Path Analysis and Structural Equation Modeling

In this section, we will introduce powerful analytical tools known as path analysis and structural equation modeling. These are well worth studying for their own sake as well as for the light they shed on multiple regression.

We could represent the relations between SO_2 content of air and the three independent variables, X_1, X_2, X_3, as shown in Figure 16.3. This representation implies that these independent variables are immediate causes of the dependent variable Y. We know that this is not true. Complicated demographic, environmental, and physicochemical factors determine the sulfur dioxide concentration in any given city, and all the cautions about cause-and-effect relationships voiced in Sections 14.6 and 15.6 apply here as well. Nevertheless, it is convenient to consider the model in this simplistic manner for the insights it gives us about the mathematical relationships between the variables.

In addition to causal relationship, we make another assumption, namely that the independent and dependent variables are related by a linear and additive model such

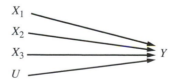

FIGURE 16.3   Path diagram showing three uncorrelated predictor (independent) variables, X_1, X_2, and X_3 and a residual variable, U, affecting one criterion (dependent) variable, Y.

as the multiple-regression equations in Expressions (16.1) and (16.2). Rather than use the terms *independent* and *dependent* variables, which carry with them definite assumptions relating to Model I regression that are generally not met in path analysis, in this section we will frequently employ the more neutral terms **predictor** and **criterion** variables, respectively—terminology that is common in multiple regression and path analysis in the social sciences.

Diagrams such as Figure 16.3 are known as **path diagrams** and obey the following conventions. Cause-and-effect relationships are depicted by one-headed arrows. The criterion variables are generally indicated as completely determined—that is, all the factors contributing to the total variation of variable Y are drawn in. Because in most instances one does not know, and has no hope of ever knowing, *all* the factors that explain the variation of a given dependent (criterion) variable, these unknown factors are lumped together as a single, unknown, residual variable (shown as U in Figure 16.3). The effect of this pooled variable on Y cannot be quantified. U is assumed to be a random normal deviate and hence could be positive or negative. However, we can ascribe to it the proportion of the total variance of Y not determined by the other causal variables. In fact, this is a familiar quantity. If we consider the total variance of Y to be 1, then the proportion of the variance due to U is r_{UY}^2, which is the coefficient of determination between Y and U. The proportion that is determined by U is the proportion not determined by X_1, X_2, and X_3. Therefore, $r_{UY}^2 = 1 - R_{Y \cdot 1,2,3}^2$, the coefficient of nondetermination of Y by X_1, X_2, and X_3.

One other assumption is implied by the diagram in Figure 16.3. Because the bases of the arrows from variables X_1, X_2, X_3, and synthetic variable U are not connected, it is implied that they are uncorrelated—that they are ***independent causes.*** Such an assumption is plausible in certain situations; in this specific case, however, it is not. Furthermore, we know empirically that there are correlations among the three predictor variables—low correlation between temperature and the others but a very high correlation (0.9553) between number of manufacturing enterprises (X_2) and population size (X_3). Thus, the model in Figure 16.3 cannot describe our data, but we will pursue this simplistic model for a little while longer to explore its consequences. Let us therefore begin by assuming that there are no correlations among X_1, X_2, X_3, and U.

In such a case, what is the independent effect of X_1 on Y? By independent effect, we mean the effect in the absence of the other predictor variables X_2 and X_3. This problem is by now familiar—such an effect can be quantified by a partial regression coefficient. Because we will want to compare the magnitudes of such effects for different factors, we will employ standard partial regression coefficients. The effect of X_1 on Y is thus expressed as $b'_{Y1 \cdot 2,3}$. The effects of the other two predictor variables can be quantified similarly by the appropriate b-primes. How do we obtain these? Remembering the normal equations of Expression (16.6), we can write in standardized form

$$b'_{Y1 \cdot 2,3} + r_{12}b'_{Y2 \cdot 1,3} + r_{13}b'_{Y3 \cdot 1,2} = r_{1Y}$$

$$r_{12}b'_{Y1 \cdot 2,3} + b'_{Y2 \cdot 1,3} + r_{23}b'_{Y3 \cdot 1,2} = r_{2Y} \qquad (16.15)$$

$$r_{13}b'_{Y1 \cdot 2,3} + r_{23}b'_{Y2 \cdot 1,3} + b'_{Y3 \cdot 1,2} = r_{3Y}$$

We do not have to use simultaneous equations or matrix methods, however, to solve these equations. We have assumed (incorrectly, but for purposes of completing our study of Figure 16.3) that the causal variables are not correlated. Lack of correlation would make $r_{12} = r_{13} = r_{23} = 0$, so all terms involving these correlation coefficients on the left-hand side of the equations vanish. We can therefore write down the solution of the equations by inspection: $b'_{Y1 \cdot 2,3} = r_{1Y}$, $b'_{Y2 \cdot 1,3} = r_{2Y}$, and $b'_{Y3 \cdot 1,2} = r_{3Y}$. Thus, the standard partial regression coefficients in this special case are equal to the correlation coefficients of the two variables concerned.

In Expression (15.7), we showed that if the simple linear regression coefficient $b_{Y \cdot X}$ is standardized by dividing by the appropriate standard deviations, we obtain the correlation coefficient r_{XY}. Thus, it follows that for the special case of Figure 16.3, $b'_{Y1 \cdot 2,3} = r_{1Y} = b'_{Y1}$; there are similar relations for the other two variables. When you think about it, this makes a good deal of sense. Because the causal variables X_1, X_2, and X_3 are here assumed to be independent, the effect of X_1 on Y, with X_2 and X_3 held constant $(b'_{Y1 \cdot 2,3})$, should be no different from the effect of X_1 on Y when X_2 and X_3 are not considered (b'_{Y1})—which, as we have seen, is the same as the correlation between X_1 and Y.

In path analysis, the standard partial regression coefficient that estimates the strength of the relationship between cause X_1 and effect Y is called a **path coefficient** and is symbolized by p_{Y1}. From what we have learned already about path analysis, we can make the first useful conclusion: In the case of independent causes, a path coefficient is equal to the correlation between the predictor and the criterion variable.

Thus, in Figure 16.3, $p_{Y1} = r_{1Y}$, $p_{Y2} = r_{2Y}$, and $p_{Y3} = r_{3Y}$. What proportion of the variance of Y is determined by X_1? Obviously, the answer is the proportion estimated by the coefficient of determination r_{1Y}^2, which equals p_{Y1}^2. Because the predictor variables are not correlated, the proportion of the total variance that each one determines is additive, and we can therefore write the following equation for the total determination of the variance of Y:

$$p_{Y1}^2 + p_{Y2}^2 + p_{Y3}^2 + r_{UY}^2 = 1 \qquad (16.16)$$

Sometimes you may encounter path diagrams for which the strength of the effect of the unknown variable U on the criterion variable Y is given as $p_{YU} = r_{YU} = (1 - R_{Y \cdot}^2)^{1/2}$. This expression is all right as long as we keep in mind that variable U does not exist as an entity but rather is a composite of all the unknown sources of unexplained variation. Therefore, any single effect ascribed to it is meaningless, except as a measure of the amount of unexplained variation.

Now we will make the model more realistic for our data—we will consider **correlated causes.** In Figure 16.4, variables X_1, X_2, and X_3 are connected by double-headed arrows. In path diagrams, such arrows indicate correlation between predictor or causal variables. What is the correlation between X_1 and Y now? In the first line of the normal equations of Expression (16.15), the correlations r_{12} and r_{13} are now not equal to zero; to obtain the standard partial regression coefficients, we have to solve the normal equations. These equations also show us, however, how to evaluate the correlation between predictor and criterion variables by means of path coefficients. If we rewrite the first normal equation in terms of path coefficients, we obtain $r_{1Y} =$

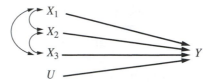

FIGURE 16.4   Path diagram showing three correlated predictor variables, X_1, X_2, and X_3, and an independent variable, U, affecting one criterion variable, Y.

$p_{Y1} + r_{12}p_{Y2} + r_{13}p_{Y3}$. This expression leads directly to the general rule for finding the correlation between two variables in a path diagram: *The correlation between two variables is the sum of the direct path between them plus the sum of the products of the chains of path coefficients or correlations along all of the paths by which they are connected.*

In evaluating the correlation between X_1 and Y in Figure 16.4, then, the first path between them is the direct one, p_{Y1}, the second is the indirect one via X_2, which is $r_{12}p_{Y2}$, while the third is an indirect path via X_3, $r_{13}p_{Y3}$. Note that you cannot consider the chain $r_{12}r_{23}p_{Y3}$ to contribute to the correlation r_{1Y} via X_3. Only the common causes behind the correlation r_{13} contribute via X_3, not the common causes behind the correlation r_{23}. The correlation r_{13} is indicated in Figure 16.4 by the outer double-headed arrow between X_1 and X_3. The following simple rule prevents such errors in evaluating correlation coefficients: *In a chain connecting two variables, no more than one double-headed arrow (correlation between predictor variables) is permitted.*

Knowing the values of the correlations represented by the double-headed arrows and of the path coefficients (the standard partial regression coefficients obtained as shown in Section 16.1), we can compute the correlation between any X and Y. Using our newly learned rules for obtaining correlations, we can write the equation for finding the correlation r_{2Y}:

$$r_{2Y} = p_{Y2} + r_{12}p_{Y1} + r_{23}p_{Y3}$$
$$= 1.7111 + (-0.1900)(-0.1807) + (0.9553)(-1.1522) = 0.6447$$

Within rounding error, this result is the same as the value of 0.6448 computed from the original data (see Box 16.1). Numerical values for the three path coefficients in the preceding equation had been obtained in the form of standard partial regression coefficients by using the regression program of BIOMstat on the air pollution data (including only the first three predictor variables X_1 through X_3) multiple.

The equation tells us something about the components of a correlation coefficient. In this particular system, the correlation between X_2 and Y is made up not only of the direct contribution of X_2 to Y (the path coefficient p_{Y2}), but also of common causes between X_2 and any other variables in the system that have a direct effect on Y. These additional contributions from common causes tend to increase the correlation between the two variables concerned, except in cases where either the paths or the correlation due to common causes are negative. In such cases, compensating effects tend to dampen the variation of the criterion variable and hence decrease correlation.

Note also that some of the path coefficients are larger than 1. The reason is that, as noted in Section 16.1, no limitation was set on the magnitude of standard partial regression coefficients. If a coefficient is larger than 1, however, the predictor variable must be connected to the criterion variable through at least one other chain of opposite sign and of a magnitude that confines the total correlation between them to the limits of -1 and $+1$ for correlation coefficients.

Next, we develop the formula for total determination of variable Y in the path diagram of Figure 16.4. Remembering the additive model for these paths, we can write $Y = X_1 + X_2 + X_3 + U$. Thus, the variance of Y must be a function of the variances of these variables. We have already examined the variance of a sum (see Section 15.4), and the formula we need now is merely an extension of Expression (15.10). We can write

$$\sum_i p_{Yi}^2 + 2\sum_{ij} p_{Yi}p_{Yj}r_{ij} + r_{UY}^2 = 1 \qquad (16.17)$$

where Σ_i is summation of all i $(1, \ldots, k)$ causal variables and Σ_{ij} is summation over all pairs of causal variables X_i, X_j, where $i < j$ and i ranges from 1 to $k - 1$, while j ranges from $i + 1$ to k. In words, this formula indicates that the total variance of Y (which in standardized scale equals unity) is composed of (1) the sum of the squares of the path coefficients along all paths leading to Y; (2) added to twice the sum of the products of the path coefficients along all paths leading to Y multiplied by their correlations for all possible pairs of predictor variables; plus (3) the coefficient of nondetermination by variable U, the collective variable representing unknown factors acting on Y. We will now compute the determination of variable Y for the example in Figure 16.4. Evaluating the formula just given, we obtain

$$p_{Y1}^2 + p_{Y2}^2 + p_{Y3}^2 + 2(p_{Y1}p_{Y2}r_{12} + p_{Y1}p_{Y3}r_{13} + p_{Y2}p_{Y3}r_{23}) + r_{UY}^2$$

$$= (-0.1807)^2 + (1.7111)^2 + (-1.1522)^2$$

$$+ 2[(-0.1807)(1.7111)(-0.1900) + (-0.1807)(-1.1522)(-0.0627)$$

$$+ (1.7111)(-1.1522)(0.9553)] + 0.38745$$

$$= 1.0001$$

The result agrees within rounding error with the correct value of 1. Note that in Expression (16.17), the terms expressing determination by known causes, $\Sigma_i p_{Yi}^2 + 2\Sigma_{ij} p_{Yi}p_{Yj}r_{ij}$, equal the coefficient of determination, R^2. This relationship is obvious, because r_{UY}^2 is the coefficient of nondetermination and $R^2 = 1 - r_{UY}^2$.

We must explore three more situations before we can apply our new knowledge of path analysis. We may wish to express the value of a path coefficient in terms of its constituent paths. An example is shown in Figure 16.5. The path from X_3 to Y goes via X_1. What is the value for p_{Y3}? We can write the prediction equation for Y

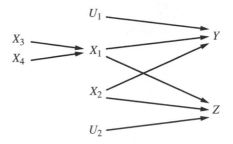

FIGURE 16.5 Path diagram showing two criterion variables, Y and Z, which are correlated by having the common predictor variables X_1 and X_2. Variable X_1 is completely determined by uncorrelated predictor variables, X_3 and X_4.

as follows: $\hat{y}' = p_{YU_1}u_1' + p_{Y1}x_1' + p_{Y2}x_2'$, using notation for standardized variables. Similarly, $\hat{x}_1' = p_{13}x_3' + p_{14}x_4'$. Substituting $\hat{x}_1'$ into the equation for $\hat{y}'$, we obtain

$$\hat{y}' = p_{YU_1}u_1' + p_{Y1}(p_{13}x_3' + p_{14}x_4') + p_{Y2}x_2'$$
$$= p_{YU_1}u_1' + p_{Y1}p_{13}x_3' + p_{Y1}p_{14}x_4' + p_{Y2}x_2'$$

Thus, the standard partial regression coefficient of Y on X_3, which is the path coefficient p_{Y3}, is shown to be the product $p_{Y1}p_{13}$. Consequently, the proportion of the variance of Y because of X_3, p_{Y3}^2, must be $p_{Y1}^2p_{13}^2$. This shows that a **compound path** is the product of the constituent paths. This is true even when paths are bidirectional, as in Figure 16.6, where one of the paths from Y to Z is via X_1 and X_2. Note that this path, symbolized as $p_{Y(12)Z}$, is not the correlation r_{YZ}; it is just one constituent in that correlation, which we will discuss presently. In such a bidirectional path, at most one two-headed arrow representing a correlation coefficient may be included.

Figure 16.5 also illustrates another important relationship—**common independent causes.** Two criterion variables are determined by two (or more) predictor variables, plus those representing unexplained variation. For the moment, we need not be concerned that one of these variables, X_1, is determined by even more remote causes,

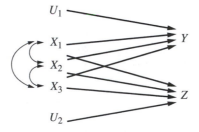

FIGURE 16.6 Path diagram showing two criterion variables, Y and Z, which are correlated by having the common correlated causes, X_1, X_2, and X_3.

X_3 and X_4. Both X_1 and X_2 affect Y and Z, and the correlation between Y and Z is due to these two common causes only. We can therefore employ the rules stated earlier that the correlation between two variables is the sum of all the paths connecting them, and we compute the two bidirectional paths $p_{Y(1)Z}$ and $p_{Y(2)Z}$ as the products of their constituent paths. Thus, $p_{Y(1)Z} = p_{Y1}p_{Z1}$ and $p_{Y(2)Z} = p_{Y2}p_{Z2}$, yielding $r_{YZ} = p_{Y1}p_{Z1} + p_{Y2}p_{Z2}$, and the general relation is

$$r_{YZ} = \sum_i p_{Yi}p_{Zi} \tag{16.18}$$

for all common independent causes, i. Because the predictors in Figure 16.5 are not correlated, the total determination of Y or Z presents no new problems. Thus, to express the total determination of Y, we can write

$$p_{Y1}^2 + p_{Y2}^2 + r_{U_1Y}^2 = 1$$

or, if we wish to express X_1 in terms of X_3 and X_4, we obtain

$$p_{Y3}^2 + p_{Y4}^2 + p_{Y2}^2 + r_{U_1Y}^2 = 1$$

Finally, we evaluate a correlation in terms of **common correlated causes** as shown in Figure 16.6. The correlation between Y and Z is still the sum of all paths connecting them, but now there are many paths—nine, to be specific. The formula for this case is

$$r_{YZ} = p_{Y1}p_{Z1} + p_{Y2}p_{Z2} + p_{Y3}p_{Z3} + p_{Y1}p_{Z2}r_{12} + p_{Y2}p_{Z1}r_{12} + p_{Y1}p_{Z3}r_{13}$$
$$+ p_{Y3}p_{Z1}r_{13} + p_{Y2}p_{Z3}r_{23} + p_{Y3}p_{Z2}r_{23}$$

The first three terms are bidirectional paths connecting Y and Z via X_1, X_2, and X_3, respectively. The next two involve paths from Y to Z via the correlation r_{12}. Note that there are two such paths: from Y through X_1 and X_2 to Z, or from Y to X_2 then to X_1 and Z. Different path coefficients are involved each time; the only common element is the correlation r_{12}. Note, however, that in any one bidirectional path the correlation occurs only once. Confirm for yourself how the remaining four terms are obtained by tracing the paths in the figure. The formula can be generalized to

$$r_{YZ} = \sum_i p_{Yi}p_{Zi} + \sum_{ij}(p_{Yi}p_{Zj} + p_{Yj}p_{Zi})r_{ij} \tag{16.19}$$

The determination of the variance of either Y or Z presents no new features.

The rules for expressing correlations and determinations by path coefficients are summarized in Table 16.2, which gives general formulas that might serve as mnemonic aides.

You now know the essential features of path coefficients and path analysis. To understand the subject fully, however, we must review some special situations—a few of which we will present later as the need arises. Now that we have achieved a reasonable mastery of the subject, though, what can we do with it? Path coefficients were first introduced by Sewall Wright in the early 1920s as a device for

TABLE 16.2 Summary Formulas for Path Coefficients

	Correlation	Determination
One criterion variable		
Independent causes[a]	$r_{jY} = p_{Y_j}$	$1 = \sum_i p_{Y_i}^2 + r_{UY}^2$
Correlated causes[b]	$r_{jY} = p_{Y_j} + \sum_{i \neq j} r_{ij} p_{Y_i}$	$1 = \sum_i p_{Y_i}^2 + 2 \sum_{ij} p_{Y_i} p_{Y_j} r_{ij} + r_{UY}^2$
Chains[c]	$r_{jY} = p_{Y_j} = p_{Y_i} p_{ij}$	
Two criterion variables		
Independent causes[c]	$r_{YZ} = \sum_i p_{Y_i} p_{Z_i}$	
Correlated causes[d]	$r_{YZ} = \sum_i p_{Y_i} p_{Z_i}$	
	$\quad + \sum_{ij} (p_{Y_i} p_{Z_j} + p_{Y_j} p_{Z_i}) r_{ij}$	

Notes:

$\sum_i$ is summation over all i $(1 \cdots k)$ causal variables.

$\sum_{i \neq j}$ is summation over all i $(1 \cdots k)$ causal variables except for j.

$\sum_{ij}$ is summation over all pairs of causal variables X_i, X_j where $i < j$, i ranges from 1 to $k - 1$, and j ranges from $i + 1$ to k.

[a]See Figure 16.3.
[b]See Figure 16.4.
[c]See Figure 16.5.
[d]See Figure 16.6.

evaluating the contributions of various factors in complicated interaction systems in animal breeding, physiology, and related subjects. Path coefficients proved useful for predicting correlations in various fields, especially in population genetics, and during the next three decades they were employed largely for predicting the consequences of breeding patterns in population and quantitative genetics. Although this use of path coefficients has diminished somewhat in recent years in favor of newer techniques, there has been a marked resurgence in the application of path analysis to problems testing structural models. Social scientists have discovered path analysis and are now among its most active users. In biology, too, there is a growing renaissance of path analysis for model testing, its original purpose.

There are three major uses of path analysis. The first is to predict correlation in systems where the strengths of component parts are known. Most such situations occur in population genetics, and we will not discuss them here. For more information on this subject, see Li (1975, 1976) and Wright (1968, 1969). We will illustrate a different application of path analysis. The example is artificial, but we chose it to illustrate an important theoretical point.

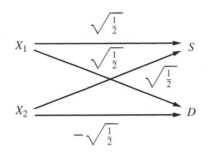

FIGURE 16.7   Path diagram illustrating computation of the correlation between sums and differences.

Suppose you wish to predict the correlation between the sums and differences of two random variables. Assume that you have sampled n pairs of numbers from a table of random numbers (such as Statistical Table **FF**) and have arrayed them into columns, X_1 and X_2. We now generate two new columns, $S = X_1 + X_2$ and $D = X_1 - X_2$. What is the correlation between variables S and D, which are both completely determined by variables X_1 and X_2? Although this problem can be solved algebraically, solving it by path coefficients is even simpler. The relationship is diagrammed in Figure 16.7. Because S and D are both completely determined by X_1 and X_2, there is no need for a residual path from unknown factors U for either variable, and we can write the following equations for the determination of the two variables, respectively: $p_{S1}^2 + p_{S2}^2 = 1$ and $p_{D1}^2 + p_{D2}^2 = 1$.

In view of the equal determination by the two predictor variables, all squared coefficients equal 0.5, and the path coefficients are $\sqrt{\frac{1}{2}}$. We make p_{D2} negative, because it is a negative contribution to D. Because X_1 and X_2 are independent random variables, they are not correlated and we can predict the correlation between S and D, using the formula for independent common causes from Table 16.2. Thus, we obtain

$$r_{SD} = p_{S1}p_{D1} + p_{S2}p_{D2} = \sqrt{\tfrac{1}{2}} \cdot \sqrt{\tfrac{1}{2}} + \sqrt{\tfrac{1}{2}} \cdot (-\sqrt{\tfrac{1}{2}}) = 0$$

The results may surprise you. The two variables S and D show no correlation. The positive contribution toward the correlation by X_1 is negated by the negative contribution because of X_2. Thus, if you can specify a model exactly (and it need not be such a simple model as this) and if you are familiar with the rules for handling path coefficients, evaluating resulting correlations is a very simple task.

This example illustrates another important point. When you observe a low or zero correlation between two variables, you cannot really be certain that they have no common causes without further study of the underlying structure of the model. Here we have a completely determined model; yet if you had no insight into the structure, you would assume that S and D had no common predictors. This notion can be extended to high positive or negative correlations as well. We do not know how many underlying common causes contribute to establishing any observed correlation; in

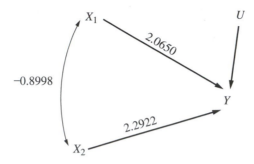

FIGURE 16.8   Path-coefficient diagram for the example presented by Hamilton (1987). (For explanation, see text.)

many cases, there may be positive as well as negative contributions, some of which will cancel each other out.

A related example, presented by Hamilton (1987), is explained here in terms of path coefficients. We encounter two independent variables, X_1 and X_2, that are only weakly correlated with a dependent variable Y. The correlation of Y with X_2 is 0.4341, but that of Y with X_1 is essentially zero (0.0025). The correlation between the two independent variables, however, is very strong and negative (-0.8998). The relations are shown in Figure 16.8. We work out the path coefficients using by now familiar procedures and arrive at $p_{Y1} = 2.0650$ and $p_{Y2} = 2.2922$. When we compute the determination of Y, we find that $p_{Y1}^2 + p_{Y2}^2 + 2p_{Y1}p_{Y2}r_{12}$ already yields an R^2 of 1, leaving essentially nothing to be determined by p_{YU}^2, the remaining unknown factors. Thus, what appeared to be a case of weak correlation between the X-variables and Y when judged by pairwise correlation is actually a case of complete determination of the criterion variable by the predictor variables.

As Hamilton (1987) points out, this example illustrates the often-counterintuitive nature of multivariate relationships. In addition, such cases almost always have negative relations in a chain of their paths, in this case in the correlation between X_1 and X_2. In this system, both predictor variables, X_1 and X_2, strongly affect the criterion variable, but because of their tight negative linkage, they dampen each other's effects, producing weak or no pairwise correlations with the criterion variable. Nevertheless, they completely determine the variation of Y.

The second use of path analysis is as an aid for illuminating certain other statistical topics where its combination of graphic representation with simple algebraic manipulation is preferable over alternative, wholly algebraic formulations of these models. Examples are partial and multiple correlations, selection of independent variables in multiple-regression analysis, and factor analysis. The first two of these topics will be discussed in later sections of this chapter; we therefore defer discussion of the relation of path coefficients to these topics until then.

The third use of path analysis, and for our purposes here its major application, is the construction and evaluation of alternative structural models—theory testing.

As an example, let us return to the sulfur dioxide pollution data analyzed in previous sections and earlier in this section in the context of learning path coefficients. In Figure 16.4, we illustrated the example using a path diagram characteristic of multiple-regression analysis: All predictor variables affect the criterion variable Y directly, but the predictors are also correlated among themselves. Is this really the most reasonable model for this example?

Remember that X_1 is average annual temperature, X_2 is number of manufacturing enterprises with 20 or more employees, and X_3 is population size of the city. Should we show each of these variables as directly affecting SO_2 content, or might one or two of them act indirectly by affecting the third variable? Temperature (X_1) is unlikely to affect population size appreciably in this data set. It is also unlikely that the number of manufacturing enterprises (X_2) is a function of temperature. Thus, it seems that temperature should continue to affect SO_2 content directly. The climate undoubtedly affects the physicochemical processes determining the degree of pollution. However, population and the number of manufacturing enterprises might form a causal chain. Do manufacturing enterprises attract their work force (i.e., increase the population), or does a population base generate its number of manufacturing enterprises? Probably there is a complex interplay of cause and effect, but a reasonable model worth investigating is one in which population size partly determines the number of manufacturing enterprises and enterprises partly determine SO_2 content.

The model in Figure 16.9a assumes that the population (X_3) does not directly engage in activities affecting the sulfur dioxide concentration in the air. Its influence is solely on increasing the number of manufacturing enterprises, which in turn release pollutants. The model as specified allows for correlation between X_3 and X_2 but not for correlation between either of these and X_1. We are encouraged in constructing this model by the observed low correlations r_{12} and r_{13}, which are -0.1900 and -0.0627, respectively. The observed correlations among the three predictor variables and of the three with Y, the SO_2 concentration are shown in the following table.

	Y	X_1	X_2	X_3
Y	1			
X_1	-0.4336	1		
X_2	0.6448	-0.1900	1	
X_3	0.4938	-0.0627	0.9553	1

We therefore wish to work out the path coefficients for the diagram in Figure 16.9a to see how well they reproduce the observed correlations. First, we write specification equations for all the path coefficients as follows:

$$p_{Y1} = r_{1Y} = -0.4336$$

$$p_{Y2} = r_{2Y} = 0.6448$$

$$p_{23} = r_{23} = 0.9553$$

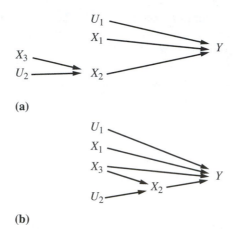

(a)

(b)

FIGURE 16.9 Alternative interpretations of the pollution data. (For an explanation, see text.)

Of the correlations of the three predictor variables with Y, two are simply equal to their corresponding path coefficients and must be correct because we designated them by the observed values. The correlation $r_{3Y} = p_{Y3} = p_{23}\,p_{Y2}$, however, is a product of two of these designated path coefficients. When we estimate r_{3Y} in this manner, we obtain $(0.9553)(0.6448) = 0.6160$. This estimate of r_{3Y} is poor, because the observed correlation is 0.4938.

We can compute the coefficient of nondetermination $r^2_{U_1Y}$ from the formula for complete determination of Y, which is

$$p^2_{Y1} + p^2_{Y2} + r^2_{U_1Y} = 1$$

Because we know the path coefficients, $r^2_{U_1Y}$ is the only remaining unknown in the equation. We estimate $r^2_{U_1Y} = 0.3962$, slightly higher than the coefficient of nondetermination implied by the multiple-regression model (using BIOMstat, we obtain $R^2_{Y·1,2,3} = 0.61255$; hence $1 - R^2_{Y·1,2,3} = 0.3874$). Thus, on the basis of reproduction of the observed correlation and by the percent variation of Y that has been determined, we conclude that the model of Figure 16.9a is not satisfactory. Perhaps we should have allowed for correlations among predictor variables, or the paths in the diagram may have been linked incorrectly.

In Figure 16.9b, we propose another model for the same example. Temperature (X_1) still independently affects SO_2 content. Population size, however, now affects pollution in two ways, indirectly through number of manufacturing enterprises, as before, but also directly. It is quite plausible that large numbers of people cause SO_2 to be liberated in the air through activities other than manufacturing enterprises. Note again that X_1 is not correlated with either X_3 or X_2. We obtain the following specification equations for the relations expressed in Figure 16.9b:

$$p_{Y1} = r_{1Y} = -0.4336$$

$$r_{3Y} = p_{23}p_{Y2} + p_{Y3}$$

$$r_{2Y} = p_{Y2} + p_{23}p_{Y3}$$

Note that p_{Y1} is the same as before, but the paths from X_2 and X_3 have changed because of the configuration of the path diagram. To solve for p_{Y2} and p_{Y3}, we solve the two simultaneous equations shown here. The results are $p_{Y2} = 1.9802$ and $p_{Y3} = -1.3979$. These values differ appreciably from those obtained for the model in Figure 16.9a because the two models differ structurally. If we try now to reconstitute correlations r_{2Y} and r_{3Y}, we find that the predictions match the observations perfectly. This, however, is no cause for satisfaction. In a triangular relationship such as that among Y, X_2, and X_3 in Figure 16.9b, we can always solve exactly for the values of the path coefficients, but we have no independent test of these estimates. However, we can estimate the amount of variance of Y determined by the model.

The equation for the determination of Y is

$$p_{Y1}^2 + p_{Y3}^2 + p_{Y2}^2 + 2p_{Y3}p_{23}p_{Y2} + r_{U_1Y}^2 = 1$$

Substituting known values of path coefficients, we obtain the coefficient of non-determination $r_{U_1Y}^2 = 0.2254$. This model thus explains even more of the variation than the standard multiple-regression model of Figure 16.4. Note that one of our predictions is not fulfilled. We thought X_3 would have an independent positive effect on pollution; people by their numbers and activities would produce more SO_2. The very opposite seems to be the case. Although people contribute to pollution through increased number of manufacturing enterprises, which in turn pollute the air, the independent effect of X_3 on Y is strongly negative.

Do people tend to clean up air? Paradoxically, yes. We are concerned with the effect of people when X_2, number of manufacturing enterprises, is held constant. Of the cities in Table 16.1, those with large numbers of people but with the number of manufacturing enterprises held constant tend to be those with large "bedroom communities," so, all other things being equal, they contribute less to pollution than do nonresidential occupants. Note also that the effects of people and manufacturing enterprises as indicated by the absolute magnitudes of those path coefficients are greater than that of temperature (-0.4336). Apparently, the lower the temperature, the more sulfur dioxide is produced. This correlation may be related to increased use of fossil fuels.

These two models are, of course, far from exhaustive, but comparing them should give you some feel for what we can do with such models. Remember that you have already observed correlations among all variables and that you are merely trying to reproduce these correlations by a heuristic path diagram. We now turn to another example to illustrate some different points.

In a laboratory study of fluctuations in housefly populations, biweekly censuses resulted in counts of live flies and counts of flies that had died during the previous two weeks. If, instead of the census of live adults, we record the change in adult

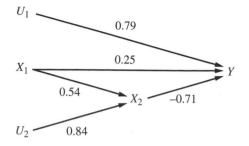

FIGURE 16.10 Path-coefficient diagram illustrating causal relationships in the housefly population study. (For explanation, see text.)

number since the last census (calling this variable Y), the density of the population at the previous census (calling it X_1), and the number of flies that died during the two-week period (calling it X_2), we can represent the interrelations by the path diagram in Figure 16.10, which represents one hypothesis. The diagram implies that density (X_1) determines mortality (X_2) in the population and that this mortality contributes to the decrease in population size. Through its effect on fecundity, however, density also affects population size directly. The observed correlations for these data are as follows: $r_{1Y} = -0.134$, $r_{2Y} = -0.573$, and $r_{12} = 0.540$; from these values, by procedures that should by now be familiar, we obtain the path coefficients that are indicated in the diagram.

From the diagram, then, density appears to affect the number of flies dying (X_2), which in turn has a strong negative effect on the population size eventually recorded. At the same time, density has a mild positive effect on the change in population size, Y. We might wish to change the model by including a variable not directly observed: the number of adult flies emerging in the population cage during the two-week period. We can argue convincingly that, except for experimental error, the change during the two weeks must reflect the balance between recruitment (newly emerged adult flies) and adult death. If we designate recruitment by X_3, we can write $Y = X_3 - X_2$, not allowing for experimental error. On this assumption, birth and death make equal contributions to the variance of Y, and we can write by definition that the paths from X_3 to X_2 and from X_3 to Y are $+\sqrt{\frac{1}{2}}$ and $-\sqrt{\frac{1}{2}}$, respectively, as they were in the path diagram of Figure 16.7.

In the path diagram for this new model (Figure 16.11), the unexplained factor completing the determination of Y has been removed, because presumably X_2 and X_3 completely determine Y. Now X_2 and X_3 are determined by unexplained factors, together with their common cause X_1 (density). For this path diagram, we lack the value for path p_{31} from density to the unrecorded recruitment variable. We can obtain this value easily, however, from the equation for the correlation, $r_{1Y} = p_{31}p_{Y3} + p_{21}p_{Y2}$. In this equation, all variables but p_{31} are known, and it is easy to solve for the path coefficient, which turns out to be 0.351. Thus, we learn that density influences the number of flies added to the population in the next two weeks positively but weakly.

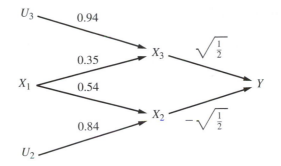

FIGURE 16.11 Path-coefficient diagram illustrating an alternative interpretation of the housefly population data. (For explanation, see text.)

This influence may be weak because the effect of density on adult recruitment is complex. The more flies there are, the more eggs are laid and the higher the input into the population. At the same time, the higher the density, the greater the crowding and the less adult input into the population (because of lower fecundities and increased egg and larval mortality). The relations may well not be linear, as assumed by the path-coefficient model, in which case further elaboration of the model would be indicated. What is important about this example is that we were able to obtain path coefficients from an observed variable to one that was not even measured (number of new flies added to the population). Using path coefficients, one can frequently construct similar models in which the effects of hypothesized intermediate variables can be estimated if there are enough equations to determine the unknown coefficients.

Another approach is to assume a reasonable experimental error in this system so that the two variables X_2 and X_3 do not completely determine Y, thus requiring an unknown factor to complete the determination of Y. In such a case, we might hesitate to postulate that the path coefficients from X_2 and X_3 were equal in magnitude. We could do a separate experiment in which the effect of density on adult production is investigated. From this experiment, we could obtain an estimate of the correlation between X_1 and X_3. We could then employ this correlation, together with the other known values, to obtain estimates of p_{Y2} and p_{Y3}.

Occasionally, situations arise in which a given unknown path coefficient is overdetermined; that is, there are more equations than unknowns. In such cases, one tries to adjust the values to effect a reasonable compromise. Estimating p_{31} from Figure 16.11 was one such example. When based on the equation for r_{1Y}, $p_{31} = 0.3504$, but when based on r_{2Y}, $p_{31} = 0.3512$. A preferable solution in such cases is SEM (discussed later), as it computes the best compromise.

We have not discussed the issue of hypothesis tests for path coefficients. Although in some cases the tests for multiple-regression analysis would apply, the main emphasis in path analysis has been on estimating magnitudes and directions of interactions rather than on the statistical tests. Remember also that underlying path analysis is linearity of the regressions. Any nonlinear effects must be accounted for in the unexplained variation or by the transformation of variables. Users of path coefficients, whose emphasis is on estimation and prediction, should be as careful as those who employ multiple-regression analysis to make certain that their data meet the assumptions and requirements of the latter.

Workers are enjoined to examine the VIF values of their data and to examine closely cases with VIFs > 10. Additionally, researchers are cautioned about using inadequate replication in view of the number of independent variables studied. Finally, the cautions voiced by us about confounding a priori with a posteriori testing of multiple hypotheses in the same data hold for hypothesis tests in path analysis as well.

Petraitis et al. (1996) have pointed out these problems and examined them in a large number of published datasets from ecology. However, in their enthusiasm for pointing out weaknesses of the path-coefficient approach, they include in their critique that more than half of the cases examined had computational errors and that they could not reproduce the results with the data furnished by the original authors. Surely, a statistical method cannot be blamed for being incorrectly calculated. Similarly, current editorial practice for most journals is to exclude voluminous original data from published papers. Authors and editors should not be held responsible for not filling up critical journal space with such material. These points and others are made in a rebuttal of Petraitis et al.'s critique by Pugesek and Grace (1998), who recommend structural equation modeling (SEM), of which path analysis is a special case. We turn to SEM immediately below and shall encounter path coefficients again later in this chapter.

Path coefficients were developed in an era when feasibility of computation necessitated relatively simple models. This simplicity extended to the number of criterion and predictor variables and to the complexity of their interactions. Furthermore, if hypothesis tests of the path coefficients were intended, the assumptions of Model I multiple regression had to be met. Finally, the number of equations established had to match the number of unknown parameters (path coefficients) to be solved.

The phenomenal increase in speed and power of computers during the latter half of the 20th century has made these constraints unnecessary and has led to a generalization of path analysis, known as **structural equation modeling, SEM.** The mathematical basis of the method was developed by Jöreskog (1970), who called it a method for the analysis of covariance structures. Since that time, it has seen wide application in the social sciences and in recent years also in biology—especially in ecological and evolutionary studies. See, for example, Shipley (2002).

Pugesek et al. (2003) is an edited volume that both introduces SEM and gives many examples of applications in ecology and evolutionary biology. In the material that follows, we shall show how SEM overcomes the limitations of path analysis and will illustrate the discussion with the already familiar dataset on city air pollution (Table 16.1). This will be one of several new parts in this edition of *Biometry* for which we cannot furnish an explicit box, because the complexity of the computations will require calculation by computer. Nevertheless, we shall furnish you enough information to give you an idea of what is being calculated.

In path analysis, the models one can consider are limited to those in which one can establish the same number of equations, as there are parameters (path coefficients) to be estimated. This is necessary in order to solve for each unknown parameter. Such models are called **just-identified** models. An **overidentified** model has more equations than parameters. Using the methods discussed earlier for path models, this would mean that one could attain different estimates of the parameters, depending on which sets of equations were used. The estimation methods used for SEM models use all of the equations simultaneously and thus obtain estimates that make use of more of the available information. Such estimates are likely to be more robust and have tighter confidence intervals. It is equally important that a wider range of models can be dealt with, as well as much larger models that would be difficult to solve by hand.

By contrast, an **underidentified** model has fewer equations than parameters and does not permit a solution. In such cases, one must either apply constraints (include additional equations) or estimate fewer parameters (by removing arrows from the model and/or by setting some of the parameters to a priori-selected numerical values).

The models for SEM can be more complex than those possible to solve by means of path analyses. As before, we distinguish between dependent and independent variables. However, in SEM, both may be subject to error. The independent variables are called *exogenous* variables because their variation is not a function of other variables in the model. The dependent variables are called *endogenous* variables because their variation is a function of independent (exogenous) variables and possibly other dependent (endogenous) variables. Neither type of variable may be directly observable due to measurement error. The unobservable (hypothetical) variables in the model are called *latent* variables. The directly observable variables (whether dependent or independent) are called *manifest* variables. The structural model describing the relationships between the m endogenous and n exogenous latent variables is formulated by the matrix algebra equation (see Appendix **B**)

$$\eta = B\eta + \Gamma\xi + \zeta$$

where η and ξ are vectors (m by 1 and n by 1) of endogenous and exogenous latent variables (η is on both sides of the equation because variation in some endogenous variables can be functions of variation in other endogenous variables), and ζ is a vector (m by 1) of structural errors (variation in the latent variables that is not predicted by other variables in the model). The matrix B (m by m) contains regression coefficients between the endogenous variables, and Γ is a matrix (m by n) of regression coefficients between the endogenous and exogenous variables. In addition, one can specify that the observed variables be measured with error and that some pairs of observed variables are correlated.

Using the path diagram in Figure 16.12a as an example, the equation for its corresponding structural model is

$$\eta = B\eta + \Gamma\xi + \zeta = \begin{bmatrix} SO_2 \\ M \end{bmatrix} = \begin{bmatrix} 0 & p_{SO_2M} \\ 0 & 0 \end{bmatrix} \begin{bmatrix} SO_2 \\ M \end{bmatrix} + \begin{bmatrix} p_{SO_2T} & p_{SO_2P} \\ 0 & p_{MP} \end{bmatrix} \begin{bmatrix} T \\ P \end{bmatrix} + \begin{bmatrix} u_1 \\ u_2 \end{bmatrix}$$

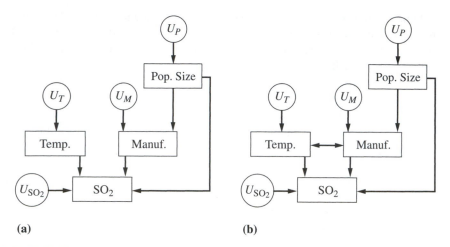

(a)   **(b)**

FIGURE 16.12   Alternative path diagrams for the air pollution data from Table 16.1. (For discussion, see text.)

Abbreviations for the variables in the preceding equation can be understood with the help of Figure 16.12**a**. As in the case of path analysis, the input data are typically the correlations among the observed variables. This means that the number of parameters to be estimated must be less than or equal to the number of unique correlations. The parameters to be estimated are just those regression coefficients corresponding to the paths that are actually included in the model. The number of possible parameters in a model could be much larger, but most are set equal to zero (these correspond to the possible paths that are not included in the model).

The actual estimation of the parameters is usually by maximum likelihood, and the fit of the model can be assessed by comparing log likelihoods (as discussed in Section 16.6 on AIC). These computations require specialized software whose details are beyond the scope of this book. Fortunately, several programs are available. The original and best-known software is LISREL (available from http://www.ssicentral.com/lisrel). Others include AMOS (http://www.spss.com/amos), EQS (http://www.mvsoft.com), the procedure CIALS within the SAS software (www.sas.com), and the sem and openmx packages for the R program library (available at http://cran.r-project.org/web/packages/sem/index.html and at http://openmx.psyc.virginia.edu). Pugesek et al. (2003) describes the method and gives many examples of applications in ecology and evolutionary biology. Some general texts include Schumacker and Lomax (2004), Kline (2004), and Raykov and Marcoulides (2006).

The model equation we furnished earlier and for the example may help to understand the SEM model, but you should know that to actually use the software one does not set up those matrices directly. One computes the correlation or covariance matrix for the observable variables before calling the sem package. The model is coded into the input for the sem software to show whether a potential path has a single-headed arrow, a double-headed arrow, or no connecting arrow at all. Many indexes are provided to measure how well the model fits. The sem software yielded the following estimates for the paths indicated in the model of Figure 16.12a.

Coefficient	Value	Standard Error
$p_{SO_2 T}$	-0.18080	0.109875
$p_{SO_2 M}$	1.70983	0.370799
p_{MP}	0.95527	0.046760
$p_{SO_2 P}$	-1.15090	0.364759
u_{SO_2}	0.38745	0.086682
u_T	1	0.223652
u_M	0.08746	0.019601
u_p	1	0.223652

The fit of the model can be assessed by how well the model can account for the correlations or covariances among the observed variables. For the path diagram in Figure 16.12a, the X^2 measure of goodness of fit was 8.8088. When compared to the χ^2-distribution with 2 degrees of freedom, it gives a P-value of 0.012224, indicating that the model did not fit satisfactorily. There were 2 degrees of freedom because there are 6 correlations among the 4 observed variables and 4 path coefficients were estimated. The mod.indices function in the sem software indicated that the greatest improvement to the fit could be made if a relationship between temperature and manufacturing were added to the model.

We show such a relation in Figure 16.12b. The input for the second model differed only by the addition of a single line indicating the presence of a correlation between manufacturing enterprises and temperature. Introducing the observed correlation of -0.130681 between temperature and manufacturing enterprises made very little difference in the estimates of the path coefficients but reduced the X^2 measure of fit to just 0.15745 with 1 degree of freedom (because another parameter was estimated from the data), yielding $P = 0.69151$. Note that because this adjustment to the model was based on an examination of five possible changes to the model, this probability does not furnish a proper test; but it is still useful as an indicator of the improvement in fit.

Factor analysis is a related method that is used to express covariation in terms of k underlying factors that explain a large part of the variance and covariance of the original variables. In this model, the variables are considered linear combinations of the underlying (latent) factors. The number of factors considered is usually much lower than the number of variables in the study. There are several methods of factor analysis, most of them somewhat complex. Introductory texts include Child (2006) and Mulaik (2009). Loehlin (2003) relates the methods of path, SEM, and factor analysis. In biology, the goals of factor analysis are (1) resolving complex relationships into the interaction of fewer and simpler factors, whether they are physiological and environmental variables underlying behavioral correlations or morphometric trends underlying morphological correlations; and (2) isolating and identifying causal factors behind biological correlations. For an example of the second, see Sokal et al. (1980).

16.4 Partial and Multiple Correlation

A **partial correlation coefficient** measures the correlation between any pair of variables when other, specified variables, are held constant. For example, the correlation $r_{13 \cdot 245}$ is the correlation between variables Y_1 and Y_3 when variables Y_2, Y_4, and Y_5 are held constant. We can keep these other variables constant experimentally or statistically, and either method should give equivalent results.

Again, we encounter the parallel approaches of experimental control and statistical control. It might not be immediately obvious why correlations should differ depending on whether other variables are free to vary or held constant. If you refer back to Figure 15.5, however, the reasons will become obvious. In Figure 15.5d, the correlation between Y_1 and Y_2 depends entirely on the common cause, Y_4. If we calculate the partial correlation $r_{12 \cdot 4}$, we should expect this to be zero because there is no reason why Y_1 and Y_2 should be correlated when Y_4 does not vary.

Keeping variable Y_4 constant in Figure 15.5e would enhance the relative correlating influence of common cause Y_5 on Y_1 and Y_2, because a larger proportion of the variation of Y_1 and Y_2 would now be accounted for by the common element Y_5. If both Y_4 and Y_5 are held constant, the partial correlation $r_{12 \cdot 45}$ would again be zero. Partial correlations are simple to compute when only three variables are involved. The formula for the partial correlation between variables Y_1 and Y_2, with variable Y_3 held constant, is

$$r_{12 \cdot 3} = \frac{r_{12} - r_{13}r_{23}}{\sqrt{(1 - r_{13}^2)(1 - r_{23}^2)}} \tag{16.20}$$

We will illustrate the computation of a partial correlation coefficient with an example from morphology based on a study of the intercorrelations among 18 morphological variables of alate aphids, *Pemphigus populitransversus* (Sokal, 1962). We compute the following three correlations for the characters forewing length (W), thorax length (T), and length of antennal segment 3 (A): $r_{TW} = 0.86$, $r_{AT} = 0.75$, $r_{AW} = 0.85$. There is a substantial correlation between wings and antennae. Some of this correlation is undoubtedly because of general size—that is, a bigger aphid will have longer wings and longer antennae. Is there also a special developmental factor, however, that tends to give aphids of the same general body length longer antennae if they have longer wings? We can investigate this question by means of partial correlations. Following Expression (16.20), we can write

$$r_{AW \cdot T} = \frac{r_{AW} - r_{AT}r_{TW}}{\sqrt{(1 - r_{AT}^2)(1 - r_{TW}^2)}} = \frac{0.85 - (0.75)(0.86)}{\sqrt{[1 - (0.75)^2][1 - (0.86)^2]}}$$

$$= 0.60736$$

The partial correlation between antennal segment 3 length and wing length, with thorax length held constant, is 0.61. Thus, although thorax length is known to be an adequate predictor of general size, it does not sufficiently explain the correlation between wing length and antennal segment 3 length. Other common causes must contribute to the correlation between these two morphological variables.

The partial correlation just computed is a first-order partial correlation coefficient (the ordinary product–moment correlation r_{12} is known as a pairwise correlation). We can also compute higher-order partial correlation coefficients. The second-order partial correlation between variables Y_1 and Y_2 with Y_3 and Y_4 held constant is

$$r_{12 \cdot 34} = \frac{r_{12 \cdot 3} - r_{14 \cdot 3} r_{24 \cdot 3}}{\sqrt{(1 - r_{14 \cdot 3}^2)(1 - r_{24 \cdot 3}^2)}} \tag{16.21}$$

From this formula we should have no difficulty constructing formulas for other second-order partial correlations or even higher-order partial correlations. However, coefficients of partial correlation of higher orders are rarely calculated, except when just one coefficient for two specific variables is the subject of interest. When a large matrix of correlation coefficients is available, methods for decomposing it—such as multiple-regression analysis, path analysis, SEM, principal-component or factor analysis—are preferable (see Section 16.11). Figure 16.13 illustrates the meaning of partial correlation graphically.

We can obtain further insight into the nature of partial correlation and the basis for Expression (16.20) by depicting the relationship by means of a path diagram. In Figure 16.14, we show the relations between the three aphid morphological variables as if thorax and wing length were causes of antennal segment 3 length. This choice is arbitrary; we could have shown wing length as the criterion with antennal segment 3 and thorax length as predictors. Given the three known correlation coefficients, we can work out the path coefficients p_{AW} and p_{AT} by the methods of Section 16.3. The path coefficients are $p_{AW} = 0.78725$ and $p_{AT} = 0.072965$, while the coefficients of nondetermination are $r_{U_A A}^2 = 0.27611$ and $r_{U_W W}^2 = r_{U_T T}^2 = 0.2604$.

We will now evaluate the partial correlation $r_{AW \cdot T}$. In this partial correlation, T (thorax length) is considered constant. By making it constant, we alter the pattern of action in the path diagram. We eliminate path p_{AT} and the resulting diagram is much simpler. The dashed lines in Figure 16.14 show paths and correlations that vanish when thorax length is assumed constant.

To obtain the partial correlation of W with A, variable T held constant, we recompute the path from W to A under the new conditions. Remember that a path coefficient is a standard partial regression coefficient. It thus measures the strength of a relationship as a proportion of the total standard deviation. In standardized variables, variances equal 1. With one variable held constant, however, the variances of neither A nor W remain at unity. The new variance of W as a proportion of the old variance is $1 - r_{WT}^2 = r_{U_W W}^2 = 0.2604$. The old variance of A was unity, and by the formula for determining the variance of a variable in a path diagram (see Section 16.3), we write

$$1 = p_{AT}^2 + p_{AW}^2 + 2p_{AT}p_{AW}r_{TW} + r_{AU_A}^2$$

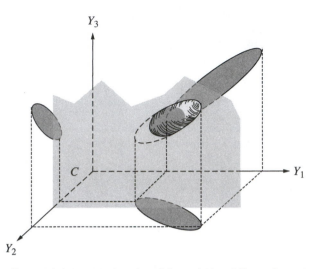

FIGURE 16.13   Geometric interpretation of partial correlation. A three-dimensional equal-frequency ellipsoid is shown at center with its two-dimensional projections (shadows) shown on each wall. Cutting through the figure at the point $Y_2 = C$ (a constant) is a plane. The trace of the ellipsoid on this plane is an equal-frequency ellipse (the central shaded ellipse). This ellipse represents the covariation in variables Y_1 and Y_3 when variable Y_2 is held constant at C. The magnitude of the partial correlation between Y_1 and Y_3 is reflected in the shape of this ellipse. The more elongate this ellipse is, the higher the partial correlation is. The shapes of the three elliptical shadows on the walls represent the ordinary correlations between the pairs of variables defining these walls.

What remains of this variance when the variation in T is held constant? All terms involving T vanish, leaving the unexplained variation $r^2_{AU_A}$ and the contribution due to the one remaining path coefficient, p^2_{AW}. But not all of p^2_{AW} remains. Remember that we have just seen that the effect of W has been diminished by the disappearance of the correlation r_{TW}. Thus, the remaining determination of A due to W is $(1 - r^2_{TW})p^2_{AW}$, and we can write the determination of A, T being held constant, as follows:

$$1 - p^2_{AT} - 2p_{AT}p_{AW}r_{TW} + r^2_{TW}p^2_{AW} = r^2_{AU_A} + (1 - r^2_{TW})p^2_{AW}$$

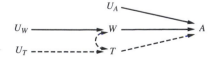

FIGURE 16.14   Path diagram illustrating partial correlation of wing length and antenna length, with thorax length held constant.

Referring back to Expression (16.3) will help you recall how standard partial regression coefficients were rescaled by dividing by the respective standard deviations of the variables. Here again we have to rescale the path coefficients in terms of the new standard deviations, which are less than unity. The appropriate scale factors are

$$p_{AW \cdot T} = \frac{p_{AW}\sqrt{1 - r_{WT}^2}}{\sqrt{r_{AU_A}^2 + (1 - r_{TW}^2)p_{AW}^2}} = \frac{0.78725\sqrt{0.2604}}{\sqrt{0.2761 + (0.2604)(0.78725)^2}} = 0.60736$$

Because the only path connecting W and A in the reduced diagram is this direct path, it is not surprising that the value of the path coefficient is equal to the value of the partial correlation coefficient $r_{AW \cdot T}$, as calculated earlier. Expressed in general terms, the adjusted path coefficient is

$$p_{Y1 \cdot 2} = p_{Y1}\frac{\sqrt{1 - r_{12}^2}}{\sqrt{1 - r_{Y2}^2}} \tag{16.22}$$

Expressed in words, it is the old path coefficient multiplied by the square root of the ratio of the remaining variation of variable X_1 to that of variable Y when variable X_2 is held constant. If we wish to express this relationship in terms of partial r's and b's, we can write

$$r_{Y1 \cdot 2} = b'_{Y1 \cdot 2}\frac{\sqrt{1 - r_{12}^2}}{\sqrt{1 - r_{Y2}^2}} \tag{16.22a}$$

When partial correlations are estimated between two criterion variables such as Y and Z in Figure 16.15, these path-coefficient formulas must be modified. Thus, to calculate the partial correlation $r_{YZ \cdot 1}$, we first calculate the correlation between Y and Z over the remaining paths, which are $p_{Y2}p_{Z2}$. This quantity must be multiplied by the square root of the proportion of the variance of X_2 remaining after X_1 has been held constant. In the example in Figure 16.15, variables X_1 and X_2 are independent, hence there is no change in the variance of X_2 when X_1 is held constant. Therefore, the proportion of X_2 remaining is 1, and $p_{Y2}p_{Z2}$ is multiplied by unity.

Just as we had to adjust for the remaining proportion of variation of variable Y in Expression (16.22), we now have to adjust for the remaining variation of both Y and Z. This adjustment signifies that a given correlating force may be more or less

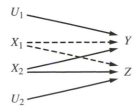

FIGURE 16.15   Path diagram illustrating partial correlation of Y and Z, with variable X_1 held constant.

important, depending on how much of the variation of Y or Z remains. Keeping X_1 constant in this instance reduces the determination of Y to $1 - p_{Y1}^2$ and that of Z to $1 - p_{Z1}^2$. Thus, we can write the formula for the desired partial correlation by inspection as follows:

$$r_{YZ \cdot 1} = \frac{p_{Y2}p_{Z2}(1)}{\sqrt{1 - p_{Y1}^2}\sqrt{1 - p_{Z1}^2}}$$

To test the null hypothesis that $\rho_{YZ \cdot 1 \ldots m} = 0$, one can proceed as in Box 15.2 for ordinary correlation coefficients, except that an adjustment must be made in the degrees of freedom for the number of variables held constant. We compute

$$t_s = r_{YZ \cdot 1 \ldots m}\sqrt{\frac{n - 2 - m}{1 - r_{YZ \cdot 1 \ldots m}^2}}$$

and test against $t_{\alpha[n - 2 - m]}$, where m is the number of variables held constant. We can also use the z-transformation (see Section 15.5), but its standard error is now modified to $1/\sqrt{n - 3 - m}$, where m is again the number of variables held constant.

Some final remarks on comparing the path-coefficient approach with the partial correlation approach: We have seen that with path coefficients, we make an explicit a-priori model of the system and then estimate the strength of the links given that model. The advantage of path coefficients is that all of the assumptions are made explicitly, and it becomes clear that inferences are a function of the underlying model. When we make different path-coefficient models of the same process, we come up with quite different coefficients. However, when computing partial correlations, some workers have been led to believe that they were carrying out "objective" and model-free computations. Yet partial correlations are also not model-free, because certain variables are isolated as the ones to be held constant. Partial correlations are a specific interpretation of relations among the variables concerned and therefore are merely a single realization of the set of possible structural relations among the set of k variables studied.

We should briefly mention $R_{Y \cdot 1 \ldots k}$, the **coefficient of multiple correlation**, which is the positive square root of the coefficient of multiple determination. Because the latter ranges from 0 to 1, the multiple-correlation coefficient can assume only positive values from zero to unity. This range distinguishes the multiple-correlation coefficient from the ordinary correlation coefficient, which ranges from -1 to $+1$. The coefficient of multiple correlation is a measure of how well the predicted values $\hat{Y}_i$ fit the observations Y_i. If we were to compute the entire array of values $\hat{Y}_i$ and calculate a correlation between $\hat{Y}_i$ and the observations Y_i, the resulting correlation coefficient $r_{\hat{Y}Y}$ would equal $R_{Y \cdot 1 \ldots k}$, which is indeed a correlation coefficient. This relation also holds for the bivariate case. Thus, $r_{\hat{Y}Y} = r_{XY}$, where $\hat{Y}$ is the estimate from the regression on X.

Our discussion of the coefficients of multiple correlation and determination must include one additional point. When $R_{Y \cdot 1 \ldots k}^2$ is estimated in the conventional way, it tends to overestimate $\rho_{Y \cdot 1 \ldots k}^2$, the parametric coefficient of determination, because the least squares estimates of the partial regression coefficients are fitted on the assumption that the variances and covariances in the normal equations [Expression (16.6)] are true values. In fact, they are only estimates. In the anova test of significance of the multiple regression (see Section 16.2), this fact is allowed for by dividing the explained and unexplained sums of squares by the appropriate degrees of freedom. $R^2_{Y \cdot 1 \ldots k}$, however, is computed from sums of squares and is not so corrected. It is therefore customary to compute an adjusted coefficient of multiple determination, as follows.

$$R^2_{Y \cdot 1 \ldots k}(\text{adj.}) = 1 - (1 - R^2_{Y \cdot 1 \ldots k})\left(\frac{n - 1}{n - k - 1}\right) \tag{16.23}$$

This value, corrected for the degrees of freedom, is always less than the $R^2_{Y \cdot 1 \ldots k}$ estimated from the sums of squares of the anova. The adjustment will be drastic when n is small and k is relatively large. By manipulating the terms in Expression (16.23), we can even show that for low values of ρ^2_Y and of n, the introduction of additional independent variables actually *decreases* R^2_Y. (adj.), although the values of estimated R^2_Y. tend to increase as the number of independent variables is augmented from k_1 to k_2. This phenomenon is the reason that many statisticians recommend that n equal or exceed $30k$, a condition not frequently met in biological research.

Box 16.2, step 14, describes the use of Statistical Table **R** to look up critical values to test whether a multiple correlation is equal to zero when $k \leq 4$. It also provides a general formula that makes use of the F-distribution.

16.5 Selection of Independent Variables

In a multiple-regression analysis, the dependent variable Y is modeled as being influenced by more than one independent variable. This makes it possible for the investigator to pursue objectives beyond the overall analyses featured in Sections 16.1 and 16.2. One goal is to find the set of variables that yield the best prediction as measured by having the smallest residual SS (and hence the largest R^2 value) for a sample. A second goal is to determine which variables are most important in the sense that their variation is the principal cause of variation in the dependent variable.

The way to reach the first goal seems clear—one should use as many variables as possible. There are, however, a number of practical problems with this approach. If the number k of independent variables is larger than the sample size n, then the SS matrix for the independent variables will be singular (see Appendix **B**) and the usual estimates for the regression coefficients cannot be computed. Even if the sample size is just somewhat larger than the number of independent variables, the ability of the regression solution to make predictions in *future* samples may be disappointing. The adjusted R^2 (used to compute the adjusted coefficient of determination in equation 16.23) takes into account the fact that larger sample sizes are needed as one includes additional independent variables.

One may find that adding further independent variables may actually decrease the adjusted R^2 values unless the new variables account for an appreciable additional proportion of the SS. A complication in such endeavors is that every time a new independent (predictor) variable is added or removed from a given dataset, many of the computations need to be repeated because the partial regression coefficient for an independent variable changes (in magnitude and sometimes even in sign) as different combinations of the other independent variables are used. Fortunately, the computations are easily accomplished nowadays with fast computers.

Another problem with including many independent variables is that it is often difficult to avoid having sets of variables that are highly correlated. As a result, the VIF values may become large and the estimates of the regression coefficients become unreliable and often overestimate the true slopes. [See Graham (2003) for a discussion of the problems that can be caused by large variance inflation factors.] Thus, it is often desirable to use a relatively small subset of the independent variables. How can one decide *which* variables to include?

The answer to this question relates to the second goal listed earlier—to determine which variables are most important in the sense that their variation is the cause of most of the variation in the dependent variable. This matter is more complex and introduces philosophical issues. In many applications, there may be variables that are highly correlated. In such studies, it is essential that one carefully define the objectives of the study and what one means by "important." The adjusted coefficient of determination given in Expression (16.23) takes into account the need for larger sample sizes as one includes additional independent variables. Whether using the adjustment or not, one usually finds that after some point, the addition of more variables tends to have a relatively small contribution to the ability to predict the dependent variable. One could consider using one of the methods described in Box 16.2 to test whether one or more independent variables account for more variation than one would expect by chance. Such tests are used in *stepwise regression procedures*, widely employed in the second half of the 20th century (and featured in detail in the third edition of this book). We give a brief description of stepwise regression below, followed by a criticism of its procedures and the reason why we no longer recommend it.

In the forward stepwise procedure, one tests whether adding variables would yield larger R^2 values than one would expect by chance. In the backward stepwise procedure, one tests whether one can leave out variables without decreasing the R^2 values appreciably. The probabilities used in such stepwise procedures are not correct probabilities because no allowance is made for the large numbers of tests carried out. Even more important is the lack of adjustments for the dependence of hypothesis tests made at one step of the selection procedure on hypothesis tests made in prior steps. The probabilities used in the stepwise procedures should be interpreted merely as parameters specifying the stringency of the criteria for adding or removing a variable from the working set.

Stepwise procedures also have the limitation that they do not necessarily find the optimal combination of variables. This is because the variables found for an optimal subset of size $m \leq k$ need not be members of a larger optimal subset. In

TABLE 16.3 Selection of Independent Variables. Summary Table Based on Exhaustive Enumeration of Subsets

Subset size m	Variables in best subset of size m	VIF_{max}	R^2	R^2_{adj}
1	2	1.000	0.41573	0.40075
2	2, 3	11.434	0.58632	0.56455
3	2, 3, 6	12.650	0.61740	0.58638
4	1, 2, 3, 5	14.642	0.63964	0.59960
5	1, 2, 3, 4, 5	14.703	0.66851	0.62115
6	1, 2, 3, 4, 5, 6	14.704	0.66951	0.61119

Air pollution data from Table 16.1 as well as Boxes 16.1 and 16.2.

Code numbers of independent variables (pollution predictors) in the best subsets (smallest residual *SS*) of the indicated size.

Table 16.3 above, we can see that for the air pollution data of Table 16.1, variable 6 is a member of the best subset of size $m = 3$ but is not a member of the best subsets of sizes $m = 4$ or $m = 5$. Unless there is a large number of variables, it is now feasible to explore all possible subsets of variables in order to be sure one has found the optimal combination of variables for a range of sizes of subsets.

For example, the air pollution data of Table 16.1 have only 6 independent variables, so it is easy to exhaustively test all possible subsets of variables taken one, two, three, ..., up to all six variables for a total of 63 subsets. For k independent variables, there will be $2^k - 1$ subsets. This quantity increases rapidly as k increases. It is over a million for $k = 20$, but such problems are still quite feasible on modern computers—especially if efficient algorithms are used and only subsets up to a specified size are considered. The results for the air pollution data are shown in Table 16.3. Note that although there is a VIF value for each variable, only the largest of these is shown in the column labeled VIF_{max} in the table.

One of the first things to notice in this table is that the results are, unfortunately, rather typical of ecological data in that there are large variance inflation factors, ≥ 10, even when only two independent variables are used. Although the R^2 and R^2_{adj} values increase as one adds more variables, one hesitates to use more variables because of the large VIF values.

Perhaps a more reasonable approach is to avoid tests of significance and explicitly seek a compromise between obtaining the smallest residual *SS* by using large numbers of variables and parsimony in which one tries to obtain simple interpretable models using as few variables as possible. The Kullback and Leibler (1951) measure of information has been suggested by several authors as an alternative. This is a distance between an estimated model and the true model. Even though one does not know the true model, one can still use this index to compare the relative closeness of different estimated models. This is the basis for what is now called the **Akaike information criterion, AIC** (Akaike, 1974).

$$\text{AIC} = -2\ln(L) + 2k \tag{16.24}$$

where L is the likelihood for the model (using parameter estimates that maximize this quantity, see Sections 13.8 and 17.1) and k is the number of parameters being fitted (the intercept and the partial regression coefficients). The $2k$ term can be interpreted as a penalty to prevent one from adding too many variables (for multivariate data an additional penalty is added). Small values of AIC indicate a better fitting model with fewer parameters. For a normally distributed dependent variable, our usual assumption, this criterion can be simplified to

$$\text{AIC} = n\left[\ln\left(\frac{\sum d_{Y\cdot1\ldots k}^2}{n}\right)\right] + 2k \tag{16.24a}$$

where $\sum d_{Y\cdot1\ldots k}^2$ is the unexplained sum of squares for the particular subset of $m < k$ independent variables chosen.

You can now easily see that it would be desirable to obtain a low value of AIC, because it would imply a low $\sum d_{Y\cdot1\ldots k}^2$ and hence high explanation by the independent variables chosen, or in other words, a good fit to the model. For small sample sizes (i.e., most practical applications), two alternative criteria have been proposed. The first is a corrected AIC,

$$\text{AIC}_c = \ln\frac{\sum d_{Y\cdot1\ldots k}^2}{n} + \frac{n+k}{n-k-2} \tag{16.24b}$$

and the second is an unbiased AIC (due to McQuarrie and Tsai, 1998) in which the residual sum of squares is divided by its degrees of freedom rather than by n:

$$\text{AIC}_u = \ln\frac{\sum d_{Y\cdot1\ldots k}^2}{n-k} + \frac{n+k}{n-k-2} \tag{16.24c}$$

Demidenko (2004) warns against the use of AIC when *multicollinearity* is present—that is, when several of the predictor variables are substantially correlated and thus the VIF value is large. Another criterion, the quasi-AIC, or QAIC, can be useful for data sets where the variance inflation factor, VIF, can become large (see Burnham and Anderson, 2002). The small sample version of this criterion is

$$\text{QAIC}_c = \frac{1}{\text{VIF}}\ln\frac{\sum d_{Y\cdot1\ldots k}^2}{n} + \frac{2k(k+1)}{n-k-1} \tag{16.24d}$$

Yet another proposed criterion is Mallows' C_p-statistic (Mallows, 1973). It is defined as

$$C_p = \frac{\sum d_{Y\cdot1\ldots k}^2}{s_{\text{resid}}^2} + 2k - n \tag{16.25}$$

where $\sum d_{Y\cdot1\ldots k}^2$ is the unexplained sum of squares using $m < k$ independent variables and s_{resid}^2 is the mean square for the residuals from the full model using all of the variables being considered. Using this criterion, one selects a combination of variables such that C_p is both small and as close to $k + 1$ as possible (the $+1$ allows for the fact that the intercept is also a parameter of the model). The decision is usually based on an examination of a plot of C_p against $k + 1$. Fortunately, these different criteria usually yield similar results. When they disagree, it is unclear which index is giving the best answer.

The results for the air pollution data are shown in the following table and in Figure 16.16 for the preceding sets of variables. For these data, the AIC_u and AIC_c criteria show minima indicating that 3 parameters (2 independent variables) should be used. These are X_2 (the number of manufacturing enterprises employing 20 or more workers) and X_3 (population size). On the other hand, the C_p-criterion indicates 4 parameters (three independent variables). Its use would have one add X_6 the average number of days with precipitation each year. The large variance inflation factors are a concern. Graham (2003) warns about using the results of analyses when the VIF values are even as small as 2.0. The $QAIC_c$ criterion takes the VIF values into account and also indicates an optimum using two variables so that seems like the best solution for these data.

Subset Size m	AIC	AIC_c	AIC_u	$QAIC_c$	C_p
1	282.78	283.10	6.8795	5.8768	21.109
2	271.69	272.34	6.6165	0.7928	5.5586
3	271.58	272.69	6.6244	1.0757	4.3610
4	272.25	273.96	6.6547	1.4778	4.0729
5	271.98	274.45	6.6658	2.0756	3.1032
6	275.05	278.44	6.7622	2.8336	5.0000

See above for definitions of the five information statistics in the column headings.

When an investigator is faced with a substantial number of putative predictor variables and has to decide which to include and which to omit, it is useful to know the consequences of the possible types of decisions. Much will depend on the correlation structure of the variables, but some general principles have been summarized by Rao (1971). In a multiple-regression situation in which the "truth" is known—that is, all the variables that contribute to an explanation of the dependent variable are known—the omission of a variable specified by the truth introduces bias to least squares estimates but reduces their variances. By contrast, including an irrelevant variable—that is, a predictor variable in addition to the truth—does not produce bias in the least squares estimates but does increase the variance and mean square error of all least squares estimates. Thus, unless a complete prediction equation is desired, omission of a variable that contributes little may be a favored procedure, because the smallest estimates of mean square error would be preferable over those that are unbiased. By contrast, adding variables that are not in the true specifications may decrease efficiency because of the increase in mean square errors.

After you have decided which subset of independent variables to employ in order to predict the dependent variable Y, you may wish to rank the independent variables in the order of their importance for predicting Y. This problem, too, does not

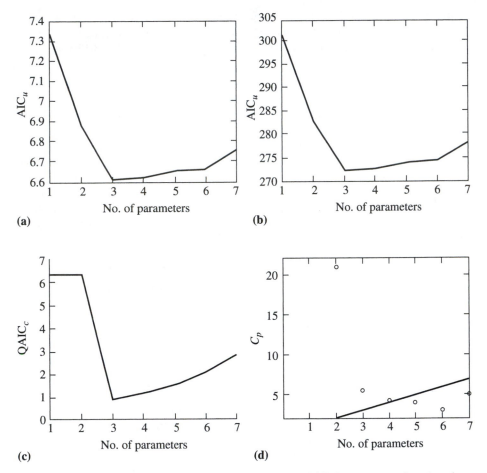

FIGURE 16.16   Plots of the **(a)** AIC_u, **(b)** AIC_c, **(c)** $QAIC_c$, and **(d)** C_p criteria, as a function of the number of parameters being fit for the city pollution data. Note that the number of parameters is the number of independent variables +1 to allow for the intercept. The line in part **(d)** connects points for which C_p is equal to the number of parameters—that is, $C_p = k + 1$.

have a unique solution. We can employ the absolute magnitude of the standard partial regression coefficients to indicate an order of importance. The complication is that when the independent variables are correlated, the relative magnitude of a coefficient depends on the other variables being considered at the same time.

The discussion in Section 16.1 about the example in Figure 16.2 illustrates how the importance of a variable depends on the other variables being held constant. Kruskal (1987) suggested using an average of squared partial correlation coefficients. He computes the average of the squared partial correlations between Y and X_i

with all possible combinations of the other independent variables held constant, e.g., $r_{Y1}^2, r_{Y1 \cdot 2}^2, r_{Y1 \cdot 3}^2, \ldots, r_{Y1 \cdot k}^2, r_{Y1 \cdot 23}^2, \ldots r_{Y1 \cdot 23 \ldots k}^2$. There are $2^{(k-1)}$ such combinations for each of the k independent variables. By this criterion, the variable with the highest average is considered the most important, that with the next highest is the second most important, and so on.

For the air pollution data of Table 16.1, we obtained the following averages (each based on 32 correlations) for the six variables: 0.16161, 0.40433, 0.25043, 0.03251, 0.04080, and 0.08159. The ranking is 3, 1, 2, 6, 5, 4. The ranking based on the b_i values is 3, 1, 2, 5, 4, 6. The ranking based on the partial correlations is 3, 1, 2, 4, 5, 6. These rankings agree only for the three most important variables. Kruskal points out that his procedure is appropriate only when the independent variables are random, a condition which actually violates the Model I regression assumption.

In practice, unbiased least squares estimates of the regression coefficients (β) are not often actually used. As mentioned earlier, whenever we use a procedure to obtain a solution with a reduced number of variables, we compromise the least squares principle by, in effect, setting some coefficients equal to zero. If the variables eliminated correspond to variables that have no effect in the true model, then we may expect this solution to be a better predictor in future samples even though we obtain a worse fit in the observed sample.

Other types of biased estimation methods—such as ridge regression, latent root (principal component) regression, and partial least-squares analysis of them—have been proposed. These methods are designed to yield more satisfactory predictive models when one is analyzing data that exhibit multicollinearity and thus have large variance inflation factors. The ridge regression method is easy to understand (once you have mastered Section 16.6). A small constant, k (the ridge coefficient, not to be confused with k, the number of independent variables), is added to the diagonal elements of the coefficient matrix on the left-hand side of the normal equations, yielding $(\mathbf{X'X} + k\mathbf{I}) = \mathbf{X'y}$. The next step is to investigate the effect of using small positive values for k ($k = 0$ yields the ordinary least squares solution).

Marquardt and Snee (1975) give an example in which the ridge regression technique yields a much better prediction equation than the one obtained by ordinary least squares. Latent root regression (also known as principal component regression) involves the regression of the dependent variable onto the principal axes of the correlation matrix for the independent variables. The partial least-squares regression method is more complex to describe, but it is potentially more powerful because it derives new variables that take into account the correlations between the independent and dependent variables. These methods should result in improved predictions in future samples when there are problems of multicollinearity.

The details of these biased regression techniques are beyond the scope of the present text, but these more advanced techniques should be considered when analyzing data sets for which the variance inflation factors are large (>10). Marquardt and Snee (1975) present an introduction to ridge regression. Latent root regression is described by Webster et al. (1974). Hocking (1976) and Draper and Smith (1998) give useful surveys of the entire field. Geladi and Kowalski (1986) is a standard introduction to partial least squares regression, though the examples are of applications in chemometrics.

IMPORTANT NOTE: Before accepting the results of a multiple-regression analysis, you should attempt to validate the adequacy of the solution. If you know enough about the phenomena under study, compare the predictions ($\hat{Y}$) and the coefficients (b_{Y_i}.) with biological theory. For example, ask whether the signs of the coefficients are reasonable. You can also check the reasonableness of extrapolations made using the model. Often one does not have sufficient theory to verify the reasonableness of the solution. The solution should then be tested using additional data. These data could be a portion of the data already available (which has not been used in the computations), or additional data could be collected specifically for testing the model.

Alternatively, one could use resampling methods such as bootstrapping. Such testing of the multiple-regression model using new data is very important because multiple-regression analysis is so powerful that it is able to capitalize on chance patterns in the observed sample to achieve a small s_{Y}^2. in the sample. Future predictions, however, may be quite poor if the data in the original sample are not representative of the population. In such cases, resampling methods will not adequately indicate the reliability of the regression equation. Validation is especially useful when the regression analysis yields very different estimates for the β's as one adds or deletes variables. The text by Burnham and Anderson (2002) provides a general discussion of these issues. They strongly emphasize the importance of careful selection of variables and discourage the use of exploratory methods such as stepwise regression or the all subsets regression. However, there is a need for exploratory studies, because biological relationships are often not sufficiently well understood.

16.6 Computation of Multiple Regression by Matrix Methods

Before we get into the material that we have to cover in this section, we have to subject you to two cautions. The first is our final reminder to those of you whose knowledge of matrix algebra is rusty (or to others unacquainted with the topic) that you should spend some time reviewing or learning the material from our précis of matrix algebra in Appendix **B.** The second caution concerns notation. Thus far, we have tried to distinguish strictly observations such as Y_i or X_i, featured as italicized *capital* letters from deviates, which are deviations from the mean and symbolized as italicized *lowercase* letters, thus: $y_i = Y_i - \overline{Y}$ or $x_i = X_i - \overline{X}$. A possible point of confusion with matrix algebra notation is that boldface capitals and lowercase letters are used to signify matrices and vectors, respectively. Therefore, **Y** represents a matrix of double subscripted observations Y_{ij}, as opposed to **y**, which is a single column of observations, shown here transposed to a row vector to save space, $\mathbf{y}' = (Y_1, Y_2, \ldots, Y_n)$. No provision is made for distinguishing deviates from direct observations; when we intend to work with the former, we shall specially indicate their nature to the reader. Contrary to our example for the vector **y,** lowercase letters are commonly used for the elements of vectors and matrices. And now, without any further cautions, let us proceed to the topic of this section.

In multiple regression, the linear model is

$$y_i = \beta_0 + \beta_1 X_{1,i} + \beta_2 X_{2,i} + \cdots + \beta_k X_{k,i} + \epsilon_i$$

As we have seen in Section 16.1, we estimate the parameters $\beta_1, \ldots, \beta_k$ and β_0, the Y-intercept α, by the solution of the set of $k + 1$ equations. These equations can be simplified greatly by stating the model in matrix notation as

$$\mathbf{y} = \mathbf{X}\boldsymbol{\beta} + \boldsymbol{\epsilon}$$

where $\mathbf{y}$ is a column vector of n observations on the dependent variable Y, $\mathbf{X}$ is an $n \times k$ matrix for the independent variables (including an initial column of all 1's if the Y-intercept β_0 is to be included in the model), $\boldsymbol{\beta}$ is a column vector of the k regression coefficients (the first one being the Y-intercept b_0 if the vector of 1's is included in the $\mathbf{X}$ matrix), and $\boldsymbol{\epsilon}$ is a column vector of random errors. Given an estimate, $\hat{\boldsymbol{\beta}}$, of the vector $\boldsymbol{\beta}$, a vector $\hat{\mathbf{y}}$ of the n predicted values of $\hat{Y}_i$ is computed as

$$\hat{\mathbf{y}} = \mathbf{X}\hat{\boldsymbol{\beta}}$$

The estimated regression coefficients are computed so as to minimize the SS of the error of prediction—that is, we minimize

$$SS_{y-\hat{y}} = (\mathbf{y} - \hat{\mathbf{y}})'(\mathbf{y} - \hat{\mathbf{y}}) = \mathbf{y}'\mathbf{y} - \mathbf{y}'\hat{\mathbf{y}} - \hat{\mathbf{y}}'\mathbf{y} + \mathbf{y}'\mathbf{y}$$

This quantity is equal to $\mathbf{y}'\mathbf{y} - 2\hat{\mathbf{y}}'\mathbf{y} + \mathbf{y}'\mathbf{y}$ because the terms are just scalars (and thus $\mathbf{y}'\hat{\mathbf{y}} = \hat{\mathbf{y}}'\mathbf{y}$). To find the vector $\mathbf{b}$ that minimizes this quantity, first replace $\hat{\mathbf{y}}$ by its estimate, $\hat{\mathbf{y}} = \mathbf{X}\hat{\boldsymbol{\beta}}$.

$$SS_{y-\hat{y}} = \mathbf{y}'\mathbf{y} - 2\hat{\boldsymbol{\beta}}\mathbf{X}'\mathbf{y} + (\mathbf{X}\hat{\boldsymbol{\beta}})'\mathbf{X}\hat{\boldsymbol{\beta}} = \mathbf{y}'\mathbf{y} - 2\hat{\boldsymbol{\beta}}\mathbf{X}'\mathbf{y} + \hat{\boldsymbol{\beta}}'\mathbf{X}'\mathbf{X}\hat{\boldsymbol{\beta}}$$

Then apply the usual procedure for finding the minimum or maximum of a function (take the derivative of $SS_{y-\hat{y}}$, set it equal to zero, and solve for $\mathbf{b}$). Using the vector and matrix derivative identities given in Appendix B, we find that the partial derivative of the residual sums of squares with respect to changes in the regression coefficients in the vector $\hat{\boldsymbol{\beta}}$ is

$$\frac{\partial SS_{y-\hat{y}}}{\partial \hat{\boldsymbol{\beta}}} = 0 - 2\mathbf{X}'\mathbf{y} + 2\mathbf{X}'\mathbf{X}\hat{\boldsymbol{\beta}}$$

Setting $\dfrac{\partial SS_{y-\hat{y}}}{\partial \hat{\boldsymbol{\beta}}}$ equal to zero makes it easy to solve for $\hat{\boldsymbol{\beta}}$. If

$$\frac{\partial SS_{y-\hat{y}}}{\partial \hat{\boldsymbol{\beta}}} = -2\mathbf{y}'\mathbf{X} + 2\mathbf{X}'\mathbf{X}\hat{\boldsymbol{\beta}} = 0$$

then $\mathbf{y}'\mathbf{X} = \mathbf{X}'\mathbf{X}\hat{\boldsymbol{\beta}}$ and thus solving for $\hat{\boldsymbol{\beta}}$ gives us the least-squares estimate of the vector of partial regression coefficients

$$\hat{\boldsymbol{\beta}} = (\mathbf{X'X})^{-1}\mathbf{X'y}$$

The vector $\hat{\boldsymbol{\beta}}$ can also be called $\mathbf{b}$ to simplify the notation. The $(\mathbf{X'X})^{-1}$ matrix is an important quantity and is often denoted $\mathbf{C}$. Its diagonal elements, c_{ii}, are called Gaussian multipliers.

It is useful to note that $\hat{\boldsymbol{\beta}} = \mathbf{b}$ is just a weighted average of the Y-values, $\mathbf{b} = \mathbf{wy}$, with the row vector of weights equal to $\mathbf{w} = (\mathbf{X'X})^{-1}\mathbf{X'}$. This allows us to easily derive a formula for its variance (the square of its standard error). If some statistic, z, can be represented as a weighted linear combination, such as $\mathbf{z} = \mathbf{wy}$, then the variance of z is $s_z^2 = \mathbf{w\Sigma w'}$, where $\mathbf{\Sigma}$ is the $n\times n$ matrix of variances and covariances for the n observations. If, as usual, one assumes the observations are independent and homoscedastic, then $\mathbf{\Sigma}$ is equal to $\mathbf{I}\sigma^2$, where σ^2 is the error variance of the observations ($\sigma_{y\cdot x}^2$ in our case) and $\mathbf{I}$ is the identity matrix. This means that the covariance matrix of the variances and covariances of the partial regression coefficients can be computed as

$$\mathbf{S_b} = \mathbf{ww'}s_{y\cdot x}^2 = (\mathbf{X'X})^{-1}\mathbf{X'X}(\mathbf{X'X})^{-1}s_{y\cdot x}^2 = (\mathbf{X'X})^{-1}s_{y\cdot x}^2$$

For a particular partial regression coefficient, say the ith, its variance is the ith diagonal element, c_{ii}, of $\mathbf{C} = (\mathbf{X'X})^{-1}$ times $s_{y\cdot x}^2$. The square roots of the diagonal elements are the standard errors of the partial regression coefficients. One can also compute the covariances among the regression coefficients using the off-diagonal Gaussian multipliers.

A similar procedure allows us to obtain the standard error of a predicted observation. The equation for $\hat{y}$ is

$$\hat{\mathbf{y}} = \mathbf{xb} = \mathbf{x}(\mathbf{X'X})^{-1}\mathbf{X'y}$$

This is again in the form of another weighted average $\hat{\mathbf{y}} = \mathbf{wy}$, where $w = \mathbf{x}(\mathbf{X'X})^{-1}\mathbf{X'}$ and $\mathbf{x} =$ a vector of the X-values used to predict Y. The standard errors are the square roots of the diagonal of the following covariance matrix:

$$\mathbf{S_{\hat{y}}} = \mathbf{ww'}s_{y\cdot x}^2 = \mathbf{x}(\mathbf{X'X})^{-1}\mathbf{X'X}(\mathbf{X'X})^{-1}\mathbf{x}s_{y\cdot x}^2 = \mathbf{x}(\mathbf{X'X})^{-1}\mathbf{x'}s_{y\cdot x}^2$$

We now present formulas for partitioning the sums of squares of the dependent variable in regression. We first carried out such partition in Chapter 14, but now we shall employ matrix notation. The total SS is $SS_{\text{Total}} = (\mathbf{y} - \bar{\mathbf{y}})^t(\mathbf{y} - \bar{\mathbf{y}})$, $df = n - 1$. The unexplained (residual) SS can be computed as $SS_{\text{Resid}} = (\mathbf{y} - \hat{\mathbf{y}})^t(\mathbf{y} - \hat{\mathbf{y}})$, $df = n - k - 1$, where k is the number of independent variables in the $\mathbf{X}$ matrix and the -1 takes into account the first column of all 1's that is used to obtain an estimate of the Y-intercept. The $\mathbf{X}$ matrix is often called a design matrix for reasons that will become obvious shortly.

The explained SS can be computed as the difference between the preceding two SS or directly as $SS_{\text{Expl}} = (\hat{\mathbf{y}} - \bar{\mathbf{y}})^t(\hat{\mathbf{y}} - \bar{\mathbf{y}})$, $df = k$, where k is the number of independent variables being tested. This quantity is usually not of direct interest. Rather, one wants to partition it to see how much of the SS can be accounted for by particular variables or sets of variables. There are two approaches for computing such quantities. In the indirect approach, one first fits a model with all of the independent variables included and notes the value for the residual SS. One then drops out the variable of interest, refits the model, and obtains a new residual SS. The difference between these two residual SS is the SS explained by the variable that was left out. If the variable was relatively unimportant, then the difference should be small. If an important variable is omitted, the residual SS should be much larger. The df of this explained SS is the number of variables that were left out.

There is also a way to compute this SS directly. One first fits the complete model with all the variables of interest. Then the SS explained by a particular set, A, of variables can be computed as $SS_{\text{exp 1}} = \mathbf{b}'_A \mathbf{C}_A^{-1} \mathbf{b}_A$, where $\mathbf{b}_A$ is the part of the vector of estimated partial regression coefficients corresponding to the variables in set A, and $\mathbf{C}_A$ is that part of the $\mathbf{C}$ matrix with elements corresponding to the variables in set A. The df are the number of variables in set A. Examples of these computations will be shown in the next section.

16.7 Solving Anovas as Regression Problems: General Linear Models

Regression methods can be used to solve for the various quantities needed in anova by expressing the linear models provided for each design using matrices and vectors. For example, for **Model I single classification designs,** the predicted value for the jth observation in the ith group would be $Y_{ij} = \mu + \alpha_i + \epsilon_{ij}$. This can be written in the form of a regression model as

$$\hat{y}_j = \mu + \alpha_1 x_{1j} + \alpha_2 x_{2j} + \cdots + \alpha_i x_{ij} + \cdots + \alpha_a x_{aj}$$

where x_{ij} is a dummy variable equal to 1 if the jth observation is in the ith group. It can be written compactly using matrix notation as $\hat{y} = \mathbf{Xb}$, where $\hat{y}$ is a vector of the predicted values of the Σn_i observations of the dependent variable, Y; $\mathbf{b}$ is a vector of the estimated coefficients; and $\mathbf{X}$ is a design matrix of dummy variables.

As an example, consider the single classification anova in Box 9.1. We use this example because it has only four groups and unequal, moderately sized sample sizes n_i, making it easy to display in this book. Note that this is explicitly a Model II anova, whose linear equation would include random effects A_i rather than treatment effects α_i. Nevertheless, we prefer to treat it here as though it were a Model I anova and the four hosts of the original example were four specially designed treatments applied to the ticks. We can define the design matrix and the matrix of regression coefficients as

$$\mathbf{X} = \begin{bmatrix} 1 & 1 & 0 & 0 & 0 \\ 1 & 1 & 0 & 0 & 0 \\ 1 & 0 & 1 & 0 & 0 \\ 1 & 0 & 1 & 0 & 0 \\ 1 & 0 & 0 & 1 & 0 \\ 1 & 0 & 0 & 1 & 0 \\ 1 & 0 & 0 & 0 & 1 \\ 1 & 0 & 0 & 0 & 1 \end{bmatrix} \text{ and } \boldsymbol{\beta} = \begin{bmatrix} \mu \\ \alpha_1 \\ \alpha_2 \\ \alpha_3 \\ \alpha_4 \end{bmatrix}$$

To save space here, just two observations in each group are shown. The actual matrix would reflect the unequal sample sizes of the four groups in this anova. $\mathbf{X}$ is called a **design matrix** because the pattern of coefficients defines the type of experimental design used. The first column of all 1's simply includes the parameter μ in all of the predictions. The other columns are dummy variables that include the proper α_i term. The first dummy variable (in the second column), for example, is defined as being equal to 1 for all observations that belong to group 1 and are equal to zero otherwise.

A problem with this formulation is that if one tries to apply the standard formula to estimate **b**, one finds that it "blows up" because the $\mathbf{X'X}$ matrix is singular and cannot be inverted. This is because the sum of the last four columns of the $\mathbf{X}$ matrix forms a unit vector that makes it equal to the first column. Thus, there is a linear dependency among the independent variables. One could have anticipated that there should be a problem because the SS for groups in a single classification anova has just $a - 1$ degrees of freedom. But in the matrix just presented, we have introduced a columns and thus a regression coefficients, corresponding to the a groups. The solution is to transform the dummy variables to get rid of this dependency by taking into account the constraint $\Sigma \alpha_i = 0$ that is part of the linear model. Perhaps the simplest transformation is to define just $a - 1$ dummy variables that are equal to 1 for observations in their group, -1 for observations in the last, ath group, and equal to zero otherwise. The matrices just presented then become

$$\mathbf{X} = \begin{bmatrix} 1 & 1 & 0 & 0 \\ 1 & 1 & 0 & 0 \\ 1 & 0 & 1 & 0 \\ 1 & 0 & 1 & 0 \\ 1 & 0 & 0 & 1 \\ 1 & 0 & 0 & 1 \\ 1 & -1 & -1 & -1 \\ 1 & -1 & -1 & -1 \end{bmatrix} \text{ and } \boldsymbol{\beta} = \begin{bmatrix} \mu \\ \alpha_1 \\ \alpha_2 \\ \alpha_3 \end{bmatrix}$$

This $\mathbf{X}$ matrix can be used to obtain the estimates, $\hat{\boldsymbol{\beta}}$, for $\boldsymbol{\beta}$. Because $\Sigma \alpha = 0$, we can, if desired, obtain an estimate of the last α_i as $\alpha_a = -\sum_{}^{a-1} \alpha_i$.

The order in which the classes or groups are listed is, of course, arbitrary. Using the classes in some other order will still result in the same SS for a given factor. Note that the dummy variables look like the vectors of contrasts described in Chapters 9 and 14. If one has $a - 1$ contrasts for the classes, they can be used in place of the arbitrary contrasts described earlier. They can be used even if they are not orthogonal—as long as they are not perfectly linearly dependent. The advantage of using contrasts is that the SS's explained by the individual dummy variables will then be of interest rather than just the SS explained by the entire set of dummy variables.

Because this is an example of a single classification anova, we only have to evaluate the *SS* among groups for a single dimension of classification. In anticipation of the two-way and multiway anovas to come, we shall call this source of variation factor *A*. The *SS* for factor *A* is the *SS* explained by the dummy variables for factor *A*. This can be computed indirectly by computing the difference in the residual *SS* for a model using the *X* matrix, including all dummy variables and a model with the dummy variables for factor *A* left out (in this case, leaving only the mean). It can also be computed directly using the following procedure.

We employ the full dataset from Box 9.1, $n = \Sigma n_i = 37$. As you may recall, this is an example of single classification anova with unequal sample sizes for the four groups. This causes no special problems with computation, so long as the matrices of dummy variables are coded correctly. To conserve space, we do not reproduce the coded input data matrix of dimension 4 by 37 here. Recall that the actual example of Box 9.1 (scutum widths of tick larvae sampled from 4 rabbit hosts) is a *Model II* anova, so that for an analysis of this dataset the *SS* among groups is the only sum of squares called for. Tests of differences between particular group means are not appropriate here. The computations shown as follows could be carried out using Excel, R, Matlab, SAS-IML, or other software that allows matrix operations.

First, compute the following matrices:

$$\mathbf{X'X} = \begin{bmatrix} 37 & 2 & 4 & 7 \\ 2 & 14 & 6 & 6 \\ 4 & 6 & 16 & 6 \\ 7 & 6 & 6 & 19 \end{bmatrix}$$

$$\mathbf{C} = (\mathbf{X'X})^{-1} = \begin{bmatrix} 0.029287 & 0.001963 & -0.004287 & -0.010056 \\ 0.001963 & 0.091787 & -0.026963 & -0.021194 \\ -0.004287 & -0.026963 & 0.079287 & -0.014944 \\ -0.010056 & -0.021194 & -0.014944 & 0.067748 \end{bmatrix}$$

$$\mathbf{b} = \mathbf{CX'y} = \begin{bmatrix} 360.8228 \\ 11.4272 \\ -6.4228 \\ -5.5151 \end{bmatrix}$$

The total *SS* is $\sum\sum(Y_{ij} - \bar{Y})^2 = (\mathbf{y} - \bar{\mathbf{y}})'(\mathbf{y} - \bar{\mathbf{y}}) = 5585.72973$, where $\bar{\mathbf{y}}$ is a column vector with all elements equal to $\bar{Y}$.

The within *SS*, $\sum\sum(Y_{ij} - \bar{Y}_i)^2$, is also the *SS* of residuals from the regression, $(\mathbf{y} - \hat{\mathbf{y}})'(\mathbf{y} - \hat{\mathbf{y}})$, where $\hat{\mathbf{y}} = \mathbf{Xb}$ and is equal to 3778.00256.

Let $\mathbf{C} = (\mathbf{X'X})^{-1}$, the inverse matrix using the **X** matrix just defined. Then let $\mathbf{C}_A$ be that part of **C** that just pertains to factor *A*—that is, the 3×3 submatrix of rows and columns 2, 3, and 4. Then let $\mathbf{b}_A$ be just the part of the vector of partial regression coefficients that pertain to factor A—that is, the last three elements in this example. The SS explained by factor A can then be computed as

$$SS_A = \mathbf{b}_A' \mathbf{C}_A^{-1} \mathbf{b}_A$$

$$= \begin{bmatrix} 11.4272 \\ -6.4228 \\ -5.5151 \end{bmatrix}' \begin{bmatrix} 13.8919 & 5.7838 & 5.6216 \\ 5.7838 & 15.5676 & 5.2432 \\ 5.6216 & 5.2432 & 17.6757 \end{bmatrix} \begin{bmatrix} 11.4272 \\ -6.4228 \\ -5.5151 \end{bmatrix}$$

$$= 1807.727$$

Anova table:

Source of variation	df	SS	MS	F_s	P
A	3	1807.72717	602.5757	5.263	0.0044
Within	33	3778.00256	114.4849		
Total	36	5585.72973			

If this example had been a Model I anova and if the dummy variables represent contrasts of particular interest, then the SS explained by each dummy variable can also be computed. For example, the SS explained by the first dummy variable for factor A (a contrast between the first and last groups) can be computed as

$$SS_{A1} = \mathbf{b}_{A1}' c_{A1}^{-1} \mathbf{b}_{A1}$$
$$= [11.4272][0.091787^{-1}][11.4272] = 1422.651$$

In a single classification anova, the results are always correct, whether sample sizes are equal or not. For that reason, it is simpler to use the conventional methods of Chapter 9 for single classification designs. That is not the case for two-way and multiway designs. For that reason, no examples with unequal sample sizes were shown in Chapter 11.

The linear model for **two-way designs** is $Y_{ijk} = \mu + \alpha_i + \beta_j + \alpha\beta_{ij} + \epsilon_{ijk}$. This model can be expressed in the form of a regression:

$$Y_j = \beta_0 + \alpha_1 X_{1j} + \cdots + \alpha_a X_{aj} + \beta_1 X_{a+1,j} + \cdots + \beta_b X_{a+b,j}$$
$$+ \alpha\beta_{11} X_{a+b+1,j} + \cdots + \alpha\beta_{ab} X_{a+b+ab,j} + \epsilon_{ijk}$$

where the X_{ij} are dummy variables equal to 1 if the jth observation is in the relevant group (and zero otherwise) and β_0 is the Y-intercept (it would be equal to μ if the independent variables were coded so that their means were all equal to zero).

For a two-way anova, say, the 2×3 anova in Box 11.2, the design matrix will again have as many rows as there are observations. The number of columns in the design matrix will be 1 plus the numbers of degrees of freedom for factors A, B, and the $A\times B$ interaction, which sum to 6 in this example. The first column is again a column of all 1's to include the mean (intercept) in the model. There are just two levels for factor A, so only one dummy variable is needed for factor A. It will contain 1 for those observations in the first level of factor A and -1 for those observations in the second level.

There are three levels for factor B, so two dummy variables must be created. The first dummy variable is 1 for observations in the first level, -1 for those in the last level (3) and 0 for all other observations (those in level 2). The second dummy variable is equal to 1 for those observations in the second level, -1 for those in the last level, and 0 for all other observations (those in level 1). Next, one must add dummy variables for the $A \times B$ interaction. There are two degrees of freedom for interaction, so there must be two dummy variables. Rather than constructing them directly, it is easiest to generate them by simply multiplying the corresponding elements of the dummy variables for factors A and B (this is why the symbol for an interaction is a multiplication sign). The one dummy variable for A times each of the two dummy variables for B will generate the required two dummy variables for the $A \times B$ interaction. The design matrix follows, but to save space only the rows for the first two observations in each cell are shown rather than all 8.

$$
\mathbf{X} = \begin{bmatrix}
1 & 1 & 1 & 0 & 1 & 0 \\
1 & 1 & 1 & 0 & 1 & 0 \\
1 & 1 & 0 & 1 & 0 & 1 \\
1 & 1 & 0 & 1 & 0 & 1 \\
1 & 1 & -1 & -1 & -1 & -1 \\
1 & 1 & -1 & -1 & -1 & -1 \\
1 & -1 & 1 & 0 & -1 & 0 \\
1 & -1 & 1 & 0 & -1 & 0 \\
1 & -1 & 0 & 1 & 0 & -1 \\
1 & -1 & 0 & 1 & 0 & -1 \\
1 & -1 & -1 & -1 & 1 & 1 \\
1 & -1 & -1 & -1 & 1 & 1
\end{bmatrix}
\quad
\mathbf{y} = \begin{bmatrix}
7.16 \\
6.78 \\
5.20 \\
5.20 \\
11.11 \\
9.74 \\
6.14 \\
3.86 \\
4.47 \\
9.90 \\
9.63 \\
6.38
\end{bmatrix}
$$

The computation of the various sums of squares, *using all 48 observations*, is carried out as follows. The total SS is $(\mathbf{y} - \bar{\mathbf{y}})'(\mathbf{y} - \bar{\mathbf{y}}) = 623.4066$.

$\mathbf{C} = (\mathbf{X}'\mathbf{X})^{-1}$

$$
= \begin{bmatrix}
0.020833 & 0 & 0 & 0 & 0 & 0 \\
0 & 0.020833 & 0 & 0 & 0 & 0 \\
0 & 0 & 0.041667 & -0.020833 & 0 & 0 \\
0 & 0 & -0.020833 & 0.041667 & 0 & 0 \\
0 & 0 & 0 & 0 & 0.041667 & -0.020833 \\
0 & 0 & 0 & 0 & -0.020833 & 0.041667
\end{bmatrix}
$$

The many zeros in the off-diagonal entries and the structured pattern are because of the equal sample sizes in this example. Solving for the regression coefficients, we obtain

$$\mathbf{b}' = [9.619583 \quad 0.588750 \quad -0.6245833 \quad -2.005833 \quad 0.9775 \quad -0.3125]$$

Because the means of the columns of $\mathbf{X}$ are equal to zero in this example, the Y-intercept, 9.619583, equals the mean, $\overline{\overline{Y}}$.

The within SS can be found as the SS of residuals from the regression, $(\mathbf{y} - \bar{\mathbf{y}}_i)'(\mathbf{y} - \bar{\mathbf{y}}_i) = 401.5213$, or most easily as the sum of the sums of squares within each cell of the two-way table.

Next, one should compute the SS for interaction, holding the main effects constant. Using the convention given earlier, the matrices $\mathbf{C}_A$, $\mathbf{C}_B$, and $\mathbf{C}_{A \times B}$ contain just the parts of matrix $\mathbf{C}$ that pertain to the variables given in their subscripts.

$$SS_{A \times B \cdot AB} = \mathbf{b}'_{AB} \mathbf{C}^{-1}_{AB} \mathbf{b}_{A \times B}$$

$$= [0.9775 \quad -0.3125] \begin{bmatrix} 32 & 16 \\ 16 & 32 \end{bmatrix} \begin{bmatrix} 0.9775 \\ -0.3125 \end{bmatrix} = 23.9262$$

Finally, we compute the SS for the main effects, holding the effects of the other main effect and interaction constant (note that in balanced designs such as this one, the results are identical whether or not one holds the other variables constant).

$$SS_{A \cdot B, A \times B} = \mathbf{b}'_A \mathbf{C}^{-1}_A \mathbf{b}_A = [0.58875][48][0.58875] = 16.6381$$

$$SS_{B \cdot A, A \times B} = \mathbf{b}'_B \mathbf{C}^{-1}_B \mathbf{b}_B$$

$$= [-0.624583 \quad -2.005833] \begin{bmatrix} 32 & 16 \\ 16 & 32 \end{bmatrix} \begin{bmatrix} -0.624583 \\ -2.005833 \end{bmatrix} = 181.3210$$

The results can be summarized in a standard anova table as follows, but note the more complicated descriptions of the sources of variation.

Anova table:

Source of variation	df	SS	MS	F	P
A adj. for B, $A \times B$	1	16.6381	16.6381	1.740	0.1943
B adj. for A, $A \times B$	2	181.3210	90.6605	9.483	0.0004
$A \times B$ adj. for A, B	2	23.9262	11.9631	1.251	0.2967
Within	42	401.5213	9.5600		
Total	47	623.4066			

These are identical to the results shown in Box 11.2 because this is a balanced design (there are equal numbers of observations in each cell of the two-way table). However, the procedures just given are also required for unbalanced designs as long as there are no empty cells. In the latter case, the $\mathbf{X}'\mathbf{X}$ matrix may be singular and special methods must be carried out. In addition, the degrees of freedom may be reduced from that indicated by the preceding formulas.

If the dummy variables represent meaningful contrasts, then the *SS* for *A*, *B*, and *A*×*B* can be partitioned into the contributions of each contrast or sets of contrasts. The sums of squares in which the *SS* for each factor is computed with all other factors held constant are called *type III sums of squares*. When the effect of interaction is assumed to be very small, some would compute the *SS* for the main effects by ignoring the interaction rather than holding it constant (this is a simple example of a type II sums of squares). To ignore a factor, we remove it from the **X** matrix before the *SS* for the other factors are computed. In the case of a balanced design (such as in the preceding example), the numerical results are identical, but they can be very different in an unbalanced design. For type I sums of squares, the factors are listed in a defined order and each *SS* holds earlier factors constant but ignores later ones—as is done in polynomial regression (see Section 16.9). Shaw and Mitchell-Olds (1993) discusses these models and their use in ecology.

The computations for multiway anova designs (three-way and higher-order factorial designs) are carried out in a similar manner. The dummy variables for the two-way interactions are as just computed for each pair of factors. The dummy variables for a three-way interaction are generated by multiplying together the dummy variables for the three factors. Once the design matrix has been constructed, the computation of the *SS* for the factors and their interactions proceeds as previously described.

The design matrix for a **nested anova design** is usually more tedious to construct because there often are more degrees of freedom and hence more dummy variables that need to be constructed. As before, if factor *A* has *a* levels, then $a - 1$ dummy variables need to be constructed. If the *i*th level of factor *A* has b_i levels of factor *B*, then $b_i - 1$ dummy variables must be constructed for each level of factor *A*. Likewise, if the *j*th level of factor *B* nested within the *i*th level of factor A has b_{ij} levels, then $b_{ij} - 1$ dummy variables must be generated. Once the design matrix is constructed, the computation of the *SS* follows the patterns just described.

This approach is very general and allows the analysis of more complex anova designs that combine the designs discussed earlier. For example, a two-way anova might be nested within a single classification anova.

16.8   Analysis of Covariance (Ancova)

An **ancova design** can be viewed as simply an anova design with the introduction of one or more independent variables for statistical control. These variables are called *covariates*. Such covariates are added to the analysis if the investigator has reason to suspect that variation in the covariates could affect the dependent variable. A classic example would be a study of the effect of some treatment such as diet, where a response such as weight gain may be influenced by the initial weight of each

TABLE 16.4 Final Weights (Y) and the Corresponding Initial Weights (X) in Bags of 10 Oysters at Five Locations with Respect to a Power Plant

Locations									
1. Intake, bottom		2. Intake, surface		3. Discharge, bottom		4. Discharge, surface		5. Bay (away from plant)	
Y	X	Y	X	Y	X	Y	X	Y	X
32.6	27.2	33.8	28.6	35.2	28.6	35.0	29.3	24.6	20.4
36.6	32.0	31.7	26.8	29.1	22.4	27.0	21.8	23.4	19.6
37.7	33.0	30.7	26.5	28.9	23.2	36.4	30.3	30.3	25.1
31.9	26.8	30.4	26.8	30.2	24.4	30.5	24.3	21.8	18.1

Data from Littell et al., 2008. Units of weight are not furnished in the original publication.

experimental animal. If it is not possible to obtain animals that are essentially identical in weight at the start of the experiment (an example of experimental control), then one could record each animal's initial weight and use ancova to correct the data for the effects of variation in initial weights (an example of statistical control).

Alternatively, one could simply perform an anova on the weight gain, but that approach does not allow for the fact that the amount of change in weight may not be the same for animals that differ in their initial weights (for instance, larger animals might gain more weight). An ancova uses regression to determine what relationship, if any, exists between the covariate (initial weight) and the dependent variable (final weight). If there is such a relationship, we allow for differences in initial weights of the animals and test whether the remaining variation of final weights differs in response to treatment (diet) differences.

As an example for single classification ancova designs, we shall use data from an experiment on the growth of oysters. The purpose of the analysis was to determine whether the elevated water temperatures near a power plant influenced the growth of oysters. Four bags of 10 oysters were placed in four locations close to a power plant, plus at a control site away from the plant. The weight of each bag was recorded at the beginning and at the end of the experiment. The initial weight was used as a covariate. We feature the original data in Table 16.4. Figure 16.17**a** shows a plot of the data with the common regression line superimposed.

The design matrix for the ancova is made up as the design matrix for an anova with five treatments but with the covariate X appended as the last column. Note: the covariate and the dependent variable Y could have been coded as deviations from their means. That would have the advantage of reducing rounding errors in the computations. To simplify the presentation, we left the variables in their original form. The design matrix, **X**, and the **y** vector follow.

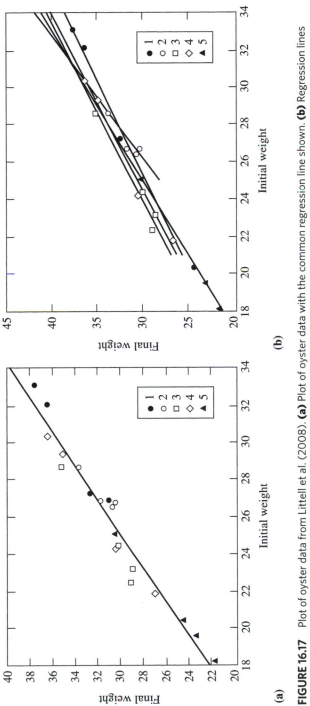

(a)

(b)

FIGURE 16.17 Plot of oyster data from Littell et al. (2008). **(a)** Plot of oyster data with the common regression line shown. **(b)** Regression lines fitted to each oyster group with the slopes allowed to vary by group.

$$\mathbf{X} = \begin{bmatrix} 1 & 1 & 0 & 0 & 0 & 27.2 \\ 1 & 1 & 0 & 0 & 0 & 32.0 \\ 1 & 1 & 0 & 0 & 0 & 33.0 \\ 1 & 1 & 0 & 0 & 0 & 26.8 \\ 1 & 0 & 1 & 0 & 0 & 28.6 \\ 1 & 0 & 1 & 0 & 0 & 26.8 \\ 1 & 0 & 1 & 0 & 0 & 26.5 \\ 1 & 0 & 1 & 0 & 0 & 26.8 \\ 1 & 0 & 0 & 1 & 0 & 28.6 \\ 1 & 0 & 0 & 1 & 0 & 22.4 \\ 1 & 0 & 0 & 1 & 0 & 23.2 \\ 1 & 0 & 0 & 1 & 0 & 24.4 \\ 1 & 0 & 0 & 0 & 1 & 29.3 \\ 1 & 0 & 0 & 0 & 1 & 21.8 \\ 1 & 0 & 0 & 0 & 1 & 30.3 \\ 1 & 0 & 0 & 0 & 1 & 24.3 \\ 1 & -1 & -1 & -1 & -1 & 20.4 \\ 1 & -1 & -1 & -1 & -1 & 19.6 \\ 1 & -1 & -1 & -1 & -1 & 25.1 \\ 1 & -1 & -1 & -1 & -1 & 18.1 \end{bmatrix} \quad \mathbf{y} = \begin{bmatrix} 32.6 \\ 36.6 \\ 37.7 \\ 31.0 \\ 33.8 \\ 31.7 \\ 30.7 \\ 30.4 \\ 35.2 \\ 29.1 \\ 28.9 \\ 30.2 \\ 35.0 \\ 27.0 \\ 36.4 \\ 30.5 \\ 24.6 \\ 23.4 \\ 30.3 \\ 21.8 \end{bmatrix}$$

We commence with a preliminary anova of the dependent variable Y, as shown in Section 16.7. We, therefore, ignore the last column of $\mathbf{X}$ for now. The total SS is 358.67 and the within treatment SS is 160.2625. The explained SS can be computed indirectly, by subtraction, or directly, as follows.

$$\mathbf{b}' = \begin{bmatrix} 30.84500 & 3.63000 & 0.80500 & 0.00500 & 1.38000 \end{bmatrix}$$

$$SS_A = \mathbf{b}'_A \mathbf{C}_A^{-1} \mathbf{b}_A$$

$$= \begin{bmatrix} 3.63 & 0.805 & 0.0045 & 1.38 \end{bmatrix} \begin{bmatrix} 0.2 & -0.05 & -0.05 & -0.05 \\ -0.05 & 0.2 & -0.05 & -0.05 \\ -0.05 & -0.05 & 0.2 & -0.05 \\ -0.05 & -0.05 & -0.05 & 0.2 \end{bmatrix} \begin{bmatrix} 3.63 \\ 0.805 \\ 0.0045 \\ 1.38 \end{bmatrix}$$

$$= 198.407$$

These results can be summarized in a standard anova table.

Source of variation	df	SS	MS	F_s	P
Among	4	198.4070	49.6017	4.6425	0.0122
Within	15	160.2625	10.6842		
Total	19	358.6695			

These results imply a substantial effect due to treatment (location). However, when we perform an ancova to take the initial weight into consideration, we obtain a very different result, as we shall soon see.

The ancova computations proceed as for an anova (description in Section 16.7), with the covariate treated simply as another dummy variable.

$$\mathbf{C} = (\mathbf{X'X})^{-1}$$

$$= \begin{bmatrix} 5.03950 & 0.77283 & 0.27407 & -0.21500 & 0.12880 & -0.19369 \\ 0.77283 & 0.31970 & -0.00755 & -0.08330 & -0.03005 & -0.03000 \\ 0.27407 & -0.00755 & 0.21505 & -0.06181 & -0.04292 & -0.01064 \\ -0.21500 & -0.08330 & -0.06181 & 0.20926 & -0.05555 & 0.00835 \\ 0.12880 & -0.03005 & -0.04292 & -0.05555 & 0.20333 & -0.00500 \\ -0.19369 & -0.03000 & -0.01064 & 0.00835 & -0.00500 & 0.00752 \end{bmatrix}$$

$$\mathbf{b'} = \begin{bmatrix} 2.94229 & -0.69189 & -0.72770 & 1.20733 & 0.65969 & 1.08318 \end{bmatrix}$$

The Y-intercept for a common regression line is 2.9423 and the slope of the regression line is 1.0832. The residual SS after fitting this model is 4.2223; thus, by subtraction from the total SS, the SS explained by the model is 354.45. This is much larger than for the ordinary anova just shown. Next, it is of interest to decompose the explained SS to give the separate contributions of the treatments and the covariate.

The SS explained by the covariate, X, with the treatments held constant is

$$SS_{X \cdot A} = \mathbf{b}_X' \mathbf{C}_X^{-1} \mathbf{b}_X = 1.08318(0.0075191)^{-1}1.08318 = 156.0402$$

The SS explained by the treatments with the covariate X held constant is

$$SS_{A \cdot X} = \mathbf{b}_A' \mathbf{C}_A^{-1} \mathbf{b}_A$$

$$= \begin{bmatrix} -0.69189 \\ -0.72770 \\ 1.20730 \\ 0.65969 \end{bmatrix}' \begin{bmatrix} 0.31970 & -0.00755 & -0.08330 & -0.03005 \\ -0.00755 & 0.21505 & -0.06181 & -0.04292 \\ -0.08330 & -0.06181 & 0.20926 & -0.05555 \\ -0.03005 & -0.04292 & -0.05555 & 0.20333 \end{bmatrix} \begin{bmatrix} -0.69189 \\ -0.72770 \\ 1.20730 \\ 0.65969 \end{bmatrix}$$

$$= 12.089$$

We can summarize these results in an ancova table as follows.

Source of variation	df	SS	MS	F_s	P
X adj A (Covariate adjusted for treatment)	1	156.0402	156.04018	517.3840	$<10^{-11}$
A adj X (Treatment adjusted for covariate)	4	12.0894	3.02234	10.0212	0.00048
Residual	14	4.2223	0.30159		
Total	19	358.6695			

An unusual property of these data is that the SS explained by the treatments without taking the covariate into account, 198.4070, is much larger than the SS explained by treatments taking X into account, 12.0894. The explanation for this anomaly (an anomaly because in an ancova one usually expects the *SS* for treatments to be larger when the confounding effects of a covariate are taken into account) can be seen by examining the data in Table 16.4 or the plots in Figure 16.17. The oysters do not seem to have been sorted evenly into the bags at the start of the experiment. The initial weights of the bags were quite different in the different treatment locations. For example, three of the bags placed in the control location (bay) were quite a bit below average in initial weight. As a result, without allowing for these differences in initial weights, there would be differences due to location even if the treatments had no effect.

An important assumption of ancova, which we have ignored so far, is that of the homogeneity of regression slopes. This assumption necessitates a test of the parallelism of the slopes of the regression lines within each treatment class. Figure 16.17**b** shows the regression lines with the slopes computed separately for each group. The slope for the second group looks distinctly steeper than that for the others. We have already encountered a test for equality of slopes in Section 14.8, but that was limited to two groups or classes; here, we study the extension of such a test to three or more classes.

The homogeneity of the slopes can be tested using a simple extension of the setup used for the ancova itself. What one does is to replace the single column in the design matrix used for the covariate with separate columns for the covariate within each treatment class. For the oyster data just previously shown, we would replace column 6 with the 5 new columns shown in the following. The resultant design matrix would then have a total of 10 columns.

$$
\mathbf{X}_{6\cdots 10} =
\begin{bmatrix}
27.2 & 0 & 0 & 0 & 0 \\
32.0 & 0 & 0 & 0 & 0 \\
33.0 & 0 & 0 & 0 & 0 \\
26.8 & 0 & 0 & 0 & 0 \\
0 & 28.6 & 0 & 0 & 0 \\
0 & 26.8 & 0 & 0 & 0 \\
0 & 26.5 & 0 & 0 & 0 \\
0 & 26.8 & 0 & 0 & 0 \\
0 & 0 & 28.6 & 0 & 0 \\
0 & 0 & 22.4 & 0 & 0 \\
0 & 0 & 23.2 & 0 & 0 \\
0 & 0 & 24.4 & 0 & 0 \\
0 & 0 & 0 & 29.3 & 0 \\
0 & 0 & 0 & 21.8 & 0 \\
0 & 0 & 0 & 30.3 & 0 \\
0 & 0 & 0 & 24.3 & 0 \\
0 & 0 & 0 & 0 & 20.4 \\
0 & 0 & 0 & 0 & 19.6 \\
0 & 0 & 0 & 0 & 25.1 \\
0 & 0 & 0 & 0 & 18.1
\end{bmatrix}
$$

One next performs the regression using the new design matrix to see how much better the prediction of Y is, using these additional variables.

$$\mathbf{b}'_{6\cdots10} = [0.982647 \quad 1.501355 \quad 1.056066 \quad 1.056925 \quad 1.223886]$$

The slope of treatments 1, 3, and 4 are rather similar. The slope of the second and fifth groups are somewhat steeper than that of the others (see Figure 16.17**b**). The residual SS from fitting this model with separate slopes is 2.834. The original residual SS was 4.2223, so the fit has improved slightly. The difference, 1.3883, is the additional SS explained by having five slopes rather than just one. This SS has 4 degrees of freedom (the difference between 5 slopes and 1) and is tested over the residual MS of the model with separate slopes. These results are summarized in the following table as well as in Figure 16.17**b**. Although suspicious, the differences among the slopes are not large enough (given the small sample sizes) to enable us to reject the null hypothesis that the slopes are homogeneous.

Source of variation	df	SS	MS	F_s	P
Residual common slope	14	4.2223	0.301594		
Residual separate slopes	10	2.8340	0.283401		
Explained by separate slopes	4	1.3883	0.347078	1.225	0.36017

Sometimes there is more than just one covariate in a study. For example, an investigator may wish to control for more than one variable such as temperature, soil moisture, concentration of nitrogen, etc. The computations are not much more difficult. One simply includes the other variables in the design matrix and then computes the sums of squares for the factors of interest, with all of the covariates held constant simultaneously. In addition, one would also have to check the validity of the assumption that the slopes are homogeneous for these additional variables.

16.9 Curvilinear Regression

In Figure 14.12, we saw that a linear regression is not always sufficient to account for differences among sample means. Often there are large deviations around linear regression. You learned to test for such deviations in Section 14.5. Figures 14.12e and 14.12f illustrate the difference in types of deviations around regression. The former appears to be a curvilinear trend; the latter seems to represent added heterogeneity without any apparent regularity, which might arise from a situation such as the following. Suppose we measure the weight of four samples of mice, each sample representing a different age group. We would expect the weights to be different and to show a positive regression on age. However, if we had also fed the four samples different rations, we could expect the sample means to deviate markedly around the regression on age because of the differences in amount or quality of food eaten.

Returning to the curvilinear trend seen in Figure 14.12e, this might best be described by fitting a **curvilinear regression,** which can be expressed as a polynomial function of the following general form:

$$\hat{Y} = a + bX + cX^2 + dX^3 + \cdots \tag{16.26}$$

or in a form more in tune with solution by matrix methods:

$$\hat{Y} = b_0 + b_1X + b_2X^2 + b_3X^3 + \cdots \tag{16.26a}$$

Such an expression uses increasing powers of X, the independent variable, and a different regression coefficient preceding each power of X. Note that this form of the equation is formally identical with the multiple-regression equation [Expression (16.1)]. The independent variables $X_1, X_2, \ldots, X_k$ are now replaced by powers of the single independent variable X, that is, $X^1, X^2, \ldots, X^k$. We will make use of this relationship in computing the coefficients of the powers of X.

A curvilinear regression equation that involves only terms of X and X^2 will yield a parabola. As the power of X increases, the curve becomes more and more complex and will be able to fit a given set of data increasingly well. However, with each added power of X, the mean square over which the regression MS must be tested loses another degree of freedom. Recall that the test for linear regression employs a mean square for deviations from regression based on $a - 2$ degrees of freedom. When there are only $a = 5$ means to be regressed, the highest-order polynomial we can fit is a cubic, because a cubic polynomial would leave us with an MS for deviations with a single degree of freedom $(5 - 2 - 1 - 1 = 1)$. Even when there are many groups, fifth and even higher powers are fitted only occasionally. For most biological work, terms of X no higher than cubic are used.

Generally, such polynomial regressions are empirical fits; that is, in most cases we cannot read structural meaning into the terms of X^2 and X^3. We simply use them to obtain a better-fitting regression line to a set of points for any of the purposes mentioned in Section 14.6. Because our aim is to find the best-fitting line, we first fit a linear regression to a set of data, and then see if we can remove an appreciable portion from the residual sum of squares by adding a quadratic term to the regression equation. We do the same for the cubic and possibly for higher terms. Curve fitting is often, therefore, a sequential addition algorithm with a hypothesis test for each increase in powers of X to find out whether an improvement in fit can be demonstrated. Of course, if we had some a priori basis for expecting, say, a quadratic relationship, then we should proceed directly to test this. As we increase the number of terms in the polynomial, the coefficients of the other powers of X also change, just as the partial regression coefficients changed as we practiced selection of independent variables in Section 16.5. To be precise in symbolism, we should write the first three polynomial regressions as follows:

$$\hat{Y} = a_0 + b_0X$$
$$\hat{Y} = a_1 + b_1X + c_1X^2$$
$$\hat{Y} = a_2 + b_2X + c_2X^2 + d_2X^3$$

where in most cases $a_0 \neq a_1 \neq a_2$, $b_0 \neq b_1 \neq b_2$, and $c_1 \neq c_2$.

Computing a polynomial regression using multiple-regression analysis is straightforward. If no special computer program for polynomial regression is available, then a multiple-regression program can be employed, with the independent variable X taking the place of X_1 in the multiple-regression analysis and with the successive powers $X^2, X^3, \ldots, X^k$ taking the place of $X_2, X_3, \ldots, X_k$, respectively. Serious computational problems can arise, however, when one goes beyond cubic regression. One way to reduce these problems is to code the X-variable so that it has a mean of zero and its sum of squares equals 1.

Box 16.4 shows the result of a multiple-regression analysis of gene frequencies in a mussel population as a function of distance along the shoreline. The data are arcsine-transformed allele frequencies (Y) regressed on distance of the sample east of Southport, Connecticut. We computed the first five powers of X. The analysis of variance and the coefficients for powers 1 through 5 (and their standard errors) are shown in Box 16.4.

BOX 16.4 Polynomial Regression

Frequency of the Lap^{94} allele (in 17 samples of the intertidal blue mussel *Mytilus edulis*) regressed on the location of the population, which is expressed in miles east of Southport, Connecticut. The allele frequencies have been transformed using the angular transformation.

Frequency of allele $Lap^{94}(Y)$	Miles east of Southport, Conn. (X)
0.155	1
0.15	9
0.165	11
0.115	17
0.11	18.5
0.16	25
0.17	36
0.355	40.5
0.43	42
0.305	44
0.315	45
0.46	51
0.545	54
0.47	55.5
0.45	57
0.51	61
0.525	67

SOURCE: Unpublished data by R. K. Koehn (graph shown in Koehn, 1978).

Box 16.4 (continued)

Testing polynomials up to the fifth power

Anova table

Source of variation	df	SS	MS	F_s	P
Explained	5	1588.6071	317.7214	28.124	6.48×10^{-6}
Linear	1	1418.3746	1418.3746	125.553	2.35×10^{-7}
Quadratic	1	57.2773	57.2773	5.070	0.04576
Cubic	1	85.1075	85.1075	7.534	0.01907
Quartic	1	11.8538	11.8538	1.049	0.32773
Quintic	1	15.9939	15.9939	1.416	0.25911
Unexplained	11	124.2671	11.2970		
Total	16	1712.8742			

$R_{Y.}^2 = 0.92745 \qquad R_{Y.} = 0.96304$

Table of coefficients

Powers of X	a or $b_{Yi.}$	SE (of a or b)
a	22.2401	4.20617
X	1.049,167,3	1.36982
X^2	$-0.151,721,4$	0.13422
X^3	0.006,555,5	5.05596×10^{-3}
X^4	$-0.000,103,313$	8.08923×10^{-5}
X^5	0.000,000,551,848	4.63791×10^{-7}

Solution of cubic polynomial

Anova table

Source of variation	df	SS	MS	F_s	P
Explained	3	1560.7594	520.2531	44.462	4.2676×10^{-7}
Unexplained	13	152.1148	11.7011		
Total	16	1712.8742			

$R_{Y.}^2 = 0.91119 \qquad R_{Y.} = 0.95456$

Table of coefficients

Powers of X	a or $b_{Yi.}$	SE (of a or b)
a	26.22325	3.07616
X	-0.94408	0.38040
X^2	0.042,145	0.01244
X^3	$-0.000,350$	0.000,170,6

Box 16.4 (continued)

Computation of cubic polynomials by matrix methods

Construct a design matrix as shown for multiple regression in Section 16.6. Instead of using independent variables X_1, X_2, X_3 as illustrated in that section, employ successive powers X^1, X^2, X^3 of the independent variable. As before, the first column corresponds to the intercept and contains the value 1 n times.

X_0	X	X^2	X^3
1	1	1	1
1	9	81	729
1	11	121	1331
1	17	289	4,913
1	18.5	342.25	6,331.6
1	25	625	15,625
1	36	1296	46,656
1	40.5	1640.3	66,430
1	42	1764	74,088
1	44	1936	85,184
1	45	2025	91,125
1	51	2601	132,651
1	54	2916	157,464
1	55.5	3080.3	170,956.7
1	57	3249	185,193
1	61	3721	226,981
1	67	4489	300,763

Except for rounding errors, the results are the same as shown above.

Note that none of the coefficients of the quintic equation has small probabilities when judged by its standard error, but overall the quintic equation removes a much larger proportion of the variation of variable Y (93%) than one would expect by chance. This result does not correspond to the results of the first anova table in Box 16.4, which tests the explained sums of squares in a successive, incremental manner—that is, whether the addition of a quartic or quintic term would explain an added proportion of the variance? By contrast, the coefficients indicate the partial contributions of the separate terms, given that a quintic equation has been fitted to the data. It appears from the anova, however, that adding powers beyond the cubic does not increase the coefficient of determination, which is 0.91119 for the cubic equation more than expected by chance. The equation was recomputed using only X, X^2, and X^3. The results are shown in the second half of Box 16.4. The polynomial regression equation is

$$\hat{Y} = 26.22325 - 0.94408X + 0.042145X^2 - 0.000350X^3$$

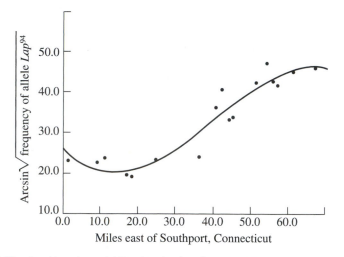

FIGURE 16.18   A cubic polynomial fitted to the data from Box 16.5 on gene frequency in the blue mussel, *Mytilus edulis,* as a function of distance from Southport, Connecticut.

Now the separate partial regression coefficients are unlikely to be zero when judged by their standard errors. The cubic equation is illustrated in Figure 16.18. The line fits the data fairly well (which we expected, because the coefficient of determination is so large, 0.91119). The considerations affecting the interpretation and choice of power of the polynomial regression are quite similar to those discussed near the end of Section 16.5.

Another useful way of examining the goodness of fit is to look at the residuals, a procedure introduced in conjunction with simple linear regression in Section 14.10. Figure 16.19 shows the standardized residuals to the linear and to the cubic fits for the example in Figure 16.18. Note that the residuals to the linear fit still show considerable trend, which is removed by fitting a third-degree polynomial.

The coefficients of the powers of the predictor variable are not tested customarily, but if desired, such a test can be carried out exactly as in multiple-regression analysis. Typically, one would retain lower powers of X up to the highest power one wishes to test. If, on the other hand, there is evidence that only a higher power of X relates meaningfully to the dependent variable Y, although lower powers have no biological meaning, then the lower powers could be omitted. We cannot easily interpret the probabilities for the coefficients for a given power of X.

Hurvich and Tsai (1990) point out that if a given dataset is used to determine a specific model (i.e., the polynomial function to be fitted), one cannot also use the same dataset to test the individual regression coefficients or to compute their confidence intervals. In a simulation study, Hurvich and Tsai showed that actual coverage rates (confidence coefficients) were substantially below nominal rates. They suggested randomly splitting the dataset into two (not necessarily equally sized) portions, determining the model in one portion and carrying out estimates and significance tests

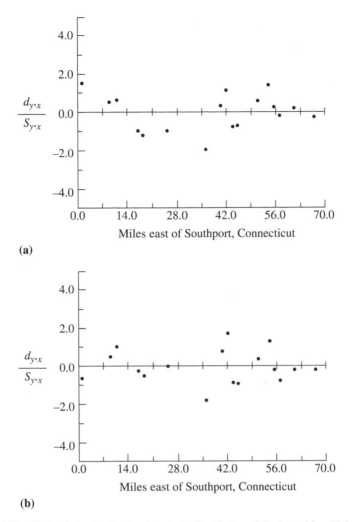

(a)

(b)

FIGURE 16.19   Plots of standardized residuals for the *Mytilus edulis* data of Box 16.4.
(a) Residuals from linear regression. **(b)** Standardized residuals from cubic regression.

using the second portion. Before accepting a given polynomial solution as anything more than a description of the observed sample, empirical testing of the coefficients as described in Section 16.5 is advisable.

Curvilinear regression by fitting a polynomial in X can be carried out by the BIOMstat computer program.

As many of our readers will have guessed by now, one can equally well solve a polynomial equation by matrix methods. One creates a design matrix with the first column, X^0, corresponding to the intercept, with all values equal to 1. The next column, X^1, has the values for the independent variable. Subsequent columns contain the values of the independent variable raised to the second, third, etc. powers. The regression coefficients are estimated from the usual equation

$$\mathbf{b} = (\mathbf{X'X})^{-1}\mathbf{X'y}$$

The main difference is that we must solve for the partial regression coefficients in a sequential manner. First, we employ just the first two columns of **X,** then the first three, first four, etc. until the desired maximum power is reached. While simple and direct, it is not computationally efficient to solve each problem from scratch. There are special algorithms that incorporate an additional variable when a solution for the previous variables already exists. An important problem is that the variance inflation factors usually become very large, and there are very serious rounding errors when more than just a few powers are used. This is because for positive values of X, the values for X and X^2 will be correlated, as will X^3, etc. The correlations between odd and even powers can be eliminated by simply coding X as deviations from its mean before computing the various powers. Unfortunately, the even powers will still be highly correlated. For evenly spaced values for the independent variable, one can use more complicated transformations such as replacing the independent variable and its various powers by the coefficients of orthogonal polynomials, which are explained immediately following.

By using **orthogonal polynomials,** we can circumvent the recomputation of coefficients as powers are added to a polynomial equation. Expression (16.26) is replaced by another polynomial regression equation,

$$\hat{Y} = A + B\xi_1 + C\xi_2 + D\xi_3 + \cdots \tag{16.27}$$

where the capital letters again are constant regression coefficients, and the ξ_j's are coefficients of special orthogonal polynomials of X; that is, the ξ_j's are complicated functions of the powers of X. These coefficients are uncorrelated with each other; by contrast, the powers of X in Expression (16.26) are clearly correlated. The orthogonal polynomials are defined so that $\Sigma\xi_j = 0$ and $\Sigma\xi_j\xi_k = 0$. As a result, the $\mathbf{X'X}$ matrix is diagonal, which makes the normal equations trivial to solve. Using the simple procedures presented earlier, orthogonal polynomials can be used only for data in which the values of the independent variable X have an equal number of replicates and are evenly spaced. Other cases will require matrix methods.

There are sets of these orthogonal polynomials for each distinct number, n, of X values and for increasing powers of X. In Statistical Table **EE**, we list orthogonal polynomials that permit the fitting of maximally third-degree regression equations to sample sizes up to $n = 12$. Coefficients for sample sizes up to 75 are found in Fisher and Yates (1963, Table XXIII). Pearson and Hartley (1958, Table 47) furnish coefficients for fitting sixth-degree equations up to sample size 52. Because the polynomial coefficients are frequently fractional, they are multiplied by the coefficients λ_j furnished underneath each column of polynomials in Statistical Table **EE**. The sum of squares of the polynomials, a convenient quantity for computation, is also provided. Remember that because the sum and hence the mean of each column of polynomial coefficients is zero, their sum of squares is equal to the sum of the squared coefficients, $\Sigma(\xi_j - \bar{\xi})^2 = \Sigma\xi_j^2$. The coefficients that have been multiplied by λ_j are generally symbolized ξ'_j. The coefficients listed in Statistical Table **EE** are not the only sets of orthogonal polynomials, but they are those commonly employed for curvilinear regression.

These orthogonal polynomials are used as dummy variables on which Y is regressed in the manner presented in Section 14.9. They are one type of coefficient of linear comparisons, whose values for different powers of X are mutually orthogonal for any given sample size n. The unweighted coefficients in the example on length of larval period in *Drosophila melanogaster* (see Box 14.8) are actually orthogonal polynomial coefficients. By successively regressing the Y-values on linear, quadratic, cubic, and higher coefficients, we can test whether the data show curvilinear regression and to what degree.

Box 16.5 illustrates such a computation for the trout data from Exercise 14.14, set up in the manner of the linear comparisons of Box 14.8. We use the coefficients

---

BOX 16.5 | **Testing for Curvilinear Regression by Orthogonal Polynomials**

This method applies to equally-spaced X-values with equal sample sizes at each X. (Trout data from Exercise 14.13; length of trout regressed on density; $a = 4$, $n = 25$.) The coefficients for the orthogonal polynomials are found in Statistical Table **EE** under sample size 4, because there are four means being regressed. These coefficients are used in linear comparisons of the type carried out in Box 14.8.

Regression		Density X				(1)	(2)	(3)
		175	350	525	700			
								$\sum \hat{y}^2 =$
	$\bar{Y}$	3.51224	3.71020	2.92680	2.53960	$\overset{a}{\sum} c\bar{Y}$	$\sum c^2$	$n(1)^2/(2)$
Linear	c_{iL}	-3	-1	1	3	-3.70132	20	17.12471
Quadratic	c_{iQ}	1	-1	-1	1	-0.58516	4	2.14008
Cubic	c_{iC}	-1	3	-3	1	1.37756	20	2.37209

Completed anova

Source of variation	df	SS	MS	F_s	P
Among densities	3	21.63688	7.212	6.941	0.000281
Linear regression	1	17.12471	17.124	16.481	0.000100
Quadratic regression	1	2.14008	2.140	2.060	0.154460
Cubic regression	1	2.37209	2.372	2.283	0.134083
Within densities	96	99.71535	1.039		

of the orthogonal polynomials given in Statistical Table **EE** for $n = 4$ points. The computation of the explained sum of squares due to the linear, quadratic, and cubic components is as shown in Box 14.8 and is not discussed further here. The completed anova is shown at the bottom of Box 16.5. It is obvious that there are differences among densities, and we can easily reject the null hypothesis for linear regression. If you carried out Exercise 14.13, however, you would have found that you were unable to reject the null hypothesis for linear regression. How can we explain this apparent discrepancy?

If we test for successively higher orders of polynomials, we use the residual MS as a denominator mean square. Because the residual MS is large in this example, we are unable to reject the null hypothesis that $\beta = 0$. If, on the other hand, we decide to test a-priori for several individual-degree-of-freedom comparisons, we assume by implication that there is no random heterogeneity around linear regression and that the departures from linearity are due to the curvilinear functional relationships. We then test over the error MS (in Box 16.5 called "Within densities"). The model changes, and the interpretation changes accordingly. Decomposing the SS for deviations around linear regression into quadratic and cubic terms showed neither term to be significant. Thus, a linear regression seems adequate to describe the relation between length of trout and density at which it has been reared.

We restrict our discussion of orthogonal polynomials in relation to curvilinear regression to just testing for the degree of curvilinearity. Although it is possible to obtain a prediction equation in terms of orthogonal polynomials as given by Expression (16.27), it is complex and subject to rounding errors.

We can also use orthogonal polynomials to decompose an interaction sum of squares into single-degree-of-freedom contrasts. An example of this application is shown in Box 16.6, which is largely self-explanatory. The analysis is a typical two-way anova with replication (see Chapter 11), with interaction between the two factors (presence or absence of auxin and number of hours in sucrose) affecting growth of pea sections. There is no overall linear regression of length of pea sections on number of hours in sucrose, but when we partition the interaction and test for the difference in linear regression on hours in sucrose between those samples run with auxin and those without auxin, we find differences in the linear component.

This difference is clearly illustrated in Figure 16.20, which shows that number of hours in sucrose increased the length of pea sections without auxin but decreased the length of those with auxin. When the data are considered together, as they would be in an overall analysis of variance, time in sucrose does not seem to affect the length of the pea sections, which is an entirely misleading finding. Thus, the careful study of interactions is frequently repaid with considerable insight into the mechanism of a given scientific process.

Remember that the method we have presented using tables of orthogonal polynomials can be applied only with equally spaced values of X and equally replicated samples of Y. When sample sizes are not equal, the same partitioning can be done, but general linear model methods have to be used to compute the sums of squares

---

BOX 16.6 | Partitioning Interaction in a Balanced Two-Way Anova by Means of Orthogonal Polynomials

Lengths of pea sections in ocular units after 20 hours of growth in tissue culture. The sections had been subjected to two treatments: (1) presence or absence of auxin and (2) various periods of the 20 hours in the presence of sucrose. Ten sections were run per treatment combination.

Subgroup means for treatment combinations based on 10 items

			Hours in sucrose			
Auxin	0	2	4	6	8	10
Absent	58.9	61.7	65.2	67.1	68.7	70.7
Present	73.5	69.8	67.4	64.7	64.2	65.0
$\overline{Y}$	66.20	65.75	66.30	65.90	66.45	67.85

SOURCE: Data from W. Purves (unpublished results).

The means are plotted in Figure 16.20. A two-way anova yielded the following results:

Anova table

Source of variation	df	SS	MS	F_s	P
Time in sucrose	5	56.5417	11.3083	0.370	0.868517
Auxin	1	126.0750	126.0750	4.120	0.044838
Time × Auxin interaction	5	1584.4750	316.8950	10.355	3.94×10^{-8}
Within subgroups	108	3304.9917	30.6018		

In view of the small probability for interaction, it is not especially interesting to partition the sums of squares for the effect of time in sucrose, but this procedure is done here because the sums of squares explained by linear, quadratic, and cubic regression are needed for subsequent computations.

Computation proceeds in the manner of Box 14.8: $n = 10$ observations per time period; $a = 6$ time periods; $b = 2$ auxin treatments.

			X = hours in sucrose				(1)	(2)	(3)
									$\sum \hat{y}^2 =$
Comparisons	0	2	4	6	8	10	$\sum\limits^{a} c\overline{Y}$	$\sum c^2$	$n(1)^2/(2)$
$\overline{Y}$	66.20	65.75	66.30	65.90	66.45	67.85			
Linear (c_{iL})	−5	−3	−1	1	3	5	9.95	70	28.2864
Quadratic (c_{iQ})	5	−1	−4	−4	−1	5	9.25	84	20.3720
Cubic (c_{iC})	−5	7	4	−4	−7	5	4.95	180	2.7225
									$\Sigma = 51.3809$

Box 16.6 (continued)

The coefficients of linear comparisons are the orthogonal polynomial coefficients from Statistical Table **EE** for sample size 6.

$$\text{SS for deviations from cubic regression}$$

$$= \text{SS (time in sucrose)} - \text{SS (linear} + \text{quadratic} + \text{cubic)}$$

$$56.5417 - 51.3809 = 5.1608$$

For none of the regression terms could we reject the respective H_0. Because the sums of squares unexplained after regression was quite small, higher polynomials were not fitted.

There is only one degree of freedom for the effects of auxin; therefore, no partitions are possible. The interaction SS has 5 df, however, so it is possible to subdivide this SS to see whether a particular aspect of the interaction is more important than others.

There are many logical ways to partition the interaction, depending on the nature of the factors. One way would be to study the differences in the linear, quadratic, cubic, and possibly higher regressions in the two treatments. To do this, perform the analysis carried out in the first part of this box on the column means separately for each row treatment as follows.

Without auxin

				X = hours in sucrose			(1) $\sum^{a} c\bar{Y}$	(2) $\sum c^2$	(3) $\sum \hat{y}^2 =$ $n(1)^2/(2)$
Comparisons	0	2	4	6	8	10			
$\bar{Y}$	58.9	61.7	65.2	67.1	68.7	70.7			
Linear (c_{iL})	-5	-3	-1	1	3	5	81.9	70	958.2300
Quadratic (c_{iQ})	5	-1	-4	-4	-1	5	-11.6	84	16.0190
Cubic (c_{iC})	-5	7	4	-4	-7	5	2.4	180	0.3200

With auxin

				X = hours in sucrose			(1) $\sum^{a} c\bar{Y}$	(2) $\sum c^2$	(3) $\sum \hat{y}^2 =$ $n(1)^2/(2)$
Comparisons	0	2	4	6	8	10			
$\bar{Y}$	73.5	69.8	67.4	64.7	64.2	65.0			
Linear (c_{iL})	-5	-3	-1	1	3	5	-62.0	70	549.1429
Quadratic (c_{iQ})	5	-1	-4	-4	-1	5	30.1	84	107.8583
Cubic (c_{iC})	-5	7	4	-4	-7	5	7.5	180	3.1250

To test whether the two treatments differ in the slope of their regression lines, add together the sums of squares explained by the separate regressions for each treatment row and subtract the sums of squares explained by the common regression line (computed from the column means). Calculated in this manner, each SS has $b - 1$ degrees of freedom.

$$SS_{\text{time}_L \times \text{auxin}} = 958.2300 + 549.1429 - 28.2864 = 1479.0865$$

$$F_s = 1479.0865/30.6018 = 48.333, P = 2.86 \times 10^{-10}$$

Box 16.6 (continued)

The slope of the linear regression of pea section length on hours in sucrose differs, depending on the presence or absence of auxin. To test whether the shapes of the two lines also differ, we perform similar operations on the sums of squares explained by higher-order regression terms (in this case, quadratic and cubic terms).

$$SS_{time_Q \times auxin} = 16.0190 + 107.8583 - 20.3720 = 103.5053$$

$$F_s = 103.5053/30.6018 = 3.382, P = 0.06866$$

$$SS_{time_C \times auxin} = 0.3200 + 3.1250 - 2.7225 = 0.7225$$

$$F_s = 0.7225/30.6018 = 0.024, P = 0.8772$$

Because the SS explained by the remaining terms is so small ($1584.4750 - 1479.0865 - 103.5053 - 0.7225 = 1.1607$), it is not worthwhile to compute the individual sums of squares for higher-order components of the time $\times$ auxin interaction ($F_s < 1$). If the null hypothesis for any of these terms were rejected, we could conclude that the lines differed in shape.

Complete anova table

Source of variation	df	SS	MS	F_s	P
Subgroups	11	1767.0917			
Time in sucrose	5	56.5417	11.3083	0.370	0.86852
$Time_L$	1	28.2864	28.2864	0.927	0.33849
$Time_Q$	1	20.3720	20.3720	0.666	0.41635
$Time_C$	1	2.7225	2.7225	0.089	0.76608
Residual	2	5.1608	2.5804	0.084	0.91920
Auxin	1	126.0750	126.0750	4.120	0.04484
Time $\times$ Auxin	5	1584.4750	316.8950	10.355	3.94×10^{-8}
$Time_L \times$ Auxin	1	1479.0865	1479.0865	48.333	2.86×10^{-10}
$Time_Q \times$ Auxin	1	103.5053	103.5053	3.382	0.06866
$Time_C \times$ Auxin	1	0.7225	0.7225	0.024	0.87717
Residual	2	1.1607	0.5804	0.019	0.98122
Within subgroups	108	3304.9917	30.6018		
Total	119				

and the SS for the individual comparisons will no longer add up to the overall SS. Unequally spaced values of X require the computation of the orthogonal polynomials—they cannot simply be looked up in a table. As discussed in Section 14.9, a common way to circumvent the computations of curvilinear regression is to transform the data to linearity.

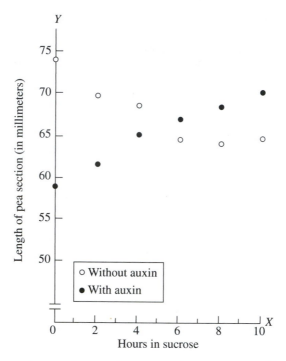

FIGURE 16.20 Mean lengths of pea sections grown with or without auxin and treated with sugar for varying periods. Data from Box 16.6.

Although the functions described in this section have been curvilinear—that is, not straight lines—they have all been of the form

$$Y = a + bX + cX^2 + dX^3 + \cdots + \epsilon \tag{16.28}$$

where the coefficients a, b, c, d, ... estimate parameters α, β, γ, δ, ... and ϵ is the error term. Although the curve described by this function is not linear, the parameters of the function are. The unknown parameters to be estimated are in a simple additive relationship. There are other equations, however, not of that form, in which the prediction equation is a nonlinear combination of the parameters. Examples are

$$Y = e^{a + bX^2 + \epsilon} \tag{16.29}$$

and

$$Y = a - bc^X + \epsilon \tag{16.30}$$

We can handle Expression (16.29) by transforming it to natural logarithms, yielding

$$\ln Y = a + bX^2 + \epsilon \tag{16.29a}$$

and solving the new equation by ordinary least squares methods. This function is therefore called an **intrinsically linear** equation.

There is no transformation, however, that will make Expression (16.30) linear. If we wish to fit this equation to a dataset by least squares for the residual terms ϵ, we must carry out complex computational procedures known as **nonlinear regression.** Why would one wish to do so? In some cases, the investigator may have direct knowledge of the true (nonlinear) structure of the process, or the theory governing it may require a nonlinear equation. In other cases, although the equation may be intrinsically linear (i.e., transformable), the error terms ϵ may be distributed normally only in the original scale. Gene exchange between populations as a function of distance, size of organisms as a function of time, or crop yield as a function of amount of fertilizer may be functions of this type. Alternatively, in many studies one would like to interpret the results in terms of the original, untransformed variables. In such cases, too, nonlinear regression is an appropriate approach.

An iterative approach (available in many program packages) is required to fit the model. There are several methods for approaching the least squares solution— the method of steepest descent and Marquardt's method, among others—which the user of the program selects. Problems that arise not infrequently with this approach include the iterative process reaching local minima and not yielding the global least squares solution, or the parameter estimates "blowing up" unless new, reduced estimates are fed to the program. An example of fitting parameters to an asymptotic regression, such as that described by Expression (16.29), is given in Snedecor and Cochran (1989, Sections 19.7 and 19.8). A practical guide for using S-plus and R is given by Huet et al. (2003) and by Ritz and Streibig (2008). For a more extensive discussion, see Motulsky and Christopoulos (2004) or Smyth (2002). An advanced treatment of nonlinear regression is found in Seber and Wild (2003).

16.10 Effect Size, Power, and Sample Size in Multiple Regression

Not surprisingly, measures of effect size for multiple regression are generalizations of those used for correlation and bivariate regression. The computations for model I regression are well-established and relatively straightforward. Methods for model II regression will be discussed briefly at the end of this section. In multiple regression, several distinct situations must be distinguished. The first situation is where one simply wishes to test whether there is any relationship between the dependent variable and a set of independent variables. In Section 15.7, r^2 was suggested as a convenient standardized measure of the strength of regression onto a single independent variable. With two or more independent variables, the corresponding measure is $R^2_{Y \cdot 1 \ldots k}$. This index can be interpreted as the proportion of variance explained, ω^2, or re-expressed as Cohen's (1988) measure of effect size, $f^2 = \dfrac{R^2_{Y \cdot 1 \ldots k}}{1 - R^2_{Y \cdot 1 \ldots k}}$.

As in Section 15.9, these measures are equivalent in the sense that one can easily convert one into another. However, a generalization of the z-transformation is not available for the multiple-correlation coefficient; therefore, tests of significance, power, and confidence intervals must be based on the central and noncentral t and F-distributions, as was done in Chapter 12 for anova. Part **1** of Box 16.7 shows the computations of the effect size for a regression in which all of the independent variables are used. An example of its application to the city air pollution data is also shown.

BOX 16.7 Effect Size and Required Sample Size, Model I Regression

Based on the air pollution data of Table 16.1 and the analyses of Box 16.2.
1. *Estimation of the overall effect size and its confidence limits.*

   Using X_1 and X_2 as independent variables, $R^2_{Y \cdot 12} = \omega^2 = 0.51611$ and thus

   $$f^2 = \frac{R^2_{Y \cdot 12}}{1 - R^2_{Y \cdot 12}} = \frac{0.51611}{1 - 0.51611} = 1.06659$$

   and

   $$E = 1 - \sqrt{1 - R^2_{Y \cdot 12}} = 1 - \sqrt{1 - 0.51611} = 0.30438$$

   The $(1-\alpha)\%$ confidence limits are computed with the aid of the *noncentral* F-distribution using a procedure similar to that described in part **2** of Box 12.1 but with the noncentrality parameter $\lambda = (\nu_1 + \nu_2 + 1)f^2$. We have seen in Section 16.10 that $\nu_1 = k$ and that $\nu_2 = n - k - 1$. Thus, $\lambda = (k + n - k - 1 + 1)f^2 = nf^2$.

   For the current example, $\lambda = 41(1.06659) = 43.73000$. We must use special software to find the noncentrality parameter, λ_1, of a noncentral F-distribution for which the observed $F_s = f^2 \dfrac{\nu_2}{\nu_1} = 20.26512$ cuts off $\alpha/2 = 0.025$ at its right tail. Then we must find the noncentrality parameter, λ_2, of the noncentral F-distribution for which the F_s cuts off $\alpha/2$ at its left tail. These values are 14.47567 and 75.67518 and are the 95% confidence limits for λ. If such software is not available, then a fairly good approximation from Patnaik (1949) can be used.

   One would search for the λ value such that $F_s = \dfrac{\nu_1 F_{\alpha[\nu_1, \nu_2]}}{\nu_1 + \lambda}$ with $\nu_1' = \dfrac{(\nu_1 + \lambda)^2}{\nu_1 + 2\lambda}$ and ν_2 degrees of freedom has a probability of $\alpha/2$ in the right tail of the central F-distribution. One would then repeat the search to find the value that has a probability of $\alpha/2$ in the left tail of the central F-distribution. This approximation is usually accurate to at least one decimal place, which may be sufficient for many practical applications.

   Confidence limits for f^2 are computed from those for λ using the relationship $f^2 = \lambda/(k + n - k - 1 + 1) = \lambda/n$ to yield the limits 0.35307 and 1.84574. These can be transformed into limits for $\omega^2 = R^2$ by using the relationship $\omega^2 = f^2/(1 + f^2)$ to yield the limits 0.26094 and 0.64860.

Box 16.7 (continued)

2. *Estimation of required sample size in order to have sufficient power to detect a specified minimum overall effect size.*

If software is available for computing probabilities for the noncentral F-distribution (as in the BIOMstat package), then the required sample size can be estimated by systematically trying different sample sizes until the desired power is achieved. Note that for a constant effect size the value of the noncentrality parameter, $\lambda = nf^2$ changes as one tries different sample sizes until the desired power is reached. For example, in order to achieve a power of at least 0.8 for detecting the effect size found earlier, $f^2 = 1.06659$, with a type I error of 0.05, a sample of only $n = 13$ would be required. The estimated power would then be 0.81973.

If such software is not available, then the approximation from Patnaik (1949) can be used. Power is estimated by comparing the quantity $F_s = \dfrac{\nu_1 F_{\alpha[\nu_1,\nu_2]}}{\nu_1 + \lambda}$ with the central F-distribution with $\nu_1' = \dfrac{(\nu_1 + \lambda)^2}{\nu_1 + 2\lambda}$ and ν_2 degrees of freedom. As just mentioned, for a fixed f^2, the value of λ changes as one tries different values of n. This approximation is usually accurate to at least one decimal place, which may be sufficient for many practical applications.

For example, we can check the results given earlier by using $n = 13$, $\nu_1 = k = 2$, $\nu_2 = n - k - 1 = 13 - 2 - 1 = 10$, and $f^2 = 1.06659$. We first compute the non-centrality parameter $\lambda = nf^2 = 13(1.06659)^2 = 13.86561$. The modified degrees of freedom and F-values are

$$\nu_1' = \frac{(\nu_1 + \lambda)^2}{\nu_1 + 2\lambda} = \frac{(2 + 13.86561)^2}{2 + 2(13.86561)} = 8.46644$$

$$F_{\alpha[k,n-k-1]} = F_{.05[2,10]} = 4.10282$$

and

$$F_s = \frac{\nu_1 F_{\alpha[k,n-k-1]}}{\nu_1 + \lambda} = \frac{2(4.10282)}{2 + 13.86561} = 0.51720$$

This F_s-value is compared to the central F-distribution with $8.46561 \approx 8$ over 10 degrees of freedom. Because this value is less than 1, it is easier to look up its reciprocal, 1.720, in Statistical Table **F** with 10 over 8 degrees of freedom. By computation, the probability is 0.18100. Because we used the reciprocal, we compute its complement to yield a power of 0.81900, which is quite close to the exact value of 0.81973.

3. *Computing the effect size and its confidence limits for a subset of variables while holding others constant.*

Cohen's (1988) measure of the effect size for a subset of one or more independent variables with the effects of the remaining independent variables held constant is

$$f^2 = \frac{\Delta R^2}{1 - R^2_{Y \cdot 1 \ldots k_2}}$$

Box 16.7 (continued)

where $\Delta R^2 = R_2^2 - R_1^2$ is the difference in squared multiple-correlation values obtained when using k_2 independent variables versus when just k_1 variables are used because a subset of the variables was not included.

For example, using the data from Table 16.1, the effect size for adding variable X_1 to a regression that already includes variable X_2 is

$$f^2 = \frac{R_{Y\cdot 12}^2 - R_{Y\cdot 2}^2}{1 - R_{Y\cdot 12}^2} = \frac{0.51611 - 0.41573}{1 - 0.51611} = 0.20744$$

Confidence limits for f^2 are computed using a procedure similar to that described earlier in part **1**, but ν_1 is now the number of variables in the subset ($\nu_1 = k_2 - k_1 = 2 - 1 = 1$ in the current example) and ν_2 is the degrees of freedom of the residual from the full model ($\nu_2 = n - k_2 - 1 = 41 - 2 - 1 = 38$ in the example). The observed noncentrality parameter is $\lambda = (n - k_1)f^2 = (41 - 1)0.20744 = 8.29775$. The lower $(1 - \alpha)100\%$ confidence limit for λ is the noncentrality parameter, λ_1, for a noncentral F-distribution for which the observed F_s (from the test for the increase in R^2 shown in Box 16.2 or computed as $F_s = f^2\dfrac{\nu_2}{\nu_1} = 7.88287$) cuts off $\alpha/2$ at its right tail. The upper confidence limit, λ_2, is the noncentrality parameter of the noncentral F-distribution for which the F_s cuts off $\alpha/2$ at its left tail. For the current example, these values are 0.52590 and 23.51617.

These are transformed into confidence limits for f^2 using the relationship $f^2 = \lambda/(n - k_1)$ to yield the limits 0.01315 and 0.58790. The latter limits can in turn be transformed into confidence limits for $\omega^2 = R^2$ by using the relationship $\omega^2 = f^2/(1 + f^2)$ to yield 0.01298 and 0.37024. Thus, although the size of the effect differs from zero (the confidence interval does not contain zero), we cannot exclude the possibility that the additional effect of adding variable X_1 to a model that already includes variable X_2 could actually be quite small (0.01298). One would need a larger sample size to reduce that uncertainty.

4. *Estimation of required sample size in order to have sufficient power to detect a specified minimum effect size for a subset of variables.*

The procedure is similar to that of part **2**, except that ν_1 equals the number of variables in the subset ($\nu_2 = k_2 - k_1$) and ν_2 is the residual degrees of freedom ($\nu_2 = n - k_2 - 1$). Different sample sizes must be tried until one is found for which power (the probability of $F \geq F_{\alpha[\nu_1, \nu_2]}$ for a noncentral F-distribution, $F_{[\nu_1, \nu_2, \lambda]}$) is at least that specified as the desired level of power. Note that for a fixed effect size, the noncentrality parameter changes as one tries different sample sizes, $\lambda = (n - k_1)f^2$.

As an example, given the effect size found in part **3**, $f^2 = 1.06659$, as a minimum effect size, the sample size needed in order to achieve a power of at least 0.8 with a type I error rate of 0.05 was found to be $n = 42$. This would give an expected power of 0.81163.

If software for the noncentral F-distribution is not available, then an approximation from Patnaik (1949) can be used for the computation of power. Power is estimated by the probability of obtaining a value of $F_s = \dfrac{\nu_1 F_{\alpha[\nu_1, \nu_2]}}{\nu_1 + \lambda}$ or greater from

Box 16.7 (continued)

the central F-distribution with degrees of freedom of $\nu_1' = \dfrac{(\nu_1 + \lambda)^2}{\nu_1 + 2\lambda}$ and ν_2. As before, $F_{\alpha[\nu_1, \nu_2]}$ is the critical value from the central F-distribution with ν_1 and ν_2 degrees of freedom. This approximation is usually accurate to at least one decimal place, which may be sufficient for many practical applications. The rest of the algorithm is unchanged. For the current example, we can try $n = 42$, $\nu_1 = k_2 - k_1 = 2 - 1 = 1$ and $\nu_2 = n - k_2 - 1 = 42 - 2 - 1 = 39$.

We first compute the noncentrality parameter:

$$\lambda = (n - k_1)f^2 = (42 - 1)1.06659^2 = 8.50520$$

The modified degrees of freedom and F-values are:

$$\nu_1' = \frac{(\nu_1 + \lambda)^2}{\nu_1 + 2\lambda} = \frac{(2 + 8.50520)^2}{2 + 2(8.50520)} = 5.01648 \approx 5$$

$$F_{\alpha[\nu_1, \nu_2]} = F_{.05[1,39]} = 4.09128$$

and

$$F_s = \frac{\nu_1 F_{\alpha[\nu_1, \nu_2]}}{\nu_1 + \lambda} = \frac{1(4.09128)}{1 + 8.50520} = 0.43043.$$

This F-value is to be compared to the central F-distribution with $\nu_1 = \nu_1' = 5$ over $\nu_2 = 39$ degrees of freedom. Because the F-value is less than one, one can look up its reciprocal, 2.32326, in Statistical Table **F** using 39 over 1 degrees of freedom. By computation, the probability is 0.17540. Because we used the reciprocal, power corresponds to its complement, 0.82460. This is fairly close to the exact value given previously.

The method for computing the confidence intervals for the overall effect size is similar to that given in Box 12.1. The noncentrality parameter, λ, for the F-distribution is estimated using the relationship $\lambda = (\nu_1 + \nu_2 + 1)f^2$, where $\nu_1 = k$ (the number of independent variables) and $\nu_2 = n - k - 1$ (the residual degrees of freedom). Thus, λ simplifies to just $\lambda = nf^2$. The lower $1 - \alpha$ confidence limit for λ is the noncentrality parameter of the noncentral F-distribution for which the observed λ is the upper $\alpha/2$ quantile. Correspondingly, the upper confidence limit is the noncentrality parameter of the noncentral F-distribution for which the observed λ is the lower $\alpha/2$ quantile.

An approximation is also provided if software for the noncentral F-distribution is unavailable. These confidence limits for λ are then converted into confidence limits for f^2 using the relationship $f^2 = \lambda/n$. In turn, these limits can be transformed into limits for ω^2 or R^2 using the relationship $\omega^2 = R^2 = f^2/(1 + f^2)$. This example illustrates that although the null hypothesis of a zero effect size can be rejected, the magnitude of the effect size is not known very precisely (the 95% confidence interval for f^2 is from 0.35307 to 1.84574).

Part 2 of Box 16.7 shows the procedure for estimating the sample size required to detect a specified minimal effect size with a given type I error rate and power. Maxwell (2004) points out the importance of making sure that a planned study has sufficiently high power. When power is low, different replications of a study can give contradictory results. For example, when there are more than just a few independent variables, there is a high probability of obtaining significant results somewhere in each analysis but a low probability of obtaining the same sets of significant independent variables in the different replicates.

As in the corresponding procedure in Box 12.1, the method is based on finding the value of λ for a noncentral F-distribution, $F_{\alpha[\nu_1, \nu_2, \lambda]}$, such that the probability of obtaining a F_s-value equal to $F_{\alpha[\nu_1, \nu_2]}$ is $\geq 1 - \beta$. An approximation is provided if software for the noncentral F-distribution is not available. In the example, a sample size of just 13 is expected to be sufficient to have a power of 0.8 to reject the null hypothesis at the 5% level if the true f^2 was as small as that observed in part **1** of that box. However, if such a small sample size were actually used the confidence intervals for the effect size would, of course, be expected to be even broader than those obtained in the example for part **1** of the box. The confidence interval would often extend downward to almost include zero. There would be a comparable increase in the upper limit. Thus, even though one would expect usually to reject the null hypothesis, one would have a much less reliable estimate of the magnitude of the effect.

For that reason, some authors, e.g., Kelley (2008) and Kelley and Maxwell (2008), advocate using a sample size that would be expected to yield a sufficiently narrow confidence interval for the parameter of interest rather than simply a sample size that just allows one to usually reject the null hypothesis. To do this, one would try larger sample sizes until the expected width of the confidence interval was sufficiently narrow.

The separate effect of one or more variables in a regression model with the effects of variation in the other variables held constant is often of interest. It can be expressed in terms of the difference, $\Delta R^2 = R^2_{Y \cdot 1 \ldots k_2} - R^2_{Y \cdot 1 \ldots k_1}$, between the squared multiple-correlation values for the complete model using all k_2 independent variables and the value obtained when only a subset of these variables are included in the regression. One can also interpret this as the effect of adding $k_2 - k_1$ variables to an initial set of k_1 independent variables. If including a subset of variables greatly improves one's ability to predict the dependent variable, then ΔR^2 should be large. However, if they provide little additional information to improve prediction, then the difference should be small.

Box 16.2 used this same quantity for tests of significance. Part 3 of Box 16.7 uses this quantity to compute the effect size for a subset of variables and its confidence limits. When the subset tested only contains a single variable, f^2 gives an indication of the importance of the contribution of just that variable with the effects of the other variables held constant. Part **4** of Box 16.7 shows a method for determining the sample size needed to detect a specified minimum effect size with a desired level of power. Note that if an investigator estimates the required sample size for each of the independent variables separately, it is likely that different sample sizes will be suggested for each variable. In that case, one should use the largest of these estimates.

Although Box 16.7 provides measures of effect size for regression and estimates of sample sizes needed for significance testing, sometimes it is more useful to have narrow confidence intervals for the partial regression coefficients themselves. This would be true if one is interested in the magnitudes of the parameters of a regression model rather than just whether or not they could be equal to zero. Even though sample sizes may be sufficient to reject the null hypothesis, the interval may be very wide, indicating the estimate for the regression coefficient is rather imprecise. For example, in part **5** of Box 16.2, the 95% confidence limits for $\beta_{Y1 \cdot 2}$ range from -1.8039 to -0.2981.

The confidence limits for partial regression coefficients can be found using the method shown in part **1** of Box 16.8. They are of the form $b_{Yj} \pm t_{\alpha[n-k-1]} s_{b_j}$ for the jth partial regression coefficient. The standard error is given in Box 16.3 as $s_{b_{Yj}} = s_Y \cdot \sqrt{c_{jj}}$ but it is convenient to rewrite it here as

$$s_{b_{Yj}} = \sqrt{\frac{1 - R^2_{Y \cdot 1 \ldots k}}{\nu} \frac{s^2_Y}{s^2_{X_j}} \text{VIF}_j}$$

where $\nu = n - k - 1$ and VIF_j is the variance inflation factor for the jth variable. The equation for the standard error can be inverted to yield a formula for the sample size needed to obtain a standard error equal to a specified value. This can then be transformed into a formula for the sample size needed to achieve a $(1 - \alpha)100\%$ confidence interval of a specified width, w, for the jth partial regression coefficient:

$$n = \left(\frac{t_{\alpha[\nu]}}{w/2}\right)^2 (1 - R^2_{Y \cdot 1 \ldots k}) \frac{s^2_Y}{s^2_{X_j}} \text{VIF}_j + k + 1$$

However, this just yields the expected sample size required. The actual width of a confidence interval for a sample could be larger or smaller than w because estimates of σ^2_Y will vary from sample to sample. Because the sample $R^2_{Y \cdot 1 \ldots k}$ is biased upwards in small samples, it should be adjusted for bias using Expression (16.23) before using it to estimate sample sizes. The following equation by Kelley and Maxwell (2003) gives an estimate of the sample size needed to have a probability, P, that the width of a $(1 - \alpha)100\%$ confidence interval for the jth partial regression coefficient will be less than or equal to w:

$$n = \left(\frac{t_{\alpha[\nu]}}{w/2}\right)^2 (1 - R^2_{Y \cdot 1 \ldots k}) \frac{s^2_Y}{s^2_{X_j}} \text{VIF}_j \left(\frac{\chi^2_{1-P[n-1]}}{\nu}\right) + k + 1$$

The solution must be iterative because the computation of the degrees of freedom depends on the sample size. The other quantities in the equation are treated as fixed. Note that large variance inflation factors result in larger sample sizes being required.

For standard partial regression coefficients, confidence limits are computed in part **2** of Box 16.8 using a very different procedure that makes use of the noncentral t-distribution. First, the noncentrality parameter is computed as $\lambda = b'_{Yj}/s_{b'_{Yj}}$,

---

BOX 16.8 Confidence Intervals and Required Sample Size for Partial and Standard Partial Regression Coefficients, Model I Regression

---

The examples are based on the air pollution data of Box 16.1 and the results of Box 16.2.

1. *Effect size and confidence intervals for partial regression coefficients.*

The measure of effect size for a partial regression coefficient, whether standardized or not, is simply the square root of the effect size given in part **3** of Box 16.7 for the effect of a single independent variable; that is, $f = \sqrt{(R^2_{Y \cdot 1 \ldots k_2} - R^2_{Y \cdot -j})/(1 - R^2_{Y \cdot 1 \ldots k_2})}$, where $R^2_{Y \cdot -j}$ is the squared multiple correlation using all of the independent variables except variable x_j.

The confidence limits for f are simply the square roots of the confidence limits for f^2 found in part **3** of Box 16.7. Confidence intervals for partial regression coefficients can be computed as shown in part **5** of Box 16.2. The limits are computed as $b_{Yj} \pm t_{\alpha[n-k-1]} s_{b_j}$.

2. *Standard partial regression coefficients.*

Confidence limits for standard partial regression coefficients are computed using a very different procedure that makes use of the noncentral t-distribution. First, the noncentrality parameter is computed as $\lambda = b'_{Yj}/s_{b'_{Yj}}$, the observed standard partial regression coefficient divided by its standard error.

Next, one must find the confidence limits for λ using the noncentral t-distribution. The lower limit, λ_1, is the noncentrality parameter for a noncentral t-distribution for which the area above the observed λ is $\alpha/2$. The upper limit, λ_2, is the noncentrality parameter of the distribution for which the area below λ is $\alpha/2$. These limits are then transformed into limits for the standard partial regression coefficient by multiplying λ_1 and λ_2 by $s_{b'_{Yj}}$, the standard error of the standard partial regression coefficient. As an example, the 95% confidence limit for $b'_{Y1 \cdot 2}$ can be found as follows:

$$\lambda = b'_{Yj}/s_{b'_{Yj}} - 1.04805/0.11494 = -9.11835$$

The noncentrality parameter for which λ cuts off 0.025 from the upper tail of a noncentral t-distribution is $\lambda_1 = -11.91946$, and the value of the noncentrality parameter for which λ cuts of 0.025 from its lower tail is $\lambda_2 = -6.25774$. These limits are then multiplied by $s_{b'_{Yj}}$ to yield the confidence limits for $b'_{Y1 \cdot 2}$:

$$L_1 = \lambda_1 s_{b'_{Yj}} = -11.91946(0.11494) = -1.37001$$

and

$$L_2 = \lambda_2 s_{b'_{Yj}} = -6.25774(0.11494) = -0.71926$$

If the adjusted R^2 value is used in the computation of $s_{b'_{Yj}}$, then the confidence interval is slightly narrower, from -1.35532 to -0.73464.

Box 16.8 (continued)

3. *Power and sample sizes for partial regression coefficients.*

The computation of power for testing whether the partial or standard partial regression coefficients are equal to zero is the same as that given in part **3** of Box 16.7 for the testing of the null hypothesis that ΔR^2 is zero for the effect of a single independent variable.

We can estimate the sample size required to have a probability, P, that the width of the $(1 - \alpha)100\%$ confidence intervals will be less than or equal to w. Using the same example, we can estimate the required sample size for a 95% confidence interval for $b_{Y1 \cdot 2}$ of width $w = 0.5$ using $\text{VIF}_1 = 1.03747$, $R^2_{Y \cdot 1 \ldots k} = 0.51611$, $s^2_Y = 550.94756$, and $s^2_{X_1} = 52.23988$. Although the required sample size must be considerably larger than in the current sample, we can still use $n = 41$ as a starting value. Then $\nu = n - k - 1 = 41 - 2 - 1 = 38$, $t_{\alpha[\nu]} = t_{0.05[38]} = 2.02439$, and $\chi^2_{1-P[n-1]} = \chi^2_{0.2[40]} = 32.34495$. The next estimate of n will then be

$$
n = \left(\frac{t_{\alpha[\nu]}}{w/2}\right)^2 (1 - R^2_{Y \cdot 1 \ldots k}) \frac{s^2_Y}{s^2_{X_j}} \text{VIF}_j \left(\frac{\chi^2_{1-P[n-1]}}{\nu}\right) + k + 1
$$

$$
= \left(\frac{2.02439}{0.5/2}\right)^2 (1 - 0.51611) \frac{550.94756}{52.23988} 1.03747 \left(\frac{32.34495}{38}\right) + 2 + 1
$$

$$
= 434.84135
$$

Rounding the result up to 435 and repeating the calculations using this new estimate results in a final estimate of $n = 354$ in just a few iterations. If the bias-adjusted estimate of the squared multiple correlation, R^2_{adj}, is used, then the estimated sample size is slightly lower: 350. The estimated sample size is rather large both because we wish to obtain a narrower confidence interval than before but also because we wish to be fairly certain ($P = 0.8$) that it will, in fact, be as narrow as desired.

4. *Power and sample sizes for standard partial regression coefficients.*

The power of a test of the null hypothesis of a standard partial regression coefficient is identical to that given in part **3** for the partial regression coefficient.

However, there is no direct formula for estimating the sample size necessary to obtain a confidence interval of a specified width. One must try the method for computing confidence intervals given in part **2** with larger sample sizes until the desired width is obtained. Using the same example, a sample size of 183 was found to be necessary for the expected width of a 95% confidence interval for $b'_{Y1 \cdot 2}$ to be less than or equal to 0.5. When the R^2_{adj} coefficient is used as the estimate, a sample size of only 148 is indicated.

---

the observed standard partial regression coefficient divided by its standard error. Next, one must find the confidence limits for λ using the noncentral t-distribution. The lower limit, λ_1, is the noncentrality parameter for a noncentral t-distribution for which the area above the observed λ is $\alpha/2$. The upper limit, λ_2, is the noncentrality parameter of the distribution for which the area below λ is $\alpha/2$. These limits are transformed into limits for the standard partial regression coefficient by multiplying them by $s_{b'_{yj}}$, the standard error of the standard partial regression coefficient.

Although there is no formula for it, the sample size necessary to give an expected width for the $(1 - \alpha)100\%$ confidence interval for the standard partial regression coefficient can be found numerically by simply trying larger sample sizes until the desired width is obtained. According to Kelley and Maxwell (2008), no analytic procedure is currently available to estimate the sample size needed in order to have a probability, P, that the width of a $(1 - \alpha)100\%$ confidence interval for the jth standard partial regression coefficient will be less than or equal to w. They suggest the use of simulations with varying values of n in order to find an appropriate sample size.

For model II regression, the computations of confidence intervals for R^2 and of samples sizes necessary to achieve sufficiently narrow confidence limits is more complex. Kelley (2008) describes a computationally intense Monte Carlo simulation approach that can be used to obtain the exact sample size, assuming that the dependent and independent variables have a joint multivariate normal distribution. He also provides tables giving n for $k = 2, 5$, and 10 for desired widths from 0.05 to 0.50. Shieh (2007) considers other possible distributions for the independent variables—they can be any discrete or continuous distribution. The computations involve a numerical integration of a function of central and noncentral F-distributions.

16.11 Advanced Topics in Regression and Correlation

As you may have gathered from the three chapters devoted to them, regression and correlation analysis are important subjects in statistics and invaluable tools in biometric work. Many other techniques had to be omitted to keep this book a reasonable length. In this section, we briefly describe several other important methods to give you some idea of their applications and refer you to suitable references where you may learn more about the techniques if you find that your work requires them.

Although developed by R. A. Fisher in the 1930s, only in the last quarter of the 20th century, as computers became widely available, has **discriminant function analysis** been applied much in biology, especially in systematics. The basic problem that this technique solves is easy to understand. Suppose we have samples from two populations representing, for example, different sexes, species, or disease states. We measure a character and find that although the means of the two distributions are not identical, the distributions overlap considerably. On the basis of this one character, then, we could not, with any degree of certainty, identify an unknown specimen as belonging to one or the other of the two populations. If a second character were measured, it might also differentiate the populations somewhat, but not absolutely.

A discriminant function is shown in Figure 16.21, in which samples of two species of limpets are plotted against two characters, X_1 and X_2, commonly used by taxonomists. On the basis of either X_1 or X_2, assigning an unknown specimen to either of the two species with any reasonable degree of certainty would be impossible. The degree of overlap of the histograms along the two coordinate axes makes this abundantly clear.

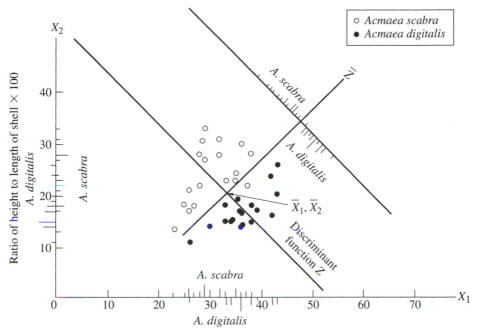

FIGURE 16.21   A discriminant function. Histograms for *Acmaea digitalis* below abscissa and to left of ordinate; those for *Acmaea scabra* above abscissa and to right of ordinate. (For explanation, see text.)

Discriminant function analysis constructs a new variable Z, which is a linear function of variables X_1 and X_2. This function is of the form $Z = \Sigma \lambda_i X_i$, which is the equation of a line cutting across the intermixed cluster of points representing the two species. This function is defined in such a way that members of one population will have high values for Z and members of the other will have low values. Figure 16.21 shows a line drawn parallel to the discriminant function on which a histogram of the Z-values (discriminant function scores) has been graphed. Note that the histograms do not overlap except for a single individual of *Acmaea scabra*. The discriminant function has served admirably to separate the two groups. Given a new, unknown specimen belonging to one of these two species, you could measure its X_1 and X_2, calculate a new value of Z employing the previously calculated coefficients λ_1 and λ_2, and assign the new individual to one of the two groups. There are also techniques for estimating the expected accuracy of such assignments.

One way of computing a discriminant function is to use multiple regression. If all members of the first group are coded $Y_i = 1$, and all members of the second group are coded $Y_j = -1$, we can compute a multiple-regression equation of Y on all independent variables X_j. The multiple-regression equation then yields the least squares best predictor of group membership, here symbolized by variable Y. As you can see, this procedure is very useful when we need to assign unknown specimens to previously recognized groups. Such situations arise in taxonomic identification where taxa have been previously established. If taxa have not been established, discriminant functions are of no assistance; other techniques have to be employed (see the end of this section).

Discriminant function analysis can save time and money. Using a complicated or expensive technique, a researcher may be able to determine definitively the sex of an immature form of an insect. Discriminant function analysis, however, enables the allocation of an immature specimen to one sex or the other simply by the measurement of several easily and inexpensively obtained external characteristics. Descriptions of discriminant functions are found in most books on multivariate analysis, including Johnson and Wichern (2007) and Morrison (2004). Computer programs for discriminant functions are widely available (e.g., in R, SAS, BMDP, SPSS, SYSTAT, NTSYSpc, and many others).

The discriminant function is a topic in the general area of **multivariate analysis,** the branch of statistics dealing with the simultaneous variation of two or more dependent variables. As computer facilities have become widely available and personal computers more powerful, these methods are ever more widely applied, and by now various types of multivariate analyses are among the arsenal of most research biologists. The degree of statistical sophistication required for work of this type is still considered relatively high, however, and is beyond the scope of this text. One technique of note is **multivariate analysis of variance** (manova for short), which carries out a generalization of an anova for cases with more than one dependent variable. Introductions to multivariate analysis with applications in biology appear in Pielou (1984), Morrison (2004), and Reyment (1991). Everitt and Dunn (2001) is a good introduction to multivariate methods. More comprehensive coverage is provided by Green (1978), Hand and Taylor (1987), Johnson and Wichern (2007), Tatsuoka (1988), Krzanowski (2000), and Jackson (2003).

Exploratory studies looking for patterns of covariation in a set of variables may start with a **correlation matrix,** which is a symmetrical table of correlation coefficients for each variable with all others (see Section 15.3).

From a matrix of correlations may emerge patterns of structure. Various methods of grouping the variables according to the magnitudes and interrelationships among their correlations have been developed. These methods are known generally as **cluster analysis.** Cluster analysis dates back to the early 1900s, but there was renewed interest in recent years. Everitt et al. (2001) gives a good introduction to cluster analysis. Comprehensive reviews are provided by Anderberg (1973), Romesburg (2004), Sneath and Sokal (1973), Hartigan (1975), and Späth (1975).

The analysis of matrices representing similarities or dissimilarities among taxonomic units developed into an extensive field of research called **numerical taxonomy.** This quantitative method is defined as the numerical evaluation of the affinity or similarity between taxonomic units and the ordering of these units into taxa on the basis of their affinities. One approach (called *phenetics*) in numerical taxonomy consists of computing similarity or dissimilarity matrices, sometimes composed of correlation coefficients, and clustering these matrices into taxonomic systems based on the values they contain. Section 18.4 gives an example of a simple application. Another approach (called *cladistics*) estimates evolutionary branching sequences from data matrices using principles such as parsimony or maximum likelihood.

All these procedures require a considerable amount of computation when more than just a few taxa are being considered. This is still a rapidly moving field. Some classic references include Sneath and Sokal (1973) and Clifford and Stephenson (1975). More recent developments are discussed in Felsenstein (2004), Jensen (2009), Gordon (1981),

Dunn and Everitt (2004), and Swofford and Olsen (1990). The current literature in the journals *Systematic Biology*, *Evolution*, *Journal of Molecular Evolution*, and *Cladistics* should also be consulted. The statistical background required for numerical taxonomy is such that readers of this book should have little difficulty in applying the methods.

EXERCISES 16

16.1 The following correlation matrix was taken from a study of the effect of growth rates, timing of breeding, and territory size on the survival of chicks of the glaucous-winged gull (Hunt and Hunt, 1976). Survival was estimated as percent chick survival to 500 g; growth rate is expressed as grams gained per day; hatching date in days; and territory size is in square meters.

		Survival	Growth rate	Hatching date	Territory size
Y	Survival	1.00			
X_1	Growth rate	0.42	1.00		
X_2	Hatching date	−0.16	−0.22	1.00	
X_3	Territory size	0.10	0.19	−0.44	1.00

$$n = 104$$

Which variable is the single best predictor of survival? Which variable has the largest standardized partial regression coefficient?

ANSWER: $b'_{Y1 \cdot 23} = 0.40536$

16.2 Use polynomial regression to predict SO_2, given population size, using the data from Table 16.1 (Y_1 and X_3). What degree polynomial should be used? How well does it fit? Are the deviations from the regression line what one would expect?

16.3 The following data are from an unpublished study by J. Levinton. This experiment measured sediment ingestion, estimated by the cumulative number (Y) of fecal pellets made of sedimentary particles by the mud snail *Hydrobia minuta*. As all of the available sedimentary grains are exhausted, the rate of accumulation slowed.

	Time in hours X	Cumulative number of pellets Y		Time in hours X	Cumulative number of pellets Y
1	6	119	10	24	774
2	6	111	11	24	896
3	6	87	12	24	670
4	12	348	13	42	896
5	12	210	14	42	729
6	12	272	15	42	886
7	18	438	16	78	763
8	18	778	17	78	954
9	18	548	18	78	954

Use polynomial regression to predict Y as a function of X. Plot the data showing the fitted linear, quadratic, cubic, and quartic curves. Plot the standardized residuals from these curves. Comment on the fit of the quartic curve to these data.

ANSWER: $R^2_{Y \cdot 1} = 0.5636$, $R^2_{Y \cdot 123} = 0.8961$, $\hat{Y} = -338.9084 + 75.7313X - 1.5446X^2 + 0.0099X^3$

16.4 The following data on growth of Hudson River striped bass are from Exhibits UT-49 and UT-50 of the 1978 Hudson River adjudicatory hearings. Perform a multiple-regression analysis of incremental growth (change in mean length from last half of July to last half of August) using the three independent variables. "Degree rise" is 4°C divided by the number of days it took the river temperature to rise from 16°C to 20°C. The variable F_7 is a standardized measure of flow in the river.

Year	Incremental growth Y	Population density X_1	Degree rise X_2	F_7 X_3
1965	23.7	1.2	0.3077	−3.2356
1966	18.3	10.0	0.4444	−1.3405
1967	27.8	4.2	0.4000	−1.5533
1968	18.5	1.3	0.1818	0.0070
1969	7.9	32.1	0.2105	−0.1987
1970	15.3	16.7	0.1820	−1.1177
1972	21.2	9.7	0.2667	4.3428
1973	21.8	28.6	0.5000	1.8497
1974	22.1	9.5	0.1540	0.8988
1975	25.2	18.3	0.4444	1.0978
1976	25.1	11.3	0.4444	5.1556

16.5 Mark et al. (1958) present the following data on the passage of thiobarbiturates from the blood plasma into the brain. Assuming the path model $X \rightarrow Y_1 \rightarrow Y_2$, estimate the path coefficients and percent determination of Y_2 by this model.

Time (in minutes) X	Log plasma concentration Y_1	Log brain concentration Y_2
3	2.3324	1.8808
3	2.2406	1.8633
30	2.1399	1.9912
30	2.1553	2.0128
60	2.0864	2.1173
60	2.0531	2.1335

ANSWER: $p_{21} = -1.0605$

16.6 Represent the data given in Exercise 16.1 in terms of a path-coefficient model. Draw the diagram and show all path coefficients. Express the correlation between survival and territory size in terms of path coefficients.

16.7 Asmundson (1931) presents the following statistics based on 707 chicken eggs.

Variable	Mean	Standard deviation
1. Length (mm)	56.93	3.02
2. Breadth (mm)	41.00	1.51
3. Maximum-minimum diameter (mm)	0.17	0.14
4. Length × 100/breadth	71.83	4.05
5. Weight of yolk (g)	16.10	2.11
6. Weight of albumen (g)	31.52	3.98
7. Weight of shell (g)	5.63	0.83

	Correlations						
	1	2	3	4	5	6	7
1	X						
2	0.208	X					
3	−0.092	0.116	X				
4	−0.758	0.431	0.166	X			
5	0.428	0.476	−0.131	−0.103	X		
6	0.582	0.726	0.105	−0.079	0.197	X	
7	0.279	0.570	0.008	0.118	0.091	0.611	X

Which variable is the best single predictor of the weight of the egg yolk (variable 5)? Develop a prediction equation for the weight of the egg yolk. Consider various subsets of the variables in order to find the smallest set of variables that enables a good prediction. Give the final prediction equation in both the standardized and the conventional form. Set 95% confidence limits to your estimates. Compute the variance inflation factors for the various models considered.

ANSWER: $r_{5.2}^2 = 0.2266$, $R_{5 \cdot 1,2,6,7}^2 = 0.6734$, $b_{5,1 \cdot 2,6,7}' = 0.81754$, VIF for variable 4 when using all 6 predictors is 16.1485.

16.8 Evaluate r_{XY} in the following path diagram.

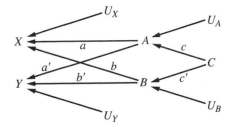

Recalculate the multiple-regression of the air pollution data of Table 16.1 after recoding X_1 as $X_1' = 10(X_1 - 40)$, X_4 as $X_4' = 10(X_4 - 5)$, and X_5 as $X_5' = 100X_5$. List the statistics that need uncoding and furnish formulas for doing so.

16.9 In a study of the numbers of species of birds on the islands off the California coast, Power (1972) presented the following data giving the number of species of birds and plants and some variables that he thought might account for the differences in numbers of species. Perform a multiple-regression analysis to determine a subset of variables most useful for predicting the numbers of species of birds. Multiple regression was also used to predict the number of plant species.

Birds	Area	Elev	Lat	Plants	I1	I2	I3	I4	I5	I6	I7	I8	I9
10	14	830	34.0	190	26	3	217.3	12	10	8	0.531	0.502	0.485
22	84	1560	34.0	340	27	3	202.4	12	8	7	0.396	0.339	0.333
32	96	2470	34.0	420	20	5	193.1	12	8	7	0.473	0.446	0.440
16	1.1	930	34.0	70	13	13	192.9	10	8	7	0.624	0.595	0.588
10	14	830	34.0	190	26	3	177.5	12	8	6	0.902	0.832	0.819
13	1.0	635	33.4	40	38	24	173.9	10	8	6	0.824	0.796	0.782
28	75	2125	33.3	392	20	20	162.6	13	8	6	0.856	0.782	0.764
23	56	1965	32.9	235	49	21	156.5	12	8	6	0.881	0.830	0.813
13	1.0	670	32.4	83	9	9	168.1	14	12	5	0.717	0.762	0.790
14	0.5	315	31.8	72	3	3	173.6	14	13	7	0.547	0.536	0.496
4	0.9	470	30.5	62	3.5	3.5	199.6	15	13	8	0.731	0.702	0.683
2	0.2	130	29.8	4	6	6	221.0	15	14	8	0.815	0.795	0.749
12	98	4600	29.0	163	165	165	243.5	16	16	9	1.000	1.000	0.964
7	2.5	660	28.3	39	44	19	278.1	14	13	10	0.873	0.896	0.910
12	134	3950	28.2	205	14	10	278.7	13	13	10	0.760	0.760	0.738
5	2.8	490	27.9	42	5	5	302.0	13	13	10	0.675	0.675	0.662

The 16 rows of this table are the 16 islands, in order: San Miguel, Santa Rosa, Santa Cruz, Anacapa, San Nicholas, Santa Barbara, Santa Catalina, San Clemente, Los Coronados, Todos Santos, San Martin, San Geronimo, Guadalupe, San Benito, Cedros, and Natividad. The first five variables are, in order: the number of breeding or summer resident bird species, island area in square miles, maximum elevation in feet, the number of native plant species, and latitude in degrees N. The remaining variables are measures of the isolation of the island. I1 is the distance in miles to nearest mainland point, I2 is distance to nearest island or mainland (whichever is shorter), I3 is the average distance to every other island and to the mainland, I4 is 16 minus the number of islands or mainland within 50 miles, I5 is 16 minus the number of islands or mainland within 100 miles, I6 is 16 minus the number of islands or mainland within 200 miles, I7 is 1 minus the sum of the reciprocals of distance to islands or mainland within 50 miles, I8 is 1 minus the sum of the reciprocals of distance to islands or mainland within 100 miles, and I9 is 1 minus the sum of the reciprocals of distance to islands or mainland within 200 miles.

Note that the isolation indexes are rather redundant and thus large variance inflation factors are to be expected. This makes it desirable to use relatively few variables in the prediction equations. Transformations of the variables should also be considered.

16.10 Data on body fat (obtained by immersion of each person in water) and three body measurements (triceps skinfold thickness, thigh circumference, and midarm circumference) on 20 healthy females aged 20 to 34 are presented in the following table. Use multiple regression to predict body fat from the body measurements. The units for the variables were not provided.

Fat	Triceps	Thigh	Midarm
11.9	19.5	43.1	29.1
22.8	24.7	49.8	28.2
18.7	30.7	51.9	37.0
20.1	29.8	54.3	31.1
12.9	19.1	42.2	30.9
21.7	25.6	53.9	23.7
27.1	31.4	58.5	27.6
25.4	27.9	52.1	30.6
21.3	22.1	49.9	23.2
19.3	25.5	53.5	24.8
25.4	31.1	56.6	30.0
27.2	30.4	56.7	28.3
11.7	18.7	46.5	23.0
17.8	19.7	44.2	28.6
12.8	14.6	42.7	21.3
23.9	29.5	54.4	30.1
22.6	27.7	55.3	25.7
25.4	30.2	58.6	24.6
14.8	22.7	48.2	27.1
21.1	25.2	51.0	27.5

Data from Neter et al. (2004).

16.11 The following data were extracted from a larger study by Castillo (1938) concerned with determining the optimum amount of fish meal to be used as a supplement in rations for young growing ducklings.

Amount of Fish Meal							
5%		10%		15%		20%	
$\bar{Y}$	X	$\bar{Y}$	X	$\bar{Y}$	X	$\bar{Y}$	X
46.5	1	67.8	2	52.2	1	100.0	2
151.6	5	142.8	4	190.8	4	249.2	4
287.1	8	365.2	7	367.2	6	525.0	6
545.9	11	637.7	10	547.8	7	942.8	9
634.1	12	855.9	12	1012.8	11	1210.2	12

X = age in weeks

$\bar{Y}$ = mean weight in grams (the original observations are not available, n varied from 11 to 46)

Does the amount of fish meal affect the rate of growth? The relationship between X and Y is obviously not linear, so the Y variable should be transformed (try $Y' = \log(Y)$). Test for differences among adjusted means. If the overall test is significant, then use the GT2-method to test differences between pairs of adjusted means.

17 Analysis of Frequencies

A lmost all our work so far has dealt with estimating parameters and testing hypotheses for continuous variables. This chapter focuses on tests of hypotheses about frequencies. Biological variables may be distributed into two or more classes, depending on a criterion such as arbitrary class limits in a continuous variable or a set of mutually exclusive attributes. An example of the former is a frequency distribution of birth weights (a continuous variable that is divided into an arbitrary number of contiguous classes); an example of the latter is a qualitative frequency distribution such as the frequency of individuals of 10 different species obtained from a soil sample. For any such distribution, we may hypothesize that it was sampled from a population in which the frequencies of the various classes represent certain parametric proportions of the total frequency.

We need a test of goodness of fit for our observed frequency distribution to the expected frequency distribution representing our hypothesis. We first realized the need for such a test in Chapters 5 and 6, where we calculated expected binomial, Poisson, and normal frequency distributions but were unable to decide whether an observed sample distribution departed more than would be expected under random sampling from the theoretical distribution. In Section 17.1, we introduce the idea of goodness of fit, discuss appropriate types of statistical tests and the basic rationale behind them, and develop general computational formulas for these tests.

Section 17.2 illustrates computations for goodness of fit when the data are arranged by a single criterion of classification, as in a one-way quantitative or qualitative frequency distribution. This test applies to cases expected to follow one of the well-known frequency distributions such as the binomial, Poisson, or normal distribution. It also applies to expected distributions following other laws suggested by the scientific subject matter under investigation, such as tests of goodness of fit of observed genetic ratios against expected Mendelian frequencies. The goodness-of-fit tests discussed in this section are the G, chi-square, and Kolmogorov–Smirnov tests.

An interesting property of the G-test statistic is its additivity: The results of several G-tests can be summed to yield meaningful results. Also, overall G-tests can be partitioned into separate G-tests representing individual degrees of freedom in a manner analogous to that practiced in anova. Such procedures are discussed in Section 17.3, which deals with pooled, total, and interaction G.

We proceed to tests of frequencies in two-way classifications—called *tests of independence*. Section 17.4 discusses various two-way classifications, including the common tests of 2 × 2 tables in which each of two criteria of classification are used to divide the frequencies into two classes, yielding a four-cell table. Procedures for analyzing such data, including the G-test, Fisher's exact test, and the traditional chi-square test are presented. We also show how to compute a measure of association (the phi coefficient) from a 2×2 table. Section 17.5 discusses procedures for testing frequency distributions classified on the basis of more than two criteria.

Section 17.6 reexamines 2×2 and $R \times C$ contingency tables from the point of view of proportions. The logit transformation of such proportions is stressed, and odds ratios and their logarithms are introduced as a way of expressing differences in proportions. This section also covers the relation of these quantities to logistic regression, which is a useful regression technique for proportions. The Mantel–Haenszel procedure for obtaining overall odds ratios from replicated 2×2 contingency tables is shown. Section 17.6 also presents methods for testing the difference between two percentages. Section 17.7 deals with goodness-of-fit tests in two-way classifications in which the two columns of a two-way table represent repeated measurements of the same individuals rather than different individuals, as in the examples from Section 17.4. This design is analogous to the randomized-blocks design of Section 11.4. Section 17.8 presents methods for estimating the strength of an effect (the magnitude of a deviation from the null hypothesis), the expected power of a planned test, and estimates of sample sizes needed in order to achieve a desired level of power.

Most of the statistical tests in this chapter involve the comparison of observed and expected frequencies. Although chi-square tests (using the X^2 statistic) are the traditional approach, we emphasize the use of the log-likelihood ratio test (using the G-statistic) throughout. As we will explain, the G statistic has theoretical advantages and is often computationally simpler. Both statistics, X^2 and G, require comparison of the test statistic with a chi-square distribution for the appropriate degrees of freedom. Computer programs yielding the correct P-value for the test statistic are now widely available. We, therefore, furnish only such "exact" values in the boxes of this chapter, in our case provided by the Statistical Calculator program of the BIOMstat program package. Readers lacking such a program can, of course, use Statistical Table **D** and express the result in the traditional manner as an inequality with critical values from the table. Although the computer-generated P-values are more precise than those looked up in the table, all conclusions based on either method of reporting the results for any one test will be identical.

17.1 Introduction to Tests for Goodness of Fit

Your by now extensive experience with statistical hypothesis testing should make the basic idea of a goodness-of-fit test easy to understand. Assume that a geneticist has carried out a crossing experiment between two F_1 hybrids and obtains an F_2 progeny of 90 offspring, 80 of which appear to be wild type and 10 of which are the mutant phenotype. The geneticist assumed dominance and expected a 3:1 ratio of the phenotypes, but the actual ratio is $80/10 = 8:1$. The parametric values assumed by the null hypothesis for p and q are $\pi = 0.75$ and $1 - \pi = 0.25$ for the wild type and mutant, respectively. Following the usual convention, we use a Greek letter here to symbolize the population parameter, even though π is the well-known symbol for the ratio of the circumference to the diameter of a circle. The context should make the meaning of the symbol clear. The *observed* proportions of these two classes are $p = 0.89$ and $q = 0.11$, respectively. The values for the proportions that are predicted by some model will be denoted $\hat{p}$ and $\hat{q} = 1 - \hat{p}$ (though $\hat{\pi}$ and $1 - \hat{\pi}$ could also be used). Note that we use the caret (generally called "hat" in statistics) to indicate the expected or predicted binomial proportions based on some model. Another way of noting the contrast between observation and expectation is to state it in frequencies: The observed frequencies are $f_1 = 80$ and $f_2 = 10$ for the two phenotypes. Assuming the true proportions are π and $1 - \pi$, the expected frequencies are $\hat{f}_1 = \pi n = 0.75(90) = 67.5$ and $\hat{f}_2 = (1 - \pi)n = 0.25(90) = 22.5$, respectively, where n refers to the total sample size, $n = f_1 + f_2$. Note that when we sum the expected frequencies, they yield $67.5 + 22.5 = 90 = n$, as they should.

The obvious question to ask when analyzing such data is whether the observed deviation from the 3:1 hypothesis is of such a magnitude as to be improbable. In other words, do the observed data differ enough from the expected values to cause us to reject the null hypothesis that $\pi = 0.75$? For this example, you already know two methods for coming to a decision about the null hypothesis. This is a binomial distribution in which π is the probability of being a wild type and $1 - \pi$ is the probability of being a mutant. We can work out the probability of obtaining an outcome of 80 wild-type phenotypes and 10 mutants, as well as all "worse" cases for $\pi = 0.75$ and $1 - \pi = 0.25$, in a sample of $n = 90$ offspring. We use the conventional binomial expression, Expression (5.9), except that p is replaced by π and q by $1 - \pi$ and k by n to yield $(\pi + (1 - \pi))^n$. In this example, we have only one sample, so what would ordinarily be labeled k in the binomial is, at the same time, n. Such a computation was illustrated in Table 5.3 and Section 5.2. We can compute the cumulative probability of the tail of the binomial distribution by the methods of Section 5.2. The result is a probability of 0.000849 for all outcomes as deviant or more deviant from the hypothesis. Note that this is a one-tailed test, the alternative hypothesis being that there are more wild-type offspring than the Mendelian hypothesis would postulate. Assuming $\pi = 0.75$ and $1 - \pi = 0.25$, the observed sample is, consequently, a very unusual outcome, and we conclude that it is highly unlikely that such a large deviation from expectation could have been obtained as a random sample from a population with a 3:1 ratio.

A less time-consuming approach based on the same principle is to look up confidence limits for the binomial proportions as we did for the sign test in Section 13.11. Interpolation in Statistical Table **P** shows that for a sample of $n = 90$, an observed proportion of 0.89 would yield approximate 99% confidence limits of 0.78 and 0.96 for the true proportion of wild-type individuals. Clearly, the hypothesized value of $\pi = 0.75$ is beyond the 99% confidence bounds.

Now, let us develop a third approach to testing the null hypothesis by a goodness-of-fit test. Table 17.1 illustrates how we might proceed. In the first column are the observed frequencies f representing the outcome of the experiment. Column (2) shows the observed frequencies as (observed) proportions p and q computed as

TABLE 17.1 Developing the *G*-Test (Likelihood Ratio Test) for Goodness of Fit

Observed and expected frequencies from the outcome of a genetic cross, assuming a 3:1 ratio of phenotypes among the offspring.

Phenotypes	(1) Observed frequencies f	(2) Observed proportions $\dfrac{f}{n}$	(3) Expected proportions $\hat{p}$ and $\hat{q}$	(4) Expected frequencies $\hat{f}$	(5) Ratio $\dfrac{\hat{f}}{f}$	(6) $f \ln\left(\dfrac{\hat{f}}{f}\right)$
Wild type	80	$p = \dfrac{8}{9}$	$\hat{p} = 0.75$	$\hat{p}n = 67.5$	0.84375	-13.59192
Mutant	10	$q = \dfrac{1}{9}$	$\hat{q} = 0.25$	$\hat{q}n = 22.5$	2.25000	8.10930
Sum	$\overline{90}$	$\overline{1.0}$	$\overline{1.0}$	$\overline{90.0}$		$\ln L = \overline{-5.48262}$

f_1/n and f_2/n, respectively. Column (3) lists the expected proportions for the particular null hypothesis being tested. In this case, the hypothesis is a 3:1 ratio, corresponding to the proportions $\pi_0 = 0.75$ and $1 - \pi_0 = 0.25$, as we have seen. Column (4) gives the expected frequencies, which we have already calculated for these proportions as $\hat{f}_1 = \pi_0 n = 0.75(90) = 67.5$ and $f_2 = (1 - \pi_0)n = 0.25(90) = 22.5$.

The log-likelihood ratio test for goodness of fit may be developed as follows. Using Expression (5.9) for the expected relative frequencies in a binomial distribution, we compute two quantities of interest to us here:

$$\binom{90}{80}\left(\frac{3}{4}\right)^{80}\left(\frac{1}{4}\right)^{10} = 0.000,551,754,9$$

$$\binom{90}{80}\left(\frac{80}{90}\right)^{80}\left(\frac{10}{90}\right)^{10} = 0.132,683,8$$

The first quantity is the probability of observing the sampled results assuming that $\pi = \pi_0 = \frac{3}{4}$, as per the Mendelian null hypothesis. The second quantity is the probability of observing the sampled results (80 wild types and 10 mutants) assuming that $\pi = p$—that is, that the population parameter equals the observed sample proportion. Note that these expressions yield the probabilities for the observed outcomes only, *not for observed and all worse outcomes*. Thus, $P = 0.000,551,8$ is less than the earlier computed $P = 0.000,849$, which is the probability of 10 *or fewer* mutants, assuming $\pi = \frac{3}{4}$ and $1 - \pi = \frac{1}{4}$.

The second probability (0.132,683,8) is greater than the first (0.000,551,754,9), because it is based on the assumption that the true proportion is identical to that of the observed data. If the observed proportion p is equal to the proportion, π_0, postulated under the null hypothesis, then the two computed probabilities will be equal and their ratio, L, will equal 1. The greater the difference between π_0 and p, the smaller the ratio will be. Thus, the ratio of these two probabilities or *likelihoods* can be used as a statistic to measure the degree of agreement between sampled and expected frequencies. A test based on such a ratio is called a **likelihood ratio test.** In our case, $L = 0.000,551,754,9/0.132,683,8 = 0.004158$. The theoretical distribution of this ratio is, in general, complex and difficult to compute. It has been shown, however, that the distribution of

$$G = -2 \ln L \qquad (17.1)$$

can be approximated by the χ^2-distribution when sample sizes are large (see Section 17.2 for a definition of "large" in this case). The appropriate number of degrees of freedom is 1. Why? As Table 17.1 shows, there are two frequency cells for these data. Together they have to add to 90. The outcome of the sampling experiment could have been any number of mutants from 0 to 90, but the number of wild types consequently would have to be constrained so that the total would add up to 90. Only one of the cells in the table is free to vary, the other is constrained. Hence, there is one degree of freedom.

In our case,

$$G = -2 \ln L = -2(-5.48262) = 10.96524$$

If we compare this observed value with a χ^2-distribution with one degree of freedom, we find that this G-value is highly unlikely ($P = 0.000{,}928$), assuming that the null hypothesis is true. We thus reject the 3:1 hypothesis and conclude that the proportion of wild types is not equal to 0.75. Because the test is two-tailed, the true proportion could be lower as well as higher, but given the observed frequencies it is more likely to be higher. The geneticist must, consequently, look for a mechanism explaining this departure from expectation.

Notation for the log-likelihood ratio test is not standardized. It is sometimes written as G^2. The symbol $2I$ is also sometimes used for G. The letter I is used because G can be interpreted as twice the amount of information present in the sample for distinguishing between the π_0 of the null hypothesis and that of the alternative hypothesis. The term "information" is used here in the sense defined in the field of information theory. Readers with a knowledge of mathematical statistics may find Kullback (1997) or Csiszár and Shields (2004) interesting in this regard. Employing the **G-test** ($= 2I$-test)—as the log-likelihood ratio test is also called—does not require any special mathematical sophistication; and, as we shall see, this method has some considerable advantages over the traditional chi-square tests.

We will now develop a simple computational formula for G. Referring to Expression (5.9), we can rewrite the two probabilities computed earlier as

$$\binom{n}{f_1}\pi^{f_1}(1 - \pi)^{f_2} \tag{17.2}$$

and

$$\binom{n}{f_1}p^{f_1}q^{f_2} \tag{17.2a}$$

then

$$L = \frac{\binom{n}{f_1}\pi^{f_1}(1 - \pi)^{f_2}}{\binom{n}{f_1}p^{f_1}q^{f_2}} = \left(\frac{\pi}{p}\right)^{f_1}\left(\frac{1 - \pi}{q}\right)^{f_2}$$

Because $\hat{f}_1 = n\pi$ and $f_1 = np$, and similarly $\hat{f}_2 = n(1 - \pi)$ and $f_2 = nq$

$$L = \left(\frac{\hat{f}_1}{f_1}\right)^{f_1}\left(\frac{\hat{f}_2}{f_2}\right)^{f_2}$$

and

$$\ln L = f_1 \ln\left(\frac{\hat{f}_1}{f_1}\right) + f_2 \ln\left(\frac{\hat{f}_2}{f_2}\right) \tag{17.3}$$

The computational steps implied by Expression (17.3) are shown in columns (5) and (6) of Table 17.1 Column (5) contains the ratios of observed over expected frequencies. These ratios would be 1 in the unlikely case of a perfect fit of observations to

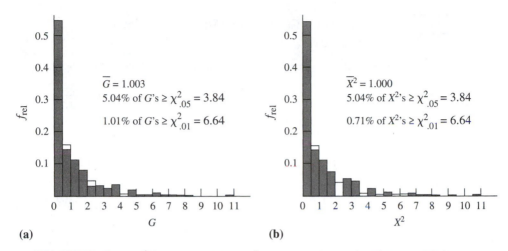

FIGURE 17.1 Expected distribution of G and X^2 for all possible samples of size $n = 90$ drawn from a binomial population with $\hat{p} = 0.75$. The shaded histograms show the distribution of the resulting values of G **(a)** and X^2 **(b)** computed for each sample on the hypothesis that $\hat{p} = 0.75$ and $\hat{q} = 0.25$. The expected $\chi^2_{[1]}$-distribution is superimposed for comparison.

the hypothesis. In such a case, the logarithms of these ratios, in column (6), would be 0, as would their sum. Consequently G, which is twice the natural logarithm of L, would be 0, indicating a perfect fit of the observations to the expectations. When the fit is not perfect, the ratio will be less than 1 and thus ln L will be negative. The negative sign in Expression (17.1) ensures that the G-value will be positive.

Next, let us study the distribution of G. Suppose our expectation of a 3:1 ratio is correct. What would be the distribution of the sample statistics G if we took a series of samples of 90 items from a population in proportions 3:1 and tested them against the hypothesis of a 3:1 ratio? If the sample came out exactly in a 3:1 ratio (mathematically impossible for 90 items), then our value of G would be zero. The greater the departure from expectation, the greater the value of G. Figure 17.1a shows the expected distribution of G's for all possible samples of 90 items from a binomial population with $\pi = 0.75$ and $1 - \pi = 0.25$. On the distribution of G we have superimposed the probability density function of the χ^2-distribution for one degree of freedom, broken up into corresponding classes for ease of comparison. There is general agreement between the two distributions. The G-values appear to be distributed approximately as χ^2 with one degree of freedom.

The test for goodness of fit can be applied to a distribution with more than two classes. Table 17.2 shows the outcome of a dihybrid cross in tomato genetics in which the expected ratio of phenotypes is 9:3:3:1. The table is set up as before, with expected frequencies calculated as $\hat{f}_i = \pi_i n$, where the values of π_i are the expected probabilities for the $a = 4$ classes. Because the probabilities of all possible outcomes sum to 1, $\Sigma^a \pi_i = 1$. The values of π_i for this example are $\pi_1 = \frac{9}{16}, \pi_2 = \frac{3}{16}, \pi_3 = \frac{3}{16},$

TABLE 17.2	The G-Test for Goodness of Fit for More Than Two Classes

Observed and expected frequencies in a dihybrid cross between tall, potato-leaf tomatoes and dwarf, cut-leaf tomatoes.

Phenotypes	(1) Observed frequencies f	(2) Expected frequencies $\hat{f}$	(3) Ratio $\dfrac{\hat{f}}{f}$	(4) $f \ln\left(\dfrac{\hat{f}}{f}\right)$
Tall, cut-leaf	926	906.1875	0.978604	−20.02752
Tall, potato-leaf	288	302.0625	1.048828	13.72996
Dwarf, cut-leaf	293	302.0625	1.030930	8.92517
Dwarf, potato-leaf	104	100.6875	0.968149	−3.36640
Sum	1611	1611.0000		$\ln L = -0.73879$

SOURCE: Data from MacArthur (1931).

and $\pi_4 = \frac{1}{16}$. Again, we calculate ratios of expected over observed frequencies in column (3) and $f \ln (\hat{f}/f)$ in column (4). The sum of the values in column (4) yields $\ln L = -0.73879$; hence $G = -2 \ln L = 1.478$. This operation can be expressed by Equation (17.4), below, whose derivation, based on the multinomial expectations (for more than two classes), is shown in Appendix A.14. Note the use of the reciprocal of the ratio defined earlier to eliminate the minus sign.

$$G = 2 \sum^{a} f_i \ln \left(\frac{f_i}{\hat{f}_i}\right) \tag{17.4}$$

Thus, the formula can be seen as the sum of the contributions of departures from expectation [$\ln (f_i/\hat{f}_i)$] weighted by the frequency of the particular class (f_i).

How can we evaluate the outcome of our test for goodness of fit? We need to know how many degrees of freedom there are in this example to be able to compare the observed value of G with the appropriate χ^2-distribution. We now have four classes constrained by a sum. Thus, any three of them can vary freely, but the fourth class must constitute the difference between the total sum and the sum of the first three. Thus, in a case with four classes we have three degrees of freedom, and in general when we have a classes, we have $a - 1$ degrees of freedom.

From Expression (17.4) and the data in Table 17.2 we arrive at $G = 2(0.73879) = 1.47758$. The P-value of a $\chi^2_{[3]}$-distribution at 1.47758 is 0.687,454. It is, therefore, quite likely that the observed frequency distribution of phenotypes could have been sampled at random from a population with a 9:3:3:1 ratio. We have no reason for rejecting the null hypothesis. In the absence of software, we can consult Statistical Table **D** (chi-square), under three degrees of freedom. We would find the probability of obtaining a $\chi^2_{[3]} > 1.478$ to be somewhat more than 50% and come to the identical conclusion. Insofar as we can tell, the dihybrid cross of Table 17.2 is consistent with the expected 9:3:3:1 ratio.

Specifying a particular alternative hypothesis is usually more difficult than specifying the null hypothesis. In some genetics experiments, there are only a limited number of possible alternative hypotheses. Thus, if a 3:1 ratio is rejected, a 1:1 ratio or a more complicated ratio might apply in a given case. In other examples, however, the alternative is just that the null hypothesis is not true. Of course, when we accept a 3:1 ratio, we still do not know that 3:1 is in fact the true ratio. All we can say is that our data are not inconsistent with the true ratio being 3:1.

In some goodness-of-fit tests, we subtract more than one degree of freedom from the number of classes, a. These are instances in which the parameters for the null hypothesis have been estimated from the sample data themselves, in contrast with the null hypotheses tested in Tables 17.1 and 17.2. In these two cases, the hypothesis to be tested was based on the investigator's general knowledge of the specific problem and of Mendelian genetics. The values of $\pi = 0.75$ and $1 - \pi = 0.25$ were dictated by the 3:1 hypothesis and were not estimated from the sampled data. Similarly, the 9:3:3:1 hypothesis was also based on general genetic theory. For this reason, the expected frequencies are said to have been based on an **extrinsic hypothesis,** a hypothesis external to the data. By contrast, consider the expected frequencies of birth weights under the assumption of normality (see Table 6.2). To compute these frequencies we needed values for μ and σ which we estimated from the sample mean $\overline{Y}$ and the sample standard deviation s of the birth weights. Therefore the two parameters of the computed normal distribution, the mean and the standard deviation, came from the sampled observations themselves. The expected normal frequencies thus represent an **intrinsic hypothesis.** In such a case, to obtain the correct number of degrees of freedom for the G-test of goodness of fit (and for the chi-square test described later in this section) we would subtract from a, the number of classes into which the data had been grouped, not only one degree of freedom for n, the sum of the frequencies, but also two additional degrees of freedom—one for the estimate of the mean and the other for the estimate of the standard deviation. Thus, in such a case a sample statistic G would be compared with chi-square for $a - 3$ degrees of freedom. Exact rules on how many degrees of freedom to subtract and computational details for such tests are given in the next section.

Now let us introduce you to an alternative technique. We must acquaint you with this method because it is the traditional approach, which you will see applied in the earlier literature and in a substantial number of current research publications. The statistic is

$$X^2 = \sum^a \frac{(f_i - \hat{f}_i)^2}{\hat{f}_i} \tag{17.5}$$

where the variables are defined as in Expression (17.4).

We turn once more to the genetic cross with 80 wild-type and 10 mutant individuals, this time laid out in Table 17.3. First we calculate $f - \hat{f}$, the deviation of

| TABLE 17.3 | Developing the Chi-Square Test for Goodness of Fit |

Observed and expected frequencies from the outcome of a genetic cross, assuming a 3:1 ratio of phenotypes among the offspring.

	(1)	(2)	(3)	(4)	(5)
Phenotypes	**Observed frequencies**	**Expected frequencies**	**Deviations from expectation**	**Deviations squared**	
	f	$\hat{f}$	$f - \hat{f}$	$(f - \hat{f})^2$	$\dfrac{(f - \hat{f})^2}{\hat{f}}$
Wild type	80	$\hat{p}n = 67.5$	12.5	156.25	2.31481
Mutant	10	$\hat{q}n = 22.5$	-12.5	156.25	6.94444
Sum	90	90.0	0		$X^2 = 9.25926$

observed from expected frequencies. Note that the sum of these deviations equals zero, for reasons very similar to those causing the sum of deviations from a mean to add to zero. In an example with two classes, then, the deviations are always equal and opposite in sign (proof of this property is developed in Appendix A.15). Following our previous approach of making all deviations positive by squaring them, we square $(f - \hat{f})$ in column (4) to yield a measure of the magnitude of the deviation from expectation. This quantity is expressed as a proportion of the expected frequency. After all, if the expected frequency were 13.0, a deviation of 12.5 would be extremely large, comprising almost 100% of $\hat{f}$, but such a deviation would represent only 10% of an expected frequency of 125.0. Each value in column (5) is the quotient of the quantity in column (4) divided by that in column (2). Note that the magnitude of the quotient is greater for the second row (mutant phenotypes), in which the $\hat{f}$ is smaller. The next step in developing the test statistic is to sum these quotients, which yields a value of 9.25926, shown at the foot of column (5).

What shall we call this statistic? We have some nomenclatural problems here. Many of you will have recognized this procedure as the so-called **chi-square test,** regularly taught to beginning genetics classes. The name of the test is too well established to make a change practical, but in fact, the quantity just computed, the sum of column (5), could not possibly be a χ^2-value. The χ^2-distribution is a continuous and theoretical frequency distribution, whereas our quantity, 9.25926, is a sample statistic based on discrete frequencies. The latter point is easily seen if you visualize other possible outcomes. For instance, we could have had as few as zero mutants matched by 90 wild-type individuals (assuming that the total number of offspring $n = 90$ remains constant), or we could have had 1, 2, 3, or more mutants, in each case balanced by the correct number of wild-type offspring to yield a total of 90. Observed frequencies change in unit increments, and because the expected frequencies remain constant, deviations, their squares, and the quotients are not continuous variables but can assume only certain values.

The reason that this test has been called the chi-square test and that many persons inappropriately call the statistic obtained as the sum of column (5) a chi-square is that the sampling distribution of this sum is approximately that of a χ^2-distribution with the appropriate number of degrees of freedom. We have previously encountered cases of this sort. The G-statistic we just studied is distributed as approximately chi-square. Similarly, Bartlett's test for homogeneity of variances (see Section 13.3) and the test for homogeneity of correlation coefficients (see Section 15.5) both employ a sample statistic distributed approximately as a chi-square. Because the sample statistic is not a chi-square, however, in these two tests we followed the prevalent convention of labeling the sample statistic X^2 rather than χ^2. We will follow the same practice here, although in one sense this is misleading. The three tests just mentioned, the homogeneity tests for variances and correlation coefficients and the goodness-of-fit test shown in Table 17.3, have only one thing in common: Their sample statistic is approximately distributed as chi-square. For this reason, we give the sample statistic a Latin letter resembling χ^2. Structurally, however, the three tests are quite different, and to be strictly correct we should provide each of these with a separate symbol to emphasize that the three sample statistics are not the same. This approach was taken in the case of the G-test, where the sample statistic is called G, although it is approximately chi-square distributed. Because we have followed a policy of conservatism on symbolism, we retain the X^2, but the reader should be clear that the sample statistic X^2 in this chapter is for the test of goodness of fit and is not the same as the X^2 for homogeneity of either variances or correlations. Some authors have called X^2 (for goodness of fit) the *Pearson statistic*.

Empirically, we can show that the distribution of the X^2-statistic is indeed close to a chi-square distribution with one degree of freedom as we did for the G-statistic. The results are graphed in Figure 17.1**b**. Note that discontinuity is more apparent in X^2 than in G. Also, G appears to follow the chi-square distribution a bit more closely. The value of $X^2 = 9.25926$ from Table 17.3, when entered in the chi-square program of BIOMstat (for 1 df) yields a $P = 0.002343$. If we employ Statistical Table **D,** our computed value of $X^2 = 9.25926$ falls between the critical χ^2-values 7.879 and 10.828 for P-values of 0.005 and 0.001, respectively. Either result will lead us to rejecting the 3:1 null hypothesis. The chi-square test is always one-tailed. Because the deviations are squared, negative and positive deviations both result in positive values of X^2. Our conclusions are the same as with the G-test. In general, X^2 will be numerically similar to G.

We can apply the chi-square test for goodness of fit to a distribution with more than two classes as well. We will practice this application on the dihybrid cross in Table 17.2. Squared deviations and their quotients over expected frequencies must now be summed for four classes. The sum of these quotients yields a value of $X^2 = 1.46872$. Again, this four-class example contains three degrees of freedom, so we compare our observed X^2 against $\chi^2_{[3]}$.

In Statistical Table **D** (chi-square), under three degrees of freedom, we find the probability of $\chi^2_{[3]} > 1.469$ to be between 90% and 50% (the computed P-value is 0.689,508). We therefore have no reason to reject the null hypothesis. As far as we can tell, the dihybrid cross of Table 17.2 follows the 9:3:3:1 ratio.

Expression (17.5) can be applied to any chi-square test of goodness of fit, although its use is somewhat cumbersome and requires the computation of deviations, their squaring, and their division by expected frequencies over all of the classes.

17.2 Single-Classification Tests for Goodness of Fit

Before discussing in detail the computational steps involved in tests of goodness of fit of single-classification frequency distributions, some remarks on the choice of a test statistic are in order. We have already stated that the traditional method for such a test is the chi-square test for goodness of fit. The newer approach by the G-test, however, has been recommended on theoretical grounds. The G-test is computationally simpler, especially in complicated designs. The BIOMstat computer program uses G exclusively for goodness-of-fit tests. Box 17.1, single-classification goodness-of-fit tests, features G-tests principally, but additionally demonstrates how to carry out a chi-square test so that you can understand published results using this statistic. The more complex designs featured in later sections employ the G-test exclusively.

The G- and chi-square tests can be applied to frequency distributions of nominal variables as well as to those of continuous variables grouped suitably. For continuous sample distributions, however, the preferred method is the Kolmogorov–Smirnov test. This simpler and more powerful procedure is described at the end of this section.

The G-tests for goodness of fit for single-classification frequency distributions are given in Box 17.1, which treats only cases in which the expected frequencies are based on a hypothesis extrinsic to the data: There are a classes, and the expected proportions in each class are assumed on the basis of outside knowledge and are not functions of parameters estimated from the sample. We start with the general case for any number of classes, where the number of classes is symbolized by a, to emphasize the analogy with analysis of variance. The data are the results of a complicated genetic cross expected to result in an 18:6:6:2:12:4:12:4 ratio. In all, 241 progeny were obtained and classified into the eight phenotypic classes. The expected frequencies can be computed simply by multiplying the total sample size n by the expected probabilities of occurrence. The overall fit to expectation is quite good. This is a test with an extrinsic hypothesis because the genetic ratio tested is based on considerations prior and external to the sample observations tested in the example.

We learned in the previous section that the probability of a computed G is obtained by comparing it to the χ^2 distribution with $a - 1$ degrees of freedom. Because the actual type I error of G-tests tends to be higher than the intended level, Williams (1976) proposed a correction for G that provides a better approximation. He divides G by a correction factor q, which is computed as $q = 1 + (a^2 - 1)/6nv$. This q should not be confused with the proportion $q = 1 - p$. Note that for the extrinsic hypothesis v will always be $a - 1$. The effect of this correction is to reduce the observed value of G slightly. When we compare our result with the χ^2-distribution, we find that the value of G obtained from our sample could readily have come from random sampling of the hypothesized population ($P = 0.269,641$). We do not have sufficient evidence to reject the null hypothesis and are led to conclude that the sample is consistent with

BOX 17.1 Tests for Goodness of Fit: Single-Classification, Expected Frequencies Based on Hypothesis Extrinsic to the Sampled Data

G-Test

1. Frequencies divided into $a \geq 2$ classes

In a genetic experiment involving a cross between two varieties of the bean *Phaseolus vulgaris*, Smith (1939) obtained the following results:

Phenotypes ($a = 8$)	Observed frequencies f	Expected frequencies $\hat{f}$
Purple/buff	63	67.78125
Purple/testaceous	31	22.59375
Red/buff	28	22.59375
Red/testaceous	12	7.53125
Purple	39	45.18750
Oxblood red	16	15.06250
Buff	40	45.18750
Testaceous	12	15.06250
Total	241	241.00000

The expected frequencies $\hat{f}_i$ were computed on the basis of the expected ratio of $18 : 6 : 6 : 2 : 12 : 4 : 12 : 4$. Compute $\hat{f}$ as $\hat{p}_i n$. Thus, $\hat{f}_1 = \hat{p}_1 n = \left(\frac{18}{64}\right) \times 241 = 67.78125$.

Employing Expression (17.4) we obtain:

$$G = 2 \sum_{}^{a} f_i \ln\left(\frac{f_i}{\hat{f}_i}\right)$$

$$= 2\left[63 \ln\left(\frac{63}{67.78125}\right) + \cdots + 12 \ln\left(\frac{12}{15.06250}\right)\right] = 8.82396$$

The degrees of freedom are $a - 1 = 7$, leading to a P-value of 0.265,543. Seeing that $P \geq 0.05$, we may consider the data consistent with Smith's null hypothesis.

Applying Williams's correction to G, to obtain a better approximation to χ^2 we compute

$$q = 1 + \frac{a^2 - 1}{6n(a - 1)} = 1 + \frac{a + 1}{6n}$$

$$= 1 + \frac{9}{6(241)} = 1.006224$$

$$G_{adj} = \frac{G}{q} = \frac{8.82396}{1.006224} = 8.76938, \quad df = 7, P = 0.269,641$$

Box 17.1 (continued)

2. Special case of frequencies divided into $a = 2$ classes

For the G-test there is no special shortcut equation for the case in which $a = 2$.

In an F_2 cross in drosophila the following 176 progeny were obtained, of which 130 were wild-type flies and 46 ebony mutants. Assuming that the mutant is an autosomal recessive, one would expect a ratio of 3 wild-type flies to each mutant fly. To test whether the observed results are consistent with this 3:1 hypothesis, we set up the data as follows.

Flies	f	Hypothesis	$\hat{f}$	Adjusted f
Wild type	$f_1 = 130$	$\hat{p} = 0.75$	$\hat{p}n = 132.0$	130.5
Ebony mutant	$f_2 = 46$	$\hat{q} = 0.25$	$\hat{q}n = 44.0$	45.5
	$n = 176$		176.0	176.0

Computing G from Expression (17.4), we obtain

$$G = 2 \sum^a f_i \ln \left(\frac{f_i}{\hat{f}_i} \right)$$

$$= 2 \left[130 \ln \left(\frac{130}{132} \right) + 46 \ln \left(\frac{46}{44} \right) \right] = 0.12002$$

$$df = a - 1 = 1, P = 0.729{,}013$$

Williams's correction for the two-cell case is $q = 1 + \dfrac{1}{2n}$, which is

$$1 + \frac{1}{2(176)} = 1.00284$$

in this example.

$$G_{adj} = G/q = 0.12002/1.00284 = 0.11968 \quad df = 1, P = 0.729{,}382$$

We clearly do not have sufficient evidence to reject our null hypothesis.

An alternative adjustment discussed in Section 17.2 is the correction for continuity. We calculate this adjustment by computing an adjusted observed frequency f_i, which is changed to reduce the difference between the observed and the corresponding expected frequencies by $1/2$, as shown in the table above. Then we employ Expression (17.4) as before to obtain G_{adj}. This correction may be applied when n is less than about 200. The computations are carried out as described above, but using adjusted f's instead of the original f's (see text).

$$G_{adj} = 2 \sum^a f_i \ln \left(\frac{f_i}{\hat{f}_i} \right)$$

$$= 2 \left[130.5 \ln \left(\frac{130.5}{132.0} \right) + 45.5 \ln \left(\frac{45.5}{44.0} \right) \right] = 0.06768$$

$$df = 1, P = 0.794{,}745$$

Note that G_{adj} is now smaller because the continuity correction results in a more conservative test than Williams's correction. One uses just one of the corrections—not both.

Box 17.1 (continued)

Chi-Square Test

1. Frequencies divided into $a \geq 2$ classes employ Expression (17.5):

$$X^2 = \sum_{}^{a} \frac{(f_i - \hat{f}_i)^2}{\hat{f}_i}$$

for the bean example featured earlier in this box:

$$X^2 = \frac{(63 - 67.78125)^2}{67.78125} + \frac{(31 - 22.59375)^2}{22.59375} + \cdots + \frac{(12 - 15.06250)^2}{15.06250} = 9.53389$$

$$df = a - 1 = 7, \quad P = 0.216{,}531$$

Because $P > 0.05$, the data may be considered consistent with the null hypothesis (as was concluded by the G-test).

2. Special case of frequencies divided into $a = 2$ classes.

First, consult the table immediately following to learn whether a correction is needed and which one to use. We are again using the drosophila data tabled earlier in this box.

Adjustment for small sample sizes in cases with $a = 2$ classes:

If	Compute
$n > 200$	X^2 by Expression (17.5), no correction needed.
$200 \geq n > 25$	X^2_{adj} (continuity correction) by Expressions (17.6) or (17.7) in Section 17.2. Note that the adjustment will result in a conservative test.
$n < 25$	Exact probabilities of the binomial as shown in Table 5.1 (Section 5.2).

We choose Expression (17.7),

$$X^2_{adj} = \frac{(|f_1 - \hat{p}_1 n| - \frac{1}{2})^2}{\hat{p}\hat{q}n} = \frac{(|130 - 132| - \frac{1}{2})^2}{0.75 \times 0.25 \times 176} = 0.068182, \quad df = 1, \quad P = 0.794{,}002$$

We do not reject our null hypothesis. In fact, these data show a remarkably good fit to the 3:1 hypothesis.

the specified genetic ratio. Especially in examples such as the one just analyzed, however, remember that we have not specified an alternative hypothesis; there are numerous alternative hypotheses that also could not be excluded if we were to carry out significance tests for them. We have not really proven that the data are distributed as specified; various other genetic ratio hypotheses could also be plausible.

No comprehensive study has determined how small a sample can be and still be suitable for the G-test of goodness of fit. The usual rule of thumb is that the smallest $\hat{f}$ should be five or more. Whenever classes with expected frequencies of less than five occur, expected and observed frequencies for those classes are generally pooled with an adjacent class to obtain a joint class with an expected frequency $\hat{f} > 5$ (such a case is illustrated later in Box 17.2). On the basis of an analysis of 126 equiprobable multinomial distributions, Conahan (1970) found that if $\hat{f}$ is > 10, then G gives a good approximation to the exact multinomial probability. She found that G was satisfactory (and better than X^2) for $a \geq 5$ and $\hat{f} \geq 3$, and she recommended the exact test (see below) when $a \geq 5$ and $\hat{f} < 3$ and for $a < 5$ and $\hat{f} < 5$.

An exact test could be carried out in a manner quite similar to working out the exact probabilities for the binomial distribution, as shown in Table 7.1. One first calculates the probability of obtaining the observed outcome on the null hypothesis. The general term for the multinomial distribution (for four classes) is given by Expression (17.11) in Section 17.4. This expression could readily be extended to any number of classes as required by the distribution to be tested. Unless the probability of the observed outcome under the null hypothesis is already greater than α, the specified type I error, the probabilities of all worse cases, must be evaluated as in the binomial case discussed in Section 7.1. A definition of "all worse cases" is not so simple in the multinomial distribution. The conventional procedure is to sum all probabilities $P_i \leq P_0$, where P_i is the probability of other outcomes under the null hypothesis and P_0 is the probability of the observed outcome under the null hypothesis. Such a computation could involve a prodigious amount of calculation in all but trivial cases. The StatXact (2010) program evaluates the exact probability of the null hypothesis in such cases.

When expected frequencies are unequal, Conahan's recommendations do not directly apply but could be interpreted conservatively by applying them to the smallest expected frequencies. Larntz (1978) compared X^2 and G at just the 5% level of significance and found that when $1.5 < \hat{f} < 4$, G rejects the null hypothesis much too often. Also when $f = 0$ or 1, G is not a close approximation to χ^2. Unfortunately this study did not consider the correction suggested by Williams (1976). Monte Carlo simulations (see Section 7.1) that we performed show that with this correction, the G-statistic approximates the χ^2-distribution more closely than it does without it.

In a goodness-of-fit analysis involving more than two classes, as in the bean example of Box 17.1, with eight classes, we may wish to probe deeper into the nature and, ultimately, reasons for departure of some classes from expectation. Such schemes are comparable to single-degree-of-freedom decompositions in analysis of variance (see Sections 9.4 and 9.5). As we did earlier, we must distinguish between planned and unplanned tests. Certain planned tests—for example, computing the goodness of fit for all single phenotypes (such as buff) together versus mixed phenotypes (such as purple–buff) also placed in one group—can be carried out simply. We evaluate their significance with one degree of freedom in the ordinary manner. By contrast, when we carry out tests suggested by the outcome of the overall analysis, we have to use the Bonferroni or Dunn–Šidák approaches learned in Section 9.4. To obtain an experimentwise error rate α, we must carry out each individual test at a critical probability of $\alpha' = \alpha/k$ (Bonferroni) or $\alpha' = 1 - (1 - \alpha)^{1/k}$ (Dunn–Šidák), where k is the number of intended tests. If we wish to test all possible sets, we undertake a simultaneous test procedure using $\chi^2_{\alpha[a-1]}$ as the critical value for all tests carried out. This test will ensure an α experimentwise error rate.

Next we consider the case for $a = 2$ cells. There is no special simplified equation for the G-test. One employs the same general equation [Expression (17.4)]. When Expression (17.4) is written explicitly for the two-cell case, it corresponds to Expression (17.3) multiplied by 2.

In tests of goodness of fit involving only two classes, the values of G and X^2, as computed from Expressions (17.4) and (17.5) typically result in type I errors at a level higher than the intended one. Williams's correction reduces the value of G and results in a more conservative test (see Box 17.1). An alternative correction that has been used widely is the **correction for continuity,** usually applied to make the value of G or X^2 approximate the χ^2-distribution more closely. This correction consists of adding or subtracting 0.5 from the observed frequencies in such a way as to minimize the value of G or X^2. In the case of the G-test, we simply adjust the f_i, changing them to reduce the difference between them and the corresponding expected frequencies by one half. Then we employ Expression (17.4) for G to obtain G_{adj}. In our experience, the correction for continuity results in excessively conservative tests.

The two-cell example in Box 17.1 is a genetic cross with an expected 3:1 ratio. The expected frequencies differ very little from the observed frequencies; it is no surprise, therefore, that the resulting value of $G = 0.12002$ yields a P-value of 0.729,013. Almost 73% of all samples from a population with the expected ratio would show as great or greater deviations than the sample at hand. The G-test could be adjusted by either Williams's correction or by the continuity correction, but not by both. The latter yields the more conservative results. However, in this case it is not worth doing it, because the G-value adjusted by either correction would only decrease in magnitude, yielding an even better fit to the null hypothesis. In Table 17.1, we feature another 3:1 genetic cross that deviates more from expectations than the one from Box 17.1. The unadjusted G, 10.96524, yields a P-value of 0.000,928; the continuity corrected $G_{adj} = 10.01214$ ($P = 0.001,555$), and Williams' correction results in $G_{adj} = 10.82987$ ($P = 0.000,999$). Thus, even after the greatest downward correction of the G_{adj}, achieved by the continuity correction, a deviation from expectation of as much as we noted, or more, will occur only roughly two out of 1000 times.

These corrections are recommended as routine by some, whereas others suggest that they are necessary only for sample sizes $n < 200$. Actually, the corrections make only a small difference in G or X^2 even when sample sizes are less than 200. We have found the continuity correction too conservative and therefore recommend that the Williams correction be applied routinely, although it will have little effect when sample sizes are large. For sample sizes of 25 or less, work out the exact probabilities as shown in Table 5.3 (see Section 5.2).

The conceptual alternative hypothesis in this genetic cross with two progeny classes was $H_2:\pi \neq \pi_0 = \frac{3}{4}$; that is, the test was two-sided, because we assumed that the proportion π of wild-type individuals could be less than or greater than the expected proportion $\pi_0 = \frac{3}{4}$. If our alternative hypothesis is that the proportion of wild types could be only less than the expected $\frac{3}{4}$, we should carry out a test of a one-sided conceptual hypothesis, and the chi-square table should be entered in the column for 2α if a type I error of α is desired. For $\alpha > 2$ classes, a simple one-sided alternative hypothesis cannot be constructed. These same considerations apply to the chi-square test with two classes, which we discuss next.

Having dealt with the problems in Box 17.1 by means of the G-test, we now turn to the *chi-square test*, which appears toward the end of that box. We employ Expression (17.5) to obtain $X^2 = 9.53389$ for the bean example with $a = 8$ classes. Again our test statistic is less than 14.067, the 5% critical value of χ^2.

The correction for continuity in the two-cell case with small sample sizes is made for X^2 by subtracting 0.5 from the absolute values of the deviation $f - \hat{f}$, yielding the following adjusted versions.

$$X^2_{adj} = \sum^2 \frac{(|f_i - \hat{f}_i| - \frac{1}{2})^2}{\hat{f}_i} \tag{17.6}$$

and

$$X^2_{adj} = \frac{(|f_1 - \hat{p}_1 n| - \frac{1}{2})^2}{\hat{p}\hat{q}n} \tag{17.7}$$

Knowledge of the size and direction of deviations from expectation aids the formulation of new hypotheses and experiments; inspection of $\hat{f}$ and of deviations $f - \hat{f}$ is therefore recommended.

Let us compare the results of all these methods for a test for goodness of fit in a small sample. We will use the example of Table 5.3: a litter of 17 offspring, of which 14 were females and 3 were males. We wish to test these data against a hypothetical sex ratio of 1:1. By the rules we have enumerated here, we should use the exact probability, which was worked out in Table 5.3 as 0.006,363,42, the probability of a deviation as great or greater in one direction from the 1:1 hypothesis. We approximate the probability of a two-tailed test by doubling this probability to yield $P = 0.01273$.

Applying the G-test to the example of Table 5.3, we obtain the following results:

G [by Expression (17.4)]

$$G = 2\left[14 \ln\left(\frac{14}{8.5}\right) + 3 \ln\left(\frac{3}{8.5}\right)\right] = 7.72303, P = 0.00545$$

G_{adj} (by Williams's correction)

$$G = \frac{7.72303}{1.02941} = 7.50237, P = 0.00620$$

G_{adj} (by the continuity correction)

$$G = 2\left[13.5 \ln\left(\frac{13.5}{8.5}\right) + 3.5 \ln\left(\frac{3.5}{8.5}\right)\right] = 6.27971, P = 0.01221$$

X^2 [by Expression (17.5)]

$$X^2 = \frac{(14 - 8.5)^2}{0.5 \times 0.5 \times 17} = 7.11765, P = 0.00763$$

X^2_{adj} [by Expression (17.7)]

$$X^2_{adj} = \frac{(14 - 8.5 - \frac{1}{2})^2}{0.5 \times 0.5 \times 17} = 5.88235, P = 0.01529$$

We computed these probabilities using the statistical calculator of BIOMstat. Without such a program, we could have used Statistical Table **A,** areas of the normal curve, by entering the table with argument $\sqrt{G}$ or $\sqrt{X^2}$ (because $\sqrt{\chi^2_{\alpha[1]}} = t_{\alpha[\infty]}$). In this instance, all of the results suggest rejection of the 1:1 hypothesis. Although the probability resulting from the G-test adjusted by the continuity correction is closest to the exact probability in this case, we would expect Williams's correction to yield probabilities closer to the exact values on the average. For both G and X^2, the adjusted value yields more nearly correct probabilities. In another experiment, carried out as described earlier for Figure 17.1 but with the continuity correction applied to both G and X^2, the percentages of type I error for the G_{adj} sampling experiment were smaller than the intended 5% and 1% values (3.84% and 0.77%, respectively); the percentages for X^2_{adj} were even smaller (2.76% and 0.71%, respectively).

Next we test the goodness of fit of frequencies arrayed in a single classification where the expected frequencies are based on a hypothesis intrinsic to the sampled data. Box 17.2 illustrates the computation using the sex ratio data in sibships of 12 introduced in Table 5.4. Only the G-test is illustrated, but the chi-square test could have been employed as well, using Expression (17.5). Recall that the expected frequencies in these data are based on the binomial distribution, with the parametric proportion of males π_δ estimated from the observed frequencies of the sample ($p_\delta = 0.519,215$). In Box 17.2, we employ the simple computational formula [Expression (17.4)] for efficient computation. The signs of the deviations are shown in column (5) of the table to indicate the pattern of departures from expectation.

Two aspects make this computation different from that in part **1** of Box 17.1. As we have explained, the G-test does not yield very accurate probabilities for small $\hat{f}_i$. The cells with $\hat{f}_i < 3$ (when $a \geq 5$) or $\hat{f}_i < 5$ (when $a < 5$) are generally grouped with adjacent classes so that the resulting $\hat{f}_i$ are large enough. The grouping of classes results in a less powerful test with respect to alternative hypotheses that differ mostly in terms of expected frequencies in the tails of the distribution. By these criteria, the classes of $\hat{f}_i$ at both tails of the distribution are too small. We group them by adding their frequencies to those in contiguous classes as shown in Box 17.2. Clearly, the observed frequencies must also be grouped to match the grouping of expected frequencies. The number of classes a is the number *after* grouping has taken place. In this case, $a = 11$.

The other new feature is the number of degrees of freedom considered for the significance test. We always subtract one degree of freedom for the fixed sum (here $n = 6115$). In addition, we subtract one degree of freedom for every parameter of the expected frequency distribution estimated from the sampled distribution. In this case, we estimated π_δ from the sample; therefore, a second degree of freedom is subtracted from a, making the final number of degrees of freedom $a - 2 = 11 - 2 = 9$.

BOX 17.2 G-Test for Goodness of Fit: Single-Classification, Expected Frequencies Based on Hypothesis Intrinsic to the Sampled Data

Sex ratio in 6115 sibships of 12 in Saxony. (Data from Table 5.4.) Column (4) gives the expected frequencies, assuming a binomial distribution. Computed originally to one decimal place, they are given here to five decimal places to give sufficient precision to the computation of G.

(1) (2)		(3)	(4)	(5)
$\male\male$	$\female\female$	f	$\hat{f}$	**Deviation from expectation**
12	0	$\left.\begin{array}{c}7\\45\end{array}\right\}52$	$\left.\begin{array}{c}2.34727\\26.08246\end{array}\right\}28.42973$	+
11	1			
10	2	181	132.83570	+
9	3	478	410.01256	+
8	4	829	854.24665	−
7	5	1112	1265.63031	−
6	6	1343	1367.27936	−
5	7	1033	1085.21070	−
4	8	670	628.05501	+
3	9	286	258.47513	+
2	10	104	71.80317	+
1	11	$\left.\begin{array}{c}24\\3\end{array}\right\}27$	$\left.\begin{array}{c}12.08884\\0.93284\end{array}\right\}13.02168$	+
0	12			
		$6115 = n$	6115.00000	

Because expected frequencies $\hat{f}_i < 3$ for $a = 13$ classes should be avoided, we group the classes at both tails with the adjacent classes to create classes of adequate size. Corresponding classes of observed frequencies f_i should be grouped to match. The number of classes after grouping is $a = 11$.

Compute G by Expression (17.4):

$$G = 2 \sum_{}^{a} f_i \ln\left(\frac{f_i}{\hat{f}_i}\right) = 2 \left[52 \ln\left(\frac{52}{28.42973}\right) + 181 \ln\left(\frac{181}{132.83570}\right) \right.$$

$$\left. + \cdots + 27 \ln\left(\frac{27}{13.02168}\right) \right] = 94.87155$$

Because there are $a = 11$ classes, the degrees of freedom are maximally $a - 1 = 10$. Because the π_i was estimated from the p_i of the sample, however, another degree of freedom is removed, and the sample G is compared to a χ^2-distribution with

Box 17.2 (continued)

$a - 2 = 11 - 2 = 9$ degrees of freedom. Applying Williams's correction to the observed G, we obtain

$$q = 1 + \frac{(a^2 - 1)}{6nv}$$

$$= 1 + \frac{(11^2 - 1)}{6(6115)(9)} = 1.000,363,4$$

$$G_{adj} = \frac{G}{q} = \frac{94.87155}{1.000,363,4} = 94.83709, P = 2.22 \times 10^{-16}$$

The null hypothesis—that the sample data follow a binomial distribution—is therefore rejected decisively.

Typically, the following degrees of freedom pertain to G-tests for goodness of fit with expected frequencies based on a hypothesis *intrinsic* to the sample data (a is the number of classes after grouping, if any):

Distribution	Parameters estimated from sample	df
Binomial	$\hat{p}$	$a - 2$
Normal	μ, σ	$a - 3$
Poisson	μ	$a - 2$

When the parameters for such distributions are estimated from hypotheses *extrinsic* to the sampled data, the degrees of freedom are uniformly $a - 1$.

These tests could also be carried out by the chi-square test, employing Expression (17.5) and using the χ^2-distribution with the same degrees of freedom as for the G-test.

We compute the probabilities for the corrected sample value of $G_{adj} = 94.83709$, using the statistical calculator of BIOMstat with χ^2 at 9 *df*. The result is $P = 2.22045 \times 10^{-16}$ and we conclude that it is extremely unlikely that the observed frequencies are a random sample from a binomial population with $\pi_\delta = 0.519,215$. Alternatively, using Statistical Table **D** for $\chi^2_{[9]}$, we find that the critical value for $P = 0.001$ is 27.877, so that our G_{adj}-value of 94.83709 would yield a very low probability ($P \ll 0.001$) assuming that the null hypothesis is correct. We therefore reject this hypothesis and conclude that the sex ratios are not binomially distributed. As is evident from the pattern of deviations, there is an excess of sibships in which one sex or the other predominates. From now on to the end of this chapter, we shall report only probability values obtained with the statistical calculator of BIOMstat. Readers will by now have realized that they could obtain critical values leading to equivalent decisions using Statistical Table **D**.

The G-test for testing the goodness of fit of a set of data to an expected frequency distribution can be applied not only to the binomial distribution, but also to the normal, Poisson, and other distributions. For a normal distribution, we customarily estimate the two parameters μ and σ from the sampled data. Hence the appropriate degrees of freedom are $a - 3$. In the Poisson distribution, only one parameter, μ, must be estimated, so the appropriate degrees of freedom are $a - 2$. Remember that the critical issue is whether the expected frequencies are based on an extrinsic or intrinsic hypothesis, not on whether the test is against a normal distribution or some other distribution. An investigator may have a hypothesis about the mean and standard deviation of a distribution and make the expected normal frequencies conform to this hypothesis. In such a case, the two extra degrees of freedom are not subtracted, so the degrees of freedom would be only $a - 1$, because estimates of the parameters were not obtained from the sample.

Conventional tests of goodness of fit specify the alternative hypothesis simply as departures from expectation, hence they are one-sided. Occasionally one may wish to test the alternative hypothesis that the observed frequencies are either too deviant or too close to the expected frequencies. In such a case, P-values resulting from the statistical calculator of the BIOMstat program should be doubled or critical values for chi-square should be looked up in columns $1 - \alpha/2$ and $\alpha/2$ of the chi-square table (Statistical Table **D**) when a type I error of α is desired. Questions of power and required sample sizes for chi-square distributed goodness-of-fit tests are treated later in this chapter (Section 17.8).

The **Kolmogorov–Smirnov test** is a nonparametric test that is applicable to continuous frequency distributions, where it has greater power than the G- or chi-square tests for goodness of fit. The Kolmogorov–Smirnov test for goodness of fit is especially useful with small samples. In such cases, it is not necessary (actually, it is inadvisable) to group the observations into classes. We have already learned the general approach of this test in the Kolmogorov–Smirnov two-sample test (see Section 13.10). As in its earlier application, this test is based on differences between two cumulative relative frequency distributions, this time it is between an observed (F) and an expected ($\hat{F}$) distribution. You will recall that in such situations we had to distinguish between cases in which the expected frequencies are based on *extrinsic* expectations—that is, models or hypotheses derived from knowledge separate from the observed sample, and those, called *intrinsic* expectations, in which the expected frequencies are based on parameters estimated from the observed sample. In the G-test for goodness of fit that we just covered in Box 17.2, we allowed for the intrinsic cases by reducing the number of the degrees of freedom of the tests. In the Kolmogorov–Smirnov test, the adjustment is made by using different tables of critical values—Statistical Table **X** for extrinsic expected frequencies and Statistical Table **Y** for intrinsic expected frequencies. Before we proceed further, we should tell you that only rarely, if ever, will you encounter a Kolmogorov–Smirnov goodness-of-fit test with extrinsic expectations. Therefore, our examples below are restricted to intrinsic tests and the use of critical values from Statistical Table **Y**. We shall briefly mention what to do, should you be faced with the much less common, extrinsic tests. Although the critical values in Statistical Table **Y** were computed from simulations using the normal distribution, the table can also be used to provide approximate tests for other distributions.

The version of the Kolmogorov–Smirnov test expounded shortly is based on Harter et al. (1984) modification of the original test with further changes and corrections by Khamis (1990, 1992, and 2000). Table 17.4 shows an example using the 12 body weights of crabs from Box 15.1. We would like to test whether the distribution of the 12 observations is consistent with having sampled them from a normal distribution with the mean and standard deviation of the observed sample. In Table 17.4, the 12 observations are ordered from smallest to largest Y_i. In column (1), we give the item numbers, i, of the ordered observations. The observations, Y_i, are shown next in column (2). The standardized deviations of each observation from the mean and their cumulative frequencies based on a normal distribution are shown in columns (3) and (4), respectively. Only relative frequencies are used in this table. Absolute frequencies would have to be divided by n. The observed cumulative frequency F_δ (not shown in Table 17.4) is given by the function $F_\delta = (i - \delta)/(n - 2\delta + 1)$, where δ is a parameter that can assume three values, 0, 1, and 0.5. Although the expected cumulative frequency $\hat{F}_i$ is continuous the observed cumulative frequency, F_δ, is a step function that increases by $1/n$ every time i increases by 1. The test statistics used by the Kolmogorov–Smirnov test is the largest absolute vertical distance between the observed and expected cumulative functions. The maximal distance is then compared with critical values furnished in Statistical Table **Y**.

However, a problem arises with how to measure the distance between the observed and expected cumulative frequencies. Seeing that the observed cumulative frequency distribution is a step function, should we measure the distance just before the step up or immediately after the step has taken place? To illustrate this problem, we show the two cumulative distributions of the crab weight data in Figure 17.2. The method chosen to overcome this difficulty has been to calculate the distance twice, once with the observed cumulative frequency distribution defined first using $\delta = 0$ and a second time with $\delta = 1$. Thus, each time F_δ is evaluated at $\delta = 0$, the formula simplifies to $F_0 = i/(n + 1)$ and consequently the test statistic $g_0 = |F_0 - \hat{F}|$, whereas for $\delta = 1$, it simplifies to $F_1 = (i - 1)/(n - 1)$ and the test statistic is $g_1 = |F_1 - \hat{F}|$. These quantities are found in columns (5) through (8) in Table 17.4. The maxima of both types of g-values are at the 6th crab weight and are 0.1800915 and 0.18708 for $\delta = 0$ and $\delta = 1$, respectively. Consulting Statistical Table **Y** for sample size $n = 12$, we find 5% critical values of 0.196 for $\delta = 0$ and of 0.220 for $\delta = 1$. By neither criterion can we reject the null hypothesis of no difference between the observed and the normalized distributions. Because the 10% critical value for $\delta = 0$ is 0.180, we can report $P \approx 0.10$ for the crab data. For $\delta = 1$, the 10% and 25% critical values are 0.201 and 0.169, so $0.25 > P > 0.10$. We are told to always use the δ-value that has yielded the lower P-value for the statistical test, so for the crab weights we would report our findings for the preceding $\delta = 0$.

The approach just learned can be used for sample sizes up to $n = 35$. For larger values of n, the δ correction has a relatively minor effect. Thus, for sample sizes from 36 to 100, critical values are tabled only for $\delta = 0.5$. We find $F_{0.5} = (i - 0.5)/n$ and $g_{0.5} = |F_{0.5} - \hat{F}|$. We then compute $d_{max} = g_{max,0.5} + 1/2n$ and compare it with the tabled critical values in Statistical Table **Y** for the appropriate n and decide on acceptance

TABLE 17.4 Kolmogorov–Smirnov Test for Goodness of Fit Using the Delta Adjustment

Cumulative expected frequencies ($\hat{F}$) of body weight in grams of 12 crabs. (Data from Box 15.2.) $\bar{Y} = 12.0475$; $s = 6.484094$.

(1) i	(2) Y	(3) $(Y - \bar{Y})/s$	(4) $\hat{F}$	(5) $F_{0.5}$	(6) $g_{0.5}$	(7) F_0	(8) g_0	(9) F_1	(10) g_1
1	1.41	−1.64055	0.05045	0.04167	0.00878	0.07692	0.02647	0.00000	0.05045
2	2.50	−1.47245	0.07045	0.12500	0.05455	0.15385	0.08340	0.09091	0.02046
3	4.19	−1.21181	0.11279	0.20833	0.09554	0.23077	0.11798	0.18182	0.06903
4	9.52	−0.38980	0.34834	0.29167	0.05667	0.30769	0.04065	0.27273	0.07561
5	11.30	−0.11528	0.45411	0.37500	0.07911	0.38462	0.06949	0.36364	0.09047
6	14.40	0.36281	0.64163	0.45833	0.18330	0.46154	0.18009	0.45455	0.18708
7	14.90	0.43992	0.67000	0.54167	0.12833	0.53846	0.13154	0.54545	0.12455
8	15.20	0.48619	0.68658	0.62500	0.06158	0.61538	0.07120	0.63636	0.05022
9	15.39	0.51549	0.69690	0.70833	0.01143	0.69231	0.00459	0.72727	0.03037
10	15.81	0.58027	0.71913	0.79167	0.07254	0.76923	0.05010	0.81818	0.09905
11	17.25	0.80235	0.78882	0.87500	0.08618	0.84615	0.05733	0.90909	0.12027
12	22.70	1.64287	0.94979	0.95833	0.00854	0.92308	0.02671	1.00000	0.05021

$$d_{max} = g_{max.0.5} + 1/2n = 0.18330 + 1/24 = 0.22496$$

$$g_{max.0} = 0.1800915$$

$$g_{max.1} = 0.18708$$

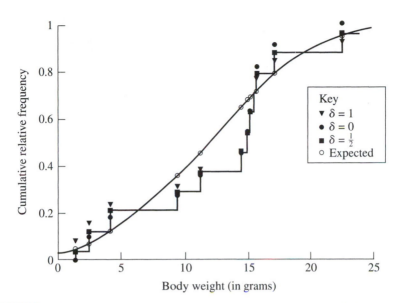

FIGURE 17.2 Observed cumulative frequency distribution of body weights of 12 crabs (stepped line). The expected cumulative distribution for these data, assuming normality, is superimposed (smooth curve). Data from Box 15.2 and Table 17.4. Deviation for the sixth variate is $g_{max,0.5}$, the largest absolute deviation in this example. The largest difference, d_{max}, between the two distributions is $g_{max,0.5} + 1/2n$.

or rejection of the null hypothesis. We actually have no such example to show you, because there are only 12 crab weights and our next example, the Chinese birth weights, has far more than 100 observations. However, those of you planning to carry out Exercise 17.9 will find a sample size of $n = 40$. We therefore show, in columns (9) and (10), the computations we would have carried out had we been faced with a sample in the middle size category.

When sample size $n > 100$, we can employ the asymptotic approximations for $P = 0.05$ and 0.01 given at the end of Statistical Table **Y** and in Box 17.3 for the critical values. In Box 17.3, we show an example of a very large sample, the Chinese birth weights from Box 6.2, whose $n = 9465$. The observations have been grouped into 15 classes to simplify computations, which are as laid out in Box 17.3. When applied to grouped data in a frequency distribution, the Kolmogorov–Smirnov test tends to be conservative; that is, the null hypothesis is rejected at a frequency less than α (D'Agostino and Noether, 1973). In the current example, the null hypothesis is rejected at $\alpha < 0.01$. The distribution of birth weights is not consistent with a normal distribution. This result was expected for this dataset because of the tests for skewness and kurtosis in Boxes 6.1 and 7.2 and a graphic test in Box 6.2. For larger samples, Gonzalez et al. (1977) describe an efficient, but complex, algorithm for the Kolmogorov–Smirnov test without the need to sort the data.

BOX 17.3 Kolmogorov–Smirnov Test for Goodness of Fit: Large Sample Sizes Grouped as a Frequency Distribution, Hypothesis Intrinsic or Extrinsic to Sampled Data

In Section 6.5 and Table 6.2, the observed frequencies of birth weights of Chinese children were qualitatively compared to the frequencies expected on the basis of a normal distribution with $\mu = \overline{Y}$ and $\sigma = s$. We test the agreement of these two distributions statistically.

1. Form cumulative frequency distributions for the observed frequencies and for the expected frequencies (based on Boxes 6.1 and 6.3). For observed cumulative frequencies, compute $F_{0.5} = (F - 0.5)/n$ for relative observed frequencies or $F_{0.5} = F - 0.5$ for absolute observed frequencies. Using absolute frequencies we obtain the following:

(1) Y (oz)	(2) $F_{0.5}$	(3) $\hat{F}$	(4) $g_{0.5} = \lvert F_{0.5} - \hat{F} \rvert$
59.5	1.5	2.8	1.3
67.5	7.5	22.7	15.2
75.5	46.5	118.3	71.8
83.5	431.5	468.5	37.0
91.5	1319.5	1368.6	49.1
99.5	3048.5	3021.2	27.3
107.5	5288.5	5184.9	103.6 ← largest difference
115.5	7295.5	7204.7	90.8
123.5	8528.5	8532.6	4.1
131.5	9169.5	9180.0	10.5
139.5	9370.5	9400.5	30.0
147.5	9444.5	9454.5	10.0
155.5	9458.5	9464.0	5.5
163.5	9463.5	9464.9	1.4
171.5	9464.5	9464.9	0.4

2. Compute $g_{0.5}$, the unsigned differences between the cumulative frequencies [these values are given in column (4)], and locate $g_{max,0.5}$, the largest difference—in this case 103.6.

For absolute frequencies, the Kolmogorov–Smirnov test statistic is

$$g_{max,0.5} = \frac{\text{largest difference}}{n} = \frac{103.6}{9465} = 0.01095$$

Had we used relative frequencies, we would not have divided by n. In such a case, the test statistic is just the largest difference.

These data have been fitted to a cumulative normal distribution based on the sample mean and standard deviation (an intrinsic hypothesis). The sample g_{max} should be tested against the adjusted critical $g_{\alpha(adj)}$ in Statistical Table **Y**. However, tabulated values are

Box 17.3 (continued)

furnished only for $n \leq 100$, too small for the present example. For larger samples, an asymptotic approximation is given in the table. For $\alpha = 0.05$ it is $0.89196/\sqrt{n} - 1/2n$ and for $\alpha = 0.01$ it is $1.0427/\sqrt{n} - 1/2n$. In the present case,

$$g_{0.05(\text{adj})} = \frac{0.89196}{\sqrt{9465}} - \frac{1}{2(9465)} = 0.00912$$

and

$$g_{0.01(\text{adj})} = \frac{1.0427}{\sqrt{9465}} - \frac{1}{2(9465)} = 0.01066$$

Because our observed $g_{\max}$ is greater than the 1% critical value, we reject the null hypothesis that the birth weights are normally distributed.

In the rarer cases of fitting observations to an extrinsic hypothesis we use the standard Kolmogorov–Smirnov critical values in Statistical Table **X** for samples up to $n \leq 100$. We would compare the value of $g_{\max}$ with the entries in Table **X**. We reject the null hypothesis if the observed $g_{\max}$ is greater than the critical value of g_α.

For larger samples, we can calculate critical values as follows:

$$g_\alpha = \sqrt{\frac{-\ln (\alpha/2)}{2n}} - \frac{1}{2n}$$

For $\alpha = 0.05$, g_α is $1.35810/\sqrt{n} - 1/2n$ as shown at the end of Statistical Table **X**.

The Kolmogorov–Smirnov test is properly applied to continuous frequency distributions only. It is conservative in the case of grouped data.

The Kolmogorov–Smirnov test is based on the assumption that the data are continuous. Pettitt and Stephens (1977) have investigated its use on discrete and grouped data and found that it provides a conservative test in such cases. They also provide a table of critical values for the case of equal expected frequencies (they suggest that it can also be used for unequal expected frequencies).

For the rare cases of fitting observations to an extrinsic hypothesis, proceed exactly as described previously but use the critical values in Statistical Table **X** instead.

The Kolmogorov–Smirnov statistic can also be used for setting confidence limits to an entire cumulative frequency distribution. By computing $f_{0.5} \pm g_{\alpha,0.5}$ for all values or classes of Y_i, we obtain $100(1 - \alpha)\%$ confidence limits. The 95% confidence limits of the cumulative distribution function of the crab body weight data are shown in Figure 17.3. We expect 95% of such sets of confidence limits to completely enclose the true cumulative frequency distribution.

The BIOMstat package of computer programs performs goodness-of-fit tests of continuous variables for normality using g_1, g_2, and the Kolmogorov–Smirnov statistic. Tests on frequency data are performed using the G-test for goodness of fit.

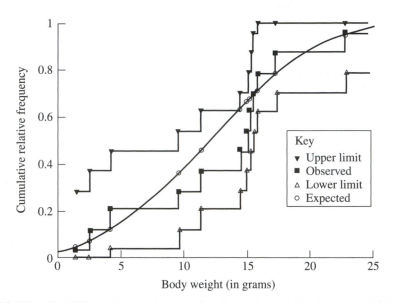

FIGURE 17.3   The 95% confidence interval for the observed cumulative frequency distribution of Figure 17.2.

17.3 Replicated Tests of Goodness of Fit

Frequently experiments leading to tests of goodness of fit are replicated. For example, the data in Box 17.4 are from a genetic experiment on houseflies in which eight replicate crosses were carried out, with the offspring expected in a ratio of 9 wild-type to 7 pale-eyed flies. Although we could simply pool the results from the eight progenies and test the hypothesis of a 9:7 ratio on the grand totals, we lose information in doing so. The tests that can be performed on such data are illustrated here in Box 17.4 (and later in Box 17.5), using the *G*-test. Such a test is called an *interaction* or **heterogeneity *G*-test.** Corresponding chi-square tests can be performed, but these have the disadvantage that the X^2 values are not completely additive, so the anova-like table given in step **6** of Box 17.4 would be only approximately correct.

The order of computations depends in part on the design of the experiment. If the replicates are obtained in sequence, with an appreciable time interval after each, the first logical step is likely to be a goodness-of-fit test of each replicate. Thus, if the progenies were obtained a week apart, the investigator usually would wish to carry out a significance test for each of the progenies as it is obtained, reserving the overall analysis for the end of the experiment. On the other hand, if the results are obtained simultaneously, the overall tests of goodness of fit may be carried out first, with individual replicates analyzed if warranted by the data and the interest of the investigator. In the example of Box 17.4, the data were collected essentially simultaneously, so we leave the analysis of the separate progenies until later in the box (step **5**). The final outcome of the analysis is identical regardless of which way the problem is approached.

BOX 17.4 Replicated Goodness-of-Fit Tests (*G*-Statistic)

Numbers of wild-type (N^+) and pale-eyed (*ge*) flies in $b = 8$ progenies of the housefly, *Musca domestica*. The expected ratio of the $a = 2$ phenotypic classes N^+ and *ge* is 9:7.

Progeny	N^+	*ge*	Σ
1	83	47	130
2	77	43	120
3	110	96	206
4	92	58	150
5	51	31	82
6	48	61	109
7	70	42	112
8	85	66	151
Σ	616	444	$1060 = n$

SOURCE: Data from R. L. Sullivan (unpublished results).

The order of computations may differ somewhat, depending on whether all replicates are obtained simultaneously or over a period of time (see text for further discussion of this issue). The resulting final analysis will be the same in any case. In this example the progenies were obtained simultaneously; we therefore proceed immediately to an overall analysis of the data, reserving tests of goodness of fit of individual progenies for step **5**.

Computation

1. The following quantities are necessary for the intended computations. They can be obtained with the help of a natural logarithm function key on a calculator.

   a. The sum of the $f \ln f$ transforms of all frequencies in the table:

   $$\sum^{a} \sum^{b} f \ln f = 83 \ln 83 + 77 \ln 77 + \cdots + 42 \ln 42 + 66 \ln 66$$
   $$= 4502.74287$$

   b. The sum of the $f \ln f$ transforms of the column sums of the table:

   $$\sum^{a} \left(\sum^{b} f \right) \ln \left(\sum^{b} f \right) = 616 \ln 616 + 444 \ln 444$$
   $$= 6663.26624$$

   c. The sum of the $f \ln f$ transforms of the row sums of the table:

   $$\sum^{b} \left(\sum^{a} f \right) \ln \left(\sum^{a} f \right) = 130 \ln 130 + \cdots + 151 \ln 151$$
   $$= 5215.20530$$

   d. $n \ln n = \left(\sum^{a} \sum^{b} f \right) \ln \left(\sum^{a} \sum^{b} f \right) = 1060 \ln 1060 = 7383.98564$

Box 17.4 (continued)

2. Are the ratios of the progenies (replicates) homogeneous? Compute G_H (G for heterogeneity).

$$G_H = 2[\text{quantity } \mathbf{a} - \text{quantity } \mathbf{b} - \text{quantity } \mathbf{c} + \text{quantity } \mathbf{d}]$$
$$= 2[4502.74287 - 6663.26624 - 5215.20530 + 7383.98564]$$
$$= 16.51394$$

The value is compared to a χ^2-distribution with $(a - 1)(b - 1) = (2 - 1)(8 - 1) = 7$ degrees of freedom. The P-value is 0.020,814. There is evidence that the ratios are not all homogeneous among the progenies.

3. A second test is for goodness of fit to the 9:7 hypothesis for the pooled data, or all eight replicates pooled together. These frequencies are $\sum^b f_1 = 616$ and $\sum^b f_2 = 444$. The test is carried out conveniently by Expression (17.4):

$$G_P = 2\left[\sum^b f_1 \ln\left(\frac{\sum^b f_1}{n\hat{p}} \right) + \sum^b f_2 \ln\left(\frac{\sum^b f_2}{n\hat{q}} \right) \right]$$

$$= 2\left[616 \ln\left(616 \div \frac{1060 \times 9}{16} \right) + 444 \ln\left(444 \div \frac{1060 \times 7}{16} \right) \right]$$

$$= 1.50039$$

This value is compared to a χ^2-distribution with $a - 1 - $ (number of parameters estimated from the sample) $= 2 - 1 - 0 = 1$ degree of freedom. Its P-value, 0.220,611, suggests that the null hypothesis for the pooled data cannot be rejected.

4. Adding the G for heterogeneity to the pooled G, we obtain a total G:

$$G_H + G_P = G_T = 16.51394 + 1.50039 = 18.01433$$

with $b(a - 1) = 8(2 - 1) = 8$ degrees of freedom. This quantity can be used to test whether the data as a whole fit the expected ratio. The P-value is 0.021,119 and we conclude that they do not. Because G_H was large and G_P was small, we know that G_T is large due to the heterogeneity. One of two analyses might next be done, depending on the interests of the investigator. One is described in step **5**; the other, an unplanned test of differences among the progenies using an STP procedure, is described in Box 17.5.

5. Partitioning the total G into contributions due to individual samples. For each sample compute G by Expression (17.4). For example, for progeny 1

$$G = 2\left[83 \ln\left(83 \div \frac{130 \times 9}{16} \right) + 47 \ln\left(47 \div \frac{130 \times 7}{16} \right) \right]$$

$$= 3.10069$$

This value has $a - 1 - $ (number of parameters estimated from sample) $= 2 - 1 - 0 = 1$ degrees of freedom. It has a probability of 0.078,259, assuming that the true ratio is 9:7. We cannot reject the null hypothesis of no difference from the 9:7 ratio

Box 17.4 (continued)

for this progeny. We can continue testing each progeny in this fashion. In doing so, we find that only for progeny 6 can we reject the null hypothesis as shown here.

$$G = 2\left[48 \ln\left(48 \div \frac{109 \times 9}{16}\right) + 61 \ln\left(61 \div \frac{109 \times 7}{16}\right)\right] = 6.53781$$

$$P = 0.010,561$$

Because the proportion of wild-type flies in progeny 6 equals $48/109 = 0.440$, we see that the discrepancy was in the direction of having too few wild-type flies ($\pi = \frac{9}{16} = 0.5625$).

As a computational check, we add the eight individual G's and note that their sum equals the total G ($G_T = 18.01433$) within rounding error.

6. We can summarize our results in the following tables

Tests	df	G	P
Pooled	1	1.50039	0.220611
Heterogeneity	7	16.51394	0.020814
Total	8	18.01433	0.021119

Progeny	df	G	P
1	1	3.10069	0.078259
2	1	3.11166	0.077734
3	1	0.67795	0.410294
4	1	1.59162	0.207095
5	1	1.19268	0.274790
6	1	6.53781	0.010561
7	1	1.80182	0.179493
8	1	0.00011	0.991632
Total	8	18.01434	0.021119

NOTE: If individual values of G have been computed first, the heterogeneity G could have been obtained simply by subtracting pooled G from the total G (the sum of the individual G's) and would not have needed to be computed independently as in step **2.**

First we test whether the outcomes of all the replicates (progenies in this case) are homogeneous—that is, whether the sampled proportions (of wild types) in the eight progenies could have come from a single population. If they are not uniform among replicates, the proportions are said to be heterogeneous. Five hypothetical heterogeneous replicates might have the following relations between the observed and expected proportions p_1 and π_1 of wild-type flies, respectively:

$$p_1 > \pi_1$$
$$p_1 > \pi_1$$
$$p_1 = \pi_1$$
$$p_1 < \pi_1$$
$$p_1 > \pi_1$$

By analogy with analysis of variance (see Chapter 11), we see immediately why such heterogeneity could also be called *interaction*.

The test is shown in step **2** of Box 17.4. G_H (heterogeneity G) is computed by a formula analogous to the interaction formula in a two-way anova: $\overline{Y} - \overline{R} - \overline{C} + \overline{\overline{Y}}$ (see Section 11.2):

$$G_H = 2[\text{sum of } f \ln f \text{ of frequencies in each cell of the table}$$
$$- \text{ sum of } f \ln f \text{ of column sums of frequencies}$$
$$- \text{ sum of } f \ln f \text{ of row sums of frequencies}$$
$$+ f \ln f \text{ of total number of items in the table}] \tag{17.8}$$

We find that we must reject the null hypothesis for the data of Box 17.4 because one or more of the progenies differ in outcome from the others. If we had simply pooled all the data and carried out an overall test, we could not, of course, have detected this heterogeneity.

The next test is carried out on the pooled frequencies, which means the eight progenies are treated as though they were a single large progeny. The test is exactly the same as that for a single classification, as shown in Section 17.2 using Expression (17.4). We find that the pooled data do not deviate from expectation more than if they had been randomly sampled. This lack of deviation might surprise you in view of the significant heterogeneity of the data, but it occurs when replicates are heterogeneous in such a way that they tend to compensate for their respective deviations from expectation.

We can illustrate this phenomenon with an artificial, deliberately exaggerated example. Suppose we test sex ratios in six bottles of *Drosophila*, three of which have only male offspring and three only female offspring. If we pooled these data (assuming that the number of male and female offspring were approximately the same), we would be unable to reject the 50:50 sex ratio hypothesis for these *Drosophila*, although there clearly is no such sex ratio in these flies. The null hypothesis of homogeneity in this example, would, of course, be rejected.

The scientific relations expressed by these tests for heterogeneity and of pooled data are of a fundamental nature and must be clearly understood. To deepen and confirm your understanding of them, we furnish, in Table 17.5, four other hypothetical examples on sex ratios in drosophila in which the differences are more subtle. We provide the P-value associated with every G-value in this table. To facilitate numerical corroboration of our verbal statements by our readers, we have boldfaced all P-values "significant" by the conventional 5% criterion.

In example A we have four progenies, each with a slight excess of females, but none with P-values ≤ 0.05 by individual G-tests using Expression (17.4). When we add these values of G together to obtain a total G, however, its P-value of 2.2% suggests rejection of its null hypothesis. When considered together, suspicious results for individual progenies yield an improbable total result. We conclude from the total G that obtaining four sample progenies, each of which is as far from expectation as these are, is unlikely. This finding is reinforced by the pooled G. Because all progenies deviate in the same direction, the pooled frequencies show this deviation clearly and, being based on a larger sample size, yield a highly improbable value

TABLE 17.5 Hypothetical Replicated Tests of Goodness of Fit Illustrating Meaning of Total, Pooled, and Heterogeneity G

Computation of G is illustrated in Box 17.4, using formulas from Box 17.1. The null hypothesis being tested is a 50 : 50 sex ratio. **Probability values $\leq$ 0.05 are shown in boldface**.

	Progenies	♀	♂	n		df	G	P
	1	59	41	100		1	3.25773	0.071,087
A	2	58	42	100		1	2.57104	0.108,836
	3	57	42	99		1	2.28150	0.130,925
	4	58	40	98		1	3.32497	0.068,235
					Total	4	11.43523	**0.022,084**
	Σ	232	165	397	Pooled	1	11.36160	**0.000,750**
					Heterogeneity	3	0.07363	0.994,802
	1	59	41	100		1	3.25773	0.071,087
B	2	42	58	100		1	2.57104	0.108,836
	3	57	42	99		1	2.28150	0.130,925
	4	40	58	98		1	3.32497	0.068,235
					Total	4	11.43523	**0.022,084**
	Σ	198	199	397	Pooled	1	0.00252	0.959,963
					Heterogeneity	3	11.43272	**0.009,602**
	1	59	41	100		1	3.25773	0.071,087
C	2	58	42	100		1	2.57104	0.108,836
	3	72	26	98		1	22.46416	**2.14 × 10⁻⁶**
	4	73	26	99		1	23.23751	**1.43 × 10⁻⁶**
					Total	4	51.53044	**1.73 × 10⁻¹⁰**
	Σ	262	135	397	Pooled	1	41.35016	**1.27 × 10⁻¹⁰**
					Heterogeneity	3	10.18027	**0.017,094**
	1	52	48	100		1	0.16004	0.689,120
D	2	36	64	100		1	7.94580	**0.004,820**
	3	52	47	99		1	0.25263	0.615,229
	4	52	46	98		1	0.36758	0.544,326
					Total	4	8.72605	0.068,324
	Σ	192	205	397	Pooled	1	0.42577	0.514,072
					Heterogeneity	3	8.30028	**0.040,197**

of G. To obtain the heterogeneity G, we subtract pooled from total G. The resulting heterogeneity G of 0.07363 with 3 df does not warrant rejection of the H_0, indicating that the deviations from expectation of the progenies are in the same direction and are not appreciably different from each other.

Example B has the same total frequencies as example A, but in progenies 2 and 4 the frequencies of males and females have been reversed. The individual values of G are necessarily the same as before, suspicious but all with P-values greater than 0.05. The total G, being the sum of the individual ones, must also be the same as before, indicating that collectively these progenies do not conform to expectation. The important difference between examples A and B lies in the pooled and heterogeneity G. Because two of the progenies in B favor females and the other two favor males, the pooled frequencies of males and females are almost identical. The pooled G is therefore negligible, 0.00252, a suspiciously good fit ($P = 0.960$), unlikely to occur in actual, experimental data. Subtracting such a low pooled G from the total G leaves a large heterogeneity G, indicating that the suspicious departures from expectation are not in a uniform direction, but are in different directions, as we know them to be.

Example C illustrates a case in which all three tests reject their null hypotheses. Although all four progenies have an excess of females, the first two have only a small excess, while the last two deviate strongly from expectation. Consequently, the total G rejects the H_0 with a near infinitesimally small P-value. The pooled G, in view of the consistent trend in favor of females, rejects its H_0 with a similarly small P-value; and the heterogeneity G rejects its H_0 as well, showing that the trend, although in all cases favoring females, is not uniform in magnitude.

Finally, in example D we show that the total G is an overall measure of the departure from expectation of the several progenies. Whereas progeny 2 has a substantially larger number of males ($P > 0.05$), progenies 1, 3, and 4 have slight, excesses in favor of females (all at $P > 0.05$). The total G, although suspiciously high, has a P-value greater than 0.05. The fit of progenies 1, 3, and 4 was good enough that the marked departure from expectation by progeny 2 was not sufficient to affect our overall judgment based on the total G. The frequencies in the progenies tended to compensate for each other, leading to acceptance of the H_0 for the pooled G, but on subtraction, the heterogeneity G has a P-value ≤ 0.05, indicating correctly that the data are heterogeneous, although we could reject neither the total nor the pooled G.

When G-values are to be summed or partitioned, neither the Williams nor the continuity correction is used because adjusted G-values are not additive.

Returning to the example in Box 17.4, we now add heterogeneity G to pooled G to obtain total G (step **4**). We find that the data as a whole do not fit the expected ratio. Occasionally G_T rejects the null hypothesis, even though G_H and G_P do not, although they are appreciable. In such a case, we would be led to conclude that the overall fit of the data to the hypothesis is poor, but we would be unable to pinpoint the exact nature of the departure from expectation.

When we compute G separately for each replicate of the study, we find only progeny 6 rejecting the null hypothesis, by having fewer wild-type and more mutant flies. The analysis can be summarized by a table similar to an anova table (step **6** in Box 17.4).

BOX 17.5 Unplanned Tests of the Homogeneity of Replicates Tested for Goodness of Fit

When an investigator concludes that there is heterogeneity among the replicates, the following techniques can be used to test whether all replicates differ from one another or whether there are homogeneous sets of replicates differing from other such sets or from single replicates. We will use the housefly progenies of Box 17.4 to illustrate the technique. For these tests, it is convenient to establish critical values of G_H for specified experimentwise comparisons α, such as 0.05 or 0.01.

When testing k, fewer than all pairwise unplanned comparisons, we can employ an experimentwise error rate α and adjust the probability using the Dunn–Šidák approach $\alpha' = 1 - (1 - \alpha)^{1/k}$ with $k < b(b - 1)/2$. Thus, let us assume we wish to make only four nonorthogonal comparisons. Then we employ $\alpha' = 1 - (1 - \alpha)^{1/k} = 1 - (1 - 0.05)^{1/4} = 0.01274$. If you have an inverse chi-square distribution program available, such as the one in BIOMstat, enter the P-value as well as the degrees of freedom $\nu = (a - 1)(b - 1)$ and obtain the critical chi-square value, 6.205. Lacking such a program, you can employ Statistical Table **E,** which gives us the critical value directly. We look up $\chi^2_{\alpha[k,\nu]}$, where a is the experimentwise type I error rate, k is the number of intended comparisons, and $\nu = (a - 1)(b - 1)$, the degrees of freedom of the test (usually this will be $a - 1$). Thus, $\chi^2_{.05[4,1]} = 6.205$, and the null hypothesis will be rejected for any of the four intended tests whose $G_H \geq 6.205$.

We may wish to test all pairs of replicates or the pairs that differ the most (which amounts to the same thing). In such a case for an experimentwise error of $\alpha = 0.05$ and for $k = b(b - 1)/2 = 28$ comparisons, we compute $\alpha' = 1 - (0.95)^{1/28} = 0.001830$. By using either the inverse chi-square function or Statistical Table **E,** we arrive at a critical G_H of 9.712. All G_H-values equal to or greater than this will lead to rejection of their associated null hypothesis.

Finally, when we wish to test all possible sets (which requires a still more conservative procedure), we carry out a simultaneous test procedure, based on the following critical value: For the heterogeneity test to be just significant at the α level, G_H must equal $\chi^2_{\alpha[(a - 1)(b - 1)]}$. This value is used as the constant critical value for G in the STP tests. In our case, $a = 2$ and $b = 8$; therefore $\chi^2_{.05[7]} = 14.067$ can be used as a critical value if a 5% experimentwise error rate is accepted. The test procedure consists of a heterogeneity test for all sets of replicates (progenies in this example) taken 2, 3, 4, . . . at a time. If there are many replicates, a computer program is useful. Usually, not all of these tests need to be carried out, however, as we will show below.

The simultaneous test procedure

1. Compute the proportion of wild-type flies, p_i, for each progeny. Reorder progenies by magnitude of p_i.

(1) Progeny	(2) N^+	(3) ge	(4) Σ	(5) p_i
2	77	43	120	0.642
1	83	47	130	0.638
7	70	42	112	0.625
5	51	31	82	0.622
4	92	58	150	0.613
8	85	66	151	0.563
3	110	96	206	0.534
6	48	61	109	0.440

Box 17.5 (continued)

2. Because progenies 2 through 4 (i.e., 1, 2, 4, 5, and 7) have similar p_i's, we will test this set first. Compute G_H using the formula in step **2** of Box 17.4. The general layout of the formula for this and subsequent values of G_H will be $G_H = 2[($ sum of $f \ln f$ for frequencies in the set $) - ($ sum of $f \ln f$ for column sums of the set $) - ($ sum of $f \ln f$ for row sums of the set $) + (f \ln f$ for total sum of the set $)]$.

$$G_H = 2[(77 \ln 77 + 43 \ln 43 + \cdots + 58 \ln 58) - (373 \ln 373 + 221 \ln 221)$$

$$-(120 \ln 120 + 130 \ln 130 + \cdots + 150 \ln 150) + (594 \ln 594)]$$

$$= 0.31210$$

which is far less than our critical $\chi^2_{.05} = 14.067$; this set of replicates therefore is considered homogeneous.

3. We now try adding progeny 8 to our set (this progeny has the next lower value of p_i). The new column totals are $373 + 85 = 458$ and $221 + 66 = 287$, and their sum is 745. Also we recompute the new sum of $f \ln f$ for the frequencies in the set by adding the $f \ln f$ for the new member to the sum. Similarly the former sum of the row sums is incremented by the $f \ln f$ of the new row.

$$G_H = 2[(77 \ln 77 + 43 \ln 43 + \ldots + 66 \ln 66) - (458 \ln 458 + 287 \ln 287)$$

$$- (120 \ln 120 + 130 \ln 130 + \ldots + 151 \ln 151) + 745 \ln 745]$$

$$= 2.43905$$

4. Progeny 3 has the next lower value of p_i. We add it to the set and recomputed G_H.

$$G_H = 2[(77 \ln 77 + 43 \ln 43 + \ldots + 96 \ln 96) - (568 \ln 568$$

$$+ 383 \ln 383) - (120 \ln 120 + 130 \ln 130 + \ldots + 206 \ln 206) + 951 \ln 951]$$

$$= 6.77410$$

Again, the set is homogeneous. It is also a maximally homogeneous set, because the test of the complete set in Box 17.4 yields a $G_H = 16.51394$ greater than the critical 14.067. The results there show that adding progeny 6 to set $\{1, 2, 3, 4, 5, 7, 8\}$ would yield a heterogeneous set.

5. We could now try to find a maximally homogeneous set involving progeny 6. Starting with progenies 6 and 3, we obtain

$$G_H = 2[(110 \ln 110 + \cdots 61 \ln 61) - (158 \ln 158 + 157 \ln 157)$$

$$- (206 \ln 206 + 109 \ln 109) + 315 \ln 315]$$

$$= 2.50317$$

This value of G_H is not greater than the critical value. Continuing in this fashion, we find that the set $\{3, 4, 5, 6, 7, 8\}$ is also maximally homogeneous because its G_H (11.80127) would increase to 14.44762 if we added progeny 1.

Box 17.5 (continued)

Thus, we may consider progenies {1, 2, 3, 4, 5, 7, 8} and {3, 4, 5, 6, 7, 8} to be homogeneous sets. Arraying the progenies by proportion of wild types (high to low), we can summarize their relations as follows:

Progeny   2 1 7 5 4 8 3 6

The next question to address is why progenies 1 and 2 (high proportion of wild types) were different from progeny 6 (low proportion of wild types).

Having determined that the replicates are not homogeneous, we now want to know the reason for this heterogeneity. Precisely which samples are homogeneous and which samples are different from the rest sufficiently to cause the heterogeneity? Although sometimes we can locate sources of heterogeneity by simple inspection, more often we need to carry out unplanned tests to accomplish this task, outlined in Box 17.5. For all pairwise (or fewer) unplanned comparisons, we can employ an experimentwise error rate α and adjust the probability per comparison (as proposed by Šidák; see Section 9.6) to $\alpha' = 1 - (1 - \alpha)^{1/k}$, making use of Statistical Table **E** and setting $k < b(b - 1)/2$. Alternatively, when comparisons will be made for all possible sets of replicates, use the simultaneous test procedure shown in Box 17.5. On carrying out the computations, we find that the significance of the heterogeneity appears to be due to differences between progenies 1 and 2 on one hand and 6 on the other. The next step might be to try to determine why these progenies gave different results. Perhaps the media in which they were reared were different, or perhaps there were biological differences among their parents.

Although the replicated goodness-of-fit tests in Boxes 17.4 and 17.5 tested a simple 9:7 ratio, these techniques can also be used for more complicated situations, such as single-classification data with more than two classes, including replicated goodness-of-fit tests to binomial or Poisson distributions.

Single-classification and replicated goodness-of-fit tests are performed by the BIOMstat package of computer programs. The unplanned simultaneous test described in Box 17.5 is also performed.

17.4 Tests of Independence: Two-Way Tables

The notion of statistical or probabilistic independence was introduced in Section 5.1, where we showed that if two events are independent, the probability of their occurring together can be computed as the product of their separate probabilities. For example, suppose that among the progeny of a certain genetic cross, the probability of a corn kernel being red is $\frac{1}{2}$ and the probability of a kernel being dented is $\frac{1}{3}$. Assuming statistical independence of these two characteristics, the probability of obtaining a dented red kernel would be $\frac{1}{2} \times \frac{1}{3} = \frac{1}{6}$.

The appropriate statistical test for the genetic problem is to test the frequencies for goodness of fit to the expected ratios of 2 (red, not dented):2 (not red, not dented):1 (red, dented):1 (not red, dented). This would be a simultaneous test of two null hypotheses: that the expected proportions are $\frac{1}{2}$ and $\frac{1}{3}$ for red and dented, respectively, and that these two properties are independent. The first null hypothesis tests the Mendelian model in general. The second tests whether these characters assort independently—that is, whether they are determined by genes located in different linkage groups. Rejection of the second hypothesis is taken as evidence that the characters are linked—that is, located nearby on the same chromosome.

In many instances in biology the second hypothesis, concerning the independence of two properties, is of great interest and the first hypothesis, regarding the true proportion of one or both properties, is of little interest. In fact, often no hypothesis regarding the parametric values $\hat{p}_i$ can be formulated by the investigator. We will cite several examples of such situations that lead to the test of independence that we will present in this section. We employ this test whenever we wish to test whether two different properties, each occurring in two states, are dependent on each other. For instance, specimens of a certain moth may occur in two color phases—light and dark. Fifty specimens of each phase may be exposed in the open, subject to predation by birds. The number of surviving moths is counted after a fixed interval of time. The proportion preyed on may differ in the two color phases. The two properties in this example are color and survival. We can divide our sample into four classes: light-colored survivors, light-colored prey, dark survivors, and dark prey. If the probability of being preyed on is independent of the color of the moth, the expected frequencies of these four classes can be computed simply as independent products of the proportion of each color (in our experiment, $\frac{1}{2}$) and the overall proportion preyed on in the entire sample. If the statistical test of independence that we explain in this section shows that the two properties are not independent, we will be led to conclude that one of the color phases is more susceptible to predation than the other. In this example, this issue is the one of biological importance; the exact proportions of the two properties are of little interest. The proportion of the color phases is arbitrary, and the proportion of survival is of interest here only insofar as it differs for the two phases.

Another example relates to a sampling experiment carried out by a plant ecologist. The first step is to obtain a random sample of 100 individuals of a fairly rare species of tree distributed over an area of 400 square miles. For each tree, the ecologist notes whether it is rooted in a serpentine soil or not, and whether the leaves are pubescent or smooth. Thus, the sample of $n = 100$ trees can be divided into four groups: serpentine–pubescent, serpentine–smooth, nonserpentine–pubescent, and nonserpentine–smooth. If the probability of a tree's being pubescent is independent of its location, our null hypothesis of the independence of these properties is true. If, on the other hand, the proportion of pubescence differs for the two types of soils, our statistical test is expected to result in rejection of the null hypothesis of independence. Again, the expected frequencies are products of the independent proportions of the two properties—serpentine versus nonserpentine, and pubescent versus smooth. In this instance, the proportions themselves may be of interest to the investigator.

The example we will work out here in detail is from immunology. A sample of 111 mice was divided into two groups, 57 that received a standard dose of pathogenic bacteria followed by an antiserum and a control group of 54 that received the bacteria but no antiserum. After sufficient time had elapsed for an incubation period and for the disease to run its course, 38 dead mice and 73 survivors were counted. Of those that died, 13 had received bacteria *and* antiserum, whereas 25 had received bacteria only. A question of interest is whether the antiserum protected the mice who received it, leading to a proportionally higher number of survivors in that group. Here again the proportions of these properties are of no more interest than they were in the first example (predation on moths).

Such data can be displayed conveniently in the form of a **two-way table** as shown later. Two-way and multiway tables (more than two criteria) are often referred to as **contingency tables.** This type of two-way table, in which each of the two criteria is divided into two classes, is known as a **2 × 2 table.**

	Dead	Alive	Σ
Bacteria and antiserum	13	44	57
Bacteria only	25	29	54
Σ	38	73	111

We can confirm from the table that 13 mice received bacteria and antiserum but died. The totals in the margins give the number of mice exhibiting any one property: 57 mice received bacteria and antiserum; 73 mice survived the experiment. Altogether 111 mice were involved in the experiment; they constitute the total sample. For discussion, we label the cells of the table and the row and column sums as follows:

a	b	$a + b$
c	d	$c + d$
$a + c$	$b + d$	n

From a two-way table one can systematically compute the expected frequencies (based on the null hypothesis of independence) and compare them with the observed frequencies. For example, using the row and column totals the expected frequency for cell d (bacteria, alive) would be

$$\hat{f}_{\text{bact,alive}} = n\hat{p}_{\text{bact,alive}} = n\hat{p}_{\text{bact}} \times \hat{p}_{\text{alive}} = n\left(\frac{c+d}{n}\right)\left(\frac{b+d}{n}\right) = \frac{(c+d)(b+d)}{n}$$

which in our case would be $(54)(73)/111 = 35.514$, a value higher than the observed frequency, 29. In general, *to compute the expected frequencies for a cell in the table, multiply its row total by its column total and divide the product by the grand total.* The expected frequencies can also be displayed conveniently in the form of a two-way table:

	Dead	Alive	Σ
Bacteria and antiserum	19.514	37.486	57.000
Bacteria only	18.486	35.514	54.000
Σ	38.000	73.000	111.000

The row and column sums of this table are identical to those in the table of observed frequencies, because the expected frequencies were computed on the basis of these same row and column totals. It should therefore be clear that a test of independence will not test whether a property occurs at a given proportion but can test only whether the two properties are manifested independently.

Before we turn to the actual statistical tests, we need to consider different models based on the design of the experiment or sampling procedure. These models relate to whether the totals in the margins of the two-way table are fixed by the investigator or are free to vary and reflect population parameters. In the example of the trees growing in two types of soils and exhibiting two types of leaves, only the total sample size n was fixed; 100 trees had been sampled. The proportion of trees growing in serpentine soils was an outcome of the sampling experiment and was not under the control of the investigator. Similarly, the proportion of leaves that were pubescent was not fixed by the investigator but was an outcome of the experiment. For a test of the null hypothesis, one should compute the probability of obtaining the frequencies observed in the 2×2 table or worse departures from independence out of all possible 2×2 tables with the same total sample size n. We will call this design the Model I for a two-way table.

There is another type of Model I, in which the same individuals are subjected to two successive tests. Again only the total sample size is fixed, but the hypotheses to be tested are different. This type of Model I is discussed in Section 17.7.

A second design, Model II, would have the margin totals for one of the two criteria fixed. An example is the immunology experiment that we just discussed in detail. The number of mice given bacteria plus antiserum and the number given only bacteria were fixed by the experimenter at 57 and 54 animals, respectively. The proportion that survived the experiment was free to vary, depending on the effect of the treatment. In testing the null hypothesis of independence, one should compute the probability of getting the observed results or worse departures from independence out of all possible two-way tables with the same marginal totals for the two treatment classes.

The moth example cited at the beginning of this section is another instance of a Model II test for independence. The investigator released 50 specimens of each color phase; hence the marginal totals for color phase were fixed by the experiment. The proportion that became prey depended on the natural forces involved and was free to vary.

In a third model for the design of a 2 × 2 table, both margin totals are fixed by the experiment. We have not yet encountered an example of this model. Assume that we wish to investigate the preference of weevil larvae for two types of beans differing in seed coats. The investigator places a fixed number of beans (100) in a jar—50 of type A and 50 of type B. Then 70 first-instar larvae are added to the jar. At these densities no more than one larva will attack any one bean. After enough time has elapsed to permit each larva to enter a bean, a count is made of how many beans of each type have been attacked by a weevil larva. Again, we have four frequency classes of beans: A–attacked, B–attacked, A–not attacked, and B–not attacked. The number of beans of both types and the total number of beans that were attacked were both fixed by the experimenter. (The number attacked was equal to the number of larvae released in the jar.) When testing the null hypothesis of independence in this model, which we call Model III, we should compute the probability of obtaining the observed frequencies and all worse departures from independence out of all possible 2 × 2 tables with the same margin totals for both criteria—bean type and presence or absence of attacking larvae.

The tests of independence proposed for these 2 × 2 tables are of three kinds: the *G*-test, an exact probability test known as Fisher's exact test, and the chi-square test. Conventional formulations of the first two tests were designed originally for specific models: The *G*-test, based on Expression (17.13) (and illustrated later in Box 17.6), was intended for Model I; Fisher's exact test [see Expression (17.14) and later in Box 17.7] was intended for Model III. The chi-square test, which has been the most common test in the past, was not designed for any particular model. The designs represented by the three models have not been clearly distinguished in textbooks of statistics, and research workers have been applying each of the tests to all three models. The criteria for choosing among the tests have not been the model of the design, but rather the sample size and the computational effort involved. It should be pointed out that the distinctions made earlier are controversial. Some workers always analyze 2 × 2 tables as if they were Model III. Their argument is that once one has the data one should compute the probability of the table conditioned on the observed marginal totals. Haviland (1990) and the commentary that follows provide a useful summary of the arguments, as does Agresti (2001). Fortunately, the different techniques seem to provide similar results even when applied to the inappropriate model. However, using Model III results in slightly larger *P*-values—that is, it results in a more conservative test.

Our recommendations for analyzing 2×2 tables are summarized in the following table:

Model	Totals	Recommended Test
I	Not fixed	G-test of independence
II	Fixed for one criterion	G-test of independence
III	Fixed for both criteria	Fisher's exact test

The **G-test of independence** tests the goodness of fit of the observed cell frequencies to their expected frequencies. The equations for the standard G-test of independence were designed for Model I and are based on a **multinomial distribution,** which is a discrete probability distribution that is a generalization of the binomial distribution to the case where an attribute has more than two classes. We do not discuss the multinomial distribution in detail; interested readers should refer to Evans et al. (2000) or to general texts on probability. The probability of observing the cell frequencies a, b, c, and d, assuming a multinomial distribution, is computed as follows:

$$\frac{n!}{a!b!c!d!}\left(\frac{a}{n}\right)^a\left(\frac{b}{n}\right)^b\left(\frac{c}{n}\right)^c\left(\frac{d}{n}\right)^d \tag{17.9}$$

If we assume that the row and column classifications are independent, this expression changes to

$$\left[\frac{n!}{a!b!c!d!}\left(\frac{(a+b)(a+c)}{n^2}\right)^a\left(\frac{(a+b)(b+d)}{n^2}\right)^b\right]$$
$$\times\left[\left(\frac{(a+c)(c+d)}{n^2}\right)^c\left(\frac{(b+d)(c+d)}{n^2}\right)^d\right] \tag{17.10}$$

If we compute $\ln L$, the natural logarithm of the ratio of these two probabilities or likelihoods, the result is

$$\ln L = \ln\left[\frac{(a+b)^{a+b}(a+c)^{a+c}(b+d)^{b+d}(c+d)^{c+d}}{a^a b^b c^c d^d n^n}\right]$$

This quantity times -2 is G and it can be simplified to

$$G = 2[a\ln a + b\ln b + c\ln c + d\ln d + n\ln n - (a+b)\ln(a+b)$$
$$- (a+c)\ln(a+c) - (b+d)\ln(b+d) - (c+d)\ln(c+d)] \tag{17.11}$$

This is approximately distributed as χ^2 with one degree of freedom. In this manner, we compute G directly without having to calculate expected frequencies ($\hat{f}$) first.

In this case also, the actual type I error of the G-test tends to be higher than the intended level. For this reason, the Williams correction is recommended (see Section 17.2). (Its formula for a 2×2 table is given in Box 17.6.) An alternative correction,

---

BOX 17.6 **G-Test of Independence**

A plant ecologist samples 100 trees of a rare species from a 400-square-mile area. He records for each tree whether or not it is rooted in serpentine soils and whether its leaves are pubescent or smooth. The results are tabled below. This represents a Model I design (only n is fixed).

Soil	Pubescent	Smooth	Totals
Serpentine	12	22	34
Not Serpentine	16	50	66
Totals	28	72	$100 = n$

The conventional algebraic representation of this table is as follows:

$$
\begin{array}{ccc}
 & & \Sigma \\
a & b & a + b \\
c & d & c + d \\
\Sigma\, a + c & b + d & a + b + c + d = n
\end{array}
$$

Compute the following quantities:

1. $\Sigma f \ln f$ for the cell frequencies

$$12 \ln 12 + \ldots + 50 \ln 50 = 337.78438$$

2. $\Sigma f \ln f$ for the row and column totals

$$= 34 \ln 34 + 66 \ln 66 + 28 \ln 28 + 72 \ln 72 = 797.63516$$

3. $n \ln n = 100 \ln 100 = 460.51702$

4. Compute G by multiplying Expression (17.11) by 2:

$$G = 2[\text{quantity } \mathbf{1} - \text{quantity } \mathbf{2} + \text{quantity } \mathbf{3}]$$
$$= 2[337.78438 - 797.63516 + 460.51702] = 1.33249$$

Williams's correction

Williams's correction for a 2×2 table is

$$q = 1 + \frac{\left(\dfrac{n}{a+b} + \dfrac{n}{c+d} - 1\right)\left(\dfrac{n}{a+c} + \dfrac{n}{b+d} - 1\right)}{6n}$$

For these data we obtain

$$q = 1 + \frac{\left(\dfrac{100}{34} + \dfrac{100}{66} - 1\right)\left(\dfrac{100}{28} + \dfrac{100}{72} - 1\right)}{6(100)} = 1.02281$$

$$G_{\text{adj}} = G/q = 1.33249/1.02281 = 1.30277, \; P = 0.253,708.$$

Based on this P-value, we accept the null hypothesis that the leaf type is independent of the type of soil in which the tree is rooted.

Box 17.6 (continued)

Yates's correction for continuity (an alternative correction)

Adjust the frequencies in the table as follows. If $ad - bc$ is positive, as it is in our example, because $(12 \times 50) - (16 \times 22) = 248$, subtract $\frac{1}{2}$ from a and d and add $\frac{1}{2}$ to b and c. If $ad - bc$ were negative, we would add $\frac{1}{2}$ to a and d and subtract $\frac{1}{2}$ from b and c. The new 2×2 table is

Soil	Pubescent	Smooth	Totals
Serpentine	$11\frac{1}{2}$	$22\frac{1}{2}$	34
Not serpentine	$16\frac{1}{2}$	$49\frac{1}{2}$	66
Totals	28	72	100

Compute the following quantities:

1. $\sum f \ln f$ for the cell frequencies

$$= 11\frac{1}{2} \ln 11\frac{1}{2} + 22\frac{1}{2} \ln 22\frac{1}{2} + 16\frac{1}{2} \ln 16\frac{1}{2} + 49\frac{1}{2} \ln 49\frac{1}{2}$$

$$= 337.54418$$

Quantities **2** and **3** are the same as those earlier in this box.

4. Compute G_{adj} by multiplying Expression (17.11) by 2:

$G_{adj} = 2[\text{quantity } \mathbf{1} - \text{quantity } \mathbf{2} + \text{quantity } \mathbf{3}]$

$= 2[337.54418 - 797.63516 + 460.51702] = 0.85208, P = 0.355,965.$

Again, we are led to accept the null hypothesis of independence between soil type and leaf pubescence.

---

that is sometimes used with small samples is Yates's correction for continuity, which, as we saw in Section 17.2, adjusts the observed frequencies by adding or subtracting 0.5 in such a way as to reduce each $|f_i - \hat{f}_i|$. In a 2×2 test of independence, we add 0.5 to cells a and d and subtract 0.5 from cells b and c when the quantity $ad - bc$ is negative. If this quantity is positive we subtract 0.5 from cells a and d and add 0.5 to cells b and c. Because this correction results in a very conservative test, however, we prefer Williams's correction, except for Model III cases, where the correction provides a better approximation to Fisher's exact test described below (Haviland, 1990).

The observed G should be compared with χ^2 for one degree of freedom for reasons that we will examine shortly.

Box 17.6 illustrates the G-test of independence (using the Williams and the Yates corrections) applied to the plant ecology sampling experiment (discussed earlier in this section) in which trees with one of two types of leaves were rooted in one of two different soils. The result of the analysis shows clearly that we cannot reject the null hypothesis of independence between soil type and leaf type. The presence of pubescent leaves is independent of whether or not the tree is rooted in serpentine soils.

The advent of easy, inexpensive computing has made it possible to solve Model I problems exactly rather than by the G-test, which is only an approximation. The program StatXact (2010) evaluates the probability of the observed frequencies a, b, c, and d and all less probable ones occurring if the row and column classifications are independent, using the multinomial distribution. When we subject the plant ecology data to this program, we obtain a probability of $P = 0.3471$ almost instantly. This exact procedure is, of course, preferable to the G-test, which yields a slightly more liberal result ($P = 0.2484$ in this case).

The method designed especially for a Model III 2×2 table is **Fisher's exact test.** The computation is based on the hypergeometric distribution (see Section 5.4) with four classes. These probabilities are computed assuming that the row and column classifications are independent (the null hypothesis) and that the row and column totals are fixed. Fisher's exact test answers the following question: Given two-way tables with the same fixed margin totals as the observed one, what is the chance of obtaining the observed cell frequencies a, b, c, and d and all cell frequencies that represent a greater deviation from expectation? As Haviland (1990) points out, this test is "exact" in the sense that it computes the exact hypergeometric probability. It is not an exact test when the marginal totals are free to vary.

The total number of ways in which a two-way table with fixed margin totals can be obtained is

$$\binom{n}{a+b}\binom{n}{a+c} = \frac{n!}{(a+b)!(c+d)!} \times \frac{n!}{(a+b)!(b+d)!}$$

This total is simply the product of the number of ways of taking $(a + b)$ items from n, multiplied by the number of combinations of $(a + c)$ items taken from n. From the coefficients of the multinomial distribution it can be shown that there are $n!/a!b!c!d!$ ways of obtaining the cell frequencies that we observed. Therefore the probability of obtaining a 2×2 table with cell frequencies a, b, c, and d is computed as the ratio of the two quantities given earlier. This expression can be simplified to

$$P = \frac{(a+b)!(c+d)!(a+c)!(b+d)!}{a!b!c!d!n!} \tag{17.12}$$

We must then compute the probability of all "worse" cases. The simplest method is to list systematically all the worse cases, compute the probability of each, and sum these probabilities. Because we are usually interested in a two-tailed test, we should consider all worse cases in both tails, which is what the BIOMstat computer program does. When the computations are carried out by pocket calculator, it may be convenient to compute the probability for the tail that contains the observed result and double this probability. Computation using a calculator can be simplified somewhat by noting that the numerator (products of factorials of the marginal totals) and $n!$ in the denominator are constant for all cases. Only the frequencies a, b, c, and d change in each table.

Fisher's exact test is illustrated in Box 17.7, which shows the results of an experiment with acacia trees and ants. All but 28 trees of two species of acacia were

BOX 17.7 Fisher's Exact Test of Independence

Invasion rate of two species of acacia trees by ant colonies. For details of this experiment see text.

Acacia species	Not invaded	Invaded by ant colony	Totals
A	2	13	15
B	10	3	13
Totals	12	16	28

The conventional algebraic representation of this table is as follows:

$$
\begin{array}{ccc}
 & & \Sigma \\
a & b & a + b \\
c & d & c + d \\
\Sigma \quad a + c & b + d & a + b + c + d = n
\end{array}
$$

Computation

We employ Expression (17.12). Many pocket and tabletop calculators now have factorial keys that compute factorials up to the capacity of the machine, usually $\leq 69!$ When the numbers in a 2×2 table are in excess of this value or when a factorial key is not available, employ a table of common logarithms of factorials, such as Table **1** in the 2nd edition of the Rohlf and Sokal *Statistical Tables*.

1. Direct computation of Expression (17.12) yields $P = 0.000,987,12$. Assuming independence of the invasion rate and the proportion of the two species of acacia trees, as well as fixed margin totals, this value is the probability of obtaining the results of the experiment by chance.

   Because factorials yield such larger numbers, however, it is often more convenient to evaluate Expression (17.12) in terms of logarithms. Most computer programs employ this device transparent to the user. Here we show how to evaluate the probability using common logarithms:

   Compute the sum of transforms of the margin totals and subtract from it the transform of the total frequency:

$$
\begin{array}{lll}
\log (a + b)! = & \log 15! = & 12.11650 \\
\log (c + d)! = & \log 13! = & 9.79428 \\
\log (b + d)! = & \log 12! = & 8.68034 \\
\log (a + c)! = & \log 16! = & 13.32062 \\
- \log n! & = - \log 28! = & - 29.48414 \\
\hline
 & & 14.42760
\end{array}
$$

2. In a manner similar to step **1** compute the sum of the transforms of the observed cell frequencies:

$$
\begin{array}{lll}
\log a! = \log 2! & = & 0.30103 \\
\log b! = \log 13! & = & 9.79428 \\
\log c! = \log 10! & = & 6.55976 \\
\log d! = \log 3! & = & 0.77815 \\
\hline
 & & 17.43322
\end{array}
$$

Box 17.7 (continued)

3. Subtract quantity **2** from quantity **1** and compute the antilogarithm of the result. The end result will be a probability (which must be ≤ 1).

$$14.42760 - 17.43322 = -3.00563$$

$$\text{antilog} - 3.00563 = 0.000,987,12$$

Next we compute the probability of all worse cases.

4. If $ad - bc$ is negative (as in the present example), decrease cells a and d by one and add unity to cells b and c. In this new 2×2 table the cell frequencies depart even more from expected frequencies (on the assumption of independence) than do the experimental results.

Acacia species	Not invaded	Invaded by ant colony	Σ
A	1	14	15
B	11	2	13
Σ	12	16	28

Note that the marginal totals have remained the same. Thus the numerator of Expression (17.12), quantity **1**, will remain unchanged and we need only compute a new value corresponding to the denominator, quantity **2**.

If $ad - bc$ had been positive, we would have decreased cells b and c by unity, and increased cells a and d by the same amount.

By direct computation we obtained $P = 0.000,038,46$. To save space we do not show steps **2** and **3** worked out in terms of logarithms.

5. Step **4** is repeated until one of the cell frequencies reaches zero. In the present example this happens on the very next iteration.

Acacia species	Not invaded	Invaded by ant colony	Σ
A	0	15	15
B	12	1	13
Σ	12	16	28

By direct computation, $P = 0.000,000,4273$.

6. Assuming independence of the invasion rate and the proportions of the two species of acacia trees, as well as fixed margin totals, the total probability of obtaining the results of our experiment and all less likely outcomes is obtained by summing the probabilities computed in steps **3**, **4**, and **5** and multiplying this sum by **2** (for a two-tailed test).

$$P = 2(0.000,987,12 + 0.000,038,46 + 0.000,000,43) = 0.002,052,02$$

The probability computed in this manner is still not "exact" because the distribution of the probabilities associated with the possible outcomes is not symmetrical. The correct but more tedious procedure is to calculate the cumulative probability for the second tail of the distribution in a manner analogous to the computation shown in this box.

Box 17.7 (continued)

We start with the most extreme case by setting to zero the smaller of the two frequencies b or c (with $ad - bc$ negative) or a or d (with $ad - bc$ positive). The other cell frequencies are adjusted to preserve the margin totals. Compute the probability of the second tail as in steps **2** and **3**. Continue the calculation by repeating step **4** until the probability of a resulting outcome exceeds that obtained in step **3** for the first tail. The probability of obtaining the observed results assuming independence (for a two-tailed test) is the sum of the probabilities obtained from steps **3** through **5** earlier, plus the sum of the newly computed probabilities for the second tail, *excluding the last one*, which exceeded the P-value for step **3** in the first tail. In this example the cumulative probability for the second tail is 0.000,598,3 and the overall probability for the two-tailed test thus is 0.000,598,3 $+ \frac{1}{2}$ (0.002,052,02) = 0.001,624,3. The discrepancy between the exact solution and the conservative value obtained by doubling the one-tailed probability increases with the difference between the two row or column totals.

Based on these results, we reject the null hypothesis that invasion rate is independent of the host species.

cleared from an area in Central America. There were 15 trees of species A and 13 of species B. These trees had been freed from ants by insecticide. Next, 16 separate colonies of ant species X, obtained from cut-down trees of species A in an area nearby, were brought in. The colonies were situated roughly equidistant to the 28 trees and were permitted to invade them. The number of trees of each species reinfested by ants was recorded. This experiment is clearly a Model III problem, because the total number of trees of the two species and the number of infested trees are both fixed. (The number of infested trees is fixed because there are only 16 ant colonies and each colony can infest only one tree; hence 12 trees must remain free of ants. No more than one colony will invade a single tree.) From the results it appears that the invasion rate is much higher for the acacia species A, in which these ants had lived before; there were $(13)(100)/15 = 86.7\%$ invasions of species A, contrasted with $(3)(100)/13 = 23.1\%$ invasions of species B. We conclude in Box 17.7 that the probability of obtaining, by chance alone, a fit to expectation as bad as shown in these data (or a worse fit) is approximately 0.002. We, therefore, reject the null hypothesis of independence and conclude that the difference in invasion rate is related to the host acacia species. Although the test is 2-tailed, we may note that species A seems to be more attractive than species B.

The computations can be long if the sample is large. When the lowest cell frequency in the table is greater than 10, the method is impractical without a computer. When no margin total is greater than 15 (and consequently the total sample size is no greater than 30), we do not have to carry out the computations but can look up the probabilities in published tables. Table 38 of the *Biometrika Tables* (Pearson and Hartley, 1958) serves this purpose; it appears in slightly simpler form as Table I of Siegel (1956). More extensive tables were published by Finney et al.

(1963). Instructions for the use of these tables, which are somewhat complicated, can be found in the cited sources.

If no computer is accessible the X^2-test with Yates's correction described later can be used to yield approximate probabilities when sample sizes are so large that calculator operation is impractical. When n is smaller, as in Box 17.7, the approximation is not so good.

During much of the 20th century, the **chi-square test of independence** was used for testing independence in 2×2 tables. To carry out such a test we could compute expected frequencies as shown earlier and apply Expression (17.5), but there is a much simpler way, adapted to the simple computational facilities of those times. The equation

$$X^2 = \frac{(ad - bc)^2 n}{(a + b)(c + d)(a + c)(b + d)} \tag{17.13}$$

yields identical results. For example, the immunology experiment with which we introduced this section; this Expression (17.13) would yield

$$X^2 = \frac{[(13 \times 29) - (44 \times 25)]^2 \times 111}{57 \times 54 \times 38 \times 73}$$
$$= 6.79555, \ P = 0.009,139$$

For comparison, the G-test for independence applied to this Model II example yields 6.87828, $P = 0.008725$.

Yates's adjustment for continuity for the chi-square test of independence is contained in the following expression:

$$X^2_{adj} = \frac{\left(|ad - bc| - \dfrac{n}{2} \right)^2 n}{(a + b)(c + d)(a + c)(b + d)} \tag{17.13Y}$$

Grizzle (1967) has shown that applying Yates's correction to chi-square tests with Models I and II almost always results in an unduly conservative test (the type I error is much lower than desired). Yates's correction therefore seems unnecessary even with quite small samples, such as $n = 20$. However, it yields a good approximation for Model III. Applying Expression (17.13Y) to the immunology experiment yields $X^2_{adj} = 5.79229$, substantially lower than 6.7955. Nevertheless, because $X^2_{adj} > X^2_{0.25[1]}$, we still conclude that mortality in these mice is not independent of the presence of antiserum. Note that the percentage mortality among animals given bacteria *and* antiserum is $(13)(100)/57 = 22.8\%$, considerably lower than the mortality of $(25)(100)/54 = 46.3\%$ among mice to whom only bacteria were administered. Clearly, the antiserum has been effective in reducing mortality. Applying this test to the Model III data in Box 17.7, we obtain $X^2_{adj} = 9.049145$, which yields a P-value of 0.002628 that is fairly close to the exact value. Without the correction, we would obtain $X^2 = 11.499145$ with a much smaller P-value, 0.000696.

Tests of independence need not be restricted to 2×2 tables. In the two-way cases considered in this section, we are concerned with only two properties, but each of these properties may be divided into any number of classes. For example, organisms may occur in four color classes and be sampled from three localities, requiring a 4×3 test of independence. Such a test would examine whether the color proportions exhibited by the margin totals are independent of the localities at which the individuals were sampled. Such tests are often called $R \times C$ **tests of independence,** R and C standing for the number of rows and columns in the frequency table, respectively. Another example, examined in detail in Box 17.8, concerns bright red color patterns found in samples of a species of tiger beetle on four occasions during spring and summer. Of interest was whether the percentage of bright red individuals changed during the time of observation. We therefore test whether the proportion of beetles colored bright red (55.7% for the entire study) is independent of the time of collection. As displayed in Box 17.8, this is a 4×2 test of independence.

In step **5** of Box 17.8, we show the following simple but general rule for computing the G-test of independence in an $R \times C$ table:

$$G = 2[(f \ln f \text{ for the cell frequencies})$$
$$- (\Sigma f \ln f \text{ for the row and column totals}) + n \ln n]$$

You can compute these quantities directly. We can also apply the Williams correction to estimates of G from $R \times C$ tables. The formula, now more complicated, is given in Box 17.8. The adjustment in G is minor when the sample is large, as in this example. When computing with a pocket calculator, the correction is necessary only when total sample size is small and the observed G-value is of marginal significance.

The results in Box 17.8 show clearly that the frequency of bright red color patterns in these tiger beetles is dependent on the season. We note a decrease of bright red beetles in late spring and early summer, followed by an increase again in late summer. To find out whether the differences between successive points in time are homogeneous, we could subject these data to unplanned tests similar to those in Box 17.5. In such an analysis, we would test the independence of selected subsets of the data.

You may have wondered about the relationship between the replicated goodness-of-fit tests in Box 17.4 and the $R \times C$ tests discussed here. The computational outline for the heterogeneity statistic, G_H, shown in step **2** of Box 17.4, is the same as that for the test statistic G in the $R \times C$ test (step **5** of Box 17.8). Are these quantities identical? Computationally speaking, they are. In fact, we could obtain the G-value of an $R \times C$ test of independence by running the data as though each row or column were one replicate of a series of replicated goodness-of-fit tests and obtaining the value of G_H. In Box 17.4, however, we were testing an *extrinsic hypothesis*, a 9:7 ratio that was applied to all eight replicates (progenies) of the experiment. In the $R \times C$ test of independence (Box 17.8), *the hypothesis is based on the totals in the margin of the table*. Thus, we are testing each row of the table as though it were a replicate against the hypothesis that 55.7% of the beetles have the bright red color pattern (and consequently 44.3% do not have it). This is clearly an *intrinsic hypothesis*.

BOX 17.8 $R \times C$ Test of Independence Using G-Test

Frequencies of color patterns of a species of tiger beetle (*Cicindela fulgida*) in different seasons.

Season	Color pattern ($a = 2$)			
($b = 4$)	Bright red	Not bright red	Totals	% Bright red
Early spring	29	11	40	72.5
Late spring	273	191	464	58.8
Early summer	8	31	39	20.5
Late summer	64	64	128	50.0
Totals	374	297	671 $= n$	55.7

SOURCE: Data from H. L. Willis (unpublished results).

Computation

1. Sum of transforms of the frequencies in the body of the contingency table

$$= \sum^b \sum^a f \ln f = 29 \ln 29 + 11 \ln 11 + \cdots + 64 \ln 64 = 3314.02462$$

2. Sum of transforms of the row totals $= \sum^b \left(\sum^a f \right) \ln \left(\sum^a f \right)$

$$= 40 \ln 40 + \cdots + 128 \ln 128 = 3760.40039$$

3. Sum of the transforms of the column totals $= \sum^a \left(\sum^b f \right) \ln \left(\sum^b f \right)$

$$= 374 \ln 374 + 297 \ln 297 = 3906.71011$$

4. Transform of the grand total $= n \ln n = 671 \ln 671 = 4367.38409$

5. $G = 2[\text{quantity } \mathbf{1} - \text{quantity } \mathbf{2} - \text{quantity } \mathbf{3} + \text{quantity } \mathbf{4}]$

$$= 2[3314.02462 - 3760.40039 - 3906.71011 + 4367.38409] = 28.59642$$

6. Williams's correction for an $R \times C$ table is

$$q = 1 + \frac{\left(n \sum^b \dfrac{1}{\sum^a f} - 1 \right)\left(n \sum^a \dfrac{1}{\sum^b f} - 1 \right)}{6n(a-1)(b-1)}$$

$$= 1 + \frac{\left[671\left(\dfrac{1}{40} + \cdots + \dfrac{1}{128} \right) - 1 \right]\left[671\left(\dfrac{1}{374} + \dfrac{1}{297} \right) - 1 \right]}{6(671)(1)(3)} = 1.01003$$

Box 17.8 (continued)

A simple (lower-bound) estimate of q is given by

$$q_{min} = 1 + \frac{(a + 1)(b + 1)}{6n}$$

$$= 1 + \frac{(2 + 1)(4 + 1)}{6(671)} = 1.00373$$

Using the more exact value for q,

$$G_{adj} = G/q = 28.59642/1.01003 = 28.312, P = 3.12 \times 10^{-6}$$

We compare this value with a χ^2-distribution with $(a - 1)(b - 1)$ degrees of freedom, where a is the number of columns and b the number of rows in the table. In our case, $df = (2 - 1)(4 - 1) = 3$.

In view of the very low P-value, we must reject the null hypothesis that frequency of color pattern is independent of season.

The three models defined earlier apply to the $R \times C$ test of independence as well. Thus, the example in Box 17.8 is Model I, because the investigator did not limit the number of beetles to be collected at any one season and had no control over the number of beetles in the two color classes. No methods have been devised especially for Models II and III; such cases are analyzed in the manner described earlier. Note, however, that the StatXact (2010) program evaluates the exact probability, or a Monte Carlo approximation thereof, for a Model I or Model III $R \times C$ test in a reasonably short time.

The *degrees of freedom for G-tests (and chi-square tests) of independence* are always the same and can be computed using the rules given earlier (see Section 17.2). There are k cells in the table, but we must subtract one degree of freedom for each independent parameter we have estimated from the data, and we must, of course, subtract one degree of freedom for the observed total sample size, n. We have also estimated $a - 1$ row probabilities and $b - 1$ column probabilities, where a and b are the number of rows and columns in the table, respectively. Thus, there are $k - (a - 1) - (b - 1) - 1 = k - a - b + 1$ degrees of freedom for the test. Because $k = a \times b$, however, by substitution this expression becomes $(a \times b) - a - b + 1 = (a - 1) \times (b - 1)$, the conventional expression for the degrees of freedom in a two-way test of independence. Thus, the degrees of freedom in the example of Box 17.8, a 4×2 case, is $(4 - 1) \times (2 - 1) = 3$. In all 2×2 cases, there is clearly only $(2 - 1) \times (2 - 1) = 1$ degree of freedom.

Another name for test of independence is **test of association.** If two properties are not independent of each other they are *associated*. Thus, in the example testing relative frequency of two leaf types on two different soils we can speak of an association between leaf types and soils. In the immunology experiment, there is a negative association between presence of antiserum and mortality. **Association** is thus similar to correlation, but it is a more general term, applying to attributes as well as to continuous variables.

The computations described in this section can be performed by the BIOMstat package of computer programs—both the Fisher's exact test and the $R \times C$ test for independence using the G-test.

Suggestions for graphic representation of two-way contingency tables have been made by Snee (1974). He depicts $R \times 2$ tables by bars representing the proportions of one of the binary properties with confidence limits superimposed on the top of each bar. The confidence limits of the proportions are evaluated as discussed at the end of Section 13.11. For $R \times 3$ tables, an approximate $100(1 - \alpha)\%$ confidence region for the parameters of the multinominal distribution of any given row is the set of points $\hat{p}_1$, $\hat{p}_2$, $\hat{p}_3$ satisfying the equation

$$\frac{p_1^2}{\hat{p}_1} + \frac{p_2^2}{\hat{p}_2} + \frac{p_3^2}{\hat{p}_3} \leq \frac{\chi^2_{\alpha[2]}}{n} + 1$$

This region can be represented on triangular graph paper.

It may be of interest to test subsets of an $R \times C$ table for homogeneity. These tests can be carried out in a manner entirely analogous to the unplanned tests in Box 17.5. We try to find the set of largest homogeneous tables, starting with various (all, if possible) smallest tables, which of course are 2×2 tables. Those that prove homogeneous are then augmented by another row or column, and testing is continued until addition of another row or column would lead to rejection of the null hypothesis of homogeneity. From the final listing of maximal homogeneous subsets we can construct a scheme of relations among the rows and columns of the overall table. Such decomposition of a sizable $R \times C$ table requires extensive computation and should be done by computer. The BIOMstat program includes an optional feature for an unplanned test of all subsets of rows and columns in an $R \times C$ contingency table.

In the tests of independence that we have discussed in this section, one way of looking for suspected lack of independence is to examine the percentage occurrence of one of the properties in the two classes based on the other property. For example, we compared the percentage of smooth leaves on the two types of soils, and we studied the percentage mortality with or without antiserum. This way of looking at a test of independence suggests another test of such data. Can we develop a test for the differences between two percentages? Several such tests are found in the statistical literature. In our experience, however, none of these tests can match the 2×2 G-test of independence in ease of computation and/or closeness of the observed type I error to the intended one. For this reason, we recommend the 2×2 G-test of independence to test the difference between two percentages or proportions.

For a divergent opinion, see D'Agostino et al. (1988). This topic has received considerable attention from statisticians in recent years (Little, 1989; Greenland, 1991), but no consensus on how to proceed has been reached. The argument pivots on the correct model of the 2×2 table to be analyzed—that is, which margin totals are fixed. With intermediate to large sample sizes, this issue is probably of only academic interest, but if n is small (say, $n < 25$), the selection of the correct model may matter.

A 2×2 contingency table can be thought of as expressing the association between the two nominal criteria that define the margins. In the immunology example with bacteria and antiserum, the table expresses the association between survival and the administration of antiserum. It is not surprising that statisticians have developed a variety of **coefficients of association** for such data. We feature only one of them, the **phi coefficient,** which is defined as

$$\phi = \sqrt{\frac{X^2}{n}} \tag{17.14}$$

where X^2 is the test statistic from a chi-square test of the 2×2 table, unadjusted for continuity, and n is the total sample size. The phi coefficient ranges between 0 and 1, and the positive or negative direction of the association can be ascertained from the sign of the determinant of the table, $ad - bc$. By substituting Expression (17.13) into Expression (17.14), we obtain a direct computational formula for phi:

$$\phi = \frac{(ad - bc)}{\sqrt{(a + b)(c + d)(a + c)(b + d)}} \tag{17.15}$$

This formulation ranges from -1 to $+1$, automatically attaching a sign to the index. Evaluated for the bacteria and antiserum example, Expression (17.15) yields

$$\phi = \frac{(13 \times 29 - 25 \times 44)}{\sqrt{57 \times 54 \times 38 \times 73}} = -0.247$$

Because of the arrangement of the table, the association is negative. The phi coefficient is only moderately strong. Specifically, the coefficient says that the administration of only bacteria is negatively related to survival in the experiment. If all antiserum-treated mice had survived and none of the bacteria-only group were left alive, there would be perfect association. Substitution of $a = d = 0$ into Expression (17.15) would yield -1.

In Section 17.8, we discuss methods for estimating effect size and sample sizes needed to be able to detect a specified effect size with a desired level of power.

Before we proceed to three-way tables, we should mention one more application of the two-way table. We showed in Chapter 13 how, if the assumptions of the analysis of variance are not met, we can carry out tests on rank—the Mann–Whitney U-test for two samples and the Kruskal–Wallis test for several samples. As we mentioned earlier, there are tests of location applied to ranked data. A (weaker) test only of medians can be constructed as follows. For two or more samples, evaluate the overall median of all samples. Then record for each sample the number of observations below and above this overall median. Omit observations lying exactly on the median. Finally, test these tables for independence by the G-test. Rejection of the null hypothesis implies that the proportions to each side of the overall median differ among the samples; hence the medians of the samples also differ. Box 17.9 shows this test applied to two examples. Note that Freidlin and Gastwirth (2000) found this test to have very low power in comparison to 6 alternative methods. They recommend that the Wilcoxon or Kolmogorov–Smirnov tests should usually be used instead.

BOX 17.9 Testing for Equality of Medians by Two-Way Tests of Independence

Two samples

Length of cheliceral base in chigger nymphs. (Data from Box 13.7.) Overall median of 26 items = 118 or rank 13.5. Two observations lie on the median and are omitted from the following table.

	Sample A	Sample B	Σ
Below median	5	7	12
Above median	9	3	12
Σ	14	10	24

A little experimentation will show that the marginal totals of this table are fixed, so we proceed to Fisher's exact test (see Box 17.7), which yields a *P*-value of 0.1069. The two medians cannot be shown to differ.

More than two samples

Length of pea sections. (Data from Box 13.6.) Overall median of 50 items = 60.5 or rank 25.5.

			Treatments			
	Control	2% glucose	2% fructose	1% glucose + 1% fructose	2% sucrose	Σ
Below median	0	8	8	9	0	25
Above median	10	2	2	1	10	25
Σ	10	10	10	10	10	

Although the marginal totals of this table are fixed (it is a Model III $R \times C$ test), we have not learned of an exact test of the probability of independence. We therefore evaluate $G = 42.797$ as an approximation. There are $(5 - 1)(2 - 1) = 4$ degrees of freedom, $P = 1.14 \times 10^{-8}$, and we reject the null hypothesis of independence. We conclude that the medians of the five samples are not the same.

There are computer programs that evaluate the exact probability of an $R \times C$ table. When we subjected the pea section table to the Fisher option of the StatXact (2010) program, we obtained $P = 0.1602 \times 10^{-9}$. The conclusion remains the same.

17.5 Analysis of Three-Way Tables

As in Chapter 11, where we proceeded from two-way classifications to three- and four-way tables, in analysis of frequencies we can turn from a test of independence in a two-way table to tests involving tables of three, four, or more dimensions. Unfortunately, the analysis of higher-dimensional tables is much more complicated than that of two-way tables, because we must test for more-complex patterns of associations among various combinations of the factors and because more tests require the iterative proportional fitting method to compute expected frequencies.

In recent years, the analysis of contingency tables has been based on **log-linear models,** which are the counterpart for attribute data to the linear model for continuous variables in analysis of variance and multiple regression. For a two-way contingency table, we have (by analogy to a two-way anova) the model

$$\ln \hat{f}_{ij} = \mu + \alpha_i + \beta_j + \alpha\beta_{ij} \tag{17.16}$$

where $\hat{f}_{ij}$ is the expected frequency in row i, column j of the two-way contingency table, μ is the mean of the logarithms of the expected frequencies, α_i and β_j are the effects of categories i and j of factors A and B, respectively, and the $\alpha\beta_{ij}$ interaction term expresses the dependence of category i of factor A on category j of factor B. For a three-way table, the log-linear model is

$$\ln \hat{f}_{ijk} = \mu + \alpha_i + \beta_j + \gamma_k + \alpha\beta_{ij} + \alpha\gamma_{ik} + \beta\gamma_{jk} + \alpha\beta\gamma_{ijk} \tag{17.17}$$

In contrast to the analysis of variance (where the main effects are of most interest), in log-linear models we are interested primarily in testing for the presence of interactions. Thus, in the two-way contingency tables of the previous section we tested the hypothesis that interaction is not present (H_0: $\alpha\beta_{ij} = 0$ for all ij), because this is a test of the independence of factors A and B. Testing for the presence of a term in the log-linear model requires, in theory, fitting two models, one with the term present and one with it omitted. The G-statistic for goodness of fit is computed for each of the two models, and the difference between the G-values is used to test for the term being left out.

In many cases, we can evaluate the G-statistics directly without having to compute the expected frequencies explicitly. However, it is good practice to compute the expected frequencies routinely, because this permits the examination of the deviations between the observed and the expected frequencies. The pattern of deviations may suggest additional models that should be tested (as we shall see later in Box 17.10).

The expected frequencies for any model are functions of the smallest set of marginal tables (one-way, two-way, and higher-dimensional tables of marginal totals) sufficient to compute these frequencies. The marginal tables in this minimal set are called configurations by Bishop et al. (1975). Thus, for example, the expected frequencies when testing for independence in a two-way table are a function of the products of the row and column sums. The test for the three-way interaction in a three-way table is a function of the AB, BC, and AC two-way tables. To find the minimal set of marginal totals needed, we examine the log-linear model to be fitted. Starting with the highest-order terms in the model, we form a set consisting of marginal tables corresponding to the terms in the model. If, for example, the term $\alpha\beta_{ij}$ is present, we include the AB two-way table in the minimal set. Such a table would contain observed frequencies f_{ij+}. This symbol stands for the observed frequencies in row i and column j summed over all the depths. If the marginal tables already in the minimal set correspond to all the terms in the model, no additional tables are needed. If there are still terms in the model that lack corresponding margin tables, however, we next consider lower-order terms and add marginal tables to the set until all terms in the model are accounted for. Note that marginal tables need not be added if they are redundant (i.e., the marginal totals can be obtained from marginal tables already in the set). For the model $\ln \hat{f}_{ijk} = \mu + \alpha_i + \beta_j + \gamma_k + \alpha\beta_{ij} + \beta\gamma_{jk}$, the AB and BC margin tables would suffice because from these we can obtain the $\alpha\beta$ and $\beta\gamma$ totals, as well as the α, β, and γ totals. If the $\beta\gamma$ term were not present in the model, the marginal tables AB and C would suffice to estimate it.

An iterative proportional-fitting algorithm can always be used to compute the $\hat{f}$. This algorithm consists simply of an iterative scaling of an initial estimate of $\hat{f}$ until its marginal totals agree with those in the minimal set of marginal totals used for the model. For some models, $\hat{f}$ can be estimated directly. In such cases, when the iterative procedure is used, it will converge after only a single iteration. In the analysis of three-way tables (the only design taken up in detail in this section), only the test for the ABC interaction requires the iterative procedure. All other estimated frequencies in a three-way table can be estimated directly. Analyses of four-way and higher-dimensional tables require iterative solutions for many of the tests. These analyses are not practical without a computer.

The example in Box 17.10 is an analysis of an experiment in which drosophila larvae were reared in medium vials containing DDT. The emerging adult drosophila were classified according to three factors: sex, whether they were healthy or poisoned (absence or presence of obvious neurological symptoms), and the location (in the medium vial) where they had pupated. It was of interest to learn whether the two sexes had different tolerances to the poison and also whether the site where the fly had pupated (in the medium, at the margin of the medium, on top of the medium, or away from the medium on the wall of the vial) affected its chance for subsequent survival.

In the analysis of a multiway table, we test a hierarchy of models, starting with the most complex. For example, as in a Model I three-way anova, in a three-way table (see Box 17.10) we first test for the three-factor interaction. This test is done by fitting a model with the $\alpha\beta\gamma$ terms deleted. This is a "partial association" model between each pair of variables, implying that there are pairwise associations between all of the factors but no joint associations between the three factors considered simultaneously. Computing $\hat{f}$ directly is not possible; an iterative proportional-fitting algorithm must be used, as shown in Box 17.10. Only five iterations were needed for this example. The expected frequencies are then compared with observed frequencies using the G-test. The very small G-value obtained (1.365, $df = 3$, $P = 0.713,759$) indicates that we have little evidence for the presence of three-factor interaction, so the $\alpha\beta\gamma$ term can safely be dropped from the model.

BOX 17.10   Analysis of a Three-Way Table Using Log-Linear Models

Emerged drosophila are classified according to three factors: pupation site, A (in the medium, IM; at the margin of the medium, AM: on the wall of the vial, OW; and on top of the medium, OM); sex, B; and mortality, C (healthy, H; poisoned, P).

Pupation site $(a = 4)$	Sex $(b = 2)$	Mortality $(c = 2)$ C		
A	B	Healthy (H)	Poisoned (P)	Totals
IM	♀	55	6	61
	♂	34	17	51
		89	23	112
AM	♀	23	1	24
	♂	15	5	20
		38	6	44
OW	♀	7	4	11
	♂	3	5	8
		10	9	19
OM	♀	8	3	11
	♂	5	3	8
		13	6	19
Totals		150	44	194

SOURCE: Data from R. R. Sokal (unpublished results).

Sex B	Mortality C		
	Healthy	Poisoned	Totals
♀	93	14	107
♂	57	30	87
Totals	150	44	194

The second table is the $B \times C$ two-way table needed for the computations below: the observed frequencies of sex against mortality summed over the four pupation sites. The other two-way tables ($A \times B$ and $A \times C$) are shown as subtotals in the first table. Numerous analyses can be performed on data such as these. Following the suggestion of Bishop et al. (1975), we will test a series of hierarchical models based on the log-linear model:

$$\ln \hat{f}_{ijk} = \mu + \alpha_i + \beta_j + \gamma_k + \alpha\beta_{ij} + \alpha\gamma_{ik} + \beta\gamma_{jk} + \alpha\beta\gamma_{ijk}$$

I. *Test for three-factor interaction (test $\alpha\beta\gamma = 0$)*

This is the only test in a three-way table for which the expected frequencies cannot be calculated directly. One therefore employs the iterative proportional-fitting algorithm to compute the $\hat{f}_{ijk}$, which are to be compared to the f_{ijk} using the G-test.

Box 17.10 (continued)

Computation

The procedure consists of adjusting an initial estimate of the $\hat{f}_{ijk}$ until their two-way margin sums $\hat{f}_{AB+}$, $\hat{f}_{A+C}$, and $\hat{f}_{+BC}$ agree with the observed two-way margin sums f_{AB+}, f_{A+C}, and f_{+BC}. The symbol f_{AB+} stands for the margin totals of an $A \times B$ table whose cell frequencies f_{ij+} for row i and column j are summed over all the depths. An $A \times B$ table can be abstracted from the three-way table given at the beginning of this box. Thus f_{12+}, the observed frequency for pupation site IM and sex ♂ summed over the two mortalities, equals 51. The margin totals for this table are 112, 44, 19, and 19 for pupation site, f_{A++}, and $(61 + 24 + 11 + 11 = 107)$ and $(51 + 20 + 8 + 8 = 87)$ for sex, f_{+B+}. The latter margin frequencies, as well as those for mortality (f_{++C}), are obtained more readily from the $B \times C$ two-way table shown below the three-way table. Here the frequencies f_{+jk} are the observed values of column j and depth k summed over the rows, and the margin totals f_{+BC} represent the totals for factors B and C, respectively. Because the computations are practical only when done by computer, we give merely a general outline of the steps.

1. Computation of the initial estimates for $\hat{f}_{ijk}$. A satisfactory initial estimate is simply

$$\hat{f}_{ijk}^{(0)} = \frac{n}{abc} = \frac{194}{4 \times 2 \times 2} = 12.125$$

for all $i, j,$ and k. The zero superscript refers to the original estimates.

2. *AB* adjustments. Form an *AB* two-way table of $\hat{f}_{ij+}$ values (based on the current estimate of $\hat{f}_{ijk}$) by pooling the $\hat{f}_{ijk}^{(0)}$ values over factor *C*. Because each cell is the sum of two values of $\hat{f}_{ijk}^{(0)}$ (for the healthy and poisoned categories of Factor *C*), pooling yields $\hat{f}_{ij+}^{(0)} = 2 \times 12.125 = 24.25$ in each cell.

	B	
A	♀	♂
IM	24.25	24.25
AM	24.25	24.25
OW	24.25	24.25
OM	24.25	24.25

New estimates, $\hat{f}_{ijk}^{(0)'}$, are then computed using the following relationship:

$$\hat{f}_{ijk}^{(0)'} = \hat{f}_{ijk}^{(0)} \frac{f_{ij+}}{\hat{f}_{ij+}^{(0)}}$$

For example,

$$\hat{f}_{111}^{(0)'} = 12.125 \frac{61}{24.25} = 30.5$$

and

$$\hat{f}_{422}^{(0)'} = 12.125 \frac{8}{24.25} = 4.0$$

Box 17.10 (continued)

These computations result in the following three-way table of estimated frequencies, positioned at the left.

		C					C	
A	B	H	P	A	B	H	P
IM	♀	30.5	30.5	IM	♀	54.632	7.163
	♂	25.5	25.5		♂	34.231	16.099
AM	♀	12.0	12.0	AM	♀	23.361	1.871
	♂	10.0	10.0		♂	14.589	4.192
OW	♀	5.5	5.5	OW	♀	6.525	2.979
	♂	4.0	4.0		♂	3.556	5.825
OM	♀	5.5	5.5	OM	♀	8.483	1.986
	♂	4.0	4.0		♂	4.623	3.883

3. Similar adjustments are made for the AC and then for the BC two-way tables.

   The resulting three-way table of estimated frequencies at the end of the first iteration is shown above, positioned at the right.

4. Test for convergence. Next the current $\hat{f}_{ijk}^{(x)}$ values are compared to those at the end of the previous iteration (or the initial estimates $\hat{f}_{ijk}^{(0)}$ the first time we pass through step **4**). If they agree to sufficient accuracy, we can stop; otherwise we must repeat the AB, AC, and BC adjustments described in steps **2** and **3**, using the current $\hat{f}_{ijk}^{(x)}$ values as input to step **2**. Note that if the estimates themselves are of special interest (rather than simply the null hypothesis test), the iterations, x, should be continued until the maximum $|\hat{f}_{ijk}^{(x)} - \hat{f}_{ijk}^{(x-1)}|$ is less than a tolerance, say 0.001.

   Another convenient measure of convergence is the change in the G-statistic computed at the end of each cycle of adjustments. We compute G in the conventional manner:

$$2 \sum^a \sum^b \sum^c f_{ijk} \ln \left(\frac{f_{ijk}}{\hat{f}_{ijk}^{(x)}} \right)$$

as shown below. When $|\Delta G|$ is less than, say 0.001, one can stop. For the present data the G and ΔG values and the maximum absolute difference between expected frequencies obtained for five iterations were as follows:

Iteration	G	ΔG	$\max\|\hat{f}_{ijk}^{(x)} - \hat{f}_{ijk}^{(x-1)}\|$
1	2.193	—	42.507
2	1.392	−0.802	0.884
3	1.366	−0.025	0.120
4	1.365	−0.001	0.023
5	1.365	−0.000	0.005

The convergence was quite rapid (as is typical). The ΔG criterion often requires fewer iterations than does the deviation between expected frequencies. Note that ΔG always decreases monotonically.

Box 17.10 (continued)

The final table of estimates $\hat{f}_{ijk}^{(5)}$ is shown here at the left; observed frequencies are in the middle, and Freeman–Tukey deviates, $\sqrt{f_{ijk}} + \sqrt{f_{ijk} + 1} - \sqrt{4\hat{f}_{ijk} + 1}$, are on the right. As a criterion for "large," the deviates should be compared to $\sqrt{\nu \chi_{.05[1]}^2 / abc} = \sqrt{3(3.841)/4(2)2} = 0.849$, where ν = the degrees of freedom of the G-test for goodness of fit (see below).

		Expected C		Observed C		Freeman–Tukey deviates	
A	B	H	P	H	P	H	P
IM	♀	54.418	6.582	55	6	0.112	−0.132
	♂	34.584	16.417	34	17	−0.057	0.201
AM	♀	22.394	1.605	23	1	0.178	−0.310
	♂	15.607	4.393	15	5	−0.091	0.376
OW	♀	7.313	3.688	7	4	−0.026	0.267
	♂	2.685	5.314	3	5	0.306	−0.032
OM	♀	8.875	2.125	8	3	−0.213	0.650
	♂	4.125	3.875	5	3	0.502	−0.330

The G-test for goodness of fit yields

$$G_{ABC} = 2 \sum^a \sum^b \sum^c f_{ijk} \ln\left(\frac{f_{ijk}}{\hat{f}_{ijk}}\right) = 1.365, P = 0.713,759$$

which has $(a - 1)(b - 1)(c - 1) = (4 - 1)(2 - 1)(2 - 1) = 3$ degrees of freedom. If P had been ≤ 0.05, we would have used the Williams (1976) correction by dividing G by

$$q_{min} = 1 + \frac{(a + 1)(b + 1)(c + 1)}{6n} = 1.03866$$

A P-value ≤ 0.05 for G would indicate the presence of a three-way interaction (e.g., the degree of association between sex and mortality differing for different pupation sites). As in a three-way Model I anova, we would then have little interest in testing the two-way interactions and the main effects.

II. *Test for two-factor interactions (conditional independence)*

Because the $\alpha\beta\gamma$ term is not significant, it is of interest to test the $\alpha\beta$, $\alpha\gamma$, and $\beta\gamma$ terms. Simple direct estimates exist for these tests.

1. Test for $\alpha\beta_{ij} = 0$, for all ij (test of independence of A and B, given the level of C). The estimates are given by

$$\hat{f}_{ijk} = \frac{f_{i+k}f_{+jk}}{f_{++k}}$$

Box 17.10 (continued)

Thus, for example,

$$\hat{f}_{111} = \frac{89 \times 93}{150} = 55.180$$

and

$$\hat{f}_{422} = \frac{6 \times 30}{44} = 4.091$$

where the f_{i+k} values are from the $A \times C$ table of marginal totals, f_{+jk} values are from the $B \times C$ marginal totals, and f_{++k} values are from the C marginal totals.

		Expected C		Freeman–Tukey deviates	
A	B	H	P	H	P
IM	♀	55.180	7.318	0.009	−0.407
	♂	33.820	15.682	0.073	0.383
AM	♀	23.560	1.909	−0.064	−0.524
	♂	14.440	4.091	0.207	0.519
OW	♀	6.200	2.864	0.395	0.707
	♂	3.800	6.136	−0.293	0.369
OM	♀	8.060	1.909	0.063	0.793
	♂	4.940	4.091	0.129	−0.435

The G-test for goodness of fit to the model is

$$G_{AB(C)} = 2 \sum^{a} \sum^{b} \sum^{c} f_{ijk} \ln \left(\frac{f_{ijk}}{\hat{f}_{ijk}}\right) = 2.869, P = 0.825,100.$$

This value has $(a - 1)(b - 1)c = (4 - 1)(2 - 1)2 = 6$ degrees of freedom. The decrease in fit due to dropping the $\alpha\beta$ terms from the model is $G_{AB(C)} - G_{ABC} = 2.869 - 1.365 = 1.504$, which has $6 - 3 = 3$ degrees of freedom, $P = 0.681,349$, and clearly leads us to accept the null hypothesis. We conclude that A and B are independent at each level of factor C. As before, the Freeman–Tukey deviates are also quite small.

Because direct estimates exist, we can also compute the G-values directly from observed frequencies and sums by using a natural logarithm function.

$$G = 2\left[\sum^{a} \sum^{b} \sum^{c} f_{ijk} \ln f_{ijk} - \sum^{a} \sum^{c} f_{i+k} \ln f_{i+k} - \sum^{b} \sum^{c} f_{+jk} \ln f_{+jk} + \sum^{c} f_{++k} \ln f_{++k}\right]$$

$$= 2[581.783 - 707.480 - 790.969 + 918.099] = 2.866, \quad P = 0.825,468$$

identical to the earlier result, except for rounding error.

2. Test for $\alpha\gamma_{ik} = 0$, for all ik (test of independence of A and C, given the level of B). The estimates are given by

$$\hat{f}_{ijk} = \frac{f_{ij+} f_{+jk}}{f_{+j+}}$$

Box 17.10 (continued)

Thus, for example,

$$\hat{f}_{111} = \frac{61 \times 93}{107} = 53.019$$

and

$$\hat{f}_{422} = \frac{8 \times 30}{87} = 2.759$$

A	B	Expected C		Freeman–Tukey deviates	
		H	**P**	**H**	**P**
IM	♀	53.019	7.981	0.302	−0.643
	♂	33.414	17.586	0.143	−0.081
AM	♀	20.860	3.140	0.506	−1.268*
	♂	13.103	6.897	0.565	−0.661
OW	♀	9.561	1.439	−0.790	1.637*
	♂	5.241	2.759	−0.955	1.216*
OM	♀	9.561	1.439	−0.436	1.133
	♂	5.241	2.759	−0.001	0.263

*indicates deviates with $P \le 0.05$ but $>$ than 0.01.   ** indicates deviates with $P \le 0.01$.

We can compute G by the test for goodness of fit or by direct computation as shown:

$$G_{AC(B)} = 2\left[\sum^a \sum^b \sum^c f_{ijk} \ln f_{ijk} - \sum^a \sum^b f_{ij+} \ln f_{ij+} \right.$$
$$\left. - \sum^b \sum^c f_{+jk} \ln f_{+jk} + \sum^b f_{+j+} \ln f_{+j+} \right]$$

$$= 2[581.783 - 673.499 - 790.968 + 888.527] = 11.684$$

This quantity has $(a - 1)b(c - 1) = (4 - 1)2(2 - 1) = 3$ degrees of freedom, yields $P = 0.069,401$, and is not quite sufficient to reject the null hypothesis.

The Williams (1976) correction is

$$q_{min} = 1 + \frac{(a^2 - 1)b^2(c^2 - 1)}{6(a - 1)b(c - 1)n} = 1.02577$$

Hence, $G = 11.684/1.02577 = 11.390$, $P = 0.077,045$. The decrease in fit due to dropping $\alpha\gamma$ from the model is $G_{A(B)C} - G_{ABC} = 11.684 - 1.365 = 10.319$ with $6 - 3$ df, $P = 0.016,040$. We can establish the decrease in fit with confidence. Note that for successive fitting, we do not use the Williams correction.

The pattern (IM plus AM versus OW plus OM) of the residuals from the model expressed as Freeman–Tukey deviates) is suspicious. To detect "large" deviates we can compare their absolute values with the rough criterion

$$\sqrt{\frac{\nu \chi^2_{.05[1]}}{abc}} = \sqrt{\frac{6(3.841)}{16}} = 1.200$$

Three of the deviates thus seem large, so it is of interest to partition $G_{AC(B)}$ even though the G-value is not quite large enough to reject the null hypothesis.

Box 17.10 (continued)

Three planned orthogonal comparisons can be made; they are shown in Box 17.11.

3. Test for $\beta\gamma_{jk} = 0$ for all jk (test of independence of B and C, given the level of A). The estimates are given by

$$\hat{f}_{ijk} = \frac{f_{ij+}f_{i+k}}{f_{i++}}$$

but to conserve space are not shown here.

The G-statistic can be computed directly using the equation

$$G_{BC(A)} = 2\left[\sum^a\sum^b\sum^c f_{ijk}\ln f_{ijk} - \sum^a\sum^b f_{ij+}\ln f_{ij+}\right.$$

$$\left. - \sum^a\sum^c f_{i+k}\ln f_{i+k} + \sum^a f_{i++}\ln f_{i++}\right]$$

$$= 2[581.783 - 673.500 - 707.480 + 806.864] = 15.338$$

$$P = 0.004{,}056$$

This quantity has $a(b-1)(c-1) = 4(2-1)(2-1) = 4$ degrees of freedom, and its P-value leads to a firm rejection of the null hypothesis. The decrease in fit due to dropping $\beta\gamma$ is $15.338 - 1.365 = 13.973$, which has $4 - 3 = 1$ degree of freedom and yields a minute P-value of $0.000{,}185$. We firmly conclude that factors B and C are not independent at each level of factor A.

III. *Test for $\alpha\beta_{ij} = \alpha\gamma_{ik} = 0$ for all ij and ik*

If two of the two-way independence terms (e.g., $\alpha\beta_{ij}$ and $\alpha\gamma_{ik}$) were deleted from the log-linear model, it would be of interest to test for the complete independence of factor A versus factors B and C; i.e., we would wish to test the model

$$\ln \hat{f}_{ijk} = \mu + \alpha_i + \beta_j + \gamma_k + \beta\gamma_{jk}$$

The estimates are given by

$$\hat{f}_{ijk} = \frac{f_{i++}f_{+jk}}{f_{+++}}$$

where $f_{+++} = n$.

		Expected C		Freeman–Tukey deviates	
A	B	H	P	H	P
IM	♀	53.691	8.082	0.211	−0.678
	♂	32.907	17.320	0.231	−0.017
AM	♀	21.093	3.175	0.455	−1.287
	♂	12.928	6.804	0.613	−0.626
OW	♀	9.108	1.371	−0.644	1.690*
	♂	5.582	2.938	−1.098	1.114
OM	♀	9.108	1.371	−0.290	1.186
	♂	5.582	2.938	−0.145	0.161

Box 17.10 (continued)

G can be computed directly by the equation

$$G_{A,BC} = 2\left[\sum^a\sum^b\sum^c f_{ijk}\ln f_{ijk} - \sum^a f_{i++}\ln f_{i++} - \sum^b\sum^c f_{+jk}\ln f_{+jk} + n\ln n\right]$$

$$= 2[581.783 - 806.864 - 790.969 + 1021.964] = 11.828$$

$$P = 0.223,180$$

This quantity has $(a - 1)(bc - 1) = (4 - 1)[(2 \times 2) - 1] = 9$ degrees of freedom. We cannot reject the null hypothesis. We would therefore conclude that factor A is completely independent of factors B and C. An examination of the residuals from the model expressed as Freeman–Tukey deviates, however, indicates that the fit is not good and one of the deviates is greater than the criterion:

$$\sqrt{\frac{\nu\chi^2_{.05[1]}}{abc}} = \sqrt{\frac{9(3.841)}{16}} = 1.470$$

The pattern of the deviates is the same as that in part **II**, step **2** of this box (apparently due to the lack of independence between the IM-plus-AM-versus-OW-plus-OM contrast and mortality).

IV. *Test for* $\alpha\beta_{ij} = \alpha\gamma_{ik} = \beta\gamma_{jk} = 0$ *for all ij, ik, and jk (test of independence of A, B, and C)*

If $\beta\gamma_{jk}$ had been found not significant in part **II**, step **3**, we would next have tested for the complete independence of all three factors. That is, we would have tested the model

$$\ln\hat{f}_{ijk} = \mu + \alpha_i + \beta_j + \gamma_k$$

The direct estimates for this model are given by

$$\hat{f}_{ijk} = \frac{f_{i++}f_{+j+}f_{++k}}{n^2}$$

G can be computed directly by the equation

$$G_{A,B,C} = 2\left[\sum^a\sum^b\sum^c f_{ijk}\ln f_{ijk} - \sum^a f_{i++}\ln f_{i++}\right.$$

$$\left. - \sum^b f_{+j+}\ln f_{+j+} - \sum^c f_{++k}\ln f_{++k} + 2n\ln n\right]$$

$$= 2[581.783 - 806.864 - 918.099 - 888.527 + 2(1021.964)] = 24.442$$

$$P = 0.006,510$$

This quantity has $abc - a - b - c + 2 = 4(2)2 - 4 - 2 - 2 + 2 = 10$ degrees of freedom. We do not test it here because we already rejected this model in part **II**, step **3**. It is given here for completeness. The Williams correction factor is

$$q_{min} = 1 + \frac{a^2b^2c^2 - a^2 - b^2 - c^2 + 2}{6(abc - a - b - c + 2)n} = 1.02010$$

$$G_{adj} = G_{A,B,C}/q_{min} = 24.442/1.02010 = 23.960, P = 0.007,707$$

Box 17.10 (continued)

V. *Final log-linear model*

The final log-linear model is

$$\ln \hat{f}_{ijk} = \mu + \alpha_i + \beta_j + \gamma_k + \alpha\gamma_{ik} + \beta\gamma_{jk}$$

The estimated values, $\hat{f}_{ijk}$, for this model are given in part **II,** step **1,** where the $\alpha\beta$ had been omitted.

We conclude that variable A (pupation site) and B (sex) are conditionally independent; that is, they are independent for each level of variable C (mortality). Both A and B are associated with variable C, however. We also found that we could simplify the model by condensing categories IM with AM and OW with OM, because the dependence of mortality on pupation site was *between* these categories rather than within them. Table 17.6 gives the estimates for the parameters in this final model. Their computation and interpretation are explained in the text.

Had we found evidence for the presence of the three-factor interaction term, the degree of association between any pair of variables would depend on the different levels of a third variable. For example, such a finding could imply that the association between factors A and B differs depending on the level of factor C (or that the association between A and C depends on B or that the association between B and C depends on A). In such a case, we would not attempt to fit any simpler models, but would make separate two-way tests of independence within each level of one of the factors.

Because there is no evidence for the three-factor interaction in the example of Box 17.10, we test for the presence of the two-factor effects. There are three possible models with one two-factor effect absent. The expected frequencies can be estimated directly for these models, so the G-test for goodness of fit can also be computed directly. We find that dropping the $\alpha\beta$ term results in $G_{AB(C)} = 2.869$, $P = 0.825,100$. The increase in G ($2.869 - 1.365 = 1.504$ with $6 - 3 = 3$ degrees of freedom yields $P = 0.681,347$) is also unlikely. The G-value from the model with the $\alpha\gamma$ term deleted is 11.684 ($P = 0.069,401$), close to but not quite at the conventional 0.05 level. However, the increase in G ($11.684 - 1.365 = 10.319$ with $6 - 3 = 3$ df) yields $P = 0.010,120$, which leads us to reject the null hypothesis of no 3-way interaction. An examination of the deviations of observed from expected frequencies reveals a suspicious pattern that supports the rejection of this model even though the G-value has a borderline probability value. The critical value used to test the deviates is a reasonable approximation. The value

$$\sqrt{\frac{\chi^2_{[\nu]}}{abc}}$$

suggested by Bishop et al. (1975) often results in a type I error much greater than the nominal α. The $\beta\gamma$ term is clearly effective. The increase in G due to the deletion of $\beta\gamma$ from the model is $15.338 - 1.365 = 13.973$ with $4 - 3 = 1$ df,

---

BOX 17.11 Decomposition of a Test for Conditional
Independence in a Three-Way Table

The drosophila data in Box 17.10 show a significant deviation from independence of
pupation site (A) and mortality (C) from sex (B). Because there are large residuals, we
want to partition the $G_{AC(B)}$ into its logical components. For these data, three comparisons (which happen to be orthogonal) are relevant.

I. *IM* + *AM versus OW* + *OM*

The expected frequencies are computed as before, but using the condensed table of
observed frequencies given in the middle.

		Expected C		Observed C		Freeman–Tukey deviates	
A	B	H	P	H	P	H	P
IM + AM	♀	73.879	11.121	78	7	0.500	−1.270
	♂	46.517	24.483	49	22	0.394	−0.460
OW + OM	♀	19.121	2.879	15	7	−0.930	1.937
	♂	10.483	5.517	8	8	−0.724	1.025

The G-value for AC, given B, is computed as before, but using only the frequencies in
the condensed table above.

$$G_{AC(B)} = 2(699.660 - 792.640 - 790.968 + 888.527) = 9.157$$

$$P = 0.010,270$$

The degrees of freedom are also computed as before but equal $(2 - 1)2(2 - 1) = 2$
because there are only two levels for factor A in the condensed table.

II. *IM versus AM*

		Expected C		Observed C		Freeman–Tukey deviates	
A	B	H	P	H	P	H	P
IM	♀	55.976	5.024	55	6	−0.097	0.502
	♂	35.197	15.803	34	17	−0.160	0.353
AM	♀	22.024	1.976	23	1	0.256	−0.570
	♂	13.803	6.197	15	5	0.376	−0.393
		$G_{AC(B)} = 1.307$		df = 2		$P = 0.520,222$	

Box 17.11 (continued)

III. *OW versus OM*

A	B	Expected C		Observed C		Freeman-Tukey deviates	
		H	P	H	P	H	P
OW	♀	7.500	3.500	7	4	−0.094	0.363
	♂	4.000	4.000	3	5	−0.391	0.562
OM	♀	7.500	3.500	8	3	0.261	−0.141
	♂	4.000	4.000	5	3	0.562	−0.391

$$G_{AC(B)} = 1.221 \qquad df = 2 \qquad P = 0.543,079$$

Because the comparisons are orthogonal, we can confirm that both G and df have been partitioned correctly: $9.157 + 1.307 + 1.221 = 11.685$ and $2 + 2 + 2 = 6$.

$P = 0.000,185$. Because of the suspicious pattern of deviates found in the test for $\alpha\gamma = 0$ and because more than one degree of freedom was involved, we partition the G-test into three orthogonal comparisons as shown in Box 17.11. We find that most of the deviation from the model is due to the differences in factor C for the first two and the last two levels of factor A.

There are also three models with two of the two-factor effects absent—for example, the model $\ln \hat{f}_{ijk} = \mu + \alpha_i + \beta_j + \gamma_k + \beta\gamma_{jk}$, which implies that variable A is completely independent of B and C, but that variables B and C are associated (i.e., not independent). Although an example of such a test is shown in Box 17.10 to illustrate the method, it is not appropriate for the current data, because we found $\alpha\gamma$ as well as $\beta\gamma$ to have notable effects.

In the final model that we can test, all two-factor effects are absent. If this model fits, the implication is that all three variables are completely independent of one another.

Figure 17.4 shows the relationships among the tests of the hierarchical models in a three-way table (using the data given in Box 17.10). The entries at the nodes of the diagram correspond to the G-values for the goodness-of-fit tests for all of the possible models. The values along the arrows represent the increase in the G-value due to the parameter being deleted when one passes from the model at the base of the arrow to the model at the head of the arrow. In practice, one would not routinely make all of these tests but would start at the top and proceed downward sequentially until reaching the simplest model that adequately fit the data. However, in BIOMstat or some similar program, it is much easier to display all the possible tests and then let the user choose appropriate results based on the hypothesis tests already performed.

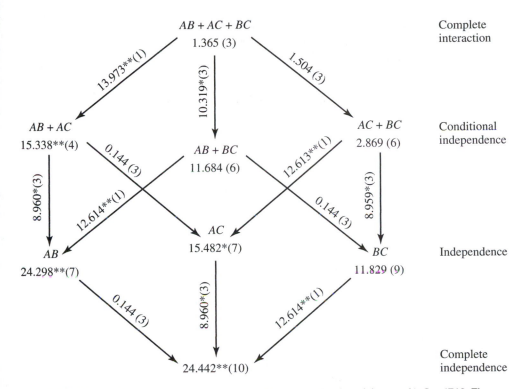

FIGURE 17.4 Diagrammatic representation of the hierarchical models tested in Box 17.10. The quantities at the nodes are G-values corresponding to the goodness-of-fit tests from the various models tested. The quantities along the arrows correspond to the differences in G-values between the model at the base of the arrow and the model at the head of the arrow. These differences correspond to the G-value explained by the parameters being deleted from the model. The values in parentheses are the degrees of freedom associated with the corresponding G-values. Single asterisks mark G-values with P-values ≤ 0.05 but > 0.01. Double asterisks mark P-values with probabilities ≤ 0.01.

The reader may already have recognized that behind the convenience of having all possible tests already performed, there lurks the danger of the multiple comparisons problem, just as it did in analysis of variance. One should not look at all of the test results but proceed in a hierarchical manner to reduce the number of tests being considered. Unless you are using an interactive program that leads the user through a hierarchic sequence of permissible tests, you are cautioned not to look at all the test results, but only at those that your research interests and prior test results warrant. The BIOMstat program carries out an analysis of all models specified in Figure 17.4 for a three-way contingency table.

If any of the marginal totals are fixed by the design of the experiment (e.g., the experimenter determines how many males and females are to be exposed to each treatment combination), then the corresponding term must be present in all models tested. If, for example, the AB table is fixed, then the $\alpha\beta_{ij}$ term must be present in all of the models—that is, it must not be dropped. The effect of $\alpha\beta$ itself is therefore never tested, because it is under the investigator's control.

An additional complication in the analysis of a multiway table is that we have tested a large number of models and, furthermore, some models have been tested only as a consequence of our having accepted null hypotheses in tests of other models. Unless one can argue that the tests are all of interest a priori, one needs to assess the overall probability of making a type I error. A simple approximate procedure has been described by Whittaker and Aitkin (1978). In Box 17.10, we tested for the presence of the $\alpha\beta\gamma$ interaction at the nominal level of $\alpha_1 = 0.05$. Because the null hypothesis that $\alpha\beta\gamma = 0$ was accepted, we then tested three models at a nominal level of $\alpha_2 = 0.05$. At this point the overall type I error rate was $\alpha = \alpha_1 + (1 - \alpha_1)\alpha_2 = 0.0975$.

If the null hypotheses of more than two of the conditional independence models had been accepted, we would have tested for complete independence at the nominal P-value of $\alpha_3 = 0.05$, when in fact this number represented an actual type I error rate of $\alpha = \alpha_1 + (1 - \alpha_1)\alpha_2 + (1 - \alpha_1)(1 - \alpha_2)\alpha_3 = 0.14263$. To avoid having such a large probability of a type I error, one may wish to use smaller values for α_1, α_2, and α_3 so that the overall type I error rate is closer to 0.05.

TABLE 17.6 Estimates of the Parameters for the Log-Linear Model $\ln \hat{f}_{ijk} = \mu + \alpha_i + \beta_j + \gamma_K + \alpha\gamma_{IK} + \beta\gamma_{JK}$ for the Data of Box 17.10

$\hat{\mu} = 1.99531$
$\hat{\alpha}_1 = 1.07331, \hat{\alpha}_2 = -0.02408, \hat{\alpha}_3 = -0.48883, \hat{\alpha}_4 = -0.56040$
$\hat{\beta}_1 = -0.06815, \hat{\beta}_2 = 0.06815$
$\hat{\gamma}_1 = 0.53032, \hat{\gamma}_2 = -0.53032$

Table of $\widehat{\alpha\gamma}$

	k	
i	1	2
1	0.16688	-0.16688
2	0.41323	-0.41323
3	-0.45702	0.45702
4	-0.12309	0.12309

Table of $\widehat{\beta\gamma}$

	k	
j	1	2
1	0.31292	-0.31292
2	-0.31292	0.31292

Up to now we have dealt only with expected frequencies and goodness-of-fit tests for the various models. It is also possible (and often of interest) to estimate the parameters themselves in the log-linear model (i.e., the α_i, β_j . . . values). For any given model, these values can easily be computed in the same way as in a corresponding anova model without replication.

For the final log-linear model in Box 17.10 let

$$l_{ijk} = \ln \hat{f}_{ijk} = \mu + \alpha_i + \beta_j + \gamma_k + \alpha\gamma_{ik} + \beta\gamma_{jk}$$

Then $\hat{\mu}$ is simply $\Sigma\Sigma\Sigma\, l_{ijk}/abc$, the mean of the l_{ijk}. The main effects are simply deviates from the mean: $\alpha_i = \bar{l}_i - \hat{\mu}$, $\beta_j = \bar{l}_j - \hat{\mu}$, and $\gamma_k = \bar{l}_k - \hat{\mu}$. The interaction terms are $\alpha\gamma_{ik} = \bar{l}_{ik} - \bar{l}_i - \bar{l}_k + \hat{\mu}$ and $\beta\gamma_{jk} = \bar{l}_{jk} - \bar{l}_j - \bar{l}_k + \hat{\mu}$. The numerical values of the estimates are given in Table 17.6. As a check one may verify that $\hat{f}_{111}$ can be computed as $\exp[1.99531 + 1.07331 + (-0.06815) + 0.53032 + 0.16688 + 0.31292] = \exp(4.01059) = 55.179$. Compare this value with 55.180 (identical within rounding error), given in the table of expected frequencies in part **II**, step **1** of Box 17.10.

As explained in the next section, log-linear models can be viewed as a special case of logistic regression models, just as we saw in Section 16.7 that anova models could be viewed as special cases of general linear models and could be solved as multiple regression problems.

17.6 Analysis of Proportions

In Section 17.4, which dealt with two-way tables, our major emphasis was on testing whether the two variables are independent. We will now present an alternative approach that can be used when one of the variables has just two states that can be interpreted as being dependent on the other variables in the study. This approach has the advantage that it can be generalized to include additional multistate or even continuous predictor variables. We shall focus on the proportions observed for the variable with just two states. Let us return to the immunology example of Section 17.4 to clarify this point. Recall that 44 of 57 mice who received bacteria *and* antiserum survived, whereas the survival rate for mice who received only bacteria was just 29 out of 54. We can express these results as proportions within the two treatment classes and arrange them in a table, as shown here:

	Dead	**Alive**	Σ
Bacteria and antiserum	$p_1 = 0.22807$	$q_1 = 0.77193$	1.0
Bacteria only	$p_2 = 0.46296$	$q_2 = 0.53704$	1.0

The proportion that survived under the treatment criterion is roughly 0.77; under the controls it is 0.54. How should we express this difference? We could describe it as a simple difference of $0.77 - 0.54 = 0.23$, but that has several disadvantages. The importance of such a difference depends on the magnitudes of the proportions being compared. An example will make this clear. Suppose the difference were only 0.01. If the survival proportion with bacteria alone were only 0.02, then an increase of 0.01 would be a change of 50%. If the survival proportion were 0.50, however, the change would be an unimpressive 2%. Perhaps a more important problem is that the standard deviation of a binomial proportion depends on the size of the proportion so that larger deviations are expected for proportions near 0.5 than for proportions near 0.02. An alternative way of expressing the difference in proportions is by stating them as **odds ratios.** If both bacteria and antiserum are administered, the odds that the mouse will survive is the ratio q_1/p_1 to 1. In our example, this is $0.77193/0.22807 = 3.38462:1$. A mouse is thus more than three times as likely to survive than to die if it receives antiserum along with its bacterial injection. The odds for survival with the treatment of only bacteria are $q_2/p_2 = 0.53704/0.46296 = 1.16000:1$. Thus, without antiserum, the odds for survival are only slightly greater than 1.

The ratio of these two odds is a useful way to compare the difference in odds. We evaluate the odds ratio, ω, as

$$\omega = \frac{q_1/p_1}{q_2/p_2} = \frac{3.38462}{1.16000} = 2.91778 \tag{17.18}$$

Thus, the odds for survival are almost three times better when antiserum is given than when it is withheld. An odds ratio of one would indicate no difference in the probabilities of survival under the two treatment conditions. Can we test the odds ratio for a departure from the null hypothesis of 1.0? Such a test is equivalent to a test of independence for a 2×2 table, so we can use the methods of Section 17.4. The natural log transform of the odds ratio, commonly called the *log-odds ratio*, is an important quantity that we will discuss shortly. Odds ratios are perhaps most meaningful in Model II two-way tables where we compare the outcomes for the two levels of the fixed treatment, but they have been used widely for the other models as well. In many applications—for example, epidemiology—both q_1 and q_2 are usually small. In such cases, $p_2/p_1 \cong 1$ and the quantity q_1/q_2, called the **relative risk,** will approximately equal the odds ratio, ω. Note that this use of the symbol ω is not the same as the measure of effect size used in Chapter 12. In Section 18.1 we will use the symbol o to distinguish a sample of the odds ratio from its parametric value. For simplicity, that is not done in Chapter 17.

In Section 14.11, we introduced the **logit transformation** of a proportion q, which is logit $q = \ln(q/p)$. We now realize that a logit is the natural logarithm of the *odds* for a given proportion. Note that when the observed $p = 0$ or 1, the logit cannot be computed. One method to avoid this problem is to use what are called **empirical logits,** $\ln[(a + \frac{1}{2})/(b + \frac{1}{2})]$, where a and b are observed frequencies that add up to n, the sample size. Another method is simply to replace the zeros by a small constant (such as $\frac{1}{2}$) and to decrease other frequencies by the same amount to maintain the original margin totals; that is, if we increase a by $\frac{1}{2}$, then we must decrease b by $\frac{1}{2}$.

Let us now examine the difference between the logit transforms of two proportions:

$$\text{logit } q_1 - \text{logit } q_2 = \ln\frac{q_1}{p_1} - \ln\frac{q_2}{p_2} = \ln\frac{q_1/p_1}{q_2/p_2} = \ln\omega$$

Note that the difference between the logits equals the logarithm of the odds ratio of the corresponding 2×2 table [see Expression (17.18)]. This relationship has made the **log-odds ratio** a very convenient statistic expressing the difference between two proportions. The odds ratio ω can be obtained from the log-odds ratio as $\omega = e^{\ln\omega}$. In the example of the mice, the odds ratio is 2.91778. The corresponding log-odds ratio is $\ln 2.91778 = 1.07082$. This quantity is also the difference between the logits of the two proportions $\ln q_1/p_1 - \ln q_2/p_2 = \ln 3.38462 - \ln 1.16000 = 1.21924 - 0.14842 = 1.07082$.

The odds ratio is easy to understand because it is in a "natural" scale. An odds ratio of 2.91778, for example, says that the odds that a mouse will survive are 2.91778 times greater if antiserum is included in their injection. If an odds ratio is less than one, then the odds of an event are reduced by that factor. By transforming the odds ratio into its logarithm, this simple interpretation is lost, but at the same time we obtain a quantity that is approximately normally distributed and that can be used in more complex models. When the proportions are equal, the odds ratio is 1 and the log-odds ratio is 0. The larger the log-odds ratio (regardless whether it is positive or negative), the greater the difference between the proportions.

For large samples, a good estimate of the standard error of the log-odds ratio can be obtained from the formula

$$s_{\ln\omega} = \left(\frac{1}{a} + \frac{1}{b} + \frac{1}{c} + \frac{1}{d}\right)^{1/2} \tag{17.19}$$

For the mouse data this value is 0.41729. Some authors prefer to add a continuity correction by adding $\frac{1}{2}$ to each of the four denominators in Expression (17.19). When we evaluate this continuity corrected standard error for the mice, we obtain the slightly smaller value 0.41190. Approximate 95% confidence limits for the log odds therefore are $1.07082 \pm 1.96(0.41190)$, or 0.26350 and 1.87814. If we back-transform these limits to the odds-ratio scale these limits become 1.30147 and 6.54135. The antiserum has clearly helped improve the odds for survival in the mice.

Another application of this standard error is to test the null hypothesis of no difference between the observed odds ratio and an established standard, ω_{st}. Suppose a veterinarian is deciding whether to inoculate the mouse colony against the bacterium in question by injecting antiserum but does not consider inoculation with antiserum worthwhile unless the odds for survival on infection with the bacterium are at least eight times those for mice not inoculated with antiserum. We compute

$$\frac{(\ln\omega - \ln\omega_{st})^2}{s_{\ln\omega}^2} = \frac{(1.07082 - \ln 8)^2}{0.41190^2} = 5.996$$

This ratio of a single squared difference over an estimated variance is distributed as chi-square with one degree of freedom. Because the odds ratio is less than 8, it may not be worthwhile to inoculate the mice even though the P-value is small, 0.014338.

For small samples, the standard error given by Expression (17.19) may not be very accurate. Various techniques exist for obtaining better estimates. All are computationally intensive. Interested readers should consult the reviews by Fleiss (1979) and Agresti (2001).

We now turn to odds ratios computed from replicated 2×2 tables, which may arise from a variety of designs. Here we will mention only two. In the first design, we consider the separate tables to be replicates sampled from the same underlying population. For example, we may wish to test the incidence of heart disease in samples from various cities. For each sample, we have also recorded whether the person was a smoker. Each 2×2 table then is smoker (yes/no) versus heart disease (yes/no) for a given city. We would like to test the homogeneity of the relation between the two variables over the cities (this is a test of conditional independence, as discussed in Section 17.5), to construct a joint estimate of the odds ratio from all the 2×2 tables (cities), and to test this estimate against the null hypothesis of an odds ratio of 1.

In the second design we suspect that the odds ratio is affected by a covariate, and we wish to control it by constructing separate 2×2 tables for different classes of the covariate. For example, the immunologist who has been inoculating mice with bacteria with or without antiserum suspects that the age of the mice plays a role in the outcome of the test. The experimenter consequently divides the test mice into three groups by age—immature, mature, and old—and divides each group roughly into two subgroups, experimentals and controls. The questions posed to the results are the same as in the first design. We will analyze an artificial example that extends the immunology experiment. The data and analysis are featured in Box 17.12.

Let us first establish the symbolism we will be using. We will designate the cells and margins in each table using the now familiar a, b, c, d, and n notation. Because there now are $k = 3$ tables, the frequencies of each table are subscripted to indicate the number $1, \ldots, k$ of the table or sampling stratum concerned. If we examine the proportion alive after treatment in the three experiments in Box 17.12, we note an apparent trend toward increasing survival in the older animals. Is this real? We first test whether there is homogeneity in the odds ratios of the three tables by calculating the odds ratio ω_i of each table using Expression (17.19) and then computing the natural logarithm of the ω_i. Heterogeneity of the odds ratios, symbolized by X_H^2, is tested with the formula

$$X_H^2 = \sum_i [w_i(\ln \omega_i - \overline{\ln \omega})^2] \tag{17.20}$$

This quantity is a weighted sum of squares that is distributed as chi-square with $k - 1$ (in this case, 2) degree of freedom. The weights, w_i, are inversely proportional to the variance of the log odds. In our example, $X_H^2 = 2.118$, $P = 0.346,802$, which does not lead to rejection of the null hypothesis despite the apparent trend toward higher

---

BOX 17.12 Test of Homogeneity of Odds Ratios for Replicated 2 × 2 Tables: The Mantel–Haenszel Procedure

Three experiments on immature, mature, and old mice, testing survival of injection with bacteria only versus injection with bacteria plus antiserum. There are $i = 1, \ldots, k$ strata. In this problem $k = 3$.

	Immature			Mature			Old		
	Dead	Alive	Σ	Dead	Alive	Σ	Dead	Alive	Σ
Bacteria only	26	29	55	43	41	84	15	23	38
Bacteria and antiserum	21	34	55	21	58	79	8	37	45
Σ	47	63	110	64	99	163	23	60	83

We examine the proportion alive in the 6 experimental groups (3 ages × 2 treatments) to obtain an overview of the results.

	Immature proportion alive	Mature proportion alive	Old proportion alive
Bacteria only	0.527	0.488	0.605
Bacteria and antiserum	0.618	0.734	0.822

Visual inspection indicates a trend toward higher survival in older mice, both with and without antiserum.

Mantel–Haenszel procedure to test odds ratios for homogeneity.

1. Compute odds ratio ω for each table. Rather than employ a formula based on proportions, such as Expression (17.18), these data can be computed most easily by the following equation:

$$\omega' = \frac{(a + 0.5)(d + 0.5)}{(b + 0.5)(c + 0.5)}$$

which is expressed in terms of cell frequencies. This formula includes continuity corrections of 0.5, which are indicated by the prime following the symbol for odds ratio. Thus,

$$\omega_1' = \frac{(26.5)(34.5)}{(29.5)(21.5)} = 1.441466$$

2. Transform the values of ω' to their natural logarithms, $\ln \omega'$. The results are the adjusted log-odds ratios. For example, $\ln \omega_1' = \ln 1.441466 = 0.365661$.

Box 17.12 (continued)

3. Compute weights, w, for the log odds. These weights are the reciprocals of the variances of the log odds, again corrected for continuity, as indicated by the prime:

$$w' = \frac{1}{\left(\dfrac{1}{a + 0.5} + \dfrac{1}{b + 0.5} + \dfrac{1}{c + 0.5} + \dfrac{1}{d + 0.5}\right)}$$

Thus

$$w'_1 = \frac{1}{\left(\dfrac{1}{26.5}\right) + \left(\dfrac{1}{29.5}\right) + \left(\dfrac{1}{21.5}\right) + \left(\dfrac{1}{34.5}\right)} = 6.796651$$

We summarize the quantities obtained in steps **1** through **3** in the following table.

	ω'	$\ln \omega'$	w'
Immature	1.441466	0.365661	6.796651
Mature	2.852059	1.048041	9.034196
Old	2.909887	1.068114	3.977993

4. Next compute the weighted mean of the adjusted log odds, $\overline{\ln \omega'}$ by the following formula:

$$\overline{\ln \omega'} = \frac{\sum\limits_i w'_i \ln \omega_i}{\sum\limits_i w'_i}$$

5. Finally, compute Expression (17.20), the weighted sum of squares of the adjusted log odds:

$$X_H^2 = \sum_i [w'_i (\ln \omega'_i - \overline{\ln \omega'})^2]$$

This quantity is distributed as chi-square with $k - 1$ degrees of freedom. When we compute this value for the data in this example, we obtain $X_H^2 = 2.118$, which for 2 degrees of freedom yields $P = 0.346,802$. We conclude that the log-odds ratios, and hence the odds ratios, are homogeneous, despite what appears to be a trend toward higher survival in older mice, and we are therefore encouraged to estimate an overall average odds ratio for our experiments. If the odds ratios are heterogeneous, separate estimates for each stratum (or at least for some strata) are indicated. To test for a trend, we can use logistic regression analysis (see text for explanation).

Mantel–Haensel Estimator

6. First compute Expression (17.21):

$$\omega_{MH} = \frac{\sum\limits_i a_i d_i / n_i}{\sum\limits_i b_i c_i / n_i}$$

Box 17.12 (continued)

which is the Mantel–Haenszel estimator of the odds ratios for the three strata. Thus,

$$\omega_{MH} = \frac{(26)(34)/110 + \cdots + (15)(37)/83}{(29)(21)/110 + \cdots + (23)(8)/83} = 2.303$$

We conclude that over the three experiments, the odds of survival for a mouse with antiserum were 2.3 times those for mice with bacteria alone.

This statistic is a function of the sums of one cell and of the margin totals of the k tables. It also contains a continuity correction. It is distributed as chi-square with one degree of freedom and tests the null hypothesis that there is no relationship between the presence of the antiserum and survival. When evaluated for these data, $X_{MH}^2 = 13.110$, which yields $P = 0.000,294$. We conclude that the two criteria are related in these three studies, and that in consequence, the joint odds ratio is greater than 1.

---

survival with age. This test does not take trend into account. To do so, logistic regression, discussed later in this section, must be used. The test of homogeneity of odds ratios is equivalent to the test of no three-factor interaction in a $2 \times 2 \times k$ three-way contingency table. For the data in Box 17.12, the methods of Box 17.10 yield $G_{ABC} = 2.161$ (2.125 with Williams's correction).

The next step is to estimate the joint odds ratio for the three experiments, using the formula

$$\omega_{MH} = \frac{\displaystyle\sum_i a_i d_i / n_i}{\displaystyle\sum_i b_i c_i / n_i} \tag{17.21}$$

The subscript MH refers to Mantel and Haenszel (1959), who in a seminal paper proposed this method of joint estimation, as well as a test for the estimate. For the three 2×2 tables in Box 17.12, the Mantel–Haenszel estimate is $\omega_{MH} = 2.303$. On the average, then, antiserum increases the odds for survival appreciably (2.303 times).

This estimate can be tested for difference from 1 (which would imply that on average factors A and B are independent within strata) by the following equation:

$$X_{MH}^2 = \frac{\left(\left| \displaystyle\sum_i a_i - \sum_i (a+b)_i(a+c)_i/n_i \right| - 0.5 \right)^2}{\displaystyle\sum_i (a+b)_i(a+c)_i(b+d)_i(c+d)_i/(n^3 - n^2)_i} \tag{17.22}$$

which is distributed as chi-square with one degree of freedom. For the data in Box 17.12, we obtain $X_{MH}^2 = 13.110$, which yields a very low P-value of 0.000,294. We conclude that there is a difference in survival on administration of the antiserum at all ages. This hypothesis can also be subjected to the test for conditional independence, $G_{AB(C)} - G_{ABC}$, which also has just 1 degree of freedom for these data. For the current example, this quantity is equal to $16.300 - 2.161 = 14.139$, $P = 0.000,170$. This implies that factors A and B are not independent within levels of factor C. One advantage of the Mantel–Haenszel test is that it can be carried out on 2×2 tables that contain very few individual observations, so long as the overall sample size, $\Sigma_i n_i$, of all strata is substantial (Agresti, 2007). It is important to note that the use of the Mantel–Haenszel estimate of the joint odds ratio assumes that the three-way interaction is not present.

Our final important topic for this section is **logistic regression.** Many situations arise in which researchers must investigate relations between a proportion and a continuous variable (with perhaps other variables taken into account). Examples are mortality as a function of dose of toxicant, morbidity as related to amount of exposure, parasitism as a function of density of intermediate hosts of the parasites, gene frequencies as affected by climatic variables, or survival as a function of age (as in the mouse example). Just as anova models can be viewed as a special cases of the general linear models (Section 16.7), log-linear models can be viewed as a special cases of logistic regression. Such an interpretation offers great flexibility in the analysis of categorical data and often permits additional insights into data, as we shall see.

Logistic regression relates the expected proportions $\hat{p}$ of a dependent variable to an independent variable X according to the following model:

$$\hat{p} = \frac{e^{a + bX}}{1 + e^{a + bX}} = \frac{1}{1 + e^{-(a + bX)}} \tag{17.23}$$

This looks more complex than it really is. It can be simplified by using the logit transform of a proportion, p, $\text{logit}(p) = \ln(p/(1 - p))$. First evaluate $1 - \hat{p}$ from Expression (17.23):

$$1 - \hat{p} = \frac{1 + e^{a + bX} - e^{a + bX}}{1 + e^{a + bX}} = \frac{1}{1 + e^{a + bX}} \tag{17.24}$$

Then

$$\frac{\hat{p}}{1 - \hat{p}} = \frac{e^{a + bX}/(1 + e^{a + bX})}{1/(1 + e^{a + bX})} = e^{a + bX} \tag{17.25}$$

Consequently,

$$\ln\left(\frac{\hat{p}}{1 - \hat{p}}\right) = \text{logit}(\hat{p}) = a + bX \tag{17.26}$$

Note that $\text{logit}(\hat{p})$ is related to X by simple linear regression (see Section 14.10). Thus, the logit is linear in its parameters, it may be continuous, and it ranges from $-\infty$ to $+\infty$. The error around the line, however, is not distributed normally.

Because the dependent variable is a proportion, the standard assumption is that it will follow the binomial rather than the normal distribution. For that reason, the general **maximum-likelihood method** is used to fit a regression line to logit-transformed data rather than the least-squares method as described in Chapters 14 and 16. Recall the notion of likelihood we encountered in Section 17.1. The maximum-likelihood method will result in the values for the parameters (a and b) that make the observed values in our data set seem most probable. Unfortunately, the solution to the nonlinear maximum-likelihood equations is complex and in practice can only be computed iteratively using a computer. We therefore do not furnish computational formulas in this book. We should point out that the maximum likelihood approach of fitting a regression line is the general method. It just so happens that for normally distributed data, likelihood is a function of the residual sum of squares so that the linear regression methods we employed in Chapters 14 and 16 would give the same results.

A typical logistic regression program, such as in the BIOMstat package, yields estimates for the intercept and regression coefficients, a and b, and their standard errors. Log-likelihood values are also usually given for models that include or exclude the regression coefficient. The difference in log likelihoods times -2 is a G-value that is approximately distributed as chi-square with as many degrees of freedom as the difference in the number of parameters in the two models being compared (one in the case of a single regression coefficient) and provides a test of the hypothesis that a parameter is equal to zero. Note that logistic regression need not be restricted to a single independent variable, as in Expression (17.26), but, as we will show later, it can be carried out as a multiple logistic regression using several independent variables.

An alternative test for a logistic regression coefficient is to compute the ratio of an estimated regression coefficient to its standard error and to compute its probability as a normal deviate (i.e., compare it to $t_{\alpha[\infty]}$). This test, known as a *Wald test*, usually gives results that are similar to those obtained with the G-test.

In the example in Box 17.13 we investigate whether variation in genotype frequencies at the ACP_1 locus among 13 villages in Sardinia can be predicted as a linear function of their elevation. A logistic regression program is essential for answering this question. A G-test of independence of the 13 by 2 contingency table indicates that we can reject the null hypothesis that the gene frequencies are homogeneous across all 13 villages. However, this test is unable to determine whether the variation in gene frequencies among villages can be accounted for by their differences in elevation. A logistic regression on elevation yields the model $\text{logit}(\hat{p}) = \hat{a} + \hat{b}X = -1.19585 + 0.00094964X$. Figure 17.5 shows an upward trend in the logit transform of the gene frequency as a function of elevation. One can test how well the model fits by comparing the observed frequencies with those predicted by this model. The expected frequencies are computed using the inverse of the logit transformation, $\hat{p} = 1/(1 + e^{-\text{logit}(\hat{p})})$, where $\text{logit}(\hat{p})$ is the logit value predicted by the regression, $\text{logit}(\hat{p}) = \hat{a} + \hat{b}X$. This yields a G-value of 19.35060 with 11 degrees of freedom (13 samples minus 2 degrees of freedom for the intercept and the regression coefficient). The P-value is 0.0551, indicating that even though we cannot reject the logistic regression model, the fit is not very impressive. Figure 17.6 shows a similar plot in terms of gene frequency. Some points look like outliers (especially the one with the highest gene frequency) but they are from especially small samples. Box 17.13 also shows the computations of confidence limits for the regression coefficient and their corresponding odds ratio.

The logistic regression coefficient is a log-odds ratio. Thus, its exponential, e^b, is an odds ratio and gives the factor by which the odds for the dependent variable

---

BOX 17.13 Logistic Regression on a Continuous Independent Variable

Data are frequencies of individuals with the A or BA phenotypes for the ACP_1 locus in 13 Sardinian villages. They were collected to determine the relationship, if any, between these phenotypes (dependent variable) and elevation (independent variable). The samples presented here are from a subset of persons bearing another genetic marker—individuals with ADA_12 or ADA_22 phenotypes for the ADA_1 locus.

(1)	(2)	(3)	(4)
	ACP_1		
Village	**A or BA**	**Other**	**Elevation in m**
1 Fonni	9	6	1000
2 Seulo	3	10	797
3 Aritzu-Belvi	3	10	796
4 Burcei	9	9	648
5 Lanusei	9	16	590
6 Bitti	4	16	550
7 Jerzu	3	7	442
8 Lode	5	2	345
9 Sedilo	1	8	228
10 Ottana	2	6	185
11 Villasimius	1	3	45
12 Tortoli	1	14	15
13 Oristano	9	17	10

SOURCE: Data from E. Bottini (unpublished results, ordered by decreasing elevation).

Analysis

1. The *G*-test for independence of village and genotype results in a *G*-value of 22.914 with 12 degrees of freedom. This yields a *P*-value of 0.028,465 so we can conclude that the gene frequency is not homogeneous—it differs among the villages. However, the purpose of the present analysis is to determine whether the differences can be accounted for by a linear function of the elevation differences among the villages.

   The maximum likelihood solution for the logistic regression of the proportion of the A or BA alleles on elevation converged quite rapidly for these data. The change in the estimated likelihood was less than 6×10^{-6} in the 4th iteration. A test of the overall fit of the logistic regression model yields $G = 19.351$ with 11 degrees of freedom (13 samples minus 2 degrees of freedom for the parameters). This corresponds to a *P*-value of 0.055,077. The logistic regression model fits the data but not very well. The log-likelihood for the model that includes estimates for both *a* and *b* was $-113.264,790,11$. The log-likelihood value for the model that only included the intercept, *a*, was $-115.046,342,64$. Their difference times 2 gives a *G*-value of

Box 17.13 (continued)

3.563 with 1 degree of freedom. This corresponds to a P-value of 0.059,081, which does not quite enable us to reject the null hypothesis that the slope is zero.

McFadden's (1974) measure of fit for a logistic regression is

$$R^{*2} = \frac{\ln L_0 - \ln L_1}{\ln L_0}$$

where $\ln L_0$ is the log-likelihood value for a logistic regression that includes only the intercept and $\ln L_1$ is the log-likelihood value for a model that includes the estimated regression coefficients for one or more independent variables. For the current example, this is

$$R^{*2} = \frac{-115.046,342,64 - (-113.264,790,11)}{-115.046,342,64} = 0.013,55$$

This is quite small, implying that relatively little of the variation has been accounted for by differences in elevation.

2. The estimated intercept and its standard error are $a = -1.195,85$ and $s_a = 0.298,39$. The Wald test for the intercept is $t_s = -1.195,85/0.298,39 = -4.007,6$ which corresponds to $P = 6.1 \times 10^{-5}$. The estimated regression coefficient and its standard error are $b = 0.000,949,64$ and $s_b = 0.000,509,831$. The regression coefficient is quite small, in part because the units are in terms of a change in just one meter of elevation. The Wald test for the regression coefficient is $t_s = 0.000,949,64/0.000,509,831 = 1.8627$ with $P = 0.0625$, not quite small enough for us to reject the null hypothesis that the slope is equal to zero. As usual, this result is similar to the test given above using likelihoods.

3. Figure 17.6 shows a plot of the data with the fitted regression line. The ordinate is the A or BA phenotype frequency expressed in logits. Figure 17.7 shows the same data with the ordinate expressed as a proportion. Using proportions, the prediction equation is computed by inverting the logit function

$$\hat{p} = 1/(1 + e^{-(a+bX)}) = 1/(1 + e^{-(-1.195,85 + 0.000,949,64X)})$$
$$= 1/(1 + 3.306,37(1.000,950,097)^X).$$

Although it is difficult to discern in the figure, the regression line is slightly curved. For each increase in elevation by 1 meter, the expected proportion of the A or BA phenotypes is increased by a factor of 1.000,9500,97 (or, more usefully, by a factor of 1.099,66 for each increase in elevation by 100 meters).

4. The confidence limits for a and b can be computed using their standard errors and the critical values from the normal distribution. The 95% confidence limits for the slope are

$$L_1 = b - t_{\alpha/2[\infty]}s_b = 0.000,949,64 - 1.960(0.000,509,831) = -0.000,049,824$$
$$L_2 = b + t_{\alpha/2[\infty]}s_b = 0.001,949,116$$

As expected, the interval includes zero because we were unable to reject the null hypothesis of a zero slope.

Box 17.13 (continued)

5. In terms of odds ratios, the limits found above can be exponentiated to yield the following 95% confidence limits for e^b

$$L_1 = e^{-0.000,049,824} = 0.99995$$

and

$$L_2 = e^{0.001,949,116} = 1.00195$$

The interval includes 1.0 because we could not reject the null hypothesis that the regression coefficient was equal to zero.

---

are increased or decreased for each unit change in the independent variable. In Box 17.13, it is for a difference in elevation of just 1 meter. This is too small to be of interest. We can multiply the regression coefficient by a scale factor, say c, to express it in more convenient units. The log-odds ratio then is cb, and the odds ratio is e^{cb} for units c times larger than the original units. For the current example, units of 100 m seem more useful. We obtained a regression coefficient of 0.000,949,64 in Box 17.13 for the $ADA_1 1$ genotype. If we multiply this by 100 we obtain a log-odds ratio of 0.094,964 for a change in 100 m. When we exponentiate this value, we obtain an odds ratio of 1.099,619. Thus, for every increase in elevation of 100 m, the expected odds for the designated ACP_1 genotype increases by a factor of almost 10%.

As in the regression analyses described in Chapters 14 and 16, analyzing the residuals from the fitted model is important. Removal of bad data points may greatly improve the fit, and the pattern of deviations may suggest other explanatory variables that can be added to the model. Note that the point at the upper left of Figure 17.5 is for the village of Lode and it seems to be an outlier. However, this point is based on a sample size of only seven, and thus large deviations are expected.

The squared multiple correlation coefficient is not appropriate here as a measure of overall fit of the regression. An alternative is McFadden's (1974) measure of fit defined as $R^{*2} = \dfrac{\ln L_0 - \ln L_1}{\ln L_0}$. We see in Box 17.13, step 1, that it is quite small, implying that relatively little of the variation in proportions has been accounted for by elevation (this is consistent with the large amount of scatter that can be seen in Figures 17.5 and 17.6). A number of other coefficients have been proposed to measure in standardized units how well a logistic regression fits the data. These are sometimes called pseudo R-squared coefficients. Another commonly used coefficient was proposed by Cox and Snell (1968) (with adjustments proposed by Nagelkerke, 1991, and Cragg and Uhler, 1970).

The independent variable X need not be a single continuous variable. There could be several continuous variables in a multiple logistic regression analysis. As in Section 16.6, the independent variables could be dummy variables that code various experimental designs such as multiway anovas and ancova. The analyses would then become the equivalent of those analyses for the analyses for proportions. As a simple

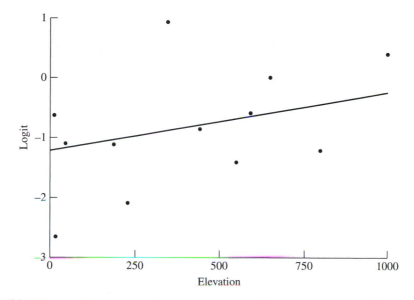

FIGURE 17.5 Logistic regression of frequency of the A or BA phenotypes of the ACP_1 locus against elevation. Data from Box 17.13. Ordinate in logit scale. Point at top corresponds to the village Lode.

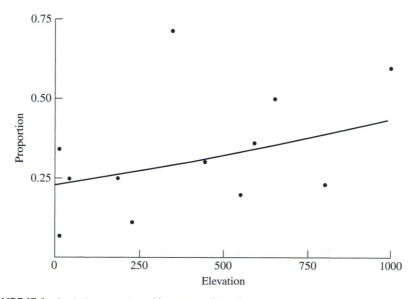

FIGURE 17.6 Logistic regression of frequency of the A or BA phenotypes of the ACP_1 locus against elevation. Data from Box 17.13. Ordinate in proportion scale. Although the regression line appears linear over the range of the independent variable shown, it is actually a curvilinear function. Point at top is for the village Lode.

example of a single-classification design with just two groups, suppose we have a sample of n individuals for whom we are interested in the proportion that have lung cancer. They are classified with respect to a single discrete variable—whether or not they are smokers. The following table relates the frequencies, f_{ij}, observed in a 2×2 contingency table to the parameters of a logistic regression.

			Smoking		
			$X = 1$	$X = 0$	
Lung cancer	$Y = 1$	$\hat{p}_1 = \dfrac{e^{a+b}}{1 + e^{a+b}}$		$\hat{p}_0 = \dfrac{e^a}{1 + e^a}$	f_{1+}
	$Y = 0$	$1 - \hat{p}_1 = \dfrac{1}{1 + e^{a+b}}$		$1 - \hat{p}_0 = \dfrac{1}{1 + e^a}$	f_{2+}
		f_{+1}		f_{+2}	$f_{++} = n$

In this table, 1 stands for presence and 0 stands for absence of the X or Y property. The values of $\hat{p}_1$ and $\hat{p}_0$ are the estimated proportions of lung cancer cases in smokers and nonsmokers, respectively. Recalling relations presented earlier in this section, we note that the odds of lung cancer are $\hat{p}_1/(1 - \hat{p}_1)$ for smokers and $\hat{p}_0/(1 - \hat{p}_0)$ for nonsmokers. The formulas in the cells are the logistic regressions, derived from Expressions (17.25) and (17.26), when X in turn assumes values of 1 and 0, respectively. Now recall that the odds ratio for a 2×2 table is $(q_1/p_1)/(q_2/p_2)$, which in this case could be rewritten as

$$\frac{\hat{p}_1/(1 - \hat{p}_1)}{\hat{p}_0/(1 - \hat{p}_0)}$$

When we substitute the logistic regression terms for the $\hat{p}$'s and cancel, we obtain $\omega = e^b$; thus,

$$\ln \omega = b$$

We have obtained the important result that for a dichotomous independent variable the logistic regression coefficient is the log-odds ratio, the difference between the two logits. This shows that when we exponentiate the logistic regression coefficient, we obtain an equation for the odds ratio for the data set in question. Knowing that the logistic regression coefficient yields the log-odds ratio for a difference of 1 in the units of the independent variable is useful for an interpretation of the results, even though it is not a practical method for computation of more complex designs.

When the independent variable is multistate categorical, with each of k states representing a separate group of the independent variable, we can obtain a solution by recoding it to $k - 1$ dummy variables for the k groups. Each dummy variable is binary and assigned a value of 1 to one of k groups and zero to all others. This differs from the coding used in Section 16.7 where one group (usually the last) was coded as -1. That coding could be used here but using just 0 and 1 leads to more convenient units. The example in Box 17.14 shows frequencies of the ACP_1 genotypes A + BA at four

BOX 17.14 | Logistic Regression on a Multistate Categorical
Independent Variable

Data extracted from the larger dataset which also served as a source for the data in
Box 17.13. These are counts of ACP_1 phenotype frequencies (in two classes, A or BA,
and Other) in four Sardinian villages. These data were taken only from individuals carrying $ADA_1 1$ phenotypes for the ADA_1 locus.

	Village				
	Fonni	Aritzu-Belvi	Lode	Oristano	Σ
A or BA	34	13	34	62	143
Other	66	94	85	123	368
Σ	100	107	119	185	511
Odds ratio, ω	1.0	0.26846	0.77647	0.97848	
Log-odds ratio, ln ω	0	−1.31505	−0.25300	−0.02176	

Computation

1. Because this table contains counts rather than proportions, we use the former to
obtain odds ratios and log-odds ratios. For example, for Aritzu-Belvi versus Fonni

$$\omega = \frac{13 \times 66}{34 \times 94} = 0.26846$$

We proceed similarly for each of the other two villages versus Fonni.

2. An alternative procedure is to subject the data to a multiple logistic regression program, by constructing three design variables as follows.

	Dummy variables		
Village	1	2	3
Fonni	0	0	0
Aritzu-Belvi	1	0	0
Lode	0	1	0
Oristano	0	0	1

The variables, serving as regressors of the proportion A or BA phenotypes, yield the
following results when subjected to a logistic regression program.

Likelihood values for models with just the constant term or the constant term plus
the dummy variables encoding differences among the villages are given below. Their
difference times 2 provides a *G*-test for the differences among the 4 villages.

Box 17.14 (continued)

Model	(1) Log likelihood	(2) df	(3) P
Intercept	$-302.923,557,22$	7	
Intercept + villages	$-292.861,082,99$	4	
Difference $\times 2 = G$	$20.124,948,46$	3	0.000,160

Note: The G-value computed above is identical to the value for a 2×4 test for independence for these data if treated as a contingency table.

The R^{*2} coefficient is relatively small,

$$R^{*2} = \frac{\ln L_0 - \ln L_1}{\ln L_0} = \frac{-302.923,557,22 - (-292.861,082,99)}{-302.923,557,22} = 0.03322$$

As shown below, logistic regression also provides estimates of the partial logistic regression coefficients and their standard errors for each dummy variable. Their ratio gives a Wald test for each coefficient. Only the first is large enough for us to reject the null hypothesis that the coefficient is equal to zero. The partial regression coefficients are log-odds ratios, and the last column gives the odds ratios. They all indicate a decrease in the odds for the A or BA phenotypes in the different villages compared to that for Fonni, but the first coefficient indicates a much stronger reduction in the odds (i.e., the expected proportion of the A or AB phenotypes will be much lower).

Village	$b_{Y_i \cdot jk}$	SE_b	$b_{Y_i \cdot jk}/SE_b$	e^b
Aritzu-Belvi	-1.31505	0.36349	-3.61785	0.26846
Lode	-0.25300	0.29281	-0.86402	0.77647
Oristano	-0.02176	0.26234	-0.08293	0.97848

The standard errors enable us to estimate confidence limits for the log-odds ratios, and, by extension, for the odds ratios. The 95% confidence limits for log-odds ratios are $b_{Y_i \cdot jk} \pm t_{.05[\infty]} SE_b$; 95% confidence limits for odds ratios can be computed as $e^{b_{Y_i \cdot jk} \pm t_{.05[\infty]} SE_b}$.

Sardinian villages. We consider the first village, Fonni, to be the referent sample, and the other three samples are evaluated with respect to it. The odds ratios are computed in a straightforward manner, following the established formula. We find that Aritzu–Belvi and Lode have substantially lower odds of possessing an A + BA genotype, but that Oristano has a genotype pattern very similar to that of Fonni. Box 17.14 also shows the design variables and the results of a multiple logistic regression of the genotype frequencies on the dummy variables. Computationally, the first approach is simpler and the overall test is identical to a test for independence, but the logistic regression enables us to perform other tests, compute odds ratios and their confidence limits, and to include continuous independent variables.

The data in Box 17.13 were the genotype frequencies from just those individuals that had the ADA_12 or ADA_22 genotypes at the ADA_1 locus. Box 17.15

BOX 17.15 | Test of Equality of Slopes in Logistic Regression Analysis

Data are ACP_1 phenotype frequencies in 13 Sardinian villages, collected to determine the relationship, if any, between genotype and elevation. Samples were from individuals who had the $ADA_1 1$ phenotype at the ADA_1 locus.

(1)	(2)	(3)	(4)
	ACP_1		
Village	**A or BA**	**Other**	**Elevation**
1 Fonni	34	66	1000
2 Seulo	27	85	797
3 Arozi-Belvi	13	94	796
4 Burcei	26	41	648
5 Lanusei	43	69	590
6 Bitti	26	93	550
7 Jerzu	16	29	442
8 Lode	34	85	345
9 Sedilo	29	75	228
10 Ottana	38	56	185
11 Villasimius	25	31	45
12 Tortoli	38	69	15
13 Oristano	62	123	10

SOURCE: Data from E. Bottini (unpublished results, ordered by decreasing elevation).

1. When a logistic regression on elevation is performed on these data using the methods described in Box 17.13, a regression coefficient of $b = -0.000,524,264$ was obtained with a standard error of $0.000,185,280$. The Wald statistic is $-2.829,569$ with a P-value of 0.0047. An obvious question to ask is whether one can reject the null hypothesis of equality with the slope obtained in Box 17.13.

2. Analogous to the procedure described in Section 16.8 to test homogeneity of slopes in multiple regression, the test for homogeneity of slopes in logistic regression first computes the log-likelihood for a model that allows the groups to have separate slopes and compares that to the log-likelihood for a model in which the groups are constrained to have the same slope. Twice the difference in log-likelihoods yields a value that can be compared to the χ^2-distribution with degrees of freedom equal to one less than the number of slopes being compared.

   The log-likelihood value for a model allowing for separate slopes is the sum of the likelihoods for the regressions fitted separately. It is found by either performing the regressions separately for the two groups and summing their log-likelihood values or by regressing on a design matrix with separate independent variables giving the

Box 17.15 (continued)

elevations for the two groups. In either case the following value would be obtained.

$$\ln L_{ADA_{,2} + ADA_{,2}} + \ln L_{ADA_{,1}} = -113.264{,}790{,}11 + (-817.205{,}051{,}08)$$
$$= -930.469{,}841{,}19$$

The log-likelihood value for a model with the two groups constrained to have the same slope is obtained by using a design matrix with only a single independent variable giving the elevation of each village. The log-likelihood value of $-934.270{,}777{,}14$ was obtained. Two times the difference between these log-likelihoods can be compared to a χ^2-distribution with one degree of freedom (one degree of freedom because just two slopes are being compared).

$$X^2 = -2(\ln L_{separate} - \ln L_{together}) = -2(-930.469{,}841{,}19 - [-934.270{,}777{,}14])$$
$$= 7.601{,}872$$

When compared to a $\chi^2_{[1]}$-distribution, a P-value of 0.005,831 is obtained, and thus we can reject the null hypothesis of equality of slopes.

provides the frequencies for individuals that had the $ADA_1 1$ genotype instead. For the combined dataset, the most obvious question to ask is whether the relationship between ACP_1 frequencies and elevation is the same as found in Box 17.13 for the other ADA_1 genotypes. The logical steps described in Box 17.15 are analogous to those described in Section 16.8 where we tested for equality of slopes using multiple regression analysis. In both cases, we compare the fit of a model with separate regression slopes within each group with a model that is constrained to force the slopes to be the same within all groups. The main difference is that in Section 16.8 the quality of fit was expressed in terms of residual sums of squares rather than log-likelihood values. Box 17.15 shows that the null hypothesis of equality of slopes can easily be rejected. Figures 17.7 and 17.8 show scatterplots of the combined data. Although there is much scatter, the differences in the trends in the two datasets is evident.

It is useful to compare these results with those of a log-linear analysis of the combined data treated as a 2 ADA_1 groups $\times$ 2 ACP_1 genotypes $\times$ 13 villages three-way contingency table. An essential difference is that in a logistic regression one factor must be defined as the dependent variable (the ACP_1 frequencies in the current example) and the others as independent variables, whereas in a log-linear analysis the variables are all treated in the same way. This means that fewer tests are appropriate in a logistic regression analysis—just those that relate the independent variables to the dependent variable and not tests of relationship just among the independent variables. The test for a three-way interaction is of interest because it can be interpreted as a test of whether the relationship between one of the dependent variables and one of the independent variables is consistent across different levels of the other

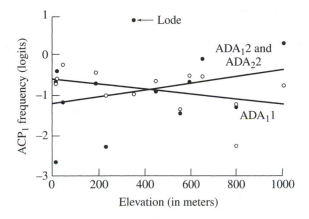

FIGURE 17.7 Logistic regression of frequencies of the A or BA phenotypes of the ACP_1 locus against elevation, separately for two ADA classes. Open circles used for $ADA_1 1$ allele and filled circles for $ADA_1 2$ and $ADA_2 2$ alleles. Data from Boxes 17.13 and 17.15. Ordinate in logit scale.

independent variable. Using the methods of Box 17.10, we find that we cannot quite reject the null hypothesis for the three-way interaction ($P = 0.0750$). This may seem inconsistent with the result in Box 17.15 that shows very different slopes. This discrepancy can be explained by the fact that the test for equality of two slopes with just 1 degree of freedom is more powerful than the test for a three-way interaction that has 12 degrees of freedom and tests for any type of deviation from parallel—not

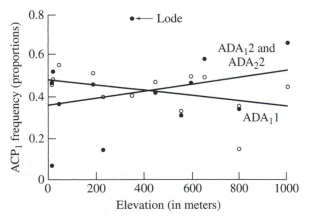

FIGURE 17.8 Logistic regression of frequencies of the A or BA phenotypes of the ACP_1 locus against locus elevation, separately for two ADA classes. Data from Boxes 17.13 and 17.15. Open circles used for $ADA_1 1$ allele and filled circles for $ADA_1 2$ and $ADA_2 2$ alleles. Ordinate in proportion scale. Although the regression lines appear linear over the range of the independent variable shown, they actually are curvilinear functions.

just linear. The log-linear analysis also tests for the overall independence of ACP_1 and ADA_1 and also within villages (conditional independence). Such tests are not part of the logistic regression analysis—just as in a multiple regression analysis one does not test whether the independent variables are correlated. However, the test for the independence of the $ACP_1 \times ADA_1$ across villages is rejected as well as the test for the homogeneity of $ACP_1 \times$ Village tables for the two ADA_1 genotypes. These results are consistent with the results of the logistic regression because the frequencies in villages differed as a function of elevation. Because it is able to take continuous covariates into account, the logistic regression approach is usually more powerful than an analysis as a three-way table when the independent variables are continuous and vary in a way that is at least approximately linear.

Now that you have learned about log odds, here is an additional method of looking at the analysis of a three-way table, such as we presented in Section 17.5. When one of the variables is of special interest and has only two categories (such as mortality: healthy versus poisoned, in Box 17.10), it is convenient to express the results in terms of the log odds of the estimated incidences as a function of the other variables. For example, the logarithmic estimated frequencies of the healthy flies are

$$\ln \hat{f}_{ij1} = \gamma_1 + \alpha\gamma_{i1} + \beta\gamma_{j1}$$

Those of the poisoned flies are

$$\ln \hat{f}_{ij2} = \gamma_2 + \alpha\gamma_{i2} + \beta\gamma_{j2}$$

Therefore, the log odds of being healthy versus poisoned are

$$\ln \frac{\hat{f}_{ij1}}{\hat{f}_{ij2}} = (\gamma_1 - \gamma_2) + (\alpha\gamma_{i1} - \alpha\gamma_{i2}) + (\beta\gamma_{j1} - \beta\gamma_{j2})$$

where the term $(\gamma_1 - \gamma_2)$ is equal to 1.06064. This value indicates that the overall odds ratio (or risk) of being in the healthy versus poisoned state is $\exp(1.06064) = 2.888$. The second term, $(\alpha\gamma_{i1} - \alpha\gamma_{i2})$, is a function of pupation site and takes on the following values: 0.33376 for IM, 0.82646 for AM, -0.95006 for OW, and -0.24618 for OM. Thus, the odds ratio is increased if the position is IM or AM and is decreased if the position is OW or OM. The last term, $(\beta\gamma_{j1} - \beta\gamma_{j2})$, gives the contribution of sex to the log odds: 0.62584 for ♀, -0.62584 for ♂ (indicating that being female increases the chances of a fly being in the healthy state). The relative risk of being healthy versus poisoned for a female pupating in the medium is then $\exp(1.06064 + 0.33376 + 0.62584) = \exp(2.02024) = 7.540$ (which agrees with the direct computation $\hat{f}_{111}/\hat{f}_{112} = 55.180/7.318 = 7.540$ using the $\hat{f}$ given in part **II**, step **1** of Box 17.10).

Section 17.8 provides methods for estimating the effect size and sample sizes needed for a desired level of power to detect a given minimum effect size.

For more information on the application of logistic regression analysis to more-complex designs (including generalizations to the case of dependent variables with more than two response categories), refer to Agresti (2002). For interpreting the results of logistic regression analyses, Hosmer and Lemeshow (1989) and Le (1998) may be useful.

In closing this section we would like to caution readers against several complications that may arise when some of the assumptions underlying tests of proportions are violated. The methods presented earlier assume that the proportions are binomially distributed but, as we saw in Section 5.2, some data seemed to have been sampled from contagious or clumped distributions. When sampling from such distributions, special methods are required that take into account the larger variances that are expected. See, for example, Le (1998).

Suppose it is suspected that a factor X predisposes toward disease A, and a retrospective study is carried out of patients with A in a hospital, investigators frequently will assemble a control group manifesting disease B. If the hospital admission rates differ for A and B, an apparent association may result between A and X, even if they are independent. This phenomenon is known as **Berkson's fallacy.** Another possible complication is that prospective studies (those that follow a cohort of individuals forward through time) and retrospective studies (those that examine events that occurred prior to a study) may yield quite different results as a function of differential duration and mortalities from disease. This case is known as **Neyman's example.** Finally, when several 2×2 tables are aggregated, an odds ratio that is similar in the 2×2 tables may become quite different when computed overall for a pooled contingency table. This case is known as **Simpson's paradox.** Workers should be aware of these (only apparent) anomalies and guard against them when designing their experiments. Goddard (1991) presents a readable discussion of these phenomena. Appleton et al. (1996) give an example where ignoring a covariate changes the direction of an association between the other two variables.

17.7 Randomized Blocks for Frequency Data

The tests of independence of Section 17.4 can be regarded as tests of the effect of treatments on the percentage of an attribute. In that section, each treatment was applied to independently and randomly selected individuals that were different for the separate treatments. Thus, in the immunology example, two samples of mice were taken and subjected to different treatments—bacteria alone, or bacteria and antiserum. These mice had presumably been selected randomly from the pool of animals available to the investigator. In the example dealing with the two leaf types found on two different types of soils, the samples were obviously separate, consisting of different groups of trees on the two soils.

Sometimes collecting data in this manner is neither possible nor desirable. For example, consider an experiment that compares the phototactic response of a species of ticks under various experimental conditions. We could take a sample of ticks, divide it into several groups, and test each group under a separate set of conditions.

What if we knew, however, that the behavior of the organisms varies, depending on factors such as age or condition of the animal? In such a case, by chance more older individuals may be placed into one group than into the other, or some other sampling accident may take place. Thus, treatment effects might be confounded with heterogeneity of the experimental subjects. There are three possible solutions to such a problem. If the factors causing heterogeneity (such as age or prior condition) are known, we can try to keep them constant. Using this approach, to obtain animals of uniform age, we may have to select from a large base population of animals of various ages, which could be difficult or expensive or both.

If the factors causing heterogeneity of response are unknown, there are two possible approaches. One is simply to take large enough samples to overcome the heterogeneity. For obvious reasons, this solution is often either not feasible or not desirable.

The alternative procedure, which is the subject of this section, is to use the same individuals for both treatments. Using this design, even if there is some heterogeneity in the inherent responses to the organisms, one can still look for and test for a significant *change* in their behavior due to the treatment. This same experimental design was discussed in Section 11.4 (randomized blocks). Whereas that section dealt with continuous variables that can be analyzed by analysis of variance, here we are considering attributes for which we can record only the frequency or percentage of individuals possessing a certain characteristic. As in its previous application, the method here is based on the correlation between treatments over blocks (which correspond to individuals in the preceding example) and assumes the absence of treatment × block interaction. Because frequently the same sample of individuals is exposed to two or more treatments, this method is also known as *repeated testing of the same individuals*. Another appropriate name for this technique is *testing of correlated proportions*.

This design is also indicated when there is some reason to believe that a prior treatment will affect the responses in a subsequent treatment. In such cases, clearly the same sample must be used because the effects of one treatment have to be compared with the effects of the prior treatment in the same individuals. An example is conditioning or learning in animals. Given a certain stimulus, individuals in a group of animals may or may not respond. If the same or a different stimulus is given at a later time, the presence or absence of a response in these animals may be conditioned by their previous response.

Our first example of repeated testing of the same individuals is the case of two treatments (only one repetition of the test). For continuous variables, this design is one of paired comparisons (see Section 11.5). This is a special case of randomized blocks in which there are only two treatments; hence each individual or block is tested twice. By analogy, for attributes we have the **McNemar test for significance of changes.** Box 17.16 shows an example in which 30 ticks were each tested twice for their phototactic response—first in a 1-inch arena and then in a 2-inch arena. The data consist of 30 paired scores. A plus was recorded if the tick left the arena toward the light, a minus if it moved away from the light. The

BOX 17.16 Randomized Blocks for Frequency Data (Repeated Testing of the Same Individuals or Testing of Correlated Proportions)

Individuals tested twice; McNemar test for significance of changes

The response of the rabbit tick (*Haemaphysalis leporispalustris*) to light was measured by placing individual ticks in arenas 1 or 2 inches in diameter and recording whether the tick left the arena on the side toward (+) or away from (−) the light. Each tick was scored first in the 1-inch arena and then in the 2-inch arena. Of interest was whether the response of an individual to light would change depending on the distance a tick had to crawl before its response was recorded.

The results can be summarized as follows:

	1-inch arena		
2-inch arena	−	+	Totals
−	8	5	13
+	9	8	17
Totals	17	13	30

SOURCE: Data from V. E. Nelson (unpublished results).

Note that the pattern of response appeared to reverse itself. In the 1-inch arena 13 out of 30 ticks turned toward the light, whereas 17 of the 30 did so in the 2-inch arena.

We label the cells of this table in the conventional manner:

	−	+	Σ
−	a	b	$a + b$
+	c	d	$c + d$
Σ	$a + c$	$b + d$	n

We wish to test whether the change in proportion between the two trials is significant. The null hypothesis is that the frequencies of the two types of "changers" are equal—that is, that the number of ticks that followed a (+) response with a (−) equaled those following a (−) response with a (+). If this condition is satisfied, the proportions in the two trials must be the same. Concisely stated, the null hypothesis is H_0: $(b − c) = 0$. The appropriate test statistic is computed by Expression (17.27):

$$G = 2\left[b \ln\left(\frac{2b}{b + c} \right) + c \ln\left(\frac{2c}{b + c} \right) \right]$$

$$= 2\left[5 \ln\left(\frac{2(5)}{5 + 9} \right) + 9 \ln\left(\frac{2(9)}{5 + 9} \right) \right] = 1.15894$$

It is advisable to apply Williams's correction, which in this case is $q = 1 + 1/2(b + c)$. Thus

$$q = 1 + \frac{1}{2(14)} = 1.03571$$

$$G_{adj} = G/q = 1.15894/1.03571 = 1.119, \quad P = 0.290,134$$

Box 17.16 (continued)

These G-values approximate the distribution of χ^2 with one degree of freedom. Clearly, there is little evidence of a significant change in behavioral response.

Individuals tested three or more times; Cochran's Q-test

The data are scores expressing the condition of 24 plants (*Acacia cornigera*) observed in three different months. All of these plants were from an experimental plot in which the growing shoots had been deprived of the ant *Pseudomyrmex ferruginea*, which normally lives within the stems. A positive rating represents a plant that has all or most of its uppermost shoot apices still intact and in which most of the foliage does not show the feeding damage of phytophagous insects. A negative rating represents a plant that has all of its growing shoot apices eaten off and in which more than half of the foliage shows severe damage. In the complete study (not shown here) there was also a control plot.

1. The data are arranged as a two-way table, using 0's for negative ratings and 1's for positive ratings. The $b = 24$ rows (blocks) of this table correspond to the individual plants scored, and the $a = 3$ columns (treatments) correspond to the repeated scoring of these plants in different months.

Individual plants (blocks)	Month in which scored (treatments)			Totals
	March	**June**	**August**	
1	0	1	0	1
2	0	0	0	0
3	0	1	0	1
4	1	0	0	1
5	1	1	0	2
6	1	0	0	1
7	0	0	0	0
8	1	1	0	2
9	1	1	0	2
10	1	0	0	1
11	1	0	0	1
12	1	1	0	2
13	0	1	1	2
14	0	1	1	2
15	0	0	0	0
16	0	1	0	1
17	0	1	0	1
18	1	0	0	1
19	1	0	0	1
20	1	0	0	1
21	1	0	1	2
22	1	1	1	3
23	1	0	0	1
24	1	1	0	2
Totals	15	12	4	31

SOURCE: Data from D. H. Janzen (unpublished results).

Box 17.16 (continued)

2. Compute the row and column sums and record them along the margins of the table.

3. Grand total of the frequencies in the table $= \sum\limits^{a}\sum\limits^{b} Y = 31$

4. Sum of the row sums squared $= \sum\limits^{b} \left(\sum\limits^{a} Y \right)^2$

$$= 1^2 + 0^2 + 1^2 + \cdots + 2^2 = 53$$

5. Sum of the column sums squared $= \sum\limits^{a} \left(\sum\limits^{b} Y \right)^2 = 15^2 + 12^2 + 4^2 = 385$

6. Test statistic [Expression (17.28)] $= Q$

$$= \frac{(a-1)\left[a\sum\limits^{a}\left(\sum\limits^{b} Y \right)^2 - \left(\sum\limits^{a}\sum\limits^{b} Y \right)^2 \right]}{a\sum\limits^{a}\sum\limits^{b} Y - \sum\limits^{b}\left(\sum\limits^{a} Y \right)^2}$$

$$= \frac{(a-1)[a \times \text{quantity } \mathbf{5} - (\text{quantity } \mathbf{3})^2]}{(a \times \text{quantity } \mathbf{3}) - \text{quantity } \mathbf{4}}$$

$$= \frac{(3-1)[3 \times 385 - 31^2]}{(3 \times 31) - 53} = 9.700$$

This value is to be compared with a χ^2-distribution with $a - 1 = 2$ degrees of freedom. The P-value is 0.007,828, and thus we reject the null hypothesis that the proportion of plants in good condition remained the same as the season progressed. It is reasonable to conclude, therefore, that when the plants are deprived of their protecting ants, their condition gradually deteriorates. Many of the plants in the treatment plot died a few months later, but the control plot showed little change.

30 paired observations can be conveniently recorded in a 2 × 2 table as shown in the box and repeated here:

	1-inch arena		
2-inch arena	−	+	Σ
−	8	5	13
+	9	8	17
Σ	17	13	30

Unlike the situation in Section 17.4, where we were interested in the proportions implied by the cells of the table, we are now more interested in the proportions in the margins. For example, it is no longer of particular interest that 8 of the 17 ticks that had a negative response in the 1-inch arena also responded negatively in the 2-inch arena. We are more interested in the fact that in the 1-inch arena 17 out of 30 ticks showed a negative reaction, whereas in the 2-inch arena 13 out of 30 ticks showed a negative reaction. We could rearrange the table as follows:

	−	+	Σ
1-inch arena	17	13	30
2-inch arena	13	17	30
Σ	30	30	60

This arrangement would imply, however, that our sample size is $n = 60$ ticks, twice the actual number of ticks involved. For this reason, we cannot use the methods of Section 17.4 on such data. The proportions within this second table are not independent of one another; they are all based on the same 30 ticks.

Inspecting the earlier table, we note that the difference between the proportions in the margins is a function only of the number of ticks that are "changers"— those that responded differently to the second test (the 2-inch arena) than they did to the first test. If there are no changers, or if the two types of changers are equally frequent, the marginal frequencies will be the same for both sets. In our example, 5 ticks changed from a positive to a negative reaction in the 2-inch arena, and 9 changed from a negative to a positive response. We may state as our null hypothesis that the proportion of individuals reacting negatively will be the same in both sizes of arenas. This statement implies that we expect the frequencies of the two types of changers to be equal. We can easily construct a G-test of this hypothesis. If we label the four cells of the table in the conventional manner (a, b, c, d), we can construct the following simple formula for G (approximating χ^2 with one degree of freedom):

$$G = 2\left[b \ln\left(\frac{2b}{b + c}\right) + c \ln\left(\frac{2c}{b + c}\right) \right] \tag{17.27}$$

This formula is derived easily from Expression (17.3). For the data in Box 17.16, $G = 1.15894$ and $P = 0.281,685$. Following our recommendation in Section 17.2 and Box 17.1, we apply Williams's correction to G, although in this case, with a P-value above the conventional 5%, the correction will not affect our decision about accepting the null hypothesis. In the box, we show that $G_{adj} = 1.118, P = 0.290,350$. In view of these results, we are unable to reject the null hypothesis that there is no change in response of the ticks between the 1-inch and the 2-inch arenas.

Though not done in Box 17.16, when samples are small $[(b + c) < 25]$, an "exact" binomial test is recommended. Such a test is carried out as follows. If we let k equal $b + c$, and let Y equal b or c, whichever is greater, then the desired probability is twice the probability of observing Y or more events of one type out of k from a binomial distribution with $p = q = 0.5$. We can employ the methods of Section 13.11 to evaluate this probability, either computing the probability or inferring it from the confidence limits for $\pi = 0.5$ in Statistical Table **P**.

In the tick experiment described in Box 17.16, all ticks were tested first in the 1-inch arena and then in the 2-inch arena. The experimenter has therefore confounded two possible effects, size of arena and order of test. The results suggest that more ticks were light-positive in the 2-inch arena (but recall that we could not substantiate this trend with our statistical test). However, this potential effect could be due to the fact that by the time ticks were placed into a 2-inch arena, they had already been tested once before. If we wish to control for the possibility of such an effect, we can subject half the ticks to the 1-inch arena first, followed by the 2-inch arena, and reverse the order of arenas for the other half (this is called a crossover design). Even though we employ equal frequencies for the two orders of treatment, however, we cannot be certain that these occur in equal frequency in the two types of changers. How can we analyze such a situation? The following table is an artificial data set intended to describe an imagined repetition of the tick experiment with a larger number of ticks, 100:

	1-inch arena		
2-inch arena	−	+	Totals
−	81	9	90
+	1	9	10
Totals	82	18	100

The data in this example are based on an experiment by Gart (1969), who developed the technique we are about to describe. In this dataset, there are only a few light-positive responses among the ticks. Of 10 changers in this table, 9 are positive in a 1-inch arena but negative in a 2-inch arena, and only 1 is negative, then positive. When we carry out McNemar's test and calculate the exact one-tailed probability of this outcome on the null hypothesis that the number of changers in each direction is equal, we obtain $P = 0.011$. Thus, there is evidence that more ticks are light-positive in the 1-inch arena than in the 2-inch arena. Until now, we have not considered the order of presentation of the arenas. The following table describes the minor diagonal of the preceding 2×2 table—that is, the changers—by tabulating the responses to first and second stimuli (treatments) against the order in which the two arenas were presented.

	Order of administration		
	First 1-inch then 2-inch	First 2-inch then 1-inch	Totals
Positive response to first treatment	5	1	6
Positive response to second treatment	0	4	4
Totals	5	5	10

Note that the total n equals 10, the number of changers. If we carry out Fisher's exact test (one-tailed) on this table (see Box 17.7), we obtain $P = 0.0238$, which suggests that the response of the first versus the second treatment is not independent of what the treatment is (i.e., 1-inch or 2-inch arena). From inspection of the table we conclude that the 1-inch arena induces more light-positive responses than one would usually expect to find by chance. Now let us rearrange this table to indicate responses to the two arenas as these relate to which one was offered first.

	Order of administration		
	First 1-inch then 2-inch	First 2-inch then 1-inch	Totals
Positive response to 1-inch arena	5	4	9
Positive response to 2-inch arena	0	1	1
Totals	5	5	10

We obtain a table with the same total counts and column totals, but the row totals have changed. A one-tailed Fisher's exact test yields $P = 0.5$. We conclude that order of administration is independent of type of treatment and plays no role in the response of the ticks.

A method for comparing the changes in two groups, each tested twice, is illustrated by Feuer and Kessler (1989). When the responses of the individuals can occur in $k = 3$ or more classes, rather than the two that we have considered, and the individuals are tested twice, the data can be arranged as a $k \times k$ square table. The diagonal frequencies, f_{ii}, of this table would list the frequencies of individuals that responded twice identically, and the off-diagonal frequencies, f_{ij}, would be those in class i for the first test and class j for the second test (or vice versa for frequencies f_{ji}). Such a design could also arise when two judges classify the same group of individuals into k classes, or when n males divided into k morphs mate with an equal number of females divided into the same k morphs. Maxwell (1970) shows how the margin frequencies of such tables can be compared and how to test the null hypothesis $H_0: f_{ij} = f_{ji}$ for all i and j.

When individuals have been tested three or more times for the presence or absence of an attribute, a test analogous to the randomized-blocks design is possible. This test is called **Cochran's Q-test.** Its test statistic is based on a model that differs from those used for previous G-tests. The original observations are replaced by ones and zeros ($Y = 1$ corresponds to the presence of the attribute; $Y = 0$ corresponds to the absence of the attribute). The resulting data are entered in a two-way table and analyzed according to the standard equations for a randomized-blocks design. For tests of significance, however, the usual test statistic is modified so that the results are expected to follow the chi-square rather than the F-distribution. If we let A stand for the columns (or the treatment effects) and B for the rows (or blocks) of the randomized-blocks design (the individuals), the test statistic Q is computed as follows:

$$Q = \frac{SS_A}{(SS_{total} - SS_B)/(df_{total} - df_B)}$$

Because the data consist only of ones and zeros, a great deal of simplification is possible, resulting in the following equation which is less intuitive but more convenient for computation:

$$Q = \frac{(a-1)\left[a\sum^a\left(\sum^b Y\right)^2 - \left(\sum^a\sum^b Y\right)^2\right]}{a\sum^a\sum^b Y - \sum^b\left(\sum^a Y\right)^2} \tag{17.28}$$

Q is expected to follow the chi-square distribution with $a - 1$ degrees of freedom (where a is the number of times each of the b individuals has been tested). There does not appear to be a G-test equivalent to the Cochran Q-test.

In the second half of Box 17.16, Cochran's Q-test is applied to data showing that acacias gradually deteriorate when an ant species that normally lives within its stems is removed. The test shows that there is a significant change in the frequency of plants in "good condition" when the plants are observed in successive months after having been deprived of the ants which normally prevent herbivores from feeding on the tender growing shoots. See Box 17.16 for details.

> The derivation of Expression (17.28) is given in Cochran (1950) who also describes an exact binomial test for Q. Patil (1975) provides a table of critical values for the case where $a = 3$ and describes the procedure for performing exact tests for $a > 3$.

17.8 Effect Sizes, Power, and Sample Sizes

There are many types of tests covered in this chapter. Thus, different procedures are needed to measure effect size and to estimate power and required sample sizes. Another complication is that the relevant distributions are discrete. Although this makes exact computations possible for small sample sizes, various approximations are usually used for medium to large sample sizes.

The simplest case is where one wishes to test whether a sample proportion could have been drawn from a population with a specified proportion, π. The procedure for estimating power for detecting a specified alternative proportion is given in Section 7.1. Table 7.1 shows the probabilities under the null and alternative hypotheses which allows one to compute power. Figure 7.3 shows these distributions graphically. To find the sample size necessary to obtain a specified power, one would increase or decrease an initial guess of the sample size until the desired power is achieved. This procedure can also be used to estimate power and sample size for the sign test

(Section 13.11) where one compares an observed proportion to the value expected under the null hypothesis, $\pi = 0.5$. When the sample size is large, computational effort can be reduced by approximating the binomial distribution by a normal distribution with the same mean and variance as the binomial.

One of the simpler cases is the **comparison of two proportions.** One could just use the difference between the two proportions, $\Delta_\pi = \pi_1 - \pi_2$ as the measure of effect size. The first part of Box 17.17 shows how this can be used to compute power and estimate sample sizes. The method is based on principles similar to those given in part **2** of Box 12.2 and discussed in Section 12.4. The formula for n was developed by Casagrande et al. (1978). It is an approximation that yields an explicit expression for the sample size, so iteration is not necessary to obtain a solution, as it was in Section 12.4. It also differs because one does not need to provide an estimate of the variance. For the binomial distribution, the variance is a function of the true proportion. This means that it is important to have a good estimate of the proportions and not just their difference. Box 17.17 also illustrates how to estimate the required sample size for a test of the difference between two proportions when one of them is a theoretical rather than a sampled value.

A statistically simpler, though a perhaps less intuitive measure of effect size, is obtained by using the difference between the arcsine transformations of these proportions—that is, $\Delta_\theta = \theta_1 - \theta_2$, where $\theta_i = \sin^{-1}\sqrt{p_i}$ for the ith observed proportion. The angles can be expressed in either degrees or radians but the formulas given in this section will assume radians because they are more commonly used, lead to simpler expressions and more convenient units. The use of arcsines takes into account that confidence intervals for proportions near 0.5 are much wider than those for proportions near 0 or 1. Cohen (1988) offers a subjective scale for the magnitudes of Δ_θ values. Values around 0.1 or less are considered small effect sizes and values around 0.4 or more are considered large effect sizes (note that Cohen's cutoff values were actually twice as large because he used arcsines multiplied by 2, to achieve effect size cutoffs about the same as for anova and correlation). The second part of Box 17.17 shows how this approximation can be used to compute power and estimate sample sizes.

Both approaches described earlier for measuring effect size divide the effect size by its standard error and treat the result as a standardized normal deviate. Power for a 2-tailed test is computed as the probability that an observation from a normal distribution will exceed the critical values of $\pm t_{\alpha[\infty]}$ for the distribution expected under the null hypothesis. Figure 7.10 can be viewed as an illustration of this method. Box 17.13 also includes the case where one of the proportions is given by theory rather than being subject to sampling error. The sign test is a special case.

The formulas for sample size find the sample sizes that correspond to confidence intervals that just barely exclude zero. In many applications, they may be so broad that many other proportions are not excluded. For that reason, it may be more useful to determine a sample size that yields a confidence interval of a specified width. There is no formula for this. One must simply try various values of n until a sufficiently narrow interval is obtained.

Different approaches are needed when there are **more than two proportions to be compared.** Several distinct cases can be distinguished. The first case corresponds

---

BOX 17.17 | Power and Sample Size Required to Detect a Given True Difference Between Two Proportions

I. *Using the Difference Between Two Proportions*

The power of a test for detecting a difference, $\Delta_\pi = \pi_1 - \pi_2$, between two true proportions π_1 and π_2 ($\pi_1 > \pi_2$) with sample sizes of n_1 and n_2 and a type I error rate of α can be computed as follows. Let

$$\bar{\pi} = \frac{n_1\pi_1 + n_2\pi_2}{n_1 + n_2}$$

$$z_1 = \frac{-t_{\alpha[\infty]}\sqrt{\bar{\pi}(1 - \bar{\pi})(n_1 + n_2)/n_1n_2} - (\pi_1 - \pi_2)}{\sqrt{\pi_1(1 - \pi_1)/n_1 + \pi_2(1 - \pi_2)/n_2}}$$

and

$$z_2 = \frac{t_{\alpha[\infty]}\sqrt{\bar{\pi}(1 - \bar{\pi})(n_1 + n_2)/n_1n_2} - (\pi_1 - \pi_2)}{\sqrt{\pi_1(1 - \pi_1)/n_1 + \pi_2(1 - \pi_2)/n_2}}$$

Power is now computed as the sum of the probabilities to the left of z_1 and to the right of z_2.

Let us apply this method to the following example: What power should be expected for detecting a true difference between the proportions, $\pi_1 = 0.65$ and $\pi_2 = 0.55$, using samples of $n = 100$ observations in each sample and a type I error rate of $\alpha = 0.05$? Because sample sizes are equal,

$$\bar{\pi} = \frac{0.65 + 0.55}{2} = 0.60$$

Then

$$z_1 = \frac{-1.95996\sqrt{0.6(1 - 0.6)(100 + 100)/100(100)} - (0.65 - 0.55)}{\sqrt{0.65(1 - 0.65)/100 + 0.55(1 - 0.55)/100}} = -3.42121$$

and

$$z_2 = \frac{1.95996\sqrt{0.6(1 - 0.6)(100 + 100)/100(100)} - (0.65 - 0.55)}{\sqrt{0.65(1 - 0.65)/100 + 0.55(1 - 0.55)/100}} = 0.51930.$$

These correspond to probabilities of 0.000,312 and 0.301,776, respectively, giving power equal to just 0.302,088.

Turning this computation around, we can estimate the sample size necessary to achieve a desired level of power to detect a specified difference, Δ_π, between two true proportions.

Let

$$\bar{\pi} = (\pi_1 + \pi_2)/2$$

$$A = [t_{\alpha[\infty]}\sqrt{2\bar{\pi}(1 - \bar{\pi})} + t_{2\beta[\infty]}\sqrt{\pi_1(1 - \pi_1) + \pi_2(1 - \pi_2)}]^2$$

Box 17.17 (continued)

where $t_{\alpha[\infty]}$ and $t_{2\beta[\infty]}$ are critical values from a two-tailed t-table. Then

$$n = \frac{A[1 + \sqrt{1 + 4\Delta_\pi/A}]^2}{4\Delta_\pi^2}$$

Applying this method to the example given above we can determine what sample size is required in each of two samples in order to obtain a power of 0.80.

$$\bar{\bar{\pi}} = \frac{0.65 + 0.55}{2} = 0.60, t_{0.05[\infty]} = 1.95996, 2\beta = 2(1 - 0.8) = 0.40,$$
$$t_{0.40[\infty]} = 0.84162$$

$$A = [1.95996\sqrt{2(0.6)(1 - 0.6)} + 0.84162\sqrt{0.65(1 - 0.65) + 0.55(1 - 0.55)}]^2$$

$$= 3.75565$$

Then

$$n = \frac{3.75565[1 + \sqrt{1 + 4(0.65 - 0.55)/3.75565}]^2}{4(0.65 - 0.55)^2} = 395.31217$$

Thus, 396 observations would be needed in each sample.

If we wish to determine the sample size when one of the proportions, say π_1, is a theoretical rather than a sample value (as in genetic research), we simply redefine the quantity A as follows:

$$A = [t_{\alpha[\infty]}\sqrt{\pi_1(1 - \pi_1)} + t_{2\beta[\infty]}\sqrt{\pi_2(1 - \pi_2)}]^2$$

For example, suppose we want to plan an experiment in which the expected ratio is 3:1 (thus, $\pi_1 = 0.75$). What sample size should we use to obtain a power of 0.80 for detecting another ratio, such as a 1:1 ratio ($\pi_2 = 0.50$), with a type I error rate of $\alpha = 0.05$?

$$A = [1.95996\sqrt{0.75(1 - 0.75)} + 0.84162\sqrt{0.5(1 - 0.5)}]^2 = 1.61163$$

Then

$$\Delta_\pi = \pi_1 - \pi_2 = 0.75 - 0.50$$

and

$$n = \frac{1.61163[1 + \sqrt{1 + 4(0.75 - 0.50)/1.61163}]^2}{4(0.75 - 0.5)^2} = 33.306$$

Therefore, 34 observations are required.

II. *Using the Arcsine Transformation of Proportions*

When the proportions are not close to 0.5, more accurate results can be obtained using arcsine transformed proportions.

$$\Delta_\theta = \theta_1 - \theta_2$$

Box 17.17 (continued)

Assuming the proportions are independent and the angles are expressed in radians, the variance of Δ_θ is

$$s_{\Delta_\theta}^2 = \frac{1}{4n_1} + \frac{1}{4n_2} = \frac{n_1 + n_2}{4n_1 n_2}$$

For equal sample sizes, this reduces to $s_{\Delta_\theta}^2 = \frac{1}{2n}$.

For a two-tailed test, power can be computed as the probability of obtaining a standardized normal deviate less than or equal to

$$z_1 = -\frac{\Delta_\theta}{s_{\Delta_\theta}} - t_{\alpha[\infty]}$$

plus the probability of obtaining a standardized normal deviate greater than or equal to

$$z_2 = -\frac{\Delta_\theta}{s_{\Delta_\theta}} + t_{\alpha[\infty]}$$

Using the example from part **I**, the power expected for detecting a difference between the proportions, $\pi_1 = 0.65$ and $\pi_2 = 0.55$, using $n = 100$, and a two-tailed type I error rate of $\alpha = 0.05$ can be computed as follows:

$$\Delta_\theta = 0.93774 - 0.83548 = 0.10226, \quad s_{\Delta_\theta}^2 = \frac{1}{2(100)} = 0.005$$

Power is the area under the normal curve to the left of $z_1 = -\dfrac{0.10226}{\sqrt{0.005}} - 1.95996 =$

-3.40618, which equals 0.00033, plus the area to the right of $z_2 = -\dfrac{0.10226}{\sqrt{0.005}} +$

$1.95996 = 0.513{,}785$, which gives a probability of 0.303701. Thus power is $0.00033 + 0.303701 = 0.30404$, a rather low value.

We can estimate the sample size necessary to achieve a desired level of power to detect a specified difference, Δ_θ, between two arcsine values using the simple formula

$$n = 0.5\left(\frac{t_{\alpha[\infty]} + t_{2(1-P)[\infty]}}{\Delta_\theta}\right)^2$$

For example, to estimate the sample size that should yield power of 0.80 to detect the Δ_θ-value of 0.10226 given above, compute

$$n = 0.5\left(\frac{1.95996 + 0.84162}{0.10226}\right)^2 = 375.3$$

This indicates that each sample should be of size 376. This is somewhat less than the earlier estimate of 396 using proportions.

to goodness-of-fit tests in which a set of observed frequencies are compared to those expected by a model—for example, genotypic ratios or expected distributions such as the normal or binomial distributions. A complication is that three or more proportions can differ in many ways. Cohen (1988) proposed the following measure of an overall effect size:

$$w = \sqrt{\sum \frac{(\pi_i - \hat{p}_i)^2}{\hat{p}_i}} \tag{17.29}$$

where the $\hat{p}_i$ are the proportions expected under the null hypothesis and they are the true proportions as specified by the alternative hypothesis. If the expected proportions are equal to the true proportions then the effect size will be zero. The expression under the square root sign resembles the expression for computing an X^2-test for goodness of fit but proportions are used rather than frequencies. Expression (17.29) can also be written as $\sqrt{X^2/n}$. Effect size can also be estimated based on the G-test

$$w = \sqrt{2 \sum \pi_i \ln\left(\frac{\pi_i}{\hat{p}_i}\right)} = \sqrt{\frac{G}{n}} \tag{17.29a}$$

This usually results in very similar values for w. Other measures of effect size may be more useful in particular applications—especially when one expects particular patterns of differences among the proportions perhaps mostly affecting a subset of the proportions.

Confidence intervals for w are found by transforming the effect size, w, to the noncentrality parameter, $\chi^2_{\alpha[\nu,\lambda]}$, for the noncentral chi-squared distribution, computing its confidence interval, and then transforming its confidence limits back to effect sizes. The parameter can be obtained from w by using the relationship

$$\lambda = nw^2 \tag{17.30}$$

Simple algebra can show that estimates for λ based on the X^2 or G-tests are just equal to those statistics themselves. For example, substituting $w = \sqrt{G/n}$ into $\lambda = nw^2$ yields

$$\lambda = n(\sqrt{G/n})^2 = G$$

Similar estimates will usually be obtained if the X^2-statistic is used instead. Confidence limits for λ are found by computing the upper and lower $\alpha/2$ quantiles for the noncentral χ^2-distribution. These are transformed back to effect sizes using the relationship $w = \sqrt{\lambda/n}$. Examples are shown in Box 17.18.

Two methods can be used to estimate the power of a test or the sample size needed in order to achieve a desired level of power. The simplest is to use the graphs in Statistical Table **JJ**. First select the graph corresponding to the degrees of freedom and type I error rate. One can then read off an estimate of power from where a value

---

BOX 17.18 | Power and Sample Size Required to Detect Specified True Differences Among Two or More Proportions

I. *Goodness-of-Fit Tests*

The data in Table 17.2, reporting the results of a genetic cross in tomatoes, can be used as an example for computing the effect size, w, its confidence interval, and the sample size needed to expect to be able to reject the null hypothesis.

The first step is to express the frequencies corresponding to the null and alternative hypotheses to proportions. First, we convert the expected 9:3:3:1 ratios in Table 17.2 to the expected proportions, π_i. We will use the observed data in that table as an example of the type of deviations from expectation that are of interest. When expressed as proportions of the total, they are denoted p_i. The effect size is then computed as

$$w = \sqrt{\sum \frac{(p_i - \pi_i)^2}{\pi_i}} = \sqrt{\frac{(0.574798 - 0.5625)^2}{0.5625} + \cdots + \frac{(0.06456 - 0.0625)^2}{0.0625}}$$

$$= 0.03019$$

This is a rather small effect size, so it is not surprising that even with a sample size of 1611 we were unable to reject the null hypothesis.

In order to compute its confidence interval we must first estimate the noncentrality parameter, λ, for the chi-squared distribution. This is

$$\lambda = nw^2 = 1611(0.03019)^2 = 1.46872$$

The effect size and the noncentrality parameter can also be estimated from the G-test for goodness of fit.

$$w = \sqrt{2\sum p_i \ln\left(\frac{p_i}{\pi_i}\right)} = \sqrt{2\sum 0.574798 \ln\left(\frac{0.574798}{0.5625}\right) + \cdots + 0.06456 \ln\left(\frac{0.06456}{0.0625}\right)}$$

$$= 0.03029$$

$$\lambda = 2n\sum p_i \ln\left(\frac{p_i}{\pi_i}\right) = n(w)^2 = 1611(0.03029)^2 = 1.47759$$

These are very similar to the values found above.

The power of the G-test given above can be computed if software for the noncentral chi-square distribution is available. Power is computed as the probability of obtaining an X^2 or a G-value equal to or larger than $\chi^2_{\alpha[\nu]}$ from the noncentral chi-square distribution with the same degrees of freedom but with the noncentrality parameter λ—that is, $power = 1 - \beta = P\{\chi^2_{\alpha[\nu]} \geq \chi^2_{\alpha[\nu,\lambda]}\}$. For the current example, this is $P\{7.81473 \geq \chi^2_{0.05[3,1.47759]}\} = 0.15120$.

Alternatively, power can be estimated using the graphs in Statistical Table **JJ**. To use these graphs, select the graph corresponding to the desired degrees of freedom and type I error rates. Then see where the curve closest to the sample size intersects the appropriate value for the effect size along the abscissa. Power is read off along the ordinate. The graph confirms that power is very low. This was expected despite the large sample size because the fit of the data to the 9:3:3:1 ratio was very good.

Although there is no direct formula for the necessary sample size to achieve a specified level of power, one can repeat the computations for power with larger sample sizes until the desired level of power is achieved. To do this, the effect size is held constant

Box 17.18 (continued)

but the noncentrality parameter is increased using the relationship $\lambda = nw^2$. For the current example and using the w-value estimated from the G-test, the sample size would have to be increased to 11,887 for power to exceed 0.80. For less extreme data, sample sizes can be estimated using Statistical Table **JJ**. One simply finds the sample size curve that comes closest to the intersection of the expected effect size along the abscissa and the desired power along the ordinate. For the current example, one can only confirm that an effect size of 0.03 and a desired power of at least 0.80 would require a sample size much larger than 2000.

Confidence limits for λ can be estimated by finding the noncentrality values, λ_1 and λ_2, such that the observed λ-value corresponds to the upper and to the lower $\alpha/2$ critical values for the noncentral chi-square distribution. By computation, using the values obtained from the G-test, these were found to be 0 and 6.68697, respectively. The lower limit is 0 because with a sample size of 1611 there is no value of λ that would make the observed noncentrality parameter correspond to an upper critical value, $\chi^2_{\alpha/2[\nu,\lambda]}$. These limits are then transformed to confidence limits for the effect size using the relationships

$$w_1 = \sqrt{\frac{\lambda_1}{n}} = \sqrt{\frac{0}{1611}} = 0 \quad \text{and} \quad w_2 = \sqrt{\frac{\lambda_2}{n}} = \sqrt{\frac{6.68697}{1611}} = 0.06443.$$

II. *Tests of Independence and Logistic Regression*

As described in the text, the methods for tests of independence and the overall fit of a logistic regression are just special cases of goodness-of-fit tests. The examples given here are just shown for completeness.

The 4×2 contingency table from Box 17.8 is a convenient example. As reported in that box, $G = 28.59642$ with 3 degrees of freedom. The total sample size is $n = 671$ and thus $w = \sqrt{G/n} = \sqrt{28.59642/671} = 0.20644$, an intermediate effect size.

The noncentrality parameter for the chi square distribution is $\lambda = nw^2 = G = 28.59642$. As before, the power expected in a new experiment with the same effect size as found in the current data is defined as $P\{\chi^2_{\alpha[\nu]} \geq \chi^2_{[\nu,\lambda]}\}$, where $\chi^2_{\alpha[\nu]} = \chi^2_{0.05[3]} = 7.81473$. Using the graph in Statistical Table **JJ** for $\nu = 3$ and $\alpha = 0.05$, power is clearly very high, almost equal to 1. By computation using the noncentral χ^2-distribution, power is equal to 0.99752. Because power is so high, it is interesting to see how small a sample size could be used while still achieving a power of, say, 0.80. Using Statistical Table **JJ**, we can see that it would be somewhat above $n = 200$. By computation, we find that a sample size of just $n = 256$ would be required on average.

Using the method described in part **I**, the 95% confidence interval for λ is from 9.88779 to 51.05902. These can be transformed to confidence limits for w using the relationships

$$w_1 = \sqrt{\frac{\lambda_1}{n}} = \sqrt{\frac{9.88779}{671}} = 0.12139 \quad \text{and} \quad w_2 = \sqrt{\frac{\lambda_2}{n}} = \sqrt{\frac{51.05902}{671}} = 0.27585$$

Although these limits give a clear indication of the likely magnitude of the effect size, they could be narrowed by increasing the total sample size.

When computing G or w for a logistic regression analysis, the p_i are the proportions expected when the alternative hypothesis is true and the π_i are the proportions predicted after fitting the logistic regression model to the p_i.

for w along the abscissa intersects a curve for sample size. Sample sizes needed to achieve a desired level of power can be found by locating the sample size curve that intersects or is slightly above the intersection of effect size along the abscissa and the desired power along the ordinate.

If the effect size or sample size is beyond the range of these graphs or if more precise results are desired (although not usually required), then an indirect computational approach must be used based on the noncentral χ^2-distribution. The approach is analogous to that described in Section 12.3 based on the noncentral F-distribution.

Power is computed by using an estimate of λ to find the probability $1 - \beta = P\{\chi^2_{\alpha[\nu]} \geq \chi^2_{[\nu,\lambda]}\}$, where $\chi^2_{\alpha[\nu]}$ is the critical value from the ordinary central χ^2-distribution and $\chi^2_{[\nu,\lambda]}$ represents values from the noncentral χ^2-distribution for the specified effect size. Although tables of $\chi^2_{\alpha[\nu,\lambda]}$ exist, the most practical method of computation is to use software such as BIOMstat. Sample sizes needed in order to achieve a desired level of power for detecting a specified effect size can then be found by trying different sample sizes until one is found for which power is sufficient. When sufficient, the graphs available in Statistical Table **JJ** require much less effort.

When the proportions being compared are not independent, as in the McNemar Cochran Q-tests (Section 17.7), different procedures are required. Lachin (1992) compares several methods estimating power and sample sizes for the McNemar test and recommends the approximation of Connor (1987) as being fairly accurate and slightly conservative. Although Patil (1975) gives the exact distribution of Cochran's Q-statistic, practical methods do not seem to have been developed for computing power and estimating sample sizes.

An $R \times C$ **test for independence** is a special case of a goodness-of-fit test. The same measures of effect size, w, can be used. Power and confidence limits for the effect sizes are computed as described for goodness-of-fit tests. One just needs to take into account that the expected proportions are computed as the product of the marginal proportions and that the numbers of degrees of freedom are computed as $\nu = (r - 1)(c - 1)$ rather than as the number of cells minus 1. The noncentrality parameter for the noncentral chi square distribution still satisfies the relationship $\lambda = nw^2$, where n is the total sample size across all cells in the table. Part **2** of Box 17.18 provides an example of the computation of effect size, power, and sample size for a contingency table.

A number of other measures of association or effect size have been proposed for contingency tables—especially for 2×2 tables. These other coefficients have more useful interpretations in particular applications. If a coefficient can be expressed as a function of w, then it may be possible to transform confidence intervals for the noncentrality parameter into confidence intervals for the coefficient as was shown earlier for w.

The computations for **logistic regression** are also just special cases of the methods presented earlier. The effect size measure, w, is used in a logistic regression analysis to indicate the strength of the overall regression as well as for different subsets of individual independent variables. The effect size for the overall regression is computed as

$$w = \sqrt{2 \sum \pi_i \ln\left(\frac{\pi_i}{\hat{p}_i}\right)}$$

where the π_i are the proportions expected if the alternative hypothesis is true and the $\hat{p}_i$ are the proportions predicted after fitting the logistic regression model to the π_i. Note that the effect size can also be computed directly from the corresponding G-test using the relationship $w = \sqrt{G/n}$. Power is computed using either the graphs in Statistical Table **JJ** or by computing power using the noncentral χ^2-distribution with $\lambda = nw^2 = G$.

As in the case of contingency tables, other measures of effect sizes may be useful in particular applications. A natural one to consider in a logistic regression is the odds ratio or its logarithm.

Additional information about this topic is available from several sources. Chapter 7 of Agresti (2002) and Agresti (2007) are helpful. Many comprehensive statistical packages contain options for computing power and estimating sample sizes for categorical data. For example, PASS (www.NCSS.com), Stata (www.stata.com), SAS (www.sas.com), SPSS (www.spss.com).

There do not seem to be any simple formulas to estimate power and sample sizes for the Kolmogorov–Smirnov test (either for extrinsic or extrinsic hypotheses). Breton et al. (2008) used Monte Carlo simulations to show that the power of the Kolmogorov–Smirnov test relative to other tests for goodness of fit to a normal distribution depends on the skewness and kurtosis of the alternative distribution. On the other hand, if the purpose of the test is to detect a difference in location for a normal distribution then Knott (1970) provides a simple approximation. Gleser (1985) describes a method for the exact computation of power when fitting to a discrete distribution but not for a practical approximation.

EXERCISES 17

17.1   In an experiment to determine the mode of inheritance of the *green* mutant, 146 wild-type and 30 mutant offspring were obtained when F_1 generation houseflies were crossed. Test whether the data agree with the hypothesis that the ratio of wild types to mutants is 3:1.

   ANSWER: $G_{adj} = 6.43715$, $X^2_{adj} = 5.523$.

17.2   Locality A has been sampled extensively for snakes of species S. An examination of the 167 adult males that have been collected reveals that 35 of these have pale-colored bands around their necks. From locality B, 90 miles away, we obtain a sample of 27 adult males of the same species, 6 of which show the bands. What is the chance that both samples are from the same statistical population with respect to frequency of bands?

17.3 Of 445 specimens of the butterfly *Erebia epipsodea* from mountainous areas, 2.5% have light color patches on their wings. Of 65 specimens from the prairie, 70.8% have such patches (unpublished data from P. R. Ehrlich). Is this difference greater than we would expect by chance? Calculate the odds ratio. If you have access to a logistic regression program, assign confidence limits to the odds ratio.

ANSWER: $G = 175.516$, $G_{adj} = 171.453$.

17.4 The following data on the sex ratios of adult long-billed marsh wrens in Seattle were presented by Verner (1964).

	Date	Males	Females
	May 6	14	11
	May 11	14	12
	May 25	14	12
1961	June 13	13	11
	June 19	13	9
	June 25	12	8
	July 8	9	6
	April 18	8	8
	May 22	6	8
1962	May 31	6	7
	June 2	6	7
	June 20	8	8
	July 23	7	5

Test whether there is a 1:1 sex ratio for each date and whether the observations on different dates are homogeneous. If heterogeneity is indicated, perform an unplanned test to obtain sets of dates homogeneous for sex ratios. Is there any evidence of heterogeneity between years?

17.5 In a study of polymorphism of chromosomal inversions in the grasshopper *Moraba scurra*, Lewontin and White (1960) gave the following results for the composition of a population at Royalla "B," Australia, in 1958.

		Chromosome CD		
		St/St	St/Bl	Bl/Bl
Chromosome EF	Td/Td	22	96	75
	St/Td	8	56	64
	St/St	0	6	6

Are the frequencies of the three different combinations of chromosome EF independent of those of the three combinations of chromosome CD?

ANSWER: $G = 7.396$, $q_{min} = 1.008,008$.

17.6 In an experiment, stem mothers of the aphid *Myzus persicae* were placed on one of three diets one day before giving birth to nymphs (data from Mittler and Dadd, 1966).

Test whether the percentage of nymphs that developed into winged forms depends on the diet of the mother.

Type of diet	% Winged forms	n
Synthetic diet	100	216
Cotyledon "sandwich"	92	230
Free cotyledon	36	75

17.7 Test agreement of observed frequencies to those expected on the basis of a Poisson distribution for the data given in (a) Table 5.5, (b) Table 5.6, (c) Table 5.7, (d) Table 5.9, and (e) Table 5.10.

ANSWER: (a) $G = 40.152$. When using more precise values of than are given in Table 5.5, one obtains $G = 40.408$.

17.8 Test agreement of observed frequencies to those expected on the basis of a binomial distribution for the data given in Tables 5.1 and 5.2.

17.9 For the pigeon data of Exercise 2.4 test the agreement of the observed distribution to a normal distribution by (a) using the Kolmogorov–Smirnov test on the ungrouped data and (b) using G-tests and chi-square tests for goodness of fit to the expected frequencies calculated in Exercise 6.1.

ANSWER: (a) $g_0 = 0.08105$ and $g_1 = 0.08168$.

17.10 Repeat Exercise 17.9 with the pigeon data as transformed in Exercise 2.6. Use the expected frequencies computed in Exercise 6.2.

17.11 Calculate expected normal frequencies for the butterfat data of Exercise 4.3. Test the goodness of fit of the observed to the expected frequencies by means of the Kolmogorov–Smirnov test.

ANSWER: $g_0 = 0.05757$ and $g_1 = 0.05709$.

17.12 If a standard drug cures disease X 60% of the time, what sample size would be required in an experiment designed to have an 80% chance of detecting (at the 5% Type I error level) a new drug that would cure disease X (a) at least 65% of the time, (b) at least 70% of the time, and (c) at least 80% of the time?

17.13 The following table of data is from a study of caste differentiation in the ant *Myrmica rubra* by Brian and Hibble (1964).

Frequencies of larvae of two sizes in three colonies at weekly intervals

	Week	Small larvae	Large larvae
Colony 1	2	12	47
	3	19	29
	4	9	52
	5	5	42
	6	8	58
	7	2	43
	8	0	35
	9	7	50
	10	9	38

Frequencies of larvae of two sizes in three colonies at weekly intervals (continued)

	Week	Small larvae	Large larvae
Colony 2	2	0	47
	3	28	36
	4	2	44
	5	1	45
	6	4	33
	7	2	46
	8	3	38
	9	3	33
	10	5	50
Colony 3	2	55	23
	3	71	4
	4	40	21
	5	29	9
	6	42	17
	7	26	8
	8	28	21
	9	29	13
	10	57	7

Are the three factors (A = larval size, B = colonies, and C = weeks) jointly independent? If not, test if the factors taken two at a time are independent. Is the degree of association between weeks and larval size homogeneous over the three colonies—that is, is there any interaction?

ANSWER: $G_{A,B,C} = 724.750$, 42 df; $G_{ABC} = 30.174$, 16 df (using the Williams correction; see Section 17.2).

17.14 In studying the effect of 1% indoleacetic acid in lanolin on fruit setting in muskmelon, Burrell and Whitaker (1939) obtained the results in the following three-way table.

		Number of flowers that did or did not set fruit			
		Treatment		Control	
		Fruit set	Fruit not set	Fruit set	Fruit not set
	1	13	10	5	23
	2	2	10	1	7
Replication	3	13	1	6	10
	4	5	6	5	9
	5	13	5	6	14

Is the effect of the treatment on fruit setting independent of the effect of different replications? Within each treatment are the replicates independent of fruit setting?

Examine the Freeman–Tukey deviates for the model that tests for conditional independence of treatment versus fruit setting, given the replicates.

17.15 A doctor had 25 patients who suffered from disease X during a recent epidemic. Although antibiotic B appeared to have no curative value against X, only 6 out of 11 patients administered B developed secondary infections, but 11 out of 14 patients not given B developed such infections. Would the doctor be justified in believing that the administration of B was beneficial to his patients? Compare the results of the G-test (using Williams's correction) with those of the chi-square test using the continuity correction. Calculate the odds ratio. If you have access to a logistic regression program, assign confidence limits to the odds ratio.

ANSWER: $X^2_{adj} = 0.716$.

17.16 Perform a logistic regression of the log odds of being male on date, using the data from Exercise 17.4. For the regression, replace the calendar dates with their equivalent number of Julian days of the year. Test whether the relationship is the same for both years.

17.17 In F_2 crosses between progeny of yellow-green and chlorophyll-deficient strains of lettuce, Whitaker (1944) obtained the following results.

Family number	Yellow-green	Chlorophyll-deficient
15487	482	156
15395	92	40
15488	28	12
15396	256	96
15397	158	47

Test agreement to a 3:1 ratio. Are the proportions of the two color types homogeneous over the families?

ANSWER: $G_H = 3.6101$, $P = 0.4613$.

17.18 In a study of the relationship between mammary tumors and dietary fiber and fat in rats, Cohen et al. (1991) obtained the following data. (Each treatment class contained 30 rats.)

Diet	No. rats with tumors
High fat	27
High fat + fiber	20
Low fat	19
Low fat + fiber	14

Compute the odds ratios of interest and their confidence limits. Test whether there is evidence of differences among the treatments. Also test whether the effect of fat content of the diet is independent of the effect of fiber in the diet.

17.19 The following results from a logistic regression of the effects of sex, race, and tar on the frequency of Kreyberg I lung cancer were extracted from the study by Harris et al. (1993). The logistic regression coefficients have been adjusted for age, year of diagnosis, alcohol consumption, body size, education, and other variables in the original study. The regression coefficients for tar exposure are relative to zero exposure.

Variable	Regression
Sex (F = 1, M = 0)	0.05
Race (black = 1, white = 0)	0.48
Cumulative (lifetime) tar exposure (in kg):	
1−4	1.94
5−8	3.02
≥8	3.76
Tar × Sex:	
1−4, F	0.23
5−8, F	0.58
≥8, F	0.58

Estimate the odds ratio for a white male who has accumulated 5 to 8 kg of tar in his lifetime. Compare this value with that of a black male who has accumulated the same amount of tar.

ANSWER: The odds for a white male with 5–8 kg of tar is 20.491 whereas for a black male it is 33.112.

17.20 The following data were taken from a case-control study by Zang and Wynder (1992) on the risk of lung cancer among cigarette smokers. Control individuals were patients admitted to the same hospital on the same day, matched as far as possible with respect to age and gender, but with different diagnoses. Adjustments for age are not included here (as they were in the published study) in order to simplify the computations.

Most recent number of cigarettes per day	Males Kreyberg I lung cancer	Control	Females Kreyberg I lung cancer	Control
Never smoked	15	822	22	941
1–10	36	136	20	118
11–20	133	328	122	193
21–40	226	311	156	115
≥41	127	91	40	17

Compute the odds ratios and log-odds ratios for having lung cancer relative to those who have never smoked (separately for each sex). Illustrate graphically. Analyze using the current level of smoking as a multistate categorical variable. Do the two sexes show the same relationships?

17.21 Reanalyze Exercise 17.20 using logistic regression analysis with the current level of smoking as an independent variable. Use the following arbitrary numerical values for this variable for the classes: 0, 5.5, 15.5, 31.5, and 45.5. Do the two sexes show the same form of relationship? Superimpose the logistic regression function on a plot of the data, as in Figures 17.5 and 17.6. Does the regression fit the data very well? Try fitting the regression with the "Never smoked" category removed. Estimate the effect sizes for the regression.

ANSWER: $b_0 = -2.8337$, $b_1 = 0.0829$, $b_2 = 0.3875$. Fit is very poor, $G = 207.714$, $df = 7$.

17.22 Using the data in Exercise 17.3, estimate the effect size of sampling from different localities. Express as both proportions and arcsines.

17.23 Estimate the effect size for the test of independence in Exercise 17.18.

18 Meta-Analysis and Miscellaneous Methods

We conclude our introduction to biometry by presenting 11 methods that do not easily fit into any of the previous chapters. All of them are useful in a variety of situations.

Six of the methods are found in Section 18.1 and deal with various approaches to the task of summarizing prior results. Methods for combining the outcome of several experiments to obtain an overall statement of the likelihood of the null hypothesis have attracted considerable attention among scientists and in news media in recent years. Such methods have received the general name of meta-analysis. In Section 18.2, we discuss runs tests, which test for randomness of nominal data in a sequence of events. In Section 18.3, we revisit the problem of a regression applied to the group means in an anova. Requiring the means to be a linear function of some independent variable is a rather strict requirement for an alternative hypothesis. A slight relaxation of this strict requirement has been named *isotonic regression*. Sections 18.4 and 18.5 complete our discussion of randomization tests. In the former section, we illustrate how we can employ randomization (in this case, by complete enumeration) to tests of unconventional, often self-devised statistics. In the latter section, we use randomization to test the association of two distance matrices, the well-known Mantel test. Finally, in Section 18.6, we discuss the outlook for future developments in biometry and introduce the concept of data analysis, which promises considerable progress in biometric research.

18.1 Synthesis of Prior Research Results: Meta-Analysis

Because statistical methodology has been the standard practice for many years in most fields of biological research, reviews of prior work are now often at least partly statistical rather than just being verbal summaries of the literature in a field. These statistical summaries of prior work are usually called *meta-analyses*, because they are statistical analyses of the results of other statistical analyses. The use of these methods has become quite standard in medical research and in the social sciences. It is now also becoming common in biological research. For example, Gurevitch et al. (2001) survey applications of meta-analysis in ecology. The analyses are used to test the consistency of prior results and whether there are convincing overall trends in the prior studies. Meta-analysis is also sometimes necessary because it may not be feasible financially or logistically to carry out a single experiment large enough to be able to detect an important but subtle effect.

In any summarization of prior research, one should be concerned that earlier studies with inconclusive results or with results that are contrary to what is expected may not have been published. In such a case, a summarization of prior results will give a biased estimate of the true mean effect and also its variance. One method aiming at detection of this problem is to construct a so-called funnel plot. It is a scatterplot of an estimate of the effect size against some measure of the precision of the estimated effect size (such as its standard error, reciprocal of its standard error, or even just its sample size). Its name comes from the fact that the scatter is expected to be funnel shaped because studies based on large sample sizes will usually have smaller standard errors and effect sizes that should be closer to their true values than studies with small sample sizes and thus larger standard errors and estimates of effect size that should vary widely. Figure 18.1**a** is an example based on a simulation using different sample sizes and a true effect size of 1. Effect sizes are expected to be symmetrically distributed around the mean effect size with the width of the scatter dependent on sample size. If, for example, the pool of published studies does not include those where a one-tailed test of the null hypothesis of zero difference was accepted, as shown in Figure 18.1**b**, then the scatter would be expected to be asymmetrical because the distribution would not include studies with small or negative effect sizes. In such a case, the average effect size would be much larger than it should be. Egger et al. (1997) proposed a test for asymmetry using a regression of the effect size on the reciprocal of its standard error. A number of modifications have been proposed since then, in part because of reports of inflated type I error rates. Stanley (2005) is a general review of this topic. Orwin (1983), Rosenberg (2005), and Rosenthal (1979) have proposed assessing the potential importance of the file drawer problem by estimating how many studies showing zero effect size would have to be found in order to reduce the estimated effect size to zero.

Several approaches have been used to synthesize the results of prior studies. The simplest is just a tabulation of how many prior studies of a certain phenomenon yielded statistically substantiated results (usually against a one-tailed alternative hypothesis). If the null hypothesis was rejected in only a few studies (say, around 5% for tests at the 5% level), then one might conclude that prior results were consistent with chance variation for a true effect size of zero. This approach is usually called *vote counting*. This is obviously a rather simplistic approach that also has a number of serious statistical problems. For example, the results of using this method can be more misleading as the number of samples increases (see, for example, Borenstein et al., 2009). It is also biased toward finding no effect (Hedges and Olkin, 1985).

Another approach is *Fisher's method for combining probabilities* (Fisher, 1954). This method uses the actual P-values from the studies being summarized. This is practical now that P-values are more commonly published. This method is useful when one wishes to test the null hypothesis that the effect sizes are zero in all of the studies being combined. An advantage of Fisher's method is that different types of statistical analyses can be combined so long as they test the same scientific hypothesis. However, a limitation of this approach is that it does not take into consideration

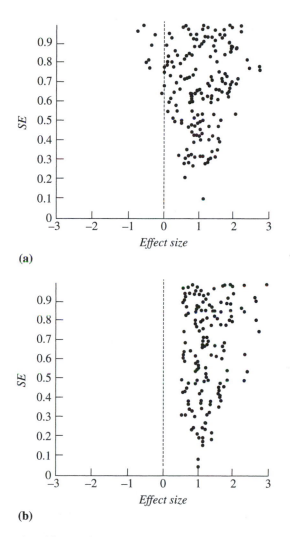

FIGURE 18.1 Examples of funnel plots (plots of the standard error of an effect size against the effect size itself). **(a)** Results of a simulation with a true effect size of 1. **(b)** Results of a simulation excluding the observed effect sizes for which the null hypothesis cannot be rejected.

estimates of the size of the effect. It cannot distinguish between large studies where only small effects were found and small studies where very large effects were detected.

Although we have used this method repeatedly with actual data, we will illustrate it with a hypothetical example to show the full range of its applicability. Box 18.1 provides the results of five experiments in which physiologists or ecologists were studying oxygen consumption in two species of crayfish. A limnologist is writing a review article

---

BOX 18.1 | Combining Probabilities from Independent Tests

Five experiments test oxygen consumption in two species of crayfish.

Experiment	P	$\ln P$
1. *t*-test (Box 9.4)		
$\quad n_1 = 14, n_2 = 12, 24\ df, t_s = 1.90$	0.069513	-2.66624
2. *G*-test for independence (Box 17.4)		
$\quad 1\ df, G = 2.95$	0.085878	-2.45483
3. Mann–Whitney *U*-test (Box 13.8)		
$\quad n_1 = 10, n_2 = 10, U_s = 72$	0.096304	-2.34025
4. Anova, single *df* comparison (Table 9.2)		
$\quad \nu_1 = 1, \nu_2 = 20, F_s = 3.522$	0.075227	-2.58725
5. t' (*t*-test with unequal variances) (Box 13.4)		
$\quad 16.2\ df, t'_s = 2.040$	0.058001	-2.84730
Total		-12.89585

$$X^2 = -2\sum \ln P = -2(-12.89585) = 25.79171, P = 0.004031$$

If all the null hypotheses in the experiments were true, then X^2 would be distributed as χ^2 with $\nu = 2k = 10$ degrees of freedom (k = the number of separate experiments and probabilities). Because X^2 corresponds to $P = 0.004031$, we can now reject the null hypothesis that there was no difference between the oxygen consumption of the two species in the five experiments.

---

and wishes to summarize the results available in the literature, none of which yielded a P-value ≤ 0.05. However, in all five experiments, species A appears to have the higher oxygen consumption than species B. The designs of the five experiments varied widely, as did the statistical tests applied to evaluate them. We shall assume that the designs chosen were correct, without describing them in detail. We shall further assume the appropriateness of the statistical tests selected to analyze the results. All the tests should by now be familiar to you.

The computation is based on the fact that $-2 \ln P$ is distributed approximately as $\chi^2_{[2]}$. By evaluating twice the negative natural logarithm of each of the five probabilities, considering each to be a $\chi^2_{[2]}$, and summing these values in the manner of Section 17.3, we obtain a total that can be looked up under 10 degrees of freedom in this example, because there are two degrees of freedom for each of the five P-values. In Box 18.1, the resulting $-2\sum \ln P$, when compared with $\chi^2_{[10]}$, yields a probability of 0.004031, leading to an overall rejection of the null hypothesis. On the basis of all these experiments together, then, we conclude that species A has a higher oxygen consumption than does species B. Details of the computation are furnished in Box 18.1.

The usefulness of combining probabilities in this instance should be apparent. This method can also be used when the separate significance tests to be combined are all of the same type but a joint overall statistical analysis is not possible. For example, suppose we measure the differences in production of a metabolite by two groups of mice, each with different diets. This experiment is repeated three times for one-week intervals. The analysis could be set up as a two-way anova with weeks as blocks and the two diets as the columns of the table. It may be, however, that the replication of this experiment is unequal, presenting the more complex problem of trying to solve a two-way anova with unequal and disproportional subclass numbers. Rather than do this, we may wish to carry out separate anovas or *t*-tests on each week, obtain the probability value for each of the tests, and combine them in the manner just shown (of course, one would not be able to test for the diet × weeks interaction by this technique).

Fisher's method for combining probabilities has been used by one of us to show that patterns of genetic variation in human populations were consistent with patterns of linguistic variation. These analyses were based on different sets of samples for different genetic loci, so they could not be combined into a single overall analysis. By obtaining a *P*-value for the test on each locus and then combining these *P*-values by Fisher's method, we were able to conclude that overall genetic variation is related to linguistic variation.

The more recent approach, collectively called **meta-analysis,** is to summarize the results of a series of individual studies by averaging their effect sizes rather than their probabilities. This allows one to estimate the strengths and direction of the relationships, test their homogeneity, and to set confidence limits rather than just accepting or rejecting the null hypothesis that there is no effect. The choice of the measure of effect size and the method of summarization depends on the type of data and hypothesis being tested. For the comparison of two means, it could be Δ, the standardized difference between two means or Cohen's *f* for differences among two or more means (see Section 12.1). Unfortunately, when summarizing a series of published studies, sufficient information may not be available to compute the desired effect size, so assumptions and approximations may be necessary in order to get the information into a comparable form. The method of analysis depends on whether one views the effect size for a series of studies as a fixed but unknown constant for the studies being summarized or as a random sample of effect sizes from a population of such studies. The fixed effects model is sometimes referred to as a *conditional model* because the inferences are limited to the studies being summarized. The random effects model is called an *unconditional model* because inferences can be made about the population of experiments that have essentially the same effect size (although that may be difficult to determine). This is not quite the same as the distinction made between model I and model II in an anova. In an anova, the question was whether the effect of a certain treatment would be the same in replications of an experiment or whether different effect sizes could be expected due to natural variation in the phenomenon being studied.

For approximately normally distributed variables and the *fixed effects* model, the estimated effect size for a difference between two means is usually transformed to

Hedges' g (Hedges, 1981). This is the standardized effect size, Δ, corrected for bias due to small sample sizes. For each individual study, g is computed as

$$g = J\frac{\bar{Y}_2 - \bar{Y}_1}{s_p} \tag{18.1}$$

where $J = 1 - 3/(4v - 1)$ is a correction for bias in small samples, v is the degrees of freedom of s_p $(n_1 + n_2 - 2$, in this case), and $s_p = \sqrt{((n_1 - 1)s_1^2 + (n_2 - 1)s_2^2)/(n_1 + n_2 - 2)}$ is the pooled standard deviation. This correction for bias could have been employed in Chapter 12, but it is not usually done because in that chapter we were just interested in the general magnitude of the effect size. However, in meta-analytic studies, the effect size is often what is of primary interest. The variance of the effect size for each individual study is approximately

$$s_d^2 = \frac{n_1 + n_2}{n_1 n_2} + \frac{g^2}{2(n_1 + n_2)} \tag{18.2}$$

A weighted average of the effect sizes is then computed to give the overall summary for a meta-analysis of the amount of difference found in a set of studies,

$$\bar{g} = \frac{\sum w_i g_i}{\sum w_i} \tag{18.3}$$

where the weights are the reciprocals of the variances, $w_i = 1/s_{g_i}^2$. The variance of the weighted average effect size is the reciprocal of the sum of the weights, $s_{\bar{g}}^2 = 1/\sum w_i$. Assuming that the effect size is normally distributed, one can use the square root of this variance as a standard error and set confidence limits to the estimated effect size. It is useful to prepare what is called a *forest plot*—a plot of the confidence intervals for each study and for their weighted average. Examples are shown in the boxes.

As a check on the assumption that all of the individual studies in a meta-analysis are estimating the same effect size, one can perform a test for homogeneity. If all k of the effect sizes are consistent with being drawn from the same population of effect sizes, then the statistic

$$Q = \sum w_i(g_i - \bar{g})^2 \tag{18.4}$$

should follow the $\chi_{[v]}^2$-distribution with $v = k - 1$ degrees of freedom.

Box 18.2 shows an example of a fixed-effects model where studies reporting on the difference between two means are divided into three classes according to habitat. In such studies, the interest is both in summarizing the size of the effect within each class (habitat) but also to compare the effect sizes in different classes.

The analysis of studies using the *random effects* model is similar but is somewhat more complex because one must allow for an added variance component for the variation in effect sizes among studies. Because of the additional source of variation,

---

BOX 18.2 Meta-Analysis of the Difference Between Two Means for a Continuous Variable: Fixed Effects Model

Data from Gurevitch and Hedges (2001) on the effects of (intraspecies) competition in three habitats. The original response variables considered were recruitment (increased numbers of individuals) or growth (increased size of individuals). They are made comparable by converting them all to effect sizes. The numbers of studies chosen for each habitat were terrestrial (19), lentic (2), and marine (22). The individual studies were restricted to a single trophic level of plants—primary producers. All the studies included in the meta-analysis were reports of field experiments, with the majority representing a decrease in the density of the hypothesized competitors. For a fixed effects model, it is assumed that there is a single true effect size within each class (habitat in the current example). For more details on the input data, see Gurevitch and Hedges (2001) and the references cited therein.

Terrestrial

(1) Sign	(2) n_c	(3) n_e	(4) $\bar{Y}_c$	(5) $\bar{Y}_e$	(6) s_c	(7) s_e
+	7	7	78.14	79.71	40.650	40.650
+	7	7	18.86	26.	9.170	9.170
−	6	6	−1.8	−2.1	0.490	0.490
−	5	5	−2.2	−2.8	0.2240	0.447
−	7	7	−2.1	−3.	0.2650	0.529
−	6	6	−2.3	−4.2	0.4900	1.225
+	3	3	85.3	285.7	115.008	153.806
+	3	3	0.	3.	0.000	2.425
+	3	3	0.	2.00	0.000	2.078
+	3	3	0.	1.67	0.000	1.732
+	5	5	17.	17.	7.603	5.367
+	5	5	47.	37.	10.286	9.391
+	4	4	87.	272.	37.712	183.532
+	18	20	−0.113	0.294	0.255	0.215
+	20	20	−0.163	0.412	0.588	0.218
+	18	20	0.140	0.632	0.380	0.359
+	20	20	−0.184	0.259	0.326	0.238
+	20	20	−0.075	0.354	0.487	0.182
+	20	20	0.147	0.541	0.340	0.299

Lentic

Sign	n_c	n_e	$\bar{Y}_c$	$\bar{Y}_e$	s_c	s_e
−	4	4	281.11	−201.03	158.038	27.520
−	4	4	187.31	−155.32	80.163	41.252

Box 18.2 (continued)

Marine

Sign	n_c	n_e	$\overline{Y}_c$	$\overline{Y}_e$	s_c	s_e
+	7	7	11.8	16.0	3.080	3.370
+	3	10	0.40	9.5	1.470	7.230
+	20	20	0	14.1	7.6030	7.603
+	20	20	0	7.1	3.130	3.130
+	20	20	0	1.4	1.789	1.789
+	10	10	82.2	94.	29.093	9.171
+	10	10	8.30	10.5	14.546	11.068
+	10	10	0	20.	42.691	42.691
+	2	2	3.63	18.5	3.352	4.257
+	2	2	0	0.25	0.000	0.354
+	2	2	3.63	2.25	3.352	0.707
+	4	4	0	34.8	0.000	58.200
+	4	4	0	25.3	0.000	35.800
+	4	4	5.40	23.6	10.880	47.000
+	4	4	1.80	10.5	5.200	24.200
+	4	4	0	10.3	0.000	17.400
+	4	4	0	8.7	0.000	17.000
+	4	4	0	5.7	0.000	14.000
+	4	4	10.8	5.4	15.800	8.800
+	4	4	21.25	37.25	9.540	22.020
+	4	4	40.25	20.25	8.780	9.000
+	5	5	15.8445	11.9533	10.787	6.240

In this box, the subscripts "e" and "c" in columns (2) through (7) are used to denote the experimental and control groups rather than referring to them more generally as "1" and "2." A "+" in column (1) indicates that the experimental group corresponds to a decrease in the density of competitors, a negative sign indicates an increase (this must be taken into account when computing an average effect size by using the difference $\overline{Y}_c - \overline{Y}_e$ rather than $\overline{Y}_e - \overline{Y}_c$), n_c and n_e are the sample sizes in the control and experimental groups, $\overline{Y}_c$ and $\overline{Y}_e$ in columns (4) and (5) are the observed means in the control and experimental groups, and s_c and s_e are the observed standard deviations in the control and experimental groups. See Gurevitch and Hedges (2001) for further explanation of these data.

1. Using the equations given in Section 18.1, compute bias-corrected effect sizes and their variances for each study.

Box 18.2 (continued)

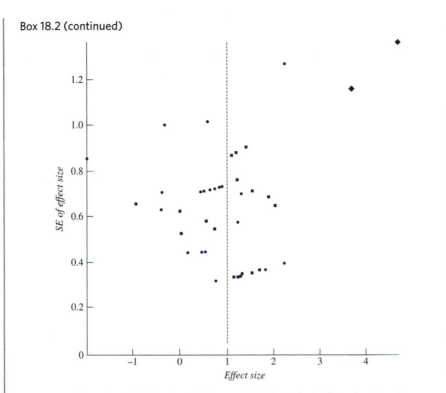

Funnel plot: For each study, the standard error of the effect size is plotted against the effect size. The terrestrial samples are shown with squares, lentic ones with diamonds, and marine ones with circles. The scatter appears more or less symmetrical, which implies that there is not likely to be a serious problem with the selection of studies to be included.

For the first study in the terrestrial group, the bias correction, J, and the pooled variance are computed using formulas immediately following Expression (18.1)

$$J = 1 - \frac{3}{4(n_e + n_c - 2) - 1} = 1 - \frac{3}{4(7 + 7 - 2) - 1} = 0.93617$$

$$s_{pooled} = \sqrt{\frac{(n_e - 1)s_e^2 + (n_c - 1)s_c^2}{n_e + n_c - 2}} = \sqrt{\frac{(7 - 1)40.650^2 + (7 - 1)40.650^2}{7 + 7 - 2}} = 40.650$$

(Because the two estimates of the standard deviation are identical for this study, the pooled standard deviation must also be the same but its computation is shown here as an example of the general case.)

Hedges' effect size, g, is then computed using Expression (18.1)

$$g = J\frac{\overline{Y}_e - \overline{Y}_c}{s_p} = (0.93617)\frac{79.71 - 78.14}{40.650} = 0.03616$$

Box 18.2 (continued)

Using Expression (18.2), the variance of this effect size is

$$s_g^2 = \frac{n_e + n_c}{n_e n_c} + \frac{g^2}{2(n_e + n_c)} = \frac{7 + 7}{7 \times 7} + \frac{0.03616^2}{2(7 + 7)} = 0.28576$$

2. The overall weighted average effect size and its variance can be computed for each of the three habitat types using Expression (18.3). The weights are the reciprocals of the variances, $w_i = 1/s_{g_i}^2$. For the terrestrial habitat, these are

$$\bar{g} = \frac{\sum w_i g_i}{\sum w_i} = \frac{3.49943(0.03616) + \cdot \cdot + 8.46121(1.20620)}{3.49943 + \cdot \cdot + 8.46121} = 1.14171$$

$$s_{\bar{g}}^2 = \frac{1}{\sum w_i} = \frac{1}{3.49943 + \cdots + 8.46121} = 0.01325$$

Assuming that $\bar{g}$ is approximately normally distributed, confidence limits can be computed using Equation (7.10). For the terrestrial habitat, the 95% lower and upper confidence limits are

$$L_1 = \bar{g} - t_{0.05[\infty]}s_{\bar{g}} = 1.14171 - 1.960\sqrt{0.01325} = 0.91609 \text{ and } L_2 = 1.36733$$

The results for all three habitats are

Habitat	$\bar{g}$	$s_{\bar{g}}^2$	L_1	L_2
Terrestrial	1.14171	0.01325	0.91609	1.36733
Lentic	4.10722	0.78444	2.37130	5.84313
Marine	0.79845	0.01521	0.55672	1.04019

By the subjective criteria discussed in Chapter 12, these average effect sizes are all quite large. In addition, all three confidence intervals exclude zero, so the null hypothesis of no effect of competition can be rejected.

3. Test of homogeneity of effect sizes within habitats.
   If the effect sizes are homogeneous, then the statistic from Expression (18.4)

$$Q = \sum w_i (g_i - \bar{g})^2$$

will follow the χ^2-distribution with $\nu = k - 1$ degrees of freedom. The sum is over all k studies within a habitat. For the three habitats, the following results were obtained.

Habitat	Q_i	df	P
Terrestrial	25.59046	18	0.10950
Lentic	0.29687	1	0.58584
Marine	43.61466	21	0.00262

Box 18.2 (continued)

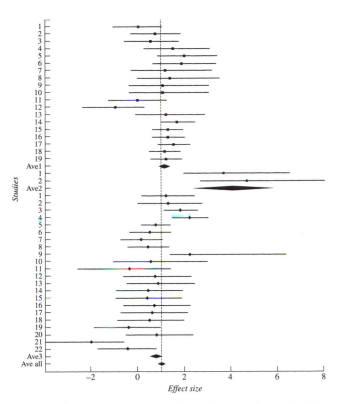

Forest plot showing effect sizes and 95% confidence limits for the studies (19 terrestrial at the top followed by the 2 lentic and 22 marine). Group mean effect sizes and their 95% confidence intervals are shown by the large diamond shapes following each group. The dashed vertical line shows the overall mean effect size, and the diamond at the bottom shows its 95% confidence interval.

While the effect sizes for the marine habitat appear heterogeneous, Gurevitch and Hedges (2001) comment that the variation is actually quite small for ecological studies. However, if one rejects the hypothesis of homogeneity for a category, then the average effect size and its confidence limits, computed previously, would not be of much interest.

4. Partitioning of variation in effect sizes.

The total variation among the effect sizes can be partitioned as was done for sums of squares in an anova (see Chapter 8). A difference is that each can be tested by comparison to a χ^2-distribution:

$$Q_{\text{within}} = \sum Q_i = 25.5905 + 0.2969 + 43.6147 = 69.5020$$

Its degrees of freedom are the sum of the degrees of freedom of the Q-values being summed. $\nu_w = 18 + 1 + 21 = 40$, which is also equal to the total number of

Box 18.2 (continued)

studies minus the number of groups. Its P-value is 0.002622, implying that we should not accept the hypothesis of overall homogeneity within groups.

The Q-value for variation among groups is a weighted sum of squares of mean group effect sizes.

$$Q_{among} = \sum_{j=1}^{a}\left(\sum_{i=1}^{\kappa_j} w_{ij}\right)(\bar{g}_j - \bar{\bar{g}})^2$$

$$= 75.4641(1.14171 - 1.0099)^2 + \cdots + 65.7382(0.79845 - 1.0099)^2$$

$$= 16.47992$$

This has degrees of freedom $\nu = a - 1 = 3 - 1 = 2$ and a P-value of 0.000264. Thus, we can reject the null hypothesis that the average effect sizes are the same within the three groups.

The within- and among Q-values can be summed to give a total Q-value, but that is usually of less interest.

confidence limits will be broader and hypothesis tests will be more conservative but in many cases more realistic.

First, the unbiased effect size and its variance are computed for each study as in a fixed-effects model.

The difference between the models is that the variance among the k studies is assumed to be of the form $\sigma^2 + c\sigma_A^2$, where the coefficient, c, is computed as

$$c = \sum w_i - \frac{\sum w_i^2}{\sum w_i} \tag{18.5}$$

The added variance component can then be estimated as

$$s_A^2 = \frac{Q - \nu}{c} \tag{18.6}$$

where the Q-statistic is computed as in Equation (18.4) for a fixed effects model, c is as computed in Equation (18.5), and ν is the degrees of freedom for Q (usually $k - 1$). Negative estimates are treated as zeros because σ_A^2 must be ≥ 0. A test of the null hypothesis H_0: $\sigma_A^2 = 0$ can be performed by comparing Q to the $\chi_{[k-1]}^2$-distribution. A related descriptive statistic is

$$I^2 = \frac{Q - \nu}{Q} \tag{18.7}$$

the proportion of the variation in the effect sizes that is due to heterogeneity rather than to sampling variation within studies. It is analogous to the intraclass correlation coefficient (Higgins and Thompson, 2002). It is sometimes multiplied by 100 to express it as a percentage.

In comparison to the fixed effects model, the variances for each study are increased by the addition of the added variance component. An asterisk is added to the symbols to indicate that the quantities are based on these inflated variances. Thus, the overall mean effect size is estimated as

$$\overline{g}* = \frac{\sum w_i^* g_i}{\sum w_i^*} \tag{18.8}$$

where the weights for each study are $w_i^* = 1/s_{g_i}^{*2}$ and

$$s_{g_i}^{*2} = s_{g_i}^2 + s_A^2 \tag{18.9}$$

As in the fixed effects model, the variance of the average effect is the reciprocal of the sum of the weights, $s_{\overline{g}*}^2 = 1/\sum w_i^*$. Assuming that $\overline{g}*$ is approximately normally distributed, confidence intervals can be computed using its square root, $s_{\overline{g}*}$, as the standard error of $\overline{g}*$.

Box 18.3 shows an example of an analysis of a random effects model for the differences in human males and females in a psychological test. In Box 18.3, the studies are not grouped into different classes. If they were, then one could have a mixed model in which the variation in effect sizes among studies within a class was

---

BOX 18.3 Meta-Analysis of the Difference Between Two Means for a Continuous Variable: Random Effects Model

Mixed Model Analysis

This analysis allows for random variation in the effect sizes in the different studies within each group (habitat in our example) rather than a fixed parameter.

The following are the results of 15 studies of gender differences in "field articulation ability" (a measure of visual-analytic spatial ability) in humans.

(1)	(2)	(3)
n	g	s_{pooled}^2
60	0.76	0.071
140	1.15	0.033
30	0.48	0.137
30	0.29	0.135
30	0.65	0.140
46	0.84	0.095
40	0.70	0.106
34	0.50	0.121
76	0.18	0.053
163	0.17	0.025
97	0.77	0.044
44	0.27	0.092
78	0.40	0.052
43	0.45	0.095

Box 18.3 (continued)

Data from Hyde (1981) and summarized by Hedges and Vevea (1998). The column for n gives the total sample size (not the sample size of each group), g is the unbiased standardized effect size for a difference between two means (males minus females), and the last column gives the pooled variance for each study.

1. *Computation of effect sizes.* In this example, the unbiased standardized effect sizes were already provided as $g = J(\overline{Y}_{males} - \overline{Y}_{females})/s_{pooled}$, where $J = 1 - \dfrac{3}{4v - 1}$

   and v is the number of degrees of freedom, which is the total sample size minus 2 in this case. The pooled variances were also provided. If not, they would be computed as in Box 18.2 or as the error MS in Chapter 9.

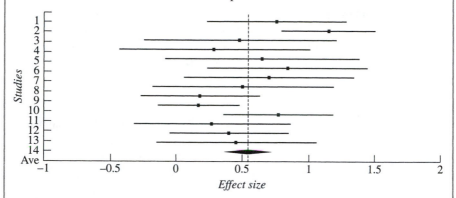

Forest plot showing the effect size and its 95% confidence interval for each study. The mean effect size and its 95% confidence interval are shown by the dashed vertical line and the diamond at the bottom.

2. *The homogeneity statistic, Q.* This is computed as in a fixed effects model (see step **3** in Box 18.2).

   For the current data, the weighted mean effect size is

   $$\overline{g} = \frac{\sum w_i g_i}{\sum w_i} = \frac{14.08451(0.76) + \cdots + 10.52632(0.45)}{14.08451 + \cdots + 10.52632} = 0.54681$$

   Then

   $$Q = \sum w_i(g_i - \overline{g})^2$$
   $$= 14.08451(0.76 - 0.54681)^2 + \cdots + 10.52632(0.45 - 0.54681)^2 = 24.10329$$

3. *Estimation of the among-studies variance component.* To compute the among-studies variance component, a coefficient is computed that is analogous to $k - 1$ times the average sample size, n_0, in a model II single-classification anova in Chapter 9.

   For the current example, the coefficient, c, is estimated as

   $$c = \sum w_i - \frac{\sum w_i^2}{\sum w_i} = 216.68366 - \frac{4615.22149}{216.68366} = 195.38431$$

Box 18.3 (continued)

The among-studies variance component can then be estimated as

$$s_A^2 = \frac{Q - (a - 1)}{c} = \frac{24.10329 - (14 - 1)}{195.38431} = 0.05683$$

Note: If the estimate is negative, then a value of zero is used.

A test of the hypothesis H_0: $\sigma_A^2 = 0$ is performed by comparing Q to the $\chi^2_{[a-1]}$-distribution. In the current example, we obtain a *P*-value of 0.0301931, so we can reject the null hypothesis at the 5% level.

4. *The overall mean effect size.* This is estimated using a new set of weights, $w_i^* = 1/s_{g_i}^{*2}$, for each study. The weights include the addition of the added variance component, $s_{g_i}^{*2} = s_{g_i}^2 + s_A^2$. For the first study, $s_{g_1}^{*2} = s_{g_1}^2 + s_A^2 = 0.07100 + 0.05683 = 0.12783$. The weight for the first study is then $w_i^* = 1/s_{g_1}^{*2} = 1/0.12783 = 7.82301$.

The weighted average effect size is then

$$\bar{g}^* = \frac{\sum w_i^* g_i}{\sum w_i^*} = \frac{7.82301(0.76) + \cdots + 6.58640(0.45)}{7.82301 + \cdots + 6.58640} = 0.54920$$

Its variance is $s_{\bar{g}^*}^2 = 1/\sum w_i^* = 1/106.49766 = 0.00939$.

Assuming that $\bar{g}^*$ is approximately normally distributed, its 95% confidence interval is

$$L_1 = \bar{g}^* - t_{0.05[\infty]} s_{\bar{g}^*}^2 = 0.54920 - 1.95996\sqrt{0.00939} = 0.35928$$

and

$$L_2 = 0.73912$$

The interval excludes zero and is in the range that Cohen (1988) called "medium to large."

5. *Grouping into classes.* If the studies had been grouped into classes as in the example in Box 18.2, then average effect sizes would be computed separately for each class. If the variation among studies could be assumed to be homogeneous across classes, then a single average variance component would be used to compute the variances, $s_{g_i}^{*2}$, for each study.

---

assumed to be random but the differences among the classes was fixed. More complex designs are possible by extending the GLM approach of Section 16.7 (see, for example, Borenstein et al., 2009, and Gleser and Olkin, 2009).

In some cases, the effect size is expressed in terms of a ratio of two means rather than of their difference. The log response ratio, *LR*, is the logarithm of such a ratio, or equivalently, the difference between the logarithms of two means. It is appropriate for variables that are always positive, and on a ratio scale, for example, log-normally distributed variables. If the original data are not available to transform the data so that the usual methods cannot be used, then a meta-analysis can be performed on LR using their variance,

$$s_{LR}^2 = \frac{s_1^2}{n_1\overline{Y}_1^2} + \frac{s_2^2}{n_2\overline{Y}_2^2} \tag{18.10}$$

where the subscripts refer to the two groups being compared (such as experimental and control) in each study. If there are more than two groups being compared, then one will need to use log-transformed data.

The methods in Boxes 18.2 and 18.3 make use of the relationship that the sum of squared standardized deviations from their mean follows the χ^2-distribution if the deviates are independently drawn from the same normally distributed population (see Section 7.8). The Q-statistics used to test homogeneity also make use of the fact that the mean of a χ^2-distribution is equal to its degrees of freedom.

As described in Chapters 14, 15, and 16, a correlation coefficient is a standardized effect size and it can also be used as effect sizes for regression coefficients (including partial regression coefficients). Box 18.4 shows the computations for both a fixed and a random effects *meta-analysis for correlation coefficients*. The method for fixed effects is identical to that shown in Box 15.3. In Box 18.4, the correlations are first transformed using Fisher's z-transformation (see Section 15.5) and then the procedures previously described for mean are used—except that the variances for each of the k z-transformed correlation are just $1/n - 3$. The average z-value and confidence limits are then back-transformed to a correlation coefficient, so it will be in familiar units. The method for random effects is slightly more complicated but follows the same form as for the mean.

Because such a diversity of methods are described in Chapter 17 for frequency data, different meta-analytic methods are needed. The method of replicated goodness-of-fit tests described in Box 17.4 can be used to summarize a series of single-classification goodness-of-fit tests if the same expected relative frequencies apply to the different studies. Much work has been done on methods for summarizing 2×2 contingency tables from a series of studies. While the Mantel–Haenszel procedure described in Box 17.12 can be used to both estimate the average odds ratio and to test the homogeneity of odds ratios for the fixed effects model, a more common and flexible approach is to compute the log-odds ratio for each 2×2 table and then to use methods similar to those previously described for Hedges' g. This approach is most appropriate when sample sizes are not small. Box 18.5 shows the procedure and an example where a series of 2×2 tables were compared for two treatment groups. When the Mantel–Haenszel method was used on these data, very similar results were obtained: The estimated odds ratios were 0.88266 for women and 0.85827 for men. The tests for homogeneity also gave similar results.

For the random effects model, the procedure is similar to that previously described for Hedges' g. The log-odds ratio is used as the effect size, and the added variance component for variation among studies is estimated as in Equation (18.6). Box 18.6 shows the details with an example. Note that in Boxes 18.5 and 18.6 the symbol o is used to represent a sample estimate of ω, the odds ratio discussed in Section 17.6.

BOX 18.4 Meta-Analysis of Correlation Coefficients: Fixed and Random Models

Fixed Effects Model

The procedure outlined in Box 15.3 is used to compute the average correlation and to test for homogeneity of the correlations found in different studies. Using the data provided in Box 15.3, the hypothesis of homogeneity was not rejected. In a meta-analysis, it will usually be of interest to also set confidence limits to the average correlation. This is done by setting confidence limits to the weighted average z-value and back-transforming them to correlation coefficients. Using the sample of 10 correlations in that box, the following results are obtained.

$$\bar{z} = 0.615268, \quad \sigma_{\bar{z}}^2 = \sum \frac{1}{n_i - 3} = 0.0025445$$

The lower 95% confidence limit for ζ is

$$L_1 = \bar{z} - t_{0.05[\infty]}\sigma_{\bar{z}} = 0.615268 - 1.95996\sqrt{0.0025445} = 0.5164008$$

and the upper limit is 0.71413503. These correspond to correlations of 0.47492 and 0.61326, respectively.

If the hypothesis of homogeneity had been rejected, then there would be interest in determining if the correlations could be classified into groups within which one expects homogeneity as shown in Box 18.2.

Random Effects Model

As usual in a random effects model, there is interest in estimating the added variance component for variation among studies.

The Q-statistic for homogeneity (called X^2 in Box 15.3) is

$$Q = \sum w_i(z_i - \bar{z})^2$$
$$= 97(0.298566 - 0.615268)^2 + \cdots + 14(0.632833 - 0.615268)^2 = 15.263517$$

It has $\nu = k - 1 = 10 - 1 = 9$ degrees of freedom. As noted in Box 15.3, its P-value is 0.08395.

The variance component is then estimated as

$$\sigma_z^2 = \frac{Q - \nu}{c} = \frac{15.263517 - 9}{337.491094} = 0.018559$$

where

$$c = \sum w_i - \frac{\sum w_i^2}{\sum w_i} = 393 - \frac{21,815}{393} = 337.491094$$

BOX 18.5 Meta-Analysis of Odds Ratios and Log Odds Ratios: Fixed Effects Model

Berger et al. (2006) published the following data from studies investigating the effects of aspirin therapy on the frequency of cardiovascular events in both women and men. Each row of this table represents a 2×2 contingency table (see the following).

Studies	Aspirin Events	Aspirin None	Control Events	Control None
Women				
1	109	4328	134	4312
2	17	1260	26	1280
3	477	19,457	522	19,420
Men				
4	289	3140	147	1563
5	173	4789	207	4738
6	307	10,730	370	10,664
7	28	921	38	925
8	228	2317	260	2280

1. *Computation of basic quantities.* Compute odds ratios, their logs, variances of the log-odds ratios, and confidence limits for each of the studies. For example, for the first study, the 2×2 contingency table is

	Aspirin	Control
Cardiovascular event	109	134
No event	4328	4312

The sample estimate for the odds ratio is then $o_1 = \dfrac{109 \; 4312}{4328 \; 134} = 0.81043$, which means that the proportion of subjects having a cardiovascular event was almost 20% less for those taking aspirin. The log-odds ratio and its variance are $\ln o_1 = \ln(0.81043) = -0.21020$ and

$$\sigma^2_{\ln o_1} = \frac{1}{n_{11}} + \frac{1}{n_{21}} + \frac{1}{n_{12}} + \frac{1}{n_{22}} = \frac{1}{109} + \frac{1}{4328} + \frac{1}{134} + \frac{1}{4312} = 0.01710$$

Assuming that the log-odds ratios are approximately normally distributed, the confidence limits can be computed as

$$L_1 = \ln o_1 - t_{0.05[\infty]}\sigma_{\ln o_1} = -0.21020 - 1.95996\sqrt{0.01710} = -0.46649$$

and

$$L_2 = \ln o_1 + t_{0.05[\infty]}\sigma_{\ln o_1} = 0.04610$$

Box 18.5 (continued)

They can be expressed in terms of the odds ratios by computing their antilogarithms. The results for all of the studies are shown in the following table. The average values for women and men and for all studies combined (see step **2**) are also shown.

Study	$\ln o_1$	L_1	L_2	o_1	L_1	L_2
1	−0.21020	−0.46649	0.04610	0.81043	0.62720	1.04718
2	−0.40913	−1.02539	0.20712	0.66422	0.35866	1.23013
3	−0.09205	−0.21778	0.03367	0.91206	0.80430	1.03425
ave ♀	−0.12452	−0.23555	−0.01349	0.88292	0.79014	0.98660
4	−0.02162	−0.22924	0.18600	0.97861	0.79514	1.20442
5	−0.19013	−0.39599	0.01572	0.82685	0.67302	1.01584
6	−0.19283	−0.34649	−0.03916	0.82463	0.70716	0.96160
7	−0.30105	−0.79765	0.19555	0.74004	0.45039	1.21598
8	−0.14743	−0.33443	0.03956	0.86292	0.71575	1.04035
ave ♂	−0.15264	−0.24325	−0.06203	0.85844	0.78408	0.93985

Forest plot showing mean effect sizes and 95% confidence intervals for the eight studies and their averages. The vertical dashed vertical line corresponds to an odds ratio of 1 (a log-odds ratio effect size of 0). Note the very wide confidence interval for the second study in the first group due to the small sample size.

2. *Tests for homogeneity among studies within a group.*
    The homogeneity of the log-odds ratios within a group can be tested using the Q-statistic.

$$Q_{\text{women}} = \sum w_i (\ln o_i - \overline{\ln o})^2$$
$$= 58.47966(-0.21020 - (-0.12452))^2 + \cdots +$$
$$243.01138(-0.09205 - (-0.12452))^2 = 1.50479$$

Box 18.5 (continued)

Under the null hypothesis, this is distributed as $\chi^2_{[\nu]}$ with $\nu = k - 1 = 3 - 1 = 2$ degrees of freedom. Its P-value is 0.471236. For men, the Q-value is 2.26598, with $\nu = k - 1 = 5 - 1 = 4$. Its P-value is 0.68697.

3. *Average odds ratios and their confidence limits.*

The average log-odds ratios for the two groups (women and men) are computed as weighted averages with the weights equal to the reciprocals of the variances, $\sigma^2_{\ln o_i}$, computed previously for each study.

$$\ln o_{\text{women}} = \frac{\sum w_i \ln o_i}{\sum w_i}$$

$$= \frac{58.47966(-0.21020) + \cdots + 243.01138(-0.09205)}{58.47966 + \cdots + 243.01138} = -0.12452$$

$$\ln o_{\text{men}} = -0.15264$$

The variance of each average is the reciprocal of the sum of the weights used to compute each average. For women, this is

$$\sigma^2_{\overline{\ln o}} = \frac{1}{\sum w_i} = \frac{1}{58.47966 + \cdots + 243.01138} = 0.0032092, \text{ and for men it is}$$

0.0021373. These can be used to compute confidence limits for the average log-odds ratios as was done for each study. The results are shown in the preceding table. It also shows these converted to odds ratios.

4. *Tests of differences among groups.*

Tests for differences among the log-odds ratios in the two groups can also be made for the Q-statistic by using the weights and average log-odds ratios for the groups.

$$\overline{\overline{\ln o}} = \frac{\sum \left(\sum w \right)_i \overline{\ln o_i}}{\sum \left(\sum w \right)_i}$$

$$= \frac{58.47966(-0.21020) + \cdots + 243.01138(-0.09205)}{58.47966 + \cdots + 243.01138} = -0.14140$$

$$Q_{\text{among}} = \sum \left(\sum w \right)_i \left(\overline{\ln o_i} - \overline{\overline{\ln o}} \right)^2$$

$$= 311.60636(-0.12452 - (-0.14140))^2$$

$$+ 467.88981(-0.15264 - (-0.14140))^2$$

$$= 0.14791$$

Under the null hypothesis that there are no differences between the two groups, Q_{among} is distributed as $\chi^2_{[\nu]}$ with degrees of freedom, ν, equal to the number of groups minus 1, 1 in the current case. The null hypothesis of equality of mean log-odds ratios is not rejected because the P-value is 0.70054.

Because there seems to be little difference between the log-odds ratios for males and females, one could pool the two groups and simply report the overall log-odds ratio (-0.14140) previously found and its 95% confidence limits (-0.21160 and -0.07120). As odds ratios, the estimate of ω is 0.86814 and the confidence limits for ω are from 0.80929 to 0.86814.

BOX 18.6 | Meta-Analysis of Odds Ratios and Log-Odds Ratios: Random Effects Model and a Test for Differences Among Classes

Ariyaratnam et al. (2007) published the following data from studies of the genetics of ischemic stroke among persons of non-European descent. Subjects were grouped by whether they possessed the insertion/deletion polymorphism of the angiotensin I converting enzyme (ACE I/D) or not.

Studies	ACE I/D		Control	
	Stroke	None	Stroke	None
Chinese				
1	21	122	28	126
2	13	31	8	54
3	21	197	54	436
4	29	17	7	36
5	26	61	48	209
6	15	50	16	101
7	26	96	214	1015
8	64	88	18	54
9	71	94	18	88
Japanese				
1	31	150	30	241
2	34	104	7	83
3	16	39	7	54
4	20	155	39	174
5	11	58	55	239
6	12	14	4	24

1. *Computation of basic quantities.* The odds ratios, logs, variances of the log-odds ratios, and confidence limits for each of the studies are computed as for a fixed effects model. For example, for the first study the 2×2 contingency table is

	ACE I/D	Control
Stroke	21	28
None	122	126

The sample odds ratio is then $o_1 = \dfrac{21}{122} \dfrac{126}{28} = 0.77459$, which means that the proportion of subjects in this first study with ACE I/D having a stroke event was about 22% less than those without this polymorphism. The log-odds ratio and its variance are

$$\ln o_1 = \ln(0.77459) = -0.25542$$

$$\sigma^2_{\ln o_1} = \frac{1}{n_{11}} + \frac{1}{n_{21}} + \frac{1}{n_{12}} + \frac{1}{n_{22}} = \frac{1}{21} + \frac{1}{122} + \frac{1}{28} + \frac{1}{126} = 0.09947$$

Box 18.6 (continued)

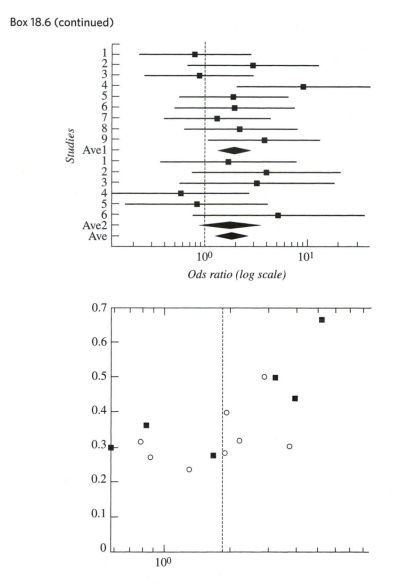

Forest plot (top) for risk of stroke in relation to the ACE I/D gene polymorphism among studies of Chinese (studies 1 to 9 at the top) and Japanese individuals (studies 1 to 6 at the bottom). The vertical dashed line corresponds to an odds ratio of 1 (a log-odds ratio effect size of 0). Diamond shapes labeled Ave1 and Ave2 show the mean odds ratios and their confidence intervals for the two groups. The diamond labeled Ave is for the overall average. Funnel plot (bottom) showing standard error of the log odds ratio plotted against the log odds ratio. The vertical dotted line shows the average log-odds ratio. Open circles correspond to Chinese studies and filled squares to Japanese. The scatter does not match the expected funnel shape. Ariyaratnam et al. (2007) reported that these data failed a test for symmetry so there is a concern that there could be a publication bias.

Box 18.6 (continued)

Assuming that the log-odds ratios are approximately normally distributed, the limits can be computed as

$$L_1 = \ln o_1 - t_{0.05[\infty]} \sigma_{\ln o_1} = -0.25542 - 1.95996\sqrt{0.09947} = -0.87356$$

and

$$L_2 = \ln o_1 + t_{0.05[\infty]} \sigma_{\ln o_1} = 0.36272$$

The confidence limits can be expressed in terms of odds ratios by computing the antilogarithms of the values given previously. The results for all of the studies are shown in the forest plot on the preceding page. The confidence intervals for the average values for Chinese and Japanese and for all studies combined are also shown. Their computations are given later in step **3**.

2. *Tests for homogeneity among studies within a group.* The homogeneity of the log-odds ratios within a group can be tested using the Q-statistic.

$$\begin{aligned}
Q_{\text{Chinese}} &= \sum w_i (\ln o_i - \overline{\ln o})^2 \\
&= 10.05363(-0.25542 - (0.51536))^2 + \cdots + \\
&\quad 10.91203(1.30635 - (0.51536))^2 \\
&= 32.51679
\end{aligned}$$

Under the null hypothesis, this is distributed as $\chi^2_{[\nu]}$ with $\nu = k - 1 = 9 - 1 = 8$ degrees of freedom. Its P-value is 0.00008. For the Japanese studies, the Q-value is 23.32296, with $\nu = k - 1 = 6 - 1 = 5$. Its P-value is 0.00029.

3. *The among-studies variance component.* To obtain this quantity, we compute a coefficient, c, analogous to $k - 1$ times the average sample size, n_0, in a model II single-classification anova in Chapter 9.

For the current example, the c-coefficient for the Chinese is

$$c = \sum w_i - \frac{\sum w_i^2}{\sum w_i} = 89.26015 - \frac{1063.30556}{89.26015} = 77.34772$$

The among-studies variance component for the Chinese studies can then be estimated as

$$s_A^2 = \frac{Q - (a - 1)}{c} = \frac{32.51679 - (9 - 1)}{77.34772} = 0.31697$$

For the Japanese studies, $c = 34.1839$ and $s_A^2 = 0.53601$. *Note:* If an estimated variance component is negative, then a value of zero is used.

A test of the hypothesis H_0: $\sigma_A^2 = 0$ is performed by the test of the Q-value described earlier in step **2**. The null hypothesis is rejected for both the Chinese and Japanese studies.

The proportion of the total variation within a group that is due to heterogeneity among the Chinese studies is

$$I^2 = \frac{Q - \nu}{Q} = \frac{32.51679 - (9 - 1)}{32.51679} = 0.75397$$

A rather similar value of $I^2 = 0.78562$ was obtained for the Japanese studies.

Box 18.6 (continued)

4. *Mean effect sizes.* Because the random effects model is being used, the mean effect size for each group is estimated using a new set of weights that include the added variance component. The weights are $w_i^* = 1/s_{\ln o_i}^{*2}$ where $s_{\ln o_i}^{*2} = s_{\ln o_i}^2 + s_A^2$. The asterisk is used to distinguish these variances and weights from those used earlier for fixed effects models.

   For the first Chinese study this is $s_{\ln o_1}^{*2} = s_{\ln o_1}^2 + s_A^2 = 0.09947 + 0.31697 = 0.41644$. The weight for the first study is then $w_i^* = 1/s_{\ln o_1}^{*2} = 1/0.41644 = 2.40134$.

   The weighted average effect size for Chinese is

   $$\overline{\ln o*} = \frac{\sum w_i^* \ln o_i}{\sum w_i^*}$$

   $$= \frac{2.40134(-0.25542) + \cdots + 2.44732(1.30635)}{2.40134 + \cdots + 2.44732} = 0.64127$$

   Its variance is $s_{\overline{\ln o}*}^2 = 1/\sum w_i^* = 1/20.58717 = 0.04857$.

   Assuming that the $\overline{\ln o*}$ are approximately normally distributed, the 95% confidence limits for the Chinese mean can be computed as

   $$L_1 = \overline{\ln o*} - t_{0.05[\infty]}s_{\overline{o}*}^2 = 0.64127 - 1.95996\sqrt{0.04857} = 0.20931$$

   and

   $$L_2 = 1.07324$$

   The corresponding values for the Japanese studies are $s_{\overline{\ln o}*}^2 = 0.11909$, $\overline{\ln o*} = 0.55284$, and the 95% confidence limits for $\overline{\ln o*}$ are -0.12354 and 1.22921. These limits are represented on the forest plot as the solid diamonds.

   Note that if one believed that the variance components within each group were estimates of the same population parameter then the estimates could be pooled (Borenstein et al., 2009) and used in step 5 below.

5. *Tests of differences among group mean effects.* Several methods have been proposed. Borenstein et al. (2009) give three equivalent ones (i.e., that yield identical probabilities). In Box 18.4, a direct approach was used. Here we compute it indirectly as the difference between the total and within Q-values.

   Compute the Q-statistic previously described separately for each group but use weights defined earlier in step 4. For the Chinese studies, this would be

   $$Q_{\text{Chinese}}^* = \sum w_i^*(\ln o_i - \overline{\ln o*})^2$$

   $$= 2.40133(-0.25542 - 0.64127)^2 + \cdots + 2.44732(1.30635 - 0.64127)^2$$

   $$= 9.38729$$

   The value for the Japanese studies is $Q_{\text{Japanese}}^* = 5.33256$. These are summed to obtain

   $$Q_{\text{within}}^* = \sum Q_i^* = 9.38729 + 5.33256 = 14.71985$$

Box 18.6 (continued)

The total Q is computed as the weighted sum of the squared deviations of the log-odds ratios from the overall weighted mean.

$$Q^*_{\text{Total}} = \sum w^*_i (\ln o_i - \overline{\ln o}^*)^2$$

$$= 2.40133(-0.25542 - 0.61565)^2 + \cdots + 1.01787(1.63761 - 0.61565)^2$$

$$= 14.76650$$

The Q-value for differences among the groups can then be computed by subtraction

$$Q^*_{\text{between}} = Q^*_{\text{Total}} - Q^*_{\text{within}} = 14.76650 - 14.71985 = 0.04665$$

This value is to be compared to the $\chi^2_{[\nu]}$-distribution, where ν is equal to the number of groups minus 1. For these data, a probability of 0.82900 is obtained, so we cannot reject the null hypothesis that the odds ratios for Chinese and Japanese are the same.

Meta-analyses of frequency data can be extended to more complex designs by using logistic regression. This allows one to adjust odds ratios for the effects of covariates. Fleiss and Berlin (2009) discuss this approach. Borenstein et al. (2009) is a general introductory text on meta-analysis. Cooper et al. (2009) is more technical but still provides a comprehensive account of meta-analysis. Gurevitch and Hedges (2001) give an overview of meta-analytic methods with an example of their application in ecology. Gurevitch et al. (2001) is a more general review of applications in ecology.

18.2 Tests for Randomness of Nominal Data: Runs Tests

In this section, we discuss ways of testing whether events occur in a random sequence or whether the probability of a given event is a function of the outcome of a previous event. These tests, which are of great usefulness, are known as **runs tests** (for reasons that will become obvious shortly). Runs tests are best introduced by an example. Consider the following familiar hypothetical case, which has undisputable biological implications. Imagine a line of 20 youngsters in front of a cinema box office, with an equal number of boys and girls. They could be lined up in many ways based on the permutations of the 10 boys and 10 girls. We will not calculate exactly how many arrangements are possible but will concentrate on extreme departures from random arrangement. If these youngsters were eight-year-olds, then a natural arrangement would be 10 boys followed by 10 girls, shown symbolically as

BBBBBBBBBBGGGGGGGGGG

Clearly, this arrangement is not random and would be quite unlikely to occur by chance. The other extreme is an arrangement that would be likely in the same group of youngsters 10 years later:

BGBGBGBGBGBGBGBGBGBG

Note the regular alternation between the sexes. How can we devise a test for these departures from randomness?

Let us call a sequence of one or more like elements preceded and followed by unlike elements a **run.** For obvious reasons, the initial and terminal sequences can only be followed or preceded, respectively, by unlike elements. In our case, the eight-year-olds comprise two runs, one of 10 boys followed by one of 10 girls. On the other hand, the 18-year-olds comprise 20 runs, because each sequence consists of a single individual preceded or followed (or both) by a member of the opposite sex.

Statisticians have worked out the expected mean and distribution of the number of runs in dichotomized samples containing n_1 individuals of one type and n_2 individuals of the other. The critical numbers of runs at both tails of the distribution of runs have been tabulated for unequal sample sizes between 2 and 20 for n_1 and n_2 and for equal sample sizes from n between 10 and 100 (see Statistical Table **AA**). Any number of runs that is *equal to or less than* the desired critical value at the left half of any row in the table or is *equal to or greater than* the desired critical value in the right half of any row leads to a rejection of the hypothesis of random arrangement. For larger samples, we use a normal approximation to the expected value and standard deviation of the distribution of runs, as shown in Box 18.7. We can then test whether the observed number of runs leads us to reject the null hypothesis by deviating from the expected value, basing our conclusion on the table of areas of the normal curve (Statistical Table **A**). For small samples, there are few computations, and even for large samples the computation is quite simple.

Box 18.7 shows a runs test for dichotomized data, such as the sexual composition of the box office line we examined earlier. There are three major applications of the runs test for dichotomized data in biometric work.

The first of these is for a dichotomy that occurs naturally, as in differences between two sexes, two color phases, two different species, and so on. The example in Box 18.7 is such a case. Among bees and wasps, fertilized eggs give rise to females, unfertilized eggs to males. Females fertilize some of these eggs by releasing sperm stored in the spermatheca during copulation, while laying others unfertilized, which result in males. Are fertilized eggs laid randomly or do they occur in batches? In Box 18.7, the sequence of male and female eggs is shown, with runs underlined. There are four runs. Consulting Statistical Table **AA**, we find that if the sequence is random, the probability of as few as 4 runs or as many as 13 runs is only 0.05 (for a two-tailed alternative hypothesis). We may therefore conclude that the two types of eggs were not laid in a random sequence but were laid in sequences of one sex. The boys and girls lined up in front of the box office, discussed earlier, are another instance of naturally occurring dichotomies.

A second application of the runs test is for studying sequences of dichotomies defined by the investigator—for example, the plus and minus signs representing

---

BOX 18.7 A Runs Test for Dichotomized Data

A wasp produced 18 offspring in the following sequence, where F stands for females (fertilized eggs) and M for males (unfertilized eggs); n_1 (males) = 6; n_2 (females) = 12; r (number of runs) = 4 (for definition of runs, see text). The runs have been underlined.

<u>F F F F F</u> <u>M M M</u> <u>F F F F F F</u> <u>M M M</u>

Critical values ($P = 0.05$) for r from Statistical Table **AA**:
  Lower bound (from column 0.025) = $r = 4$
  Upper bound (from column 0.975) = $r = 13$

We conclude that the two types of eggs were not laid in random sequence. The eggs are laid in sequences of one sex. A sample value of $r = 13$ would have indicated an alternating sequence of male and female eggs.

When one of two unequal sample sizes is greater than 20, we use a normal approximation and test

$$t_s = \frac{r - \mu_r}{\sigma_r} = \frac{r - [2n_1n_2/(n_1 + n_2)] - 1}{\sqrt{[2n_1n_2(2n_1n_2 - n_1 - n_2)]/[(n_1 + n_2)^2(n_1 + n_2 - 1)]}}$$

where μ_r is the expected number of runs and σ_r its standard deviation. This formula seems formidable, yet is easy to work out because all the quantities in it are integer numbers. We calculated the P-value for the dataset of this box, despite the fact that its sample size does not meet the requirements stated in this box. We obtained $t_s = 2.755$, which yields $P = 0.005{,}869$, a substantially lower value than the 5% critical value that we could claim by using Statistical Table **AA**. However, we cannot rely on this P-value, because our data do not satisfy the minimum sample size requirement of 20 for the larger of the two samples. For $n_1 = n_2 = n$ this formula simplifies to

$$t_s = \frac{r - \mu_r}{\sigma_r} = \frac{[r - n - 1]}{\sqrt{n(n - 1)/(2n - 1)}}$$

We reject the null hypothesis at the 5% level if t_s is greater than 1.960 (using Statistical Table **A**). For equal sample sizes, however, Table **AA** furnishes critical values for r up to $n = 100$.

---

deviations from expectation in Table 5.4, the sex ratios in 6115 sibships of 12. Recall that we fitted expected binomial frequencies to these data, discovering that observed frequencies were higher than expected at the tails and deficient at the center of the distribution. The G-test for goodness of fit discussed in Section 17.2 led us to reject the null hypothesis and substantiated the departure from expectation. By looking only at the signs of the deviations, we are ignoring their magnitude. Thus, it is conceivable (but improbable) that we might have a sequence of positive deviations for the first half of the distribution, followed by a sequence of negative deviations, with all of these deviations being quite minute and the observed data fitting their expectations quite faithfully. In cases where the fit to expectation is satisfactory, however, the pattern of pluses and minuses among the deviations is much more likely to be random, and any departure from randomness, especially in the direction of long sequences, will be matched by large deviations from expectation.

Let us look at the pattern in column (6) of Table 5.4. There are nine plus signs and four minus signs, and only three runs in the data. We can reject the null hypothesis at $P = 0.025$. We conclude that the sequence of signs of the deviations is not random. Of course, this test does not prove that these data do not fit the binomial expectations. This was not the hypothesis being tested. The test does, however, by an extremely simple method, suggest that all is not as expected in this instance and leads us to further tests of various hypotheses.

The third major application of the runs test for dichotomized data, is the so-called **runs test above and below the median.** This test is especially useful in testing the random sequence of a series of observations. As you will recall, we have stressed that independence of the observations is a fundamental assumption of analysis of variance, without which the value of any statistical test is put into question (see Section 13.2). In Section 13.2, you were referred to a test to be introduced later. This is the runs test above and below the median. We find the median item in a given sample and label all items above the median item + and all those below it – (any observations equal to the median must be ignored). Then we array these items in their natural order in the sample and undertake a runs test of the pluses and minuses. If individual observations are independent of their predecessors—that is, the data are in random order—the values above and below the median (the plus and minus signs) should be in random sequence. If, on the other hand, the data lack independence, there may be either more or fewer runs than expected.

It is obvious how a sequence over an area such as a transect across a field might lack independence. Several plots might occur on a particularly rich patch of ground followed by other plots on relatively infertile soil. A similar lack of independence can occur in a temporal sequence. For example, a technician weighing specimens for a given analysis might commit various types of errors. He or she might have a bias for high readings, and would periodically overcorrect these by deliberate low readings, creating an unusual alternation of high and low readings—that is, a greater number of runs than expected. Or the balance used for weighing may gradually go out of adjustment during the experiment, yielding a trend in which all the later-weighed specimens would tend to be lighter than the earlier-weighed specimens. This phenomenon would tend to place earlier-weighed specimens above the median and later-weighed ones below the median. A similar trend would occur if the specimens had been dried in an oven from which they were all removed before weighing, permitting them to absorb water from the atmosphere before being weighed. Specimens weighed last would have had the greatest opportunity to absorb water and would likely be heaviest. In a perfect linear trend, the plus or minus deviations from the median would be in two groups to each side of the median. In such a case, the runs test provides a simple, distribution-free test of regression with a null hypothesis simpler than that of the ordering test discussed in Section 14.11. Neither test can substitute for regression analysis, of course, because it does not fit a line or allow one to predict the dependent variable.

We shall briefly consider a second type of runs test called **runs up and down** that is especially suited to trend data. If n items are ordered in their natural sequence of occurrence and the sign of the difference from the previous value is recorded for each item except the first, then we can carry out a runs test on the resulting sequence of $n - 1$ signs. These signs would all be alike if the data were monotonically increasing or decreasing. Cyclical data would show more than the number of runs expected in a random sequence of values. The expected distribution of runs has been tabulated for sample sizes up to $n = 25$ (see Statistical Table **BB**). The normal approximation shown in Box 18.8 is adequate for larger samples.

BOX 18.8 A Runs Test for Trend Data (Runs Up and Down)

Percent survival to the pupal stage in 32 generations of a line of *Drosophila melanogaster* selected for central pupation site. The data are graphed in Figure 18.1. From the differences between successive generations, the following set of signs is obtained (the runs have been underlined): $(+)$ and $(-)$ indicate an increase or a decrease, respectively, from the previous generation.

Generation	1	2	3	4	5	6	7	8	9	10	11
Increase or decrease		−	+	−	+	−	+	+	−	−	+

Generation	12	13	14	15	16	17	18	19	20	21	22
Increase or decrease	−	+	−	+	−	−	−	−	−	+	−

Generation	23	24	25	26	27	28	29	30	31	32
Increase or decrease	+	−	+	−	−	+	−	−	+	−

SOURCE: Data from Sokal (1966).

Number of points or elements, $n = 31$. Number of runs (sets of like signs preceded or followed by unlike signs), $r = 23$. Since $n > 25$, we test r by means of a normal approximation:

$$t_s = \frac{r - \mu_r}{\sigma_r} = \frac{r - [(2n - 1)/3]}{\sqrt{(16n - 29)/90}}$$

where μ_r is the expected number of runs and σ_r is its standard deviation. The numbers in this expression are constants independent of the specific problem. In this example,

$$t_s = \frac{23 - [61/3]}{\sqrt{[16(31) - 29]/90}} = 1.171, \quad P = 0.241,560$$

We accept the null hypothesis of a random trend, because 24% of the number of runs obtained at random would differ from the expected number of runs by as much or more than our observed difference. We cannot show any departure from a random trend in survival over the 32 generations tested.

Box 18.8 (continued)

Rejecting the null hypothesis at the 5% level would require the number or runs to exceed $\mu_r \pm 1.960\sigma_r$ runs, which works out to $20.333 \pm 1.960(2.278) = 15.87$ or 24.80 runs. Thus, with 15 or fewer runs, we would have concluded that one or more systematic trends were exhibited by the data. With 25 or more runs, we would be led to suspect a more or less regular alternation of differences, as might be caused by a cyclical phenomenon.

For sample sizes $n \leq 25$ the critical number of runs for several probabilities can be looked up directly in Statistical Table **BB**.

Box 18.8 illustrates a test for runs up and down. The data are from a selection experiment in which survival to the pupal stage was a measure of fitness, possibly correlated with the progress of selection. According to the simple analysis of this test, the results do not lead to rejection of the null hypothesis of no trend in successive differences. Visual inspection of the data, however, indicates an apparent downward trend followed by a recovery after generation 20 as Figure 18.2 shows. This discrepancy illustrates two points: the need for careful statistical testing of what may appear obvious, and that all runs tests apply to attributes only. Runs tests work with dichotomized attributes, such as pluses and minuses or males and females, but they take neither absolute magnitude nor order of magnitude into consideration. Therefore, a small positive deviation counts as much as a large

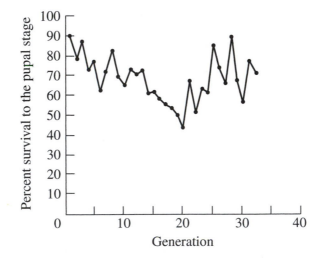

FIGURE 18.2   Percent survival to pupal stage in the CP line of *Drosophila melanogaster* selected for central pupation site. (Data from Sokal, 1966.)

positive deviation. Had the experiment of Box 18.8 and Figure 18.2 been terminated at generation 20, then we could have demonstrated a downward trend using ordinary regression analysis, even though this trend cannot be shown by the runs test, which uses less information from the data. With this caution, we readily recommend runs tests as "quick and dirty" methods for assessing trends and randomness. Runs tests are also useful when the investigator is unwilling to make assumptions necessary for more elaborate tests, such as a regression analysis. When ranks or actual measurements are not at hand, runs tests are often the only practical alternatives.

18.3 Isotonic Regression

In Section 14.5, we learned that a single-classification anova, whose groups correspond to a measurement variable X, can be tested against a stricter alternative hypothesis than that applied in the anova. The conventional alternative hypothesis of anova is that at least one group mean differs from the others. The stricter hypothesis relates to regression and states that the group means are a linear function of the variable X. Many situations arise in science, however, where the investigator is not prepared to posit such a strong alternative hypothesis. He or she may be prepared to say only that the means are in monotonic order, or, weaker yet, that they are in isotonic order, meaning that designated means are (1) greater than or equal to or (2) less than or equal to each other. These situations include not only cases in which the category differentiating the groups is a quantitative variable but also those in which it is a qualitative variable—for example, the effects of different drugs for which the experimenter has a clear notion of the ordering of their anticipated effects. In Sections 7.7 and 9.4, we mentioned that when there are only two groups, you can increase the power of the test considerably if you anticipate the direction of the differences and can carry out a one-tailed test. There is no direct analogue of a one-tailed test in an analysis of variance with many groups, but by specifying an alternative hypothesis with ordered expectations, one can increase the power of the analysis considerably. Gaines and Rice (1990), who have brought this approach to the attention of biologists, have referred to it as **isotonic regression.**

In Box 18.9 we analyze the dataset from Exercise 14.13, which gives the weight of brown trout at four densities in increasing order. When we analyze these data by analysis of variance, we are led to reject the null hypothesis of no differences among densities, but if we carried out Exercise 14.13, we could not demonstrate linear regression. The data do not conform very well to a linear regression model. In the second density, 350 fish/container, the weight of the fish was higher than at the preceding lower density and at the succeeding higher density. Could we have altered our conclusions had we had relaxed our alternative hypothesis, asking only for isotonicity of the means? Note that for analysis of variance, any inequality among the means, such as the ones observed in this data set, enlarges the mean square among groups,

---

BOX 18.9 Isotonic Regression

Effect of crowding on weight of *Salmo trutta*, the brown trout; X = number of fish per container; Y = weight in grams; n = 25 (observations per group); a = 4 groups. Original data are not available.

	Density = X			
	175	350	525	700
$\bar{Y}_i$	3.51224	3.71020	2.92680	2.53960

SOURCE: Unpublished data of J. Calaprice.

Completed anova table with regression

	Source of variation	df	SS	MS	F_s	P
$\bar{Y} - \bar{\bar{Y}}$	Among densities	3	21.63688	7.21229	6.944	0.000,800
$\hat{Y} - \bar{\bar{Y}}$	Linear regression	1	17.12471	17.12471	7.590	0.110,377
$\bar{Y} - \hat{Y}$	Deviations from regression	2	4.51218	2.25609	2.172	0.119,518
$Y - \bar{Y}$	Within groups	96	99.71534	1.03870		
$Y - \bar{\bar{Y}}$	Total	99	121.35222			

Computation

1. The null hypothesis of the anova is $H_0: \mu_1 = \mu_2 = \cdots = \mu_a$. The alternative hypothesis of the ordinary least squares regression as computed in the anova is $H_1: \mu_Y = a + bX$. The simplest alternative hypothesis of isotonic regression as employed here is $H_1: \mu_1 \geq \mu_2 \geq \cdots \geq \mu_a$, with at least one inequality retained. This alternative hypothesis is appropriate for this example of density effects with a negative regression slope. In the case of positive regression, the inequality signs would be reversed.

2. We make the data conform to the specifications of the alternative hypothesis. Since $\mu_{175} < \mu_{350}$, we amalgamate these two groups and compute their weighted mean:

$$\frac{1}{50}[(25 \times 3.51224) + (25 \times 3.71020)] = 3.61122$$

This procedure is repeated (always starting at the same end of the array of means) until the specifications are met. In our example, a single amalgamation, as shown, is sufficient. We call the means of the new amalgamated sample $\bar{Y}_i^*$.

Box 18.9 (continued)

3. Compute a new test statistic, $E^2 = SS_{among}/SS_{total}$, where SS_{among} is based on the $m \leq a$ amalgamated means $\bar{Y}_i^*$:

$$E^2 = \frac{\sum\limits_{}^{m} n_i(\bar{Y}_i^* - \bar{\bar{Y}})^2}{\sum\limits_{}^{m}\sum\limits_{}^{n_i}(Y - \bar{\bar{Y}})^2} = \frac{21.14703}{121.35222} = 0.17426$$

4. The general method for evaluating the probability of obtaining an E^2 value equal to or greater than the observed value is to perform a sampling experiment, since available tables are limited to equal sample sizes and only a few groups. For the present problem we drew 4 random samples of size 25 from the same normal distribution 10,000 times and then counted how often we obtained an E^2 value that was equal to or greater than the observed value. Thus we assume normality and homoscedasticity of the data. None of the samples from these data resulted in an E^2 value larger than the observed, so the probability is $\leq 10^{-4}$.

thereby decreasing the P-value and firming up the rejection of the null hypothesis. But, if the hypothesis that the investigator wishes to test is either linear regression or isotonic regression, then the excessive magnitude of the second group should not contribute to the rejection of the null hypothesis because it counters the expectations of the alternative hypothesis.

Box 18.9 illustrates isotonic regression analysis. In step **1**, we specify an alternative hypothesis. We describe only the simplest isotonic regression model; other models are given in Gaines and Rice (1990). In step **2**, we make the group means conform to the expectations of the alternative hypothesis that the means stay the same or decrease as density increases. Because $\mu_{350} > \mu_{175}$, we average μ_{350} with its predecessor, making the remaining three groups $\mu_{175+350}$, μ_{525}, and μ_{700}. If this were insufficient to achieve the expected relationship, then further merging would be done until the array of means was isotonic. We now arrive at $m < a$ merged classes. In our case, $m = 3$. In step **3** of Box 18.9, we compute the sum of squares of the m groups and divide it by the total sum of squares of the data set. The ratio between these two quantities is our test statistic, E^2. The distribution of E^2 is a complicated function and has not been tabled, except for the case of a few small equal-sized samples. Gaines and Rice (1990) suggest that one determine the probability by a sampling experiment as described in the box.

Note that the outcome of the isotonic regression test rejects the null hypothesis at a minute $P < 0.0001$, because none of the 10,000 random trials resulted in an E^2-value as large as that for the observed sample. Thus, although we cannot claim a linear regression for the means, we can at least claim an isotonic decrease in the means in response to increases in density. A randomization test to determine the probability of the observed E^2-value would also be possible. We would perform a randomization test if we were concerned about the assumption of normality and homoscedasticity for these data (but such a test would require access to the original data, which were not available in this case).

18.4 Application of Randomization Tests to Unconventional Statistics

An important property of randomization tests is that they can be applied to unconventional statistics that seem appropriate in a particular application. To illustrate this, we have taken a problem from numerical taxonomy. In this example, we compute the probability of an observed sample under the null hypothesis for an unconventional statistic. Table 18.1 is a half-matrix of correlation coefficients indicating the degree of similarity among 10 species of bees whose numerical code names are shown at the margins of the matrix. The correlation coefficients are computed from a set of observations representing the characters of these species. The matrix enables us to

TABLE 18.1 Similarity Coefficients (Correlation Coefficients) Among 10 Species Selected from the *Hoplitis* Complex of Bees

Species code numbers are shown at the margins of the half-matrix.

					Species					
Species	**4**	**5**	**8**	**26**	**35**	**36**	**40**	**50**	**67**	**68**
4	—									
5	.65	—								
8	.70	.84	—							
26	.40	.43	.49	—						
35	.41	.45	.49	.60	—					
36	.41	.42	.48	.52	.94	—				
40	.17	.39	.44	.57	.42	.29	—			
50	.35	.48	.52	.45	.41	.33	.51	—		
67	.36	.37	.41	.20	.25	.29	.20	.49	—	
68	.33	.36	.43	.18	.25	.29	.19	.46	.96	—

SOURCE: Data from Sokal (1958).

look up the similarity between any two species. Thus, the similarity between species 26 and 36 is 0.52.

Conventional taxonomic work has resulted in species 4, 5, and 8 being assigned to a single genus, *Proteriades*. We will now justify this judgment by developing a criterion for distinctness of a group. By distinctness, we mean a measure of homogeneity or cohesion of the members of a group relative to their similarity with other species. A very simple and intuitively obvious method for determining distinctness is to take average similarity among the members of the group (that is, the average similarity between species 4 and 5, 4 and 8, and 5 and 8) and subtract from it the average similarity of each of the species 4, 5, and 8 with all the other species that are not in this group. This calculation yields a positive or negative value that indicates the distinctness of the original group. The higher the positive value, the more distinct the group; that is, similarity is high within the group and low with species outside the group. Conversely, if the distinctness value is negative, the average similarity with species outside of the group is higher than that within the group; hence there would be little justification for its formal recognition. When we calculate the distinctness value for the group containing species 4, 5, and 8, we obtain $1/3(0.65 + 0.70 + 0.84) - 1/21(0.40 + 0.41 + 0.41 + \cdots + 0.43 + 0.45 + \cdots + 0.49 + \cdots + 0.41 + 0.43) = 0.730 - 0.409 = 0.321$, indicating that this group is markedly distinct from the other species.

How can we compute the probability of obtaining such a value, or a more extreme one, by chance alone? The similarity coefficients (in this case correlation coefficients) in the half-matrix of Table 18.1 have an unknown distribution, and the distribution of the distinctness values is even less evident. The problem was solved by computing distinctness values for all possible samples of 3 out of the 10 species. Our null hypothesis is that all groups of 3 species from among the assemblage of 10 bee species equal each other in their distinctness values. In such a case, we could not justify retaining the genus *Proteriades*. There are $\binom{10}{3} = 120$ different ways of making sets of 3 out of 10 species. The distribution of the 120 distinctness values is shown in Figure 18.3. We find that 5% of the distinctness values lie beyond -0.148 and 0.210 at the two tails of the distribution. Thus, the distinctness value 0.321 of the group containing species 4, 5, and 8 leads us to reject the null hypothesis. By contrast, an artificial group made up of species 4, 26, and 50 (which have been placed in different genera) forms a group with little cohesion (distinctness $= -0.039$, and we cannot reject the null hypothesis according to Figure 18.3). Had this problem included more species (hence led to more possible outcomes), we might have decided to carry out a sampled randomization test (discussed in Section 7.1) by taking several hundred samples and computing distinctness values for these, rather than for all outcomes. In a similarity matrix among 100 species, there are 161,700 possible sets of 3; we might not want to compute a distribution of distinctness values for all of them.

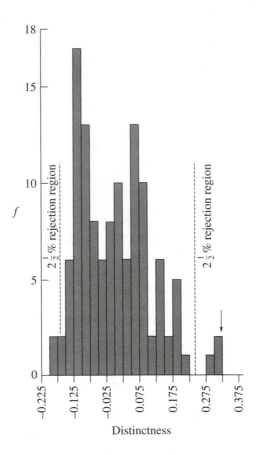

FIGURE 18.3 Frequency distribution of distinctness values for all sets of 3 species from 10 species of bees (see Table 18.1). The broken lines mark off $2\frac{1}{2}\%$ rejection regions. The arrow identifies the distinctness value of the group being tested, that containing species 4, 5, and 8.

18.5 The Mantel Test of Association Between Two Distance Matrices

A test that has enjoyed considerable popularity in recent years is the **Mantel test** (Mantel, 1967; Sokal, 1979). It is a randomization technique that applies to data expressed as dissimilarities. In evolutionary biology and ecology, dissimilarity coefficients are frequently used to measure the degree of difference between individuals, populations, species, or communities. We have not yet encountered such coefficients in this text. Sometimes distance coefficients are used (actual geographical distances or multidimensional distances). The correlations between species of bees, discussed in the previous section, can easily be turned into dissimilarities by subtracting them from 1. Then instead of indicating greater similarity, higher values of the coefficient would indicate greater dissimilarity. Dissimilarities can be computed from paired strands of DNA in different species, from gene frequencies that describe different population samples, or from morphological characters measured for individuals within a sample or among an array of samples. For purposes of learning the Mantel test, we will not be concerned with different methods for computing dissimilarities between a set of objects ($=$ individuals, populations, species, communities), but we will assume these as givens. Readers specializing in the fields employing dissimilarities will know how to compute them. If dissimilarities are computed between all pairs of n entities, then these values can be arranged in an $n \times n$ dissimilarity matrix that, because it is symmetric, can be displayed as a half-matrix of $n(n - 1)/2$ elements.

A Mantel test is used to estimate the association between two putatively independent dissimilarity matrices describing the same set of entities and to test whether the association is stronger than one would expect from chance. For example, we may be interested in learning whether genetic distances between populations are related to their morphological dissimilarity. We would construct a genetic-distance matrix based on sequence similarity of nuclear or mitochondrial DNA or on gene frequencies for various loci estimated for these populations. We would also construct a dissimilarity matrix based on morphological measurements. We would then ask whether larger genetic distances are associated with larger morphological dissimilarities. The Mantel test posits the null hypothesis that there is no association between the elements in one matrix and those in the other. Alternative scenarios include tests of whether differences in song patterns, measured spectrographically in local populations of birds, are related to genetic differences between these populations; whether language differences in human populations are related to genetic distances between them; and whether distances based on species composition of different habitats are related to microclimatic differences between these habitats. As we will see, the Mantel test is a very powerful procedure that can test other biological questions that do not appear at first sight to relate to dissimilarities but can be formulated to make them amenable to analysis by this test.

The example we will use to illustrate the Mantel test concerns genetic distances and geographic distances (in kilometers) between all pairs of 10 human populations. Are the genetic distances associated with the geographic distances? This is a question of spatial differentiation. If these populations had differentiated in situ, then it would stand to reason that geographically closer populations would also be closer genetically. If, on the other hand, the populations had differentiated elsewhere, and then migrated into their current locations, no such association need occur. The data for this example, shown in Box 18.10, are from a study (Sokal et al.) of the Yanomama Indians, who live in a border area between Brazil and Venezuela and who have been studied very intensively by J. V. Neel and his associates. These Amerindians are relatively untouched by Western civilization. All together, 50 populations

BOX 18.10 Mantel Test of Association Between a Genetic
and a Geographic Distance Matrix

Genetic and geographic distances computed for 10 villages of the Yanomama Amerindians.

Genetic distances ($\times 10^5$):

Village	3G	8L-11P	8F	8ABC	11YZ	11D	11X	3RS	80	15L
3G	0000									
8L-11P	2040	0000								
8F	4143	3944	0000							
8ABC	1213	2549	5285	0000						
11YZ	1246	1362	4473	1682	0000					
11D	2868	3908	4109	2195	4362	0000				
11X	3393	3347	3313	4774	4928	4567	0000			
3RS	3023	2816	8473	2804	2605	5730	5661	0000		
8O	3784	3229	3276	2445	2767	2986	4330	4805	0000	
15L	5698	6308	7491	3722	3377	6550	10242	5331	3661	0000

Geographic distances (in km):

Village	3G	8L-11P	8F	8ABC	11YZ	11D	11X	3RS	80	15L
3G	0.0									
8L-11P	38.7	0.0								
8F	254.4	220.5	0.0							
8ABC	43.8	81.3	286.3	0.0						
11YZ	116.2	147.6	368.1	102.3	0.0					
11D	86.2	88.9	216.0	89.0	190.5	0.0				
11X	86.4	57.1	168.5	118.8	201.9	72.4	0.0			
3RS	174.2	203.3	349.6	136.1	203.6	141.0	211.1	0.0		
8O	118.9	83.3	137.3	155.3	230.8	109.0	39.0	249.3	0.0	
15L	303.4	298.3	249.2	299.9	401.6	217.2	252.9	250.5	263.8	0.0

SOURCE: Data extracted from Sokal et al.

Computation

1. The Mantel coefficient Z is computed using Expression (18.1):

$$(0.02040 \times 38.7) + (0.04143 \times 254.4) + \cdots + (0.03661 \times 263.8) = 365.03379$$

The null hypothesis of no association between the elements of the two distance matrices is evaluated by random permutation of rows and columns of the geographic-distance matrix (see Table 18.2), followed by a recomputation of the Mantel coefficient. This procedure is repeated many times, until a reference distribution of Z-values has been established. The observed Z-value (in this case 365.03379) is then compared with this reference distribution.

Box 18.10 (continued)

2. If the number of replicated randomizations is N, the cumulative short tail of the distribution is evaluated as $(n_T + 1)/(N + 1)$, where n_T is the number of randomized Z-values equal to or above (or equal to or below) the observed value. In our case $n_T = 1, N = 249$, and the right-hand tail of the distribution is $(1 + 1)/(249 + 1) = 0.008$.

3. Alternatively, the correlation between the two matrices can be computed: $r = 0.51840$. The significance of this correlation can also be tested by random permutation of rows and columns of the geographic-distance matrix and recalculation of the correlation coefficient. If the same rows and columns are permuted as in step **1**, the numerical result will be identical. Otherwise, the results are likely to differ slightly.

---

were studied, but for this example we selected at random one sample from each of 8 village clusters and 2 samples from the Parima village cluster, making a total of 10 samples. We matched their genetic distances (based on 15 independent allele frequencies) with geographic distances between any two villages computed from their geographical coordinates.

As we show in Box 18.10, the Mantel statistic Z is computed as

$$Z = \sum_{i=1}^{n-1} \sum_{j=i+1}^{n} X_{ij} Y_{ij} \tag{18.11}$$

where X_{ij} and Y_{ij} are elements of the two distance matrices **X** and **Y**, respectively. This is called the *Hadamard product* of two matrices. It is the sum of the products of corresponding off-diagonal elements in the two matrices. Thus, the computation proceeds as shown in step **1** of Box 18.10, yielding 365.03379. If large genetic distances X_{ij} match large geographic distances for corresponding elements Y_{ij}, then the value of Z will be larger than average. Conversely, if there is a negative association—that is, if large values of X_{ij} match small values of Y_{ij} (and vice versa)—then the quantity Z will be small. To test whether the Z is an unusually deviant value, we can carry out a randomization test. Whenever application of the Mantel test involves a substantial number of entities (more than eight), we carry out a sampled permutation test in which the elements of one matrix are randomly rearranged. We could permute the elements of one of the matrices (it does not matter which one) randomly, each time calculating Z for, say, 499 randomized values, making the observed Z the 500th. Then we could evaluate the probability of the observed value of Z by noting its position in the distribution of randomized outcomes. We must permute the n rows (and columns) of the matrix. Therefore, if we interchange entities 1 and 3, for example, these two rows and their corresponding columns in the matrix must be interchanged in their entirety. We show such an

TABLE 18.2 Random Permutation of Rows and Columns of a Distance Matrix

A schematic 5 × 5 distance matrix is shown here, followed by a random permutation of its rows and columns.

Distances

	1	2	3	4	5
1	0				
2	20	0			
3	41	39	0		
4	12	25	53	0	
5	13	14	45	17	0

A random permutation of rows and columns

	2	1	5	4	3
2	0				
1	20	0			
5	14	13	0		
4	25	12	17	0	
3	39	41	45	53	0

interchange for a 5 × 5 matrix in Table 18.2. Because an 8 × 8 matrix will have 8! = 40,320 permutations, you can easily see why we will usually take a sample of these permutations rather than enumerating them completely. Usually a few hundred to a few thousand permutations are sufficient to obtain a good enough estimate of the probability.

The Z-statistic is expressed in arbitrary units, and its implications are hard to understand without an examination of the frequency distribution of the outcomes. For this reason, workers often employ the standardized Mantel coefficient, which is the product–moment correlation between the elements of the two dissimilarity matrices. However, one cannot use conventional probability distributions for correlation coefficients (such as are shown in Statistical Table **R**), because the elements of the matrices are not independent bivariate observations, as is assumed by correlation theory. Instead we must enumerate the many permutations of rows and columns of one of the matrices, computing a correlation coefficient each time, and compare the observed coefficient against the reference distribution. A product–moment correlation coefficient can be computed automatically by a Mantel program if the matrices are first transformed by a suitable linear transformation. In consequence, the P-value of a Mantel coefficient will be exactly the same as that for the standardized coefficient. The application of the Mantel test requires a computer.

For the example in Box 18.10, we obtained a standardized Mantel statistic of $r = 0.51840$ for the association between genetic and geographic distances. The probability of this observed value was estimated as $P = 0.008$. (We carried out 249 random permutations of the rows and columns of the geographic-distance matrix. When the resulting correlation coefficients were arrayed in a frequency distribution, the observed value, the 250th item in the sample, ranked second from the right tail, yielding a probability of $P = 2/250 = 0.008$.) Spatial distance clearly is correlated with genetic distance.

When the size of the distance matrix is substantial (say, of dimension greater than 100×100), the computer time required for computation of a sampled permutation test begins to be prohibitive. Mantel (1967) worked out an approximate standard error for Z, which is asymptotically normally distributed. We find that this standard error gives good approximations to the sampled randomization probability values, even when matrices as small as 25×25 are analyzed. The formula for the standard error is complex and is not given here, but obtaining the outcome of a Mantel test by an asymptotic approximation is an option in most computer programs for the Mantel test. When the asymptotic approximation was applied to our 10 Yanomama samples, we obtained $t_{[\infty]} = 2.21862$, which yields a probability of $P = 0.01326$. Although not as small as that obtained with the random permutation test, we are still led to reject the null hypothesis by conventional criteria.

We alluded in the preceding discussion to the wide applicability of the Mantel test. We have found that it can be used as an excellent nonparametric equivalent of analysis of variance as follows. Visualize a simple single-classification anova, with $a = 3$ groups and $n = 5$ replicates per group. Create a dissimilarity matrix between all pairs of 15 observations. The elements can be as simple as absolute differences between the observations. If the dissimilarity matrix is arranged as shown in Figure 18.4a, you will notice that all within-group dissimilarities are found in triangular submatrices, whereas all the between-group dissimilarities are shown as square submatrices. (They would be rectangular if sample sizes in the groups were unequal.) Next, we construct a design matrix that matches the structure of the dissimilarity matrix. We place 0's in the triangular within-group submatrices and 1's in the square between-group submatrices (Figure 18.4b). We now subject these two matrices to a Mantel procedure. If differences between groups are greater than those within groups (rejection of the null hypothesis in an anova-like design), then the 1's in the design matrix will be associated with the largest differences. When we randomly permute rows and columns of one of these matrices, we find that the 1's no longer are necessarily associated with differences between groups, and the randomized Z-values are lower than the observed ones. Sokal et al. (1993) have shown that the power of such a test is nearly the same as that of an anova. By constructing various types of design matrices, a variety of hypotheses can be

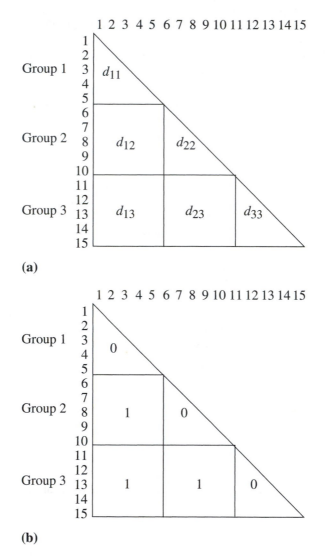

FIGURE 18.4 Schematic distance **(a)** and design **(b)** matrices for carrying out a nonparametric analysis of variance by means of a Mantel test. The data consists of 15 observations grouped into 3 groups of 5 each. The d_{ii} symbolize distances within groups, the d_{ij} distances between groups. The design matrix replaces within-group distances by zeros, between-group distances by ones.

tested. For an illustration of the construction of such design matrices, see Livshits et al. (1991).

Pairwise association between matrices has been extended to include three or more matrices by Smouse et al. (1986); see also Oden and Sokal (1992). For a discussion of distance coefficients see Sneath and Sokal (1973) and for a more up-to-date account with special reference to ecology consult Legendre and Legendre (1998).

18.6 The Future of Biometry: Data Analysis

Although we indicated in Chapter 1 that biometry is a rapidly changing field and that it would be risky to try to anticipate its future development, we will nevertheless comment on one current trend. Our discussion of randomization tests in Chapter 7 and the current chapter lead us naturally into an area of great promise known as **data analysis**. Data analysis is defined as the systematic search through a set of data to reveal information and relationships of interest. The procedures of data analysis, although essentially numerical, are also experimental. In the previous two sections, we were, in a manner of speaking, carrying out experiments with a set of data. Data analysis consists of such techniques as transformations, to see what effect they have on the distribution of a sample or on the relationships between two variables in a sample; robust estimation, by minimizing the effects of a certain percentage of the extreme observations; the examination of residuals and influence functions; and especially graphic techniques, to suggest interesting patterns in the data. In essence, these techniques are nothing new. They have been carried out by perceptive research workers and statisticians for many years. The development of high-speed computation—and especially of interactive data processing—however, permits an almost instantaneous reply to any query an investigator wishes to ask of a set of data. Therefore the investigator can ask many more questions, in the hope that some of them will yield insights through results that can then be followed up with additional questions or with subsequent experiments to elaborate and confirm new results.

Why should there be this emphasis on analyzing the actual research data instead of computing statistics that are abstractions or summaries of these data? More and more data are being obtained because of the enormous expansion of scientific activity and the development of automatic data-gathering devices. Research workers in many disciplines are confronted with masses of data the like of which have never been reckoned with before. Digestion of these data presents a serious challenge to the meaningful interpretation of underlying factors and regularities. In addition, there is an ever-increasing tendency toward quantification in all fields of human knowledge, as numerical methods enter fields previously untouched by quantification. Most important, however, is the revolution in the speed and capabilities of computing devices and in visual and graphic display units for the results of computation. At a computer terminal or workstation, it is now easy to truncate a sample and automatically recompute analyses or transform a variable and study its new distribution curve, with all of these computations being carried out essentially instantaneously. Many attempts at data analysis, which previous researchers might wistfully have thought of undertaking but abandoned because of the amount of labor involved, can now be easily accomplished. Clearly, much of this analysis will be useless and wasted, but enough new information should emerge to give the overall approach great promise.

We can view data analysis as the systematic and largely automated scrutiny of sets of data to yield both summaries of and fresh insights into the relationships in a given study. Tukey (1977) places special emphasis on the analysis of residuals when carrying out an anova or a regression analysis. In data analysis, as much attention is paid to the residual deviations as to the main effects and interactions. An unusually large or small deviation is examined critically to discover reasons for such a deviation. Only with automatic computational devices is it feasible to inspect, routinely and systematically, all residual deviations. Because these deviations represent the remaining variability, they are, of course, the object of inquiry for further scientific research. General rules and procedures for data analysis cannot yet be given, inasmuch as the hardware and software necessary to carry out such work are not yet sufficiently standardized. It is quite clear, however, that data analysis will continue to be a major aspect of biometric work. One of the problems encountered in data analysis is that the longer one spends adjusting the data and checking various models, the less reliable the final statistical inferences become for those data due to the need to correct for multiple comparisons. Some workers unthinkingly apply various statistical tests to the same data and then choose the test results that suit them best. Such a procedure is analogous to selecting data to fit models and then presenting the spectacularly accurate regression equation as if it represented a genuine random sample.

EXERCISES 18

Part I. Exercises 18.1 through 18.5 relate to the subject matter of Chapter 18.

18.1   An investigator believes that an increase in temperature speeds development of the immature insects she is studying. This is a one-tailed hypothesis; the null hypothesis is that high and low temperatures produce equal developmental rate; the alternative hypothesis is that high temperatures produce more rapid rates. In the first experiment, 25 insects in individual containers are placed in each of two temperature chambers at low and high temperatures, respectively. The mean developmental time at the low temperature is 11.6 days, at the high temperature 7.4 days. Analysis of variance results in a value of $F_s = 2.94$. In a second experiment, the investigator constructs an ingenious device in which the rank order of pupation in two experiments is recorded without the individuals being inspected twice a day, as had been necessary in the first experiment. With these rank orders of 50 pupae in each of two temperature conditions, she is able to carry out a Mann–Whitney U-test of the observations, which suggests again that the individuals at the higher temperature pupated earlier. The test statistic $t_s = 1.55$. Finally, the investigator wished to test so many individuals that neither of the previous two experimental setups would have been practical. She therefore arbitrarily started 500 larvae at low temperature and 495 larvae at high temperature. Seven days later, 203 individuals in the low-temperature chamber had pupated, and 241 individuals of the high-temperature culture had pupated.

Do the individual tests each support acceleration of development at higher temperatures? Do the three tests jointly show this?

ANSWER: $-2\sum \ln P = 16.738, 6\,df, P < 0.025$ (using G with Williams's correction for the third experiment).

18.2   As part of a larger study, Strum et al. (2010) published the following data on two types of plasma cholinesterase activities (AChE and BChE; units mL^{-1} plasma) for shore birds captured in South America. Is there evidence for an overall difference between the reference and agricultural sites? Analyze and prepare forest plots separately for the two types of cholinesterase activity.

Species	ChEtype	Reference Sites			Agricultural Sites		
		n	Mean	SD	*n*	Mean	SD
American golden plover	AChE	27	0.37	0.13	18	0.40	0.07
	BChE	27	1.08	0.30	18	1.14	0.33
White-rumped sandpiper	AChE	64	0.59	0.19	16	0.57	0.19
	BChE	64	2.32	0.87	16	2.34	0.61
Pectoral sandpiper	AChE	3	0.52	0.03	39	0.39	0.14
	BChE	3	1.84	0.32	39	2.09	0.53
Buff-breasted sandpiper	AChE	15	0.41	0.16	8	0.26	0.13
	BChE	15	1.79	0.28	8	1.23	0.66
South American	AChE	3	0.44	0.17	2	0.47	0.11
painted snipe	BChE	3	0.89	0.17	2	0.58	0.10

18.3 In addition to the data shown in Box 18.5, Berger et al. (2006) also published the following data on hemorrhagic stroke in relation to aspirin therapy. Analyze and interpret.

Studies	Aspirin		Control	
	Events	None	Events	None
Women				
1	0	1277	2	1304
2	51	19,891	41	19,901
Men				
3	13	3416	6	1704
4	23	11,014	12	11,022
5	2	947	1	962
6	12	2533	25	16,222

18.4 Test the fit of the linear, quadratic, and cubic regression analyses performed in Exercise 16.3 using a runs test on the signs of the deviations from regression.

18.5 Test for a trend in the flow variable F_7 in Exercise 16.4 using the runs up and down test and the runs above and below the median test. Comment on why they yield different results for these data.

ANSWER: Seven runs up and down; two runs above and below median.

Part II. The following exercises range over the entire subject matter of the book. These problems do not require any computation, but serve as review problems for the student, who is required to indicate the appropriate method for a solution. The following steps should be taken for *each* of the exercises that follow:

a. State the statistical method appropriate for providing the solution, and why.
b. Outline the setup of the data.
c. Give formulas for computations.
d. State degrees of freedom; if none applicable, state "none."
e. Draw attention to any requirements to which the data must conform to be tested by the method you have chosen, as well as to any corrections that might need to be applied.
f. Enumerate the specific answers that could emerge as the result of your method.
g. List any alternative method applicable.

18.6 A normal probability curve has been fitted to certain data. There are 15 classes. Are the deviations such as could be expected from accidents of sampling?

18.7 The amount of inorganic phosphorus in the blood plasma of 10 unrelated men is determined from specimens taken every morning for two weeks. Are there differences among individuals or does the amount vary from morning to morning? Would you expect the interaction to be negligible? If not, how would you design the experiment to demonstrate interaction?

18.8 An ecologist claims that there is a narrow optimum size for a burrowing animal, above or below which the animal is at a disadvantage. How could you investigate this hypothesis in a frequency distribution of 300 body diameters of this species?

18.9 You have data on the physical characteristics of 1221 sets of 4 siblings each. Among other traits, you know whether each individual's eye color is blue or not. Without studying any pedigrees, how could you demonstrate that blue eyes are controlled wholly or in part by genetic factors? (If we can prove that sibs are more like each other than are randomly chosen individuals, we will consider this sufficient evidence, although theoretically sibs could be more alike because of common environment as well.)

18.10 We are measuring tarsal claw length of samples of male lice of species P taken from 8 different individuals of host species X. Is there an added variance due to differences among hosts? Remember that we are unlikely to obtain the same number of lice from each host individual.

18.11 Counts are made of the number of mites found on blades of grass species G in the sunlight and in the shade. Twenty blades of grass in the sun and 31 in the shade were examined. The number of mites per blade varied from 3 to 32. Are there more mites per blade in the shade?

18.12 An animal behaviorist is studying mating behavior in animal species X. Females of equal ages are introduced singly into a mating cage, where they are exposed to several males of known and demonstrable aggressiveness. In species X, however, the female can accept or reject copulation at will. Thus, the number of copulations per unit time is an index of the female's submissiveness or sexual appetite. The investigator tests 100 females, and they range from 0 to 11 matings per unit time, with a mean of 3.2 matings. His hypothesis is that the females can be subdivided into at least two races or groups differing in submissiveness. How can he support his findings with an analysis of this experiment? Another way of looking at this analysis is to ask what results the scientist would expect if copulations were independent of the genetics of the female— that is, if they were merely a randomly varying variable.

18.13 A study of respiration under the influence of drugs in salamanders yields data on seven animals for each of five drugs and seven control animals. The observations are volume of oxygen consumed, and a preliminary investigation shows that the variances for each drug are anything but homogeneous. Before the investigator attempts to equalize variances by transformation, how could she test for differences among the drugs? If such differences are not present, she might not think it worthwhile to spend time looking for a suitable transformation.

18.14 An investigator studying weight gain in rats on different diets is comparing two groups of 20 rats each. Not only are the variances of the two samples different but also the observations are clearly not distributed normally. No transformation known to the investigator appears able to take care of the difficulty. Suggest a method of testing for differences in weight gain between the two samples.

18.15 Samples of moths were collected from four localities. The percentage of melanistic individuals varied from sample to sample. Can the four samples be considered as coming from a homogeneous population with regard to incidence of melanistic forms? (You possess the original data from which the percentages were calculated.) If not, are there any homogeneous subsets of localities?

18.16 An investigator wishes to demonstrate normality in a great number of frequency distributions. Computer facilities are not available. None of the frequency distributions has more than 20 items.

18.17 The lengths of 50 skulls of a certain species of field mouse from each of two localities are measured. Can you demonstrate a difference in average length? (Variances of the two samples do not appear to differ greatly.)

18.18 You want to determine whether variations in body weight and amount of bile pigment in urine are due to common causes. You have data on 1243 male white Americans, ages 45 to 50.

18.19 The number of nests of bird species A per square mile of a certain forested area has been determined on the basis of a study of 60 square miles. This bird is known to be territorial (that is, nonrandomly distributed). How can you show this from your data, and would data based on square mile units necessarily show the phenomenon?

18.20 The mean number of dorsocentral bristles in dichaete *Drosophila melanogaster* is determined for flies emerging on each of five two-day intervals. Separate readings are obtained for each of 12 replicate bottles. Can the whole set of data be treated as if from a homogeneous population?

18.21 A biologist collects 357 specimens of a species of lizard in one locality. She measures tail length on all of them. How many must she measure in the future from any other locality so that a 5% difference of the current mean tail length of her lizards will lead her to reject the null hypothesis of no difference?

18.22 The variation "banded" is found in the dragonfly species L. Both males and females show this variation. During an afternoon's intensive collection in a rice paddy in the South, students in a field zoology class collect 117 mating pairs (in copula). As each pair is collected, the students classify it into one of the following four categories:
a. both sexes common (unbanded) type
b. both sexes banded
c. ♀♀ common, ♂♂ banded
d. ♀♀ banded, ♂♂ common

We wish to test whether mating was at random or whether these dragonflies prefer to mate with their own type.

18.23 Sixty-seven field mice and 23 pocket gophers were collected from a certain area. Femur length is measured for both species. Is length of the femur more variable in the mice?

18.24 Seven doses of chemical A (1, 2, 4, 8, 16, 32, and 64 mg) are administered to 3-day-old chicks, 15 chicks per dose. All 105 chicks are caged together and distinguished by head stains denoting the dose. At the end of 3 weeks, the chicks are killed, and the glycogen content of the liver is determined separately for each chick. Did changes in the dose have effects on glycogen content, and can this relation be quantified? Is a linear equation adequate to describe the effect of a dose of chemical A on glycogen content?

18.25 A botanist collects five samples of herbs of species B from different localities. The sample sizes vary. The correlation between height of plant and number of flowers has been calculated for all the samples, and the null hypothesis of no correlation can be rejected for every one. Could the correlation between height and flower number be the same for the five localities? If so, what estimate can you give of ρ?

18.26 You would like to predict the total length of a cat from the length of its femur. What are the confidence limits of your predicted value for a given femur length?

18.27 An investigator wishes to establish a correlation between number of ovarioles and fecundity in *Drosophila*. He first raises the female under optimal conditions, counting her egg output over a 2-week period. Then he removes her ovaries by dissection, sections them, and counts the ovarioles. He has counted fecundity on 100 individuals but has sectioned only 25 ovaries to date. Before proceeding with his work, he would like to have reassurance that correlation occurs in his data. What preliminary test would you suggest on the 25 paired readings amassed to date?

18.28 An ecologist watches a bird nest to record the food brought in by the female for her young. He can distinguish two size classes of insect prey, big and small. Reviewing a sequence of 30 consecutive observations taken in the course of one afternoon and containing 20 big and 10 small insects, he wonders whether the sequence of size classes is random.

18.29 To test one aspect of a new theory, an ecologist wished to compare two populations for an unconventional statistic that was unlikely to be distributed normally and for which the standard error was unknown.

18.30 The progeny of a cross was expected to yield dominants, heterozygotes, and recessives in a 1:2:1 ratio. Separate replicates were reared under five environmental conditions. Do the data fit the expected 1:2:1 ratio? Is there any evidence of differential survivorship in the different environments?

18.31 A pharmacologist wishes to compare the effects of two drugs on the longevity of hamsters, mainly to determine which drug has the greatest effect in shortening the life span. The variable recorded (length of life in days) is not distributed normally. The investigator wishes to complete the experiment in as short a time as possible, but a few animals of each group live a very long time.

18.32 An immunologist compared the strength of two antigens by two separate tests. The first test induced an all-or-nothing skin reaction and was analyzed by a 2×2 test for independence. The second was a precipitin test, and the difference in turbidity was analyzed by a *t*-test (after the observations had been transformed). Both tests were inconclusive. How could an overall test of the relevant null hypothesis based on both tests be constructed?

18.33 A physiologist wishes to study the effect of different levels of a new drug on mice. The response of the mice (gain in weight after two weeks) is expected to be influenced somewhat by their initial weights.

18.34 An ecologist would like to determine which of a series of environmental variables are the best predictors of the abundance of a species of plant.

18.35 We want to fit a linear model to a set of data in which it is assumed that variables A and B determine (cause) a variable C, which in turn directly affects another variable, D, which is also affected by variable A. How important is the indirect effect on B on variable D? The correlations between all pairs of variables have been computed.

18.36 You have a sample of 35 measurements of the height of a plant. You have concocted an unusual statistic describing the bimodality of the distribution. You would like to assign confidence limits to this statistic but do not know how to construct a formula for a standard error. How would you proceed?

18.37 Refer to Exercise 18.22. Assume that you have shown that mating pairs are preferentially of the same type. The data were obtained at the height of the season. Suppose you also have similar data for the early and the late season. Are the outcomes homogeneous? If so, how can we estimate the preference jointly from the three data tables?

18.38 An investigator has studied mating behavior in an insect species. This behavior consists of a complex repertoire, which the ethologist has decomposed into some 20 steps. Samples from populations at 10 localities show differences in these steps, and the investigator is able to devise a distance measure indicating the differences in the repertoires of individuals from all pairs of the 10 localities. She would like to know whether these behavioral distances are related to the geographic distances of these populations.

18.39 Thirty patients with disease X were given a course of injections of drug D. Another sample of 25 with the same disease were given a placebo injection. The numbers of patients cured were recorded for both groups. How much more likely is a patient to be cured by drug D than by the placebo? Set confidence limits to this estimate.

18.40 Injection of sex hormone T brings on premature development of the reproductive organs in an animal species. An investigator would like to investigate the relation between dose of hormone and magnitude of effect (expressed as length of the reproductive organ) in samples of 20 animals per dose. He employs 10 doses ranging from low to high. It is obvious, after the results are inspected, that the function is not linear, and an alternative hypothesis of a linear ordering of the responses to increasing doses cannot be supported. Attempts to transform the variables are not successful. The investigator would be satisfied with a simpler demonstration. Can it be shown that organ size is a nondecreasing function of dose?

18.41 An insecticide is administered to 8 batches of 50 insects for 7 increasing doses and one control. Calculate a regression equation for proportion mortality on dose and set confidence limits to your slope estimate.

18.42 Samples of fish were collected from three lakes. Do the relative proportions of the six most common species differ among the lakes?

18.43 A researcher plants seedlings of a species of plant in two types of soil. One soil has a high proportion of clay and the other a high proportion of sand. Within each soil type, plants are watered at three levels (2-day intervals, 4-day intervals, 6-day intervals). Each watering regime by soil-type treatment combination has 25 plants. Vegetative biomass is measured at the end of the summer. The researcher would like to know if the frequency of watering had a greater effect on plant size in soil with high clay content or high sand content.

18.44 Snakes of species X are suspected to have higher rates of evaporative water loss (EWL) than snakes of species Y. Furthermore, rates of EWL increase linearly with the log of body size. You measure EWL in the two species under controlled conditions, but for both species you have individuals of varying body sizes. Does species X, in fact, show higher rates of EWL than species Y?

18.45 A clinician is evaluating the efficacy of a new drug for treating students suffering from test anxiety. She hopes the drug will improve their chances of passing a critical exam. A sample of n_1 students is given the drug and n_2 are given a placebo. The students then take a biometry exam and are scored as either passing or failing. Is the proportion of students passing the exam independent of whether they are given the drug or a placebo? Quantify the odds of failing the exam given either the drug or a placebo.

18.46 After an extensive search of the literature, you have data from many studies comparing the mean biomass of plants grown under high or low levels of CO_2. How can the results of these studies be summarized (you have the means, standard errors, and sample sizes for all of the studies)?

18.47 You want to know whether graduate students who have taken a biostatistics course appropriately analyze biological data faster than graduate students who have not taken such a course. You assign a problem set to a sample of 15 new and 15 experienced students and record the identity (new vs. experienced) of the first 15 students to correctly finish the problem set. Students who finish but do not have the correct answers are required to redo all problems until the entire problem set is correct. Does student experience influence the speed of correctly analyzing data?

18.48 A plant ecologist samples the abundance of an invasive plant species from twenty sites in the pine barrens of Long Island. At each site, 12 environmental variables are measured. The researcher would like to know if any of these environmental factors might be facilitating the establishment of this plant species.

18.49 You want to predict the number of eggs in an ovulating frog based on its body size. You suspect that the total number of eggs is also influenced by species identity. You have body-size and egg data for the females of five species of frog. How can you develop a predictive equation? Assume all females are reproductively mature.

18.50 One hypothesis for the importance of spicy food is that compounds in chilies might have antimicrobial properties. Thus, you suspect that spicy food might be more frequent in warmer climates. You have latitudinal data on a sample of 40 nations in the Northern Hemisphere. You also have data for each of these nations on the proportion of meals that have chilies in the ingredients. Do you find that warmer (closer to the equator) nations have a higher frequency of spicy food? Assume that you can ignore spatial nonindependence among nations.

18.51 You have devised an index to score the degree of aggressive response a male fish uses to defend its territory. The index is rather ad hoc and not likely to be normally distributed. How can you test for differences in aggressiveness in two different species?

18.52 A researcher would like to determine if the degree of aggression in species X (measured in an arbitrary scale developed for this study) differs depending on food abundance in two locations. The index is not likely to be normally distributed. Using the same observational effort in each site, he used 15 plots in each of the two sites.

18.53 A behavioral ecologist runs 50 trials in which an individual mouse is placed into a cage that contains two petri dishes filled with equal amounts of birdseed. One dish is placed under an overhanging branch (sheltered), and the other is in the open (unsheltered). Each time, a different mouse is used. The researcher notes which dish the mouse feeds from first. In half of the trials, a replica of a predatory bird is placed on a perch at the top of the cage; in the other half of the trials, no simulated predator is present. The researcher would like to know if the apparent presence of a predator affects the foraging behavior of this species of mouse.

18.54 An entomologist is interested in whether a particular herbivorous insect species prefers to feed on plant species A or B. He captures a large number of these insects, places them individually in a chamber with a choice of either plant to feed on, and notes which plant they choose. Does the insect prefer either species A or B? How strong is the preference?

18.55 A researcher wishes to compare the sodium content in the plasma of young southern fur seals with the level of sodium in the milk of their mothers. Sodium content in plasma and milk of 10 randomly selected pairs of seals was measured. Is there evidence that a difference in sodium levels exist?

18.56 You have data on the relationship between color (light, dark) and predation pressure (high, low) for moths from five separate studies conducted in different cities of England. Are the chances of experiencing high predation pressure different for light and dark moths? Are these odds the same in all five cities?

18.57 After surveying the published results of studies of the effect of competition on a certain species of insect, you are concerned that many of the experiments that failed to substantiate the competitive effect might not have been published. Hence your summary might not accurately represent the effects of competition. How can you check for such a bias?

18.58 In a study of the frequency of stomach cancer the number of cancer cases observed in all the hospitals of a city over a 10-year period as well as the total number of patients seen at the hospitals was recorded. The frequencies were grouped into three racial groups and by sex. Do these factors have an effect on cancer frequency and are these effects independent of each other?

18.59 You have seven strains of a species of *Drosophila* in which you want to test a 9:7 ratio for some trait.

18.60 An ecologist studying an insectivorous bird species suspects that the swarms of insects that are the bird's prey persist long enough for a bird to collect insects, feed them to its young at its nest, and then return to the same swarm to collect more. She makes observations on the success (returning with insects) or failure (returning without insects) of individual birds on each successive foraging trip. She predicts that successful foraging trips should be grouped together in time if birds are able to return to forage in a particular insect swarm several times before it disappears. Other studies suggest that birds initially locate insect swarms by a random search and that the probability that a bird locates a swarm remains constant over time.

18.61 A geneticist believes that gene frequency in samples of a certain locus should vary as a function of mean annual temperature. How can he test this hypothesis for the 10 samples of a species of mosquito that he has collected at different locations?

18.62 A class of students is given a biometry final exam, but not all were able to pass the test. The instructor makes them attend a review session and gives them the same final again. Has the performance of the class improved, deteriorated, or remained the same? (Everyone retook the final, and no new students were allowed to take the exam.).

18.63 A geneticist is interested in the mutation rate of a particular gene in humans. She sequences alleles for 500,000 sets of 2 parents and 1 child and identifies all the mutations of the gene in the children by comparing their sequences with those of both their parents. The children are observed to possess between 0 and 4 mutations in the gene, with most children carrying no mutations. Estimate the mutation rate and suggest a probability model for studying the mutation process and evaluating the fit of the model to the data.

18.64 The diameters of 20 shells of a species of mollusk from each of 15 localities were measured. Can the whole population be considered homogeneous?

18.65 We want to test the effects of 4 germicides. Bacterial counts were taken after the treatments, each of which was performed in 10 replicates, values range from in the hundreds to values in the millions. How do we analyze the data? (Besides the four germicides groups, there is also a control group.)

Appendix A

Mathematical Proofs

A.1 Demonstration that the sum of the deviations from the mean is equal to zero.

We must learn two common rules of statistical algebra. The first is that we can open a pair of parentheses with a Σ sign in front of them by treating the Σ as though it were a common factor:

$$\sum_{i=1}^{n}(A_i + B_i) = (A_1 + B_1) + (A_2 + B_2) + \cdots + (A_n + B_n)$$
$$= (A_1 + A_2 + \cdots + A_n) + (B_1 + B_2 + \cdots + B_n)$$

Therefore

$$\sum_{i=1}^{n}(A_i + B_i) = \sum_{i=1}^{n}A_i + \sum_{i=1}^{n}B_i$$

The second rule is that $\sum_{i=1}^{n} C$ developed during an algebraic operation, where C is a constant, can be computed as follows:

$$\sum_{i=1}^{n}C = C + C + \cdots + C \quad (n \text{ terms})$$
$$= nC$$

Because in a given sample a mean is a constant value, $\sum^{n}\overline{Y} = n\overline{Y}$. In the subsequent demonstration and others to follow, whenever all summations are over n items, we have simplified the notation by dropping subscripts for variables and superscripts above summation signs.

We wish to prove that $\Sigma y = 0$. By definition,

$$\Sigma y = \Sigma(Y - \overline{Y})$$
$$= \Sigma Y - n\overline{Y}$$
$$= \Sigma Y - \frac{n\Sigma Y}{n} \quad \left(\text{because } \overline{Y} = \frac{\Sigma Y}{n} \right)$$
$$= \Sigma Y - \Sigma Y$$

Therefore $\Sigma y = 0$.

A.2   Demonstration of the effects of additive, multiplicative, and combination coding on means, variances, and standard deviations.

For this proof, we must learn one more convention of statistical algebra. In Section A.1, we saw that $\Sigma C = nC$. When the Σ precedes both a constant and a variable, however, as in ΣCY, the constant can be placed before the Σ, because

$$\sum_{i=1}^{n} CY_i = CY_1 + CY_2 + \cdots + CY_n$$

$$= C(Y_1 + Y_2 + \cdots + Y_n)$$

$$= C\left(\sum_{i=1}^{n} Y_i\right)$$

Therefore

$$\sum_{i=1}^{n} CY_i = C\sum_{i=1}^{n} Y_i$$

Thus $\Sigma CY = C\Sigma Y$, $\Sigma C^2 y^2 = C^2\Sigma y^2$, and $\Sigma 2\bar{Y}Y = 2\bar{Y}\Sigma Y$ (because both 2 and $\bar{Y}$ are constants).

Means of Coded Data

ADDITIVE CODING.   The variable is coded $Y_c = Y + C$, where C is a constant, the additive code. Therefore

$$\sum Y_c = \sum(Y + C) = \sum Y + nC$$

and

$$\bar{Y}_c = \frac{\sum Y_c}{n} = \frac{\sum Y}{n} + \frac{nC}{n} = \bar{Y} + C$$

To decode $\bar{Y}_c$, subtract C from it and you will obtain $\bar{Y}$; that is, $\bar{Y} = \bar{Y}_c - C$.

MULTIPLICATIVE CODING.   The variable is coded $Y_c = DY$, where D is a constant, the multiplicative code. Therefore

$$\sum Y_c = D\sum Y$$

and

$$\bar{Y}_c = \frac{\sum Y_c}{n} = D\frac{\sum Y}{n} = D\bar{Y}$$

To decode $\bar{Y}_c$, divide it by D and you will obtain $\bar{Y}$; that is, $\bar{Y} = \bar{Y}_c/D$.

COMBINATION CODING.   The variable is coded $Y_c = D(Y + C)$, where C and D are constants, the additive and multiplicative codes, respectively. Therefore

$$\sum Y_c = D\sum(Y + C) = D\sum Y + nDC$$

and

$$\bar{Y}_c = \frac{\Sigma Y_c}{n} = D\frac{\Sigma Y}{n} + \frac{nDC}{n} = D\bar{Y} + DC$$

To decode $\bar{Y}_c$, divide it by D, then subtract C and you will obtain $\bar{Y}$; that is,

$$\bar{Y} = \frac{\bar{Y}_c}{D} - C$$

Variances and Standard Deviations of Coded Data

ADDITIVE CODING.   The variable is coded $Y_c = Y + C$, where C is a constant, the additive code. By definition, $y = Y - \bar{Y}$, and

$$\begin{aligned} y_c &= Y_c - \bar{Y}_c \\ &= [(Y + C) - (\bar{Y} + C)] \quad \text{(as shown for means above)} \\ &= [Y + C - \bar{Y} - C] \\ &= [Y - \bar{Y}] \\ &= y \end{aligned}$$

Therefore $\Sigma y_c^2 = \Sigma y^2$, and $\Sigma y_c^2/(n - 1) = \Sigma y^2/(n - 1)$.

Thus additive coding has no effect on sums of squares, variances, or standard deviations.

MULTIPLICATIVE CODING.   The variable is coded $Y_c = DY$, where D is a constant, the multiplicative code. By definition, $y = Y - \bar{Y}$, and

$$\begin{aligned} y_c &= Y_c - \bar{Y}_c \\ &= DY - D\bar{Y} \quad \text{(as shown for means above)} \\ &= D(Y - \bar{Y}) \\ &= Dy \end{aligned}$$

Therefore $y_c^2 = D^2y^2$, $\Sigma y_c^2 = D^2\Sigma y^2$, $\Sigma y_c^2/(n - 1) = D^2[\Sigma y^2/(n - 1)]$, and $s_c^2 = D^2s^2$.

Thus, when data have been subjected to multiplicative coding, a sum of squares or variance can be decoded by dividing it by the *square* of the multiplicative code; a standard deviation can be decoded by dividing it by the code itself, that is, $s^2 = s_c^2/D^2$ and $s = s_c/D$.

COMBINATION CODING.   The variable is coded $Y_c = D(Y + C)$, where C and D are constants, the additive and multiplicative codes, respectively. By definition, $y = Y - \bar{Y}$, and

$$\begin{aligned} y_c &= Y_c - \bar{Y}_c \\ &= [D(Y + C) - (D\bar{Y} + DC)] \quad \text{(as shown for means above)} \\ &= [DY + DC - D\bar{Y} - DC] \\ &= D[Y - \bar{Y}] \end{aligned}$$

Therefore $y_c = Dy$, as before.

Thus, in combination coding only the multiplicative code needs to be considered when decoding sums of squares, variances, or standard deviations.

A.3 Demonstration that the sum of squares within groups and the sum of squares among groups add to the total sum of squares in an analysis of variance (see Section 8.5).

By definition,

$$SS_{\text{total}} = \sum_{i=1}^{a} \sum_{j=1}^{n_i} (Y_{ij} - \overline{\overline{Y}})^2$$

Add and subtract $\overline{Y}_i$ within the brackets:

$$= \sum_{i}^{a} \sum_{j}^{n_i} [(Y_{ij} - \overline{Y}_j) + (\overline{Y}_i - \overline{\overline{Y}})]^2$$

Square and then remove brackets:

$$= \sum_{i}^{a} \sum_{j}^{n_i} (Y_{ij} - \overline{Y}_i)^2 + 2 \sum_{i}^{a} \sum_{j}^{n_i} (Y - \overline{Y})(\overline{Y} - \overline{\overline{Y}}) + \sum_{i}^{a} n_i (\overline{Y}_i - \overline{\overline{Y}})^2$$

$$= SS_{\text{within}} + 2 \sum_{i}^{a} \sum_{j}^{n_i} (Y_{ij} - \overline{Y}_i)(\overline{Y}_i - \overline{\overline{Y}}) + SS_{\text{among}}$$

The middle term is equal to zero, because $\sum^{n_i}(Y_{ij} - \overline{Y}_i)$ is zero within each group. Therefore, $SS_{\text{total}} = SS_{\text{within}} + SS_{\text{among}}$.

A.4 Derivation of simplified formulas for the standard error of the difference between two means.

The standard error squared, from Expression (9.2), is

$$\left[\frac{(n_1 - 1)s_1^2 + (n_2 - 1)s_2^2}{n_1 + n_2 - 2} \right] \left(\frac{n_1 + n_2}{n_1 n_2} \right)$$

When $n_1 = n_2 = n$, this expression simplifies to

$$\left[\frac{(n - 1)s_1^2 + (n - 1)s_2^2}{2n - 2} \right] \left(\frac{2n}{n^2} \right) = \left[\frac{(n - 1)(s_1^2 + s_2^2)(2)}{2(n - 1)(n)} \right] = \frac{1}{n}(s_1^2 + s_2^2)$$

When $n_1 \neq n_2$, but each is large enough so that $(n_1 - 1) \approx n_1$ and $(n_2 - 1) \approx n_2$, the square of the standard error of Expression (9.2) simplifies to

$$\left[\frac{n_1 s_1^2 + n_2 s_2^2}{n_1 + n_2} \right] \left(\frac{n_1 + n_2}{n_1 n_2} \right) = \left[\frac{n_1 s_1^2}{n_1 n_2} + \frac{n_2 s_2^2}{n_1 n_2} \right] = \frac{s_1^2}{n_2} + \frac{s_2^2}{n_1}$$

A.5 Demonstration that t_s^2 obtained from a test of significance of the difference between two means (as in Box 9.4) is identical to the F_s-value obtained in a single-classification anova. This derivation is for the simple case of two equal-sized groups (as in Table 9.2).

$$t_s = \frac{\bar{Y}_1 - \bar{Y}_2}{\sqrt{\frac{1}{n(n-1)} \left(\sum^n y_1^2 + \sum^n y_2^2 \right)}} \quad \text{(from Box 9.4)}$$

$$t_s^2 = \frac{(\bar{Y}_1 - \bar{Y}_2)^2}{\frac{1}{n(n-1)} \left(\sum^n y_1^2 + \sum^n y_2^2 \right)} = \frac{n(n-1)(\bar{Y}_1 - \bar{Y}_2)^2}{\sum^n y_1^2 + \sum^n y_2^2}$$

In the two-sample anova,

$$MS_{means} = \frac{1}{2-1} \sum^2 (\bar{Y}_i - \bar{\bar{Y}})^2$$

$$= (\bar{Y}_1 - \bar{\bar{Y}})^2 + (\bar{Y}_2 - \bar{\bar{Y}})^2$$

$$= \left(\bar{Y}_1 - \frac{\bar{Y}_1 + \bar{Y}_2}{2} \right)^2 + \left(\bar{Y}_2 - \frac{\bar{Y}_1 + \bar{Y}_2}{2} \right)^2 \quad \text{(because } \bar{\bar{Y}} = (\bar{Y}_1 + \bar{Y}_2)/2)$$

$$= \left(\frac{\bar{Y}_1 - \bar{Y}_2}{2} \right)^2 + \left(\frac{\bar{Y}_2 - \bar{Y}_1}{2} \right)^2$$

$$= \frac{1}{2}(\bar{Y}_1 - \bar{Y}_2)^2 \quad \text{(because the two terms are identical)}$$

Then

$$MS_{among} = n \times MS_{means} = n \left[\frac{1}{2}(\bar{Y}_1 - \bar{Y}_2)^2 \right]$$

$$= \frac{n}{2}(\bar{Y}_1 - \bar{Y}_2)^2$$

$$MS_{within} = \frac{\sum^n y_1^2 + \sum^n y_2^2}{2(n-1)}$$

$$F_s = \frac{MS_{among}}{MS_{within}}$$

$$= \frac{\frac{n}{2}(\bar{Y}_1 - \bar{Y}_2)^2}{\left(\sum^n y_1^2 + \sum^n y_2^2 \right) \Big/ [2(n-1)]}$$

$$= \frac{n(n-1)(\bar{Y}_1 - \bar{Y}_2)^2}{\sum^n y_1^2 + \sum^n y_2^2}$$

$$= t_s^2$$

A.6   Derivation of the expected error mean square of a completely randomized design estimated from the mean squares of a randomized-complete-blocks design.

Randomized-complete-blocks design

Source of variation	df	MS	Expected MS
Blocks	$b-1$	MS_B	$\sigma^2_{(RB)} + a\sigma^2_B$
Treatments	$a-1$	$MS_{A(RB)}$	$\sigma^2_{(RB)} + \dfrac{b}{a-1}\sum \alpha^2$
Error	$(a-1)(b-1)$	$MS_{E(RB)}$	$\sigma^2_{(RB)}$
Total	$ab-1$		

Completely randomized design

Source of variation	df	MS	Expected MS
Treatments	$a-1$	$MS_{A(CR)}$	$\sigma^2_{(CR)} + \dfrac{b}{a-1}\sum \alpha^2$
Error	$a(b-1)$	$MS_{E(CR)}$	$\sigma^2_{(CR)}$
Total	$ab-1$		

If we assume that the total SS of the two designs is the same, we can write the following identity where each side represents the total SS:

$$(b-1)MS_B + (a-1)MS_{A(RB)} + (a-1)(b-1)MS_{E(RB)}$$

$$= (a-1)MS_{A(CR)} + a(b-1)MS_{E(CR)}$$

Rewriting the identity in terms of the variance components of the expected means squares, we obtain

$$(b-1)(\sigma^2_{(RB)} + a\sigma^2_B) + (a-1)\left(\sigma^2_{(RB)} + \frac{b}{a-1}\sum \alpha^2\right) + (a-1)(b-1)\sigma^2_{(RB)}$$

$$= (a-1)\left(\sigma^2_{(CR)} + \frac{b}{a-1}\sum \alpha^2\right) + a(b-1)\sigma^2_{(CR)}$$

$$[(b-1) + (a-1) + (a-1)(b-1)]\sigma^2_{(RB)} + a(b-1)\sigma^2_B + b\sum \alpha^2$$

$$= [(a-1) + a(b-1)]\sigma^2_{(CR)} + b\sum \alpha^2$$

$$(ab-1)\sigma^2_{(RB)} + a(b-1)\sigma^2_B = (ab-1)\sigma^2_{(CR)}$$

$$\sigma^2_{(CR)} = \sigma^2_{(RB)} + \frac{a(b-1)}{ab-1}\sigma^2_B$$

We rewrite this formula for $\sigma^2_{(CR)}$ in terms of the mean squares. Because $\sigma^2_{(RB)} = MS_{E(RB)}$ and $\sigma^2_B = (MS_B - MS_{E(RB)})/a$, we obtain

$$MS_{E(CR)} = MS_{E(RB)} + a(b-1)\frac{MS_B - MS_{E(RB)}}{a(ab-1)}$$

$$= MS_{E(RB)} + (b-1)\frac{MS_B}{ab-1} - (b-1)\frac{MS_{E(RB)}}{ab-1}$$

$$= [(ab-1) - (b-1)]\frac{MS_{E(RB)}}{ab-1} + (b-1)\frac{MS_B}{ab-1}$$

$$= \frac{b(a-1)MS_{E(RB)} + (b-1)MS_B}{ab-1}$$

A.7 Demonstration that t_s^2 obtained from a paired-comparisons test (as in Box 11.6) is identical to the F_s-value for treatments obtained in a two-way anova without replication (same box).

We will use the simpler $\sum^b(Y_1 - Y_2)$ in place of $\sum^b_{i=1}(Y_{i1} - Y_{i2})$. Also, remember that in such an anova $a = 2$.

$$t_s = \frac{\overline{D} - (\mu_1 - \mu_2)}{s_{\overline{D}}} \qquad \text{(from Box 11.6)}$$

where $\overline{D} = \overline{Y}_1 - \overline{Y}_2$, the difference $\mu_1 - \mu_2$ is hypothesized to equal zero, and

$$s_{\overline{D}} = \sqrt{\frac{\sum^b[(Y_1 - Y_2) - (\overline{Y}_1 - \overline{Y}_2)]^2}{(b-1)b}}$$

By squaring and rearranging a little, we can rewrite the expression for t_s as

$$t_s^2 = \frac{b(b-1)(\overline{Y}_1 - \overline{Y}_2)^2}{\sum^b[(Y_1 - Y_2) - (\overline{Y}_1 - \overline{Y}_2)]^2}$$

In a randomized-complete-blocks anova with $a = 2$ groups

$$F_s = \frac{MS_{among}}{MS_{AB}}$$

In our case, because we have only two groups,

$$MS_{among} = b\sum^a(\overline{Y} - \overline{\overline{Y}})^2 = \frac{\left(\sum Y_1 - \sum Y_2\right)^2}{2b} = \frac{b}{2}(\overline{Y}_1 - \overline{Y}_2)^2$$

$$MS_{AB} = \frac{\sum^a_i \sum^b_j (Y_{ij} - \overline{Y}_i - \overline{Y}_j + \overline{\overline{Y}})^2}{b-1}$$

Here $\bar{Y} = (\bar{Y}_1 + \bar{Y}_2)/2$ because of equal sample size, and likewise $\bar{Y}_j = (Y_{1j} + Y_{2j})/2$. Because there are only two groups, we may drop the summation over a and write out the two squared terms for each pair of observations:

$$MS_{AB} = \frac{1}{b-1} \sum_{j}^{b} \left\{ \left(Y_{1j} - \bar{Y}_1 - \frac{Y_{1j} + Y_{2j}}{2} + \frac{\bar{Y}_1 + \bar{Y}_2}{2} \right)^2 \right.$$

$$\left. + \left(Y_{2j} - \bar{Y}_2 - \frac{Y_{1j} + Y_{2j}}{2} + \frac{\bar{Y}_1 + \bar{Y}_2}{2} \right)^2 \right\}$$

With a little rearrangement we find that

$$MS_{AB} = \frac{1}{b-1} \sum_{j}^{b} \left\{ \left[\left(Y_{1j} - \frac{1}{2}Y_{1j} - \bar{Y}_1 + \frac{1}{2}\bar{Y}_1 \right) + \left(-\frac{Y_{2j}}{2} + \frac{\bar{Y}_2}{2} \right) \right]^2 \right.$$

$$\left. + \left[\left(Y_{2j} - \frac{1}{2}Y_{2j} - \bar{Y}_2 + \frac{\bar{Y}_2}{2} \right) + \left(-\frac{Y_{1j}}{2} + \frac{\bar{Y}_1}{2} \right) \right]^2 \right\}$$

$$= \frac{1}{b-1} \sum_{j}^{b} \left\{ \left[\frac{1}{2}(Y_{1j} - \bar{Y}_1) - \frac{1}{2}(Y_{2j} - \bar{Y}_2) \right]^2 \right.$$

$$\left. + \left[-\frac{1}{2}(Y_{1j} - \bar{Y}_1) + \frac{1}{2}(Y_{2j} - \bar{Y}_2) \right]^2 \right\}$$

The two terms in square brackets above are identical except for the sign, which is not important because the terms are squared. The constant $\frac{1}{2}$ may also now be factored out (becoming $\frac{1}{4}$). We can therefore simplify this expression as follows:

$$MS_{AB} = \frac{1}{b-1} \frac{1}{4} \sum_{j}^{b} 2[(Y_{1j} - \bar{Y}_1) - (Y_{2j} - \bar{Y}_2)]^2$$

$$= \frac{1}{2(b-1)} \sum_{j}^{b} [(Y_{1j} - Y_{2j}) - (\bar{Y}_1 - \bar{Y}_2)]^2$$

F_s may now be written as follows:

$$F_s = \frac{\left[\frac{b}{2} (\bar{Y}_1 - \bar{Y}_2)^2 \right]}{\frac{1}{2(b-1)} \sum_{j}^{b} [(Y_{1j} - Y_{2j}) - (\bar{Y}_1 - \bar{Y}_2)]^2}$$

After canceling the two 2's and moving $(b-1)$ into the numerator, the expression for F_s is identical to our expression for t_s^2 given above.

A.8   Derivation of the formulas for b and a in regression. This derivation requires a knowledge of calculus.

We define

$$\sum d_{Y \cdot X}^2 = \sum (Y - \hat{Y})^2$$

$$= \sum [Y - (a + bX)]^2$$

$$= \sum (Y - a - bX)^2$$

To minimize $\sum d_{Y \cdot X}^2$ with respect to the two parameters (a and b), we must first take partial derivatives with respect to each:

$$\frac{\partial \sum d_{Y \cdot X}^2}{\partial a} = 2\sum (Y - a - bX)(-1)$$

$$= -2\left[\sum Y - na - b\sum X\right]$$

$$\frac{\partial \sum d_{Y \cdot X}^2}{\partial b} = 2\sum (Y - a - bX)(-X)$$

$$= -2\left[\sum XY - a\sum X - b\sum X^2\right]$$

Next we set each partial derivative equal to zero and solve the pair of simultaneous linear equations for a and b.

$$\begin{cases} 0 = \sum Y - na - b\sum X \\ 0 = \sum XY - a\sum X - b\sum X^2 \end{cases}$$

If we multiply the first equation by $\sum X/n$ and then subtract it from the second equation, we cancel out the term involving a:

$$0 = \sum XY - \frac{\sum X \sum Y}{n} - b\sum X^2 + \frac{b\left(\sum X\right)^2}{n}$$

Thus

$$b = \frac{\sum XY - \dfrac{\sum X \sum Y}{n}}{\sum X^2 - \dfrac{\left(\sum X\right)^2}{n}}$$

The numerator of this expression

$$\sum XY - \frac{\sum X \sum Y}{n} = \sum XY - \bar{X}\sum Y \quad \text{or} \quad \sum XY - \bar{Y}\sum X$$

Add and subtract $\bar{Y}\sum X$:

$$= \sum XY - \bar{X}\sum Y - \bar{Y}\sum X + \bar{Y}\sum X$$

Factor out the summation signs and substitute $\sum \bar{X}\,\bar{Y}$ for $\bar{Y}\sum X$:

$$= \sum \left[(XY - \bar{X}Y) - (X\bar{Y} - \bar{Y}\bar{X})\right]$$

$$= \sum \left[(X - \bar{X})Y - (X - \bar{X})\bar{Y}\right]$$

$$= \sum (X - \bar{X})(Y - \bar{Y})$$

Because $x = X - \overline{X}$ and $y = Y - \overline{Y}$,

$$\sum (X - \overline{X})(Y - \overline{Y}) = \sum xy$$

By analogous derivation, the denominator of this expression, $\sum XY - (\sum X)^2/n$, can be shown to equal $\sum (X - \overline{X})(X - \overline{X})$ which equals $\sum x^2$. Therefore,

$$b = \frac{\sum xy}{\sum x^2}$$

Given b, we can solve for a from the first equation:

$$a = \frac{\sum Y}{n} - \frac{b \sum X}{n}$$

$$= \overline{Y} - b\overline{X}$$

A.9   Derivation of the formulas for b and a in regression. This derivation requires algebra only.

We need to minimize $\sum d^2_{Y\cdot X}$, the unexplained sum of squares

$$\sum d^2_{Y\cdot X} = \sum (Y - \hat{Y})^2$$

$$= \sum (Y - a - bX)^2 \quad (\text{since } \hat{Y} = a + bX)$$

$$= \sum Y^2 + na^2 + b^2 \sum X^2 - 2a \sum Y - 2b \sum XY + 2ab \sum X$$

Add and subtract the following three terms from this expression (not changing its value in the process):

$$\frac{(\sum Y)^2}{n}, \quad \frac{2b \sum X \sum Y}{n}, \quad \text{and} \quad \frac{b^2 (\sum X)^2}{n}$$

Rearranging the terms, we obtain

$$\sum d^2_{Y\cdot X} = \left[\sum Y^2 - \frac{(\sum Y)^2}{n} \right] - 2b \left[\sum XY - \frac{\sum X \sum Y}{n} \right]$$

$$+ b^2 \left[\sum X^2 - \frac{(\sum X)^2}{n} \right] + \frac{(\sum Y)^2}{n} - \frac{2b \sum X \sum Y}{n}$$

$$+ \frac{b^2 (\sum X)^2}{n} + na^2 - 2a \sum Y + 2ab \sum X$$

The first three terms can be rewritten as sums of squares (see Section A.8) or sums of products (see Section 14.4). For the remaining six terms, we rewrite ΣX and ΣY in terms of $\bar{X}$ and $\bar{Y}$, respectively. We obtain

$$\Sigma d_{Y \cdot X}^2 = \Sigma y^2 - 2b \Sigma xy + b^2 \Sigma x^2 + n\bar{Y}^2 - 2bn\bar{X}\bar{Y} + b^2 n\bar{X}^2$$
$$+ na^2 - 2an\bar{Y} + 2abn\bar{X}$$

Factoring out n for the last six terms, we discover that they can be collected as

$$n(\bar{Y} - a - b\bar{X})^2$$

Therefore

$$\Sigma d_{Y \cdot X}^2 = \Sigma y^2 - 2b \Sigma xy + b^2 \Sigma x^2 + n(\bar{Y} - a - b\bar{X})^2$$

By adding and subtracting $(\Sigma xy)^2/\Sigma x^2$ we obtain

$$\Sigma d_{Y \cdot X}^2 = \Sigma y^2 - \frac{(\Sigma xy)^2}{\Sigma x^2} + \frac{(\Sigma xy)^2}{\Sigma x^2} - 2b \Sigma xy + b^2 \Sigma x^2$$
$$+ n(\bar{Y} - a - b\bar{X})^2$$

$$= \Sigma y^2 - \frac{(\Sigma xy)^2}{\Sigma x^2} + \Sigma x^2 \left[\frac{\Sigma xy}{\Sigma x^2} - b \right]^2 + n(\bar{Y} - a - b\bar{X})^2$$

If we wish to minimize $\Sigma d_{Y \cdot X}^2$ (barring trivial solutions), the two terms involving b as well as both b and a must equal zero. Hence, when $\Sigma d_{Y \cdot X}^2$ is minimal, $b = \Sigma xy/\Sigma x^2$ and $a = \bar{Y} - b\bar{X}$.

Also because the last two terms must equal zero for minimal $\Sigma d_{Y \cdot X}^2$ we have obtained as a by-product a derivation of an alternative formula:

$$\Sigma d_{Y \cdot X}^2 = \Sigma y^2 - \frac{(\Sigma xy)^2}{\Sigma x^2} \qquad \text{[Expression (14.5)]}$$

A.10 Demonstration that the sum of squares of the dependent variable in regression can be partitioned exactly into explained and unexplained sums of squares, the cross products canceling out.

By definition (see Section 14.3),

$$y = \hat{y} + d_{Y \cdot X}$$
$$\Sigma y^2 = \Sigma (\hat{y} + d_{Y \cdot X})^2 = \Sigma \hat{y}^2 + \Sigma d_{Y \cdot X}^2 + 2 \Sigma \hat{y} d_{Y \cdot X}$$

If we can show that $\sum \hat{y} d_{Y \cdot X} = 0$, then we have demonstrated the required identity. We have

$$\sum \hat{y} d_{Y \cdot X} = \sum bx(y - bx) \qquad \begin{array}{l} \text{[because } \hat{y} = bx \text{ from Expression (14.3) and} \\ d_{Y \cdot X} = Y - \hat{Y} = (Y - \bar{Y}) - (\hat{Y} - \bar{Y}) = \\ y - \hat{y} = y - bx] \end{array}$$

$$= b \sum xy - b^2 \sum x^2$$

$$= b \sum xy - b \frac{\sum xy}{\sum x^2} \sum x^2 \qquad \left(\text{because } b = \sum xy / \sum x^2 \right)$$

$$= b \sum xy - b \sum xy$$

$$= 0$$

Therefore $\sum y^2 = \sum \hat{y}^2 + \sum d_{Y \cdot X}^2$ or, written in terms of original observations,

$$\sum (Y - \bar{Y})^2 = \sum (\hat{Y} - \bar{Y})^2 + \sum (Y - \hat{Y})^2$$

A.11   Derivation of the standard error of a regression coefficient.

$$b = \frac{\sum xy}{\sum x^2}$$

By factoring out y, we express the regression coefficient as a weighted sum of the deviates $Y - \bar{Y}$:

$$b = \sum \left(\frac{x}{\sum x^2} \right) y$$

$$= \sum w_i y \qquad \left(\text{where } w_i = \frac{x_i}{\sum x^2} \right)$$

Then (assuming that the deviations from regression are uncorrelated, and with the same variance, $s_{Y \cdot X}^2$) we obtain

$$s_b^2 = \sum w_i^2 s_{Y \cdot X}^2$$

$$= \left[\frac{1}{\left(\sum x^2 \right)^2} \sum x_i^2 \right] s_{Y \cdot X}^2$$

$$= \frac{s_{Y \cdot X}^2}{\sum x^2} \qquad \text{(as shown in Box 14.2)}$$

A.12 Derivation of the variance of $\hat{Y}$, the estimated value of Y for a given value X. In the equations below, an X without a subscript signifies the given value of X for which we wish to obtain $\hat{Y}$, the predicted value of Y. Values of X and x with a subscript represent the observed values and deviates of the sample.

First we must prove the identity $\sum xy = \sum xY$. By substitution,

$$\sum (X - \bar{X})(Y - \bar{Y}) = \sum (XY - \bar{X}Y - X\bar{Y} - \bar{X}\bar{Y})$$

$$= \sum [(X - \bar{X})Y - \bar{Y}(X - \bar{X})]$$

$$= \sum xY - \bar{Y}\sum (X - \bar{X}) \quad \text{(but because } \sum (X - \bar{X}) = 0)$$

$$= \sum xY$$

Then

$$\hat{Y} = a + bX$$

$$= \bar{Y} - b\bar{X} + bX \qquad \qquad \text{(because } a = \bar{Y} - b\bar{X})$$

$$= \bar{Y} + b(X - \bar{X})$$

$$= \frac{\sum Y}{n} + \frac{\sum xy}{\sum x^2}(X - \bar{X}) \qquad \left(\text{because } \bar{Y} = \frac{\sum Y}{n} \text{ and } b = \frac{\sum xy}{\sum x^2} \right)$$

$$= \sum \left(\frac{1}{n} + (X - \bar{X})\frac{x_i}{\sum x^2} \right) Y_i \qquad \left(\text{because } \sum xy = \sum xY \right)$$

$$= \sum w_i Y_i \qquad \left(\text{where } w_i = \frac{1}{n} + (X - \bar{X})\frac{x_i}{\sum x^2} \right)$$

Assuming that the deviations from regression are uncorrelated, and with the same variance $s_{Y \cdot X}^2$, we obtain

$$s_{\hat{Y}}^2 = \sum w_i^2 s_{Y \cdot X}^2$$

$$= \sum \left(\frac{1}{n} + (X - \bar{X})\frac{x_i}{\sum x^2} \right)^2 s_{Y \cdot X}^2$$

$$= \sum \left[\frac{1}{n^2} + \frac{2(X - \bar{X})x_i}{n \sum x^2} + (X - \bar{X})^2 \frac{x_i^2}{\left(\sum x^2 \right)^2} \right] s_{Y \cdot X}^2$$

$$= \left[\frac{1}{n} + \frac{2(X - \bar{X})}{n \sum x^2} \sum x_i + \frac{(X - \bar{X})^2}{\sum x^2} \right] s_{Y \cdot X}^2 \qquad \begin{array}{l}\text{(because } \sum x_i = 0, \text{ the}\\ \text{middle term in brackets}\\ \text{vanishes)}\end{array}$$

$$= \left(\frac{1}{n} + \frac{(X - \bar{X})^2}{\sum x^2} \right) s_{Y \cdot X}^2 \qquad \text{(as shown in Box 14.2)}$$

A.13 Proof that the variance of the sum of two variables is

$$s_{(Y_1+Y_2)}^2 = s_1^2 + s_2^2 + 2r_{12}s_1s_2$$

where s_1 and s_2 are standard deviations of Y_1 and Y_2, respectively, and r_{12} is the correlation coefficient between Y_1 and Y_2.

If $Z = Y_1 + Y_2$, then

$$s_Z^2 = \frac{1}{n}\sum(Z - \bar{Z})^2 = \frac{1}{n}\sum\left[(Y_1 + Y_2) - \frac{1}{n}\sum(Y_1 + Y_2)\right]^2$$

$$= \frac{1}{n}\sum\left[(Y_1 + Y_2) - \frac{1}{n}\sum Y_1 - \frac{1}{n}\sum Y_2\right]^2 = \frac{1}{n}\sum[(Y_1 + Y_2) - \bar{Y}_1 - \bar{Y}_2]^2$$

$$= \frac{1}{n}\sum[(Y_1 - \bar{Y}_1) + (Y_2 - \bar{Y}_2)]^2 = \frac{1}{n}\sum[y_1 + y_2]^2$$

$$= \frac{1}{n}\sum[y_1^2 + y_2^2 + 2y_1y_2] = \frac{1}{n}\sum y_1^2 + \frac{1}{n}\sum y_2^2 + \frac{2}{n}\sum y_1y_2$$

$$= s_1^2 + s_2^2 + 2s_{12}$$

Because $r_{12} = s_{12}/s_1s_2$, however, we have

$$s_{12} = r_{12}s_1s_2$$

Therefore,

$$s_Z^2 = s_{(Y_1+Y_2)}^2 = s_1^2 + s_2^2 + 2r_{12}s_1s_2 \qquad \text{[Expression (15.10)]}$$

Similarly,

$$s_D^2 = s_{(Y_1-Y_2)}^2 = s_1^2 + s_2^2 - 2r_{12}s_1s_2 \qquad \text{[Expression (15.10a)]}$$

A.14 Proof that the general expression for the G-test can be simplified to expression (17.4).

In general, G is twice the natural logarithm of the ratio of the probability of the sample (with all parameters estimated from the data) and the probability of the sample (assuming the null hypothesis is true). Assuming a multinomial distribution, this ratio is

$$L = \frac{\dfrac{n!}{f_1!f_2!\cdots f_a!}\,\pi_1^{f_1}\pi_2^{f_2}\cdots\pi_a^{f_a}}{\dfrac{n!}{f_1!f_2!\cdots f_a!}\,p_1^{f_1}p_2^{f_2}\cdots p_a^{f_a}}$$

$$= \prod_{i=1}^{a}\left(\frac{\pi_i}{p_i}\right)^{f_i}$$

where f_i is the observed frequency, p_i is the observed proportion, and π_i the expected proportion for class i (assumed by the null hypothesis), while n is the sample size, the sum of the observed frequencies over the a classes.

$$G = -2 \ln L$$

$$= -2 \sum^a f_i \ln\left(\frac{\pi_i}{p_i}\right)$$

$$= 2 \sum^a f_i \ln\left(\frac{p_i}{\pi_i}\right)$$

Because $f_i = np_i$ and $\hat{f}_i = n\pi_i$,

$$G = 2 \sum^a f_i \ln\left(\frac{f_i}{\hat{f}_i}\right) \qquad \text{[Expression (17.4)]}$$

A.15 Proof that in a goodness-of-fit test with only two classes the deviations are always equal in magnitude but opposite in sign.

Let f_1 and f_2 be the two observed frequencies $(f_1 + f_2 = n)$, and let π be the expected probability of the event that has an observed frequency of f_1. The expected frequencies $\hat{f}_1$ and $\hat{f}_2$ would then be

$$\hat{f}_1 = n\pi$$
$$\hat{f}_2 = n(1 - \pi) = n - n\pi$$

The deviations from expectation are

$$f_1 - \hat{f}_1 = f_1 - n\pi$$

and

$$f_2 - \hat{f}_2 = f_2 - n + n\pi$$
$$= (n - f_1) - n - n\pi \qquad (\text{because } f_1 + f_2 = n, \ f_2 = n - f_1)$$
$$= -f_1 + n\pi$$
$$= -(f_1 - n\pi) \qquad (\text{because } n\pi = \hat{f}_1)$$
$$= -(f_1 - \hat{f}_1)$$

Appendix B

Introduction to Matrices

The use of vector and matrix notation is convenient for describing more complex statistical analyses (especially those involving several variables) because it usually allows the necessary operations on the data to be expressed more succinctly. For this reason, it is worth taking a small digression to learn the necessary notation. This short introduction (that can also serve as a review) covers only the matrix operations that are actually used in Chapter 16. For a more general account, see texts such as Gentle (2010), Graybill (2001), Harville (2009), or Searle (2006). An understanding of the implications of the various vector and matrix operations is usually facilitated by the fact that most of the operations have a geometric interpretation. Carroll et al. (1997) emphasize this aspect of the subject.

Definitions

Scalar: a single numerical value, such as 2.5. It is symbolized using a lowercase letter in italics such as a.

Vector: a list of numerical values given in some particular order. It is symbolized by a lowercase boldfaced letter such as $\mathbf{a}$. It is important to distinguish vectors with the elements written vertically (called *column vectors*), such as $\mathbf{a} = \begin{bmatrix} 1 \\ 2 \\ 3 \\ 4 \end{bmatrix}$, from those written horizontally (*row vectors*), such as $\mathbf{b} = [1, 2, 3, 4]$. Unless specified otherwise, we will assume vectors are written vertically. A particular element of the vector is symbolized using an italic letter with a subscript to indicate its position in the vector. For example, $a_2 = 2$ in the vector given above. Certain vectors have special names. A *null vector*, $\mathbf{0}$, contains only zeros, and a *unit vector*, $\mathbf{1}$, contains only 1s. It is often helpful to visualize vectors graphically as a directed line from the origin with each element interpreted as the coordinates on the tip of the vector (usually shown with an arrowhead). Figure B.1 shows three vectors, $\mathbf{a}$, $\mathbf{b}$, and $\mathbf{c}$, emanating from the origin.

Vector transpose: this operation, symbolized by a superscript t, indicates that a column vector should be turned on its side and written as a row vector and vice versa. Thus, for the vectors given above, one could write that $\mathbf{b} = \mathbf{a}^t$.

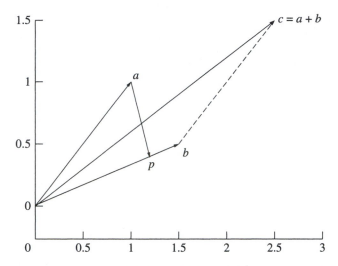

FIGURE B.1 Example of vectors **a** = (1,1), **b** = (1.5,0.5), and their sum **c** = (2.5,1.5). The projection of **a** onto **b** is a vector from the origin ending at **p**. The dashed line shows the effect of appending vector **a** to the end of vector **b** to form the sum **a** + **b** geometrically. It also shows the vector **p**—that is, the projection of vector **a** onto the vector **b**.

Matrix: a two-dimensional rectangular array of numerical values. It is symbolized by a boldfaced capital letter such as **A**. An example would be $\mathbf{A} = \begin{bmatrix} 1 & 1 & 1 \\ 2 & 3 & 4 \\ 3 & 5 & 7 \\ 4 & 7 & 10 \end{bmatrix}$.

Each element of a matrix is symbolized with a lowercase italic letter with subscripts indicating its row and column position. For example, $a_{2,3} = 4$ in row 2 and column 3 of matrix **A**. The rows or the columns are often interpreted as vectors (as a special case, a matrix could consist of just a single row or column). As with many topics in statistics, there is no general agreement about the meaning of the rows and the columns of a data matrix. Some authors let the columns correspond to the variables and the rows to the observations and others will employ the opposite convention. For that reason, one must make sure one understands the conventions used by an author before one can apply a particular matrix equation.

As is the case for vectors, certain matrices have special names. A null matrix, **0**, contains only zeros. It is often important to distinguish matrices of different "shapes." A matrix with the same number of rows and columns is called a *square matrix*. A square matrix of all zeros except for values down the main diagonal (the a_{ii} values going down from the upper left to the lower right) is called a *diagonal matrix*. A diagonal matrix with all of the elements along the diagonal equal to unity is called an *identity matrix* and given the special symbol **I**.

Matrix transpose: this operation is symbolized by a superscript t (many authors use a prime symbol, but we prefer to use primes for other purposes). It results in rewriting the matrix with the rows as columns and the columns as rows. Using the matrix $\mathbf{A}$ above, $\mathbf{A}^t = \begin{bmatrix} 1 & 2 & 3 & 4 \\ 1 & 3 & 5 & 7 \\ 1 & 4 & 7 & 10 \end{bmatrix}$. A square matrix that is identical before and after transposition is called a *symmetric matrix* (if a matrix $\mathbf{A}$ is symmetric then $a_{ij} = a_{ji}$).

Arithmetic Operations on Vectors and Matrices

As in ordinary (scalar) algebra, one can define various arithmetic operations such as addition, subtraction, and multiplication on vectors and matrices. There are, however, some constraints on when and how these operations can be applied.

Vector and matrix addition and subtraction are defined element by element for vectors and matrices. Thus, the sum, $\mathbf{c}$, of two column vectors $\mathbf{a}$ and $\mathbf{b}$ has elements $c_i = a_i + b_i$, and the sum, $\mathbf{C}$, of two matrices $\mathbf{A}$ and $\mathbf{B}$ has elements $c_{ij} = a_{ij} + b_{ij}$. Subtraction is also defined element by element. An obvious constraint is that the vectors or matrices being added or subtracted must have the same number of rows and the same number of columns so that there are corresponding elements to add. For example, $\begin{bmatrix} 1 \\ 2 \\ 3 \\ 4 \end{bmatrix} + \begin{bmatrix} 1 \\ 3 \\ 5 \\ 7 \end{bmatrix} = \begin{bmatrix} 2 \\ 5 \\ 8 \\ 11 \end{bmatrix}$.

Figure B.1 shows an example of adding two vectors.

Vector multiplication is defined in a less obvious way and there are two distinct types of products. The equation $c = \mathbf{a}^t\mathbf{b}$, for $\mathbf{a}$ and $\mathbf{b}$, both column vectors of the same length, is known as an *inner product*, also called a *scalar product* because the result, c, is a single numerical value. In this operation, the corresponding elements of the two vectors are multiplied together and then summed—that is, $c = \sum a_i b_i$. For example, $\begin{bmatrix} 1 & 2 & 3 & 4 \end{bmatrix} \begin{bmatrix} 1 \\ 3 \\ 5 \\ 7 \end{bmatrix} = 1 + 6 + 15 + 28 = 50$.

The sum of a vector of data values can be computed as $\sum Y_i = \mathbf{1}'\mathbf{y}$, where $\mathbf{1}$ is a unit vector with n elements. Multiplying the result by the scalar $1/n$ would then give the mean. This operation can also be used to compute the length of a vector, $length = \sqrt{\mathbf{a}'\mathbf{a}}$.

The second type of multiplication, $\mathbf{C} = \mathbf{a}\mathbf{b}^t$, called an *outer product* or matrix product, is somewhat less common. The matrix $\mathbf{C}$ will have as many rows as there are elements in **a** and as many columns are there are elements in **b**. The elements of matrix

$$C \text{ are defined as } c_{ij} = a_i b_j. \text{ For example, } \begin{bmatrix} 1 \\ 2 \\ 3 \end{bmatrix} \begin{bmatrix} 1 & 3 & 5 & 7 \end{bmatrix} = \begin{bmatrix} 1 & 3 & 5 & 7 \\ 2 & 6 & 10 & 14 \\ 3 & 9 & 15 & 21 \end{bmatrix}$$

Matrix multiplication can be defined in terms of inner products of vectors. For the product $C = AB$, each element of C is computed as $c_{ij} = A_i B_j$, where A_i is the ith row of matrix A and B_j is the jth column of matrix B. In order for this to work, the number of columns of matrix A must be the same as the number of rows of matrix B. For example,

$$\begin{bmatrix} 1 & 2 & 3 \\ 4 & 5 & 6 \end{bmatrix} \times \begin{bmatrix} 1 & 3 & 5 & 7 \\ 2 & 4 & 6 & 8 \\ 3 & 5 & 7 & 9 \end{bmatrix}$$

$$= \begin{bmatrix} 1+4+9 & 3+8+15 & 5+12+21 & 7+16+27 \\ 4+10+18 & 12+20+30 & 20+30+42 & 28+40+54 \end{bmatrix} = \begin{bmatrix} 14 & 26 & 38 & 50 \\ 32 & 62 & 92 & 122 \end{bmatrix}$$

Unlike scalar multiplication, the order of operations is important. In general, $AB \neq BA$; that is, it does not have the commutative property of scalar algebra. In fact, multiplication in a different order will not even be defined unless the number of columns in the matrix on the left happens to match the number of rows for the matrix on the right. Matrix multiplication does have the associative property: $A(B + C) = AB + AC$ and $(A + B)C = AC + BC$. Another useful relationship is that $(AB)' = B'A'$ (i.e., the transpose of a product is the product of the transposes of the matrices written in the reverse order).

Multiplication of a vector or a matrix by a scalar simply multiplies every element of the vector or the matrix by the scalar. In particular, multiplication by -1 changes the sign of every element. For example, if we multiply a unit vector by the mean we generate a column vector with every element equal to the mean, $\bar{\mathbf{y}} = \bar{Y}\mathbf{1}$. This enables the following convenient formula for the sum of squares: $(\mathbf{y} - \bar{\mathbf{y}})'(\mathbf{y} - \bar{\mathbf{y}})$.

The **projection** of one vector onto another is often needed and is related to other useful quantities. The vector, **p**, that is the projection of a vector **a** onto another vector **b** is computed as $\mathbf{p} = \dfrac{\mathbf{a}'\mathbf{b}}{\mathbf{b}'\mathbf{b}}\mathbf{b}$. An example of a projection is given in the Figure B.1. The **length of the projection vector** is $p = \mathbf{a}'\mathbf{b}/\sqrt{\mathbf{b}'\mathbf{b}}$, and the **cosine of the angle between the two vectors** is $\cos\theta = \dfrac{\mathbf{a}'\mathbf{b}}{\sqrt{\mathbf{a}'\mathbf{a}}\sqrt{\mathbf{b}'\mathbf{b}}}$. The slope of the regression of Y on X can be interpreted as the projection of the vector corresponding to the dependent variable onto the vector of the deviations of the independent variable from its mean, $b = \dfrac{\mathbf{y}'(\mathbf{x} - \bar{\mathbf{x}})}{(\mathbf{x} - \bar{\mathbf{x}})'(\mathbf{x} - \bar{\mathbf{x}})}$ (note that it is not necessary to also express the dependent variable as deviations from its mean). Thus, $\hat{\mathbf{y}} = b(\mathbf{x} - \bar{\mathbf{x}})$, the vector of predicted deviations of **y** from its mean, is a projection of **y** onto the vector $\mathbf{x} - \bar{\mathbf{x}}$; the explained SS is the square of the length of the projection; and the cosine of the angle between **y** and $\mathbf{x} - \bar{\mathbf{x}}$ is the correlation between X and Y.

A **matrix inverse** can be defined simply as the matrix that yields an identity matrix as the result when multiplied by the original matrix (in either order). Thus, $\mathbf{AA}^{-1} = \mathbf{A}^{-1}\mathbf{A} = \mathbf{I}$, where $\mathbf{A}^{-1}$ is the inverse of matrix $\mathbf{A}$ and $\mathbf{I}$ is the identity matrix with the same number of rows and columns as $\mathbf{A}$. The matrix inverse corresponds to the reciprocal in scalar arithmetic, as we have seen in Expressions (16.8) and (16.9).

For an inverse of a matrix to exist, it must obey certain constraints. First, the matrix must be square. An important requirement for a reciprocal to exist is that the matrix must not be singular. This is a generalization of the requirement that a scalar must not be equal to zero in order for its inverse to exist. A singular matrix is one in which one or more rows or columns are linear combinations of the other rows or columns. It may be difficult to detect this just by inspection. However, computer algorithms for constructing an inverse of a matrix (of which there are many) will detect this condition and provide a warning. Most numerical methods for constructing an inverse of a matrix are straightforward but tedious and best left for computer software and hence not shown here. See the books cited at the beginning of this appendix.

An alternative way of handling the problem of inverting singular matrices is to make use of what are called **generalized inverses**. These matrices have many but not all of the properties of the usual inverse. There are several types of generalized inverses, and they are described in the books cited at the beginning of this appendix. The use of generalized inverses can provide an elegant approach to the solution of regression models because their use can deal with the problems of singular design matrices automatically.

It is sometimes necessary to be able to compute the **derivative of a vector or a matrix**. These operations are defined element-wise—for example, the derivative of a vector by a scalar variable is just a vector of the derivatives of each element by that scalar. The following identities are used in Section 16.6. The partial derivative symbols are used because the vector $\mathbf{b}$ corresponds to more than one scalar variable.

The derivative of a scalar constant (i.e., a scalar that is not a function of $\mathbf{b}$) by the vector $\mathbf{b}$ is the scalar zero: $\dfrac{\partial c}{\partial \mathbf{b}} = 0.$

The derivative of a scalar constant times a vector by that vector is just the scalar constant: $\dfrac{\partial c\mathbf{b}}{\partial \mathbf{b}} = c.$

The product $\mathbf{b}'\mathbf{Ab}$, where $\mathbf{b}$ is a vector and $\mathbf{A}$ is a square symmetric matrix that is not a function of $\mathbf{b}$ and has as many rows and columns as there are rows in $\mathbf{b}$, appears often. Its derivative with respect to the vector b is $\dfrac{\partial \mathbf{b}'\mathbf{Ab}}{\partial \mathbf{b}} = 2\mathbf{Ab}.$

There are, of course, many other identities for other combinations of vectors and matrices, but the ones given above are the ones needed in Chapter 16. This appendix only provides a very brief introduction to matrices. Many important topics such as eigenvalues and eigenvectors are left for the more comprehensive accounts in the books cited at the beginning of this appendix.

Bibliography

Agresti, A. 2001. Exact inference for categorical data: Recent advances and continuing controversies. *Stat. Med.* **20**: 2709–2722.

Agresti, A. 2002. *Categorical Data Analysis.* 2nd ed. Wiley, New York. 734 pp.

Agresti, A. 2007. *An Introduction to Categorical Data Analysis.* 2nd ed. Wiley, New York. 400 pp.

Akaike, H. 1974. A new look at the statistical model identification. *IEEE Trans. Auto. Contr.* **19**: 716–723.

Allee, W. C., and E. Bowen. 1932. Studies in animal aggregations: Mass protection against colloidal silver among goldfishes. *J. Exp. Zool.* **61**: 185–207.

Allee, W. C., E. S. Bowen, J. C. Welty, and R. Oesting. 1934. The effect of homotypic conditioning of water on the growth of fishes, and chemical studies of the factors involved. *J. Exp. Zool.* **68**: 183–213.

Altman, D. G. 1978. Plotting probability ellipses. *J. Royal Stat. Soc., Series C (Appl. Stat.)* **27**: 347–349.

Anderberg, M. R. 1973. *Cluster Analysis for Applications.* Academic Press, New York. 359 pp.

Anderson, V. L., and R. A. McLean. 1974. Restriction errors: Another dimension in teaching experimental statistics. *Amer. Stat.* **28**: 145–152.

Anscombe, F. J. 1948. The transformation of Poisson, binomial and negative binomial data. *Biometrika* **35**: 246–254.

Appleton, D. R., J. M. French, and M. P. J. Vanderpump. 1996. Ignoring a covariate: An example of Simpson's paradox. *Amer. Stat.* **50**: 340–341.

Archibald, E. E. A. 1950. Plant populations. II. The estimation of the number of individuals per unit area of species in heterogeneous plant populations. *Ann. Bot., N.S.* **14**: 7–21.

Asmundson, V. S. 1931. The formation of the hen's egg. *Sci. Agric.* **11**(9): 50 pp.

Balandra, K. P., and H. L. MacGillivray. 1988. Kurtosis: A critical review. *Amer. Stat.* **42**: 111–119.

Bancroft, T. A. 1964. Analysis and inference for incompletely specified models involving the use of preliminary test(s) of significance. *Biometrics* **20**: 427–442.

Banta, A. M. 1939. *Studies on the Physiology, Genetics, and Evolution of Some Cladocera.* Carnegie Institution of Washington, Dept. Genetics, Paper 39. 285 pp.

Barnes, H., and F. A. Stanbury. 1951. A statistical study of plant distribution during the colonization and early development of vegetation on china clay residues. *J. Ecol.* **39**: 171–181.

Barnett, V., and T. Lewis. 1994. *Outliers in Statistical Data.* 3rd ed. Wiley, Chichester, UK. 604 pp.

Bartels, R. H., S. D. Horn, A. M. Liebetrau, and W. L. Harris. 1977. *A Computational Investigation of Conover's Kolmogorov–Smirnov Test for Discrete Distributions.* Johns Hopkins Univ. Dept. Math. Sci. Tech. Report No. 260. 14 pp.

Benjamini, Y. 1988. Opening the box of a box plot. *Amer. Stat.* **42**: 257–262.

Bhalla, S. C., and R. R. Sokal. 1964. Competition among genotypes in the housefly at varying densities and proportions (the *green* strain). *Evolution* **18**: 312–330.

Bishop, Y. M. M., S. E. Fienberg, and P. W. Holland. 1975. *Discrete Multivariate Analysis: Theory and Practice.* MIT Press, Cambridge, MA. 557 pp.

Blakeslee, A. F. 1921. The globe mutant in the jimson weed (*Datura stramonium*). *Genetics* **6**: 241–264.

Bliss, C. I. 1967. *Statistics in Biology.* Vol. 1. McGraw-Hill, New York. 550 pp.

Bliss, C. I. 1970. *Statistics in Biology.* Vol. 2. McGraw-Hill, New York. 639 pp.

Bliss, C. I., and D. W. Calhoun. 1954. *An Outline of Biometry.* Yale Co-op. Corp., New Haven, CT. 272 pp.

Bliss, C. I., and R. A. Fisher. 1953. Fitting the negative binomial distribution to biological data and note on the efficient fitting of the negative binomial. *Biometrics* **9**: 176–200.

Block, B. C. 1966. The relation of temperature to the chirp-rate of male snowy tree crickets, *Oecanthus fultoni* (Orthoptera: Gryllidae). *Ann. Entomol. Soc. Amer.* **59**: 56–59.

Boardman, T. J. 1974. Confidence intervals for variance components—a comparative Monte Carlo study. *Biometrics* **30**: 251–262.

Borenstein, M., L. V. Hedges, J. P. T. Higgins, and H. R. Rothstein. 2009. *Introduction to Meta-Analysis.* Wiley, New York. 450 pp.

Bortkiewicz, L. von. 1898. *Das Gesetz der kleinen Zahlen.* Teubner, Leipzig. 52 pp.

Box, G. E. P., and D. R. Cox. 1964. An analysis of transformations. *J. R. Stat. Soc.*, Ser. B **26**: 211–243.

Bradley, R. A., and S. S. Srivastava. 1979. Coding in polynomial regression. *Amer. Stat.* **33**: 11–14.

Breiman, L., and J. H. Friedman. 1985. Estimating optimal transformations for multiple regression and correlation. *J. Amer. Stat. Assn.* **80**: 580–598.

Breton, M. D., M. D. Devore, and D. E. Brown. 2008. A tool for systematically comparing the power of tests for normality. *J. Stat. Comp. Simul.* **78**: 623–638.

Brian, M. V., and J. Hibble. 1964. Studies of caste differentiation in *Myrmica rubra* L. 7. Caste bias, queen age and influence. *Insectes Sociaux* **11**: 223–228.

Brower, L. P. 1959. Speciation in butterflies of the *Papilio glaucus* group. I. Morphological relationships and hybridization. *Evolution* **13**: 40–63.

Brown, B. E., and A. W. A. Brown. 1956. The effects of insecticidal poisoning on the level of cytochrome oxidase in the American cockroach. *J. Econ. Entomol.* **49**: 675–679.

Brown, F. M., and W. P. Comstock. 1952. Some biometrics of *Heliconius charitonius* (Linnaeus) (Lepidoptera, Nymphalidae). *Amer. Mus. Novitates* **1574**: 53 pp.

Buley, H. M. 1936. Consumption of diatoms and dinoflagellates by the mussel. *Bull. Scripps Inst. Oceanography, Tech. Ser.* **4**: 19–27.

Burdick, R. K., C. M. Borror, and D. C. Montgomery. 2005. *Design and Analysis of Gauge R&R Studies: Making Decisions with Confidence Intervals in Random and Mixed ANOVA Models.* SIAM: Philadelphia. 201 pp.

Burdick, R. K., and F. A. Graybill. 1992. *Confidence Intervals on Variance Components.* Marcel Dekker, New York. 230 pp.

Burdick, R. K., J. Quiroz, and H. K. Iyer. 2006. The present status of confidence interval estimation for one-factor random models. *J. Stat. Plan. Infer.* **136**: 4307–4325.

Burnham, K. P., and D. R. Anderson. 2002. *Model Selection and Inference—A Practical Information-Theoretic Approach.* 2nd ed. Springer Verlag.

Burr, E. J. 1960. The distribution of Kendall's score S for a pair of tied rankings. *Biometrika* **47**: 151–171.

Burrell, P. C., and T. W. Whitaker. 1939. The effect of indol-acetic acid on fruit-setting in muskmelons. *Amer. Soc. Hort. Sci.* **37**: 829–830.

Burstein, H. 1973. Close approximation of binomial confidence limits at 10 confidence levels. *Commun. Stat.* **2**: 185–187.

Cade, B. S., and B. R. Noon. 2003. A gentle introduction to quantile regression for ecologists. *Frontiers in Ecology and the Environment* **1**: 412–420.

Carroll, J. D., P. E. Green, and A. Chaturvedi. 1997. *Mathematical Tools for Applied Multivariate Analysis*. Academic Press, New York. 376 pp.

Carroll, R. J., and D. Ruppert. 1996. The use and misuse of orthogonal regression in linear errors-in-variables models. *Amer. Stat.* **50**: 1–6.

Carter, G. R., and C. A. Mitchell. 1958. Methods for adapting the virus of rinderpest to rabbits. *Science* **128**: 252–253.

Casagrande, J. T., M. C. Pike and P. G. Smith. 1978. An improved approximate formula for calculating sample sizes for comparing two binomial distributions. *Biometrics* **34**: 483–486.

Castillo, J. V. 1938. Determination of the optimum amount of fish meal in rations for growing ducklings. *Philippine Agric.* **27**: 18–30.

Chernick, M. R. 2007. *Bootstrap Methods: A Guide for Practitioners and Researchers*. 2nd ed. Wiley, New York. 369 pp.

Clark, F. H. 1941. Correlation and body proportion in mature mice of the genus *Peromyscus*. *Genetics* **26**: 283–300.

Clifford, H. T., and W. Stephenson. 1975. *An Introduction to Numerical Classification*. Academic Press, New York. 229 pp.

Clifford, P., S. Richardson, and D. Hémon. 1989. Assessing the significance of the correlation between two spatial processes. *Biometrics* **45**: 123–134.

Child, D. 2006. *Essentials of Factor Analysis*. 3rd ed. Continuum, London. 192 pp.

Cochran, W. G. 1950. The comparison of percentages in matched samples. *Biometrika* **37**: 256–266.

Cochran, W. G., and G. M. Cox. 1957. *Experimental Designs*. 2nd ed. Wiley, New York. 611 pp.

Cohen, A. C., Jr. 1960. Estimating the parameter in a conditional Poisson distribution. *Biometrics* **16**: 203–211.

Cohen, J. 1988. *Statistical Power Analysis for the Behavioral Sciences*. 2nd ed. Lawrence Erlbaum Associates, Hillsdale, NJ. 567 pp.

Cohen, L. A., M. E. Kendall, E. Zang, C. Meschter, and D. P. Rose. 1991. Modulation of N-nitrosomethylurea-induced mammary tumor promotion by dietary fiber and fat. *J. Nat. Cancer Inst.* **83**: 496–501.

Conahan, M. A. 1970. The comparative accuracy of the likelihood ratio and X-squared as approximations to the exact multinomial test. Ed.D. diss., Lehigh University 64 pp.

Connor, R. J. 1987. Sample size for testing difference in proportions for the paired-sample design. *Biometrics* **43**: 207–211.

Cook, P. A., and C. P. Wheater. 2000. *Using Statistics to Understand the Environment*. Routledge, New York. 272 pp.

Cooper, H., L. V. Hedges, and J. C. Valentine (eds.). 2009. *The Handbook of Research Synthesis and Meta-Analysis*. Russell Sage Foundation, New York. 615 pp.

Copenhaver, M. D., and B. Holland. 1988. Computation of the distribution of the maximum Studentized range statistic with application to multiple significance testing of simple effects. *J. Stat. Comp. Simul.* **30**: 1–15.

Cowan, I. M., and P. A. Johnston. 1962. Blood serum protein variations at the species and subspecies level in deer of the genus *Odocoileus*. *Syst. Zool.* **11**: 131–138.

Cox, D. R., and N. Reid. 2000. *The Theory of the Design of Experiments*. Chapman and Hall, New York. 336 pp.

Cox, D. R., and E. J. Snell. 1968. A general definition of residuals. *J. Roy. Stat. Soc., Series B—Statistical Methodology* **30**: 248–275.

Cragg, J. G., and R. S. Uhler. 1970. The Demand for Automobiles. *The Canadian Journal of Economics / Revue canadienne d'Economique* **3**: 386–406.

Crow, E. L. 1956. Confidence intervals for a proportion. *Biometrika* **43**: 423–435.

Crow, E. L., and R. S. Gardner. 1959. Confidence intervals for the expectation of a Poisson variable. *Biometrika* **46**: 441–453.

Csiszár, I., and P. Shields. 2004. *Information Theory and Statistics: A Tutorial.* Now Publishers Inc., Boston. 124 pp.

D'Agostino, R. B., W. Chase, and A. Belanger. 1988. The appropriateness of some common procedures for testing the equality of two independent binomial populations. *Amer. Stat.* **42**: 198–202.

D'Agostino, R. B., and G. E. Noether. 1973. On the evaluation of the Kolmogorov statistic. *Amer. Stat.* **27**: 81–82.

D'Agostino, R. B., and E. S. Pearson. 1973. Tests for departure from normality: Empirical results for the distribution of b_2 and $\sqrt{b_1}$. *Biometrika* **60**: 613–622.

D'Agostino, R. B., and G. L. Tietjen. 1973. Approaches to the null distribution of $b1$. *Biometrika* **60**: 169–173.

Davis, E. A., Jr. 1955. Seasonal changes in the energy balance of the English sparrow. *Auk* **72**: 385–411.

Davison, A. C., and D. V. Hinkley. 1997. *Bootstrap Methods and Their Application.* Cambridge University Press, London. 592 pp.

Demidenko, E. 2004. *Mixed Models: Theory and Applications.* Wiley-Interscience, New York.

Denny, M., and S. D. Gaines. 2002. *Chance in biology: Using probability to explore nature.* Princeton Univ. Press: Princeton. 416 pp.

Devlin, S. J., R. Gnanadesikan, and J. R. Kettenring. 1975. Robust estimation and outlier detection with correlation coefficients. *Biometrika* **62**: 531–545.

Dixon, P. M. 2001. The bootstrap and the jackknife: Describing the precision of ecological indices. Pages 267–288 in *Design and Analysis of Ecological Experiments* (S. M. Scheiner, and J. Gurevitch, eds.). Oxford University Press, New York. 432 pp.

Dixon, P. M. 2002. Bootstrap resampling. Pages 212–219 in *Encyclopedia of Environmetrics* (A. H. El-Shaarawi, and W. W. Piegorsch, eds.). John Wiley and Sons, New York.

Dixon, W. J. 1950. Analysis of extreme values. *Ann. Math. Stat.* **21**: 488–506.

Dixon, W. J., and F. J. Massey, Jr. 1969. *Introduction to Statistical Analysis.* 3rd ed. McGraw-Hill, New York. 638 pp.

Draper, N. R., and H. Smith. 1998. *Applied regression analysis.* 3rd ed. Wiley: New York. 736 pp.

Dunn, G., and B. S. Everitt. 2004. *An Introduction to Mathematical Taxonomy.* Dover Publications, New York. 160 pp.

Dunn, O. J., and V. A. Clark. 1974. *Applied Statistics: Analysis of Variance and Regression.* Wiley, New York. 387 pp.

Dunnett, C. W. 1980. Pairwise multiple comparisons in the homogeneous variance, unequal sample size case. *J. Amer. Stat. Assn.* **75**: 789–795.

DuPraw, E. J. 1965. Non-Linnean taxonomy and the systematics of honeybees. *Syst. Zool.* **14**: 1–24.

Dwass, M. 1960. Some k-sample rank-order tests. Pp. 198–202 in *Contributions to Probability and Statistics, Essays in Honor of Harold Hotelling* (I. Olkin et al., eds.) Stanford University Press, Stanford, CA.

Efron, B. 1987. *The Jackknife, the Bootstrap, and Other Resampling Plans.* Society for Industrial Mathematics, Philadelphia. 100 pp.

Efron, B., and G. Gong. 1983. A leisurely look at the bootstrap, the jackknife, and cross-validation. *Amer. Stat.* **37**: 36–48.

Efron, B., and R. J. Tibshirani. 1994. *An Introduction to the Bootstrap.* Chapman and Hall, London. 456 pp.

Eisenhart, C. 1947. The assumptions underlying the analysis of variance. *Biometrics* **3**: 1–21.

Evans, M., N. Hastings, and B. Peacock. 2000. *Statistical Distributions.* 3rd ed. Wiley: New York. 221 pp.

Everitt, B. S., and G. Dunn. 2001. *Applied Multivariate Data Analysis.* 2nd ed. Wiley, New York. 342 pp.

Everitt, B. S., S. Landau, and M. Leese. 2001. *Cluster Analysis.* 4th ed. Wiley, New York. 320 pp.

Falconer, D. S. 1981. *Introduction to Quantitative Genetics.* 2nd ed. Longman Press, New York. 340 pp.

Falconer, D. S., and T. F. C. Mackay. 1996. *Introduction to Quantitative Genetics.* Prentice Hall, New York. 464 pp.

Fawcett, R. F. 1990. The alignment method for displaying and analyzing treatments in blocking designs. *Amer. Stat.* **44**: 204–209.

Federer, W. T., and F. King. 2007. *Variations on Split Plot and Split Block Experiment Designs.* Wiley, New York. 270 pp.

Felsenstein, J. 1985. Confidence limits on phylogenies: An approach using bootstrap. *Evolution* **39**: 783–791.

Felsenstein, J. 2004. *Inferring Phylogenies.* Sinauer, Sunderland. 644 pp.

Feltz, C. J. and G. E. Miller. 1996. An asymptotic test for the equality of coefficients of variation from k populations. *Statistics in Medicine* **15**: 646–658.

Ferreira, D. F., C. G. B. Demetrio, B. F. J. Manly, and A. de Machado. 2007. Quantiles from the maximum studentized range distribution. *Rev. Mat. Estat., Sao Paulo* **25**: 117–135.

Feuer, E. J., and L. G. Kessler. 1989. Test statistic and sample size for a two-sample McNemar test. *Biometrics* **45**: 629–636.

Finney, D. J. 1971. *Probit Analysis.* 3rd ed. Cambridge University Press, London. 333 pp.

Finney, D. J., R. Latscha, B. M. Bennett, and P. Hsu. 1963. *Tables for Testing Significance in a 2 × 2 Contingency Table.* Cambridge University Press, London. 102 pp.

Fisher, R. A. 1954. *Statistical Methods for Research Workers.* 12th ed. Oliver & Boyd, Edinburgh. 356 pp.

Fisher, R. A., and F. Yates. 1963. *Statistical Tables for Biological, Agricultural and Medical Research.* 6th ed. Oliver & Boyd, Edinburgh. 138 pp.

Fleiss, J. L. 1979. Confidence intervals for the odds ratio in case-control studies: The state of the art. *J. Chronic Diseases* **32**: 69–77.

Fligner, M. A., and G. E. Policello. 1981. Robust rank procedures for the Fisher–Behrens problem. *J. Amer. Stat. Assn.* **76**: 162–168.

Freeman, M. F., and J. W. Tukey. 1950. Transformations related to the angular and the square root. *Ann. Math. Stat.* **21**: 607–611.

Freidlin, B. and J. L. Gastwirth. 2000. Should the median test be retired from general use? *Amer. Stat.* **54**: 161–164.

French, A. R. 1976. Selection of high temperatures for hibernation by the pocket mouse, *Perignathus longimembris*: Ecological advantages and energetic consequences. *Ecology* **57**: 185–191.

Fröhlich, F. W. 1921. Grundzüge einer Lehre vom Licht- und Farbensinn. *Ein Beitrag zur allgemeinen Physiologie der Sinne*. Fischer, Jena. 86 pp.

Gabriel, K. R. 1964. A procedure for testing the homogeneity of all sets of means in analysis of variance. *Biometrics* **20**: 459–477.

Gabriel, K. R. 1978. A simple method of multiple comparisons of means. *J. Amer. Stat. Assn.* **73**: 724–729.

Gaines, S. D., and W. R. Rice. 1990. Analysis of biological data when there are ordered expectations. *Amer. Nat.* **135**: 310–317.

Galambo, J. 1984. *Introduction to Probability Theory*. Marcel Dekker, New York. 200 pp.

Games, P. A., and J. F. Howell. 1976. Pairwise multiple comparison procedures with unequal N's and/or variances: A Monte Carlo study. *J. Educ. Stat.* **1**: 113–125.

Gans, D. L. 1991. Preliminary test on variances. *Amer. Stat.* **45**: 258.

Gart, J. J. 1969. An exact test for comparing matched proportions in crossover designs. *Biometrika* **56**: 75–80.

Gartler, S. M., I. L. Firschein, and T. Dobzhansky. 1956. Chromatographic investigation of urinary amino-acids in the great apes. *Amer. J. Phys. Anthropol.* **14**: 41–57.

Gaylor, D. W., and F. N. Hopper. 1969. Estimating the degrees of freedom for linear combinations of mean squares by Satterthwaite's formula. *Technometrics* **11**: 691–705.

Geissler, A. 1889. Beiträge zur Frage des Geschlechtsverhältnisses der Geborenen. *Z. K. Sächs. Stat. Bur.* **35**: 1–24.

Geladi, P., and B. Kowalski. 1986. Partial leastsquares regression: A tutorial. *Analytica Chimica Acta* **185**: 1–17.

Gentle, J. E. 2010. *Matrix Algebra: Theory, Computations, and Applications in Statistics*. Springer, New York. 530 pp.

Gideon, R. A., and D. E. Mueller. 1978. Computation of two-sample Smirnov statistics. *Amer. Stat.* **32**: 136–137.

Glass, D. J. 2006. *Experimental Design for Biologists*. Cold Spring Harbor Laboratory Press, New York. 206 pp.

Gleser, L. J. 1985. Exact power of goodness-of-fit tests of Kolmogorov type for discontinuous distributions. *J. Amer. Stat. Assn.*, **80**: 954–958.

Gleser, L. J., and I. Olkin. 2009. Stochastically dependent effect sizes. Pages 357–376 in *The Handbook of Research Synthesis and Meta-Analysis* (H. Cooper, L. V. Hedges, and J. C. Valentine, eds.). Russell Sage Foundation, New York.

Goddard, M. J. 1991. Constructing some categorical anomalies. *Amer. Stat.* **45**: 129–134.

Gonzalez, T., S. Sahni, and W. R. Franta. 1977. An efficient algorithm for the Kolmogorov–Smirnov and Lilliefors tests. *ACM Trans. Math. Software* **3**: 60–64.

Good, P. 2000. *Permutation Tests: A Practical Guide to Resampling Methods for Testing Hypotheses*. Springer-Verlag, New York. 270 pp.

Gordon, A. D. 1981. *Classification: Methods for Exploratory Analysis of Multivariate Data*. Chapman and Hall, New York. 193 pp.

Gould, S. J. 1966. Allometry and size in ontogeny and phylogeny. *Biol. Rev.* **41**: 587–640.

Gould, S. J. 1975. Allometry in primates, with emphasis on scaling and the evolution of the brain. *Contrib. Primatol.* **5**: 244–292.

Govindarajulu, Z. 2000. *Statistical Techniques in Bioassay*. 2nd ed. S. Karger AG, Switzerland. 234 pp.

Graybill, F. A. 2001. *Matrices with Applications in Statistics*. 2nd ed. Duxbury, Belmont, CA. 480 pp.

Green, P. E., with contributions by J. D. Carroll. 1978. *Analyzing Multivariate Data*. Dryden Press, Hinsdale, IL. 519 pp.

Greenland, S. 1991. On the logical justification of conditional tests for two-by-two contingency tables. *Amer. Stat.* **45**: 248–251.

Greenwood, M., and G. U. Yule. 1920. An inquiry into the nature of frequency distributions of multiple happenings. *J. R. Stat. Soc.* **83**: 255–279.

Greig-Smith, P. 1964. *Quantitative Plant Ecology*. 2nd ed. Butterworths, Washington, DC. 256 pp.

Grizzle, J. E. 1967. Continuity correction in the X^2-test for 2×2 tables. *Amer. Stat.* **21**(4): 28–32.

Groggel, D. J., and J. H. Skillings. 1986. Distribution-free tests for main effects in multifactor designs. *Amer. Stat.* **40**: 99–102.

Grubbs, F. E. 1969. Procedures for detecting outlying observations in samples. *Technometrics* **11**: 1–21.

Gumbel, E. J. 1954. *Statistical Theory of Extreme Values and Some Practical Applications*. U.S. Govt. Printing Office, Washington, D.C. 51 pp.

Gurevitch, J., and L. V. Hedges. 2001. Meta-analysis: Combining the results of independent experiments. Pages 347–370 in *Design and Analysis of Ecological Experiments* (S. M. Scheiner and J. Gurevitch, eds.). Oxford University Press, New York.

Gurevitch, J., P. S. Curtis, and M. H. Jones. 2001. Meta-analysis in ecology. *Advances in Ecological Research* **32**: 199–247.

Gurland, J., and R. C. Tripathi. 1971. A simple approximation for unbiased estimation of the standard deviation. *Amer. Stat.* **25**: 30–32.

Haine, E. 1955. Biologisch-ökologische Studien an *Rhopalosiphoninus latysiphon* D. Landwirtschaftsverlag, Hiltrup bei Münster (Westf.). 58 pp.

Haltenorth, T. 1937. Die verwandtschaftliche Stellung der Grosskatzen zueinander. *Z. Säugetierk.* **12**: 97–240.

Hamilton, D. 1987. Sometimes $R^2 > r_{yx_1}^2 + r_{yx_2}^2$ correlated variables are not always redundant. *Amer. Stat.* **41**: 129–132.

Hand, D. J., F. Daly, A. D. Lunn, K. McConway, and E. Ostrowski (eds.). 1994. A Handbook of small Data Sets, New York: Chapman & Hall. 474 pp.

Hand, D. J., and C. C. Taylor. 1987. *Multivariate Analysis of Variance and Repeated Measures: A Practical Approach for Behavioural Scientists*. Chapman and Hall, New York. 262 pp.

Hanna, B. L. 1953. On the relative importance of genetic and environmental factors in the determination of human hair pigment concentration. *Amer. J. Hum. Genet.* **5**: 293–321.

Harris, R. E., E. A. Zang, J. I. Anderson, and E. L. Wynder. 1993. Race and sex differences in lung cancer risk associated with cigarette smoking. *Internat. J. Epidem.* **22**: 592–599.

Harter, H. L., H. J. Khamis, and R. E. Lamb. 1984. Modified Kolmogorov–Smirnov tests for goodness of fit. *Comm. Stat. Simul. Comput.* **13**: 293–323.

Hartigan, J. A. 1975. *Clustering Algorithms*. Wiley, New York. 351 pp.

Hartley, H. O. 1950. The maximum *F*-ratio as a short cut test for heterogeneity of variances. *Biometrika* **37**: 308–312.

Hartley, H. O. 1962. Analysis of variance. Pages 221–230 in *Mathematical Methods for Digital Computers*, Vol. 1 (A. Ralston and H. S. Wilf, eds.). Wiley, New York.

Harville, D. A. 2009. *Matrix Algebra from a Statistician's Perspective*. Springer, New York. 636 pp.

Hasel, A. A. 1938. Sampling error in timber surveys. *J. Agric. Res.* **57**: 713–737.

Hastie, T. J., and R. J. Tibshirani. 1990. *Generalized Additive Models*. Chapman and Hall, New York. 352 pp.

Haviland, M. G. 1990. Yates's correction for continuity and the analysis of 2×2 contingency tables. *Statistics in Medicine* **9**: 363–367.

Hazelton, M. L. 2003. A graphical tool for assessing normality. *Amer. Stat.* **57**: 285–288.

He, X. 1997. Quantile curves without crossing. *Amer. Stat.* **51**: 186–192.

Hedges, L. V. 1981. Distribution theory for Glass's estimator of effect size and related estimators. *Journal of Educational Statistics* **6**: 107–128.

Hedges, L. V., and I. Olkin. 1985. *Statistical Methods in Meta-Analysis*. Academic Press, New York. 369 pp.

Hinkelmann, K., and O. Kempthorne. 2005. *Design and Analysis of Experiments*, Vol. 2, *Advanced Experimental Design*. Wiley-Interscience, New York. 780 pp.

Hinkelmann, K., and O. Kempthorne. 2007. *Design and Analysis of Experiments*. Vol. 1, *Introduction to Experimental Design*, 2nd ed. Wiley-Interscience, New York. 672 pp.

Hinkley, D. V. 1978. Improving the jackknife with special reference to correlation estimation. *Biometrika* **65**: 13–21.

Hoaglin, D. C., and R. E. Welsch. 1978. The hat matrix in regression and ANOVA. *Amer. Stat.* **32**: 17–22.

Hochberg, Y. 1974. Some generalizations of the T-method in simultaneous inference. *J. Multivar. Anal.* **4**: 224–234.

Hochberg, Y. 1976. A modification of the T-method of multiple comparisons for one-way layout with unequal variances. *J. Amer. Stat. Assn.* **71**: 200–203.

Hochberg, Y., and A. C. Tamhane. 1983. Multiple comparisons in a mixed model. *Amer. Stat.* **37**: 305–307.

Hocking, R. R. 1976. The analysis and selection of variables in linear regression. *Biometrics* **32**: 1–49.

Holm, S. 1979. A simple sequentially rejective multiple test procedure. *Scand. J. Stat.* **6**: 65–70.

Hosmer, D. W., and S. Lemeshow. 1989. *Applied Logistic Regression*. Wiley, New York. 307 pp.

Hotelling, H. 1953. New light on the correlation coefficient and its transforms. *J. R. Stat. Soc., Ser. B* **15**: 193–232.

Huet, S., A. Bouvier, M.-A. Poursat, and E. Jolivet. 2010. *Statistical Tools for Nonlinear Regression: A Practical Guide with S-PLUS and R Examples*. 2nd ed. Springer, New York. 246 pp.

Hunt, G. L., and M. W. Hunt. 1976. Gull chick survival: The significance of growth rates, timing of breeding and territory size. *Ecology* **57**: 62–75.

Hunter, P. E. 1959. Selection of *Drosophila melanogaster* for length of larval period. *Z. Vererbungsl.* **90**: 7–28.

Hurvich, C. M., and C.-L. Tsai. 1990. The impact of model selection on inference in linear regression. *Amer. Stat.* **44**: 214–217.

Huxley, J. 1932. *Problems of Relative Growth*. Methuen & Co.: London. Reprinted 1972, Dover Publications: New York. 312 pp.

Hyde, J. S. 1981. How large are cognitive gender differences? A meta-analysis using w^2 and d. *Amer. Psych.* **36**: 892–901.

Jackson, J. E. 2003. *A User's Guide to Principal Components*. Wiley-Interscience, New York. 592 pp.

Jensen, R. J. 2009. Phenetics: revolution, reform or natural consequence? *Taxon* **58**: 50–60.

Johnson, N. K. 1966. Bill size and the question of competition in allopatric and sympatric populations of dusky and gray flycatchers. *Syst. Zool.* **15**: 70–87.

Johnson, N. L., and S. Kotz. 1969. *Distributions in Statistics: Discrete Distributions,* Vol. 1. Houghton Mifflin, Boston. 328 pp.

Johnson, R. A., and D. W. Wichern. 2007. *Applied Multivariate Statistical Analysis.* 6th ed. Prentice Hall. Upper Saddle River, New Jersey. 800 pp.

Johnston, R. G., S. D. Schroder, and A. R. Mallawaaratchy. 1995. Statistical Artifacts in the Ratio of Discrete Quantities. *American Statistician,* **49**: 285–291.

Jolicoeur, P. 1968. Interval estimation of the slope of the major axis of a bivariate normal distribution in the case of a small sample. *Biometrics* **24**: 679–682.

Jolicoeur, P. 1975. Linear regressions in fishery research: Some comments. *J. Fish. Res. Board Can.* **32**: 1491–1494.

Jöreskog, K. G. 1970. A general method for analysis of covariance structures. *Biometrika* **57**: 239–251.

Karten, I. 1965. Genetic differences and conditioning in *Tribolium castaneum. Physiol. Zool.* **38**: 69–79.

Kelley, K. 2008. Sample size planning for the squared multiple correlation coefficient: Accuracy in parameter estimation via narrow confidence intervals. *Multivariate Behav. Res.* **43**: 524–555.

Kelley, K., and S. E. Maxwell. 2003. Sample size for multiple regression: Obtaining regression coefficients that are accurate, not simply significant. *Psych. Methods* **8**: 305–321.

Kelley, K., and S. E. Maxwell. 2008. Power and accuracy for omnibus and targeted effects: Issues of sample size planning with applications to multiple regression. Pages 166–192 in *Handbook of Social Research Methods* (P. Alasuuta, L. Bickman, and J. Brannen, eds.). Sage, Newbury Park, CA.

Kempthorne, O. 1975. Fixed and mixed models in the analysis of variance. *Biometrics* **31**: 473–486.

Kendall, M. G., and J. D. Gibbons. 1990. *Rank Correlation Methods.* 5th ed. Edward Arnold, London.

Kendall, M. G., and A. Stuart. 1961. *The Advanced Theory of Statistics,* Vol. 2. Hafner, New York. 676 pp.

Kermack, K. A., and J. B. S. Haldane. 1950. Organic correlation and allometry. *Biometrika* **37**: 30–41.

Khamis, H. J. 1990. The δ-corrected Kolmogorov–Smirnov test for goodness of fit. *J. Stat. Plan. Infer.* **24**: 317–335.

Khamis, H. J. 1992. The δ-corrected Kolmogorov–Smirnov test with estimated parameters. *Nonparametric Stat.* **2**: 17–27.

Khamis, H. J. 2000. The two-stage δ-corrected Kolmogorov–Smirnov test. *J. Appl. Stat.* **27**: 439–450.

Khuri, A. I. 1995. A measure to evaluate the closeness of Satterthwaite's approximation. *Biometrical J.* **37**: 547–563.

King, J. A., D. Maas, and R. G. Weisman. 1964. Geographic variation in nest size among species of *Peromyscus. Evolution* **18**: 230–234.

Kline, R. B. 2004. *Principles and Practice of Structural Equation Modeling.* 2nd ed. The Guilford Press, New York. 366 pp.

Klingenberg, C. P. 1996. Multivariate allometry. Pages 23–49 in *Advances in Morphometrics* (L. F. Marcus, M. Corti, A. Loy, G. J. P. Naylor, and D. E. Slice, eds.). Plenum Press, New York.

Knott, M. 1970. Small sample power of one-sided Kolmogorov tests for a shift in location of normal distribution. *J. Amer. Stat. Assn.* **65**: 1384–1394.

Koehn, R. K. 1978. Physiology and biochemistry of enzyme variation: The interface of ecology and population genetics. Pages 51–72 in *Ecological Genetics: The Interface* (P. F. Brussard, ed.). Springer-Verlag, New York.

Koenker, R. 2005. *Quantile Regression*. Cambridge University Press, Cambridge. 368 pp.

Kotz, S., and D. F. Stroup. 1983. *Educated Guessing: How to Cope in an Uncertain World*. Marcel Dekker, New York. 187 pp.

Kouskolekas, C. A., and G. C. Decker. 1966. The effect of temperature on the rate of development of the potato leafhopper. Empoasca fabae. (Homoptera: Cicadellidae). *Ann. Entomol. Soc. Amer.* **59**: 292–298.

Kraemer, H. C., J. Mintz, A. Noda, J. Tinklenberg, and J. A. Yesavage. 2006. Caution regarding the use of pilot studies to guide power calculations for study proposals. *Arch. Gen. Psych.* **63**: 484–489.

Krebs, C. J. 1989. *Ecological Methodology*. Harper and Row, New York. 654 pp.

Kruskal, W. 1987. Relative importance by averaging over orderings. *Amer. Stat.* **41**: 6–10.

Kruskal, W. H., and W. A. Wallis. 1952. Use of ranks in one-criterion variance analysis. *J. Amer. Stat. Assn.* **47**: 583–621.

Krzanowski, W. J. 2000. *Principles of Multivariate Analysis*. Oxford University Press, Oxford. 608 pp.

Kuhry, B., and L. F. Marcus. 1977. Bivariate linear models in biometry. *Syst. Zool.* **26**: 201–209.

Kullback, S. 1997. *Information Theory and Statistics*. Wiley, New York. 416 pp.

Kullback, S., and R. A. Leibler. 1951. On information and sufficiency. *Ann. Math. Stat.* **22**: 79–86.

Kvånalseth, T. O. 1985. Cautionary note about R^2. *Amer. Stat.* **39**: 279–285.

Lachin, J. M. 1992. Power and sample size evaluation for the McNemar test with application to matched case-control studies. *Stat. Med.* **11**: 1239–1251.

Lande, R. 1977. On comparing coefficients of variation. *Syst. Zool.* **26**: 214–217.

Larntz, K. 1978. Small-sample comparisons of exact levels for chi-squared goodness-of-fit statistics. *J. Amer. Stat. Assn.* **73**: 253–263.

Le, C. T. 1998. *Applied Categorical Data Analysis*. Wiley: New York. 287 pp.

Legendre, P., and L. Legendre. 1998. Numerical ecology, 2nd. English edition. Elsevier Science. 853 pp.

Leggatt, C. W. 1935. Contributions to the study of the statistics of seed testing. *Proc. Internat. Seed Testing Assn.* **5**: 27–37.

Levine, M., and M. H. H. Ensom. 2001. Post hoc power analysis: An idea whose time has passed? *Pharmacotherapy* **21**: 405–409.

Lewontin, R. C. 1966. On the measurement of relative variability. *Syst. Zool.* **15**: 141–142.

Lewontin, R. C. 1974. The analysis of variance and the analysis of causes. *Amer. J. Hum. Genet.* **26**: 400–411.

Lewontin, R. C., and J. Felsenstein. 1965. The robustness of homogeneity tests in 2 × n tables. *Biometrics* **21**: 19–33.

Lewontin, R. C., and M. J. D. White. 1960. Interaction between inversion polymorphisms of two chromosome pairs in the grasshopper, *Moraba scurra. Evolution* **14**: 116–129.

Li, C. C. 1975. *Path Analysis—A Primer*. Boxwood Press, Pacific Grove, CA. 346 pp.

Li, C. C. 1976. *First Course in Population Genetics*. Boxwood Press, Pacific Grove, CA. 631 pp.

Li, X. and G. Li. 2007. Comparison of confidence intervals on the among group variance in the unbalanced variance component model. *J. Stat. Comp. Simul.* **77**: 477–486.

Littell, R. C., W. Stroup, and R. Freund. 2008. *SAS System for Linear Models*. 4th ed. Wiley-Interscience, New York. 504 pp.

Little, R. J. A. 1989. Testing the equality of two independent proportions. *Amer. Stat.* **43**: 283–288.

Littlejohn, M. J. 1965. Premating isolation in the *Hyla ewingi* complex. *Evolution* **19**: 234–243.

Livshits, G., R. R. Sokal, and E. Kobyliansky. 1991. Genetic affinities of Jewish populations. *Amer. J. Hum. Genet.* **49**: 131–146.

Loehlin, J. C. 2003. *Latent Variable Models: An Introduction to Factor, Path, and Structural Equation Analysis*. 4th ed. Psychology Press, Sussex, UK. 336 pp.

Lunt, H. A. 1947. The response of hybrid poplar and other forest tree species to fertilizer and lime treatment in concrete soil frames. *J. Agric. Res.* **74**: 113–132.

MacArthur, J. W. 1931. Linkage studies with the tomato. III. Fifteen factors in six groups. *Trans. R. Can. Inst.* **18**: 1–19.

Mallows, C. L. 1973. Some comments on C_p. *Technometrics* **15**: 661–675.

Manly, B. F. J. 1977. A further note on Kiritani and Nakusuji's model for stage frequency data including comments on the use of Tukey's jackknife technique for estimating variances. *Res. Pop. Ecol.* **18**: 177–186.

Manly, B. F. J. 1997. *Randomization, Bootstrap and Monte Carlo Methods in Biology*. Chapman and Hall, New York. 399 pp.

Manly, B. F. J. 2004. *Multivariate Statistical Methods: A Primer*. 3rd ed. Chapman and Hall: London. 225 pp.

Mantel, N. 1967. The detection of disease clustering and a generalized regression approach. *Cancer Res.* **27**: 209–220.

Mantel, N., and W. Haenszel. 1959. Statistical aspects of the analysis of data from retrospective studies of disease. *J. Nat. Cancer Inst.* **22**: 719–748.

Mark, L. C., J. J. Burns, L. Brand, C. I. Campomanes, N. Trousof, E. M. Papper, and B. B. Brodie. 1958. The passage of thiobarbiturates and their oxygen analogs into brain. *J. Pharmacol. Exp. Therap.* **123**: 70–73.

Markowski, C. A., and E. P. Markowski. 1990. Conditions for the effectiveness of a preliminary test of variance. *Amer. Stat.* **44**: 322–326.

Marquardt, D. W. 1970. Generalized inverses, ridge regression, biased linear estimation and non-linear estimation. *Technometrics* **12**: 591–612.

Marquardt, D. W., and R. D. Snee. 1975. Ridge regression in practice. *Amer. Stat.* **29**: 3–20.

MathWorks, The. 2009. MATLAB 7.8. The MathWorks, Natick, MA (www.mathworks.com).

Maxwell, A. E. 1970. Comparing the classification of subjects by two independent judges. *Brit. J. Psych.* **116**: 651–655.

Maxwell, S. E. 2004. The persistence of underpowered studies in psychological research: Causes, consequences, and remedies. *Psych. Methods* **9**: 147–163.

Mayer, L. S., and M. S. Younger. 1976. Estimation of standardized regression coefficients. *J. Amer. Stat. Assn.* **71**: 154–157.

Mayr, E., and P. D. Ashlock. 1991. *Principles of Systematic Zoology*. 2nd ed. McGraw-Hill, New York. 475 pp.

McFadden, D. 1974. Conditional logit analysis of qualitative choice behavior. Pages 105–142 in *Frontiers in Economics* (P. Zarembka, ed.), Academic Press, New York.

McGill, R., J. W. Tukey, and W. A. Larsen. 1978. Variations of box plots. *Amer. Stat.* **32**: 12–16.

McQuarrie, A. D. R., and C.-L. Tsai. 1998. *Regression and Time Series Model Selection*. World Scientific, Hackensack, NJ. 455 pp.

Miller, R. 1968. Jackknifing variances. *Ann. Math. Stat.* **39**: 567–582.

Miller, R. G. 1974. The jackknife—a review. *Biometrika* **61**: 1–15.

Millis, J., and Y. P. Seng. 1954. The effect of age and parity of the mother on birth weight of the offspring. *Ann. Hum. Genet.* **19**: 58–73.

Mittler, T. E., and R. H. Dadd. 1966. Food and wing determination in *Myzus persicae* (Homoptera: Aphidae). *Ann. Entomol. Soc. Amer.* **59**: 1162–1166.

Montgomery, D. C. 2004. *Design and Analysis of Experiments.* 6th ed. Wiley, New York. 660 pp.

Morrison, D. F. 2004. *Multivariate Statistical Methods.* 4th ed. McGraw-Hill, New York. 469 pp.

Mosimann, J. E. 1970. Size allometry: Size and shape variables with characterizations of hte lognormal and generalized gamma distributions. *J. Amer. Stat. Assn.* **65**: 930–945.

Motulsky, H., and A. Christopoulos. 2004. *Fitting Models to Biological Data Using Linear and Nonlinear Regression: A Practical Guide to Curve Fitting.* Oxford University Press, Oxford. 352 pp.

Mulaik, S. A. 2009. *The Foundations of Factor Analysis,* 2nd ed. Chapman and Hall, New York. 548 pp.

Mullett, G. M. 1972. Graphical illustration of simple (total) and partial regression. *Amer. Stat.* **26**(5): 25–27.

Nagelkerke, N. J. D. 1991. A note on a general definition of the coefficient of determination. *Biometrika* **78**: 691–692.

Nelson, V. E. 1964. The effects of starvation and humidity on water content in *Tribolium confusum* Duval (Coleoptera). Ph.D. diss., University of Colorado. 111 pp.

Nemenyi, P. B. 1963. Distribution-free multiple comparisons. PhD thesis, Princeton University. 1408 pp.

Neter, J., M. H. Kutner, C. J. Nachtsheim, and W. Wasserman. 2004. *Applied Linear Statistical Models,* 5th ed. McGraw-Hill, Boston. 1396 pp.

Newman, K. J., and H. V. Meredith. 1956. Individual growth in skeletal bigonial diameter during the childhood period from 5 to 11 years of age. *Amer. J. Anat.* **99**: 157–187.

Noether, G. E. 1985. Elementary estimates: An introduction to nonparametrics. *J. Educ. Stat.* **10**: 211–222.

O'Connell, C. P. 1953. *The Life History of the Cabezon Scorpaenichthys marmoratus (Ayres).* Calif. Div. Fish Game, Fish Bull. 93. 76 pp.

Oden, N. L., and R. R. Sokal. 1992. An investigation of three-matrix permutation tests. *J. Classif.* **9**: 275–290.

Olson, E. C., and R. L. Miller. 1958. *Morphological Integration.* University of Chicago Press, Chicago. 317 pp.

Onwuegbuzie, A. J., and N. L. Leech. 2004. Post hoc power: A concept whose time has come. *Understanding Statistics* **3**: 201–230.

Orwin, R. G. 1983. A fail-safe *N* for effect size in meta-analysis. *J. Educ. Stat.* **8**: 157–159.

Ostle, B. 1963. *Statistics in Research.* 2nd ed. Iowa State University Press, Ames, IA. 585 pp.

Owen, D. B. 1962. *Handbook of Statistical Tables.* Addison-Wesley, Reading, MA. 580 pp.

Page, E. B. 1963. Ordered hypotheses for multiple treatments: A significance test for linear ranks. *J. Amer. Stat. Assn.* **58**: 216–230.

Park, W. H., A. W. Williams, and C. Krumwiede. 1924. *Pathogenic Micro-Organisms.* Lea & Febiger, Philadelphia. 811 pp.

Patil, K. D. 1975. Cochran's Q test: Exact distribution. *J. Amer. Stat. Assn.* **70**: 187–189.

Pearce, S. C. 1965. *Biological Statistics.* McGraw-Hill, New York. 212 pp.

Pearson, E. S., and H. O. Hartley. 1958. *Biometrika Tables for Statisticians*. Vol. 1. 2nd ed. Cambridge University Press, London. 240 pp.

Petraitis, P. S., A. E. Dunham, and P. H. Niewiarowski. 1996. Inferring multiple causality: The limitations of path analysis. *Functional Ecology* **10**: 421–431.

Pettit, A. N., and M. A. Stephens. 1977. The Kolmogorov–Smirnov goodness-of-fit statistic with discrete and grouped data. *Technometrics* **19**: 205–210.

Phillips, J. R., and L. D. Newsom. 1966. Diapause in *Heliothis zea* and *Heliothis virescens* (Lepidoptera: Noctuidae). *Ann. Entomol. Soc. Amer.* **59**: 154–159.

Pielou, E. C. 1977. *Mathematical Ecology*. Wiley, New York. 385 pp.

Pielou, E. C. 1984. *The Interpretation of Ecological Data: A Classification and Ordination*. Wiley, New York. 263 pp.

Power, D. M. 1972. Numbers of bird species on the California islands. *Evolution* **26**: 451–463.

Powick, W. C. 1925. Inactivation of vitamin A by rancid fat. *J. Agric. Res.* **31**: 1017–1027.

Prange, H. D., J. F. Anderson, and H. Rahn. 1979. Scaling of skeletal to body mass in birds and mammals. *Amer. Nat.* **113**: 103–122.

Price, R. D. 1954. The survival of *Bacterium tularense* in lice and louse feces. *Amer. J. Trop. Med. Hyg.* **3**: 179–186.

Pugesek, B. H. and J. B. Grace. 1998. On the utility of path modeling for ecological and evolutionary studies. *Functional Ecology* **12**: 853–856.

Pugesek, B. H., A. Tomer, and A. von Eye (eds.). 2003. *Structural Equation Modeling: Applications in Ecological and Evolutionary Biology*. Cambridge University Press, Cambridge. 336 pp.

Quenouille, M. H. 1952. *Associated Measurements*. Academic Press, New York. 242 pp.

Quinn, G. P., and M. J. Keough. 2002. *Experimental Design and Data Analysis for Biologists*. Cambridge University Press, Cambridge. 520 pp.

R Development Core Team. 2011. *R: A Language and Environment for Statistical Computing*. R Foundation for Statistical Computing, Vienna, Austria. www.r-project.org.

Raktoe, B. L., A. Hedayat, and W. T. Federer. 1981. Factorial Designs. Wiley, New York. 209 pp.

Ramsey, P. H., P. P. Ramsey, and K. Barrera. 2010. Choosing the best pairwise comparisons of means from non-normal populations, with unequal variances, but equal sample sizes. *J. Stat. Comp. Simul.* **80**: 595–608.

Rao, P. 1971. Some notes on misspecification in multiple regressions. *Amer. Stat.* **25**(5): 37–39.

Raykov, T., and G. A. Marcoulides. 2006. *A First Course in Structural Equation Modeling*. 2nd ed. Lawrence Erlbaum Associates, Mahwah, NJ.

Reyment, R. A. 1991. *Multidimensional Paleobiology*. Pergamon Press, New York. 377 pp.

Ricker, W. E. 1954. Stock and recruitment. *J. Fish. Res. Board Can.* **11**: 559–623.

Ricker, W. E. 1973. Linear regression in fishery research. *J. Fish. Res. Board Can.* **30**: 409–434.

Ritz, C., and J. C. Streibig. 2008. *Nonlinear Regression with R*. Springer: New York. 148 pp.

Robertson, J. L., R. M. Russell, H. K. Preisler, and N. E. Savin. 2007. *Pesticide Bioassays with Arthropods*. 2nd ed. CRC Press, Boca Raton, FL. 224 pp.

Rohlf, F. J. and R. R. Sokal. 1981. *Statistical Tables*. 2nd ed. W. H. Freeman, San Francisco. 219 pp.

Romesburg, C. 2004. *Cluster Analysis for Researchers*. Lulu.com. 340 pp.

Rosenberg, M. S. 2005. The file-drawer problem revisited: A general weighted method for calculating fail-safe numbers in meta-analysis. *Evolution* **59**: 464–468.

Rosenthal, R. 1979. The "file drawer problem" and tolerance for null results. *Psychol. Bull.* **86**: 638–641.

Ross, S. M. 1985. *Introduction to Probability Models.* 3rd ed. Academic Press, Orlando, FL. 502 pp.

Ruxton, G. D. 2006. The unequal variance *t*-test is an underused alternative to Student's *t*-test and the Mann–Whitney U test. *Behavioral Ecology* **17**: 688–690.

Ruxton, G. D., and N. Colegrave. 2006. *Experimental Design for the Life Sciences.* Oxford University Press, Oxford. 184 pp.

Ryan, T. P. 1997. *Modern Regression Methods.* Wiley, New York. 515 pp.

Ryan, T. P. 2007. *Modern Experimental Design.* Wiley-Interscience, New York. 593 pp.

Sahai, H., and M. M. Ojeda. 2004. *Analysis of Variance for Random Models: Theory, Methods, Applications and Data Analysis.* Vol. I: *Balanced Data.* Birkhauser, Boston. 484 pp.

Sahai, H., and M. M. Ojeda. 2005. *Analysis of Variance for Random Models: Theory, Methods, Applications and Data Analysis.* Vol. II: *Unbalanced Data.* Birkhauser, Boston. 480 pp.

SAS Institute. 1988. *SAS/STAT™ User's Guide.* Release 6.03 ed. SAS Institute, Cary, NC. 1028 pp.

Scheirer, C. J., W. S. Roy, and N. Hare. 1976. The analysis of ranked data derived from completely randomized factorial designs. *Biometrics* **32**: 429–434.

Schumacker, R. E., and R. G. Lomax. 2010. *A Beginner's Guide to Structural Equation Modeling.* 3rd ed. Routledge Academic, Mahwah, NJ. 536 pp.

Scott, A., and C. Wild. 1991. Transformations and R^2. *Amer. Stat.* **45**: 127–129.

Searle, S. R. 1989. Statistical computing packages: Some words of caution. *Amer. Stat.* **43**: 189–190.

Searle, S. R. 2006. *Matrix Algebra Useful for Statistics.* Wiley, New York. 472 pp.

Seber, G. A. F. 1977. *Linear Regression Analysis.* Wiley, New York. 465 pp.

Seber, G. A. F., and C. J. Wild. 2003. *Nonlinear Regression.* Wiley, New York. 792 pp.

Shapovalov, L., and H. C. Taft. 1954. The life histories of the steelhead rainbow trout (*Salmo gairdneri gairdneri*) and silver salmon (*Oncorhynchus kisutch*) with special reference to Waddell Creek, California, and recommendations regarding their management. *Calif. Div. Fish Game, Fish Bull.* **98**: 375 pp.

Shapiro, S. S., and M. B. Wilk. 1965. An analysis of variance test for normality (complete samples). *Biometrika* **52**: 591–611.

Shaw, R. G., and T. Mitchell-Olds. 1993. Anova for unbalanced data: An overview. *Ecology* **74**: 1638–1645.

Shieh, G. 2007. A unified approach to power calculation and sample size determination for random regression models. *Psychometrika* **72**: 347–360.

Shipley, B. 2002. *Cause and correlation in biology: A user's guide to path analysis, structural equations and causal inference.* Cambridge University Press: Cambridge, UK. 332 pp.

Shoemaker, L. H. 1995. Tests for differences in dispersion based on quantiles. *Amer. Stat.* **49**: 179–182.

Siegel, S. 1956. *Nonparametric Statistics for the Behavioral Sciences.* McGraw-Hill, New York. 312 pp.

Siegel, S., and N. J. Castellan, Jr. 1988. *Nonparametric Statistics for the Behavioral Sciences.* 2nd ed. McGraw-Hill, New York. 339 pp.

Simpson, G. G., A. Roe, and R. C. Lewontin. 1960. *Quantitative Zoology.* Rev. ed. Harcourt, Brace, New York. 440 pp.

Sinnott, E. W., and D. Hammond. 1935. Factorial balance in the determination of fruit shape in *Cucurbita*. *Amer. Nat.* **64**: 509–524.

Skinner, J. J., and F. E. Allison. 1923. Influence of fertilizers containing borax on growth and fruiting of cotton. *J. Agric. Res.* **23**: 433–445.

Smith, F. L. 1939. A genetic analysis of red seed-coat color in *Phaseolus vulgaris. Hilgardia* **12**: 553–621.

Smith, R. J. 1993. Logarithmic transformation bias in allometry. *Amer. J. Phys. Anthropol.* **90**: 215–228.

Smithson, M. 2003. *Confidence intervals.* Sage Publications, Newbury Park, CA. 90 pp.

Smouse, P. E., J. C. Long, and R. R. Sokal. 1986. Multiple regression and correlation extensions of the Mantel test of matrix correspondence. *Syst. Zool.* **35**: 627–632.

Smyth, G. K. 2002. Nonlinear regression. Pages 1405–1411 in *Encyclopedia of Environmetrics* (A. H. El-Shaarawi and W. W. Piegorsch, ed.). Wiley, Chichester, UK.

Sneath, P. H. A., and R. R. Sokal. 1973. *Numerical Taxonomy.* W. H. Freeman, San Francisco. 573 pp.

Snedecor, G. W., and W. G. Cochran. 1989. *Statistical Methods.* 8th ed. Iowa State University Press, Ames, IA. 593 pp.

Snee, R. D. 1973. Some aspects of nonorthogonal data analysis. Part I. Developing prediction equations. *J. Quality Tech.* **5**: 67–79.

Snee, R. D. 1974. Graphical display of two-way contingency tables. *Amer. Stat.* **28**: 9–12.

Snee, R. D. 1977. Validation of regression models: Methods and examples. *Technometrics* **19**: 415–428.

Sokal, R. R. 1952. Variation in a local population of *Pemphigus. Evolution* **3**: 296–315.

Sokal, R. R. 1958. Quantification of systematic relationships and of phylogenetic trends. *Proc. Tenth Internat. Congr. Entomol.* **1**: 409–415.

Sokal, R. R. 1962. Variation and covariation of characters of alate *Pemphigus populitransversus* in eastern North America. *Evolution* **16**: 227–245.

Sokal, R. R. 1965. Statistical methods in systematics. *Biol. Rev.* (Cambridge) **40**: 337–391.

Sokal, R. R. 1966. Pupation site differences in *Drosophila melanogaster. Univ. Kansas Sci. Bull.* **46**: 697–715.

Sokal, R. R. 1967. A comparison of fitness characters and their responses to density in stock and selected cultures of wild type and black *Tribolium castaneum. Tribolium Inf. Bull.* **10**: 142–147.

Sokal, R. R. 1979. Testing statistical significance of geographic variation patterns. *Syst. Zool.* **28**: 227–231.

Sokal, R. R., and C. A. Braumann. 1980. Significance tests for coefficients of variation and variability profiles. *Syst. Zool.* **29**: 50–66.

Sokal, R. R., and P. E. Hunter. 1955. A morphometric analysis of DDT-resistant and nonresistant housefly strains. *Ann. Entomol. Soc. Amer.* **48**: 499–507.

Sokal, R. R., and I. Karten. 1964. Competition among genotypes in *Tribolium castaneum* at varying densities and gene frequencies (the black locus). *Genetics* **49**: 195–211.

Sokal, R. R., and N. L. Oden. 1978a. Spatial autocorrelation in biology. 1. Methodology. *Biol. J. Linn. Soc.* **10**: 199–228.

Sokal, R. R., and N. L. Oden. 1978b. Spatial autocorrelation in biology. 2. Some biological implications and four applications of evolutionary and ecological interest. *Biol. J. Linn. Soc.* **10**: 229–249.

Sokal, R. R., N. L. Oden, B. A. Thomson, and J. Kim. 1993. Testing for regional differences in means: Distinguishing inherent from spurious spatial autocorrelation by restricted randomization. *Geogr. Anal.* **25**: 199–210.

Sokal, R. R., and R. C. Rinkel. 1963. Geographic variation of alate *Pemphigus populitransversus* in eastern North America. *Univ. Kansas Sci. Bull.* **44**: 467–507.

Sokal, R. R., F. J. Rohlf, E. Zang, and W. Osness. 1980. Reification in factor analysis: A plasmode based on human physiology-of-exercise variables. *Mult. Beh. Res.* **15**: 181–202.

Sokal, R. R., P. E. Smouse, and J. V. Neel. 1986. The genetic structure of a tribal population, the Yanomama Indians. XV. Patterns inferred by autocorrelation analysis. *Genetics* **114**: 259–287.

Sokal, R. R., and P. A. Thomas. 1965. Geographic variation of *Pemphigus populitransversus* in eastern North America: Stem mothers and new data on alates. *Univ. Kansas Sci. Bull.* **46**: 201–252.

Sokoloff, A. 1955. Competition between sibling species of the *Pseudoobscura* subgroup of *Drosophila. Ecol. Monogr.* **25**: 387–409.

Sokoloff, A. 1966. Morphological variation in natural and experimental populations of *Drosophila pseudoobscura* and *Drosophila persimilis. Evolution* **20**: 49–71.

Späth, H. 1975. *Cluster-Analyse-Algorithmen.* R. Oldenbourg Verlag, München. 217 pp.

Spjøtvoll, E., and M. R. Stoline. 1973. An extension of the *T*-method of multiple comparison to include the cases with unequal sample sizes. *J. Amer. Stat. Assn.* **68**: 975–978.

SPSS. 1990. *SPSS Reference Guide.* SPSS, Chicago. 949 pp. (www.spss.com).

StatXact. 2010. Version 9. Cytel Software Corp., Cambridge, MA. (www.cytel.com).

Steel, R. G. D., and J. H. Torrie. 1980. *Principles and Procedures of Statistics.* 2nd ed. McGraw-Hill, New York. 633 pp.

Student (W. S. Gossett). 1907. On the error of counting with a haemacytometer. *Biometrika* **5**: 351–360.

Sullivan, R. L., and R. R. Sokal. 1963. The effects of larval density on several strains of the housefly. *Ecology* **44**: 120–130.

Sullivan, R. L., and R. R. Sokal. 1965. Further experiments on competition between strains of houseflies. *Ecology* **46**: 172–182.

Swanson, C. O., W. L. Latshaw, and E. L. Tague. 1921. Relation of the calcium content of some Kansas soils to soil reaction by the electrometric titration. *J. Agric. Res.* **20**: 855–868.

Swofford, D. L., and G. J. Olsen. 1990. Phylogeny reconstruction. In *Molecular Systematics* (D. M. Hillis and C. Moritz, eds.). Sinauer Assoc., Sunderland, MA. pp. 411–501.

Tate, R. F., and G. W. Klett. 1959. Optimal confidence intervals for the variance of a normal distribution. *J. Amer. Stat. Assn.* **54**: 674–682.

Tatsuoka, M. M., with contributions by P. R. Lohnes. 1988. *Multivariate Analysis: Techniques for Educational and Psychological Research.* 2nd ed. Macmillan, New York. 479 pp.

Teissier, G. 1948. La relation d'allometrie: Sa signification statistique et biologique. *Biometrics* **4**: 14–48.

Teissier, G. 1960. Relative growth. Pages 537–560 in *The Physiology of Crustacea,* Vol. 1 (T. H. Waterman, ed.). Academic Press, New York.

Thomas, J. D., and R. A. Hultquist. 1978. Interval estimation for the unbalanced case of the one-way random effects model. *Ann. Stat.* **6**: 582–587.

Thomson, G. H. 1951. *The Factorial Analysis of Human Ability.* 5th ed. Houghton Mifflin, New York. 383 pp.

Ting, N., R. K. Burdick, F. A. Graybill, S. Jeyaratnam, and T-F. C. Lu. 1990. Confidence intervals on linear combinations of variance components that are unrestricted in sign. *J. Stat. Comp. Simul.* **35**: 135–143.

Tukey, J. W. 1949. One degree of freedom for non-additivity. *Biometrics* **5**: 232–242.

Tukey, J. W. 1977. *Exploratory Data Analysis.* Addison-Wesley, Reading, MA. 688 pp.

Underwood, A. J. 1997. *Experiments in Ecology: Their Logical Design and Interpretation Using Analysis of Variance.* Cambridge University Press, Cambridge. 522 pp.

Ury, H. K. 1976. A comparison of four procedures for multiple comparisons among means (pairwise contrasts) for arbitrary sample sizes. *Technometrics* **18**: 89–97.

Utida, S. 1943. Studies on experimental population of the Azuki bean weevil. *Callosobruchus chinensis* (L.) VIII. Statistical analysis of the frequency distribution of the emerging weevils on beans. *Mem. Coll. Agric. Kyoto Imp. Univ.* **54**: 1–22.

Velleman, P. F., and D. C. Hoaglin. 1981. *Applications, Basics, and Computing of Exploratory Data Analysis.* Duxbury Press, Boston. 354 pp.

Verner, J. 1964. Evolution of polygamy in the long-billed marsh wren. *Evolution* **18**: 252–261.

Vollenweider, R. A., and M. Frei. 1953. Vertikale und zeitliche Verteilung der Leitfähigkeit in einem eutrophen Gewässer während der Sommerstagnation. *Schweiz. Z. Hydrol.* **15**: 158–167.

Wainer, H. 1974. The suspended rootogram and other visual displays: An empirical validation. *Amer. Stat.* **28**: 143–145.

Wallis, J. R., N. C. Matalas, and J. R. Slack. 1974. Just a moment! *Water Resources Res.* **10**: 3–11.

Warton, D. J., and N. C. Weber. 2002. Common slope tests for bivariate errors-in-variables models. *Biometrical J.* **44**: 161–174.

Warton, D. J., I. J. Wright, D. S. Falster, and M. Westoby. 2006. Bivariate line-fitting methods for allometry. *Biol. Rev.* **81**: 259–291.

Watson, R. A., and D. B. Tang. 1980. The predictive value of prostatic acid phosphatase as a screening test for prostatic cancer. *New England J. Med.* **303**: 497–499.

Webber, L. G. 1955. The relationship between larval and adult size of the Australian sheep blowfly *Lucilia cuprina* (Wied.) *Austral. J. Zool.* **3**: 346–353.

Webster, J. T., R. F. Gunst, and R. L. Mason. 1974. Latent root regression analysis. *Technometrics* **16**: 513–522.

Weir, B. S. 1990. *Genetic Data Analysis: Methods for Discrete Population Genetic Data.* Sinauer Assoc., Sunderland, MA. 377 pp.

Weir, J. A. 1949. Blood pH as a factor in genetic resistance to mouse typhoid. *J. Inf. Dis.* **84**: 252–274.

Welch, B. L. 1951. On the comparison of several mean values: An alternative approach. *Biometrika* **38**: 330–335.

Welsch, R. E. 1977. Stepwise multiple comparison procedures. *J. Amer. Stat. Assn.* **72**: 566–575.

Whitaker, T. W. 1944. The inheritance of chlorophyll deficiencies in cultivated lettuce. *J. Hered.* **35**: 317–320.

Whittaker, J., and M. Aitkin. 1978. A flexible strategy for fitting complex log-linear models. *Biometrics* **34**: 487–495.

Whittaker, R. H. 1952. A study of summer foliage insect communities in the Great Smoky Mountains. *Ecol. Monogr.* **22**: 1–44.

Wilkinson, L. 1989. *SYSTAT: The System for Statistics.* SYSTAT, Evanston, IL. 638 pp.

Williams, C. B. 1964. *Patterns in the Balance of Nature.* Academic Press, London. 324 pp.

Williams, D. A. 1976. Improved likelihood ratio tests for complete contingency tables. *Biometrika* **63**: 33–37.

Willis, E. R., and N. Lewis. 1957. The longevity of starved cockroaches. *J. Econ. Entomol.* **50**: 438–440.

Wolberg, J. 2006. *Data Analysis Using the Method of Least Squares: Extracting the Most Information from Experiments*. Springer, New York. 250 pp.

Wolfram Research. 2008. Mathematica version 7. Wolfram Research, Inc., Champaign, IL (www.wolfram.com).

Woodson, R. E., Jr. 1964. The geography of flower color in butterflyweed. *Evolution* **18**: 143–163.

Wright, S. 1968. *Evolution and the Genetics of Populations*. Vol. 1, *Genetic and Biometric Foundations*. University of Chicago Press, Chicago. 469 pp.

Wright, S. 1969. *Evolution and the Genetics of Populations*. Vol. 2, *The Theory of Gene Frequencies*. University of Chicago Press, Chicago. 511 pp.

Wuhrmann, K., and H. Woker. 1953. Über die Giftwirkungen von Ammoniak und Zyanidlö-sungen mit verschiedener Sauerstoffspannung und Temperatur auf Fische. *Schweiz. Z. Hydrol.* **15**: 235–260.

Young, L. C. 1941. On randomness in ordered sequences. *Ann. Math. Stat.* **12**: 293–300.

Zang, E. A., and E. L. Wynder. 1992. Cumulative tar exposure: A new index for estimating lung cancer risk among cigarette smokers. *Cancer* **70**: 69–76.

Zar, J. H. 2010. *Biostatistical Analysis*. 5th ed. Prentice-Hall, Englewood Cliffs, NJ. 944 pp.

Author Index

Subject Index

KEY TO STATISTICAL METHODS

	Design or Purpose	Measurement Variables	Ranked Variables	Attributes
1 variable 1 sample	Examination of a single sample	Procedure for grouping a frequency distribution, Box 2.1; stem-and-leaf display, Section 2.5; testing for outliers, Section 13.4; boxplots, Section 7.11 Test for serial independence, Box 13.1 Computing median of frequency distribution, Section 4.3 Computing arithmetic mean and standard deviation: unordered sample, Box 4.1; frequency distribution, Box 4.2 Confidence limits of coefficient of variation by jackknife, Box 7.6 Setting confidence limits: mean, Box 7.3; variance, Box 7.4 Computing g_1 and g_2; Box 6.1; Standard errors, Box 7.1		Confidence limits for a percentage, Section 17.1 Confidence limits for binomial and Poisson parameters, Box 7.5 Runs test for randomness in dichotomized data, Box 18.7
	Comparison of a single sample with an expected frequency distribution	Normal expected frequencies, Section 6.2 Goodness-of-fit tests: parameters from an extrinsic hypothesis, Box 17.1; from an intrinsic hypothesis, Box 17.2 Kolmogorov–Smirnov test of goodness of fit, Box 17.3 Shapiro–Wilk test for normality, Box 13.5 Graphic "tests" for normality: normal quantile plots and rankits, Box 6.2 Test of sample statistic against expected value, Box 7.2		Binomial expected frequencies, Section 5.2 Poisson expected frequencies, Section 5.3 Goodness-of-fit tests: parameters from an extrinsic hypothesis, Box 17.1; from an intrinsic hypothesis, Box 17.2
1 variable ≥2 samples	Single classification	Single-classification anova, Box 9.1 Planned comparison of means in anova, Box 9.5; single degree of freedom comparisons of means, Box 9.6, Box 14.8; Dunn–Šidák and Bonferroni methods, Box 9.6 Unplanned comparison of means: T-method, equal sample sizes, Box 9.7; T', GT2, and Tukey–Kramer, unequal sample sizes, Box 9.7; Welsch step-up, Box 9.7; STP test, Box 9.9; contrasts using Scheffé, T, and GT2, Box 9.8; multiple confidence limits, Section 9.5 Estimate variance components: unequal sample sizes, Box 9.2; equal sample sizes, Box 9.3 Confidence limits to a variance component, unequal n, Box 9.3 Tests of homogeneity of variances, Box 13.2 Tests of equality of means when variances are heterogeneous, Box 13.3, Box 13.4 Required sample size and power, Box 12.2; effect size, Box 12.1 Nonparametric test of equality of medians, Box 17.9	Kruskal–Wallis test, Box 13.7 Unplanned comparison of means by a nonparametric STP, Box 13.9	G-test for homogeneity of percentages, Boxes 17.6 and 17.8 Replicated goodness-of-fit test, Box 17.4; unplanned analysis of replicated tests of goodness of fit, Box 17.5
	Nested classification	Two-level nested anova: equal sample sizes, Box 10.1, 10.2; unequal sample sizes, Box 10.4 Three-level nested anova: equal sample sizes, Box 10.3; unequal sample sizes, Box 10.5		
	Two-way or multi-way classification	Two-way anova: with replication, Box 11.1, Box 11.2; without replication, Box 11.4 Randomized blocks, Box 11.5 Multiple comparisons, Box 11.3 Three-way anova, Box 11.7 Partitioning interaction, Box 16.6 More-than-three-way classification, Section 11.8 Test for nonadditivity in a two-way anova, Box 13.6	Friedman's method for randomized blocks, Box 13.11 Two-way replicated anova, Box 13.13	Three-way log-linear model, Box 17.10 Randomized blocks for frequency data (repeated testing of the same individuals), Box 17.16